MULTICHIP MODULES
Systems Advantages, Major Constructions, and Materials Technologies

EDITED BY

R. Wayne Johnson
Assistant Professor of Electrical Engineering
Auburn University

Robert K. F. Teng
Associate Professor of Electrical Engineering
California State University at Long Beach

John W. Balde
Senior Consultant
Interconnection Decision Consulting, Inc.

**A volume in the IEEE PRESS Selected Reprint Series,
prepared under the sponsorship of the IEEE Components, Hybrids,
and Manufacturing Technology Society.**

IEEE
PRESS

The Institute of Electrical and Electronics Engineers, Inc., New York

IEEE PRESS
445 Hoes Lane, PO Box 1331
Piscataway, NJ 08855-1331

ISBN 0-87942-267-X

IEEE Order Number: PC0260-0

Printed in the United States of America

10 9 8 7 6 5 4 3 2 1

Cover photo: Courtesy of Rockwell International Corp.

Library of Congress Cataloging-in-Publication Data

Multichip modules : systems advantages, major constructions, and
materials technologies / edited by R. Wayne Johnson, Robert K. F.
Teng, John W. Balde.
     p.   cm. — (A volume in the IEEE Press selected reprint
series)
  Includes index.
  ISBN 0-87942-267-X
  1. Electronic packaging. 2. Microelectronic packaging.
I. Johnson, R. Wayne, 1957–   . II. Teng, Robert K. F. III. Balde,
John W., 1923–   . IV. Series: IEEE Press selected reprint series.
TK7870.15.M85   1991
621.381′046—dc20   90-20827

Contents

Contents

Contents

Contents

Contents

Preface

MULTICHIP packaging technology has received widespread attention in the electronics industry. The use of multichip modules promises to increase circuit density and increase circuit performance beyond that otherwise possible through the use of VLSI and surface mount technology. Because this technology is so new and growing so fast, there is no comprehensive book on the subject. Chapters of handbooks and textbooks presently available cover only selected aspects of the technology.

As with any such fast-moving technology, the best way to capture the available information is to publish a selection of seminal papers. Some of the information was first published years ago, but much is in very recent conference proceedings. In very few cases have the papers reached the archival technical publications. A selection of these conference papers provides the best available information today. This book should be the primary source of information for engineers working in this area, students and professors in the packaging-related courses at colleges and universities, and perhaps even for the managers who need to make the hard technical choices about the technology for their new systems.

The original idea for this book came from some discussions during the 38th ECC in Los Angeles, 1988. Dr. Leo Feinstein, IEEE CHMT Society President, and Dr. Paul Wesling, Vice-President for Publications, both urged submission of a proposal for this book to the IEEE Press. Approval was received in December 1989. In keeping with the need to make this information relevant, Dr. Wayne Johnson of Auburn University and Jack Balde of Interconnection Decision Consulting were asked to select the papers and organize the text. Both have been particularly active in the newly developing area.

I wish to acknowledge the continuing support of Paul Wesling and the editorial contributions of Carrie Briggs of the IEEE Press and of Iwona Turlik and John Segelken, who both offered many useful suggestions.

ROBERT K. F. TENG
University of California,
Long Beach

General Concepts

IN THE EARLY 1970s, Rex Rice in the IEEE VLSI Tutorial stated the case for high density silicon, by showing that the cost of interconnection on silicon was lower than the cost on any other interconnection level. That realization, and the expectation that circuit performance and reliability could be improved by packing more circuit elements on each chip led to a drive for greater and greater density. The fact that solid-state technology was taught at all the universities and colleges also led to a major trend to higher and higher density silicon circuits, and larger and larger chips. There was no education base devoted to increasing the density of interconnections between chips. The result was a major increase in chip size and circuits.

Interconnection on silicon greatly exceeded the interconnection capability of the hybrid, the printed circuit board, and the backplane. The interconnection density on the chip increased 10:1 in a few years, but in the same period there was an almost trivial increase in the circuit density of the printed circuit interconnection board. Many speakers noted the trend with concern, including Ray Rinne of IBM at the Computer Packaging Committee Workshop in 1972, Fred Buelow of Amdahl in 1975 at the introduction of the 470 Computer, and more recently Barry Whalen of MCC at the Materials Research Society meeting, Boston 1988. It was obvious that a means had to be found to increase the density of interconnection wiring.

The first to accept an interconnect medium with higher interconnection density were the mainframe manufacturers. IBM was first in 1980, and then NEC and Mitsubishi in 1984, to pioneer the use of multilayer cofired alumina ceramic multichip interconnects. While the technology was expensive initially (greater than $1.00 per layer per square inch of substrate), the reduced size of the assemblies permitted a major increase in operating speed, and a reduction of cycle times. Thus the first extensive use of multichip technology with many chips was for computer mainframes. Some of the descriptions of these applications state the fundamental rationale for increasing density and decreasing size to achieve improved performance.

Concurrently, there was some modest use of small multichip modules with only a few chips for increased density provided through the reduction of the board space by elimination of part of the packaging borders. John Bauer of RCA Moorestown was its principal advocate, and produced considerable speed increases and size reduction by the use of smaller multichip packages. One figure[1] from his 1980 reports and articles shows the density increase using three- and four-chip multichip modules.

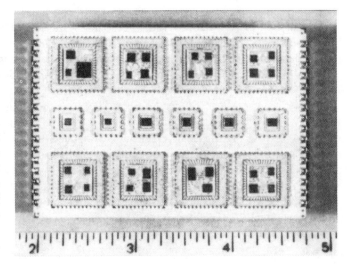

The mainframe multilayer cofired assemblies were not notable for their cost reductions; the improvement in performance was the expected benefit, even if costs were greater. The costs eventually lowered somewhat. With the small multichip assemblies from RCA it started to become apparent that the use of multichip modules should be considered for cost reasons also. There was still the feeling that multilayer ceramic was higher cost than printed circuit board technology, and that the yield problems of small multichip modules would limit their use. There was the feeling that only special situations could justify the use of such technology.

When Goddard of Bell Laboratories plotted the densities and costs of interconnections, it suddenly became evident to Knausenberger and Schaper that the normalized costs of all interconnect was identical for identical lengths. If silicon interconnection was cheaper, it was only because the line lengths were shorter. The costs at the next interconnect level would be less, and the performance would increase if line lengths were similarly reduced.

Connecting discrete chips in discrete packages forced the cost to be large, because costs were dependent on the area of the interconnect board. It made little difference if the wiring density on the chips was great if the wiring density between the chips was low. The use of multichip modules improved the density both by removing the cases from each chip, and by using wiring technology with closer lines and spaces.

The ultimate criteria of density is the percentage of the total area that is occupied by functioning circuit elements, i.e., active silicon. The density of active devices per unit area determines the amount of functional activity possible in that defined area. Pack things closer together by whatever means possible—using dense chips on open PC boards, or low density chips on high-interconnect-density substrates.

If the substrate is to be high density, it must have the ability to interconnect the required circuits. The metric that

[1]Reproduced with permission from *Electronic Packaging and Production*, vol. 20, no. 5, May 1980. © 1980 Cahners Publishing Co., Inc.

Knausenberger and Schaper chose was inches of wiring conductors crossing a square inch of surface area. This "inches per square inch" measure is now almost universally quoted to assess the interconnection capability of a substrate.

The circuit density of an assembly can be expressed as the percentage of the surface area covered by active or working silicon, or often more simply by the density of silicon die area per unit area of substrate. It is the accepted criteria for efficient packaging, a legacy of this new idea. Suddenly the issue was clearly defined—increasing the density of an interconnect will improve the wiring efficiency, even if the chips are identical.

Implementations Needed

One way to increase the packaging efficiency is to reduce the number of packages. Even simple four- and six-chip hybrids could reduce the size of the interconnect board, and improve speed and performance. The problem with that approach in the mid 1980s was that the silicon companies did not want to provide the multichip assemblies or the bare chips, so the idea was only useful to vertically integrated companies such as Hitachi. Still, this was a good idea waiting for wider acceptance.

Some of these general concepts are presented in Chapter 2. The ideas were there quite early, but the implementation was not.

Multilayer Cofiring Ceramic Hybrids

The early acceptance of multichipping by IBM was possible because, as a vertically integrated company, they could obtain uncased chips and could design chip sets that could be mounted in the regular arrays needed for the cofired ceramic technology. With up to 35 layers of metallization on this doctor-bladed "green" ceramic tape, assemblies could be laminated and cofired to produce the desired dense interconnect assemblies.

Indeed, the chips were placed on the surface with minimal space between chips, but the price was an expensive substrate, with costs for a 4″ × 4″ module running from $1000 to $10,000. Though the system cost was comparable to other solutions, the cost and complexity of this cofired ceramic solution was only possible for the mainframe companies: IBM, NEC, and Mitsubishi. The perception was that this solution was only for the high-performance mainframes, not really acceptable for cost performance applications, or even the military computers. Historically important, and still useful today, this technology is described in Chapter 3.

Again, a better way was needed. One of the problems with the ceramic multilayer technologies was that the dielectric constant of alumina is over 9, and there was the realization that the one way to increase the density per layer was to reduce the dielectric constant. Some companies had this insight as early as 1984, but it was still news in the presentations in 1987 at the IPC and the MRS.

How so? If the lower dielectric constant permits decreasing the thickness of the interlayer dielectric, then the ground planes can be brought closer to the signal planes. With closer ground planes, the fields around the lines are made smaller, and the effect of one field on an adjacent line is reduced proportionally. Lower dielectric constant, decrease the dielectric thickness, lower crosstalk, and smaller spacing is possible. Suddenly the race was to utilize lower dielectric constants.

Silicon-on-Silicon with Silicon Dioxide Dielectric

The first technology to use integrated-circuit (IC) processing technology as an alternative to multilayer ceramic employed deposited layers of silicon dioxide and aluminum metallization to produce an interconnect substrate on a silicon "mother board." Silicon-on-silicon investigations were under way at many companies, with Hitachi using the technology for small multichip modules, and with Honeywell and eventually Mosaic, N-Chip, and IBM exploring this approach.

Ten-micron lines and spaces were easy for the silicon manufacturers. Silicon dioxide/aluminum had many advantages: low cost, availability, and proven technology available from multiple sources. But silicon dioxide has a number of deficiencies, also. Multiple thick layers of silicon dioxide are difficult to produce, yet they are needed for controlled impedance. The higher resistance of aluminum limits high-speed performance. For multichip modules, die attachment may be by reflow soldering, which is more difficult with aluminum than with copper.

Silicon as a substrate also poses difficulties. It is not strong enough mechanically to be the total package, but must be mounted in a ceramic package to provide protection. If the silicon "mother board" is of considerable size, bonding that silicon to a package must either be done with a *very* resilient adhesive (Hitachi uses silicone rubber, AT&T uses silicon gel) or the package support must be a very close expansion match to the silicon, or both. This forces the uses of aluminum nitride (Rockwell) or silicon carbide (Hitachi). Alternatively, the die bond adhesive can be thick and resilient (DEC), but for good heat transfer it requires loading the adhesive with thermally conductive filler. Unfortunately, thermal fills only help conductivity to a limited extent.

Multilayer Interconnection with Organic Insulation

Polymer dielectric layers have material properties that are better than silicon dioxide in many ways. They can be used instead of silicon dioxide to avoid the vacuum steps, and permit easier use of copper instead of aluminum. Organic dielectric coatings make additional layers easier, with copper conductors having lower circuit resistance. Copper could be solder-plated or tinned, permitting the use of solder-bonded TAB connections for easier repair or replacement. The developments through the use of polyimide or equivalent dielectric are discussed in Chapter 4.

Though these dielectrics are somewhat lower in dielectric constant, there was considerable variation among products of different manufacturers. New materials such as benzocyclobutane (BCB) and polyphenyl quinoxalene (PPQ) offer advantages of lower dielectric constant and lower moisture absorption and greater leveling. Initial work with polymer dielectrics was performed with silicon substrates.

MULTILAYER INTERCONNECTIONS DIRECTLY ON CERAMIC AND METAL SUBSTRATES

When the silicon substrate is housed in a protective package, the match of the substrate expansion requires the use of aluminum nitride or silicon carbide. But if the substrate matches the temperature coefficient expansion (TCE) of the silicon substrate, it also matches the TCE of the silicon chips themselves. It suddenly became evident: if the only thing that the silicon was doing was to hold up the polyimide or other organic interconnect, why not put the interconnect directly on the ceramic? With a match of temperature coefficient, the active silicon chips could be attached directly to the packaging ceramic through openings in the dielectric. If silicon is retained, it becomes just a support substrate for the circuits, unnecessary if the chips can be bonded directly to the substrate through openings or holes. Once polyimide or other organic was used for the insulation, the continued use of silicon as the substrate was no longer necessary for many applications.

The resultant multichip assembly with organic insulation on dimension-controlling substrates is now being called "deposited" multichip module technology, or MCM-D. The dimensional control and the heat removal is supplied by the rigid silicon, ceramic, or perhaps metal substrate, and the interconnect circuits are aluminum or copper on a low-dielectric-constant insulating deposited film layer. Sometimes power circuits are provided in thick-film technology on the base ceramic, but in many cases the passive devices are bonded on the surface.

Eliminating the silicon does have some disadvantages. Silicon-based depletion layer capacitance, dual dielectric capacitors, or other silicon-based passive components are no longer possible, but there are alternatives. Thin depositions of high dielectric insulation can provide the capacitance, and either thin- or thick-film resistors can be deposited on the ceramic.

Chapter 5 presents papers that advocate the deposition of the circuits directly on the ceramic substrate. Though most of the dielectric is deposited in place on the substrate, it is also possible to form the interconnect circuit separately and then transfer the circuit to the rigid substrate. That approach is also presented in papers from the companies that create the interconnect circuits by lamination, including the use of laser patterning of the circuits.

Some new considerations were discovered during the shift to deposited dielectrics on silicon or ceramic. The first is that most of the dielectric materials deposited on the ceramic were conformal to an excessive extent. Parylene is nearly 100% conformal—cover a 5-μ path with parylene, and there will be a replicate 5-μ ridge on the top of the dielectric layer. Polyimide similarly replicates the patterns of the metallization layer, though only 80% of the metal stripe height is replicated to the surface. Both are unacceptable for subsequent layer processing, because photolithographic patterning will have defocusing of the light, making the patterns fuzzy.

The costly solution is to "planarize," or polish, the surface after each layer of deposition. Alternatively, new dielectrics like benzocyclobutane have greater flow characteristics and little conformality print-through to the top surface. If the surface is planar within 1 μ, the polishing step can usually be eliminated.

ALL-CERAMIC MULTICHIP MODULES ARE NOT DEAD

No new technology replaces one previously dominant without a struggle. If lower dielectric constant is a good idea, why not seek a lower dielectric ceramic? If copper metallization is better electrically and for assembly, why not change the firing process to permit copper metallization? Why not, indeed, produce circuits of the same density as were now possible with polyimide or benzocyclobutane? Chapter 6 explores these possibilities.

There are problems. The newer organic dielectrics have permittivities of 2.8 or less, but the borosilicate glasses cannot go below 5. Will foam ceramic do the job? Generally they are difficult to metallize and produce processing problems, and have poor thermal conductivity.

Will these new low-dielectric ceramics be widely accepted? Well, possibly—but the technology is a little late, and highlights a major problem with the low-dielectric constant ceramics—they are great insulators! No low-dielectric-constant ceramic has a thermal performance greater than 5 W/m K, a poor performance compared to aluminum nitride at 140 or possibly 260, or silicon carbide at 150 to 270 W/mK. Alternatively, the dielectrics for MCM-D can be deposited on metal or metal laminates, such as copper–molybdenum–copper. Never mind that the organics are even poorer thermal conductors; they can be bypassed using thermal vias or holes or square "well" openings in the interconnect layers for the mounting of the chips. The lower interconnect density per layer for cofired ceramics requires more layers. Thermal vias or wells would further reduce the wiring density for this technology. It is the thermal considerations that determine the acceptance of a substrate.

The low-temperature cofired ceramic technology can be used in low-power applications. Also, it is still of importance for mainframes. The multilayer cofired ceramic multichip constructions are acceptable as the interconnect substrates, because mainframes have used top heat removal in the past and can continue to do so.

DIELECTRIC MATERIALS

There are different questions about the proper materials to be used for the dielectric for the organic or oxide insulation between the signal planes of the MCM-D constructions. The earlier oxide insulation gives way to polyimide, and the polyimide is now giving way to benzocyclobutane or PTFE Teflon, or some of the many modified polyimides. Dielectric constant is important, but so is planarization, and so is the material compatibility with copper. It does little good to have a proper TCE or tensile modulus if there is no adhesion to copper, or if the materials corrode copper and destroy the circuit. And some of the materials do just that. Is it better to precoat the polyimide with chromium and nickel before the copper, or is it better to use another material? These and other considerations are explored in Chapter 7.

There is another problem not well reported. Matching the

TCEs of the dielectric and the substrate is not the whole story. A dielectric material that has a high tensile modulus can produce substantial bending of the silicon or ceramic substrate as the temperature is reduced from the coating temperature to the temperature used in photolithography. Typical bending of a 5-in. wafer of silicon is about 50 μ at the center for the usually unmodified polyimides. Reduce the TCE and increase the tensile modulus, and there is no improvement. Reduce the tensile modulus alone, and the substrate remains planar. The papers in Chapter 7 present part of the story, but the reader must realize that all the data are not yet reported.

THERMAL CONSIDERATIONS

As the density of the active silicon increases, the density of the heat generation does also. Not only are individual chips increasing from 10 to 40 watts, but the assemblies increase to 25 or more watts per square inch. In one assembly presented, there are 25 40-watt chips on a 4″ square interconnect module. Dissipating 1000 watts is not trivial, so that the earlier top-contacting assemblies of IBM are giving way to liquid cooling, either on the surface by boiling heat transfer, impingement cooling, or sample immersion. Water cooling can also be performed by introducing cooling channels in the ceramic substrate, which produces quite acceptable values for temperature rise. This whole area is moving rapidly, and the papers of Chapter 8 explore some of the proposals. Unfortunately, some of the most recent ideas in this area are not yet in print, but there will be time for updates and new developments later.

INSPECTION AND ELECTRICAL TEST

The yield for such a complicated multilayer assembly is going to be proportional to the product of the yields for each layer. Going the whole way to the final assembly before test will almost assure zero yield. Each layer must be tested and repaired before the next layer is deposited. Furthermore, process inspection is necessary for good quality control and yield improvement. Automatic optical inspection, acoustic microscopy, scanning electron microscopy, X-ray and laser scan technologies have been proposed for inspection. Electrical test using boundary scan techniques, primarily developed for testing of the silicon die, can be adapted to test the functional assemblies. Each of these subjects has insufficient published literature available relevant to the multichip; we can select only the few papers available at this time.

ELECTRICAL CONSIDERATIONS

All of these constructions are of course only useful if they meet the electrical needs of the densely packaged circuits.

Crosstalk must be controlled, impedances must be managed, and resistive losses must be controlled. Current surge losses, or dI/dt noise must also be manageable, even in the face of huge currents. Consider the 25-chip assembly with each chip dissipating 40 watts. That 1000 watts must come from the 5-volt power source, requiring 200 amperes average, and over 200,000 amperes surge current at the time of clock-cycle coordination. Certainly 2 μ of copper deposited per layer in the interconnect substrate will be inadequate. We will shortly see 10 mil copper for the ground planes, and distributed capacitances to permit the necessary decoupling of the effects of the current draw on the power planes. Some of these electrical considerations are discussed adequately in the papers of Chapter 9, but other work in progress must further develop the electrical needs of these modules.

OVERVIEW

It is clear that multichip modules will have a major impact on all portions and aspects of electronic systems design. Multichip technology offers advantages for all types of electronic assemblies. Mainframes will need organic-on-ceramic to interconnect the high numbers of custom chips needed for the new systems. Cost-performance systems will use the high-density interconnect to assemble a few new chips with a collection of currently available chips to achieve high performance without time-consuming custom silicon, allowing quick time to market.

Military and automotive applications will prefer multichip modules for the small size and weight. Telecommunications will be in favor of the increased capability at low cost. Consumer electronics applications are now beginning to appear, as the realization that this technology lowers costs becomes more widely recognized. All levels of technology benefit from this technology.

The benefits are great. There is such a reduction in size, number of levels, and complexity of the interconnection interfaces that the assemblies will clearly be lower in cost as well as smaller in size and higher in performance. Virtually every major company is either actively using or looking hard at the advantages. The multichip revolution in the 1990s will have an impact on electronics as great as or greater than surface mount technology had for the 1980s. We hope that this collection of fundamental papers will provide a useful base from which to understand the excitement and the potential of the next advance in electronic packaging.

J. W. BALDE
Interconnection Decision Consulting
R. WAYNE JOHNSON
Auburn University

BECAUSE multichip packaging is a rapidly growing technology, there are many technical and strategic issues that are to be considered in future developments. The historical analysis of the advantages of multichip modules and the analyses of the advantages of the multichip technology in system interconnection are presented in the following pages.

John Bauer of RCA stated the advantages of small multichip modules for military applications, and showed assemblies of three- and four-chip modules, but stated the advantages orally, never in writing. The Amey–Balde papers state the case for the economic advantages of small multichip modules, while the Bauer emphasis had previously been on performance alone. Both results stem from the increase in interconnection density. Jensen, Rubin, and later Neugebauer show the advantages for even higher numbers of chips, giving data on size and density advantages.

The theory of the advantages of multichip modules did not mature until after the Goddard paper on costs of interconnect in various technologies, and the Knausenberger–Schaper insight that the normalized cost of the interconnection circuit determines total cost. Suddenly, density of the *assembly* was understood to be the appropriate criteria.

The impact of multichip modules on system costs were explored by Sage *et al.* of BPA and George Messner of PCK. Here, the advantages of the improved interconnection substrate are stated in comparison with alternative interconnection methods. For the first time, the differences of ceramic multichip module capability and deposited dielectric and metal technology were presented in cost/density charts.

Finally, an overview paper stating the materials requirements is offered. It has been said that multichip technology is really silicon processing technology with a great increase in materials knowledge required. This paper provides an introduction to that material information.

New Packaging Strategy to Reduce System Costs

JOHN W. BALDE

Abstract—Packaging density for mid-sized computers, personal computers, and other cost-performance equipment has been limited by the need to use available interface or "glue" circuit chips. This makes the system and circuit boards DIP dominated, with a continuing lower circuit density than achieved by bigger main frame systems. Using the percent active silicon area yardstick, various multichip package strengths are explored, showing that there is an increase in board efficiency for easily achievable package constructions that are possible with open plastic premolded packages. Packages are shown that demonstrate that this approach is available whenever the systems manufacturers and the integrated circuit (IC) houses decide to set up the necessary standard packages.

INTRODUCTION

PACKAGING COSTS are very greatly dependent on the PC board area and the number of layers of multilayer boards required for interconnect. Larger high performance mainframe systems have often made a design commitment for a highly efficient design using much VLSI technology, but the cost-performance machines have had to settle for VLSI for only the logic chips. The remaining "glue" chips (hex convertors, line drivers, etc.) are usually implemented in DIP's, with their attendant considerable board area and continuing through-hole assembly techniques.

When a computer system central processor is converted to VLSI, only a few chips might then constitute 90 percent of the total semiconductor circuitry, and those few chips may be packaged in new surface mounting packages, or in some other square packages. Square packages are needed because the VLSI chips usually have I/O requirements from 68 to 164, and the rectangular DIP or SOIC packages cannot be used.

THE GLUE CIRCUIT PROBLEM

Many cost-performance systems have these few VLSI packages, constituting perhaps 90 percent of the working system, mounted on a PC board otherwise occupied with DIP's. The interface circuits, hex convertors, line drivers, etc., are still SSI and MSI chips, and DIP mounted units are readily available. The resultant system board is dominated by these "glue" circuit chips. Fig. 1 shows an NCR computer with the few VLSI circuits hardly noticeable. No wonder that pin grid array packages, or pin-mounted sockets for packages that are meant for surface mounting are used so that the board remains in through-hole pin-mounted package technology. The board area and the system packaging costs are dominated by the older DIP technology, and no effective use of the new surface mounting techniques is being made in the cost-performance system area.

Manuscript received July 11, 1984.
The author is with Interconnection Decision Consulting, Route 4, Flemington, NJ 08822.

Fig. 1. The NCR four board minicomputer, showing only four VLSI packages.

Somehow the glue chips must be more closely packed to reduce the board area. Since board area for the glue chips may constitute as much as 95 percent to 98 percent of the board and system area, it is more important to find a new strategy for reducing the area occupied by the support chips than it is to look for further increases in VLSI density as a means of reducing system costs. For a given system with the glue chips occupying 95 percent of the board area, a 10 fold increase in logic capability would reduce system size and board costs by 4 percent at most, and probably not at all because of the PC board layout constraints.

ALTERNATIVE STRATEGIES USING MULTICHIP PACKAGES

Let us examine the packaging efficiency gain to be expected if the mounting of the glue circuits were to be changed.

If we use the percent active silicon area criteria for packaging efficiency proposed by Schaper and Knausenberger [1], we see that close packed arrays of 28 lead chip carriers produce an active silicon area of less than one percent. Actual values are quite dependent on the standard board size chosen for production reasons. For the example chosen here, the active area was only 0.83 percent. If we could increase the package size to the use of all 68 lead chip carriers, there would be an increase in packaging efficiency to 4.1 percent. Great—a 4.9 fold or almost 5 fold increase, but out of the question for cost-performance systems with tight design intervals and limited resource investment.

How can one get some of the area efficiency of the all 68 lead package solution for the glue or interface circuits? The use of multichip hybrids, with TAB or ceramic chip carriers on a ceramic substrate has often been used in higher performance systems. The Amdahl four chip hybrid for the cache memory of the 580 systems is an example (Fig. 2). It

Reprinted from *IEEE Trans. Components, Hybrids, Manuf. Technol.*, vol. CHMT-7, no. 3, pp. 257–260, September 1984.

Fig. 2.   Amdahl four chip carrier hybrid for the cache memory.

Fig. 3.   AMP JEDEC leaded 'A' premolded package with four mounted chips.

Fig. 4.   Silicone gel filled pin grid array package, showing the Hitachi aluminum dam and aluminum lid seal.

Fig. 5.   Hitachi version of premolded package.

requires a shift to ceramic technology and ceramic packages with their higher cost, both of which are unacceptable to many of the producers of cost-performance systems. The increase in packing density can be achieved more economically with the use of multichip mounting in plastic packages, such as the recently described AMP or Hitachi packages of Figs. 3–5 [2], [3].

What happens when these packages are used for the glue circuits? In a system with 90 percent of the circuitry in VLSI, 90–95 percent of the chips are SSI and MSI. Using 90 percent, a 180 chip system would have 20 VLSI chip packages, and 40 4-chip multichip packages. Roughly 2/3 of the packages on a board would be multichip, and about 1/3 VLSI. Look at the packaging efficiency (Table I). Over half of the efficiency of the all 68 lead VLSI packages was achieved, and almost 3X the packaging efficiency of an all 28 lead package system.

What happens if advancing technology produces VLSI logic chips with such I/O requirements that 84 pin packages are required? It looks as if the chip size will not require a larger cavity, so that most of the industry is looking at chip carrier or pin grid packages with higher density leads. If we look at the 84 lead package with leads on 25 mil centers, it can be used not only for the VLSI, but also for 4-chip multichipping. The packaging density increases from 2.3 to 3.5 percent, or another 50 percent increase. Significant—but what happens next?

Look at the same calculations with the use of 164 lead packages (Table I, column 5). Again assuming no basic change in the need for SSI and MSI glue circuits, the larger packages with 164 I/O can accommodate perhaps 8 SSI/MSI chips. Is this worth doing? No. The density on the board is so

decreased by the use of the larger packages that the advantage in silicon density is lost. (Only $4 \times 9 = 36$ packages instead of $7 \times 13 = 91$, and only 2 percent silicon instead of 3.5 percent). There may be advantages to the system through the use of more powerful VLSI chips, but that is another matter.

Rather the use of 164 lead packages will work to the reduction of board area and increase in silicon efficiency only if there are more efficient interface chips. If, for example, there were interface chips 300 mils square with I/O require-

TABLE I
COMPARISON OF PERCENT ACTIVE SILICON FOR VARIOUS PACKAGING STRATEGIES

I/O	28	68	68	84	164	164
Package Size	0.450"	0.950"	0.950"	0.650"	1.150"	1.150"
Chip Size	0.100 MSI	0.400 VLSI	0.400" VLSI / 0.100" MSI	.400" VLSI / 0.100" MSI	.400" VLSI / .100" MSI	.600" VLSI / .300 MSI
Array Size	12 x 20	6 x 10	6 x 12	7 x 13	4 x 9	4 x 9
Silicon VLSI 10% / SSI/MSI 90%	240 x .01 = 2.4	60 x .16 = 9.6	20 x .16 = 3.2 / 40 x 4 x .01 = 1.6	31 x .16 = 3.2 / 60 x 4 x .01 = 2.4	18 x .16 = 2.9 / 18 x 8 x .01 = 1.4	18 x .36 = 6.5 / 18 x 4 x .09 = 6.5
Total Silicon	2.4	9.6	4.8	7.4	4.3	13.0
Package Spacing	0.75"	1.3"	1.3"	1.0"	1.3"	1.3"
Usable Board Area	7" x 12" = 84"	7" x 12" = 84"	7" x 12" = 84"	7" x 12" = 84"	7" x 12" = 84"	7" x 12" = 84"
% Silicon	2.4/84 = 2.8%	9.6/84 = 10.2%	4.8/84 = 5.7%	7.4/84 = 8.6%	4.3/84 = 5.1%	13/84 = 15%
% Active Si	2.8% x .4 = .83%	10.2% x .4 = 4.1%	5.7% x .4 = 2.3%	8.6% x .4 = 3.5%	5.1% x .4 = 2.0%	15% x .4 = 5.9%

ments of 44, four of these could be combined in a 164 lead package, and intermixed with VLSI chips. That possibility produces a packaging efficiency of almost 6 percent active silicon, a density that has been expected only through the use of wafer scale integration!

Can these packages be interconnected with reasonable boards? The interconnection requirements may be expected to fall in accordance with the trend line of the Schaper-Knausenberger plot, as shown in Fig. 6, which has each of the densities in the table indicated. Actually, specific interconnection capabilities of a PC board structure often make it possible to get interconnection more efficiently than in the curve, and this is indicated at the off-line points "0," plotted from the interconnection analysis of Wulf Knausenberger [4].

THE MULTICHIP PACKAGE TYPES

Is close packing chip carriers an alternative? Not if they are close packed on the board surface. Tightly packed packages are difficult to inspect, difficult to replace or repair, and boards with closed spaced packages are particularly difficult to change if change in circuit is required. Certainly the military systems suppliers do close pack and do make changes, but careful analysis must be made to be sure that the solution is cost effective for the commercial market.

How about the close spaced carriers of the Amdahl type? Yes the density increases, and the risk to the board during repair or replacement is reduced, but the cavities are quite small, and the cost high. Large cavity plastic packages are much more economical.

Is it possible to obtain such packages in a post molded package from a semiconductor supplier? No, becuase the cost of such post molded packages would be greater because of the burn in and test losses. Hermetic sealed packages are a

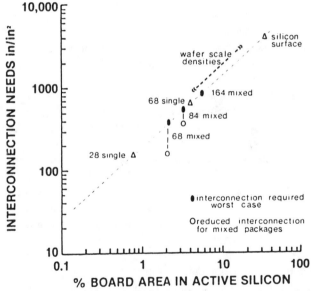

Fig. 6. Plot of the packing densities of Table I, and expected or realizable interconnection density requirements.

possibility, but the premolded or open package is the near-perfect solution, so that the package is not sealed until after it is established to be a good unit.

The premolded packages by AMP or Hitachi show the feasibility of such multichip packaging. Chip protection uses silicon gel to protect the chip from surface contamination and subsequent corrosion. Recent advances in gel technology make this possibility much more likely [5].

STANDARDIZATION NEEDED

Using the surface mounted chip carrier packages reduces the interconnection costs, but does not make the package costs low if the packages are of many different types. If multichip packaging is to work, the internal connections that are always certain, *no* internal connections should be used. Instead *all* I/O connections from the chip should be brought out to the edge connections of the package [6].

PIN GRID PACKAGING AS A POSSIBILITY

Pin grid packages can also be suitable for multichipping, and can also be available in open packages, gel filled. Are both pin grid array and surface attach packages equally feasible? Yes, if full array pin grid array packages are used, and if only occupied board area is considered. But most pin grid array packages use only the peripheral pins, because of both the area required for the lid and the heat sink, and the need to reduce the interconnection demands of the PC board. Such pin grid array packages use *more* board area than the equivalent surface mounted chip carriers [7]. Through-hole packages require more interconnection layers because of the reduced wiring channels between the pin attach holes [8]. This greater interconnection costs reduces the advantage of multichip pin grid array packages to improve packaging costs.

MULTICHIP OR NEW IC CHIPS

If condensing the glue circuits is so useful, why not produce those circuits in multiple on IC chips and avoid the use of hybrids? Of course, but the whole premise is that the designers of cost-performance systems have neither the clout or the time to try to get special chips that do not now exist. The reason for a multichip module is the same as the time-honored reasons for using a hybrid—it enables one to mount as a unit and purchase as a unit the function that is not now available on a single chip.

What happens when the IC industry decides that some of these multichip glue circuit packages are becoming so popular that single chip production is justified? Does one abandon the approach—is all of this strategy of just passing interest? Not so. If combined glue circuits are available on a 300 mil square chip, then they could be combined with the VLSI chips in the 164 lead package that we just examined. With the 4 such combined support circuit chips in the package, and packaging densities over 9 percent, that would produce pressure again for the production of an even larger support chip.

The problem with expectations of IC chip solutions is that each combined function chip gets more specialized, and is apt to be produced in smaller and smaller numbers. The point of hybrids and multichip assemblies is that the chips inside are

standard, and the package is standard, though the assembly may be special. The trick is to make the assembly as standard as possible, though the combination of the contents may be unique.

Hybrid technologies are always being pushed away by silicon integration, but have you noticed—there always seem to be more hybrids rather than less. There will always be a place and a rationale for combining chips of the next level of complexity into manageable assemblies. Furthermore, such hybrid assemblies are at least as plentiful as the single chip modules.

CONCLUSION

If plastic body multichip packages are available to the designers of cost-performance systems, the advantages of surface mounting can be made available to personal computers, minicomputers, and the office terminal machines. One important consideration—these small multichip packages will only be available to the designers of cost-performance systems if they are low cost and standard. It should be possible to specify standard production chips in one of a very few high volume standard packages, so that delivery is fast and costs low. Some systems designers are looking this way—it should be a major alternative.

REFERENCES

[1] L. W. Schaper and W. H. Knausenberger, "The myth of cheap wire," oral presentation at NEPCON West, Feb. 1984; *IEEE Trans. Components, Hybrids, Manuf. Technol.*, vol. CHMT-7, no. 3, Sep. 1984, *this issue*.
[2] J. W. Balde, "As packaging density increase, focus shifts to SSI and MSI chips," *Electronics*, vol. 57, no. 11, pp. 103–105, May 31, 1984.
[3] K. Otsuka, Y. Shirai, and K. Okutani, "A new silicone gel sealing mechanism for high reliability encapsulation," *IEEE Trans. Components, Hybrids, Manuf. Technol.*, vol. CHMT-7, no. 3, Sept. 1984, *this issue*.
[4] W. H. Knausenberger, AT&T Bell Laboratories, unpublished data.
[5] K. Otsuka, Y. Shirai, and K. Okutani, "Sealing mechanism of a silicon jelly encapsulation with high reliability," *Proc. ECC*, pp. 88–94, May 1984.
[6] D. I. Amey and J. W. Balde, "Circuit-board packaging considerations for optimum utilization of chip carriers," *IEEE Trans. Components, Hybrids, Manuf. Technol.*, vol. CHMT-3, no. 1, pp. 105–110, Mar. 1980.
[7] E. T. Lewis, "The VLSI package–An analytical review," *IEEE Trans. Components, Hybrids, Manuf. Technol.*, vol. CHMT-7, no. 2, June 1984.
[8] W. H. Knausenberger and N. A. Teneketges, "High pinout IC packaging and the density advantages of surface mounting, *IEEE Trans. Components, Hybrids, Manuf. Technol.*, vol. CHMT-6, no. 3, pp. 298–304, Sept. 1983.

Interconnection Costs of Various Substrates—The Myth of Cheap Wire

WULF H. KNAUSENBERGER, MEMBER, IEEE AND LEONARD W. SCHAPER

Abstract—Silicon has always been considered a dense and cheap medium for making wires, especially when compared with traditional interconnection media on a per interconnection basis. However interconnection costs in all technologies are essentially invariant on a per unit length basis. The advancing scale of integration has allowed active devices to be smaller and closer together, so wiring, both on chip and in systems, has been cheaper because there has been less of it. It is shown that substrate interconnection capability (in inches of wire per square inch of substrate) is the critical factor in achieving compact low-cost systems characterized by high interconnection density. A relationship between the active silicon area of devices on a substrate and the interconnection capability of the substrate is presented.

I. INTRODUCTION

THE CONCEPT of decreasing interconnection cost as the interconnection medium approaches the chip level has become well-known and generally accepted. The authors will examine this concept and show that this decrease in interconnection cost is entirely due to decreases in interconnection lengths. Interconnection cost is essentially invariant on a per unit length basis.

Fig. 1 is the late C. T. Goddard's compilation of interconnection cost as a function of interconnection capability, which was originally proposed by D. E. McCumber [1]. It gives the impression that silicon technology, besides being a very powerful (in terms of density) way to make wires, is a very cheap one. Unfortunately, this plot confuses two issues—the cost of making wires on a particular substrate, and the length of the wires required to connect devices on that substrate.

This paper provides a more accurate rationale for continued progress in the scale of integration (SOI) of integrated circuits, as well as a demonstration that new interconnection approaches are needed to deliver the system advantages which those silicon advances promise.

II. HIERARCHY

Consider a hierarchy of interconnection substrates used for connecting smaller devices. For the purpose of this paper, we consider the chip itself as a substrate used to connect polycells, or other cell-like elements. The hybrid connects bare chips, and the printed wiring board connects packaged IC's. For all "substrates," we are concerned with the interconnection capability, and the interconnection cost.

Manuscript received September 1984.

W. H. Knausenberger is with AT&T Bell Laboratories, Whippany, NJ 07981.

L. W. Schaper is with AT&T Bell Laboratories, 600 Mountain Avenue, Murray Hill, NJ 07974.

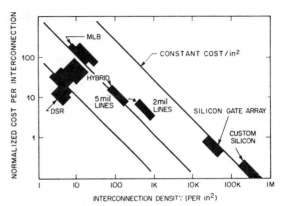

Fig. 1. Cost and density of interconnections.

III. INTERCONNECTION CAPABILITY

Goddard measured interconnection density in interconnections per square inch. Unfortunately, this figure of merit confuses two things: the intrinsic interconnection capability of a substrate, and the size of the devices or packages which are to be connected. Small devices (i.e., polycells) will obviously have many interconnections per square inch: the wires will be short as well as dense. Conversely, large packages such as DIP's will require longer wires. Thus, simple interconnection density is not a good measure of substrate capability.

We choose to use the figure of merit of inches of interconnection available per square inch of substrate as a more valid measure of interconnection capability. This measure is directly related to substrate technology, not to the size of the devices to be connected. As we shall show later, however, there is a strong correlation between the two.

IV. COST

Goddard's cost measure was cents per interconnection. Again, this figure is strongly related to device size, since size will influence interconnection length. We use as cost measure the cost per inch of available wire as a way to determine the relative cost of making wires in the various technologies.

V. COMPARISONS

We compare five interconnection technologies: double-sided rigid printed wiring boards (DSR), multilayer printed wiring boards with surface mounted (MLB-SM), and thru-hole mounted components (MLB-TH), ceramic hybrids, and 2.5 μm CMOS chip technology (2.5 μm). The assumptions used in the comparison are listed in Table I. The area derating term is the fraction of substrate area available for wiring. Packages or

Reprinted from *IEEE Trans. Components, Hybrids, Manuf. Technol.*, vol. CHMT-7, no. 3, pp. 261–267, September 1984.

11

TABLE I
COMPARISON ASSUMPTIONS

Substrate	Area inch²	Design Rule and Layers	Potential inch/inch²	Area derating factor*	Actual inch/inch²	Relative Cost
2.5μm	0.056	3μm Line + Space-2 Layer	8466	0.65	5503	1.1
HIC	5.76	5 mil Line + Space/2 Sides	200	~0.9	180	1.7
MLB-SM	104	8 mil Line/9 mil Space 4 signal layer + pad layer 3 track, 30 mil via lands	235	0.8	188	0.8
MLB-TH	104	8 mil Line/9 mil Space 4 signal layer 2 track, 60 mil lands	235	0.6	140	1
DSR	104	12 mil Line + Space 2 sides, 1 track 60 mil lands	83	0.6	50	0.5

* The fraction of substrate area which is available for placement of wiring tracks. (Based on the authors' best estimate.)

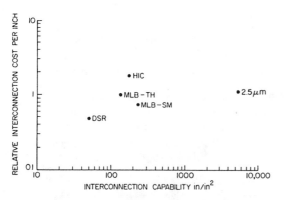

Fig. 2.   Relative interconnection costs.

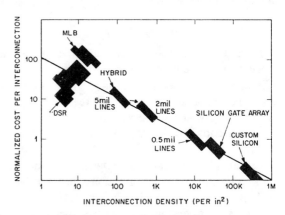

Fig. 3.   Constant cost per inch trend line.

chips block areas or interfere with wiring tracks; thus, the difference between potential and actual available wiring density.

Fig. 2 plots relative cost per inch of wire against the capability, in inches of wire/inch² of substrate. Even though the capability range is over two orders of magnitude, the range of relative costs is very small. In effect, wire costs the same per unit length, whether on printed wiring boards or on silicon. The importance of high density wiring is that it allows the total length of wire required to interconnect the active devices in the system to decrease; cost is directly proportional to wire length.

Reexamination of the plot in Fig. 1 reveals that all the data follow a trendline with a slope of 1/2 (see Fig. 3). The dimensions of this line make it clear that this is again a statement that interconnection costs are constant on a per unit length basis.[1]

VI. POTENTIAL VERSUS USABLE WIRE

For economic reasons, there must be a match between any substrate's wiring capability and the interconnection requirements of the packages placed on that substrate. If packages or

[1] Observation by D. E. McCumber.

devices with high interconnection demands are placed on a low capability substrate, routing will be impossible, or the devices will have to be spread out over a much greater substrate area.

Conversely, putting packages with low interconnection demand on a high density substrate is uneconomic, because expensive substrate area (therefore a great deal of wiring capability) is not being used.

Optimal packaging strategies match the interconnection capability of the substrate to the wiring demand placed on that substrate by the devices mounted thereon.

VII. A DEMAND MEASURE

A useful measure of wiring demand is what we call the active fraction F, defined as the fraction of the substrate area occupied by active devices (active areas on the chips). Thus we relate wiring demand to, in effect, total gate area over an entire substrate. This should hold for random logic, interconnect dominated systems.

Table II gives the values used to compute F for the technologies considered. These are plotted in Fig. 4, which shows the high correlation between wiring demand and the appropriate interconnection capability to support that demand. (The HIC point lies far off the line because the particular HIC used for comparison mounts power IC chips,

TABLE II
FRACTION OF SUBSTRATE AREA OCCUPIED BY ACTIVE DEVICES

Substrate	Subs. Area inch^2	Devices	Typical Number of Devices	Area/Device inch^2	Total Si Area	Silicon F	Total Active Area inch^2	Overall F
2.5μm	0.056	—	—	—	0.056	0.35	0.0195	0.35
HIC	5.76	Power device in chip carrier	16	0.0164	0.262	0.9		
		latch	8	0.0064	0.051	0.5	0.267	0.0464
		buffer	8	0.0009	0.0072	0.7		
MLB-SM	104	28 lead chip carrier MSI & SSI	240	0.0125	3.00	0.4	1.2	0.0115
MLB-TH	104	DIPS MSI & SSI	144	0.0125	1.8	0.4	0.72	0.0069
DSR	104	DIPS—SSI	60	0.0034	0.204	0.6	0.122	0.0012

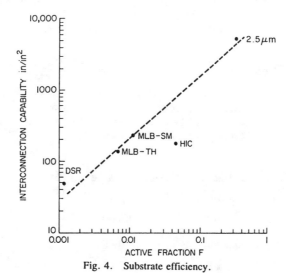

Fig. 4. Substrate efficiency.

with an active fraction of ~0.9 on both sides of the ceramic substrate, doubling the effective density of devices.)

VIII. CONCLUSION

We conclude that no fabrication technology offers cheap wiring; this is a myth. Interconnection costs are essentially invariant on a per unit length basis. Denser technologies, however, can lead to the use of shorter wires, and thus real savings. The tremendous gap in capability between printed wiring boards and silicon argues for the intensive development of higher capability substrates to accommondate the ever increasing densities available on silicon devices.

REFERENCE

[1] C. T. Goddard, "The role of hybrids in LSI systems," *IEEE Components, Hybrids, Manuf. Technol.*, vol. CHMT-2, p. 367, Dec. 1979.

FUTURE OF MULTICHIP MODULES IN ELECTRONICS

BY

DR MAURICE G SAGE
BPA (TECHNOLOGY & MANAGEMENT) LTD
DORKING, SURREY, UK

Existing IC packaging and interconnection is creating a barrier for advances taking place in IC technology. Wafer scale integration (WSI) is being proposed as the next major step in electronics, bringing with it the removal of a large number of these barriers. However there are still major problems with WSI technology that are unlikely to be solved before the late nineties. In the meantime a new packaging and interconnection technology will be introduced - the multichip module (MCM), which will act as a hybrid for WSI. The MCM will not make the PWB obsolete, but act as an important space transformer between the IC die and the next level of interconnection. This paper looks at the importance and place of the MCM, predicting typical costs, applications and demand.

THE MOVE TO MULTICHIP MODULES (MCMs)

Existing packaging and interconnection technology is not complimenting the advances being made in the IC. In particular the trend towards higher pin-count is demanding new forms of IC packaging which result in ever larger footprints. This is illustrated in Table 1 below.

Table 1

FOOTPRINTS OF VARIOUS IC PACKAGES

Type	Dimensions inches	Area inches	Pin Spacing
20 pin DIP	0.4 x 1.0	0.40	100 mil
64 pin DIP	1.0 x 3.0	3.00	100 mil
64 pin LCC	0.5 x 0.5	0.25	25 mil
132 pin LCC	1.22 x 1.22	1.50	25 mil

MS 402

From Table 1 the move from a 20 pin DIP to a 64 pin DIP increases the footprint by 7.5 times. This situation is eased by a factor of twelve by moving to the LCC (pin spacing 25 mil) but as soon as the pin-count goes to 132 pins there is an increase in the footprint of 6 times. The continuing growth in pin-count has created a move to surface mount technology (SMT) which is even now pointing towards the use of unpackaged ICs, many of which will be used in thin film multilayer interconnect structures known as multichip modules (MCMs).

Is this intermediate step necessary, why not move straight to wafer scale integration (WSI)? The move to WSI is still some way off, since a number of major technical problems have yet to be solved and the move to MCMs, is essentially that of a hybrid for WSI. It will perform a similar function to the original thick film hybrid ie a collection of die substituting for a single IC.

The MCM uses thin film multilayer technology utilising organic dielectrics such as polyimide. Manufacture involves a number of process techniques borrowed from the integrated circuit industry. Line widths will be reduced to typically 10 microns with thin dielectrics that will enable the necessary controlled impedance value to be obtained, as well as increased density of interconnection with reduced crosstalk.

Figure 1 below shows the BPA world forecast of demand for different IC packages through to 1997. Essentially there are three generic types ie packages for insertion, packages for SMT and a third sector covering TAB, chip on board and flip chip.

<u>Figure 1</u>

WORLD IC PACKAGE TYPES 1987-1997

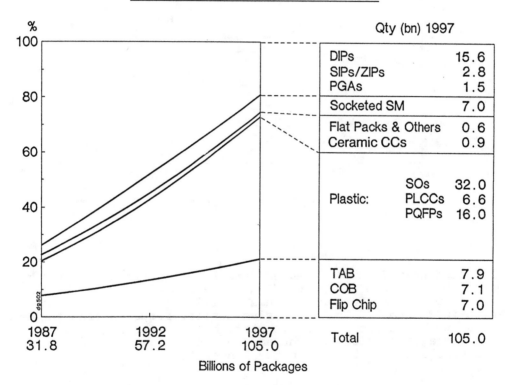

Billions of Packages

These three distinct sectors are plotted in Figure 2 with the lower curve indicating the area in which the MCM will emerge as a major and new IC packaging and interconnection technology.

Figure 2

IC PACKAGING SUBSTITUTION CURVE

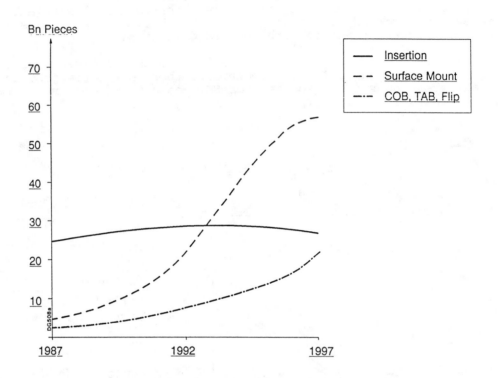

The need to meet the increased density of interconnection that will be required between the advancing IC and the PWB will be met by the MCM acting as a space transformer and this is illustrated on the cost/density axis shown in Figure 3. The increasing density required from the PWB is giving rise to higher costs, but is still holding its own against the MCM. However advantages of size, weight and reliability, even the for the type 1 MCM, will ensure MCM viability in the longer term.

Figure 3

PRICE DENSITY RELATIONSHIPS

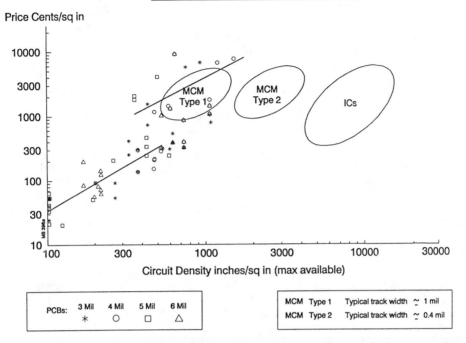

Note: An additional diagram has been added to the original paper covering density and numbers of layers, see Appendix 1.

A figure of merit is derived for various technologies using the propagation speed in inches/p sec times the interconnection density, typical figures are given in Table 2. The figure of merit for VLSI will need to be partially adjusted (down) to take into account line resistance and capacitance which form a CR network. This effect is not significant in MCMs due to higher line conductivity, lower drive currents and lower output capacitance from the drivers.

Table 2

FIGURE OF MERIT SPEED - DENSITY

Technology	Figure of Merit (ins/p sec . density ins/sq in)
VLSI	28.0
MCM	14.6
PWB	2.2

Advantages of Multichip Modules

The advantages of MCMs are significant and listed below:

. Smaller size   -   between a 5 to 10 reduction in area
                              against SMT solutions.

. Higher reliability   -   a vital objective due to the increasing
                              functionality of electronic products.

. Improved electrical   shorter lines between ICs and less heat
  performance   -   dissipation

The cost/square inch of a 5 metal layer MCM is not yet competitive against advanced PWBs. However cost savings in MCMs are obtained from factors such as improved reliability, size, performance. As the MCM is a new technology and development is mainly concentrated on the product, lower costs will not occur until the this effort moves to process development, with further reductions obtained from the experience curve.

The Demand for Multichip Modules

The MCM is a new technology and is not simply an extension of thick film hybrids or simply the application of well proven IC processes. It will take time for the technology of MCM to be accepted in the market place, but when established it is most likely to be in use for twenty years or more. It is interesting to reflect on the fact that thick film hybrids and PWBs have been in existence for nearly forty years.'

The current worldwide usage of circuitised substrates for MCMs is about $50m. Companies currently using MCMs in their own products or marketing them externally have all been involved in the development of this technology for about five years - leadtimes are not short. The current product development activity in MCMs is considerable, with over 100 companies worldwide participating. The next stage after product development is process development and this has yet to take place on any large scale.

In terms of applications, MCMs will find use across the whole electronics industry for example:

Military   -   the advantages of reduced size, weight,
                      improved performance and reliability are
                      significant

Consumer   -   in Japan the use of MCMs are being developed
                      to replace thick film hybrids. Size
                      reductions of around 10 to 1 in area, with
                      lower costs than functionally equivalent and
                      existing thick film hybrids are reported. In
                      the auto industry the need for higher
                      reliability combined with CPUs running at
                      100MHz in the mid nineties is creating
                      considerable interest in the MCM.

Computers	-	higher performance in CPUs and cache memory are prompting considerable development effort. Japanese computer companies are already using the technology in their products.
Telecomm/Comms	-	MCMs will find applications in Main Switches and Transmission Multiplexers. The demand is very dependant on the introduction of ISDN particularly the broadband services and this will vary throughout the world, with Japan and W Europe most possibly leading the way. Opportunities exist for MCMs in hand-held and pocket communication products.
Integrated Circuits	-	the yield problems of large die from probed wafers is creating a move to partitioning and rebuilding on silicon MCMs. Some companies have looked at restricting die size to 8mm/side with a view to adopting a flexible hybrid approach via the MCM.

Using a variety of forecasting approaches the MCM market for circuitised substrates in 1997 is estimated at $7bn - the PWB market alone at this time will be in excess of $38bn.

A New Technology

The MCM offers significant performance and cost advantages. However it is a new technology and its full introduction will take time, with take off forecast for the early nineties. A list of advantages and cautions are given in Table 3 below. With all new technologies there will be plenty of 'good reasons' to downgrade it as many of the vacuum tube manufacturers were eventually to find to their cost in avoiding solid state.

Table 3

MCMs - ADVANTAGES AND CAUTIONS

Advantages	Cautions
• Smaller size	• A new technology
• Reduced weight	• Design and test problems
• Increased interconnection density	• Material and process problems
• Lower heat dissipation	• Not simply IC technology
• Improved reliability	• Connector problems
• Ability to mix technologies	• PCBs will compete for many applications
• Repairability	• Not an extension of thick film hybrids

INTERCONNECTION DENSITY VERSUS LINE TECHNOLOGY

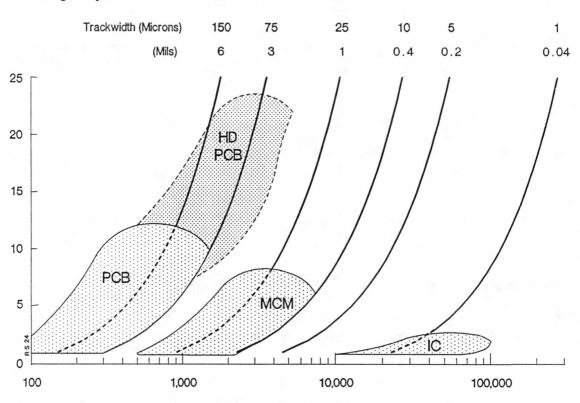

Interconnection Density inches/sq in (max available)

Cost–Density Analysis of Interconnections

GEORGE MESSNER, MEMBER, IEEE

Abstract—A very rapid increase in the conductor density of various interconnecting substrates makes it desirable to have a uniform method for evaluation of their relative density capabilities. The derivation of one such method of evaluation, based on conductor density per unit of total area of the substrate, is made. This method permits a uniform density analysis of the entire interconnection spectrum from printed to integrated circuits. It is also useful in derivation of a price–density analysis and other figures of merits for assisting in the comparison of various packaging techniques. A description of a generalized approach for development of such a graphic price–density analysis is provided.

INTRODUCTION

WITH THE very rapid increase in the number and density of active elements placed on a single chip, there is a corresponding increase in the number of input/output (I/O) terminals needed for their operation. This has resulted in an unabating pressure to increase the pathway density on the substrates which interconnect these complex large-scale integration (LSI) and very large-scale integration (VLSI) chips. Terminal counts on these chips are today reaching 200 and are expected shortly to go beyond 400.

There were several approaches which attempted to provide a yardstick to evaluate, on a uniform basis, the performance capabilities of various interconnection techniques used to interconnect these high I/O IC packages. This paper describes a derivation of one such potentially universal yardstick, which is based on connectivity or on conductor density per unit area. This method permits a uniform density analysis spanning and uniting the interconnection technologies from printed circuit boards through integrated circuits themselves.

DERIVATION OF UNIFORM CONNECTIVITY MEASUREMENT

There is a constant need to compare in a uniform manner the interconnection density provided by various substrates or to estimate the wiring lengths needed to interconnect such multi-I/O packages. As shown in Fig. 1, there have been a number of attempts made to define such packaging or interconnection (*P/I*) substrate connectivity.

Additional examples could be given, but these will suffice to show a variety of approaches, all of which interestingly result in numerically close values for selected examples of connectivity.

The purpose of this paper is to propose as a universal definition of connectivity a much simpler and a more

Don Frank (IBM):

$$\text{PC complexity} = \frac{(\text{Lines/Chan.})\,(\text{Layers})}{(\text{Hole c-c})\,(\text{Line Width})}$$

Wm. Loeb:

$$\text{Density factor} = \frac{(1 + \text{Lines/Chan.})\,(\text{layers})}{(\text{grid})}$$

Petschauer (Sperry):

$$\text{Tracks/in} = \frac{(K)\,(\text{Pins/Chip})\,(Q \text{ of Chips})^2}{\text{Spacing}}$$

Fig. 1. Connectivity definitions [1]–[4].

Tracks/in = Lines/in =
= in of cond/in² of substr. area
Lines/cm = cm of cond/cm² of area.

Fig. 2. Definition of connectivity.

straightforward approach. As will be shown, this connectivity measure also provides a greater number of opportunities for the evaluation of performance and rating of various *P/I* substrates. The proposed definition is given in Fig. 2.

This connectivity definition describes the total signal wire channel capacity provided by a given substrate type. It is very easy to establish: it only requires that one count the numbers of conductors located in a 1-inch length on a single level of substrate. Since these conductors could come on various levels from different directions, the number of 1-inch segments of these wiring channels are multiplied by the number of levels to finally obtain the total connective capacity of a given system, expressed in inches of wiring per square inch of substrate. (In metric countries, centimeters/centimeters² can be substituted for the inch values by dividing them by 2.5.)

In the printed wiring board (PWB) industry, for holes on 0.100 centers, it is customary to talk about two/between and three/between technologies as shown in Fig. 3. In such cases, a single level will have a corresponding 20 or 30 inch/inch² wiring and, depending on the number of signal wiring levels, the total connectivity of a multilayer board (MLB) structure will be a multiple number of these layers. Table I gives examples of wiring capacity or the connectivity analysis for various PWB structures (PTH and SMT) based on the use of uniform conductor and space widths.

DERIVATION OF MEASURE FOR REQUIRED WIRING LENGTH

This measure of channel capacity or available wiring density is simple and quite useful. It does not, however, answer the second need: to establish an approximation to the working length required for the interconnection of various

Manuscript received November 21, 1986; revised March 9, 1987. This work will be presented at the Printed Circuit World Convention IV, Tokyo, Japan, June 1987.

The author is with PCK Technology, 322 L.I.E. South Service Road, Melville, NY 11747.

IEEE Log Number 8714729.

Reprinted from *IEEE Trans. Components, Hybrids, Manuf. Technol.*, vol. CHMT-10, no. 2, pp. 143–151, June 1987.

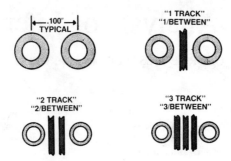

Fig. 3. "Tracks between" technologies for holes on 0.100 centers.

TABLE I
CONNECTIVITY OF PWB*

Line/space width, mils	Hole dia: 0.030 inch PTH Hole c-c: 0.100 inch	0.013 inch PTH (SM) 0.100 inch	0.050 inch	Via = cond
0.012 inch	10	20		30
0.010 inch	10	30		40
0.008 inch	20	40		50
0.007 inch	20	40	20	60
0.006 inch	30	50	40	70
0.005 inch	40	60	40	90
0.004 inch	50	70	60	120
0.004 inch (wire dia.) Multiwire	60	80		

* Number of conductors/inch/level.

electronic components and devices. Such a derivation (based on empirical analysis) has been made by Dr. D. Seraphim of IBM [5]. His equation is shown below and provides a value for the required length of wiring L per chip site:

$$L = 1.5P \left(1.5 \frac{N_T}{2} \right) \qquad (1)$$

where

L total wiring length/site;
P pitch between packages;
N_T total number of I/O pins.

The derivation of this equation is rather simple and logical. If you analyze Fig. 4, you see that the centrally located component has connections going to neighboring locations for an average distance P, or to the next one beyond, for an average Manhattan distance of $2P$. Averaging the two gives the $1.5P$ of the above equation. Since in a number of cases, the average net has three connections to four terminals, the average number of connections will be 3/4 N_T or 1.5 $N_T/2$, as given in (1).

The above equation, therefore, provides the necessary information about the required length of wiring for the interconnection of these devices. In his paper [5], however, Dr. Seraphim has observed that it is impossible to fill with interconnections every available wiring channel. He has empirically established that, at best, there is a 50 percent efficiency in actual wiring. Thus the maximum available wiring capacity (connectivity) must be at least twice the length of the necessary or actually required wiring (wiring demand), as established by (1). In other words, for $P = 1$, the length of

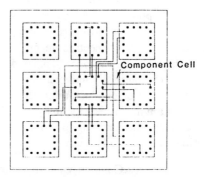

Fig. 4. Wiring density derivation.

required interconnections is $Lr = 1.125 Nt$. To achieve this, the wiring capacity (or connectivity) must be $2L$ or $Lc = 2.25 Nt$ to ensure that all of the required wiring demand will be accommodated. Therefore, attention must always be paid as to which type of density a reference is made: is it the provided *wiring capacity*, (connectivity), or is it the *wiring demand*, to interconnect the selected component (which is at best half of the first)?

USE OF THE WIRING DEMAND AND WIRING CAPACITY VALUES

Through the application of either of the two values, or both, it is now possible to perform a variety of useful design and trade-off tasks. You can establish the wiring necessary to be provided for interconnection of selected device packages on a substrate. You can evaluate if the selected packaging method is suitable to accommodate the necessary wiring demand. You can compare complexity of interconnecting substrates, and so on.

TABLE II
CONNECTION DENSITY [6]

Pads	DIP[a]	Chip Carriers[a]	TAB[a]
(#/comp.)	in² of substr.		
16	40	80	200
44	43	123	309
68	44	139	347
156	—	160	399

[a] Inches of wire per one square inch of substrate.

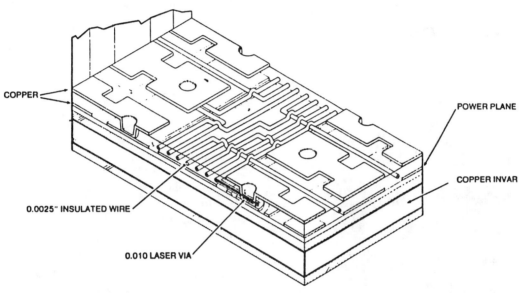

Fig. 5. Microwire.

For example, for the specific cases of square packages with peripheral terminals, normalized to 1 square inch, J. Welterlen [6] has extended the Seraphim equation as

$$L/in^2 = \frac{2.25N}{\frac{NS}{4} + 5S}.$$ (2)

where

S is the terminal spacing

Since he uses a value of 2.25, it indicates that (2) provides a value for the wiring capacity needed to interconnect the selected chip carriers. Based on the above equation, he has established a table of wiring channel (connectivity) requirements for a variety of IC package types, based on their terminal pitch. These are given in Table II.

Using this approach, one can now approximate the interconnection densities needed to accommodate a variety of packages.

Table III shows how (1) is used at PCK Technology, where a discrete wiring product called Microwire™ is manufactured (see Fig. 5). The table is constructed using the Seraphim equation, but Nt is derated to account for power and ground terminals.

TABLE III
INTERCONNECTION DENSITY REQUIREMENTS[a]

Packages per in²	Wiring Density	Required Circuit Capacity
20 Pad LCC 0.050 in Pitch	20.25 in/in²	40.5 in/in²
40 Pad LCC 0.050 in Pitch	40.5 in/in²	81 in/in²
124 Pad LCC 0.025 in Pitch	126 in/in²	252 in/in²

[a] Assume that 10% of the total I/O are used for power and ground.

It has been our experience (at PCK Technology) that the wiring lengths and densities which have been achieved by Microwire for interconnecting various chip carrier designs agree very closely with precalculated values using (1) and (2). For instance, Fig. 6 shows the top layer of a typical Microwire board and its interconnecting pattern. As calculated, the wiring density of this part was supposed to be 95 inch/inch²; the actual wiring density measured on the finished board was 93 inch/inch², a fairly close correlation.

Therefore, it could now be postulated that use of this connectivity analysis method can result in a useful, quick, and practical analysis of wiring requirements for selected packages or give reasonably accurate estimates of the signal layer and

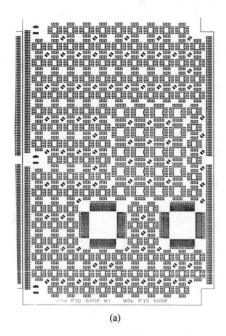

(a)

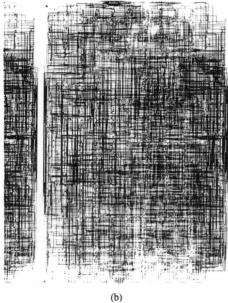

(b)

Fig. 6. Typical Microwire board. (a) Top layer. (b) Interconnecting patterns.

channel capacity required to accommodate specific wiring densities. It can also serve as a simple universal yardstick to compare various packaging and interconnection (P/I) substrates, as shown in Table IV which is based on the assumption of equal widths for conductors and spaces.

EXTENSION OF THE CONNECTIVITY CONCEPT

This particular figure of merit can also be combined with other elements to provide a much broader analysis or a better yardstick for a specific purpose.

Today, for instance, the question of electrical performance is very important for the operation of high-speed systems. It is the desire of packaging engineers to preserve through all packaging levels the high-speed capabilities of the IC's used in the system. One of the basic parameters which affects signal

TABLE IV
CONNECTIVITY OF VARIOUS INTERCONNECTS[a]

Conductor Width		Connectivity	
mils	microns	Level	Technique
10	250	50	Normal Thick Film Hybrid
5	125	100	Best Thick Film Hybrid
4	100	50–70	Best PTH PCB's
4[b]		60–80	Multiwire
3	75	165	Ceraclad
2.5[b]		160	Microwire
2	50	250	Thin Film on Polyimide

[a] Number of conductors/inch/level.
[b] Measured in diameter.

propagation through a system is the dielectric constant of the substrates, which are related by

$$Vp = \frac{C}{\sqrt{K}} \text{ m/s} \qquad (3)$$

where

Vp propagation velocity
C speed of light (3×10^8 m/s)
K relative dielectric constant
  for air $K = 1$
  for epoxy glass $K = 4$–5
  for ceramic $K = 8$–10
Vp (air) 0.3 m/ns = 0.98 ft/ns.

For instance, P. McEnroe from Trilogy [7] has used connectivity (lines/centimeters) divided by signal delay (pS per centimeter) to arrive at a figure of merit useful for the cases where speed of operation is of importance. Using this method, he had rated various interconnection approaches as shown in Table V. Similar derivations for other useful figures of merit are possible.

Another set of useful information based on connectivity analysis has been developed by W. H. Knausenberger and L. W. Schaper of AT&T [8] in a paper given at NEPCON 1984, wherein they have plotted connectivity versus cost of interconnection per inch as shown in Fig. 7.

From an analysis of this graph, they came to the conclusion that the cost per inch of interconnection—on a multilayer board, on a hybrid, or on an IC chip—is basically the same, if analyzed per unit length. It was a very intriguing, but as you will see later, a somewhat premature conclusion.

In their further work [9], they have defined a new term, *substrate efficiency*, in which they were attempting to determine how much active silicon could actually be placed on a given substrate system. Their figure of merit F is given by

$$F = \frac{\text{active silicon area (cell area)}}{\text{total substrate area}}. \qquad (4)$$

Using that active fraction F as the x-axis and moving the interconnection capability (or wiring capacity) onto y, they have devised a new graph (Fig. 8), which ranks various interconnection approaches based on substrate efficiency.

TABLE V
INTERCONNECT FIGURE OF MERIT [7][a]

PC Board	48/80 = 0.6
VLSI Chip	1250/1000 = 1.25
MLC	160/125 = 1.28
Trilogy Thin Film	200/100 = 2.0

[a] (lines per centimeter)/pS per centimeter).

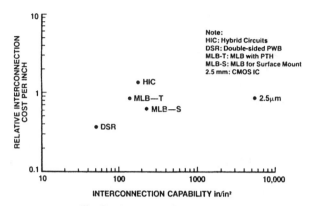

Fig. 7.   Interconnection measures.

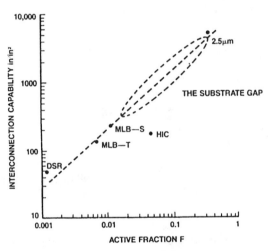

Fig. 8.   Substrate efficiency.

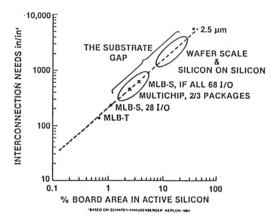

Fig. 9.   Substrate efficiency for multichip models and interconnections and silicon wafers.

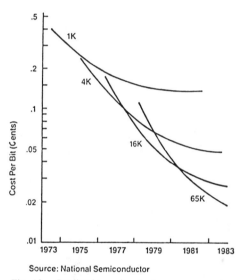

Fig. 10.   Decline in semiconductor memory costs.

As seen from Fig. 8, most of the interconnection substrates fall on a straight line. What becomes immediately obvious from this graph is that between the interconnections made on silicon chips and interconnections made on complex multilayer boards, there are no intermediate interconnection methods available, i.e., they have established the "substrate gap." It is a quite powerful concept, and today a great deal of activity is concentrated in development of interconnection systems which can fill this gap. We will return to a discussion of candidates to fill this gap toward the end of my presentation, but Jack Balde in his presentation [10] at the World Printed Circuit Conference in 1984 has already indicated two candidates for that region: multichip modules and interconnections on silicon wafers, as shown in Fig. 9.

FUNCTIONAL PERFORMANCE OF IC'S VERSUS PWB'S

The methodology described above has permitted us to take a much more general and universal view of interconnections through the application of channel connectivity measurements and has opened opportunities for wider areas of comparisons.

It has been a unanimous observation that the wide penetration of IC's into electronic equipment is due to the unrelenting progress that has been made in improving their cost–function performance. That situation is exemplified in Fig. 10, which shows price–function relationships of memory chips between 1973 and 1983.

Such performance was achieved by constant technical improvements of the processes, which kept reducing the IC feature sizes, and by manufacturability improvements, which continuously increased yields on wafers.

But as shown in Fig. 11 [11], [12], the rate of feature size reduction in the manufacture of IC's was not significantly better than the reduction rate of conductor widths on printed boards. Granted, the features are 50 times greater on circuit boards, but the board sizes and other conditions are certainly different from silicon wafers. Still, the technical progress of the PWB industry, as indicated by the typical conductor widths manufacturable at leading-edge facilities, has been quite respectable during the past 20 years.

It has always been assumed that the least costly interconnec-

25

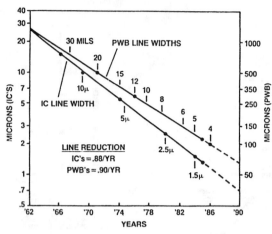

Fig. 11. Rate of feature size reduction in the manufacture of IC's and PWB's.

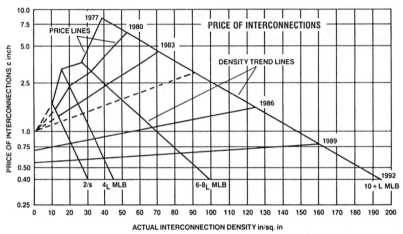

Fig. 12. Plot of typical price–density analysis of PWB's.

tions are those made on the chip, becoming so because of the continuous technological and manufacturing progress of IC's. But as we just saw, the technological progress in PWB technology is comparable.

This leads to a legitimate question: if the rate of technological progress is not that different between IC's and PWB's, is progress in the cost–function of PWB's also similar to that for IC's?

PRICE–DENSITY ANALYSIS OF PWB'S

In an attempt to develop an answer to the above question and establish a more universal price-function relation for PWB's, the approach of Knausenberger and Schaper, shown in Fig. 7, to plot wiring density on one axis and price per unit length of conductors on the other, is extremely interesting and has been pursued further by this author. The graph in Fig. 12 shows a plot of a typical price–density analysis. Such a plot has enabled us to derive a more structured and systematic view of usually chaotic price–density relationships in the printed wiring area.

What emerges first from this graph is recognition that there are four basic PWB types, differentiated by their price–density relationships. The values for these four typical PWB types were arrived at by analyzing the general average conductor density they have actually provided, and the price levels typical for each board type when produced in large production

volumes in the years 1977, 1980, and 1983. The plot also indicates that, while technical improvements kept increasing the density capabilities of these four distinct types of PWB's, the learning curve experience was simultaneously instrumental in the gradual price reduction of these board types, when measured in cents per inch of wiring.

This price movement was apparently hinged at the fulcrum located around the 1¢/inch price of interconnections. That pivoting effect is quite rational, since for lower board densities, where the material content is dominant, changes in pricing and technical progress is more limited and slower than in higher density boards where technological improvements have a multiplying effect.

Therefore, following the price situation between 1977 and 1983, it was prudent to expect that further movement in prices would occur around that pivot, as indicated by the dotted line in Fig. 12. Unfortunately, the price dislocation of 1985 has moved the pivot itself downward, and it is not yet clear where its downward movement will stop. The 1986 prices per inch of interconnects could be plotted easily because the slope of the price line did not change appreciably—only the pivot moved. This pricing line represents a level reached in the U.S. at the beginning of 1986 in the period of greatest price erosion. However, no price rebound is yet evident in the U.S. due to continuously depressed sales levels.

26

Since the 1985 price situation had not materially changed the technological factors, the projections of these trends could be plotted as straightline extensions of the 1977–1983 lines. When that is done, as indicated on the graph in Fig. 12, an interesting situation emerges: at some future time, arbitrarily set here in 1992, there could be a condition where the price for one inch of interconnect will be uniform regardless of on what type of PW board it is located. In the above graph this occurs at the point of 0.4¢/inch of interconnect. It is not a definite law but a very intriguing possibility and 0.4¢/inch represents a very interesting number, as you will see later. It also indicates that the statement made by Knansenberger and Schaper in 1984 about the uniformity of interconnect price per unit length may become true sometime in the early 1990's.

Note also that the *x*-axis in this plot represents the wiring demands (not the wiring capacity). Also, you can see that densities for four types of PWB's, as plotted for the year 1986, can be easily accommodated by the wiring capacity (connectivity) provided by three/between technology, and density plotted for 1992 by five/between technology. The three/between wiring capacity is a reality today and five/between probably will be by 1992. Thus these plots seem to produce a reasonably accurate technological forecast.

If we run a vertical line on this graph at any desired density, intersecting the plots of the year lines of the above graph (and suitably multiply the values), we can replot the information on a different graph which will relate the price per square inch of printed board substrates versus their density for any given year. Such a plot is given below in Fig. 13.

In this graph, the *y*-axis has the more familiar units which are dollars per square inch of PWB substrate but there are two values for *x*. The top one is the same as in the previous graph and shows the wiring demand in inch/inch² of a board. The line below is doubled in density and represents the necessary wiring capacity which must be provided on a given board to ensure that the desired wiring demand (as given by the top line) is achievable. Thus this bottom line is the measure of connectivity.

Analysis of the above graph does not lead to many surprises. The gap between the 1983 and 1986 price plots represents the price erosion of 1985. The 1989 and 1992 plots are somewhat arbitrary since, as mentioned above, all future pricing depends on the movement of that pivot. The 1992 curve represents the specific case of uniform price at 0.4¢/inch of actual wiring (or 0.2¢/inch of provided channel capacity) as discussed above.

The people who have analyzed the U.S. MLB pricing and have compared it to the above 1986 price plot agree that it generally reflects the existing U.S. pricing structure in 1986. It should be remembered, however, that this graph is a generalization based on a derivation and therefore is only an approximation. But still it gives a reasonably good ballpark pricing level for 1986.

Derivation of Universal Price–Density Plots

As indicated in the beginning, there has always been a strong desire to establish a system which could provide a clear ranking of various interconnection technologies and also point

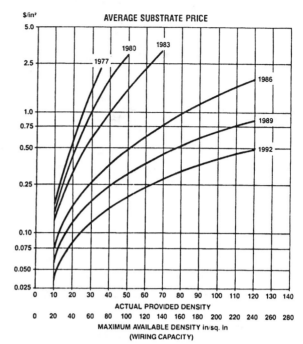

Fig. 13. Plot of the price per square inch of printed board substrates versus density for any given year.

out trends and directions of electronic packaging developments. For that purpose, and partly influenced by the previous price–density plot, BPA, a British consulting firm, has constructed an interesting graph locating various types of interconnection structures on a similar price–density plot, mainly using 1984–1985 worldwide pricing levels, as shown below in Fig. 14 [13].

In the plot of Fig. 14, in addition to ranking the various levels of complexity for PWB's, MLB's, and hybrids, we also see a reappearance of the "substrate gap," as well as the location of connectivity requirements for various types of IC packages as shown previously in Table II as derived by J. Welterlen.

In order to put the entire spectrum of interconnections, starting from simple print and etch one-sided boards and going to sophisticated integrated circuits, into a single graph, the above BPA graph is replotted on log/log paper, which also straightens the lines. The solid line on the graph of Fig. 15 is the replotted BPA data from Fig. 14. The two dashed straight lines are replots of my cost curves for 1986 and 1992 from Fig. 13. This last one represents that specific case when all actual interconnections will be uniformly priced at 0.4¢/inch of their length, regardless of on which type of PWB they are located. (Note again that 0.4¢/inch of actual wiring is equal to 0.2¢/inch of available channel density.)

The location of two emerging technologies destined to bridge the "substrate gap" should be noted, as well as the price–density area, where the IC chip interconnects are operative.

The *x*-axis here, as in Fig. 14, is connectivity, i.e., maximum available wiring capacity.

An examination of the above graph shows that this 0.2¢/inch line reaches directly into the IC region. Calculating, we find that 2.5 μm design rules and two metallization layers

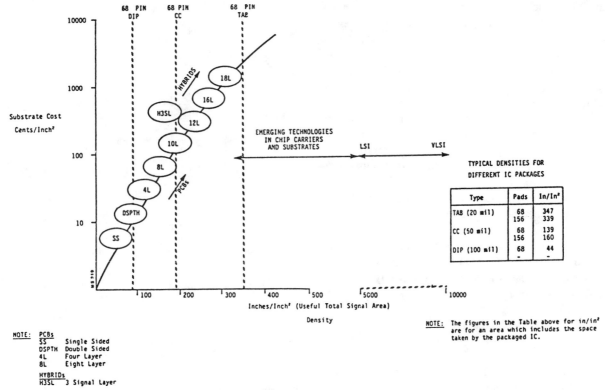

NOTE: PCBs
SS   Single Sided
DSPTH   Double Sided
4L   Four Layer
8L   Eight Layer

HYBRIDs
H3SL   3 Signal Layer

NOTE: The figures in the Table above for in/in²
are for an area which includes the space
taken by the packaged IC.

Fig. 14.   Cost versus interconnection density.

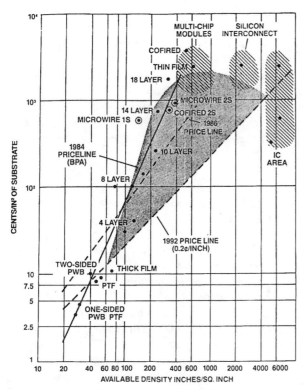

Fig. 15.   Price–density relationships.

the low-density area, the one-sided and two-sided printed board area. Accordingly, 25 inch/inch² of conductors on a one-sided board should be costing 5¢/inch²; on two-sided boards the 50 inch/inch² wiring capacity should be sold for 10¢/inch². Both channel densities are achievable by three/between technology, which is already here. The indicated pricing levels for such boards, if you examine the graph, already exist in various parts of the world, but possibly not yet in the U.S.

We see, therefore, that this uniform cost line of 0.2¢/inch is actually being experienced today in the lowest—and highest—density elements of the interconnection spectrum. In other words, if you interconnect an uncased chip on a single-sided or two-sided PW board, you probably are achieving the lowest packaging cost possible with present technology. This explains the great interest in such forms of component packaging for the low-cost mass-production consumer market.

As one climbs in density levels from two-sided boards toward more complicated interconnections, a "price bulge" appears which progressively raises the prices of more complex interconnections above the 0.2¢/inch levels (the shaded area in Fig. 15). For four-layer MLB's, it is twice or 0.4¢/inch; for 10–12 layers, it is around 1¢/inch. On 14-layer boards, interconnecting cost is an order of magnitude higher (i.e., 2¢/inch), and so on. It is interesting to note the levelling of this price rise in the area of multichip modules, where IC manufacturing technologies are being employed for the next level of interconnects.

The reason for the low pricing levels at the extremes of the density spectrum is mass production, where the cumulative experience levels and the use of advanced and highly

would provide 10 000 inch/inch² connectivity or interconnection channel density. At 0.2¢/inch this will be priced at $20/inch², which is exactly the current price of IC's. Therefore, the plot is right at that point of the density scale.

If we look at the other end of this line, we see it is located in automated manufacturing technology have resulted in the lowest possible cost levels. With an increase in interconnection density, the complex interconnect structures call for the introduction of specialized technologies which, in conjunction with an accumulation of manufacturing problems, tend to progressively reduce the yields and thus increase the relative price. Another element, reduced size lot processing, becomes a significant factor here because at very high densities, customization and specialization is much more dominant. It is difficult to assess how much each of these elements contributes to this "price bulge," but they all contribute to the obvious deviation from this 0.2¢/inch "baseline" pricing.

As long as this price bulge exists, the market forces will continuously exert pressure to reduce it and bring the prices of more complex *P/I* structures and smaller lot manufacturing costs closer to this "baseline." The technological *fan out* effect, described in Fig. 12, will certainly be a major factor contributing to the shrinking of this bulge. It is interesting, however, to see that the location of the pivot on the *y*-axis of the graph of Fig. 12 already appears to be too conservative, and should probably be moved downward. But since the purpose of this paper is not to provide exact price numbers but to describe a useful methodology where trends can be observed, from which everyone can construct his own set of graphs based on his local price–density situation, it is not important to place that pivot more accurately now.

Another general observation which can be made from an analysis of the above graph is the questionable economic viability of extending PW technologies beyond 500–600 inch/inch^2 densities. Such multilayer structures can be built, and are. For example [14], [15], Fujitsu's 42 layer multilayer board, built for their M-780 computer, provides close to 800 inch/inch^2 density on 18 signal layers using 2.4 mil (60 μm) wide lines, but it costs tens of thousands of dollars to produce. This chart shows that multichip modules using thin film deposits on polyimide or on silicon could provide comparable or even higher densities at lower cost levels. This probably explains why there are currently very intensive engineering efforts mounted by all major OEM's in this area of technology and why the "substrate gap" (as defined by Knausenberger and Schaper in 1984) is presently being quickly filled in. The graph of Fig. 15 shows such price, density, and technology tradeoffs quite clearly and can be a useful tool for the tradeoff analysis of various new or existing interconnection technologies.

CONCLUSION

This paper has attempted to derive a universal and useful yardstick which will permit comparative ratings of various interconnecting substrates. Such a measure is the concept of *connectivity*, which provides information about the total signal wiring channel capacity per unit area of a selected substrate system.

An extension of this method allows analysis of the entire interconnection spectrum through historical or technological comparison of interconnection costs per unit length or per unit of substrate area. Through the use of this tool, a variety of studies and tradeoff comparisons can be made to evaluate the current state or even to project future developments in the electronic packaging field.

This price–connectivity analysis provides an indirect answer to the question posed earlier about the cost–function progress of PWB technology. The indications are that unit length prices of PWB interconnections, when manufactured by the most-advanced and automated mass manufacturing methods, are equal to interconnection prices on IC chips. Therefore, the general PWB cost–function performance improvement rate has been similar to that of integrated circuit technology. This is the surprising but inescapable conclusion derived from this analysis.

REFERENCES

[1] D. Frank, "Computer system performance and technology trends," IBM Corp., Endicott, NY, presented at the IEEE Computer Society Packing Committee Workshop, Oiso, Japan, Jan. 1986.

[2] G. L. Ginsberg, "A look at developments in multilayer fabrication," *Electron. Packaging Prod.*, pp. 34, June 1986.

[3] W. Loeb, "Printed circuits mirror integrated circuits," *Electron. Packaging Prod.*, pp. 168–170, Feb. 1986.

[4] R. J. Petschauer, "Evolution of high performance computer packaging," in *Proc. WESCON*, Nov. 1985.

[5] D. P. Seraphim, "Chip–module-package interfaces," in *Proc. Electron. Insulation Conf.*, Sept. 1977.

[6] D. Welterlen, "Trends in interconnect technology," in *Proc. 1980 ISHM Annu. Conf.*

[7] F. C. Chong and P. McEnroe, "A high density multichip memory module," in *Proc. WESCON*, Nov. 1985.

[8] W. H. Knausenberger and L. W. Schaper, "The myth of cheap wire," presented at NEPCON West, Anaheim, CA, Feb. 1984.

[9] ——, "Interconnection costs of various substrates—The myth of cheap wire," *IEEE Trans. Components, Hybrids, Manufact. Technol.*, vol. CHMT-7, no. 3, Sept. 1984.

[10] J. Balde, "Interconnection electronics in the mid-80's: A summary and overview," in *Proc. Printed Circuit World Conv. IV*, Washington, D.C., May 1984.

[11] M. P. Cassidy, B. C. Fichter, and A. C. Lubowe, "Trends in Western Electric circuit pack manufacture," PC Fab., Jan. 1984.

[12] R. Scott, "Dry etch systems," *Semiconductor Int.*, Oct. 1985.

[13] M. Sage, "Electronic interconnections and packaging—Now a critical technology," in *Proc. Internepcon Japan*, Jan. 1987.

[14] M. Nishihara, "A new concept for performance circuit boards," presented at the IEEE Computer Society Packing Committee Workshop, Oiso, Japan, Jan. 1986.

[15] H. Nakahara, "High performance MLB's for Japanese general purpose computers," *Electron. Packaging Prod.*, Feb. 1987.

Comparison of Wafer Scale Integration with VLSI Packaging Approaches

CONSTANTINE A. NEUGEBAUER, SENIOR MEMBER, IEEE, AND RICHARD O. CARLSON, SENIOR MEMBER, IEEE

Abstract—A comparison is made of various high-density packaging approaches, including printed wiring board, thick-film hybrids, and wafer scale integration (WSI). Criteria include power dissipation, density, delays, and cost. It is concluded that thin-film hybrids using state-of-the-art VLSI chips have the potential for WSI density and performance. The requirement for fault tolerance, additional levels of metallization, excess power dissipation, process conservatism to achieve finite yield, and the nonoptimum nature of the Al/SiO_2 transmission line for cross wafer communication have made WSI noncompetitive.

I. INTRODUCTION

WAFER SCALE integration requires leap-frogging in chip dimensions by several orders of magnitude. This can be accomplished in one of three ways.[1] The first of these involves the interconnection of only the "good" circuits, determined by previous testing, by use of a discretionary interconnection pattern made custom for each wafer. The second involves interconnecting such "good" circuits by use of programmable interconnections which, in turn, requires a matrix of fuses and "anti-fuses." The third is based on massive redundancy, such as triple redundancy. Beyond this, new schemes using radically new architectures and involving self-test, self-diagnostics, and self-repair are being postulated for much more efficient approaches to wafer scale integration in the future.

In addition to much higher circuit density and lower cost or circuit function, wafer scale integration can also lead to much higher performance. This is because the line delays associated with the ordinary printed wiring board or conventional hybrid packaging approaches are largely absent, such as bonding pads, wires, package conductor runs, and conductor runs between packages. Similarly, the output capacitances would be expected to be much less, leading to reduced power dissipation and smaller output buffer requirements.

It is the intent of this paper to compare various high-density packaging approaches, either presently available or projected into the future, to wafer scale integration. We conclude that advanced hybrid packaging approaches using pretested and replaceable chips can achieve circuit densities, performance, and costs comparable to wafer scale integration. The big advantage over wafer scale integration is the much lower risk,

which would be essentially limited to packaging issues such as chip testing, replacement, and substrate rewiring. But the chips used could be state of the art, they can be any technology (CMOS, ECL, TTL, etc.), and they can be obtained from any source. If, in addition to that, silicon is used as a substrate for the hybrid circuit, further advantages accrue. Among these are the matching thermal expansion coefficients between chip and substrate, the excellent thermal conductivity of silicon ($5\times$ better than alumina), and the ability to pattern a multilayer fine-line interconnect pattern on the silicon using well-developed silicon processing techniques.

II. COMPARISON OF ADVANCED VLSI PACKAGING APPROACHES TO WAFER SCALE INTEGRATION

The authors have chosen five VLSI packaging approaches to compare to wafer scale integration. These are the printed wiring board approach, using discrete (ceramic or plastic) packages; the thick-film multilayer hybrid on ceramic; the ceramic multilayer hybrid; and the thin-film multilayer hybrid populated with chips on one side, and also on both sides. These approaches will now be briefly described. Table I summarizes the most important assumptions.

III. DESCRIPTIONS OF APPROACHES

In the printed wiring board approach, a multilayer board is used with four to eight x and y conductor planes appropriately spaced between ground planes. The conductor run width is 0.005 inch. The chips are packaged in chip carriers which, in turn, are mounted on the board surface. The board, in turn, has a plated-through-hole (PTH) grid of 0.100 inch center-to-center. A mix of VLSI chips (0.300 × 0.300 inch) to SSI/MSI "glue" chips of 1:1 was assumed. The board is terminated in edge connectors on two edges.

The thick-film multilayer hybrid consists of many alternate layers of thick-film conductors and thick-film dielectrics screened on a sintered ceramic substrate. Bare VLSI chips (0.300 × 0.300 inch) are mounted on this substrate face-upward, and wire bonds are used to bond the chip terminations to the appropriate bonding pads on the substrate. The conductor width is 0.010 inch, and the dielectric thickness is 0.0015 inch.

The ceramic multilayer hybrid differs from thick-film multilayer hybrid in that the dielectric layers are cofired ceramic layers of the order of 0.020 inch in thickness. Refractory metal thick film x and y interconnection planes, as well as ground planes and power distribution planes, are screened on "green tape" into which holes have been punched

Manuscript received October 31, 1986; revised March 12, 1987. This paper was presented at the IEEE Computer Packaging Workshop, Split Rock Lodge, PA, May 1986.

The authors are with General Electric Company, Bldg. KW, Room C-1603, P.O. Box 8, Schenectady, NY 12301.

IEEE Log Number 8715550.

[1] Panel discussion, 1983 IEEE International Solid State Circuit Conference.

Reprinted from *IEEE Trans. Components, Hybrids, Manuf. Technol.*, vol. CHMT-10, no. 2, pp. 184–189, June 1987.

TABLE I
ASSUMPTIONS FOR VLSI PACKAGING COMPARISONS

VLSI chip (or circuit) size	0.300 inch × 0.300 inch
Number of output buffers	30/chip
Number of input buffers	60/chip
Capacitance of output driver gate, C_g	3.0 pF for $R_0 \leq 50\ \Omega$ 1.0 pF for $R_0 = 150\ \Omega$ 0.3 pF for $R_0 = 450\ \Omega$
Input capacitance	1 pF
Output buffer delay, $t_{O.B.}$	3 ns for $R_0 \leq 50\ \Omega$ 2 ns for $R_0 = 150\ \Omega$ 1 ns for $R_0 = 450\ \Omega$
VLSI chip pin count	132
Voltage swing, V	5
Internal chip power (exclusive of buffers)	0.5 W

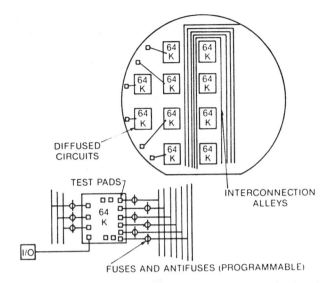

Fig. 1. Model for wafer scale integration: wire-in "good" circuits only.

for vias. This stack of ceramic layers is then fired at high temperatures. Flip-chips are mounted on this substrate using solder bumps.

The thin-film multilayer hybrid consists of layers of thin-film conductors, such as aluminum or copper, separated by thin-film layers of a dielectric-like silicon dioxide or polymeric material like polyimide. Line widths are much smaller than the previous approaches, typically of the order of 0.001 inch or less, and the dielectric thickness is typically several micrometers. This allows all interconnections between chips to be accomplished in two or, at most, three layers of metal. Bare chips, 0.300 × 0.300 inch, using solder bumps are mounted using flip-chip techniques.

Finally, for the last packaging approach, both sides of such a substrate are populated with chips and communication between the two sides is established through via hole connections. This can give very short interconnections between chips, depending on the fanout pattern.

To compare these approaches to wafer scale integration, the authors have picked the relatively simple case of identical circuits diffused in a silicon wafer, in our case 64K memory circuits (Fig. 1). These circuits are endowed with test pads which allow identification of "good" circuits. The circuits are separated by interconnection alleys, which are 0.100 inch wide and which contain power, ground address, and I/O lines. These lines can be connected or disconnected to each 64K memory circuit separately by means of floating gate transistors, which can be individually accessed to program them "on" or collectively discharged to program them "off". Using a circuit size of 0.300 × 0.300 inch, the footprint for each circuit is 0.16 inch². For comparison purposes we assume that each circuit is good (100-percent yield).

The comparison below between these approaches consists of estimates of the packaging density, the package delay, the power requirements at 100 MHz, the relative packaging cost at 100 MHz, and a comparison of several packaging figures of merit.

IV. PACKAGING DENSITY

The packaging density is calculated by dividing the chip area by the chip footprint, which consists of the substrate overhead area assigned to this chip and is due to the PTH grid, bonding pad periphery, engineering change pads, or other design rule determined parameters. This is shown for the various VLSI packaging approaches in Table II. It can be seen that the spread for the packaging density is a factor of 15. In particular, it is the thin-film approaches (including wafer scale integration) which are particularly dense, being on the average 10× more dense than the printed wiring board approach.

V. PACKAGE DELAY

While the total signal delay is the sum of the chip and the package delays, the authors are interested here only in the latter. Package delay (see Fig. 2) was estimated from the output capacitance C_0, which the package presents to the output driver and the on-resistance of the output driver R_0 (the line resistance is assumed to be negligible), and the buffer delay $\tau_{O.B.}$, by taking

$$\text{package delay} = \tau_{O.B.} + 3R_0C_0\ \text{ns}.$$

Here $\tau_{O.B.}$ is taken to be 3 ns for a large output driver ($R_0 = 150\ \Omega$), 2 ns for a medium-sized output driver ($R_0 = 150\ \Omega$), and 1 ns for a small output driver ($R_0 = 450\ \Omega$). C_0 is the sum of the package and line capacitances being driven, as well as the sum of the input capacitances of the I/O's of the chips being driven.

The authors have taken a time constant of 3 R_0C_0 to allow for adequate time to discharge a node below the threshold voltage. The use of the RC product to estimate the package delay is an over-simplification and assumes that the short line lumped parameter equivalent circuit model is adequate and that transmission-line effects are negligible. While this is not always the case in practical cases, this simplification appears nevertheless justified for the purposes of this comparison.

The comparison of package delays for the various approaches is given in Table III. The first three columns give the line capacitance per centimeter of line length, the average chip-to-chip line length required, and the output capacitance C_0, respectively. In calculating C_0, a fanout of 2 was assumed, which approximately doubles the line capacitance. The next

TABLE II
COMPARISON OF PACKAGING DENSITY

Packaging Approach	Footprint Required for Chip of Dim. 0.300×0.300 inch^2	Footprint Area Is Limited by	Packaging Density A(Si)/A(F.P.) Percent
Printed Wiring Board	1.35×1.35	PTH grid required by thermal vias to achieve 20 deg. C/W	5
Thick-Film Multilayer Hybrid on Ceramic	0.85×0.85	Limited by pad periphery 0.25 C-C around chip	12
Ceramic Multilayer Hybrid	0.60×0.60	Engineering change pads periphery around chip	25
Thin-Film Multilayer Hybrid on various substrates, populated on one side	0.50×0.50	Engineering change periphery 0.015 C-C around chip	36
Wafer Scale Integration	0.40×0.40	Requires 0.100 inch routing alleys between circuits	56
Thin-Film Multilayer Hybrid on various substrates, populated on both sides	0.35×0.35	$2 \times$ density of one sided Si substrates	72

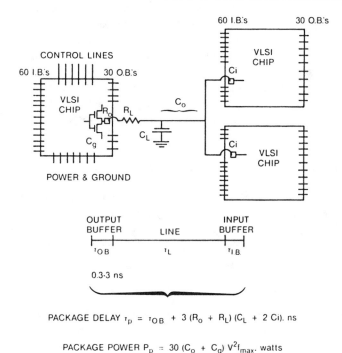

PACKAGE DELAY $\tau_p = \tau_{OB} + 3 (R_o + R_L) (C_L + 2 C_i)$. ns

PACKAGE POWER $P_p = 30 (C_o + C_g) V^2 f_{max}$. watts

Fig. 2. Estimating package delay and power for VLSI packaging approaches.

four columns give the package delay for various assumed values of the output driver on-resistance (line resistance was ignored as negligible). The last column then gives the value of the output driver on-resistance required at a 100-MHz clock rate.

This resistance is important since the silicon area requirement of the output buffer increases sharply as the required on-resistance decreases. Since there is a large number of such output buffers per VLSI chip, typically 30–60, this translates into much larger chip sizes.

The benefit which is obtained from the use of the thin-film packaging technologies is again visible in Table II. These are shorter conductor line lengths and reduced output buffer requirements, or conversely, a smaller package delay.

VI. PACKAGE POWER REQUIREMENTS

The "package" power differs from the "chip" power in that it is the power dissipated in the output buffers in driving the package capacitance C_0, while the latter is dissipated internal to the chip in switching logic nodes. The power/chip is then the sum of the two.

The power requirements at 100 MHz are compared for the various packaging technologies in Table IV. Here we have taken a constant value of 0.5 W/chip for the internal "chip" power dissipation, while the "package" power per chip is given by

$$\text{Package power/chip} = 30(C_g + C_0)V^2 f$$

where the factor 30 reflects our estimate of 30 output buffers per chip, C_g is output driver (gate) capacitance, V is the voltage swing in charging and discharging the output capacitance C_0, and $f \equiv 100$ MHz. This is shown in the first column. The second column is the power density, obtained by driving the power/chip by its package footprint. The third column expresses a (power $\times$ delay) figure of merit for the particular packaging approach, given by the power/chip times the $R_0 C_0$ time constant.

It can be seen that for each of the packaging approaches the package power exceeds the internal chip power, in some cases severalfold. This is particularly true for the printed wiring board and the thick-film multilayer approaches where the package power exceeds the internal chip power by a factor of 5.

The reduced power requirement for the thin-film approaches again becomes evident. This is, of course, because of the shorter distances and smaller package capacitances involved. However, this reduced power dissipation/chip is bought at the expense of an increased power density (column 2) indicative of the higher packaging density associated with these approaches. This means that more efficient cooling techniques are required, even though the total power dissipation is down.

The package power $\times$ delay product, given in the third

TABLE III
COMPARISON OF PACKAGE DELAYS

Packaging Approach	Line Cap., pF/cm	Ch.-Ch. Line Length, cm	Out. Cap. C_0, pF	Package Delay, nS, for Various Output Buffers				Requ. R_0, Ω At 100 MHz
				25	50	150	450	
Printed Wiring Board	1.0	15	32	5.4	7.8	15.4	43.5	25
Thick-Film Multilayer Hybrid on Ceramic	2.0	8	34	5.5	7.8	16.4	44.2	25
Ceramic Multilayer Hybrid	2.5	6	17	—	5.6	9.7	24.0	50
Thin-Film Multilayer Hybrid on various Substrates, populated on one side	1.0	5	12	—	4.8	7.4	17.2	50
Wafer Scale Integration	1.0	3	8	—	4.2	5.6	11.8	150
Thin-Film Multilayer Hybrid on various Substrates, populated on both sides	1.0	2.5	7	—	4.1	5.2	10.5	150

TABLE IV
COMPARISON OF POWER REQUIREMENTS AT 100 MHz

Packaging Approach	Power per Chip $= \dfrac{0.5 + 30(C_g + C_0)}{V^2 f}$ W	Power Density = Power/Chip Area, Footprint W/cm^2	Power $\times$ Delay $= P \times R_0 C_0$ ($R_0 = 50\ \Omega$) nJ
Printed Wiring Board	3.1	0.27	5.00
Thick-Film Multilayer Hybrid on Ceramic	3.2	0.66	5.40
Ceramic Multilayer Hybrid	2.0	0.85	1.70
Thin-Film Multilayer Hybrid on various Substrates, populated on one side	1.6	1.00	0.96
Wafer Scale Integration	1.2	1.10	0.48
Thin-Film Multilayer Hybrid on various Substrates, populated on both sides	1.1	1.35	0.39

column, is a figure of merit which can be used to compare various packaging approaches in the accustomed manner, with the lowest of these being the most favorable [2]. Thus the power $\times$ delay product of a particular chip technology, when packaged with an approach with a large package power $\times$ delay, would experience a relatively severe degradation. Again the thin-film packaging approaches show the most favorable power $\times$ delay, being roughly $10 \times$ smaller than the printed wiring board approach.

VII. Cost

This comparison is made for a comparable state of maturity of the various packaging technologies, i.e., the packaging yield is assumed to be high and comparable and, thus not a factor in estimating cost. We start with the observation, based on past experience, that for packaging on printed wiring boards, the chip cost, package cost, and package overhead cost are about 1/3 each of the total cost. Here the package cost includes the ceramic package, wiring cost, and that of the printed wiring board, and the overhead cost includes power

supplies and power removal equipment, such as heatsinks and fans. The authors, therefore, take the relative cost of the printed wiring board approach to be 1. They further assume that the chip cost decreases as the output buffer required for a 100-MHz operation decreases in area. They have taken a relative chip cost of 0.33 for $R_0 = 25\ \Omega$, 0.25 for $R_0 = 50\ \Omega$, and 0.20 for $R_0 = 150\ \Omega$.

We have taken the package and overhead cost to be proportional to the power dissipated per chip. This appears to be a reasonable assumption since power supplies and heat removal equipment dominate the cost. Thus, if the power per chip is, say 1/2 that of the printed circuit board approach, the package + overhead cost of this approach was taken to be 1/2 for this approach. Cost comparisons are shown in Table V. It can be seen that the estimated costs of the various approaches are not greatly different from each other.

VIII. Package Figures of Merit

The three most important parameters which must be kept at a minimum for any given packaging approach are the

TABLE V
COMPARISON OF RELATIVE PACKAGING COSTS AT 100 MHZ

Packaging Approach	(Chips) + (Package & Overhead) × Power/3.1		
	Chips	Package & Overhead	Total
Printed Wiring Board	0.33	0.67	1.00
Thick-film Multilayer Hybrid on Ceramic	0.33	0.69	1.02
Ceramic Multilayer Hybrid	0.25	0.40	0.65
Thin-Film Multilayer Hybrid on various Substrates, populated on one side	0.25	0.35	0.60
Wafer Scale Integration	0.20	0.26	0.46
Thin-Film Multilayer Hybrid on various Substrates, populated on both sides	0.20	0.24	0.44

TABLE VI
COMPARISON OF PACKAGE FIGURES OF MERIT

Packaging Approach	Power × Delay (Normalized)	Size or Weight (Normalized)	Cost (Normalized)	Overall Figure of Merit
Printed Wiring Board	1.00	1.00	1.00	1.00
Thick-Film Multilayer on Ceramic	1.08	0.42	1.02	0.46
Ceramic Multilayer Hybrid	0.34	0.20	0.65	0.044
Thin-Film Multilayer Hybrid on various Substrates, populated on one side	0.19	0.14	0.60	0.016
Wafer Scale Integration	0.10	0.09	0.46	0.0041
Thin-Film Multilayer Hybrid on various Substrates, populated on both sides	0.08	0.07	0.44	0.0025

performance (power × delay), the size or weight, and the cost. The authors compare these parameters for the various packaging approaches in Table VI, normalized to 1 for the printed wiring board approach. It can be seen that they cover a range of a factor of 12 in the power × delay product, a factor of 14 in size or weight, and a factor of 2 in cost. Overall, therefore, these approaches differ by as much as a factor of almost 400 when these figures of merit are multiplied together. This shows that there is still considerable room left for improvement in packaging techniques and that by applying an advanced packaging approach, particularly a thin-film interconnection approach of the type described, densities, performance, and cost comparable to that achievable by wafer scale integration could be achieved.

IX. THE THIN-FILM HYBRID APPROACH TO WAFER SCALE INTEGRATION

It is clear from the above comparison that the thin-film interconnection approaches give a particularly favorable figure of merit. This is for the following reasons.

1) Thin-film interconnections can have a line width comparable to that used on the chip itself, thus giving highest density and minimum capacitance.

2) The number of interconnection layers between VLSI chips can be as low as two because of the very high line density achievable by the thin-film technology.

3) Many of the materials and processes developed for IC fabrication are also useful here, such as Al/polyimide multilayering and even the use of the silicon as a substrate.

Perhaps the principal disadvantage of the thin-film multilayer packaging approach is that the line resistances are appreciable when the line length exceeds a few centimeters. This would cause additional delays and power dissipation. It is thus desirable to use copper as the conductor material, along with a low dielectric constant dielectric layer, such as SiO_2 or a polymer. Ho et al. [3] have discussed the merits of thin-film packaging in high performance computers.

The VLSI and glue chips can be pretested prior to assembly in the hybrid and can be replaced at least a few times even after incorporation in the hybrid. Testability of individual sections of the hybrid is thus mandatory for fault isolation. Power delivery and removal are difficult problems, particularly when a high power density is involved. In the limit of very high power density levels, liquid cooling will become necessary, perhaps by direct immersion. Particularly attractive might be cooling by nucleate boiling by direct immersion of the flip-chip hybrids in Freon where the backside of the chips could be the boiling interface by adding a Linde surface[2] on which the nucleation barrier for boiling is lower.

X. 3-D PACKAGING

Regardless of which approach is used to implement wafer scale integration, interconnections between wafers (or substrates) would still present a major packaging problem. Such a wafer would be expected to have hundreds or even thousands of terminations. Further, if all those terminations were placed on the wafer periphery, then, in a worst-case situation, a circuit located in the interior of one wafer, in order to communicate with a circuit located in the interior of another

[2] Trademark of Union Carbide.

wafer, would have to drive a considerable line length with its attendant delay. It is almost inevitable that interconnections will have to be provided from wafer to wafer in the Z-direction between wafers in a stack, and not just at the wafer peripheries, but in the wafer interior. A possible version of 3-D packaging has recently been proposed by Grinberg *et al.* [3] using microspring contacts between wafers in a stack.

It should also be pointed out that power voltage dips due to simultaneous switching and coupled noise are both reduced for the higher density packaging approaches, where the average line length is shorter.

REFERENCES

[1] R. W. Keyes, *IEEE J. Solid State Circuits*, vol. SC-13, p. 265, 1978.
[2] C. W. Ho, D. A. Chance, C. H. Bajorek, and R. E. Acosta, *IBM J. Res. Develop.*, vol. 26, no. 3, p. 286, 1982.
[3] J. Grinberg, G. R. Nudd, and R. D. Etchells, *IEEE Trans. Computers*, vol. C-33, p. 69, 1984.

TECHNOLOGY AND DESIGN FOR HIGH SPEED DIGITAL COMPONENTS IN ADVANCED APPLICATIONS

R.P. Vidano, J.P. Cummings, R.J. Jensen, W.L. Walters, M.J. Helix
Honeywell Corporate Technology Center
10701 Lyndale Avenue South
Bloomington, Minnesota 55420

SUMMARY

GaAs digital circuitry poses challenges that must be met in materials/process development and in design of 1st and 2nd level packaging structures. The development of controlled characteristic impedance chip carriers (C^2IC^2) which surface mount with compatible printed wiring boards (CPWB) have several promising features to solve the electrical, thermal, and mechanical issues in high speed digital systems. We have used a methodology employing computer aided design (CAD) techniques to arrive at efficient interconnection schemes, good cooling techniques, and a stable mechanical outline suitable for a wide range of advanced applications. Excellent correlation has been found between electrical simulation and actual circuit operation. We show that controlled impedances are mandatory at the 1st level of packaging for > 1 Gbit/sec logic ICs and have demonstrated processes/materials which offer multilayer, fine lines, high resolution, controlled characteristic impedance interconnections accommodated within the C^2IC^2 component. Finally, we illustrate and discuss the necessity to design C^2IC^2/(CPWB) hardware in accordance with system requirements with respect to interconnection issues, thermal management, along with static and dynamic forces.

INTRODUCTION

The combination of very high speed and high circuit density for SSI-MSI-LSI GaAs digital circuitry presents unique challenges for interconnection and packaging. GaAs digital circuitry with rise times of less than 150 pico-seconds and clocking frequencies of 1-4 GHz offers a five to six fold potential increase in performance over current high performance ICs. These performance benefits will be realized provided that materials/process developments and design advances are attained for 1st and 2nd level packaging structures. The drive towards MSI/LSI in GaAs circuitry will be rapid because losses and parasitics in multi-chip SSI circuit modules can be expected with even the best packaging technologies[1]. These MSI/LSI circuits will require high I/O accommodation at all levels of packaging. A recent study has shown that there is not currently available a 1st level package in excess of 6-8 I/O that will provide performance for high speed digital (4 GHz) applications[2]. The development of hermetic chip carriers with efficient heat sinking and controlled characteristic impedance I/Os can meet the packaging needs of GaAs ICs[3,4]. These controlled characteristic impedance chip carriers (C^2IC^2) may be single or multichip depending upon application, performance requirements, and circuit partitioning. The development of C^2IC^2 components will offer the possibility of incorporating a large number of controlled impedance I/O ports for chip to chip interconnection along with efficient electrical/mechanical partitioning.

APPROACH

At Honeywell, we have been using Computer Aided Design (CAD) to evaluate alternative interconnection and packaging schemes. Figure 1 shows the methodology we use to combine system requirements, packaging technology and circuit options to obtain various package designs. Each of these options must be assessed with respect to electrical, thermal, and mechanical characteristics.

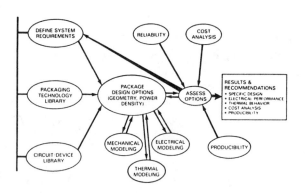

Fig. 1   Methodology for Assessing Packaging Technology

First, because of the short rise times and high clocking frequencies transmission line models are used along with appropriate driving, receiving, and loading circuitry. Relationships between materials parameters and electrical performance is established using a combination of analysis and experimental verification.

Second, thermal management of high speed GaAs ICs for military missions takes into account possible advanced applications in satellites, aircraft, missiles and ground based vehicles. Thermal analysis using CAD techniques can address both the component and system levels. Direct heat sinking of the GaAs ICs using high thermal conductivity materials is mandatory. Efficient system cooling techniques such as thermal planes and heat pipes are being investigated and developed.

Surface mounted components have special thermal stress compatibility problems with respect to multilevel wiring boards. C^2IC^2 surface mounted components being developed are designed taking into account thermal cycling, vibrational spectra, shock, and acceleration. The effects of large thermal excursions are modeled using CAD analysis and these computer predictions are being related to experimental data.

In many applications, GaAs ICs will be subjected to adverse environmental conditions such as high humidity, contamination, and temperature. Therefore, these circuits will be packaged in a hermetic, moisture free enclosure. Furthermore, due to storage requirements as well as projected usage in advanced applications, the long term reliability of the package and life time of the circuit is of foremost importance.

MECHANICAL OUTLINES

Hermetic, surface mounted components (C^2IC^2) which are the 1st level of packaging interface with compatible multilayer wiring boards, the 2nd level of packaging. The concept for C^2IC^2 components with integrated heat sinks and controlled impedance I/Os is shown in Figure 2 for single chip components and Figure 3 for multi-chip components. A multilayer technology must be accommodated at the 1st level of packaging that has high tolerance, controlled impedance microstrip or stripline interconnections. Furthermore, termination resistors, provisions for decoupling capacitors, and direct grounding must be made within the first level of packaging.[5]

Reprinted from *Proc. 33rd Electron. Components Conf. (ECC)*, pp. 334–343, 1983.

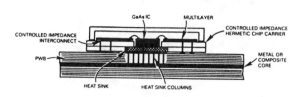

Fig. 2. Single Chip C²IC² Component for GaAs ICs

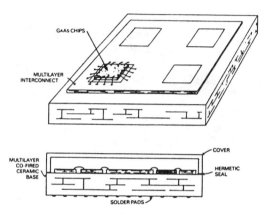

Fig. 3. Multi-Chip C²IC² Component for GaAs ICs

There are two options for the mechanical outline footprint. The C²IC² may utilize a JEDEC standardized periphery footprint or it may use a pad grid array footprint. Either option will allow surface mounting with the next level of interconnection, the multilevel wiring board (Figure 4).

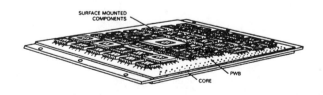

Fig. 4. High Speed Digital System with C²IC² Surface Mounted Components and Compatible PWBs

C²IC² SUBSTRATE TECHNOLOGY DEVELOPMENT

The combined demands of increased circuit density and speed requires new advances in the materials and processes used in interconnection and packaging. The accommodation of high I/O count chips requires narrow conductor linewidths and reasonable spacing to reduce crosstalk while short delay times and low resistive losses necessitate low resistivity conductor materials with large cross sections and hence, large aspect ratios. Relatively thick dielectric layers are required for high characteristic impedances (Z_0) and low noise coupling, and a low dielectric constant material is needed for reduced propagation delays and high Z_0. Finally, routing flexibility and transmission line structures require multilayer metallization with accurate control over dielectric thickness and planarity for acceptable yields.

Table 1 shows line width, thickness, and dielectric thickness data to obtain 50 ohm characteristic impedance interconnections with three different technologies; Cu thick film/thick film dielectric, Cu/polyimide and co-fired ceramic. The Cu thick film/thick film dielectric technology has been researched extensively at this laboratory[6]. Recently, it has been shown that thick film technology may be employed in microstrip interconnections by using a technique of multiple dielectric screen printings to obtain the required dielectric thicknesses[7]. The Cu/polyimide system is currently being researched at this laboratory[8]. Polyimide dielectrics with copper conductors have also been developed for usage in commercial applications[9]. Co-fired ceramic technology has been available for years and is often designed and fabricated into controlled characteristic impedance interconnections[10].

TABLE I

(MICROSTRIP TRANSMISSION LINE)

$$Z_0 = \frac{87}{\sqrt{\varepsilon_r + 1.41}} \ln\left[\frac{5.98\ h}{.8\omega + t}\right]$$

Technology	ε_r	$\omega(\mu m)$	$h(\mu m)$	$t(\mu m)$	Z_0
Cu Thick Film	8	100	90	12	50
Cu/Polyimide	3.5	50	30	12	50
Co-fired Ceramic	9	100	100	12	50

(STRIPLINE TRANSMISSION LINE

$$Z_0 = \frac{60}{\sqrt{\varepsilon_r}} \ln\left[\frac{4b}{.67\pi(.8\omega + t)}\right] \quad b = 2h$$

Technology	ε_r	$\omega(\mu m)$	$b(\mu m)$	$t(\mu m)$	Z_0
Cu Thick Film	8	100	500	12	50
Cu/Polyimide	3.5	25	80	12	50
Co-fired Ceramic	9	100	600	12	50

Although the current needs of high speed digital systems may be met by standard thick film processing techniques using inorganic dielectrics, these methods are limited in resolution capability. Current state of the art for both thick film and co-fired ceramic technologies is about 100 micron line widths. A thin film multilayer technology employing an organic dielectric appears promising in terms of resolution and multilayer capabilities, and dimensional control. We have therefore devoted considerable effort to the development of a controlled characteristic impedance multilayer technology employing copper conductors, polyimide dielectrics, and thin film processing techniques.[8] The Cu sputtered film technology is applicable for both organic and inorganic dielectric systems.

Copper is chosen as a conductor material for its low electrical resistivity (1.6 micro-ohm-cm) low cost, and solderability. Polyimide is an advantageous dielectric for a number of reasons. The polymer precursor, polyamic acid, is a viscous liquid which can be applied at the desired thickness of 15 to 50 microns by spinning or spraying providing partial planarization of conductor grids and accurate thickness control. The fully cured polyimide is chemically inert, mechanically tough, and thermally stable up to 400°C, and has favorable electrical properties, i.e. low dielectric constant (3.5-4.0), low dissipation factor over a wide frequency range, and high breakdown voltage.

We have fabricated copper conductor lines with 10 micron thicknesses, widths, and spacing by two techniques: (1) DC magnetron sputtering followed by photolithography and ion milling, and (2) selective plating into thick photoresist patterns[8]. We have also developed reactive ion etching processes for patterning very high aspect ratio features in polyimide films[8].

Multilayer Cu/PI structures have been fabricated in our laboratory by a variety of techniques involving different sequences of etching and deposition processes. These techniques can be divided into two basic approaches, additive and subtractive metallization. In the additive metallization approaches, a blanket coating of polyimide is first deposited and patterned by wet or plasma etching, then conductor lines or vias are metallized by sputtering or plating into the holes or troughs in the polyimide, and finally unwanted conductor material is removed by wet etching or lift-off. In the subtractive approaches, the conductor layer is first sputter deposited and patterned by wet etching or ion milling, then polyimide is coated over the conductor, and finally the structure is planarized by mechanical or plasma etching techniques.

The feasibility of component packaging using Cu/PI multilayer technology was demonstrated by designing and fabricating a 3 layer interconnect structure on a pad grid array ceramic substrate. The experimental package is shown in Figure 5 and a cross section of the Cu/PI layers is shown in Figure 6. The ceramic substrate contains plated through via holes. Signal lines on the first layer are 25 microns wide and 7 microns thick. Exposed metal lines and contact pads are gold plated for reliability. The ground plane is in the middle layer, and the power lines are on the top layer. Time Domain Reflectometry (TDR) using 35 pico-second risetimes of the controlled impedance signal lines showed a characteristic impedance of 55 ohms. The package outline provides for direct grounding and for the accommodation of terminating resistors and decoupling capacitors. Therefore, this C^2IC^2 component which has been developed using the Cu/PI system for controlled impedance interconnects is suitable as a 1st level packaging technology for GaAs ICs.

Fig. 5. C^2IC^2 Component using Cu/PI Technology for Controlled Impedance Interconnections

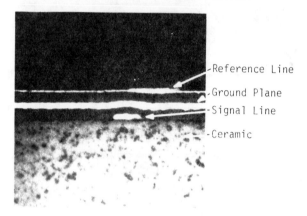

Fig. 6. Cross-Section of Multilayer Cu/PI Controlled Characteristic Impedance Interconnections

INTERCONNECTION REQUIREMENTS

The interconnection issue in high speed digital logic is similar to the impedance matching problem in microwave systems. Common geometric configurations for high speed interconnection are microstrip, stripline, coplanar and hybrid[11]. Microstrip is the term used for a conductor strip on the surface of a dielectric sheet which separates the strip from the ground plane. The characteristic impedance of the microstrip transmission line is given by[12]:

$$Z_0 = \frac{87}{\sqrt{\varepsilon_r + 1.41}} \ln\left[\frac{5.98\ h}{0.8\omega + t}\right] \qquad (1)$$

Where h = dielectric thickness, ω = conductor width, and τ = conductor thickness.

Stripline is the technique used on the interior layers of a multilayer interconnection where a metal strip is symmetically placed between two ground planes. The characteristic impedance of the stripline is given by[12]

$$Z_0 = \frac{60}{\sqrt{\varepsilon_r}} \ln\left[\frac{4b}{.67\pi(0.8\omega + t)}\right] \qquad (2)$$

Where b = dielectric thickness (b=2h), ω = conductor width, and t = conductor thickness.

Low impedance transmission lines ($Z_0 < 50$ ohms) are undesirable because they require large drivers which occupy considerable chip area, dissipate large amounts of power, and are slow due to their large geometries. High impedance transmission lines ($Z_0 > 80$ ohms) have long RC time constants and more attenuation than low impedance transmission lines. Furthermore, high impedance lines are extremely difficult to fabricate due to the large dielectric thicknesses required.

Synchronous logic is dependent upon logic and clock signal arriving at the predicted time but it is difficult to achieve this when the clock signal is distributed to many chips in high speed systems. To overcome this difficulty a very low interchip signal delay must be achieved by a combination of high packing density and low media delay. Media delay introduced between interconnecting chips can result from parasitic capacitances, excessive line resistances, as well as wave propagation through the dielectric. The propagation delay is a function of dielectric constant for a TEM wave propagating through a "lossless" microstrip structure. The propagation delay for such a TEM wave can be given by[12]:

$$(\tau_{pd})_{TEM} = 33\sqrt{\varepsilon_r} \text{ ps/cm} \qquad (3)$$

where ε_r is the relative dielectric constant of the media. Clearly, low dielectric constant materials are desirable for the minimization of propagation delay.

If the propagation delay path for an interchip network is longer than 1/3 to 1/2 of the signal rise or fall time then characteristic impedance transmission line technology must be utilized. This may be expressed as a critical path length given by[13]:

$$L_c = \frac{t_r \text{ or } t_f}{3\tau_{pd}} \qquad (4)$$

where L_c = critical path length, t_r = rise time, t_f = fall time, and τ_{pd} = propagation delay.

Therefore, for GaAs circuitry with rise and fall time of less than 150 pico-seconds, a transmission line structure must be used for interconnection lengths in excess of 0.9cm with PI dielectrics and 0.5cm with ceramic dielectrics.

Because of the short rise and fall times associated with high speed digital GaAs circuitry, the transmission line structures must have large line bandwidths. The line bandwidth is given by[14]:

$$f = \frac{k}{t_r} \qquad (5)$$

where $k = 0.37$, and t_r = risetime

Therefore, for rise times of less than 150 pico-seconds the transmission line bandwidth must be essentially from DC to greater than 2.5 GHz.

Microstrip and stripline losses originate with two sources; ohmic losses in the conductor and ground plane, and dielectric losses in the substrate[15]:

$$\alpha = \alpha_c \text{ (ohmic)} + \alpha_D \text{ (dielectric)} \qquad (6)$$

Therefore, it is important to consider the properties of the conductor with respect to attenuation and dielectric with respect to loss tangents when choosing materials systems for high speed interconnections. Furthermore, design considerations such as line

discontinuities (vias, bends, impedance mismatches) cause radiation losses and must be considered in efficient interconnections[16].

An illustration why controlled characteristic impedance interconnections in the Gbit/sec region are mandatory at the first level of packaging is shown in Figure 7. A 4:1 MUX GaAs circuit was packaged into a standard JEDEC Type C 44 I/O chip carrier. JEDEC leadless chip carriers (LC3) have favorable electrical performance characteristics compared to other standard packaging technologies[17] due to short lead length but they do not possess controlled characteristic impedances interconnections. Figure 7 shows the 4:1 MUX output at 500 MHz and 1 GHz clock input. At 1 Gbit/sec operation, the package parasitics are substantial and cause signal degradation. Therefore, these types of 1st level packaging structures would be unsuited for Gbit/sec logic circuitry. However, the 4:1 multiplexer circuit has operated up to 1.9 GHz clock input in a microstrip test fixture with replacable controlled impedance single layer alumina chip carriers with integrated heat sinks[18]. Because of the relatively low I/O requirements of the 4:1 MUX, conventional thin film processing techniques could be used for controlled impedance lines and to incorporate termination resistors and decoupling capacitors within the 1st level of interconnection. The waveforms from the microstrip test fixture are shown in Figure 8 at 1.0 and 1.5 GHz clock input with 0001 multiplexed outputs.

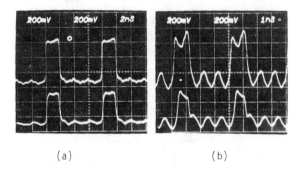

(a)   (b)

Fig. 7.   4:1 GaAs MUX Out and Sync Out at (a) 500 MHz and (b) 1 GHz Clock Input when Packaged in JEDEC Type C 44 I/O LC3 Component

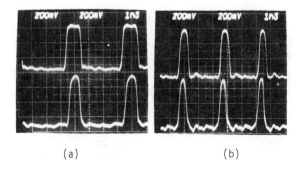

(a)   (b)

Fig. 8.   4:1 GaAs MUX Out and Sync Out at (a) 1 GHz and (b) 1.5 GHz Clock Input with Package in Microstrip Test Fixture

ELECTRICAL SIMULATION AND VERIFICATION

For interconnections which are close to the ground plane a lumped element model is used (Figure 9). In these cases the capacitance of the signal line over the ground plane is given by(19).

$$Co = \frac{\epsilon A}{h} \quad \epsilon = \epsilon_r \epsilon_0 \tag{7}$$

where ϵ_r = relative dielectric constant of the medium, ϵ_0 = permittivity of free space, A = signal line area, and h = dielectric thickness.

The line inductance of a signal run is calculated by(19)

$$Lo = Zo^2 Co \tag{8}$$

The line resistance is calculated according to the relationship:

$$R = \frac{\rho L}{A} \tag{9}$$

where ρ = resistivity of the line, A = signal line cross section, and L = signal line length.

Media delay introduced by the substrate technology can limit the performance and is therefore a very critical parameter. The lumped element parameters capacitance, inductance, and resistance all per unit length are calculated from the material parameters and line dimensions and used to model the media delay properties of a transmission line. This type of analysis has been applied previously to high speed interconnections(20).

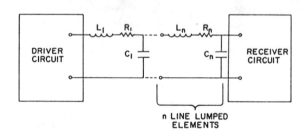

Fig. 9.   Lumped Element Transmission Line Model for Interconnections

The high speed circuit families which we have investigated are silicon based emitter coupled logic (ECL) and GaAs based Schottky Diode FET (SDFL). The basic circuit used for the ECL line drivers and with current mode logic (CML) line receivers is shown in Figure 10(21). The basic circuit used for SDFL multiplexers is shown in Figure 11(18).

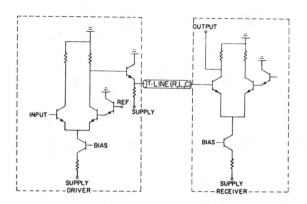

Fig. 10.   Basic Model Used for HSPICE Simulation of ECL Driver Lumped Element Transmission Line, and Receiver (Ref. 21).

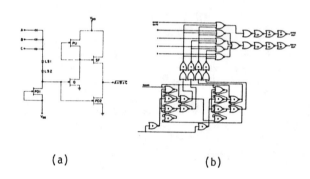

(a)   (b)

Fig. 11.   Basic Model Used for HSPICE Simulation of (a) SDFL NOR Gate and (b) 4:1 Multiplexer (Ref. 18).

Table 2 lists data for several types of 1st and 2nd level packaging and interconnection technologies. For the co-fired ceramic and Cu/PI technologies, dimensions to yield a controlled characteristic impedance of 80 ohms was assumed. The PWB was assumed to be lossless with Z_0=80 ohms and a propagation delay of 60 ps/cm. The lumped element data for a typical JEDEC chip carrier was inferred from the literature(17).

TABLE 2

Interconnection Technology	R_L	C_L	L_L
JEDEC 44 I/O LC^3	.1 Ω	.1 Pf	2 NH
Co-fired Ceramic w = 100µm t = 12µm h = 300µm	1 Ω/cm	.27 Pf/cm	1.7 NH/cm
CU/PI w = 50 µm t=12µm h = 65 µm	.25 Ω/cm	.25 Pf/cm	1.5 NH/cm

Note: PWB assumed Z_0 = 80Ω, τ_D = 60 ps/cm "Lossless" T. Line

For a GaAs 4:1 multiplexer to operate at 1.5 Gbit/sec, it must have 4 parallel ECL 375 Mbit/sec inputs assuming rise and fall times of 500 pico-seconds with a total period of 2.67 nano-seconds. Using HSPICE, a serial chip to chip interconnection comparing multi-chip C^2IC^2 components to single chip JEDEC LC^3 components was performed. It is assumed that 4 separate chip to chip serial interconnections would be inputs for the 4:1 GaAs MUX circuit. In this case, ECL driver/CML receiver models were employed with 500 ps rise time digital input. The chip to chip interconnection consisted of a serial net of 2 circuits, each circuit consisting of a CML buffer, ECL driver, and CML receiver. Figure 12 shows the waveform for the 2 circuit serial net using JEDEC LC^3 components. From Figure 12, circuit 2 switched after 6 ns including a total media delay of 1.25 ns due to the PWB connections inherent with any single chip component technology. Figure 13 shows the waveform for the 2 circuit net using multi-chip C^2IC^2 components. The last circuit in the net switched after 5 ns with a total media delay of .25 ns. The primary reason for the reduced media delay in this comparison was the high packing density achieved with the multi-chip components. In this example the ECL inputs to a GaAs multiplexer were shown to have reduced media delay in a multi-chip component and therefore, it is more likely that proper timing of the clock and input signal may be achieved. This HSPICE comparison illustrates the importance to consider, when appropriate, multi-chip packaging methods to realize high performance.

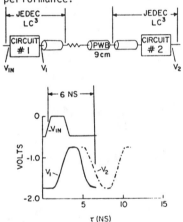

Fig. 12. HSPICE Simulation of ECL Serial Network Using JEDEC LC^3 Components

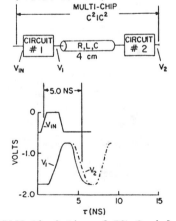

Fig. 13. HSPICE Simulation of ECL Serial Netowrk Using Multi-Chip C^2IC^2 Components

The entire 4:1 multiplexer GaAs IC was modeled using HSPICE simulation techniques. Appropriate lumped element interconnect data was used on the input and output ports of the replacable microstrip chip carrier and associated test fixture. Figure 14 compares the simulated HSPICE with actual waveforms obtained for the 4:1 MUX circuit at 1.3 Gbit/sec operation. There is excellent correlation between the HSPICE simulation and actual waveforms.

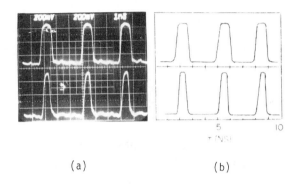

(a)   (b)

Fig. 14. Comparison of 4:1 MUX Out and Sync Out at 1.3 GHz Clock Input for (a) Actual Waveform and (b) HSPICE Simulated Waveform in Microstrip Test Fixture with Appropriate I/O Models

A copper/polyimide (Cu/PI) multilayer structure fabricated using controlled impedance microstrip interconnections was evaluated by computer simulation and experimental measurement. The dimensions of the structure were chosen to yield a characteristic impedance greater than 50 ohms. Transient electrical analysis of the structure used HSPICE simulation. The capacitance was experimentally determined using a Wayne-Kerr capacitance bridge and was used to calculate the inductance from expression[8]. The resistance of the signal line was also experimentally determined. A 0.5 volt square wave pulse with a rise time of 2 ns was used.

The results of the simulation and experimental tests are given in Figure 15 showing excellent correlation between simulated and actual waveforms. The rise times and reflections are largely dependent upon the pulse generator characteristics, line discontinuities, and impedance mismatch due to termination effects. There is no reflection or ringing in the simulated case at the input since the computer model mandates a pure wave at the input. The attenuation and propagation delay are related to the metal/dielectric system. Both electrical properties are dependent upon the R, L, and C values for the system which relate back to fundamental material properties ρ, μ, ϵ for the copper, and PI respectively. The propagation delay was found to be 69 ps/cm by computer analysis and 66 ps/cm by experimental test. Similarly, the attenuation was found to be 5.4 mv/cm by computer analysis and 5.2 mv/cm by experimental test.

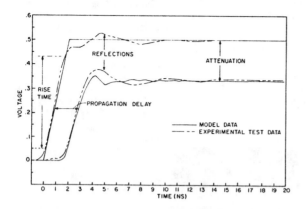

Fig. 15. Comparison of Actual Versus HSPICE Simulation for 2NS Pulse Using Cu/PI Transmission Line

THERMAL MANAGEMENT

Power densities for GaAs SDFL is in the 500 mW/mm² range when optimum gate packing densities are assumed. Therefore, when considering MSI-LSI circuitry, it is probable that power levels of 2-5 watts per chip will be realized.

For surface mounted C^2IC^2 components, the thermal impedance is a function of the chip carrier dimensions and material properties along with the GaAs chip dimensions and material properties. Figure 16 shows the coefficient of thermal conductivity for several materials. Compared to silicon, GaAs exhibits thermal conductivity values of about half. As a result, GaAs suffers from a disadvantage with respect to spreading resistance of junction generated heat. This means that thermal management techniques developed for silicon circuitry should not be automatically used for GaAs.

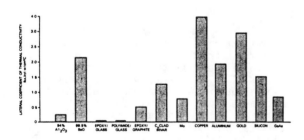

Fig. 16. Coefficient of Thermal Conductivity for Materials Used in C^2IC^2/CPWB Systems

The spreading resistance between the GaAs chip and substrate is given by[22]:

$$R_s = \frac{1}{\pi a k} \ln 4 \, (a/b) \tag{10}$$

where a,b = chip dimensions and k = substrate thermal conductivity.

Ideally, it would be desirable to have a large chip bonded to a high thermal conductivity substrate. To date, the largest GaAs ICs have been about 2.5 mm by 2.5 mm. Although materials such as copper exhibit

excellent thermal conductivity, for reasons of mechanical compatibility it is best to die bond the GaAs chip to materials such as Mo, Al₂O₃, or BeO. By proper material choices, one major source of thermal resistance, the substrate spreading resistance may be contained.

Heat transfer from the chip to the final heat sink can be accomplished by a variety of means. Which technique is used is a matter of system application. For example, ground based equipment may accommodate drastically different cooling methods than airborne equipment. It is important, however, that the first level (C^2IC^2) and second levels (PWB) of interconnection and packaging are constructed with materials and designs amenable to efficient cooling.

The conductive heat flow from the chip junction to the heat sink is given by:

$$Q = \frac{T_j - T_c}{\Theta_{JX}} \tag{11}$$

where Q = power input, T_j = junction temperature, T_c = case temperature (at point x), and Θ_{JX} = thermal impedance ($^oC/W$)

The thermal impedance is given as:

$$\Theta = \frac{L}{KA} \tag{12}$$

where K is the thermal conductivity of the material, L is the distance of the thermal path, and A is the cross sectional area of the thermal path.

When using surface mounted components such as chip carriers, one of the best methods of accomplishing short, efficient thermal paths is by direct heat sinking into the second level of packaging[20].

Using finite element techniques (ANSYS) the thermal impedance of several material combinations were calculated. Figure 17 shows the C^2IC^2 surface mounted component populating both sides of the compatible PWB with direct heat sinking. The dimensions of the GaAs chip was assumed to be 2.5 mm x 2.5 mm x 0.5 mm and the C^2IC^2 component was assumed to be 12.5 mm x 12.5 mm with a substrate thickness of 0.1 mm. A C^2IC^2/CPWB component "site" was assumed of 17.5 mm x 17.5 mm. The PWB thickness assumed was 1 mm and the core thickness assumed was 1 mm for a total compatible PWB thickness of 3 mm.

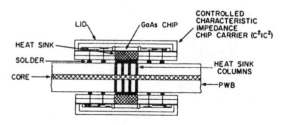

Fig. 17. Direct Heat Sinking Model of Surface Mounted C^2IC^2 Component

Since the C^2IC^2 mounted on the PWB has mirror symmetry, a two-dimensional axisymmetric solution was used. A 1 watt heat source across the surface of the GaAs chip was assumed with 66% of the power concentrated at the center and 33% distributed at the periphery. The temperature rise from point x, the edge of the PWB core to the junction was calculated. Since the power (P) and temperature rise (ΔT) are known, a thermal impedance Θ_{JX} value could be computed. Temperature rises and corresponding thermal impedances for an entire board were not calculated since this study was to determine the relative merits of various materials choices at the C^2IC^2/PWB component site. However, once a system is identified, it is mandatory to calculate the temperature rise data for the entire populated board using appropriate boundary conditions.

Table 3 lists the Θ_{JX} values for the cases considered. It is apparent that when directly surface mounting upon a PWB, efficient thermal paths must be designed. Without direct heat sinking into an efficient thermal transfer PWB core, thermal impedance values would be prohibitively high. However, when direct heat sinking using high thermally conductive materials from the chip to the core, reasonable thermal impedance values are obtained allowing the designer flexibility in component placement. The usage of BeO C^2IC^2 components offers marginally better results than alumina components with integrated heat sinks.

MECHANICAL

There are two methods to interconnect surface mounted C^2IC^2 components, (1) upon thick film ceramic daughter boards, and (2) direct attachment to the PWB. In each case the component is relow soldered to the circuit board. The direct attachment method employs reflow solder surface mounting of the component to the next level of interconnection, the PWB. Therefore, the direct attachment method eliminates the costly intermediate level of interconnection; the ceramic multilayer circuit board.

The primary problem with the direct attachment method is the lack of acceptable compatible printed wiring board (CPWB) materials. Differences in coefficient of thermal expansion and stiffness mismatch of the ceramic compared to commonly used PWB materials create solder joint failures during thermal cycling. Other effects such as moisture cause PWB material property changes which also play a role in solder joint reliability.

During the past few years there has been extensive industry-wide investigations concerning which type of compatible PWBs and solder materials should be used with surface mounted components(24,25,26 - list is only representative). Furthermore, there have been investigations in the fundamental mechanisms for solder joint failures(27) due to power cycling and studies relating fatigue failure to the number of thermal cycles(28). Honeywell is currently engaged in the development of reliable compatible PWB structures for surface mounted components. As this work progresses complete experimental data will be collected and related to finite element analysis (ANSYS).

COMPONENT RELIABILITY

Hermetically sealed packages protect against moisture and contamination. It is important, however, to utilize hermetic sealing processes whereby impurities and moisture are not trapped within the cavity. Furthermore, when using organic dielectric systems such as PI, it is necessary to use low temperature sealing processes (< 300° C). Trapped impurities along with residual moisture can result in migrative resistance shorts or electrolytic corrosion of the GaAs IC metallization. Hermetic sealing is readily accomplished by a process utilizing low temperature, vitreous, solder sealing glass. These types of low temperature sealing processes such as "hot cap" sealing have been developed under government MMT contracts(29) and are widely used in the semiconductor industry.

TABLE 3

Case	Chip Carrier	Chip Carrier Heat Sink	PWB Heat Pipes	CORE	PWB	Θ_{JX} (°C/W)
A	Al_2O_3	Mo	Cu	Alloy 42	Glass/Epoxy	36.3
B	Al_2O_3	Mo	Cu	Cu Clad Invar	Glass/Epoxy	13.3
C	Al_2O_3	Mo	Al_2O_3	Al_2O_3	Al_2O_3	17.5
D	Al_2O_3	Al_2O_3	Al_2O_3	Al_2O_3	Al_2O_3	16.0
A'	BeO	BeO	Cu	Alloy 42	Glass/Epoxy	32.3
B'	BeO	BeO	Cu	Cu Clad Invar	Glass/Epoxy	12.8
C'	BeO	BeO	Al_2O_3	Al_2O_3	Al_2O_3	11.1

CONCLUSIONS

The development of hermetic controlled characteristic impedance chip carrier (C^2IC^2) components which surface mount with compatible printed wiring board (CPWB) structures can meet the stringent performance and reliability challenges posed by GaAs digital ICs. These C^2IC^2/CPWB systems must be developed with a global view for interconnection requirements, thermal management, and application/environment by using CAD for electrical, thermal, and mechanical design.

We have demonstrated materials/processes advancements at the 1st level of packaging (C^2IC^2) which will deliver controlled characteristic impedance interconnections suitable for MSI, LSI GaAs circuitry. Metal-dielectric systems such as Cu/PI can satisfy multilayer, fine line, high resolution controlled impedance interconnection requirements. These controlled impedance interconnections have been incorporated into mechanical outlines which include termination resistors, decoupling capacitors, direct grounding, and hermeticity.

Single chip conventional packaging such as JEDEC LC^3 may be utilized below 500 Mbit/sec for digital technologies such as ECL. However, multi-chip C^2IC^2 technologies offer media delay reductions due to higher packing densities and therefore the increased liklihood of attaining proper timing of clock and data inputs when interfacing with GaAs multiplexers.

Our HSPICE electrical simulations have been experimentally verified at ECL type performance ($t_r \sim 2$ ns) and GaAs SDFL performance ($t_r \sim 150$ ps). Thermal design using finite element analysis (ANSYS) has shown efficient thermal management for GaAs SDFL power densities in C^2IC^2/CPWB systems by proper materials/design tradeoffs.

ACKNOWLEDGEMENTS

The authors wish to acknowledge the following individuals for their contributions to this work:

Processing: D. Paananen, A. Rau, T. Gruchow,
A. Fraash, J. Smeby
CAD: D. Kompelein, T. Heisler
Testing: S. Parsons, S. Dice

REFERENCES

1. "III-V Semiconductor Integrated Circuits - A Perspective"; R.P. Mandal, Solid State Technology, January 1982, pp . 94-103.

2. "Package Study for High Speed (GHz) Commercial GaAs Products", D. Wilson, N. Frick, J. Kwiat, S. Lo, J. Churchill, and J. Barrera, 1982 GaAs IC Symposium, Extended Abstracts, November 1982, New Orleans, La., pp. 13-16.

3. "Special Considerations in the Probing and Packaging of High Speed Digital ICs", J.A. Higgins, M.K. Kilcogne, Semiconductor International, May 1981, pp. 75-85.

4. "Tools and Technologies Required for the Design and Fabrication of Compact, High Performance Digital Signal Processors using Gallium Arsenide Integrated Circuits", B.K. Gilbert, T.M. Kinter, D.J. Schwab, B.A. Naused, L.M. Krueger, K.M. Rice, A. Firstenberg, Abstracts; The 1983 Space Electronics Conference, pp. 15-16, January 1983, Los Angeles, Ca.

5. "A New Chip Carrier for High Performance Applications Integrated Decoupling Capacitor Chip-Carrier 'IDCCC'", C.M. Val, J.E. Martin, Proceeding of the Technical Program - International Electronics Packaging Society, Nov. 1982, pp. 143-159.

6. "Status of Copper Thick Film Hybrids", D.E. Pitkanen, J.P. Cummings, C.J. Speerschneider, Solid State Technology, October 1980, pp. 141-146.

7. "An All Thick-Film Stripline Construction", M. Ahmad, S.M. Riad, A.A.R. Riad, and F.W. Stephenson, IEEE Trans, CHMT, September 1982, Vol. CHMT-5, No. 3 pp. 294-296.

8. "Technology Base for High Performance Packaging", J. P. Cummings, R.J. Jensen, D.J. Kompelien, T. J. Moravec, Proceedings, 32nd Electronics Components Conference, May 1982, San Diego, Ca. pp. 465-468.

9. "A Method of Manufacturing High Density Fine Line Printed Circuit Multilayer Substrates Which Can Be Thermally Conductive", S. Lebow, Proceedings, 30th Electronics Components Conference, May 1980, San Francisco, Ca., pp. 307-309.

10. "Thermal Conduction Module: A High Performance Multilayer Ceramic Package", A.J. Blodgett, D.R. Barbour, IBM J. Res. Develop. 26, 30-36 (1982).

11. "High Speed Digital Packaging", J. Vidich, Proceedings of the "80 ERADCOM Hybrid Microcircuit Symposium, June 1980, Fort Monmouth, NJ, pp. 84-97.

12. "Characteristics of Microstrip Transmission Line", H.R. Kaupp IEEE Trans on Computers, Vol. EC-16, No. 2, April 1967, pp. 185-193.

13. "High-Density High-Impedance Hybrid Circuit Technology for Gigahertz Logic", E.T. Lewis, IEEE Trans. CHMT, Vol. CHMT-2, No. 4, December 1979, pp. 441-450.

14. "High Speed Digital Logic for Satellite Communications", O.G. Gabbard, Electro-Technology, April 1969, pp. 59.

15. "Losses in Microstrip", R.A. Pucel, D.J. Masse, C.P. Hartwig, IEEE Trans MTT, Vol. MTT-16, No. 6, June 1968, pp. 342-350.

16. "Losses of Microstrip Lines", E.J. Denlinger, IEEE Trans MTT, Vol. MTT-28, No. 6, June 1980, pp. 513-522.

17. "Chip Carriers as a Means for High Density Packaging", J. S. Prokop, D.W. Williams, IEEE Trans CHMT, Vol. CHMT-7, No. 3, September 1978, pp. 297.

18. "Improved Logic Gate With a Push-Pull Output for GaAs Digital ICs", M. Helix, S. Jamison, S. Hanka, R. Vidano, P. Ng, C. Chao, Proceedings of the 1982 GaAs IC Symposium, Nov., 1982, New Orleans, La. pp. 108-112.

19. "Large Scale Digital Hybrid Microcircuits for New Military Systems", D.R. Kling, R.W. Ilgenfritz, IEEE Trans CHMT, Vol. CHMT-2, No. 4, December 1979, pp. 372-377.

20. "Hybrid Technology for VHSIC", E.T. Lewis, J. Ciccio, R.E. Thun, Proceedings of the "80 ERADCOM Hybrid Microcircuit Symposium, June, 1980, Fort Monmouth, NJ, pp. 48-83.

21. "CML Scraps Emitter Follower for ECL Speed, Lower Power", D. Buhanan, Electronics, Vol. 55, No. 22, Nov. 3, 1982, pp. 93-96.

22. R.C. Buchanen, M.D. Reeber, Solid State Technology, Feb. 1973.

23. "Advances in Microelectronic Subsystems with Chip Carriers", J.S. Prokop, D.W. Williams, R.P. Vidano, Insulation Circuits, August, 1979.

24. "Hermetic Chip Carrier Implementation", J.F. Fenimore, Elec. Pack. and Prod., May 1981, pp. 172-186.

25. "Clad Metal Circuit Board Substrates for Direct Mounting of Ceramic Chip Carriers", F.J. Dance, J.L. Wallace, Elec. Pack. and Prod., January 1982, pp. 228-237.

26. "A New Family of Microelectronic Packages for Avionics", R.E. Settle, Sol. State Tech. Vol. 21 No. 6, June 1978, pp. 54-58.

27. "Effects of Power Cycling on Leadless Chip Carrier Mounting Reliability and Technology", W. Engelmaier, Proceedings, 2nd International Electronics Packaging Conference, Nov. 1982, San Diego, Ca., pp. 15-23.

28. "Predicting Fatigue Life of Leadless Chip Carriers Using Manson-Coffin Equations", J.K. Hagge, Proceedings, 2nd International Electronics Packaging Conference, Nov. 1982, San Diego, Ca. pp. 199-208.

29. "Manufacturing Technology for Low Cost Hermetic Chip Carrier Packaging", J.S. Prokop, R.D. Vernon, R.P. Vidano, USAF-ML/LTE Contract F33615-78-C-5147 4th Quarterly Report IR-537-8-4.

MULTICHIP MODULE DESIGNS

FOR

HIGH PERFORMANCE APPLICATIONS

by

C. Neugebauer, R.O. Carlson, R.A. Fillion, T.R. Haller

INTRODUCTION

Much of the improvement in system performance promised by the ever increasing semiconductor device performance has not been realised because of the performance barriers inherent with today's packaging and interconnection approaches. Through semiconductor materials, process and equipment improvements, today's VLSI devices are faster, more complex and more efficient. However, these on chip improvements have not been matched at the package and board levels. Because of the rapidly increasing pin count of today's complex devices, VLSI packages and their multilayer boards (MLB's) have generally not reduced interconnect distances or loading and for the more complex devices, both the package and board interconnect distances have increased, thereby limiting the system performance.

To reverse this effect, chip to chip interconnections must be shortened, must have reduced loading and must have reduced circuit noise from reflections, losses and cross talk. One approach to achieve this is to eliminate chip packages and to make direct chip to chip interconnections, thereby eliminating completely the package delays and minimizing the chip to chip interconnect delays.

SEMICONDUCTOR TECHNOLOGY TRENDS

The key to semiconductor device improvements is the shrinking feature size, i.e., the minimum gate or line width on a device. Feature sizes have gone from 4 microns or larger for MSI devices, to 2 microns or larger for LSI devices and to 1 micron or larger for VLSI or VHSIC devices. The next generation devices, ULSI (Ultra Large Scale) or VHSIC-II, will have submicron features in the 0.5 to 0.8 micron range. The shrinking feature size provides increased gate density, increased gates per chip and increased clock rates. These benefits are offset by an increase in the number of I/O's and in the chip power dissipation. The increased clock rate is directly related to device feature size. With reduced feature sizes, each on-chip device is smaller and therefore has reduced parasitics allowing for faster switching. Furthermore, the scaling has reduced on-chip gate to gate distances and, therefore, interconnect delays.

PACKAGING AND INTERCONNECTION TECHNOLOGY TRENDS

Electronic packaging and interconnection approaches have had to evolve and improve to handle these increasingly more complex semiconductor devices and to meet increasingly more demanding system performance requirements. The thru-hole packaging and assembly of the MSI and LSI era were based upon individual IC's packaged into ceramic or plastic carriers called dual-in line-packages (DIP's). These had leads on 100 mil centers on two sides of a rectangular package. These leads were bent down and were soldered into a plated hole that went through the printed circuit board. With MSI devices with up to 20 leads, the DIP footprint (required board mounting area) was 0.4 inch by 1.0 inch. The footprint increased to 1.0 inch by 3.0 inch for the largest DIP's, the 64 pin LSI devices, a seven fold increase. Thru-hole assemblies are giving way to Surface Mounted Assemblies (SMA's) with devices mounted, not to thru holes but to surface pads, to reduce package footprints and to improve performance.

SMA's eliminate the need for large diameter plated-thru-holes, permitting finer pitch packages and increased routing density. Leaded and leadless chip carriers are available in quad packages with leads on 50, 40 or 25 mil centers. This reduces package foot prints, decreases chip-to-chip distances and permits higher pin count ICs. A 64 pin leadless chip carrier (LCC) requires only a 0.5 inch by 0.5 inch footprint with a 25 mil pitch. This represents a 12 fold density improvement on a 4 fold reduction in interconnect distances over DIP assemblies.

For the more complex VLSI devices, with 120 to 196 I/O's, even surface mounted approaches become inefficient and limit system performance. A 132 pin device in a 25 mil pitch carrier, requires a 1.0 to 1.5 inch^2 footprint. This represents a 4 or 6 fold density loss and a 2x increase in interconnect distances verses a 64 pin device. As more and more higher pin count devices are designed into a system, the loss of board density and operating speed becomes unacceptable, even with surface mounting. A higher performance packaging and interconnection approach is necessary to achieve the performance improvements promised by VLSI and VHSIC technologies.

MULTICHIP MODULE TECHNOLOGIES

An alternate approach to the interconnect and packaging limits of conventional chip carrier/MLB assemblies, is to insert another packaging level between the chip (level 0) and the MLB (level 2), namely the multichip module (MCM) (level 1.5). The MCM eliminates the single chip package (level 1) and instead, mounts and interconnects the chips directly onto a higher density, fine pitch interconnection substrate that provides all of the chip to chip interconnections within the MCM. Because the chips are only one tenth of the area of the packages, they can be placed closer together providing for both higher density assemblies and shorter/faster interconnects. As with each innovative packaging approach, MCM's have some drawbacks, particularly in the areas of test, repair and yield. These include bare chip testing, chip handling damage, burn-in yield, parts availability, customer acceptance, repair processes and engineering change capability.

In the multichip module (MCM), bare chips are interconnected on a substrate with substantially finer conductor lines, smaller dielectric thicknesses, and denser via grid than the board, and thus is not subject to the conventional PWB design rules and assembly restriction. At first glance, it appears easy to place bare chips closer and closer together. There are, however, limits to how close chips can be placed together on a substrate. There is, for example, a certain peripheral area around the chip which is normally required for bonding, engineering change pads and chip removal and replacement. With some substrate materials, such as thick film ceramic, the pad or routing density or the chip power dissipation limits the chip placement.

The interconnect technology of bare chips with vacuum deposited thin films or with screened-on thick films dates back many years. Fine line geometries are possible with thin film gold with minimum space or weight needs for space operation or for microwave circuitry. The thick film approach using metal/glass pastes has typically 10 mil minimum widths but can build up many signal layers using dielectrics of ceramic and devitrifying glasses. It is attractive for analog circuitry because laser trimmable resistors can be fabricated from screened on pastes, but is suitable for high performance circuitry only if many interconnect layers are employed and line impedance is controlled.

Table I summarizes some high performance interconnect technologies for bare chips taken from the recent literature. The predominant dielectric material is polyimide with aluminum or copper metallization processed on a silicon or alumina substrate. Chip attach methods include wirebond, TAB, solder bump and two direct on chip metallizations. Polyimide offers a low dielectric (~ 3.5) material that is readily applied by spraying, spinning or film lamination and readily patterned for vias and interconnection runs. The GE approach uses an overlay interconnect geometry that eliminates wirebonds, allows for direct chip-to-chip interconnect and requires no masks for patterning using a computer driven, laser controlled lithography technique. (This novel process is described in paper #2 in session 15, "A High Density Copper/Polyimide Overlay Interconnection").

The interconnect issues which all of these high density approaches using bare chips must face are listed in Table II below. In many cases, there are conflicting demands such as line density, cross talk, yield, ease of fabrication, ease of repair and circuit changes, and cooling difficulty. Thus, the systems designer must understand the tradeoffs involved in committing a circuit design to a module layout, working closely with the circuit designer to understand systems requirements such as chip placement, I/O layout to the next level of packaging, etc.

All of these high performance multichip module interconnect technologies hold the promise of overcoming the chip-to-chip interconnect and the package density barriers facing single chip packaging. However, the vast capacity of these approaches to assemble in a relatively small area, 10 to 20 times more function, leads to interconnect problems at the higher packaging levels. This can be viewed as a series of interconnect barriers along a long obstacle course one must take in order to efficiently package todays more complex devices.

As long as carriers were limited to a 40 or 50 mil pitch, a 20 mil pitch MLB was not a barrier. However, with the availability of 25 mil carriers, the 20 mil pitch MLB was a barrier to efficient packaging. The 200 pin MLB connector was not a barrier, until carrier and MLB barriers were overcome, allowing more function on each MLB.

RENT'S RULE VS. SYSTEM ARCHITECTURES

Rent's Rule, which relates the number of I/O required for a block of logic elements, has withstood many generations of semiconductor device technology improvements. The rule has been modified and adjusted at times, but it is still generally valid.   Random logic devices such as gate arrays follow the original curve. Microprocessors, with their more structured architectures, fall on a curve with less pins per gate. Finally, memory devices, which are an array of elements, are on the lowest curve with more than an order of magnitude fewer pins than a microprocessor with an equivalent gate count. Generally, the more structured a device is the fewer I/O pins it needs to function effectively[1].

The basic equation for Rent's Rule is:

$$P = A (G) \text{ Exp. } B,$$

where P is the number of signal I/O, G is the number of gates, and A and B are the coefficient and exponential factors, respectively, that vary with the type of logic being considered. Table III shows the Rent's constants for several types of logic functions.

TABLE I

HIGH PERFORMANCE MULTICHIP MODULE INTERCONNECT TECHNOLOGIES

USER	SUBSTRATE	CONDUCTOR W,T,P	INSULATOR	CHIP ATTACH	MODULE I/O	ADVANTAGES	DISADVANTAGES
MOSIAC (2)	SILICON	ALUMINUM 11,2,22	SILICON DIXIDE	WB	WB	VENDOR/SI CHIP TECHNOLOGY/TCE MATCH	HIGH CAPACITANC HIGH RESISTANCE LINES
IBM (3)	ALUMINA	COPPER 8,6,25	POLYIMIDE	SB	PGA	LOW ATTACH IN-DUCTANCE SUBSTRATE FOR PWR/GND	HEAT REMOVAL
HONEYWELL (4)	ALUMINA	COPPER 50,5,125	POLYIMIDE	WB/T	WB	SUBSTRATE FOR PWR/ GND	
ATT (5)	SILICON	COPPER 25,5,50	POLYIMIDE	SB	PGA	LOW ATTACH INDUCT-ANCE/TCE MATCH	HEAT REMOVAL
RAYCHEM (6)	SILICON	ALUMINUM 40,5,100	POLYIMIDE	WB	WB	VENDOR TCE MATCH	
GE (7)	ALUMINA	COPPER 25,5,75	POLYIMIDE	OVERLAY	WB	NO MASKS/FAST PROTO-TYPE/LOW ATTACH INDUCTANCE	MUST REMOVE OVERLAY TO REPLACE CHIPS
LIVERMORE (8)	SILICON	VARIOUS	VARIOUS	BEVELED CHIP EDGE		HANDLE HIGH PIN COUNT TCE MATCH	BEVELING STEP LIMITED LINE ROUTING

W = WIDTH	WB = WIRE BOND	NOTE:	CONDUCTOR DIMENSIONS IN MICRONS.
T = THICKNESS	T = TAB BOND		NUMBERS LISTED ARE CONSERVATIVE
P = PITCH	SB = SOLDER BUMP		REPORTED VALUES, NOT MINIMUM
	PGA = PIN GRID ARRAY		VALUE.

TABLE II INTERCONNECT BARRIERS FACING LSI AND VLSI

BARRIER	SOLUTION
o DIP Pin Limitation: 64 pins	o Surfaced Mounted Chip Carriers
o MLB Routing Density: 100 lines/inch	o Finer Pitch lines/Buried Vias
o MLB Connector Limits: 100 pins	o Finer Pitch/Multiple-Row
o Back Plane Routing Limits: 100 lines/inch	o More layers/Buried Vias/Finer Pitch
o LCC Pin Limitations: 200 pins	o High Density Multichip Module
o Large LCC Foot Print: 1 to 1.5 inch^2	o High Density Multichip Module
o Large MCM Foot Print: 25 mil pitch	o Finer Package Pitch - 20/15 mils
o MLB PTH Density: 100 mil grid	o Finer Holes/Tighten Grid - 75/50 mils
o MLB Routing Density: 200 lines/inch	o ?
o MLB Connector Limits: 200 pins	o ?
o Backplane Routing Limits: 200 lines/inch	o ?

TABLE III

Rent's Constants for Various Logic Types

LOGIC TYPE	A	B
Static RAM	6.0	0.12
Microprocessor	0.8	0.45
Gate Array	1.9	0.45
Random Logic	1.4	0.63

A new aspect to Rent's Rule has developed with the advent of distributed processing. In conventional processing systems a central processing element such as a computer, processor or microprocessor is involved in virtually all operations. To increase throughput, additional processing capability is added to the central processor, such as increasing the number of internal registers, the width of the data bases or the operating speed. In distributed processing, though, additional processing capability is provided by adding additional processing elements. In some cases, it will be a mix of specialized processors, with different operations performed by different processors. But in an increasing number of cases, the multiple processing elements are identical. The increased functional throughput is achieved by having more than one processor either solving different parts of the same problem or by solving different problems at the same time.

There are a host of new distributed architectures being championed, each with its unique advantages and its restrictions. These generally fall into three families; parallel processors, pipeline processors and array processors. In parallel processors, multiple processors operate on different parts of the same problem or similar problems. In pipeline processing, each processor performs one step of the operation, then passes the partially processed data to the next element, forming a serial chain or even a loop of processors. Finally, array processors, combine the operations of parallel and pipeline elements by having a block of elements working in parallel for one step then passing the partially processed data to the next bank of elements for the next stage of the operation.

Although these architectures offer the system's designer many new ways to solve problems, all of these distributed approaches strain the limits of our packaging and interconnection technologies. Distributed processing requires a large number of wide data buses going between the distributed elements. The number of interconnections varies a great deal depending upon which of the three types of architectures, parallel, pipeline or array, are involved.

For example, if we have four microprocessors each with 100 I/O pins (32 Data In, 32 Data Out plus controls), we can configure them into any one of these three architecture types. Let's then calculate the number of interconnects between the four chips and the number that go external to the cluster. In the parallel configuration, almost all of the control lines and the datalines are unique and therefore are external. There would be be only 40 internal interconnects and 360 external interconnects. In the pipeline configuration, the data out of the first processor feeds the second processor and so on. There would be about 140 external interconnects and about 260 internal interconnects. Finally, in the array processor, where there is both parallel and pipeline operations, there would be about 200 external interconnects and 200 internal.

The array configuration follows Rent's Rule for processors (P = 0.95 x G Exp. 0.45) with a 2x increase in pins for a 4x increase in gates (four chips vs. one chip). The parallel configuration on the other hand, has many more pins than predicted by Rent's Rule, while the pipeline configuration has many less. In this example, we showed that with the same device type, a 100 pin microprocessor, there are three Rent's Rule curves needed for the three distributed processing architectures described.

Table **IV** summarizes the Rent's constants for the various system architectures.

TABLE **IV**

RENT'S CONSTANTS FOR VARIOUS SYSTEM ARCHITECTURES

ARCHITECTURE	A	B
High Speed Computers	82	0.25
Parallel Processing	0.005	0.92
Array Processing	0.82	0.45
Pipeline Processing	8.0	0.25

BREAKING RENT'S RULE

The systems designer is normally constrained to present a relatively small interface to the outside world. Thus, a state of the art printed circuit board today is typically limited to a 200 to 400 pin interface to the next packaging level, depending on board size and edge connector technology. On the other hand, the number of pins which can be accommodated on the board itself is of the order of several thousand, depending on the PTH grid, the wirability, and the number of interconnect layers. If the board is to be effectively utilized, the systems designer must look for ways to drastically reduce the pin count to conform to the edge connector limit. For example, for an array processor design using a mix of 160 VLSI logic and memory chips for a total of 7M gates, 11920 signal I/O and power/ground pins are presented to the designer on the chip level, which he must now reduce to a few hundred by proper interconnect design. Only perhaps 400 of these can be brought out through the board edge connectors to the backplane. The remaining pins must be interconnected on the board. The difficulty is that, in order to achieve this pin count reduction, even the modified Rent's Rule must be broken.

For the array processor in our example, the form of Rent's Rule which applies is given approximately by P=0.82(G)exp0.45 (1), where P is the pin count (including power and ground pins) associated with a system of G gates. According to this rule, the number of edge connector pins on the board should be more than 1100, if the entire system is to reside on one board. But only 400 are available. The systems designer must therefore achieve a pin count reduction in excess of that predicted by Rent's Rule.

The system designer achieves this pin reduction by innovation and multiplexing. Various data buses are multiplexed onto a common set of I/O pins on a time shared basis. This reduces the number of pins but at a cost of reduced performance, speed and programmability. In higher performance systems, these options are unacceptable and other means must be employed to break Rent's Rule.

Rent's Rule can be broken at any level of packaging. The microprocessor chip, with its predominantly structured architecture, is an example of breaking Rent's Rule at the chip level. High performance multichip modules can get around the pin limits at the package to board interface. It is advantageous to push the Rent's Rule restrictions to the higher interconnect levels, allowing parallel data paths between chips and thus higher system performance and greater architectural flexibility.

A COMPARISON OF HIGH DENSITY INTERCONNECTION APPROACHES

It is instructive to compare four high density packaging approaches for implementing a hypothetical array processor system containing 6.98 gates consisting of 160 VLSI chips. They are:

- 160 SCM's containing 1 chip each mounted directly on a double sided PWB

- 16 MCM's containing 1 array processor (10 VLSI chips) each mounted on a PWB

- 4 MCM's containing 4 array processors (40 VLSI chips) each mounted on a PWB

- 1 MCM containing 16 array processors (160 VLSIchips)

This array processor system contains 16 array processor chip sets with ten VLSI chips per chip set. There are four 100 pin, 40,000 gate microprocessors, four 25 pin, 64,000 gate RAM's, one 220 pin, 20,000 Gate Array are one bus driver glue chip. Although there are 745 I/O in the ten chip set, the high degree of chip to chip interconnect reduces the number of I/O for the array processor to only 200 pins, consistant with Rent's curve for microprocessors.

A mix of VLSI chips as shown in Table V is assumed.

TABLE V   CHIP MIX FOR 7MGATE ARRAY PROCESSOR SYSTEM

Chip Type	No. of Chips	No. of Signal I/O's Chip	Gate/Chip
Microprocessor	64	100	40,000
RAM	64	25	64,000
Gate Array	16	220	20,000
Glue	16	25	100
TOTAL	160	11920	$6.98.10^6$

(1840 for power and ground)

High density packaging techniques are postulated, and include:

- The minimum allowed pin pitch on packages is 0.025 inches.

- The PTHs grid on the PWB are on a 0.100 inch grid.

- The wirability of the PWB is 40 lines/inch on each layer.

- The wirability of the MCM is 100 lines/cm on each layer.

- There are 2 signal layers on the MCM.

- There are no via limitations on the MCM.

- Custom MCM packages of various sizes are available.

- The PWB presents 200 pins to the backplane per side.

- VLSI chips are CMOS and have an average fan-out of 2.

- Rent's curve for array processors will hold.

A comparison of packaging density and performance related parameters for these four packaging approaches is given in Table VI. The most striking result is the dramatic gains in both functional density and performance which are achievable by use of the MCM. A total of 11900 signal I/Os and 1840 power and ground pins must be interconnected for this system. If implemented conventionally by mounting SCM's directly on the PWB, 5760 I/Os must be interconnected on each of the PWBs. On the other hand, if 16 MCM's with one array processor in each, mounted instead, only 1608 I/Os must be interconnected on each of the PWB, while the rest (745 per MCM) are interconnected on the MCMs themselves. Because of the MCM's superior wirability, the ten chips in the MCM can be interconnected in a much smaller area then on the PWB. A density increase of 3x is thus possible, with the atttendant speed benefits of 2:1 and a speed, power, density figure of merit improvement of 15x.

An even higher packing density and performance are possible if more of the I/O interconnection effort is thrown into the MCM, although this is somewhat less dramatic. In our example, this implies interconnection of four array processors (40 chips) in one single MCM and mounting four MCMs onto the PWB. A packing density increase of 1.3x, a speed increase of 1.3x, and a power dissipation decrease of 1.2x are thus achievable over the 10 chip MCM version. Finally, the highest packing density and performance would be possible if all the I/O interconnection effort is thrown into the MCM. In this version, the PWBs would be completely eliminated, all chip to chip interconnection of the 160 chips in the 16 processor array would be within the high performance, high density MCM substrate. The key issue, though, is whether a single MCM can be fabricated with 160 complex ICs with any reasonable yield or with any practical test and fault isolation methodology. For most applications, modules with up to 30 or 40 chips would be practical if the individual ICs are fully tested prior to module fabrication and if the interconnection substrate is high yielding or readily repairable. With lower chip yields and/or interconnect yield, then modules may be limited to 10 or 20 devices.

The packaging figure of merit is the ratio of the assembly gate density, and the interconnect power-delay product. With the standard 160 single chip, PWA approach normalized to 1, the 10 chip, 16 module MCM approach shows a 15 fold improvement. The 40 chip MCM approach showed another factor of 2x (30x total) improvement. The additional 1.6x improvement in the 160 chip MCM approach is marginal at best and more than offset by lower module yields.

In each of these configurations, the package footprints are limited by chip or substrate area, package I/O pitch and PWA PTH Grid. First, the package must be large enough to contain the single chip or the multichip substrate. Second, the package perimeter must be at least equal to the numer of I/O pins times the pitch. (25 mil Pitch). Finally, the board surface area for PTH interconnection must be at least equal to the number of I/O pins times the square of the PTH Grid (100 mil grid).

TABLE VI

COMPARISON OF INTERCONNECTIONS FOR THREE HIGH DENSITY
PACKAGING APPROACHES FOR 100 CHIP, 4.36 MGATE ARRAY
PROCESSOR SYSTEM

	Packaging Parameter	160 SCMs on PWB	16 MCMs on PWB	4MCMs on PWB	Single MCM
Density	Total Package Foot-print Area, cm^2	1202 cm^2	370 cm^2	280cm^2	207 cm^2
	Functional Density, Gates/cm^2	6K Gates/cm^2	19 K Gates/cm^2	25 K Gates/cm^2	34 K Gates/cm^2
I/Os	No. of Interconn, on MCM	-	8384(72%)	9900(84%)	11500(97%)
	on PWB	11500(97%)	3216(25%)	1600(13%)	-
	on Backplane	400(3%)	400(3%)	400(3%)	400(3%)
Inter-Connections	Total Interconn. length, meters	1750m	883m	684m	644m
	Ave. Signal Path Length between I/Os, cm	15.2 cm	7.68 cm	5.95 cm	5.6 cm
	MCM Area	-	2.25 inch^2	9.0 inch^2	36 inch^2
	PWB Area	210 inch^2	64 inch^2	49 inch^2	-
	MCM - Layers	-	2	2	2
	PWB - Layers	6	6		4
Package Performance	Ave. propagation delay between I/Os, nS	1.08	0.48		0.34nS
	CV^2 power dissipated in ave. signal path, mW	25mW	13mW	10mW	9mW
	Power x Delay for average signal path, pJ	27pJ	6pJ	4pJ	3pJ
Figure of Merit	Density x Performance gates/cm^2 pJ	220 gates/cm^2 pJ	6250 gates/cm^2 pJ	11300 gates/cm^2 pJ	pJ
	(normalized)	(1)	(15)	(30)	(50)

POWER LIMITATIONS

Only geometric factors have been considered in estimating the achievable packaging density. In so doing, we have assumed use of CMOS technology and modest (50MHz) clock rates, so that conventional air cooling approaches still apply. However, it should be recognized that we are very close to the ragged edge of power removal capability, and that drastically improved cooling approaches, such as liquid immersion, nucleate boiling, or direct fluid cooling of the grooved chip backsides must soon be employed if the functional and power densities are expected to improve much in the future.

Power delivery is not much easier. Because of ever dropping power supply voltages in the face of increasing power requirements, inordinate current density requirements can be avoided only by use of local power supply modules dispersed among the digital circuitry itself. This allows power distribution to be largely by a higher voltage, and avoids unmanageable ΔI noise.

CONCLUSIONS

1. Systems density, speed, and power dissipation can be greatly improved by the use of advanced packaging techniques.

2. The most dramatic improvements can be obtained by placing the bulk of the interconnection effort into the lowest possible packaging level, namely the chip itself (level 0), and the MCM (level 1.5), where the wirability is high, as opposed to the PWB (2nd level) or the backplane (3rd level), where the wirability is low.

3. For systems of moderate pin intensity, such as array processors, most of the improvements can be made with MCM's of moderate chip count. In systems of higher pin intensity the optimum module size is expected to be larger.

4. Pin-out limitations in the higher packaging levels require that the systems designer achieves substantial pin count reductions in the MCM packaging level; i.e., the number of pins per gate being packaged on the MCM and the PWB must decrease below the value predicted by Rent's Rule for VLSI chips.

5. Systems with pipeline or array processing will have inherently lower module and PWB pin counts than predicted by Rent's Rule, while parallel processing systems will have much higher pin counts.

REFERENCES:

1. J.D. Meindl, "Opportunities for Giga scale Integration", Solid State Technology, Dec. 1987, pg. 85.

2. H. Stopper, "A Wafer with Electrically Programmable Interconnects", in Tech. Dig. IEEE Int. Solid State Circuits Conf., paper FAM18.5, pp. 268-269, Feb. 1965.

3. C.S. Ho, D.A. Chance, C.H. Bajorek, and R.A. Acosta, "The Thin Film Module as a High Performance Semiconductor Package", IBM J. Res. Develop., Vol. 36, No. 3, pp. 286-296, May 1982.

4. R.J. Jansen, J.P. Cummings, and H. Vora, "Copper/Polyimide Materials System for High Performance Packaging", IEEE Trans. Comp., Hybrids, and Manuf. Tech. Not. Vol. CHMT-7, pp. 384-393, Dec. 1984.

5. C.J. Bartlett, J.M. Segelken, and N.A. Teneketges, "Multichip Packaging Design for VLSI-Based Systems", IEEE Trans. Comp., Hybrids, and Manufac. Tech., Vol. CHMT-12, No. 4, pg. 647 (1987).

6. E. Bogatin, "Beyond Printed Wiring Board Densities: A New Commercial Multichip Packaging Technology", Proceedings of Nepcon/East '87, pp. 218-226, June 1987.

7. C.W. Eichelberger, R.J. Wojnarowski, R.O. Carlson, and L.M. Levinson, "HDI Interconnects for Electronic Packaging", SPIE Symposium on Innovative Science and Technology, Paper 877-15, Jan. 1988.

8. D.W. Tuckerman, "Laser Patterned Interconnect for Thin-Film Wafer Scale Circuits, "IEEE Electron Device Letters, Vol. EDL-8, NO. 11, pp. 540-543, Nov. 1987.

Multichip Packaging and the Need for New Materials

J. W. BALDE

Interconnection Decision Consulting
Flemington, New Jersey 08822

The sole determiner of speed, of cost, and of possible system size is the interconnection density of the circuits. The best way to increase the interconnection density is to reduce the dielectric constant, and there are new materials that offer almost a 2:1 reduction.

Designing new modules with these low dielectric constant materials and with larger chips increases the wattage and increases the wattage density. Ceramic substrates must be used to dissipate the heat of the new large higher wattage chips without destroying them. This requires the use of new ceramic materials. A choice must be made between the leading contenders. A new multichip module technology is being defined, and will result in a major packaging change. The processes and the materials choices are presently overwhelming, but there are great advantages to those who identify the best solutions.

Key words: Multi-chip packaging, packaging materials, benzocyclobutene, AIN, SiC

INTRODUCTION

We live in a time of change, and the electronics industry sees more change than any other. Semiconductor packaging is changing from DIPs to Chip Carriers, and now to Multichip Packaging. Each change brings a reduction in size, an increase in circuit speed, and eventually a decrease in cost. Each change also brings an increase in complexity, and diversity of choices, and an increasing risk. The risk comes from the choice of the wrong material, the wrong assembly, or the commitment to a technology that is not ready. This paper will attempt to discuss the new technologies needed and the choices that must be made.

We have learned in the past few years that the original perception by Rex Rice and others[1] was only a part of the story. They said correctly that interconnections are cheaper on the silicon wafer than on a Hybrid, and that PC boards, backplanes and cabinet interconnects were each greater in cost by a factor of 10. That led to larger and larger chips, bigger and more complex multichip hybrids and higher density printed circuit boards. But it was only part of the story.

Knausenberger and Schaper[2] normalized the costs as a function of the wire length, and found that interconnections on the wafer surface were no cheaper PER INCH than any other wiring; an inch of cable or an inch of PC board costs the same as an inch of interconnect on the silicon wafer.

The result of that difference is emphasis is that the only cost criteria of ultimate interest is the size of the system; an increase in density reduces costs whether that increase in density is due to more interconnect on a chip, or more interconnect on the substrate board of ceramic or plastic. If the total size is reduced by using a more efficient interconnect media, it is possible to reduce costs and increase speed. Semiconductor chips of ordinary density can then be used because the interconnection is more efficient.

How does one do this? By understanding the way materials affect the interconnection density. It is the change of materials that makes possible the higher density electronic assemblies and the high density modules.

INCREASING DENSITY THROUGH LOW DIELECTRIC THICKNESS

There have been discussions elsewhere on the effect of low dielectric constant.[3,4] If density is to be increased the adjacent lines cannot be brought closer together in interconnect structures of the same materials and structures without incurring excessive inter-circuit crosstalk. The coupling between the fields of adjacent wiring traces is just too great. What are the possibilities for decreasing the coupling?

The first and earliest was to substitute silicon dioxide for epoxy-glass for the interconnect insulation. Both dielectrics have about the same dielectric constant, about 4.5 to 5. The capacitance of a wiring trace on the surface of these insulating materials must be equal to the thickness of the dielectric to achieve a characteristic impedance of 50 ohm. If the epoxy-glass is 7 mil thick, the printed board lines are 7 mil wide; if the lines are reduced to 3 or 4 mil, the epoxy glass must then be 3.5 mil. If silicon dioxide is used as the insulation, a 10 micron metallized line on 10 micron of dielectric also produces the same characteristic impedance.

For ECL circuits which are terminated in characteristic impedance loads, 50 ohm is about as low as one can risk going. In a paper by Evan Davidson,[5,6] lower impedances produce greater noise, mostly from dI/dT noise. Furthermore, lowering the line impedance will require greater power consumption and heat dissipation problems, and increase the I/O leads to all devices to cope with the dI/dT problems.

Even for CMOS circuits, low line impedance pro-

(Received June 7, 1988; revised July 18, 1988)

duces greater skew delay, and forces greater power consumption as the circuits are required to charge lines of higher capacitance. No, the ratio of line width to dielectric thickness must be about unity *provided the dielectric constant remains the same.*

For these materials, the closest line spacing is determined by the permissible crosstalk coupling between adjacent lines. For an acceptable level of unwanted signal pickup of about 10%, the nearest approach of a quiet line to a active pulse-carrying line is 3:1. The printed circuit board industry has been producing 4 mil line and 12 mil space boards, 3 mil line and 9 mil space, and most recently 1 mil line and 3 mil space in the new thin film interconnect. Even the reduction in size of the interconnect in this technology has not yet reduced the relationship of line and space. The only difference is that there is recognition that the interconnect capability is approaching that of the silicon chip, so that the dimensions are quoted in microns instead of mils. But 25 micron line and 75 micron space is the same three to one ratio.

What then is the advantage of using polyimide insulation on ceramic or on the surface of a silicon wafer? It is the reduction in size through the reduction of the dielectric thickness, and this reduction in scale brings to the electronics systems designer an alternative to placing the interconnect wiring on the same silicon semiconductor chip with the active ICs. With only a modest fine line capability, say 10 micron or even 25 micron, the interconnections between chips have been fully allocated in just 2 layers of interconnect. No workers in this "new" multichipping technology have had to go to any more than a single pair of *x-y* signal planes to perform all the required interconnect.[7-15] By using polyimide of 15 micron thickness, interconnect traces of 15 micron can have acceptable capacitance.

But there is a problem. Fifteen micron traces of aluminum provide appreciable resistance for the longest paths of 8 inches or more. Copper would be better, and 25 micron lines would be better yet. Multilayer copper traces in silicon dioxide insulation is difficult, but is readily achievable in multilayer polyimide construction. Many companies are developing interconnect structures using copper and organic films; the NEC and NTT structures are typical (Fig. 1, Refs. 7–8). Increasing the stripe width, however, requires increasing the dielectric thickness. Present technology of spin coating cannot satisfactorily go above 15 micron per coating layer. Increasing thickness to 25 micron doubles the cost.

INTRODUCING LOW DIELECTRIC CONSTANT DIELECTRICS

The introduction of a lower dielectric constant organic coating can have two significant effects of this problem of increasing density and increasing performance. The first is on the permissable nearest approach of a second line. If the dielectric constant is reduced to one half, the fringing fields around an

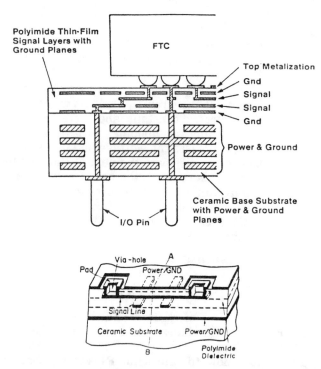

Fig. 1 — The Two Main Approaches to Organic Insulated Signal Interconnections Organic insultated interconnection on co-fired Ceramic Substrate (the NEC structure, as shown in the '85 ECC Proceedings) and the Organic Thin-Film Interconnect Layers on solid substrate (The NTT structure, as shown in the '87 IEMT Proceedings)

Reproduced with permission IEEE IEMT '87 and IEEE ECC '85.

active line are also halved, because the ground plane can be nearer. If the characteristic impedance is kept the same, then a reduction in dielectric constant can permit a reduction of the thickness of the dielectric.

If the fields are compressed because of the closer approach of the signal traces to the ground plane, then the nearest trace can also approach by that same ratio. The new Dow Chemical Polycyclobutene[16] with its dielectric constant of 2.6, about half of the 5.0 of polyimide under non-hermetic use conditions, permits the reduction of the space to the nearest line from 75 micron to 40. A 3 trace wiring "street," 450 micron or 18 mils wide can be replaced by a 6 trace street in the same width. That is double the wiring density for the same line width (Fig. 2).

There is more to this than doubling the wiring density. It has accomplished this without needing to reduce the stripe or trace width. There is no need to reduce the path below 1 mil, which not only has problems of higher resistance, but is extremely difficult to produce without pinholes.

Reducing the dielectric constant for additive dielectric thin film circuits for the "new" multichipping can double wiring density, permit thinner dielectrics, and retain wiring path width all at the same time. Costs of thin film dielectrics are proportional to dielectric thickness, because thicker layers are more difficult to coat. Costs are also proportional to dielectric constant, because the line widths can be greater, and therefore cheaper to produce with good yield.

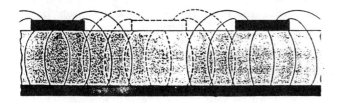

THIN AND LOW DIELECTRIC CONSTANT SUBSTRATE
PROVIDES HIGHER WIRING DENSITY

Fig. 2 — Reduced Dielectric Constant Permits Reduced Dielectric Thickness. With thinner dielectric of lower dielectric constant, field spreading and lateral capacitance if reduced. The result is higher wiring density with no increase in crosstalk.

J. W. Balde, IDC Corporation, reproduced with permission of IPC World.

Why then is this material not used? First, it is a new material, just available. Second, the circuit producers have had no experience with it, and worry that there might be some hidden problem. This is a reasonable worry–there have been plenty of materials that have promised much and delivered little, or with a major flaw. For example, even polyimide has continuing problems, moisture absorption and change in dielectric constant, and incompatibility with copper. One has to use a barrier of chrome-nickel, and use the material only in hermetic sealed packages, both of which are possible, but a pain.

Either polycyclobutene or some other material is needed for the new multichipping to bring it to full potential. In particular, abandonment of hermetic sealing as a means of protecting the semiconductor chips will increase the pressure for the abandonment of polyimide. If a silicon gel is used to prevent corrosion of the semiconductors[17] then the polyimide surface will be wet continuously, because of the ability of silicon gels to transmit water to the interconnection media surface. This is a truly great problem for the polyimides and other interconnection dielectrics that change dielectric constant with absorbed water.

USING LOW DIELECTRIC CONSTANT MATERIALS IN THE PRINTED CIRCUIT BOARD

The printed circuit board used for the interconnection of the semiconductor packages can also benefit from the use of a lower dielectric constant material. Just as in the thin organic film dielectric case, a reduction of the dielectric constant permits the closer approach of the adjacent line. A circuit of 3 mil trace and 9 mil space can be reduced to a circuit of 4 mil trace and 4 mil space. This would also double the wiring density, permitting more traces within

the via pad metalization. It also would permit increasing the stripe width to 4 mil, much easier to do.

The problem here is different from that of the additive dielectric. Additive dielectric is difficult to make THICK, but epoxy-glass or other glass reinforced printed circuit board laminates are difficult to make THIN. The dielectric constant can be made less using expanded PTFE for the reinforcing fibers,[18] but the thickness cannot be reduced below that required for the overlaping intersection of the reinforcing fibers, coated with a little dielectric.

Here also the solution is a new material. Rogers Corporation (Fig. 3), with their new RO 2800 material,[19] has produced a stable printed circuit board laminate with PTFE as the matrix, a ceramic plus chopped fiberglass for the filler. The PTFE is a low dielectric material, and the silicon dioxide-based ceramic filler keeps it low. The chopped fiberglass fibers are small (10 micron) and in such sufficiently low quantities that they do not adversely affect the dielectric constant, but do produce the necessary board integrity and stability.

What of this material? It CAN be made thin, and it has a lower dielectric constant of about 2.8. It can double the wiring density, and therefore reduce the number of layers of signal wiring needed for a high performance circuit. Typically a 12 to 15 layer board with 6 signal layers can be reduced to a 4 layer board with only 2 signal traces. At the least, the number of layers can be reduced by 2:1. Since board costs are more than proportional to the number of layers, the board costs can be reduced by 80% or more. In a recent study by the author, a multichip modules with organic insulation for the interconnection layers were 5 to 10 times higher in cost than printed circuit boards, but about 1/10 to 1/5 of the cost of an implementation in multilayer co-fired ceramic.

Why is this technology not widely used? Same answer. A single supplier of the material, and the need to change such standard practices as the lamination temperature, and the required etch of the lami-

RO2800™ Fluoropolymer Composite

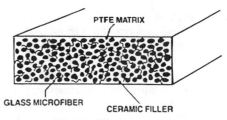

- **Highly** Filled
- Reinforced with Randomly Oriented Microfibers

Fig. 3 — The Rogers RO 2800 Board Construction, showing how the low dielectric constant matrix and fillers can be produced in reduced thickness.

Dave Arthur, Rogers Corporation, reproduced with permission of the IPC.

nated before metalization. There ARE new processes to be learned, but the benefits are there for the company willing to change.

USING NEW CERAMICS IN THE NEW MULTICHIPPING APPLICATIONS

Increasing the density of circuits requires that the individual IC chips be liberated from their customary individual packages and grouped into larger packages with many chips. Multichip packages have been with us for many years, and the Hybrid industry, with $9 Billion in sales, is as large as the printed circuit board industry, and almost as large as the semiconductor industry itself. High density multichip packages have been used by IBM and Mitsubishi.[20-21] They are no longer the proper solution for new systems packaging. The size of the chips is increasing rapidly, and the temperature coefficient of expansion (TCE) differences require a better match between the chip and the package than possible using alumina as the dielectric. Furthermore, the high dielectric constant and the poor thermal properties of alumina are limiting the possible high performance applications.

Ceramic manufacturers can make the dielectric constant lower,[22-24] but at the expense of even poorer thermal performance. Heat must be removed from the larger die expected in the future, and the multiple chips mounted in the new multichipping packages make for total wattages often above 300 W per package and 10 W per chip. Taking the heat away from the chips from the top of the chips is possible, but extremely costly. Only the high performance manufacturers can use multiple plunger heat removal. Even providing heat caps with massive fins and close approach to the die back surfaces is limited. A ceramic interconnection that does not permit efficient heat transfer is a problem.

Substrate materials are available with better heat transfer, but those such as Boron Nitride, Beryllium Oxide, Graphite and others that do not match the TCE are not usable. The die must be mounted to the substrate with a very thin layer of bonding material, because adhesives are very poor thermally, and produce great thermal impedances unless they are very thin. But if the adhesive is to be thin, the bigger die required for future designs must be mounted to a substrate with a very close match in TCE or die breakage will occur. (The alternative, using a very thick compliant die bonding adhesive is only useful if the die is low wattage, and can tolerate the poor thermal conductivity of thick die bonding adhesive.)

The best substrate material for direct die bond must have a TCE through the room temperature-to-85° C range of 2.5 or so. There are only three materials that presently qualify: mullite, aluminum nitride and silicon carbide. There are important differences between those materials.

The first resolution criteria is thermal conductivity. Mullite is quite inferior to the other two, even though it does match the TCE and affords a good dielectric constant.[24] For high density, high performance custom chips, heat spreading conductivity is of marginal interest. As Avram Bar-Cohen has pointed out,[25] if the adjacent packages also dissipate approximately equal heat, the only thermal impedance of importance are those in the path directly down. The poor thermal performance of the adhesives for the die bond and the heat sink bond make the requirement of the ceramic modest. That is not the general case however.

For cost-performance and military packaging only a few of the chips in the package generate appreciable heat. These hot chips can transmit heat to the ceramic substrate, and that substrate can conduct heat laterally under the lower-power chips (Fig. 4). The better the thermal performance of the ceramic substrate, the greater is the heat spreading. For silicon carbide or aluminum nitride with a thermal conductivity of 270 W/m-K heat spreading to the adjacent portions of the interconnect area can transfer about 75% of the heat away from the area directly under the chip. The heat THEN transfers to the heat radiator.[26] The heat flow through a unit area of the adhesive is reduced. The result is an apparent increase in the thermal conductivity of the adhesive to the heat sink because of the greater area for the heat flow. The adhesive between the ceramic base and the heat sink must be thick because of the great TCE difference between the ceramic and the aluminum of the heat sink. The reduction of the thermal impedance between the ceramic and the aluminum finned radiator makes an important differences.

The adhesive can also be reduced in thickness by using heat sinks of Copper-Tungsten. Matching the

WHY GOOD HEAT SPREADING ?

Main Frame case didn't need that

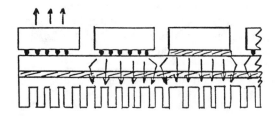

But "NEW" Multichipping does

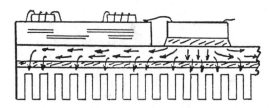

Fig. 4 — Cross-section diagram of a multichip module showing lateral heat spreading in modules with mixed die sizes.

TCE makes no sense—so much tungsten is needed that the thermal performance is reduced from the 360 W/m-K of copper to nearly 140, worse than aluminum nitride, and the TCE can not be reduced below 5.

Another approach is to use a solid copper heat sink, and connect the large high-wattage die with 98/2 lead tin solder. That material is so compliant that it does not break the die, even for the high power industrial semiconductors that dissipate 100 W. But the electrical conductivity of solder limits the multichip-applications to those which can tolerate common base connections. Coating the copper with deposited thin film diamond, metalizing the die sites, and then soldering is possible, but this technology is very new and not yet recognized. Clearly the ceramics are more acceptable.

How does one choose between the remaining contenders: Silicon Carbide and Aluminum Nitride? Let us list some of the advantages and disadvantages of each.

Aluminum Nitride: Advantages:
  Cheaper
  More suppliers
  Lower dielectric constant—can have signal pads
  Can be laminate co-fired

Aluminum Nitride: Disadvantages:
  Inconsistent characteristics
  Dissolves in alkaline solutions
  270 w/m-K non-available in big sizes
  Surface decomposition to Aluminum Oxide
  May be unsuitable for non-hermetic packaging

Silicon Carbide: Advantages
  Physically and chemically strong
  4 inch square substrates possible
  Consistent in material characteristics
  Hitaceram offers 270 W/m-K

Silicon Carbide: Disadvantages
  High dielectric constant, requires insulation
  As-fired surface is rough; 16 microns
  Requires polishing or insulation metallization
  Procurement is difficult

Both these substrates can be metallized, but require either polishing or the ceramic itself, or polishing of the first coating material. Coating with dielectric first may be easier, because the organic only has to be made planar and smooth. Subsequent metallization provides ground or reference plane for the interconnect.

ACHIEVING LID HERMETICITY

Many applications for these assemblies are military, and the acceptance of silicone gel protection for the bare chips is some time away. Lids of aluminum nitride or silicon carbide are just too expensive, and the TCE of Kovar does not match. The solution is mullite, or even a mullite with a little silicon carbide powder added. The problem is then the sealing of the package with organic material inside. Many sealing glasses are low melting point materials, but low melting point glasses often have low dielectric constant, which causes capacitative crosstalk between signal traces, requiring interposed ground lines, and greater lead density.

This is another area of research effort on materials. One solution is to used the co-fired aluminum nitride, bringing the signal traces down to a lower level at the seal area. Expensive. Another is to continue to use the multilayer co-fired material for the power and ground traces, and use a layer of buried signal traces to get to the pin grid array pins. The capacitance and poor characteristic impedance throws away much that has been gained using the organic dielectric.

A better solution is to use surface metalization, and protect the surface traces with silicon dioxide over the conductors and below the low melting point glass sealant to the mullite cap (Fig. 5). Research effort is about to begin, sponsored by a few of the companies moving to the multichip technology.

MATERIAL PROBLEMS IN COATING CERAMIC WITH DIELECTRIC

Electrical Problems Requiring Planarity

All dense circuits need to have the signal wiring sandwiched between power and ground planes. If the circuits are ECL or bipolar, the ground planes establish the characteristic impedance, and make the signal line into mini-transmission lines. That completely avoids reflections if the end of the line is terminated, and makes the transmission delay independent of voltage.

If the circuits are CMOS or other high impedance driving semiconductors, the ground plane is still needed to control the crosstalk. From the viewpoint of the CMOS driver, it would be better if that were not so—it takes energy and current flow to charge the capacitances, and that takes time, so there is a delay that is similar to the transmission line delay, but now dependent on the driving voltage.

LID SEALING ON A SILICON CARBIDE SUBSTRATE

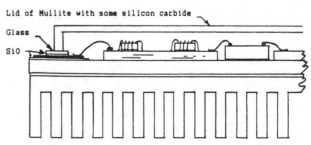

Fig. 5 — One possible construction that permits lead exit under edge sealing glass, avoiding the need for via holes or co-fired constructions.

This skew delay is small if the distances are small, and therefore is reduced due to the space compression of the new higher density interconnect. More important, the ends of the line cannot be terminated, because the termination resistance would cause a continuous draw of current from the CMOS driver, making the power consumption excessive. There will then be reflections of the signal pulse from this unterminated transmission line. If the reflection of the leading edge of the pulse comes back so soon that the outgoing pulse has not yet reached the top of its risetime climb, then the distortion of the outgoing signal occurs during the rise time, and becomes a sort of hiccup in the climb to the maximum pulse amplitude.

Round trip return pulse time of 15% is obviously negligible—that was the standard for many years. Now 25% is considered safe, and there are some designers considering 50% permissible time delay for the return pulse. That may be acceptable, but it depends on the returning pulse shape. If it is just the leading edge that causes the outgoing pulse disturbance, well fine. If much of the returning pulse echoes the outgoing pulse, the returning pulse can disturb the top edge of the outgoing pulse and cause difficulty.

In my opinion, acceptance of a returning pulse delay up to 50% of the rise time is more a statement of acceptance of the capability of present circuit fabrication limitations than a statement of really acceptable performance. Try for 25% first by increasing the density through multichip modules; go to terminated transmission lines beyond that acceptable unterminated distance.

If the first metalization layer of a multichip is a ground plane, it need not be insulated from a substrate of otherwise unacceptable dielectric constant. Why not metalize the ceramic directly, as in Fig. 1? There are two reasons why this might be a problem.

PROBLEMS OF DIRECT CERAMIC COATING

If the ceramic is to be directly metallized, the as-fired surface roughness of the ceramic is a problem. Roughness measurements of 16 micron can occur. It is difficult to plate or even sputter metal coatings that can be subsequently plated with a surface so rough. The solution is to polish the surface to 3 micron, and then the sputtering and plating becomes possible. Titanium palladium and gold platings are applied to permit eutectic die bonding, and copper is more difficult.

Polishing is not a good solution. Polishing a sintered granular material often pulls out grains from the substrate, leaving holes or craters. Sure, most of the surface is smooth, but these random pits can be a disaster if they occur at the location of a future feature of the signal traces. The variation in the surface can cause problems with the future circuit stripe.

The alternative is to coat the ceramic surface with an organic. Think epoxy for the moment, though many organics can be used.

This coating fills in the valleys of the ceramic landscape, if the organic material is chosen with care, and the right additives and technique are used to permit wetting of the rough terrian, without any "tenting over" of the low places. Not an easy trick. When the coating is applied properly, it is rough at the top, but now polishing is in organic material, and it is easier to get a smooth surface. That is the message of the Microtek process, which has polished the layers to get planarity of the organic before metalization for some years.

Adherence problems, compatibility with the plastics, etc. are all materials problems to be addressed. For example, if polyimide is the dielectric, direct contact with copper causes an adverse reaction, so that chromium or chromium-nickel must be used as a barrier.

There is a more severe problem that this polishing may also fix. Ceramics are not flat. After firing, there is a bow or camber that may be as much as 0.3% and not less than .03% of the width of the substrate piece. If a 4" square multichip substrate is proposed, that camber or bow may be as much as 300 micron, and not less than 30 micron. If any of the process steps tend to coat the material and produce a planar top layer, metalization of the ceramic substrate directly can produce severe variations in the capacitance and the characteristic impedance.

For example, if the ceramic were directly metallized, and the dielectric were to be applied by spin coating, producing a near planar top surface, then the distance from the ground plane to a subsequent signal path could vary from 15 micron to possibly 30! Well, things are not that bad, spin coating does not produce a planar top surface—it is not like the top of the water surface in an overflowing bathtub. Rather it is a surface that is not as flat as one would wish, but it is also not uniformly thick over the metallized ceramic.

Coating the ceramic first, and the polishing can improve the planarity. Producing the interconnection circuits elsewhere, and then transferring them to the metallized ceramic can assure parallelism to the ceramic metallized surface. Not straight-flat parallelism, but conformity to the curvature of the surface that it finds.

There are materials problems that must be solved. And they must be solved for the organic materials that one wishes to use for the circuit insulation layers.

PROBLEMS WITH THE ORGANIC DIELECTRIC AND THE SIGNAL METALIZATION PROCESSES

There are other material problems in the metallization of the dielectric for the signal traces and the vias. If the vias are laser cut, the material melted by the laser drops back on the surface, unless an Excimer laser is used. If the organic is etched, the materials must permit that etching, another material consideration.

Then there is the adherence of copper metalization to the dielectric chosen. We have mentioned earlier the incompatibility of polyimide with copper, and the need for chrome-nickel interfaces, top and bottom around each copper line. Any other dielectric must either permit direct copper contact, or have a similar interface treatment established and proven.

Signal metalization can be sputtered in the vias, but that does not fill the vias. This is a problem for the planarity of the subsequent organic dielectric. Does one overcoat and polish again, or tolerate the depression caused by the unfilled via? Perhaps it is better to use a process that fills the via.

That process is electroplating, or even electroless plating of copper. If one plates enough, the vias fill, but then the metal paths are very thick. If one masks properly, the vias can be filled, but that requires masking steps. Another material choice must be made.

Alternatively if electroless copper deposition is to be used, perhaps the dielectric itself can contain the necessary catalyzing material, so that just an exposure step can define the path, eliminating one application of a mask material that must be subsequently removed. This is another area of material development. I cannot list all the activity in these areas; much of it is proprietary development not yet publicly discussed. There are many research projects in process at this time.

SUMMARY AND CONCLUSIONS

The next major thrust in semiconductor packaging will be the increasing use of multichip modules. These modules will have a few layers of organic film applied to ceramic substrates, or in some cases, silicon substrates mounted on ceramic. They will be used for mainframe and other large scale computers with many custom chips, but they will also be used for military and cost-performance computers and other electronic equipment. The advantages are the increase in speed, reduction in size and weight, and reduction in design time. As the technology matures, there will be reductions in cost also.

There are many materials problems to be solved, and the ferment today is great, with over 30 companies committed to this new technology. Of the many alternatives, quite a few will survive; there are indeed choices to be made with advantages and disadvantages to each.

New materials and processes are needed in the area of low dielectric organic dielectrics suitable for copper metalization and use in non-hermetic environments. There are new high thermal conductivity ceramics needed that are stable, and that can be coated and metallized and survive high temperature moist environments. There are problems of lid sealing for hermetic devices, and package constructions and materials for non-hermetic.

There is a rich stew of material development activity that will impact this market. The size and performance advantages, and ultimately the cost advantages of this approach are so great as to insure a boil of activity for some time.

REFERENCES

1. R. Rice, "Tutorial on VLSI Packaging," IEEE Computer Society Press, 1980.
2. W. H. Knausenberger and L. W. Schaper, IEEE Trans. Components, Hybrids, Manuf. Technol. CHMT-7, 261 (1984)
3. J. W. Balde, "Low Dielectric Constant-The substrate of the Future," Printed Circuit World Convention IV Technical Paper WCIV-59, June 2-5, 1987.
4. J. W. Balde, "Small Dimensions or Low Dielectric Constant-The competing approaches to High Density Interconnect," the Mat. Res. Soc. 1987 Fall Meeting, paper I 2.1, Proc. publication in process.
5. E. Davidson, IBM J. Res. Dev. 26, 349 (1982).
6. E. Davidson, IEEE Trans. Components, Hybrids, Manuf. Technol, CHMT-6, 272 (1983).
7. T. Watari and H. Murano, "Packaging Technology for the NEC SX Computer" IEEE ECC Proc., pp. 192-198.
8. T. Ohsaki et al, "A Fine-Line Multilayer Substrate with Photosensitive Polyimide Dielectric and Electroless Copper-Plated Conductors," IEEE-CHMT Int. Electron. Manuf. Tech. Symposium," 1987, pp 178-183.
9. A. A. Evans and J. K. Hagge, "Advanced Packaging Concepts—Microelectronic Multiple Chip Modules Using Silicon Substrates," SAMPE Electron. Mater. and Packaging Conf. June 1987, pp 37-45.
10. P. L. Young, et al, "Thin Film Multi-chip Module," IEEE/NBS VLSI Packaging Workshop, September 1985.
11. E. Bogatin, "High Density Interconnect: An Enabling Technology for Multichip Packages," IEE/NBS VLSI & GaAs Packaging Workshop, September 1987.
12. A. Wayne Johnson, et al, "Thin-Film Silicon Multichip Technology," IEEE ECC May 1988, pp 267-275.
13. R. J. Jensen, "Recent Advances in Copper/Polyimide Thin film Multilayer Interconnect Technology for IC packaging," Materials Research Society Fall Meeting, December 1987, Paper I 2.2.
14. J. H. Hagge, "Ultra-Reliable Packaging for Silicon-On-Silicon WSI," 1988 IEEE ECC Proceedings, pp 282-292.
15. B. Nelson and T. Pan, "Reliability Issues in Copper Polyimide Systems," Engineering Foundation Workshop on Advanced Materials and Processes for High Density Interconnect, March 1988.
16. R. W. Johnson et al, "Thin-Film Silicon Multichip Technology," IEEE ECC 1988, pp 271.
17. C. P. Wong, "High Performance Silicon Gel as IC Device Chip Protection," Mat. Res. Soc Mtg, Dec 1987, paper I 5.7.
18. J. Mosko, "Printed Wiring Boards Using Mixed Dielectrics" Int. Electron. Packaging Conf., 1988, pp 161-172.
19. D. J. Arthur, "Electrical and Mechanical Characteristics of Low Dielectric Constant Printed Wiring Boards," IPC Conference, Boston, 1986.
20. H. Kohara et al, "High-Thermal Conduction Module" IEEE Elec. Comp. Conf, 1985, p 180-186.
21. G. G. Werbitsky and F. W. Haining, "Circuit Packaging for Large Scale Integration," Ibid., pp 187-192.
22. Y. Shimada et al, "Low dielectric Constant Multilayer Glass-Ceramic Substrate with Ag-Pd Wiring for VLSI Package, IEEE ECC 1987, pp 398-405.
23. D. L. Shealy et al, "Use of Cordierite and Copper as Advanced Packaging Materials," IEEE-CHMT/NBS VLSI & GaAs Packaging Workshop, 1987.
24. M. Horiguchi et al, "New Mullite Ceramic Packaging and Substrates," IEEE-CHMT 1988, pp 574-583.
25. A. Bar-Cohen, "Thermal management of Air and Liquid-Cooled Multichip Modules," IEEE CHMT Trans., June 1987, pp 159-175.
26. Hiroshi. Shibata, Mitsubishi, Oral Presentation, IEEE VLSI Workshop, Santa Clara 1984.

ANALYSIS OF MATERIALS AND STRUCTURE TRADEOFFS IN THIN AND THICK FILM MULTI-CHIP PACKAGES

J.P. Krusius[1] and W.E. Pence
Cornell University, Electrical Engineering,
119 Phillips Hall, Ithaca, New York 14853

Abstract

System simulation, a new design methodology for electronic packages, has been used to study multi-chip module tradeoffs. Modules are described with a set of structural, electrical and materials related model parameters. Parameter ranges cover wide ranges including multi-layer ceramic, thin film and polyimide on ceramic technologies. 1.5 μm CMOS chip technology and 14 mil line width, 100 mil grid space printed wiring board technology are used. Modules have been optimized for three cases using the simulated annealing technique: maximum module clock frequency, maximum board level clock frequency, and optimum system performance. Many similar mathematical solutions are typically found, an indication of the complexity of the response surface. These examples demonstrate the power of package system simulation for identifying, quantifying, and predicting technology design, development, and tradeoff issues and solutions for multi-chip modules.

Introduction

The design of advanced IC packages for electronic systems is complicated by many tradeoffs resulting from conflicting design constraints and performance goals. The ability to optimize a packaging structure for each system application requires considering a large number of parameters and their dependencies simultaneously together with the appropriate figures of merit. This is often difficult to handle with confidence using ad hoc methods. Package system simulation, a new design methodology, has been created to solve such problems. While package system simulation is in many ways analogous to circuit simulation, it relies on a mixture of physical and statistical models as a way to describe the much higher complexity of multi-level packaging hierarchies.

Package system simulation is based on a large system model, which spans the range from individual devices and gates, to circuit boards and sub-assemblies. The intention is to create a model, which is accurate enough to allow for detailed tradeoff studies early in the design stage, permitting consideration of a number of competing IC, packaging, and system partitioning approaches in all combinations. In this way, the merits of a

particular packaging technology, or hierarchy of technologies, can be measured objectively.

Since the details of the approach and the current set of physical and architecture models have been described elsewhere [1,2,3], only a brief outline will be given here. A subsystem (logic, memory) is specified in terms of the total circuit count, the device technology (circuit family, feature sizes, electrical characteristics, interconnects), the chip technology (integration level, metal levels, interconnects, I/O drivers), the packaging technology (single/multi-chip, chip count, materials, signal planes, interconnect geometry), and the board (grid, signal planes, interconnect structure). These models and parameter values for specific technologies constitute the input technology database. An algorithm has been devised to compute the characteristics of the packaging hierarchy described with a descriptive set of output parameters, such as chip size, I/O count, power dissipation, switching noise, package size, lead count, crosstalk, board size, and signal delays. Changing any single parameter value in the input database alters the output parameter set, resulting in a system with different characteristics.

Three primary objectives can be pursued using package system simulation: (1) to analyze fundamental limits and trends, and predict technologies required for future systems, (2) to compare the performance leverage of competing technologies, and (3) to determine the structure and technologies required for optimal systems. The third goal is the most intriguing, since it couples packaging and functional architectures. Questions of optimum integration levels, both on chip and across the packaging hierarchies, i.e. system partitioning, have been difficult to address in the past.

In this paper we report on an attempt to explore multi-chip module tradeoff issues using package system simulation. We have included three levels of the system hierarchy: chip, module, and board. The results will uncover the complex response surface of the multi-chip module design space.

Design Goals and Ground Rules

Multi-chip packaging is clearly superior to conventional packaging (DIP, PGA, SMT) in terms of density and speed. Higher integration levels in integrated circuits also provide enhanced performance, provided on-chip delays remain small.

1 On sabbatic leave at IBM T.J. Watson Research Center, Yorktown Heigths, New York

Reprinted from *Proc. 39th Electron. Components Conf. (ECC)*, pp. 641–646, 1989.

It is well known that as chips grow in size (> 1 cm) and circuit count (> 10^4), on-chip interconnects present a problem due to large RC delays in the on-chip thin-film interconnects. Assuming that it is undesirable to increase the geometrical sizes of on-chip interconnect lines and spaces, the question of an optimum level of integration arises. Specifically, given a multi-chip module with a fixed target size, what combination of chip integration level, chip count, and multi-chip packaging technology (MLC, Cu-polyimide, etc.) results in optimal performance? Performance can be defined in many ways, for example path delays, power density, noise, and functional density. This study will focus on the question of optimal multi-chip module designs for logic and memory applications, as determined by gate density, module clock rate, and system clock rate as figures of merit.

The chip technology employed in this analysis is specified in Table 1. The level of integration (number of circuits per chip) for logic and memory sub-systems, is a variable subject to optimization.

parameter	value
circuit technology	CMOS
gate length [μm]	1.5
linewidth [μm]	4.0
# metal levels:	
logic	2
memory	5
typical integration:	
logic	10^4
memory	10^6
rent coefficient (logic)	0.50
rent exponent (logic)	0.51
data bus width (memory)	16 bits
size at typical integration level:	
logic	6.13 mm
memory	3.70 cm

Table 1. Definition of selected logic and memory chip technology.

The technology employed in the module is described with a few primary design parameters. The interconnect geometry is fixed by the via grid spacing (in mils) and the number of interconnect lines per channel. Other variables include the number of signal planes, interlayer dielectric thickness, metal line thickness, and the dielectric constant. The range of these design parameters is set to represent technologies from multilayer ceramics (MLCs) [20 mil grid], to a thin-film, Cu-polyimide on ceramic technology [3 mil features]. Unorthodox combinations of design parameters, such as 20 layers with 3 mil feature sizes and 2-layers with a 20 mil grid, are obviously permitted. Table 2 lists the important parameters in the description of the multi-chip packaging technology,

together with their assigned ranges.

In the following sections, the design space spanned by the variables in Table 2 will be explored. The first design goal is to determine the combination of design parameters values, which results in maximum functional density in a module which is confined to 3" in size. There are clearly two competing extreme solution: a large number of low-integration chips, and a small number of high-integration chips. The second design goal is to determine the combination of parameters which results in minimum path delays in the module. This clearly involves a tradeoff between the level of integration on the chip and the distribution of signal paths (i.e. how much of the critical path is on-chip, as opposed to in the module, as a function of integration level). The third design goal is to determine the combination of parameters which results in minimum system delays, for systems which contain more than one module, and which are fixed in size (i.e. circuit count). The fourth and final goal is to find the optimum partitioning subject to maximizing the board level clock rate.

parameter (unit)	range
Via grid spacing (mils)	3-50
Lines per channel	0-2
Signal planes	2-20
Line width (mils)	3-50
Line thickness (mils)	0.5-5
Dielectric thickness (mils)	1-10

Table 2. Multi-chip module design parameters ranges.

(1) Module Circuit Density

Fig. 1 shows a plot of the simulated size (in cm) of a multi chip module as a function of the number of chips it contains, and the number of logic circuits contained on each chip. The chip technology parameters are described in Table 1. The parameters describing the module are: 20 mil via grid spacing, 1 line per channel, and 16 signal planes.

Increasing the number of circuits per chip increases chip I/O (Rent's rule), and the size of the chip. The increase in module size is a result of increasing wiring requirements with a fixed number of wiring planes. The maximum circuit count in a 3" module is represented by that point in Fig. 1 which lies below a line at 3" (7.62 cm) and maximizes total circuit count (number of chips per module x number of circuits per chip). To locate that point, the simulated annealing algorithm has been used. The results for the case described above are given in Tables 3 and 4 for logic and memory sub-systems. Three runs with different seeds to start the optimization) have been performed. There is no guarantee that any of these

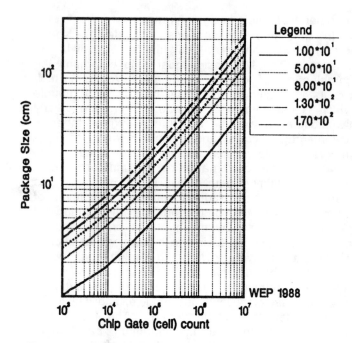

Legend

——	$1.00*10^1$
········	$5.00*10^1$
········	$9.00*10^1$
——	$1.30*10^2$
— ·	$1.70*10^2$

WEP 1988

Figure 1. Simulated multi-chip module size as a function of gates/chip with chips/module as the parameter (logic).

maxima located by simulated annealing constitutes the absolute maximum. Examining the three different solutions e.g. in Table 3 one finds that the results listed in Table that they correspond to similar module circuit counts but different integration levels. Thus the circuit counts per chip and chips counts per module differ greatly.

parameter	run 1	run 2	run 3
circuits/module	4.4×10^6	2.13×10^6	2.44×10^6
chips/module	2	56	42
circuits/chip	2.2×10^6	3.8×10^4	5.8×10^4
chip I/O	858	108	134
chip size (cm)	7.58	1.07	1.27
chip power (W)	5.4	.106	.153
module grid (mils)	3	3	10
lines/channel	0	0	0
signal planes	20	20	20
module clock F (MHz)	57	53	54

Table 3. Results of simulated annealing runs to determine maximum module circuit count for the logic chip set (Table 1).

High integration levels ($> 10^4$) for the ICs result in maximum circuit density in the module, although high module circuit counts can also be obtained with low integration level chips, if the packaging technology is stressed. Memory and logic are different in terms of I/O growth. Lower I/O requirements for the memory chips allows more of them to be mounted and interconnected in the module, with fewer signal wiring planes, than the logic chips. Signal delays in the module for the

logic and memory modules, as expressed with critical path delays and memory access delays, respectively, are assessed in the next section to determine optimum levels of integration for maximum speed.

parameter	run 1	run 2	run 3
bits/module	9.8×10^6	8.3×10^6	5.3×10^6
chips/module	10	200	108
bits/chip	9.7×10^5	4.2×10^4	9.1×10^4
chip I/O	32	27	28
chip size (cm)	2.35	.544	.731
chip power (W)	2.14	.094	.202
module grid (mils)	3	9	6
lines/channel	0	0	0
signal planes	2	20	5
module clock F (MHz)	24	22	23

Table 4. Results of simulated annealing runs to determine maximum module bit count for the memory chip set (Table 1).

(2) Maximum Module Clock Frequency

The simulated multi-chip module clock rate in MHz for logic as a function of chip circuit count and module chip count is shown in Fig. 2. The case for memory chips results in a plot with a similar form. Module designs, which allow for the highest clock frequency are again found using simulated annealing. In this optimization the module size is constrained to be greater than 3", thus excluding trivial solutions such as single chip or 1 circuit.

The results from the simulated annealing runs for both logic and memory are given in Table 5. Two runs with two different seeds were made again for each case.

parameter	logic		memory	
	run 1	run 2	run 1	run 2
module clock (MHz)	42	51	23	25
circuits/chip	8.6×10^5	5.6×10^3	3.4×10^5	9×10^6
chips/module	200	29	43	3
circuits/module	1.7×10^8	1.6×10^5	1.5×10^7	3×10^8
chip I/O	532	40	30	35
chip size (cm)	4.6	.52	1.3	7.8
chip power (W)	2.1	.021	.740	20.0
via grid (mils)	3	50	50	50
lines/channel	0	0	0	0
signal planes	20	14	20	10
module size (cm)	64.7	10.1	9.4	15.6

Table 5. Results of simulated annealing runs to determine maximum module clock rate for logic and memory.

62

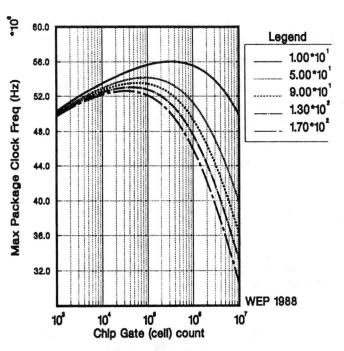

Fig. 2. Simulated module clock rate as function of chip circuit count with the module chip count as the parameter for logic.

An examination of the results in Table 5 again reveals that very different combinations of module chip count, chip integration level, and module design can produce modules with similar clock rates. While the obtained solutions obtained were constrained to a module size larger than 3", the optimization algorithm does not necessarily converge to solutions with minimum module sizes. The runs listed in Table 5 assume a logic depth of 12 stages. The module clock rate is determined from the critical path delay (1 off-chip net, 12 on-chip stages), plus a clock skew term. Note that in none of the solutions the via grid limits the module wireability (0 lines/channel).

(3) Maximum System Clock Rate

For large processors or memory boards which may contain more than one of the multi-chip modules examined in previous sections, it is desirable to minimize the processor cycle time or memory sub-system access time. With the same assumptions made above, a system, consisting of a number of circuits larger than one module can accomodate, is modeled. The board technology is a two-layer printed wiring board (permitivity = 4.8) with a 100-mil via grid and 3 lines per channel (14 mil nominal linewidth). The entire system is assumed to contain 1.024×10^8 circuits and to consist on one board only. Simulated annealing techniques were again employed to find maxima on the system clock response surface. The results are given in Table 6 for logic for three solutions with different seeds. The logic chip circuit count,

module chip count, and module design parameters vary again from one optimum to another. Table 7 lists corresponding results for memory. In both cases, the module size was again constrained to be larger than 3".

parameter	run 1	run 2	run 3
board clock (MHz)	8.2	8.2	23.7
circuits/chip	1014	7×10^6	5.2×10^5
chips/module	82	3	200
circuits/module	8.3×10^4	1.8×10^7	1.1×10^8
chip I/O	17	1543	413
chip size (cm)	0.3	13.8	3.6
chip power (W)	0.006	16.7	1.24
via grid (mils)	50	46	11
lines/channel	0	0	0
signal planes	10	12	20
module size (cm)	8.97	123.1	50.3

Table 6. Simulated annealing results for maximum board clock rate for logic.

In Table 6 the logic depth is assumed to be 12 stages. An additional board clock skew term is added to compute the board clock rate. The low board clock frequencies are a result of the large planar board size. Fig. 3 shows the simulated system clock rate for the logic system. Discontinuities in the curves result from the discrete number of modules permitted in the solution. The modules are identical and may be only partly populated. Plots such as Fig. 3 graphically illustrate the complexity and local characteristics of the response surface for this input technology parameter set. Fig. 3 also gives justification for using simulated annealing techniques rather than classical techniques for non-linear optimization. The discontinuity in the maximum frequency results from the increasing chip integration level and small module count.

parameter	run 1	run 2	run 3
board clock rate	10.1	11.7	11.8
bits/chip	10^7	10^6	3.4×10^5
chips/module	2	14	33
bits/module	2×10^8	1.4×10^7	1.1×10^7
chip I/O	35	32	30
chip size (cm)	8.3	2.4	1.4
chip power (W)	22.1	2.3	0.74
via grid (mils)	3	11	50
lines/channel	0	0	0
signal planes	4	12	20
module size (cm)	8.3	9.8	8.

Table 7. Results of simulated annealing runs to determine maximum system clock rate for memory.

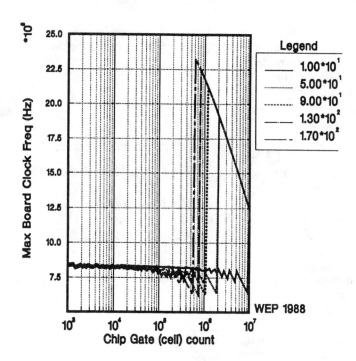

Figure 3. Simulated system clock rate as a function of chip circuit count and the module chip count as a parameter for logic.

parameter	logic value	memory value
board clock (MHz)	8.4	11.8
circuits/chip	1.7×10^3	6.3×10^5
chips/module	11	18
circuits/module	18.6×10^3	1.1×10^3
chip I/O	23	31
chip size (cm)	0.36	1.86
chip power (W)	.090	1.37
via grid (mils)	35	3
lines/channel	0	0
signal planes	3	2
module size (cm)	10.2	22.3

Table 8. Optimum multi-chip module results for logic and memory like systems with maximum clock frequency. Unconstrained simulated annealing has been used.

(4) Optimum Module Designs

The size of the multi-chip module was in the above sections required to be larger than 3". The removal of this constraint allows the consideration of a wider range of module designs, which maximize system clock rate. Table 8 lists the results of two unconstrained simulated annealing runs, one for logic and the other for memory, to locate optimal multi-chip module designs. The fixed chip and board technologies were specified as before, but integration levels (circuits/chip and chips/module) were allowed to vary. Due to the higher I/O counts in the logic chips, more chips and lower chip integration levels emerges as an possible optimum solution. Additional runs with different seeds reveal as before that combinations of high integration levels and low chip counts also produce similar board speeds. Similar observations can be made from results for the memory like system. The fact that a clear optimum does not emerge from these simulations is indicative of (1) the complex nature of the design space, and (2) the fact that one technology (e.g. packaging) may compensate for deficiencies in another technology (e.g. chip), and vice-versa.

Discussion

Previous sections have shown that optimum module designs (in terms of circuit count, module clock rate, and system clock rate) can be obtained with very different combinations of design parameters and integration levels. This is not surprising considering the nature of the module design problem. An analogous situation exists in circuit design.

Table 9 list the results of the simulated system performance measured with the board clock rate for the module designs found in earlier sections. These module designs (chips/module, circuits/chip, and module design parameters), which resulted in maximum module circuit count and maximum module clock rate, are used to simulate systems containing 10^7, 10^8, and 10^9 circuits. An increasing system size (board size) results in a slower system clock rate. Table 9 shows that the module designs, which maximized circuits/module or the module clock rate, result in very similar system clock rates (Table 9). This indicates that the design space contains widely separated points with similar performance (board clock rate). A large number of simulated annealing runs have been made with random seeds in order to map out the details of the response surface. While details of the response surface will be reported elsewhere, we have found both widely different and closely similar local minima in the design space. Package system simulation models and algorithms seem to be sufficiently accurate to capture the complexity and rich tradeoff potential characteristic of this complex response surface.

While a detailed understanding of the multi-chip module design space will result from systematic and quantitative design techniques, such as package system simulation, other design criteria, availability of technologies, manufacturability, and cost issues will need to be specified in order to discriminate among several possible solutions. We thus again invoke the analogy to circuit simulation, where the designer himself has the task to perform the final selection. Remaining with this analogy, it is the goal of the present authors to make package system simulation one of the most significant tools for this decision process.

circuits/system	module 1	module 2
10^7	15 , 12	16 , 10
10^8	8 , 12	8 , 10
10^8	3 , 12	3 , 10

Table 9. Simulated maximum system clock rate (MHz) for module 1 (maximum number of circuits), and module 2 (maximum module clock rate). In each module column the results for logic and memory are given in that order.

Acknowledgements

This work has been supported by Semiconductor Research Corporation and IBM. One of us (W.E.P) is grateful for continued fellowship support from IBM-Endicott. The authors have benefited from fruitful technical discussions with Drs. D. Seraphim, W. Chen, R. Guernsey and W. Vilkelis of IBM.

References

[1] W.E. Pence and J.P. Krusius, "Fundamental Limits for Electronic Packaging," Proceedings of the Spring Meeting of the Electrochemical Society, p. 60, May 1988.

[2] W.E. Pence and J.P. Krusius, "Computer Packaging Tradeoffs," IEEE 1988 Pocono Computer Packaging Workshop, May 1988, unpublished.

[3] W.E. Pence and J.P. Krusius, "System Simulation: A New Method for the Design of Electronic Packages", Technical Digest of TECHCON'88, October 1988, pp. 102-105.

A COMPARISON OF THIN FILM, THICK FILM, AND CO-FIRED HIGH DENSITY CERAMIC MULTILAYER WITH THE COMBINED TECHNOLOGY: T&T HDCM (THIN FILM AND THICK FILM HIGH DENSITY CERAMIC MODULE)

Dr. M. Terasawa and S. Minami
Kyocera Corporation
Advanced Package Division
52-11 Inoue-Cho, Higashino, Yamashinaku, Kyoto, 607, Japan

and

J. Rubin
Kyocera International, Inc.
6211 Balboa Ave., San Diego, CA, 92123

ABSTRACT

A new ceramic packaging technology has been developed for advanced high speed VLSI. The technology combines thin film multilayer technology; thick film multilayer technology; and co-fired, high density, multilayer ceramic (MLC) technology. This new technology is called T&T HDCM (THIN and THICK FILM HIGH DENSITY CERAMIC MODULE).

The choices of materials for the various technologies are discussed. Polyimide and silica-fused glass are preferred as low dielectric constant insulating materials for the thin and thick film technologies respectively. Copper is selected as the low resistance conductor for both thin and thick film with aluminum being an alternate choice for thin film.

Thin film photolithography is used for generating the necessary fine patterns. High precision patterns can be achieved with both thin and thick film processing.

The advantages of, and the design guidelines for the thin film method are discussed, as well as the characteristics of, and the design guidelines for the thick film method. The characteristics of, and the design guidelines for the advanced co-fire multilayer ceramic (MLC) are also discussed. This technology is known as HIGH DENSITY CERAMIC MODULE (HDCM). The process flow charts of the three technologies are given.

A comparison of the three basic technologies: thin film, thick film, and MLC are made and the advantages and disadvantages of each are discussed.

It is shown that best answer to advanced high speed VLSI packaging is achieved by combining the three technologies to produce a new package type known as T&T HDCM (THIN AND THICK FILM HIGH DENSITY CERAMIC MODULE).

These T&T HDCMs are now commercially available. Several examples are shown.

INTRODUCTION

Rapid progress in the improvement of VLSI technology has brought about a requirement for new packaging technologies which do not compromise the properties of advanced VLSI devices. Conventional co-fired alumina multilayer packages, which traditionally have had the highest reliability among the various packaging technologies, may cause problems for VLSI packaging.

VLSI is characterized by high integration and high speed. Requirements for VLSI packages are shown in Table I. High integration brings about a remarkable increase of input-output (I/O) counts that leads to the reduction of lead pitch and an increase in chip size. In the case of logic circuits, the relationship between number of gates and I/O counts can be expressed by the following equation[1]:

$$N_T = \alpha\, N_C^{\beta}$$

where N_T and N_C are I/O count and gate number, respectively and α and β are constants. For a large computer system, it has been reported that $\alpha = 2.5$, $\beta = 2/3$. On the basis of the data, it is understood that single chip systems and multichip systems require 200 to 400 I/Os and 1000 and up I/Os, respectively. This trend can be seen in Figure 1.

Conventional multilayer ceramic (MLC) technologies use a "silk screen" printing method for the formation of traces. Consequently, the MLC can not meet the requirements for the formation of fine patterns. Minimum line width is about 100 microns (0.004") for the printing method. Therefore, it seems highly improbable that this printing method can be applied for VLSI packages having 200 and up I/Os, as shown in Figure 2. Furthermore, the difficulty in using co-fired technology is aggravated by the large dimensional tolerance created by the large shrinkage which occurs during sintering. That is to say, methods which can produce fine and precise patterns should be used to replace the printing method.

Reprinted with permission from *The Int. J. for Hybrid Microelectron.*, vol. 6, no. 1, pp. 607–615, 1983. Copyright © 1983 by International Society for Hybrid Microelectronics.

Therefore, we can conclude that MLC technology by itself is not enough to meet the requirements for advanced VLSI packages. Therefore, a new packaging technology which meets the requirements, but is not expensive, and is also commercially available, should be developed as early as possible. The technology must be compatible with the following factors:

1. Low Dielectric Constant Insulator.
2. Low Resistance Conductor.
3. Fine Pattern Process.
4. High Precision Process.
5. High Thermal Conduction.

This paper details a new packaging technology developed by Kyocera, which meets the above factors and is feasible for practical manufacturing.

CHOICE OF MATERIALS

As shown in Table II, there is an extremely large choice of insulator materials which are applicable for IC packages. Low dielectric constant materials, which could replace the alumina, used in conventional ceramic packages, were investigated. Among the materials listed in Table II, quartz glass and polyimide have the lowest dielectric constants, about 3.8 and 3.2, respectively. These two materials were selected for use as the insulators in the new packaging technology.

Table III shows various metallizations . At Kyocera, tungsten is used as a conductor material in MLC. Tungsten has a relatively high electrical resistivity, which is not desirable for meeting the requirements for VLSI.

Gold is the best material in electrical conductivity, but its cost is very high, and if possible, its use should be limited.

Copper and aluminum have excellent electrical conductivity and are not very expensive. Copper can be used in two processes, thick film and thin film, both of which can produce fine precise patterns. Aluminum is used in thin films. On the basis of the above, copper and aluminum have been selected as the metallic conductors.

Thick film copper is inexpensive and has excellent electrical characterictics, but it must be fired in neutral or reducing atmosphere to be protected from oxidation. This enables the copper to be used with MLC because the tungsten used in MLC must also be fired in an inert or reducing atmosphere.

The increase of I/Os in VLSI requires finer and higher precision patterns. In the very best case of conventional MLC, dimensional tolerances resulting from shrinkage during sintering are 0.5 per cent.

The next feature is the remarkable increase of switching speeds. The relationship between access time and power is illustrated in Figure 3. Currently available commercial devices are based on silicon and gallium-arsenide. Advanced devices have an access time in sub-nano seconds.

In the packaging of such high speed devices, the following factors must be strictly controlled:

1. Noise.
2. Impedance matching.
3. Electrical resistivity.
4. Temperature rise.

Propagation delay of a package can be minimized by reducing the dielectric constant of an insulator and the length of conductor traces. It is generally known that the propagation delay (T_{PD}) obeys Kaupp's equation[2]

$$T_{PD} = 0.0333 \sqrt{\varepsilon_r}$$

The reduction of dielectric constant follows the reduction of capacitance, which solves noise problems. Alumina used in MLC has a dielectric constant of 8.5 to 10. By considering the two factors mentioned above, the dielectric constant of alumina is considered to be too high.

Also, the use of low electrical resistivity conductors is very important. Refractory metals such as tungsten and molybdenum-manganese are used as conductors in MLC. These metals have relatively high resistivity, so they must be replaced by metals having excellent electrical conductivities. Furthermore, it is preferred that the replacement metals have high skin effects.

The high dimensional tolerances have caused the main problems in application of MLC for VLSI. A high precision, fine pattern process can be achieved by the use of thin film and thick film technologies.

In order to meet the requirement for VLSI packages, the following materials and processes have been selected for the new packaging technologies.

1. Low Dielectric Constant Material:
   Polyimide and Quartz glass.

2. Low Resistance Conductor:
   Copper and Aluminum.

3. Fine Pattern Process:
   Thin Film Photolithography.

4. High Precision Process:
   Thin Film and Thick Film.

THIN FILM MULTILAYER

A typical thin film process is shown in Figure 4. The characteristics of a typical polyimide are shown in Table IV. This polyimide has the lowest value of dielectric constant and, also, it has a low dissipation factor over a wide frequency range. Furthermore, it is chemically inert, very pure, and thermally stable up to 400°C. The good heat resistance enables the polyimide to be operated under general conditions of wire bonding, die bonding, sealing, etc.

The maximum thickness of polyimide formed at one operation is about 20 microns (0.0008"). Via holes are formed by wet photolithography and the minimum size is about three times the thickness of the polyimide.

In the thin film, both copper and aluminum can be used as a conductor. Copper is mainly used in the case of multilayer systems.[3] The metals are ion-plated by the use of electron beam deposition in order to enhance the conductor adhesion. Chromium, whose thickness is less than $1000\overset{\circ}{A}$, is used as the substrate metal. The adhesion strength is about 2 Kg/mm^2 (2850 psi).

The sheet resistance of the thin film copper deposit is shown as a function of its thickness in Figure 5. Experimental results and theoretical values are shown in dotted line and solid line, respectively. The observed values are 1.6 times the theoretical value.

The main reason for the use of polyimide as an insulator is to minimize propagation delay (T_{PD}). The interesting phenomena of the change in propagation delay with the change in characteristic impedance is shown in Figure 6. The delay was measured in a strip line structure by TDR.

It is interesting to note that if the data for either of the two insulators, i.e. alumina and polyimide are connected, they show a straight line. In other words, the propagation delay is only a function of dielectric constant and can be given by the equation, $T_{PD} = 0.0333\sqrt{\varepsilon_r}$ ns/cm, where ε_r is the relative dielectric constant of the media.

It is clear that low dielectric constant materials, e.g., polyimide in this case, are desirable to minimize propagation delay.

Figure 7 shows the typical design guideline for the thin film multilayer. For convenience, the data is shown in both microns and inches. This example employs a co-fired alumina/tungsten MLC coated with three layers of polyimide. The conductors are chromium/copper.

The design for the traces is shown in Figure 8.

The design guidelines for the overall package are schematically shown in Figure 9.

THICK FILM MULTILAYER

Conventional thick film systems may use various metals such as gold, silver-palladium, and copper. In this development, copper has been selected for the following reasons:

1. Copper is one of the best materials for its electrical characteristics.

2. Copper thick film system must be fired under a nitrogen atmosphere.

3. Copper system is relatively inexpensive.

The second reason is especially significant. Since the thick film multilayer is combined with co-fired multilayer ceramics, the thick film must be fired under a nitrogen or reducing atmosphere in order to protect against the oxidation of the co-fired conductors, viz., tungsten.

As shown in Figure 10, the thick film copper process is very simple. Copper conductor and glass dielectric are fired at approximately 900°C under a pure nitrogen atmosphere.

The material characteristics are listed in Table V. The sintered conductor has a sheet resistance of approximately 2 Ω/sq at 20 microns (0.0008") thickness.

The glass dielectric material is based on silica. If it is pure silica, it should have a theoretical dielectric constant around 3-4. However, the silica glass has a higher dielectric constant of 6-7. The result depends on the use of alumina as a filler. However, this value is a significant improvement when compared with the alumina system.

The metallization patterns are produced by screening. Therefore, the patterns are not as fine as those which can be generated by the thin film or co-fired systems.

On the other hand, the thick film system has a high dimensional tolerance and geometrical precision very similar to the thin film system.

Figure 11 illustrates a cross section of the thick film multilayer technology and is utilized as a design guideline for the construction of the package. As was previously shown, dimensions are given in both microns and inches for convenience.

A plan-view of the package is shown in Figure 12 and is used also as a design guideline for the conductor traces and pads.

The generalized outline of the package is shown in Figure 13 and gives the design guideline for the total package.

ADVANCED CO-FIRED MULTILAYER CERAMICS

At Kyocera, the most advanced co-fired multilayer ceramic technology is named HDCM, which stands for HIGH DENSITY CERAMIC MODULE. The basic technology is well-known, as it is used for practically all co-fired alumina ceramic modules. The Kyocera technology, HDCM, is comparable to the IBM technology being used in the production of IBM 3081 TCM (Thermal Conduction Module).[4]

Features of HDCM are as follows:

1. Fine via and line geometries.
2. Multiple layers and ultra thin tape.
3. Precise layer registration.
4. High pin count capability.

In Figure 14, it can be seen that the minimum via hole diameter is 100 microns (0.004") and the minimum line width and spacing are both 100 microns (0.004").

Figure 15 illustrates chip contact pads for TAB and wire bonding, as well as pads for flip chip bonding.

Tape layer thickness can range from 100 microns (0.004") to 250 microns (0.010") and with 30 and up layers being attainable (Figure 16).

Geometrical tolerances are shown in Table VI. The X-and Y-axis tolerances are excellent for co-fired ceramic packages. Unfortunately, these values are still too high to meet the tight dimensional tolerance requirement of VLSI packages.

Since alumina is used as the insulator, the dielectric constant is 8.5 to 10. When considering increased device speed, this high dielectric constant is considered untolerable.

For the two reasons mentioned above, the current multilayer ceramics are considered to be limited for advanced VLSI packaging technology.

COMPARISON OF THREE BASIC TECHNOLOGIES

The comparison of the three basic technologies mentioned in this paper, viz., thin film, thick film, and HDCM, is very useful to seek a solution to the problem of packaging advanced VLSI. The comparisons can be seen in Table VII.

Polyimide, silica glass, and alumina are employed as the insulators for thin film, thick film, and HDCM, respectively. A dielectric constant of 3.2 and, therefore, the propagation delay is best for the thin film system.

The conductor metal is copper for both the thin film and thick film, which results in an electrical conductivity which is superior to that of co-fired system using tungsten. Since the pattern generation method in thin film is photolithography, this system can produce the finest patterns among the three systems. Minimum via grid size for the thin film is one-half the size for the co-fired system. Therefore, the wiring density of the former becomes four times that of the latter.

Currently, the maximum number of layers available are five for both thin film and thick film. This number is not enough to meet the requirements for VLSI multichip modules. Consequently, further development work is being continued.

On the other hand, the co-fired system has no limitations in numbers of layers. Kyocera has even built packages with up to 150 layers.

Another unique factor is in I/O pin attachment. Co-fired ceramic is metalized by high temperature refractory metals and is, therefore, capable of being brazed with molten metals such as silver-copper or silver. This point has a great advantage in pin attachment reliability. Unfortunately pins can only be attached to the thin and thick film by soldering.

In dimensional tolerance, however, the co-fired system is very inferior to the other two systems. Even in the best process, the tolerance cannot be reduced to less than one-half per cent. This disadvantage causes a great difficulty in the application of co-fired ceramics to large size VLSI packages or modules with fine patterns.

Pattern modification is relatively easy in the thin film and thick film systems as it is just simply a matter of mask or pattern change .

THIN FILM AND THICK FILM HDCM

As mentioned above, the three methods presented here have their individual advantages and disadvantages. When taking these many aspects into consideration, if the three technologies are applied independently, it is very difficult to meet the requirements for VLSI packages and/or modules. The most desirable and probable method to meet these requirements is to combine their individual advantages. In other words, THIN FILM and THICK FILM HIGH DENSITY CERAMIC MODULE, T&T HDCM, seem to be the best technology to overcome the problems presented in applying conventional packaging technologies to advanced VLSI packages.

Structures of T&T HDCM can be seen in Figures 17 and 18. It is preferable that the base ceramic contain ground planes, power planes, and common lines and has brazed pins.

As examples of the T&T HDCM concept, two demonstration samples are shown. Figure 19 is a top and bottom view of a 400 pin flat package which has copper thin film metallization pattern on the top surface with a polyimide layer between the metallization and alumina substrate. The details of this package are seen in Figures 20 and 21. Figure 20 shows a diagrammatical cross-section of the package. Figure 21 is a detail of the cavity and adjacent area. Table VIII shows electrical characteristics of this package.

Figure 22 is a copper thick film substrate. Although this particular sample has no co-fired metallization nor brazed leads, such combination has been produced at Kyocera. Copper thick film HDCM appears to be the most suitable for surface mounting application because of its excellent solder leach resistance and geometric capability. When brazed I/O pins are required and low electrical resistivity are required this system is especially advantageous.

Thus, the combination technology, T&T HDCM, is commercially available on a production basis to solve the complex problems associated with the packaging of advanced high speed VLSI devices.

SUMMARY

Marriage of thin film technology with thick film technology on a MLC base has now provided an extremely flexible and excellent packaging technology for VLSI. It is believed that the T&T HDCM technology is the trend of the near future and will solve many of the packaging problems which were not solvable by standard MLC approaches.

REFERENCES

1. D. Seraphim, "Chip-Module Package Interface", IEEE Transactions on Components, Hybrids, and Manufacturing Technology, Vol. CHMT-1, pp. 305-309, 1978.

2. H.R. Knaupp, "Characteristics of Microstrip Transmission Line", IEEE Transactions on Computers, Vol. EC-16, No. 2, pp. 185-193, 1967.

3. C.W. Ho et.al., "The Thin-Film Module as a High-Performance Semiconductor Package", IBM J. Res. Develop., Vol. 26, No. 3.
   and
   J. P. Cummings, et. al., "Technology Base for High Performance Packaging", Proc. 32nd Electronics Components Conference, pp. 465-468, May, 1982.

4. B.T. Clark, "Design of the IBM Thermal Conduction Module", Proc. of the 31st Electronics Components Conference, 1981.

Table I
Advanced VLSI Packages

Table II
Material Characteristics — Insulator

Material	Dielectric Constant (1 MHz)	Volume Resistivity (Ω·cm)	Thermal Conductivity (25°C) (cal/cm·sec·°C)
Silicon Carbide	40	Semiconductive	0.16 ~ 0.2
• Alumina	8.5 ~ 10.5	>10^{14}	0.03 ~ 0.04
Zircon	7.8	>10^{14}	0.012
Silicon Nitride	7.5	>10^{14}	0.03 ~ 0.04
Beryllia	6.9	>10^{14}	0.55 ~ 0.65
Forsterite	6.5	>10^{14}	0.008
Mullite	6.0	>10^{14}	0.016
Glass·Epoxy	5.0	>10^{13}	0.0007 ~ 0.0015
Borosilicate Glass	4.1	10^{14}	0.002
• Quartz Glass	3.8	10^{14}	0.003 ~ 0.004
• Polyimide	3.2	10^{16}	0.0005

Table III
Conductor Metalization

Material	Sheet Resistance (mΩ/sq.)	Application (atm)	X-Y Tolerance
Tungsten (W)	8 ~ 15	Co-fire with alumina (H_2)	± 0.5%
Mo-Mn	—	Post-fire on alumina (H_2)	Negligible
Au	2 ~ 5	Thick Film (Air)	Negligible
Ag/Pd	15 ~ 35	Thick Film (Air)	Negligible
Cu (t = 20 μm) (.0008")	1.6 ~ 2.3	Thick Film (N_2)	Negligible
Cu (t = 10 μm) (.0004")	3	Thin Film	Negligible
Al (t = 10 μm) (.0004")	4.2	Thin Film	Negligible

Copper: Lowest resistance and best for high frequency

Table IV
Characteristics of Polyimide

Characteristic	Item	Value
Composition	Impurity	Na 0.4 ppm K <0.1 ppm Ca 0.3 ppm Cu <0.1 ppm Cr <0.1 ppm Fe 0.1 ppm U <0.3 ppb
Mechanical	Tensile Strength	20.000 psi (14.2 Kg/mm²)
Thermal	Melting Point Decomposition	>500°C
Electrical	Dielectric Constant	3.2 (1KHz, 25°C)
	Dielectric Loss Angle	0.0018 (1KHz, 25°C)
	Volume Resistivity	1.3×10^{16} Ω·cm
	Dielectric Strength	7.8 KV/mil (307 KV/mm)

Table V
Thick Film — Material Characteristics

● Dielectric (SiO_2 Glass)

Volume Resistivity	(100 VDC)	>10^9 Ω
Dielectric Constant	(100 KHz, 25°C)	6-7
Dielectric Loss Angle	(100 KHz, 25°C)	0.0017
Voids in Glass		Minimum

● Conductor (Copper)

Metalization Thickness	.0008" to .001" (20-25 μm)
Sheet Resistance (mΩ/Sq.)	1.6 to 1.8
Adhesion to Al_2O_3	4,000 psi (2.8 Kg/mm²)

Table VI
Size (O.D.) & Tolerance

- Max. Size: 6" × 6" (150 × 150 mm)
- Tolerance (X & Y): ± 0.5% NLT ± .002"
- Tolerance (Z): ± 5% NLT ± .001"
- Flatness: .0015" - .002"/inch NLT .001"

Table VII
Comparison of 3 Basic Technologies

Item	Thin Film	Thick Film	Co-Fired
Dielectric	Polyimide	Silica Glass	Alumina
Dielectric Constant	3.2	6 - 7	9.5
Dielectric Loss Angle	0.0018	0.0017	0.0003
Propagation Delay nsec/inch (nsec/cm)	.15 (0.06)	0.22 (0.085)	.25 (0.10)
Thickness - inch (µm)	.0004'' - .0008'' (10 - 20 µm)	.0016'' (40 µm)	.004'' up (100 µm up)
Conductor	Cu	Cu	W
Sheet Resistance	3mΩ/Sq.	2mΩ/Sq.	8mΩ/Sq.
Line Width	.0012'' (30µm)	.006'' (150µm)	.004'' (100 µm)
Via Hole Size	.0016'' (40µm)	.012'' (300 µm)	.004'' (100 µm)
Min Via Grid	.005'' (125µm)	.025'' (625 µm)	.010'' (250µm)
Maximum Layers	5	5	30 up
Conductor	Ion Plate & Photolithography	Screen	Screen
Dielectric & Via	Coat & Photolithography	Screen	Tape & Punch
Firing (Curing)	(400°C)	900°C	1.600°C
Firing Atmosphere	—	Nitrogen	Reducing
Pin Attach	Soldering	Soldering	Brazing
Plating	If Required	If Required	Yes
Via & Pattern Dim. Tolerance	Negligible ↑	Negligible ↑	± 0.5% ↑
∵ (Base Ceramic)	(Sintered)	(Sintered)	(Co-fired)
Pattern Modification	Fast With Low Retooling Cost	Fast With Low Retooling Cost	Hi Volume Oriented
∵ (Tool Change)	(Photo-mask)	(Screen Pattern)	(Pattern & Punch)

T&T HDCM → HYBRID TECHNOLOGY combines best of all technologies for high performance, high density and practical packages.

Table VIII
400 Pin Flat Package Electrical Characteristics

Polyimide Base	With	Without
Capacitance (pF)	1.21	2.07
Characteristic Impedance (Ω)	95 - 110	100 - 110
Propagation Delay (nsec)	0.20	0.24
Inductance (nH)	11 - 15	21 - 25

$L = Z_0^2 C$

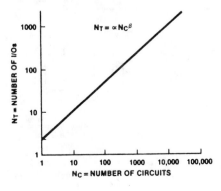

Figure 1

TREND OF INPUT/OUTPUT TERMINALS VS CIRCUITS ON MODULES

$N_T = \propto N_C^\beta$

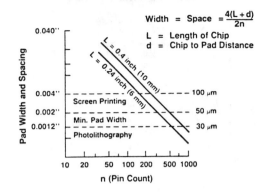

Figure 2

Packaging Density Relationship
Bonding Pad Density (W/B and TAB)

$$\text{Width} = \text{Space} = \frac{4(L + d)}{2n}$$

L = Length of Chip
d = Chip to Pad Distance

Figure 3

Device Technology

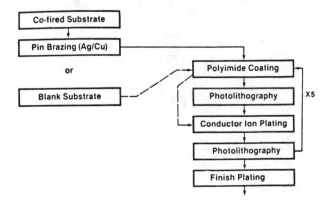

Figure 4

Thin Film Process Flow Chart

Figure 5
Sheet Resistance vs. Conductor Thickness

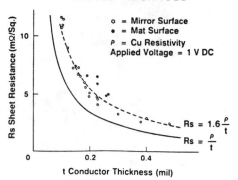

o = Mirror Surface
● = Mat Surface
ρ = Cu Resistivity
Applied Voltage = 1 V DC

$Rs = 1.6 \dfrac{\rho}{t}$

$Rs = \dfrac{\rho}{t}$

Rs Sheet Resistance (mΩ/Sq.)

t Conductor Thickness (mil)

Figure 6
Propagation Delay

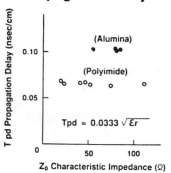

(Alumina)

(Polyimide)

$Tpd = 0.0333 \sqrt{\varepsilon r}$

T pd Propagation Delay (nsec/cm)

Z_0 Characteristic Impedance (Ω)

Figure 7
Thin Film HDCM — Design Guidelines

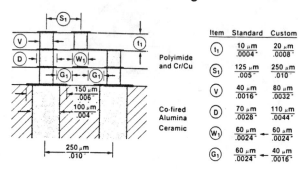

Item	Standard	Custom
Polyimide and Cr/Cu		
t_1	$\dfrac{10\ \mu m}{.0004"}$	$\dfrac{20\ \mu m}{.0008"}$
S_1	$\dfrac{125\ \mu m}{.005"}$	$\dfrac{250\ \mu m}{.010"}$
V	$\dfrac{40\ \mu m}{.0016"}$	$\dfrac{80\ \mu m}{.0032"}$
D	$\dfrac{70\ \mu m}{.0028"}$	$\dfrac{110\ \mu m}{.0044"}$
Co-fired Alumina Ceramic		
W_1	$\dfrac{60\ \mu m}{.0024"}$	$\dfrac{60\ \mu m}{.0024"}$
G_1	$\dfrac{60\ \mu m}{.0024"}$	$\dfrac{40\ \mu m}{.0016"}$

Cr/Cu/Cr: t = 5 ~ 10 µm (200 ~ 400 µ") Common to Both

Figure 8
Thin Film HDCM — Design Guidelines

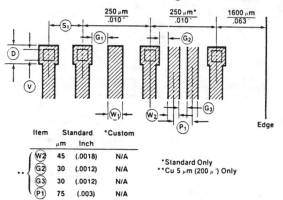

Item	Standard		*Custom
	µm	Inch	
W_2	45	(.0018)	N/A
G_2	30	(.0012)	N/A
G_3	30	(.0012)	N/A
P_1	75	(.003)	N/A

*Standard Only
**Cu 5 µm (200 µ") Only

Figure 9
Thin Film HDCM — Design Guidelines

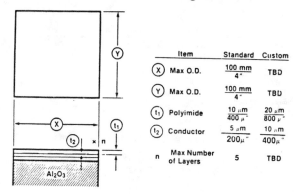

Item	Standard	Custom
X Max O.D.	$\dfrac{100\ mm}{4"}$	TBD
Y Max O.D.	$\dfrac{100\ mm}{4"}$	TBD
t_1 Polyimide	$\dfrac{10\ \mu m}{400\ \mu"}$	$\dfrac{20\ \mu m}{800\ \mu"}$
t_2 Conductor	$\dfrac{5\ \mu m}{200\ \mu"}$	$\dfrac{10\ \mu m}{400\ \mu"}$
n Max Number of Layers	5	TBD

Figure 10
Copper Thick Film — Process Flow Chart

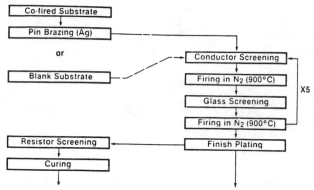

Co-fired Substrate

Pin Brazing (Ag)

or

Blank Substrate

Conductor Screening

Firing in N₂ (900°C)

Glass Screening

Firing in N₂ (900°C)

X5

Resistor Screening

Finish Plating

Curing

Figure 11
Copper (Cu) Thick Film — Design Guidelines

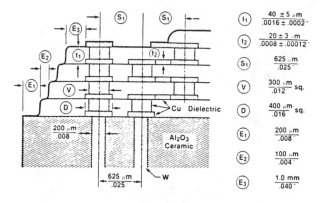

Cu Dielectric

Al₂O₃ Ceramic

Item	
t_1	$\dfrac{40 \pm 5\ \mu m}{.0016 \pm .0002"}$
t_2	$\dfrac{20 \pm 3\ \mu m}{.0008 \pm .00012"}$
S_1	$\dfrac{625\ \mu m}{.025"}$
V	$\dfrac{300\ \mu m}{.012"}$ sq.
D	$\dfrac{400\ \mu m}{.016"}$ sq.
E_1	$\dfrac{200\ \mu m}{.008"}$
E_2	$\dfrac{100\ \mu m}{.004"}$
E_3	$\dfrac{1.0\ mm}{.040"}$

Figure 12
Copper Thick Film — Design Guidelines

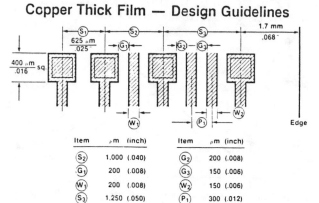

Item	µm	(inch)	Item	µm	(inch)
S_2	1.000	(.040)	G_2	200	(.008)
G_1	200	(.008)	G_3	150	(.006)
W_1	200	(.008)	W_2	150	(.006)
S_3	1.250	(.050)	P_1	300	(.012)

Figure 13
Copper Thick Film — Design Guidelines

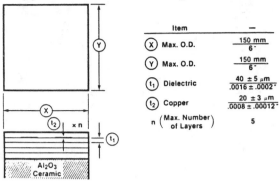

Item	—
X Max. O.D.	$\dfrac{150\ mm}{6"}$
Y Max. O.D.	$\dfrac{150\ mm}{6"}$
t_1 Dielectric	$\dfrac{40 \pm 5\ \mu m}{.0016 \pm .0002"}$
t_2 Copper	$\dfrac{20 \pm 3\ \mu m}{.0008 \pm .00012"}$
n (Max. Number of Layers)	5

Figure 14
Via Holes & Lines

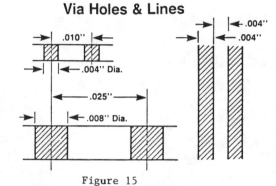

Figure 15
Chip I/O Contact Pads

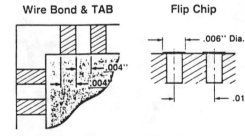

Figure 16
Tape Layer Thickness

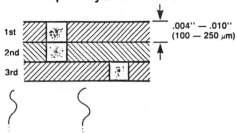

Figure 17
Thin Film HDCM Concept

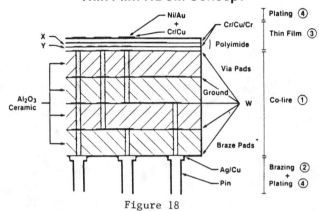

Figure 18
Copper Thick Film and Co-fire Concept

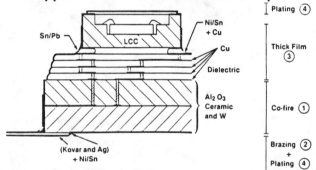

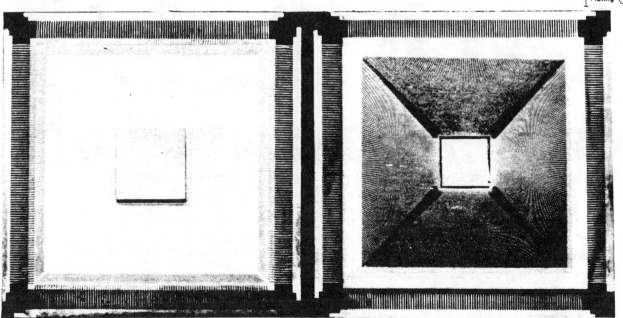

Figure 19

Figure 20
400 Pin Flat Package (Thin Film)

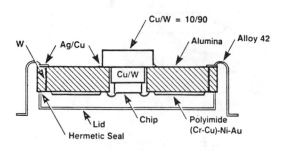

Figure 21
400 Pin Flat Package (Thin Film)

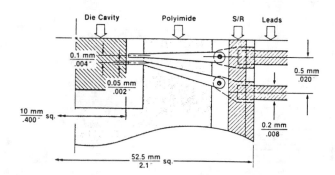

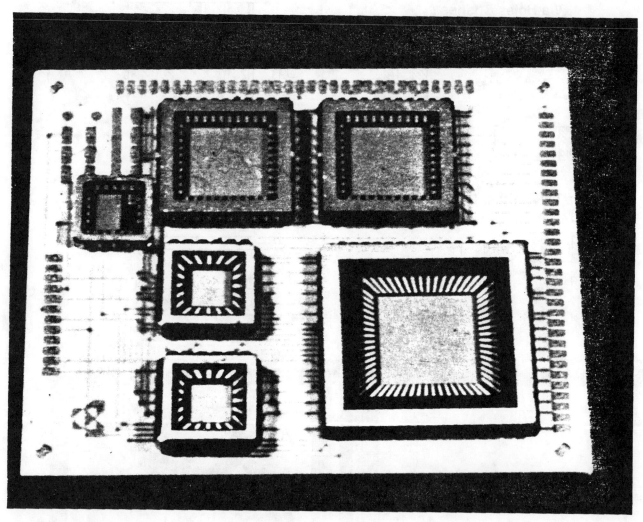

Figure 22

Part 3
Cofired Ceramic Technology

COFIRED multilayer ceramic technology was originally developed to produce chip capacitors, and then used for single chip packaging. Its capability to make high density packages was soon realized in the application of multichip modules. The first explosive increase in interconnection density came with the selection by IBM of the alumina ceramic cofired technology for their mainframe computers. While there are a number of IBM articles by Yates and others, the review articles in the *IEEE Transactions on Components,* *Hybrids, and Manufacturing Technology* and *Scientific American* (Blodgett) are excellent, and the Schwartz overview is most useful. The NTT paper by Kishimoto and Ohsaki provides a description of the alternative base heat removal to the previous heat removal methods of substrates from the backs of the die. The paper by Shibata *et al.* of Mitsubishi provides a similar substrate, with an alternate heat removal method.

BERNARD SCHWARTZ
IBM Corporation
T. J. Watson Research Center
Yorktown Heights, N. Y. 10598

ABSTRACT

Multilayer ceramics (MLC) technology is the basis for several major products today, which include capacitors and packages for integrated circuits. This technology was originally developed to produce miniature capacitors for the micro-module program. When these goals were achieved, new applications in packaging were found. The MLC technology is now the building block in IBM's medium and large computers. The current state of this technology will be described. It's advancement will require improved understanding of the sciences involved in each of the process steps. Several of these technical areas will be discussed, particularly the relationships among the electrical properties of the materials and the performance of the package. As the dimensions become smaller, more concern will have to be shown for the defect levels. This implies improved control of material preparation and processing.

INTRODUCTION

Multilayer ceramics, (MLC) are monolithic structures of personalized and interconnected ceramic and metal layers. The individual layers can vary from less than one mil to over twenty mils. The materials in greatest use today are titanate dielectrics with palladium metal electrodes for MLC capacitors and a variety of alumina ceramics with either molybdenum or tungsten conductors for MLC substrates. A short history of these developments [1,2,3] will be presented which will be followed by a discussion of the future requirements.

HISTORY OF MLC

The MLC technology started in 1958 with the Radio Corporation of America (RCA) and U.S. Signal Corps Micromodule Project. At that time I was hired by RCA to develop a family of miniature capacitors, with the requirements shown in Table I. The calculation shows that eleven layers of one-mil thickness would be needed in order to achieve these values.

All of the thick and thin film technologies being used at that time were reviewed. The doctor-blading method was finally selected. Fortunately, the paint industry was developing many new binder systems for their products. A plasticized copolymer of polyvinyl chloride acetate was found, which could be heavily loaded with titanate powders, with the aid of suitable solvents and deflocullants[4]. The initial casting was done on glass plates, but a plastic carrier was found to be more suitable for production, similar to the method described by J. L. Park[5].

The next problem was the selection of a metallurgy. After testing many noble metals pastes, palladium was selected because it could be co-fired with the titanates in air. Also it was a relatively inexpensive and reliable material. Screening provided adequate resolutions. The lamination process was the most difficult to develop. One afternoon I went to the Chemical Show in New York City, where I saw a demonstration of a photograph being laminated between two vinyl sheets in a heated press. The press was purchased and in a short time the lamination conditions were established. After lamination, the individual capacitors were punched or cut out of the laminated stack and then sintered. The complete process is shown in Table II.

Because of the success of this project, the Signal Corps requested two additional proposals for: (1) a high temperature, radiation resistant capacitor and (2) a micromodule wafer package for a quartz crystal oscillator. A fifty-one layer alumina capacitor was built, which led to the concept of MLC alumina substrates. This package was the first MLC substrate built. It was a three layer structure, as shown in Fig. 1. Mo-Mn metallization was used to provide the hermeticity. A metal cap was ultrasonically soldered to the metallized sealing ring on the top layer. This package passed all the MIL specifications.

TABLE I. Capacitor Requirements

1958

Micro-Module Program Needed, in a 0.3 In. by 0.3 In. Ceramic Wafer Form, the Following:

1. Precision Capacitors with up to 3000 pf and 50 Vdc Ratings
2. General Purpose Capacitors with up to 300,000 pf and 50 Vdc Ratings

$$C = \frac{.224\ KAn}{T}$$

C = Capacitance (pf)
K = Dielectric Constant
A = Electrode Area (Sq. In.)
N = Number of Layers
T = Layer Thickness (In.)

For (1) K$\approx$30
A$\approx$.04
T$\approx$.001
N$\approx$11

For (2) K$\approx$3000
A$\approx$.04
T$\approx$.001
N$\approx$11

Figure 1. Quartz Crystal Package

Up to this time, the vertical connections could only be made at the edges of the various packages. The next major innovation was the internal via [6]. A via is a hole punched through the ceramic green sheet and subsequently filled with a metal paste. It is used to make an electrical connection from one layer to another. The problem of bridging the grids from the chips to the board could then be solved, thereby eliminating the card, as represented in Fig. 2.

The flow diagram for the alumina MLC process is given in Table III. A modern doctor blading system used in IBM in shown in Fig. 3. As noted in Table III, molybdenum pastes are used in IBM MLC substrates. They provide excellent adherence to the ceramic and an acceptable electrical conductivity, see Table IV. Several of the patterns used in the IBM Thermal Conduction Module are shown in Fig. 4.

A schematic of this module, see Fig. 5, shows several key features: the "microsocket" area array connections for the chips on the top layer, the engineering change capability on the top surface where the discrete wires are bonded to the surface pads, the redistribution layers, the X and Y signal layers, the power distribution and voltage reference layers. The pin brazing pads are on the bottom surface. Fig. 6 is a schematic of the substrate, chip, cooling design, and sealed enclosure.

The module pins provide the electrical connections to the board for power and I/O connections. The heat is removed from the chip directly by the piston and helium gas in the hermetically sealed reservoir to a heat sink in the cover of the module.

The MLC module technology has enabled IBM to make significant increases in circuit density and speeds, as shown in Fig. 7. The specific module functions are shown in Fig. 8 and will be discussed in the next section.

Table II. MLC Capacitor Process

1. **Ball Mill Slip**
   - **Ceramic Dielectric (Titanates) Powder**
   - **Thermo-Plastic Binder and Plasticizer**
   - **Deflocculant**
   - **Solvents**
2. **Doctor Blade a Tape on a Glass or Plastic Substrate**
3. **Dry, Strip, and Cut Process Sheets**
4. **Screen Noble Metal Paste Electrodes and Dry**
5. **Laminate Sheets to Design**
6. **Cut Units to Size**
7. **Connect Alternate Layers at Edges (Notches)**
8. **Sinter**
9. **Tin Terminals**
10. **Test**

Table III. MLC Substrate Process

1. **Ball Mill Slip (Aluminas)**
2. **Doctor Blade Tape on Mylar**
3. **Dry, Strip and Reel**
4. **Blank**
5. **Punch Holes (Location, Vias, Etc.)**
6. **Screen Mo (or W) Pastes and Dry**
7. **Stack and Laminate**
8. **Size (Cut)**
9. **Sinter**
10. **Plate**
11. **Braze (Pins, Flange)**
12. **Mount Chips**
13. **Seal**
   (Testing Steps Omitted)

The problem:

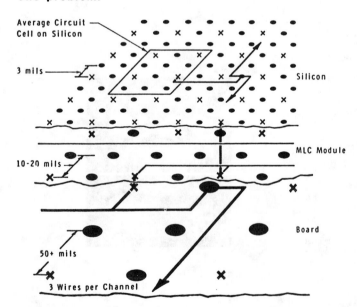

Figure 2. Bridging the Grids

Figure 3. Modern Doctor-Blading System

Table IV. Module Metallurgies

Metal	Melting Point (°C)	Electrical Resistivity ($10^{-8}\Omega$-m)	Coefficient of Thermal Expansion (10^{-7}/°C)	Thermal Conductivity (W/mK)
Ag	960	1.6	19.7	418.4
Au	1063	2.2	14.2	297.06
Cu	1083	1.7	17.0	393.29
Pd	1552	10.8	11.0	71.13
Pt	1774	10.6	9.0	71.13
Mo	2625	5.2	5.0	146.44
W	3415	5.5	4.5	200.83
Ni	1455	6.8	13.3	92.05
Cr	1900	20	6.3	66.94

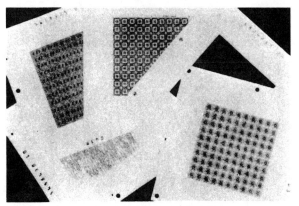

Figure 4. Screened Mo Patterns

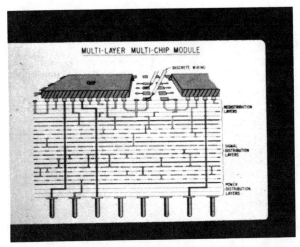

Figure 5. Schematic Drawing of a MLC Substrate.

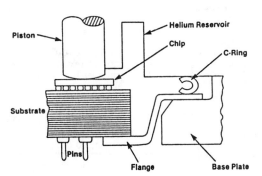

Figure 6. TCM Module Cross Section.

FUTURE DIRECTIONS

An important property of the ceramic is its dielectric constant because it controls the time delay (Td) of the signals according to the equation:

$$T_d = \frac{\sqrt{K}\,l}{C}$$

where K is its dielectric constant, l the distance the signal travels and C the speed of light. Typical values for several ceramic substrate materials are listed in Table V. These values may seem very small, but for many megabits of information the totals can be significant. It is desirable, therefore, to reduce the dielectric constant as much as possible. The characteristic impedance (Z_O) also depends on the dielectric constants of the ceramic, as well as the geometry of the wiring, according to the following equation:

$$Z_O = \sqrt{L/C}$$

Where L is the inductance and C is the capacitance. Reducing K will increase Z_O, unless the dielectric thickness is also reduced. The problem associated with a smaller module, is that there is less space for the conductors. As a result metals with higher electrical conductivities will have to be used. NEC [7] is now producing a glass-ceramic MLC module with gold paste conductors. Their glass is a lead-borosilicate, which has to be fired in air. If lower firing ceramics can be developed which can withstand reducing atmospheres, then copper may be used. Controlling the adherence of copper to ceramics is still a problem.

Lead reduction is the ratio of the total chip connections to module pins, and is controlled by the product designs. A TCM module can have a lead reduction ratio up to 8.

The chip joining method changed when IBM went from single layer SLT, MST, and MC substrates to the MLC construction. The "microsocket" permits many more connections per unit area, with the use of redistribution patterns. As the chips become larger, the thermal expansion of the ceramic will have to match that of the silicon. The intermittent heating and cooling of the chip induces a stress in the solder pads, based on their thermal expansion differences, which could lead to fatigue failures. To reduce this problem, the ceramic and semiconductor should have closely matched thermal expansions.

Power distribution could become the next limiting factor in module design. As each chip increases its circuit density, the current requirements become greater. Some of the proposed solutions include metals with higher electrical conductivity, chip capacitors on the module or direct current bussing. These ideas are still being tested. The other functions listed in Fig. 8 can be accommodated through design changes.

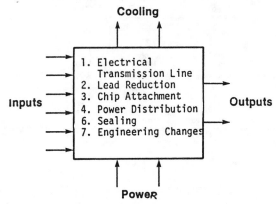

Figure 8. Module Functions

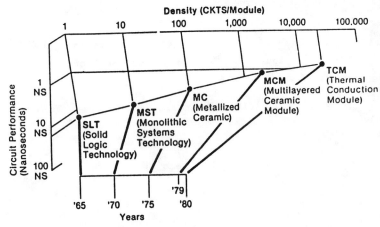

Figure 7. Improvements in Performance

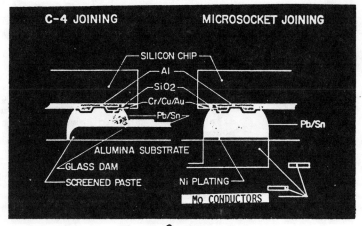

Figure 9. Chip Joining

Table V. Ceramic Substrate Materials

MATERIALS	DIELECTRIC CONSTANT	DELAY TIME $(10^{-9}sec/in)$	THERMAL EXPANSION $(10^{-6}/°C)$	STRENGTH (KPSI)	THERMAL CONDUC- TIVITY (CGS)
96% Al_2O_3	9.3	0.26	6.4	46.0	0.06
92% Al_2O_3	8.5	0.25	6.5	48.0	0.04
ELECTRICAL PORCELAIN	5.5	0.20	4.4	13.0	0.004
SiO_2	3.8	0.17	0.6	8.0	0.005
Si_3N_4	6.0	0.21	3.0	85.0	0.08
Al N[8]	8.8	0.25	4.5	53.3	0.02
Glass- Ceramic[7]	7.5	0.23	4.2	42.6	0.01

The real problems will be in the control of materials and processes as dimensions become smaller. Current use of natural materials will be replaced with chemically synthesized powders. Some of the processes in development include: Sol-Gel synthesis, co-precipitation, metal powder oxidation and solution drying. Each of these processes can provide very pure powders with well controlled particle sizes. The application of these powders are still in the early development stages. Fabricating processes will have to be carried out in clean room environments in order to avoid defects in the ceramics. As the structures become smaller, the microstresses will increase. These conditions will require stronger ceramics. Several methods being studied are: zirconia toughening, surface compression layers, reduced defect levels and sizes and improved densities and microstructures.

Ceramic packaging technology is still in a very early stage. In this brief review, a few new directions were proposed. The big objective ahead of us will be to achieve higher circuit densities in the new computers.

References

1. B. Schwartz, "Microelectronics Packaging," Electronic Ceramics: Amer. Ceram. Soc. Publ. No. 3, pp 12-25 (3 May 1969).

2. B. Schwartz, "Microelectronics Packaging II," Amer. Ceram. Soc. Bull., 63(4) 577-581 (1984).

3. B. Schwartz, "Review of Multilayer Ceramics for Microelectronics Packaging," J. Phys. Chem. Solids, (to be published).

4. R. E. Mistler, D. J. Shanefield and R. B. Runk, "Tape Casting of Ceramics," Ceramic Processing Before Firing, (Ed. by G. Y. Onoda, Jr. and L. L. Hench) pp. 411-448, Wiley, NY (1978)

5. J. L. Park, Jr. U.S. Patent No. 2,966,719 (3 Jan. 1961).

6. H. Stetson, U.S. Patent No. 3,189, 978 (22 June 1965).

7. Y. Shimada et. al., "Low Firing Multilayer Glass-Ceramic Substrate," "IEEE Trans. Comp. Hybrids, Mfg. Tech.," Vol. CHMT-6, No. 4, 382-388 (Dec. 1983).

8. N. Iwase, et. al., "Development of A High Thermal Conductive Substrate Technology," IMC 1984 Procedings, pp. 180-185 (May 21-23, 1984).

Microelectronic Packaging

Methods of housing, cooling and interconnecting the silicon chips in a digital computer have an important bearing on the machine's performance. The aim is to fit the most chips in the least volume

by Albert J. Blodgett, Jr.

The packaging of the microelectronic silicon chips in a high-speed digital computer might seem only incidental to the machine's design. One could argue that the actual processing of information is done entirely by the circuitry on the chips themselves; the functions of the packaging are simply to interconnect the chips and other devices, to distribute electric power and to provide a suitable operating environment. Clearly, every computer needs packaging of some kind, but the nature of the packaging would not appear to have much influence on the functioning of the machine. The facts are otherwise. In many high-speed data-processing units packaging technology is the factor that determines or limits performance, cost and reliability.

One reason packaging has become so important is the imperative to make the central elements of a computing system exceedingly compact. Improvements in the design and fabrication of microelectronic devices have greatly increased the number of logic functions that can be put on a chip as well as the speed at which arithmetic operations are performed. As a result a major source of delay in the central processing unit of many computers now is the time needed for a signal to pass from one chip to another. In order to reduce the delay the chips must be put closer together. Fitting many chips, each with many terminals, into a small volume challenges packaging technology in several ways. First, there is little space available for the thousands of conductors needed to distribute electric power and information-bearing signals among the chips. Furthermore, the electrical properties of this network of conductors must be designed to minimize the distortion of signals, which becomes more difficult as switching speeds are increased and dimensions are reduced. Finally, a dense array of chips gives off a fair amount of heat, which must somehow be removed if the circuits are to operate properly. In many instances the thermal problem is the most challenging one.

The designers of high-performance digital computers have devised a number of ingenious and quite varied solutions to these problems. Here I shall discuss some general considerations that go into the design of any packaging technology. I shall also describe in some detail the system developed for one high-performance system, the model 3081 processor of the International Business Machines Corporation. It is large "mainframe" computers of this kind that make the severest demands on packaging methods and materials. The technology I shall describe is capable of interconnecting more than 100 chips—which might incorporate circuitry equivalent to the entire central processing unit of an earlier computer—in a module that can be held in one hand.

For more than a decade the hierarchy of packages in a typical mainframe computer has had three levels. Each chip is permanently mounted on an individual chip carrier, a plastic or ceramic housing with a dozen or more metal leads. A number of chip carriers, along with a few discrete components such as resistors and capacitors, are mounted on a card made of fiberglass impregnated with an epoxy resin. Metal conduction paths are printed on the surface of the card, and in some cases the card is a laminate with multiple layers of printed wiring. Several cards are mounted through connectors along one edge to a larger board whose printed circuitry establishes connections between the cards. The entire assembly of chip carriers, cards and main board is tied to the rest of the system (including other assemblies of the same kind) by multiconductor cables.

The three-level hierarchy of packages is still the standard in many computers, most notably the ones called minicomputers. In both the lowest- and the highest-speed systems, however, technological advances have made it possible to dispense with one level in the hierarchy. In the low-performance realm of the microcomputer each chip is still housed in a separate carrier; owing to the dramatic increase in the number of circuits per chip, however, the number of chips has been reduced to the point where they all fit on one board. In some of the latest large mainframe computers another approach has been taken: many chips are bonded to a single substrate, forming a multichip module that plugs directly into a large board. Thus in both cases the card level is eliminated.

The evolution of microelectronic devices and that of packaging technology are strongly interdependent. For example, putting more circuits on a single chip generally requires adding more electrical contacts, but that is feasible only if the package can accommodate additional connections and signal paths. Similarly, chips with more circuits as a rule consume more power; they can be incorporated into a system only if the package can safely remove the addition-

LACEWORK OF CONDUCTORS interconnects 100 silicon chips in the central processing unit of a "mainframe" computer. Each of the patterns on the opposite page forms one layer of a ceramic substrate that can have as many as 33 layers. At the upper left is the top layer, to which the chips are bonded, establishing electrical connections through a dense array of contact pads in the middle of each chip site. At the upper right is one of five redistribution layers, which carry signals from the chip contact pads to engineering-change pads that surround the chip site. The two layers in the middle are signal planes that provide for communication between chips; on a given plane all the conductors are oriented parallel to either the x or the y axis. A complete substrate has 16 signal planes. Between each pair of signal planes is a voltage-reference plane, such as the one at the lower left, that helps to define the electrical properties of the signal lines. The layer at the lower right is one of three that distribute power to the chips. In the finished substrate the layers are stacked, laminated and fired, so that they fuse into a tilelike ceramic. The patterns are shown at their approximate actual size after firing.

83

al heat. Because of these relations the constraints that apply to packaging design today can be fully understood only by first considering the nature of the chips that go into the packages.

The crucial event in the evolution of semiconductor technology was the introduction of the integrated circuit, which combines a number of transistors and other basic circuit elements on a single piece of silicon. The first integrated circuits, in the early 1960's, included only from six to eight transistors, diodes and resistors, enough to implement a "gate," or simple logic function. A typical gate receives two or three input signals and issues a single output; the pattern of the input signals determines whether the output is on or off. Several hundred gates might be needed to build a functional unit of a computer such as an adder.

The early integrated circuits were built on a silicon chip with a surface area of a few square millimeters, and the smallest features in the pattern that defined the circuitry were from 10 to 20 micrometers across. The chip communicated with the rest of the system through eight or 10 terminals. The switching delay, which represents the time from the arrival of an input signal to the production of an output signal, was generally between 20 and 40 nanoseconds, or billionths of a second.

Advances in the art of semiconductor manufacture over the past 20 years have been dramatic. The maximum available surface area on a chip has increased by more than 10 times (to about 50 square millimeters), and the minimum feature size has decreased by a similar factor (to less than 1.5 micrometers). High-speed logic chips have been built with more than 40,000 transistors and other circuit elements, organized into 5,000 gates that switch in about two nanoseconds. A chip of this kind can have 200 electrical contacts.

The integrated circuits employed in a high-performance mainframe computer are built out of basic circuit elements called bipolar transistors. Chips based on another kind of transistor, the field-effect transistor, can fit many more circuit elements in a given area. For example, microprocessors and related devices have been built on a single chip with more than 100,000 field-effect transistors. Why, then, do mainframe computers employ bipolar transistors? The reason is that they are faster: as a rule they have a shorter switching delay.

The chips with the highest transistor count are random-access memory arrays, whose function is to store information. Early semiconductor memory devices, introduced in 1971, had 128 cells; each cell was capable of holding one bit (binary digit) of information. Today memory chips with a capacity of 256,000 bits are in production.

For the packaging designer the logic chips of the central processor present a greater challenge than memory chips. In general logic chips dissipate more power and hence give off more heat; they also require more electrical contacts for signal communication. Similarly, a chip based on bipolar transistors consumes more power than a chip with equivalent functions made with field-effect transistors. It follows that packaging design is most critical in the central processor of a high-speed mainframe computer.

Packaging begins where the chip ends: at the metallic pads on the surface of the chip where contact is made with the external wiring. The length of that wiring is among the packaging designer's primary concerns.

In the analysis of many electrical circuits it is reasonable to assume that a voltage applied at one end of a wire appears simultaneously at all points along the length of the wire. Actually the speed of a voltage signal along a wire is finite; it is determined by a property of the insulator surrounding the wire, namely its relative dielectric constant. If the insulator is air (or a vacuum), the relative dielectric constant is 1 and the signal travels with the speed of light in free space, which in units appropriate to this discussion is about 30 centimeters per nanosecond. In other insulators the dielectric constant is larger and the speed is reduced by a factor proportional to the square root of the dielectric constant.

For a fiberglass printed-circuit board the dielectric constant is approximately 4, and so the propagation speed is reduced by a factor of two; in other words, signals travel through the conductors in the board at about 15 centimeters per nanosecond. Because a signal may have to go appreciably farther than 15 centimeters to get from one chip to another, propagation delays can exceed a nanosecond. In slower digital devices a delay of this magnitude is insignificant because the switching delays of the logic gates are tens or hundreds of nanoseconds. In a computer built out of devices that switch in a nanosecond, however, propagation delays clearly have a major influence on the overall speed of operations. It is for this reason that minimizing wire lengths and maximizing circuit density are of critical importance in the design of packaging.

When propagation delays are important, the wire that carries a signal cannot be considered a simple conductor but instead must be treated as a transmission line. The signal is represented as a wave propagating along the transmission line, and the voltage at any point along the conductor depends on both distance from the source and time since the signal was emitted.

In a transmission line the electrical resistance of the conductor is not the only property that affects the propagation of a signal. It is also important to know the inductance, which determines the amount of energy stored in the magnetic field set up by a passing current, and the

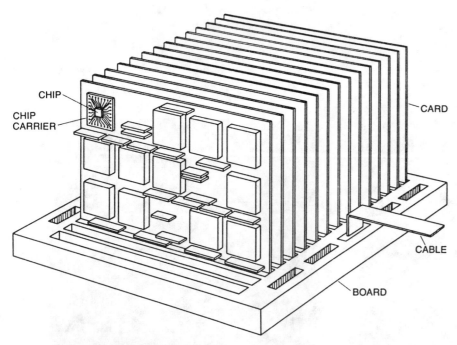

HIERARCHY OF PACKAGES in a large digital computer has traditionally had three levels. Each chip is housed in an individual carrier; a number of carriers and other components are attached to a printed-circuit card; several cards are in turn mounted on a larger printed-circuit board. The assembly is connected to the rest of the system by cables. With the recent introduction of a module capable of holding more than 100 chips it has been possible to eliminate the card level and thereby to reduce the number of interconnections and the total wiring length.

CHIP

CHIP CARRIER

CARD

CABLE

BOARD

capacitance, which determines the energy stored in the corresponding electric field. The inductance and the capacitance depend on the geometry of the transmission line and on electrical and magnetic properties of the materials it is made from; together they define the impedance of the line. For a low-resistance transmission line the impedance is equal to the square root of the ratio of the inductance per unit length to the capacitance per unit length. It is measured in ohms, the same unit employed for resistance, but its effects on a propagating signal are more complicated than the effect of resistance on a steady current.

One characteristic of all waves is that they can be reflected. Similarly, a digital signal can be partially reflected from a discontinuity in the transmission line or from the end of the line. The reflection coefficient, which gives the fraction of the signal reflected, is determined by the impedance and by the load resistance that terminates the line. Suppose a given transmission line has an impedance of 100 ohms. If the load resistance is also 100 ohms, the signal is totally absorbed by the load and none of it is reflected back into the line; this is the ideal situation. If the load resistance is 200 ohms, however, a third of the signal is reflected and adds to the initial signal on the line. A load resistance of 50 ohms also yields a reflection coefficient of one-third, but the reflected signal is subtracted from the initial one. It is clear that such reflections must be controlled to prevent switching errors.

Reflections are only one of several ways the electrical design of a package can modify a signal or inject "noise" into a circuit. For example, two adjacent conductors can be coupled through their mutual inductance and capacitance, so that a signal sent down one line may also appear on the other. Such "crosstalk" must be avoided if the behavior of the system is to be predictable.

In a high-performance package the basic method of controlling the characteristics of the transmission lines is to separate layers of signal wires with conductive sheets called voltage-reference planes. The reference planes can also provide a path for return currents. Each plane is at a uniform electric potential, either zero volts (ground voltage) or one of the supply voltages needed by the chips and other components. (Hence the planes can also be employed to distribute power.) A signal line encased in an insulating medium and sandwiched between two such planes makes a transmission line whose properties can be calculated. The planes give the line a uniform and well-defined impedance and also inhibit crosstalk between lines in adjacent layers.

The design of a transmission line begins with the specification of its direct-

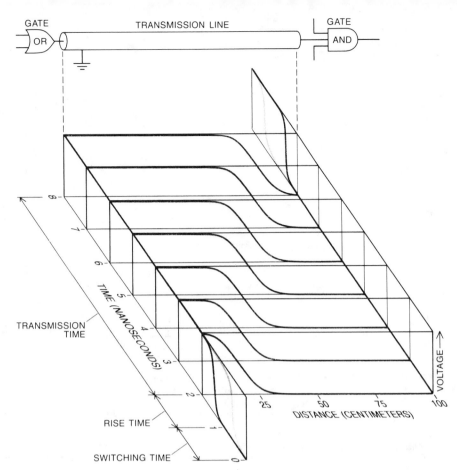

SIGNAL DELAY in a digital computer limits the maximum speed of operation. The delay has three components: the rise time of a voltage applied to a logic gate, the switching time of the gate itself and the transmission time needed for the signal to reach the next gate. When the signal must pass from one chip to another, the transmission delay, which is determined by the packaging technology, is often the longest of the three. Here the rise time and the switching time are each about a nanosecond, but transmitting the signal over a distance of roughly a meter takes six nanoseconds. The signal propagates as an electromagnetic wave in a transmission line; the wave is represented by a graph of voltage as a function of time and distance.

current resistance. The resistance must be small compared with the load resistance or the input voltage will be seriously attenuated when it reaches the output of the line. The resistance per unit length is determined by the resistivity of the material and the cross-sectional area of the conductor; once the material is chosen only the latter property can be altered by the designer. For printed circuits and conductive traces fabricated by similar techniques the cross section is a flattened rectangle.

Given the dimensions of the conductor, the line impedance is determined by two additional factors: the dielectric constant of the insulating medium and the distance between the voltage-reference planes. For a particular insulating material the distance between reference planes is adjusted to achieve the desired impedance. The design value depends

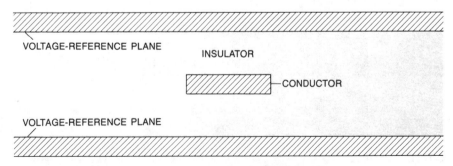

TRANSMISSION LINE consists of a conductor embedded in an insulator and sandwiched between two conductive planes. The electrical characteristics of the line depend on the dimensions and the physical properties of the component parts. For example, the speed of a wave propagating through the line is determined by the dielectric constant of the insulator. The parallel conductive planes reduce coupling, or crosstalk, between lines in adjacent layers.

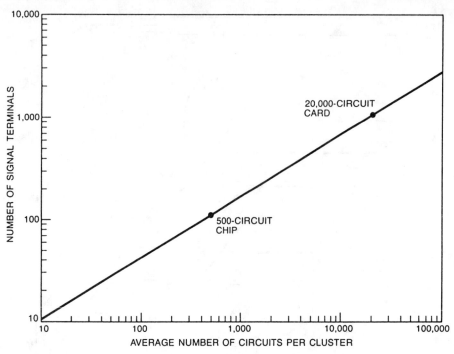

NUMBER OF TERMINALS needed for communication with a cluster of logic circuits can be estimated from an empirical relation called Rent's rule. In a mainframe computer a cluster of C logic circuits needs $2.5C^{.61}$ signal terminals. Thus a chip with 500 circuits would need about 110 terminals, and a card with 20,000 circuits would need about 1,000 signal terminals.

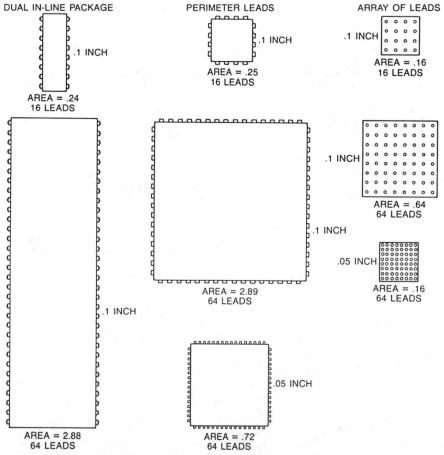

"FOOTPRINT" OF A CHIP CARRIER, or the area it occupies on a card or board, depends on the spacing and arrangement of the terminals. The dual in-line package, with two rows of pins .1 inch apart, is efficient for a chip with 16 leads but not for one with 64. A square package with leads at the perimeter saves no space if the distance between terminals remains .1 inch, but it covers only one-fourth the area when the leads are on .05-inch centers. The most efficient practical configuration for a chip carrier is a full grid of terminals. Indeed, a square carrier with 64 leads on a .05-inch grid takes up less space than a 16-pin dual in-line package does.

on many factors, including the electrical properties of the chips in the system, the dimensions and other specifications of the package and the amount of power available to drive the transmission lines. Typically the impedance is in the range from 50 to 100 ohms.

A conductor between two reference planes can only approximate an actual transmission line. In practice a signal line connecting two chips may follow a tortuous route, threading from one layer of wiring to another. At transitions between packaging levels—such as those from the chip to the carrier or from the card to the board—the electrical properties of the wiring depart significantly from the ideal. As noted above, such discontinuities can cause reflections. They also introduce additional delays, proportional to their capacitance and inductance. The extra delays must be added to the basic propagation delay of the circuit to determine the total delay introduced by the packaging system.

A single transmission line extending between two isolated terminals could be given precisely defined characteristics. The actual task of the packaging designer, however, is to create a network of intermeshed pathways to interconnect thousands of device terminals. The topological complexity of the network is formidable.

The number of signal terminals needed to establish connections at any given level in the packaging hierarchy can be estimated from an empirical relation called Rent's rule, developed in 1960 by E. F. Rent of the International Business Machines Corporation. The rule, which can be applied to chips, multichip modules, cards or boards, gives the approximate number of terminals as a function of the number of logic circuits, C. A form of the rule based on data obtained by analyzing large data-processing systems indicates that the average number of terminals is equal to $2.5C^{.61}$. Thus the number of terminals needed to support 100 circuits is about 40 and the number needed for 1,000 circuits is almost 170.

There are certain limitations to the scope of Rent's rule. First, it can be applied only to the logic elements of a processing system; arrays of memory cells need far fewer terminals. Second, each cluster of circuits must be a small, "random" subset of the entire logic complex. If a cluster constitutes a complete functional unit of the computer, then again fewer terminals suffice. Third, it is assumed that information passed between packages is not specially encoded for serial transfer, a technique that can reduce the number of terminals, although only at the cost of slower operation. The clusters of logic circuits for the mainframe computers being considered here meet these conditions.

Consider the packaging of a functional unit with 10,000 logic circuits, built up out of chips that have a maximum of 25 circuits each. The chips are housed in carriers with 14 signal pins and mounted on boards with 100 signal terminals. At first it might seem that a 10,000-circuit unit could be assembled out of 400 of the 25-circuit chips. Rent's rule shows, however, that a package with 14 signal pins can support an average of only 17 circuits; hence the actual number of chips needed is 10,000 divided by 17, or 588. (Almost a third of the capacity of the chips remains idle, suggesting that a chip carrier with more than 14 signal pins would be more efficient.)

A second application of Rent's rule indicates that the 100 signal terminals on each board provide enough communications capacity for 424 circuits. The number of boards needed is therefore 10,000 divided by 424, or 24, and each board can hold an average of about 25 chip carriers.

Rent's rule estimates only the number of packages needed at a given level in the hierarchy; the actual size of a card or board must be determined from the details of the physical design. Much depends on the lowest-level package, which is generally the single-chip carrier. The factors that need to be considered in card design include the configuration of the terminals on the carrier, the maximum density of the signal paths, the electrical performance required, the power requirements of the chips and the cooling capacity of the overall packaging design.

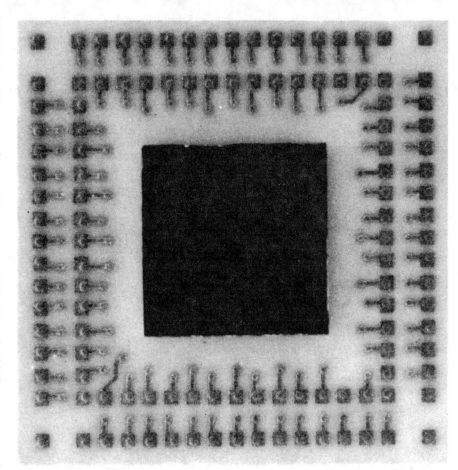

MULTICHIP CARRIER can greatly reduce the average distance between chips and hence the average transmission delay. Here a single chip is shown mounted face down on a multilayer ceramic substrate that can accommodate about 100 chips. Both power and signals pass through conductors buried within the substrate. The double row of pads surrounding the chip gives access to the signals, so that the assembly can be tested and connections can be altered.

A chip carrier in its simplest form is a space transformer. It serves as a bridge between the small and closely spaced contact pads on the surface of the chip and the larger wiring network on the card or board. One of the commonest chip carriers is the dual in-line package, a rectangular enclosure of plastic or ceramic with a row of leads along each of its two longer sides. The leads are arranged on .1-inch centers and are inserted into holes in the printed-circuit board. The dual in-line package makes fairly efficient use of space on the card for chips that have comparatively few terminals (say from 10 to 20), but it becomes unwieldy as the number of leads increases. For a 64-pin chip not only is the package greatly lengthened but also its width must be increased to allow space inside the package for connections between the leads and the chip contact pads.

The industry standard is now shifting from the dual in-line package to a square chip carrier with leads, or contacts, on all four sides. The leads are on .05-inch centers and are soldered to corresponding pads on the surface of the printed-circuit card. The design significantly reduces space requirements. A square chip carrier with 64 perimeter leads on .05-inch centers occupies only a fourth the area taken up by a 64-pin dual in-line package.

The densest terminal configuration for a chip carrier is a full array of pins or contact pads. Sixty-four terminals in an eight-by-eight array on .05-inch centers take up less than a fourth the space needed for the same number of leads in a square perimeter arrangement at the same spacing. The advantage of the array format becomes even greater as the number of terminals increases. Making connections to the denser array of pins or contacts, however, requires a printed-circuit board that is more complex and hence more expensive.

As the number of terminals per chip and the number of chips per unit area increase, the major limiting factor becomes the capability of the board or multichip module to provide interconnections. The simplest printed-circuit boards have only one layer of signal lines, and so the wiring must be laid out with careful attention to an obvious constraint: no two conductors can be allowed to cross. Multiple layers of wiring can eliminate this problem, particularly if all the conductors in a given layer are oriented along one major axis, either x

or y. On the other hand, with a multilayer board space is needed for vias, or vertical pathways, that connect the layers. A signal crossing the board diagonally would be directed through a via to an x plane, where a signal line would take it along one edge of the board, through another via to a y plane, with a signal line perpendicular to the first one, and finally through a third via to the surface of the board at the destination.

The layout of a multilayer board is governed by design rules that specify the dimensions of the conductors and where they can be placed. For example, the vias are generally arranged in a grid; not every intersection of the grid lines necessarily has a via, but vias can be installed only at the grid points. The signal lines must run between the rows of vias. Inserting multiple conductors between adjacent vias can increase the overall wiring density, but it also complicates the design of the board since a conductor may be blocked from a particular via by another conductor.

The design rules establish the maximum wiring capacity of a packaging technology. The maximum length of wiring per unit area of board surface is equal to the number of signal layers multiplied by the number of signal lines between adjacent vias and divided by the grid spacing of the vias. For example, a board with two signal planes, two wires between vias and vias on .1-inch centers has a wiring capacity of 40 inches per square inch. Both the wiring capacity and the efficiency with which it can be put to use are strongly influenced by the via size and the availability of vias for communication between signal planes.

A technology that packs many chips into a small volume is futile if the heat generated by the chips cannot be safely carried off. The property that determines the cooling capacity of a chip carrier or module is its thermal resistance, which is closely analogous to electrical resistance. According to Ohm's law, the resistance between two points in an electrical circuit is equal to the voltage difference between the points divided by the current flowing between them. Similarly, the thermal resistance of a package is the temperature difference between the heat source (the chip) and the ultimate heat sink (the ambient air) divided by the heat flux passing through the package. When the system is in a steady state, the heat flux must be equal to the power being dissipated by the chip. The thermal resistance can be expressed in units of degrees Celsius per watt.

The thermal path from the chip to the air can be considered in two parts. First, heat is conveyed, usually by conduction, from the semiconductor junctions—the areas within the structure of a chip where most of the heat is generated—to the surface of the package. The thermal resistance of this part of the path is called the internal resistance and depends mainly on the geometry of the chip carrier and the thermal conductivity of the materials. Second, heat is removed from the package itself, in most cases by moving air; the mechanism of heat transfer is either forced or natural convection. The external thermal resistance is a complicated function of many factors, including the area and thermal emissivity of the module and the velocity and turbulence of the airstream.

The overall thermal resistance of a typical plastic dual in-line package in moving air is about 50 degrees C. per watt. Thus a chip dissipating .5 watt would undergo a 25-degree rise in temperature. The internal resistance can be reduced by replacing the plastic package with a ceramic one, which has a higher thermal conductivity, or by enlarging the area over which the chip is bonded to the carrier. Such measures can reduce the thermal resistance by half and thereby double the power capacity. The external resistance can also be reduced, for example by adding a heat sink to increase the effective area of the package or by increasing the velocity of the air, but it is very difficult to reduce the overall resistance by more than another factor of two.

For a package with a given thermal resistance the maximum power dissipation depends on the temperature difference available for cooling. Although some silicon devices can operate at a junction temperature exceeding 150 degrees C., the maximum operating temperature of digital circuits is usually

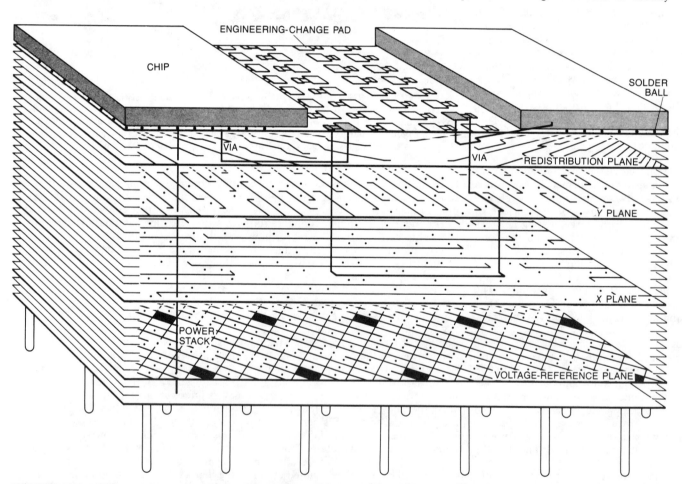

MULTILAYER CERAMIC SUBSTRATE can accommodate 320 centimeters of wiring per square centimeter of surface area. Here a small region of the substrate is shown in a schematic and greatly magnified cross section. A typical signal path proceeds from a chip down a "via" that passes between the layers of the structure. A conductor on one of five redistribution layers carries the signal to another via that returns to the surface at an engineering-change pad. The signal then passes through a metallic link along the surface and into a third via, which takes it to a signal plane deep within the substrate. A conductor oriented along the x axis leads to still another via and a y-oriented conductor; a final via returns to the surface at another chip site to complete the signal path. Some signal paths go to pins on the bottom of the substrate, where connections are made to other multichip modules. Power is distributed by three planes at the bottom of the substrate. The pattern of connections can be modified by cutting the link at an engineering-change pad and bonding a new wire to the pad.

limited to between 75 and 85 degrees to ensure reliability and uniformity of electrical properties from chip to chip. The inlet temperature of the cooling airstream can be as high as 30 degrees, and the air heats up (by as much as 10 to 15 degrees) as it passes through the package. Hence the maximum available temperature difference is less than 50 degrees. These calculations imply that the power-handling capacity of a ceramic dual in-line package is less than two watts. In general the maximum heat flux with air cooling is about two watts per square centimeter at the chip level and about .5 watt per square centimeter at the module level. (The values can be increased by using special heat sinks and high-velocity chilled air.) These limits represent important constraints on circuit power and circuit density and therefore on performance.

Some designers of high-speed computing systems have explored a number of alternatives to packaging technology based on the air-cooled chip carrier. Here I shall describe one high-performance technology, which I believe has extended the state of the art. It was developed for a new series of mainframe computers by scientists and engineers at the IBM facilities in East Fishkill, Endicott and Poughkeepsie, N.Y., and in Sindelfingen, West Germany. Two fundamental design objectives were to reduce the number of interconnections between packaging levels and to reduce the total wiring length. These goals offered three potential benefits: higher speed, lower cost and improved reliability. The package that resulted from these efforts has two main components: a multilayer ceramic substrate on which the chips are mounted and through which all interconnections are made, and a module assembly that provides a direct thermal path from the back of the chips to a water-cooled heat sink.

The ceramic substrate compresses a wiring network of extraordinary complexity into a square, tilelike object 90 millimeters on a side and roughly five millimeters thick. On the top surface are sites for between 100 and 133 high-speed chips, with a total of more than 12,000 chip contact pads. On the bottom surface of the substrate are 1,800 pins that supply power to the chips and route signals to and from the next level in the packaging hierarchy. Both the chip terminals and the module terminals are organized as two-dimensional arrays to minimize the space required. The substrate itself has 33 layers of conductors, which are interconnected by more than 350,000 vias.

Sixteen of the layers in the substrate are x or y planes of signal wiring. The design rules for these layers allow vias to be placed on grid points with a spacing of .5 millimeter. Just one signal line

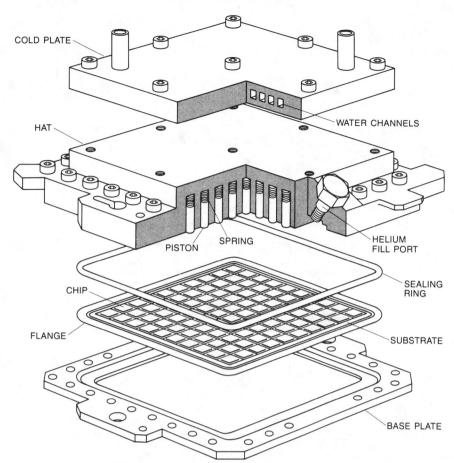

THERMAL-CONDUCTION MODULE houses the multilayer substrate and cools the 100 or more chips mounted on its surface. The substrate is clamped between a baseplate and a "hat" equipped with spring-loaded metal pistons. Each piston presses against the back of a chip, conducting heat to a cold plate bolted on top of the hat. The cold plate in turn gives up the heat to chilled water that is pumped through channels in the plate. The module's ability to dissipate heat is enhanced by filling the internal volume with helium, which has a higher heat conductivity than air. The rated power-handling capacity of the thermal-conduction module is 300 watts.

is allowed between adjacent vias, eliminating the possibility of conflict for access to a via. Thus the 16 signal layers have a maximum wiring capacity of 320 centimeters per square centimeter of substrate. A typical substrate has 130 meters of wires on these planes. A voltage-reference plane is inserted between each pair of x and y layers to control the impedance of the signal lines. The characteristic impedance is 55 ohms.

The top five layers of the substrate have the most vias. Here the vias are on a .25-millimeter grid to match the area array of 120 contact pads on each chip. These layers are employed to redistribute the signal lines (96 per chip site) from the chip pads to a set of contact pads that surround each site. From there the signal lines return to the interior of the substrate. The surface pads allow the module to be tested with the chips in place. Furthermore, if changes in wiring are needed, a small link on the surface can be severed to isolate any signal line from the internal wiring of the substrate. A new connection can then be made by bonding a fine wire to the metal pad and laying it in the channels between the

chips. The ability to make such engineering changes is particularly important during the development of a new product. There are also methods for replacing an individual chip because of either a design change or the failure of a component.

Power is distributed by three planes at the bottom of the substrate; two of the planes carry the voltages required by the chips and the third is at ground potential. The planes themselves are supplied in parallel by an array of pins distributed uniformly across the bottom surface; 500 of the 1,800 available pins are used for this purpose. From the power planes the current flows directly to the chip power pads through parallel vertical stacks of vias. The substrate is designed to supply up to four watts to each chip site, although not all the chips draw the maximum current. The module as a whole is limited to a total of 300 watts, or an average of about three watts per chip. The power-supply voltage loss due to resistance in the package is less than 15 millivolts.

As might be expected, some compromises had to be made in matching the design to the materials for such a complex packaging technology. The principal component of the substrate is alumina, or aluminum oxide, a ceramic that was chosen because of its superior mechanical properties and its ability to withstand the various chemical and thermal processes employed in manufacturing the substrate and module. The principle drawback of alumina is its relatively high dielectric constant of 9.4. As a result the propagation speed of signals in the module is lower than it would be in a fiberglass printed-circuit board; since chips can be mounted directly on the ceramic substrate, however, the average distance between chips is much smaller than it would be with card-on-board technology, and so the total propagation delay is significantly less. Another compromise was made in choosing the material of the conducting paths inside the substrate. Because the ceramic must be fired at 1,500 degrees C., the conductors must be made out of a re-fractory metal; other metals have higher electrical conductivity, but their melting point is too low. (Copper, for example, melts at 1,083 degrees C.) The metal chosen is molybdenum, whose bulk resistivity is approximately three times that of copper. This constraint is compensated for by the high density of vias in the design; large currents are supplied directly to the chips by stacks of vias.

The fabrication of the multilayer ceramic substrate begins with the casting of the individual layers. Ceramic and glass powders are mixed with an organic binder and solvent to form a slurry, which has the consistency of paint. The slurry is deposited on a moving plastic belt and passes under a blade that sets the thickness of the layer. A long drying oven drives off the solvent, leaving a cohesive but still flexible material that looks rather like thick paper. Square blanks called green sheets are then cut from the web. ("Green" is the ceramist's term for unfired material; the sheets are actually white.) Alignment holes are punched in the corners of each sheet to aid in subsequent operations.

The next step is the punching of via holes. It is done under computer control by a high-speed multiple-punch-and-die machine. In the top layer of a 100-chip module, for example, every chip site has an identical array of holes. One hundred punches are therefore mounted in a grid with a spacing equal to the spacing of the chip sites. Each operation of the tool punches a single hole at the same relative position in each chip site; the entire sheet is then moved slightly and the next set of 100 holes is punched.

The metal patterns are laid down on the green sheets by a process similar to stenciling or silk-screen printing. A paste of molybdenum in a binder and solvent is extruded through a metal mask that has the pattern of the wiring cut into it. (The pattern itself is generated automatically by data from the computer-aided-design system.) The paste is applied under pressure so that the punched via holes are also filled. The metallized sheets are then dried and inspected. The detection of flaws in individual sheets before they are combined with other sheets to form a complete substrate is important to maintaining the overall yield of the manufacturing process.

The sheets that pass the inspection are stacked in the appropriate sequence and laminated under high pressure at 75 degrees C. Because vias only 120 micrometers in diameter must be continuous from one layer to the next, control of dimensions and alignment is critical. The green laminate is trimmed to size and subjected to a long firing cycle in which the peak temperature of more than 1,500 degrees is reached in a hydrogen atmosphere. At lower tempera-tures the organic material decomposes and becomes volatile, and at the higher temperatures the ceramic and the metal sinter into a monolithic structure.

The rate of heating must be carefully controlled; if the temperature were to rise too quickly, the organic binder would volatilize faster than it could diffuse to the surface, causing the substrate to delaminate. During the sintering process the substrate shrinks by approximately 17 percent in each dimension, for a total volume reduction of about 40 percent. Given the tight dimensional tolerances of the final assembly, it is clear that the amount of shrinkage must be known precisely when the patterns are first inscribed on the green sheets. After firing, the substrate has the size, shape and characteristic hardness of a ceramic tile; if it is struck, it rings.

The exposed metal areas on both surfaces of the completed substrate are plated with nickel and then with gold. An extensive electrical test is done by an automated tester that again utilizes data from the computer-aided-design system to verify the correct pattern of interconnections. The machine must confirm that each pad is connected to other pads as specified by the design; in addition the machine must make certain there are no extra, improper connections. When the testing is completed, the 1,800 pins are brazed to the bottom surface; a metal flange is attached, also by brazing, in the same operation.

The integrated circuits are attached to the substrate by a method developed at IBM for an earlier generation of computers. First a tin-lead solder is evaporated through a metal mask onto the contact pads on the surface of the chip. The chip is then heated in an inert atmosphere to melt the solder, which, under the influence of surface tension, forms a spherical droplet on each terminal. The solder is allowed to harden again, and the chip is inverted over the substrate with the contact pads aligned. When all the chips are in place, the assembly is once more heated to the melting point of the solder; each solder pad assumes the shape of a truncated sphere, connecting the chip and the substrate terminals electrically but still holding the chip above the surface. The mounted chips can be tested through the surrounding engineering-change pads. Any surface wires needed are then attached by ultrasonic bonding.

The multilayer ceramic substrate, with its flange and pins, forms the base of the thermal-conduction module, which has two other major components, called the hat assembly and the cold plate. The hat is clamped to the substrate flange, compressing a pliable ring with a C-shaped cross section and thereby sealing the internal volume. The cold plate is bolted on top of the hat. Within

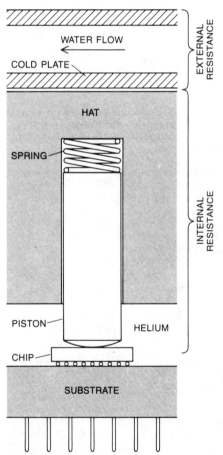

THERMAL RESISTANCE of the module determines the maximum temperature rise of a chip with a given power dissipation. The resistance of the heat path from the chip through the piston (and the helium atmosphere) to the hat is defined as internal; the resistance from the cold plate to the flowing water is external. The total resistance of the package is some 11 degrees Celsius per watt per chip site. Therefore a chip that dissipates four watts undergoes a rise in temperature of 44 degrees above the water temperature.

the hat is an array of spring-loaded aluminum pistons with the same pattern as that of the chips on the substrate. Each piston presses against the back of a chip, conducting the heat upward to the cold plate. The cold plate in turn is cooled by water chilled to 24 degrees C. flowing through internal channels at a rate of 40 cubic centimeters per second. The thermal properties of the module are enhanced further by filling the sealed volume with helium, which at room temperature has a much higher thermal conductivity than air. The helium reduces the internal thermal resistance by more than half.

The assembled module has an internal thermal resistance (measuring from the chip up to the cold plate) of nine degrees per watt per chip site and an external resistance of two degrees per watt per chip site. The module is conservatively specified for use at a maximum of four watts per chip and a total of 300 watts. From the thermal resistance it can be calculated that a chip dissipating four watts should reach a temperature of 68 degrees, which is well below the maximum operating temperature of the circuits. The heat flux is about 20 watts per square centimeter at the chip level and four watts per square centimeter at the module level, an order of magnitude greater than the heat flux in a typical air-cooled package.

The thermal-conduction module represents only one level out of three levels in an advanced-technology central processing unit. The first level is that of the chips themselves, which were developed at the IBM laboratory in East Fishkill. The logic chips all have the same underlying structure of 704 basic logic cells; over this structure, however, there are three levels of metallic conductors that are customized to create various configurations of gates and other devices. The metal layers also supply power to the circuits and distribute signals between the cells and the contact pads. The bipolar logic circuits on the chips have a switching delay of 1.1 nanoseconds.

The third component of the system, which represents another major advance in packaging technology, is the large printed-circuit board that interconnects up to nine thermal-conduction modules. The board, developed at IBM's Endicott facility, has 20 layers, including six impedance-controlled signal planes. A nine-module board measures 60 by 70 centimeters and provides connections between the 16,200 pins of the modules and more than 2,000 additional terminals for cables leading to other subassemblies. The board also supplies up to 600 amperes of current to the modules.

The thermal-conduction module and the associated technologies were introduced in 1981 in the IBM 3081 high-

BED OF NAILS on the underside of the thermal-conduction module is made up of 1,800 connecting pins. Five hundred of them supply power to the module; the rest are available for communication with other modules and with other components of the computing system.

performance data processor, developed at the Poughkeepsie laboratory. In a typical module in the 3081, 52 of the chip sites are occupied by logic chips. There are also 34 array chips, a form of high-speed semiconductor memory employed for data and instructions that must be immediately available to the central processing unit, and five terminator chips, which are arrays of resistors used to match the impedance of long, off-chip transmission lines. A module with this complement of chips holds more than 25,000 logic circuits,

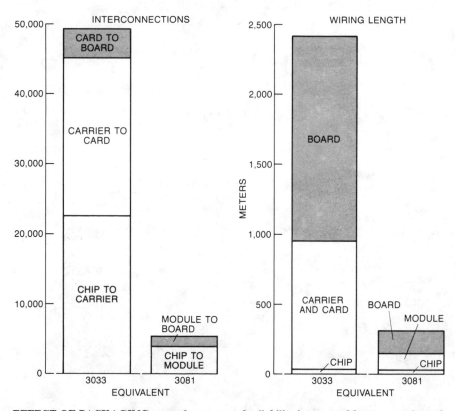

EFFECT OF PACKAGING on performance and reliability is suggested by a comparison of the packaging technologies employed in two mainframe computers made by the International Business Machines Corporation. The IBM 3033 is built with individual chip carriers, cards and boards; the 3081 employs the thermal-conduction module. Here the comparison is made between a single thermal-conduction module and a set of 3033 components with the same number of logic circuits. The number of connections between packaging levels in the thermal-conduction module is smaller by a factor of 10; furthermore, most of the remaining connections are solder joints between the chip and the substrate, which are more reliable than mechanical connections. The signal wiring length in the thermal-conduction module is one-eighth that in the 3033 equivalent, leading to a commensurate reduction in signal-transmission time.

65,000 array bits and almost 500 terminating resistors.

It is instructive to compare the packaging technology of the 3081 with that of another high-performance IBM computer, the 3033, which is built with a more conventional packaging technology. In the 3033 each chip has an average of 12 logic circuits and is housed in an individual chip carrier. The carriers in turn are mounted with terminating resistors and other components on printed circuit cards, which connect along one edge to larger circuit boards. If the functions of the typical thermal-conduction module described above were implemented in the 3033 technology, 1,880 chip carriers would be needed for the logic circuits, another 80 chip carriers

for the memory arrays and additional components for the line terminators. To mount and interconnect all these components would take 52 printed-circuit cards, four larger boards and the cabling required for connections between the boards. Note that these components would be needed to reproduce the functions of a single module; a full-featured 3081 system has 26 modules.

Three effects of the packaging technology based on the thermal-conduction module are particularly noteworthy. First, the large reduction in packaging hardware results in a significantly lower cost. Second, the system has improved reliability. The main sites of failure in electronic assemblies are the connections between package levels. The

thermal-conduction module has eliminated one level of packaging entirely and has reduced the number of logic signal connections between levels by a factor of almost 10. Furthermore, most of the remaining connections are the chip-to-module solder joints, which are inherently more reliable than mechanical connections at higher packaging levels.

The third effect is improved performance. The total length of logic wiring in the 3081 processor is roughly one-eighth what it would be in equivalent 3033 technology. The result is a fourfold reduction in the total packaging delay of the central processing unit, including the modules, the boards and the cables; this in turn allows a twofold decrease in processor cycle time.

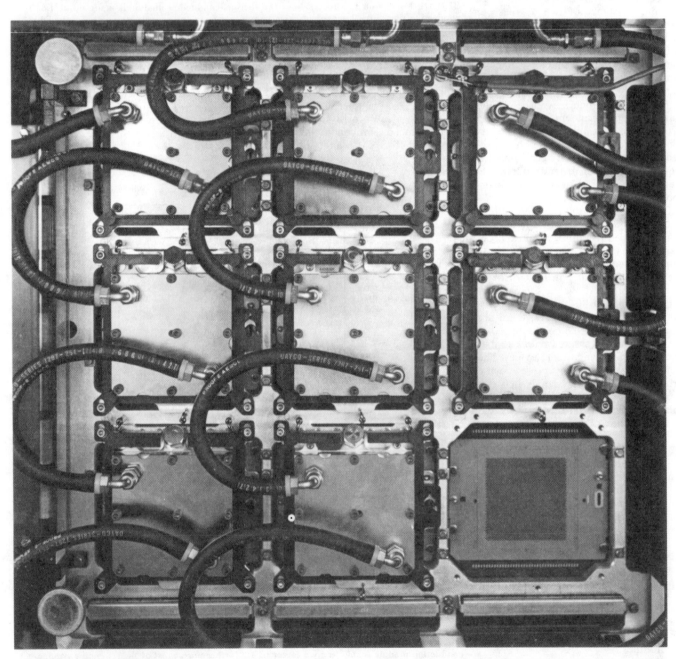

PRINTED-CIRCUIT BOARD mounted on a massive steel frame represents the next level (after the thermal-conduction module) in the packaging hierarchy of the 3081 computer. The board accepts nine modules. Here eight modules have been installed and connected to the water-cooling system; the ninth position is left vacant, exposing the connectors that receive the 1,800 pins on each module.

A Multilayer Ceramic Multichip Module

ALBERT J. BLODGETT, JR., MEMBER, IEEE

Abstract—The increased performance and circuit/bit densities of today's semiconductor chips require corresponding technological advancements in chip packaging. More interconnections and increased heat dissipation must be accommodated in a package compatible with the higher performance of these chips. In the packaging approach described, increased package performance and reliability were achieved by the use of a multilayered ceramic substrate upon which nine chips can be mounted. Conductors for ground, signal, and power lines are interlayered in the ceramic with connections to the top surfaces and to pins at the bottom surface for connection to external circuitry. This multilayer multichip module significantly advances the state of the art; it provides higher circuit density and performance with a ten-fold increase in circuit density over previous logic systems hardware.

I. INTRODUCTION

THE EVER-INCREASING performance and circuit/bit densities of today's semiconductor chips present a variety of challenges to the packaging engineer. The chip package must deliver an increased amount of power and an increased number of interconnections to chips in a manner which allows removal of the increased heat dissipated and which is compatible with the higher performance proffered by the chip.

This paper describes a multichip module which uses multilayer ceramic technology [1], [2] to achieve these goals. The design approach requires a high density interconnection capability at the substrate level. These modules are currently in production for use in IBM 4331 and 4341 processors [3].

II. MULTICHIP MODULE DESIGN

The objectives for the package design are simply stated: provide the requirements of large-scale integrated (LSI) chips

Manuscript received February 15, 1980; revised July 18, 1980. This paper was presented at the 30th Electronic Components Conference, San Francisco, CA, April 28–30, 1980.

The author is with the IBM Research Center, P. O. Box 218, Yorktown Heights, NY 10598.

Fig. 1. Uncapped 50-mm square multichip module.

at a reduced cost with increased performance and with improved reliability. A design approach to achieve these objectives is equally simple: reduce the number of interconnections and reduce total wiring length. This leads to a multichip module design.

A. General Features

An uncapped 50-mm square multichip module (MCM) is shown in Fig. 1 against a background of the 11.6-mm square logic modules used in IBM System/370 (S/370). By utilizing high density logic devices (up to 704 circuits per device) and a multilayer ceramic (MLC) substrate, this module replaces approximately 670 of the S/370 modules and associated printed circuit cards and hardware. The back side of the 50-mm MCM (Fig. 2) is populated with a 19 × 19 array of pins (361 pins total) on 2.54-mm centers for interconnection to a multilevel glass epoxy printed circuit card. A smaller 35-mm square MCM, which has a 14 × 14 array of pins (196 pins total), is used when fewer connections are required.

Both versions, 35 and 50 mm, interconnect a maximum of nine chips each with 120 terminals utilizing the flip-chip solder

Reprinted from *IEEE Trans. Components, Hybrids, Manuf. Technol.*, vol. CHMT-3, no. 4, pp. 634–637, December 1980.

Fig. 2.   Capped 50-mm square multichip module.

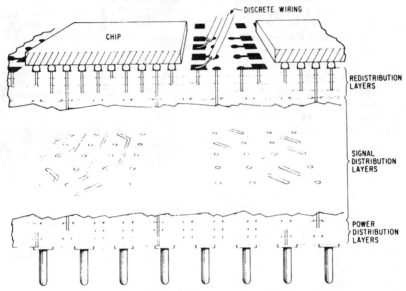

Fig. 3.   Sketch of multilayer ceramic multichip module.

reflow technology first introduced in 1969 [4]. The chip-to-substrate interconnections are arranged in an 11 × 11 array on 0.25-mm centers. Each chip site on the top surface is surrounded by a double row of pads which are used for testing and, in some designs, engineering changes.

B. Multilayer Ceramic Substrate

Twenty-three molybdenum-metalized alumina layers are used in the MCM substrates to provide the design features sketched in Fig. 3. The total substrate thickness is approximately 4 mm.

The top six layers are used to redistribute the signal lines from the 0.25-mm chip pad grid, first to the top surface, and then to the 0.5-mm internal wiring grid. This allows electric test and engineering change capabilities. A wiring change can be made by deleting an internal wire net and replacing it with discrete surface wires. Individual chips can be removed and replaced.

The center layers of the substrate contain the eight X-Y wiring planes. Voltage reference layers are interspersed for impedance control. The vias between layers are on a 0.5-mm grid allowing true three-dimensional wiring. A typical substrate contains 5 m of signal wiring. The bottom five layers are used for power distribution and signal redistribution to the 2.54-mm pin grid. A simplified 17-layer design is utilized when the engineering change feature is not required.

C. Thermal and Electrical Parameters

The LSI logic chips require 1.5 W to power the full 704 available circuits. This power is distributed in parallel paths directly from the power planes to the chip power pads with a maximum voltage drop of 15 mV. These direct paths are clearly visible in the cross section of an MLC substrate shown in Fig. 4. The main thermal path for the power dissipated in the chip is through the 120 solder pad connections to the substrate and through the substrate to its surface and the metal cap. The thermal conductivity of the high alumina content (90 percent) substrate is approximately 21 W/m/°C. Forced air cooling allows a total power dissipation of 9 W.

The dielectric constant of the ceramic is 9.4, and the resistivity of the sintered molybdenum metalization is 10 $\mu\Omega \cdot$ cm. The electric design of the substrate achieves a 55-Ω characteristic impedance for the signal lines.

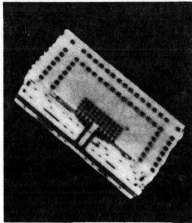

Fig. 4.   Cross section of multilayer ceramic.

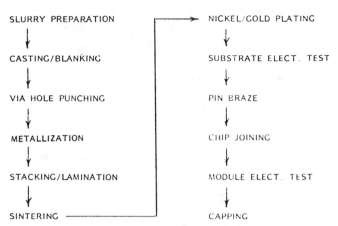

Fig. 5.   MLC module process flow.

D. Comparison with System/370 Packaging

A typical 50-mm MCM contains six LSI logic chips utilizing a total of 4000 circuits. This corresponds to the circuit count of a logic board fully populated with 18 circuit cards in a System/370 Model 148.

The S/370 technology averages six logic circuits per chip and utilizes 16-pin single chip modules. It requires 670 modules with a total of 10 720 chip-to-substrate interconnections, 10 720 module-to-card interconnections, and 1720 card-to-board interconnections.

The 50-mm MCM requires only 720 (6 × 120) chip-to-substrate and 361 module-to-card interconnections. This 20× reduction in interlevel connections plays a major role in improving circuit reliability.

The 5 m of wiring in the MCM coupled with a similar length of wiring on the semiconductor replaces 300 m of wiring required for the same function in S/370 technology. This density improvement is the major factor in increased package performance.

III. MULTILAYER CERAMIC TECHNOLOGY

A. Ceramic Greensheet Fabrication

The process flow for fabricating MLC modules is outlined in Fig. 5. The process begins with mixing alumina powder and glass frit in a 90:10 ratio and blending it with an organic binder (polyvinyl butyral) and solvents. The resulting slurry is cast on a moving plastic web which carries it through a series of drying ovens that drive off the solvents. The thickness of the cast sheet is determined by the distance between the web and doctor blade assembly. The dried, green (unfired) ceramic material is then separated from the plastic web.

The green ceramic material is next inspected and blanked into individual 185-mm square greensheets in preparation for personalization. The greensheet is made in two thicknesses: 0.2 and 0.28 mm.

B. Greensheet Personalization

Alignment holes are mechanically punched in four corners of each greensheet for use in the personalization tooling. Next, a precision cluster punch is used to punch the desired via pattern into the blank greensheet. The nominal diameter of the

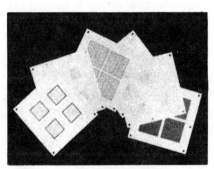

Fig. 6.   Metalized ceramic greensheets.

via holes is 0.13 mm, and a greensheet may have as many as 29 000 via holes.

Metalization is accomplished by extruding a molybdenum paste through a wide nozzle as the nozzle traverses a metal mask resting on the greensheet. The punched vias can be metalized at the same time the surface is defined by using the appropriate etched pattern on the metal masks.

A variety of personalized greensheets are shown in Fig. 6. Note that the pattern is repeated on each layer. The pattern is repeated in a 2 × 2 array for 50-mm products and in a 3 × 3 array for 35-mm products.

C. Substrate Sintering and Finishing

After careful inspection, the personalized greensheets are stacked in the desired sequence using the alignment holes and then laminated together at 75°C with a pressure of approximately 25 MPa. The green laminates are then cut into pieces corresponding to the final substrates prior to firing. Sintering is performed in a kiln in hydrogen. The organic binder is first volatilized at low temperatures, then the molybdenum and ceramic are sintered in a temperature cycle which peaks at 1550°C. The linear shrinkage of parts is approximately 17 percent.

The MLC substrates are then processed through a series of plating steps to provide the required nickel and gold metalization for subsequent brazing, chip joining, and capping operations. After electrical test, nickel plated Kovar pins are brazed to the input/output (I/O) pads using a low temperature noble metal braze.

D. Multichip Module Assembly

The LSI logic chips, which are terminated in lead tin solder pads [5], are placed with the pads down on the substrate. The assembly is then reflowed in a furnace. The molten solder wets the corresponding plated vias on the substrate surface providing the required connection of hundreds of terminals in a single furnace reflow process.

After electric test, a plated Kovar cap is seam welded to the metalized band which surrounds the chip site area on the top surface (Fig. 1). The resulting hermetic seal provides protection from the environment.

IV. CONCLUSION

An advanced package design which interconnects and distributes power to a number of LSI semiconductor chips is described. High interconnection density between the chip and substrate is achieved by using flip-chip, solder-reflow technology.

The module utilizes a multilayer ceramic substrate which significantly advances the state of the art. The substrate, which is fabricated by cofiring up to 23 metalized ceramic layers, provides high density, high performance wiring to exploit the performance and reliability proffered by today's LSI devices.

The result is a 20-fold reduction in the number of packaging interconnections and a ten-fold increase in circuit packaging density over earlier logic systems hardware.

REFERENCES

[1] B. Schwartz and D. L. Wilcox, "Laminated ceramics," in *Proc. Electronics Components Conf.*, 1967, pp. 17–26.
[2] H. D. Kaiser, F. J. Pakulski, and A. F. Schmeckenbecher, "A fabrication technique for multilayer ceramic modules," *Solid State Technol.*, pp. 35–40, May 1972.
[3] E. Bloch, "Very large scale integration for the 1980's," presented at the 1979 Electronics Components Conf.
[4] L. F. Miller, "Controlled collapse reflow chip joining," *IBM J. Res. Develop.*, vol. 13, pp. 239–250, 1969.
[5] P. A. Totta and R. P. Sopher, "SLT device metallurgy and its monolithic extension," *IBM J. Res. Develop.*, vol. 13, pp. 226–238, 1969.

VLSI Packaging Technique Using Liquid-Cooled Channels

TOHRU KISHIMOTO AND TAKAAKI OHSAKI

Abstract—A new packaging technique which employs innovative indirect liquid cooling is described. The technique involves mounting very large-scale integrated (VLSI) chips on a multilayered alumina substrate which incorporates very fine coolant channels. In particular, an investigation into the optimal structure for the cooling section by computer simulation and by experiment involving the physical implementation of this structure is discussed. The numerical solution of the coolant flow distribution obtained ensures that the coolant distributor and collector structure dimensions can be determined to meet the uniform velocity distribution condition. Additionally, the channel cross section is designed to be 800 μm wide $\times$ 400 μm high to achieve a lower thermal resistance. An outline of an indirect liquid cooling package fabricated based on the results of these structures is presented. The package mounts a 5×5 array of 8-mm^2 VLSI chips on a substrate measuring 85 mm $\times$ 105 mm. The substrate features 29 very fine coolant channels, six conductor layers, and 900 input/output (I/O) pins. The technique permits the realization of an allowable heat dissipation higher than 400 W per package at a flow rate of 1.0 l/min. Furthermore, since the thickness of the cooling section is smaller than 1.0 mm, the volume power density increases 17 kW/l or more. This cooling capability is tenfold greater than that obtained by conventional indirect water cooling.

INTRODUCTION

CONTINUAL ADVANCES in the speed and integration scale of the integrated circuits used in high-performance systems have created even greater demands for higher density packaging to ensure reduced wiring delay for improved electrical performance. In high-performance electronic equipment, effectively removing the considerable amount of heat generated is important for increasing the circuit speed and ensuring the high reliability of an electronic circuit.

At present, most of the high-performance systems use forced air convection cooling or indirect liquid cooling, of which the latter promises higher power dissipation and more compact packaging. Recently, the use of indirect water cooling for the central processing unit of a high-performance computer system has been reported [1], [2].

Although quite suitable for two-dimensional packaging, conventional liquid cooling systems are not well suited to three-dimensional or high-density stack packaging. This is because the application of these systems results in complicated and bulky modules as well as inevitably increasing the interconnection length for high-density stack packaging. Therefore, it is essential to develop a liquid-cooled package

Manuscript received March 22, 1986; revised July 14, 1986. This paper was presented at the 36th Electronic Components Conference, Seattle, WA, May 5-7, 1986.
The authors are with the Electrical Communications Laboratories, Nippon Telegraph and Telephone Corporation, 3-9-11, Midoricho, Musashimo-shi, Tokyo 180, Japan.
IEEE Log Number 8611125.

having a higher cooling capability and a better suited high-density stacked package.

In line with this, the present paper describes a new packaging technology which employs innovative indirect liquid cooling. The technique involves the mounting of very large-scale integrated (VLSI) chips on a multilayered alumina substrate which incorporates very fine coolant channels comprising a cross section of less than 1 mm^2. This paper first describes the design procedure for the innovative liquid cooling technique, that is, the fabrication of very small cross-section channels which achieve lower thermal resistance. It also discusses the design for the coolant distributor and collector structure which work together to realize a uniform liquid flow distribution. It then takes a look at the actual fabrication process for the newly proposed package. Finally, it presents an evaluation of cooling capability for the fabricated package based on experimental and numerical methods.

STRUCTURE

Fig. 1(a) shows the cross-sectional arrangement of the newly proposed indirect liquid cooling package, and Fig. 1(b) illustrates the high-density stacked system conception. Very fine coolant channels having a cross section of less than 1 mm^2 were formed into the lower portion and between the buried via-holes of a multilayered alumina substrate. Interconnection pads were fabricated on the front and rear surfaces of the substrate to permit connection of the substrate to the wiring boards which were located on three sides of the package. The coolant distributor and collector were formed in two opposite edges of the multilayered alumina substrate. These were fabricated using the same punching process as is employed in via-hole formation.

In the cooling process itself the coolant is supplied from the coolant inlet port and distributed to each coolant channel by the coolant distributor, thereby ensuring a uniform velocity profile. The heat generated in the chips is conducted to the coolant channels by way of the alumina substrate, where it is transferred to the liquid flow. The warmed coolant is subsequently collected by the coolant collector after flowing through the coolant channels. It then flows out through the coolant outlet port. The upper and lower stack packages are connected in series by piping.

This new cooling technique offers four principal structural advantages. First, the cooling channels and the conductor layer sections can be formed into a single substrate by the same punching process as is used in via-hole formation. Second, a high cooling capability is realized by fabricating channels just inside the multilayer alumina substrate. Such a

Reprinted from *IEEE Trans. Components, Hybrids, Manuf. Technol.*, vol. CHMT-9, no. 4, pp. 328–335, December 1986.

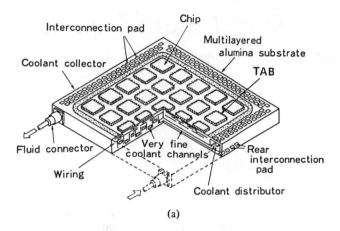

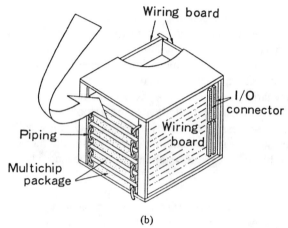

Fig. 1. (a) Schematic of new VLSI package using innovative indirect liquid cooling technique. (b) High-density stack-packaged system.

structure affords a lower thermal resistance because of the very short thermal path and the high heat transfer coefficient obtained through miniturized coolant channels. Third, high-density stack packaging can be realized because of the minimal thickness of the cooling channels. Finally, when replacement or inspection of VLSI chips is needed, performance tests can be carried out under the same conditions as those of an actual operating system because the cooling section is formed within the same substrate [3].

Using this innovative indirect liquid cooled package enables the realization of an attractive cooling system. Furthermore, such a package is suitable for high-density stack-packed electronic equipment which uses a large number of high-power dissipation chips.

DESIGN

Cooling capability depends on the thermofluid dynamic characteristics of very fine coolant channels, as well as on coolant distributor and collector structures. In this section, we present the design procedure for the cooling section in which the channel cross section was determined based on the decision to attempt to achieve a substantially lower thermal resistance. Additionally, the coolant distributor and collector structures were determined to satisfy the condition of a uniform velocity distribution for the liquid flow through the channels.

Channel Cross Section

To obtain the optimal channel cross section, we evaluated the thermal resistance R_s as a function of channel height, channel width, and pressure drop along the cooling channel. We define R_s as a reciprocal of the product of the cooling channel's surface area S and the average heat transfer coefficient h_m. We evaluated the thermal resistance based on the pressure drop because a marked increase is indicated under the constant flow rate condition as the channel cross section gets smaller. In general, conventional pumps are ineffective for driving coolants through very narrow channels.

The average liquid velocity V, related to the pressure drop ΔP from the inlet to the outlet channel, is given by Poiseuille's and Blasius's formulas. Using V, the average heat transfer coefficient can be derived from empirical equations (1) and (2) below.

When the liquid flow is laminar, the average heat transfer coefficient h_m is written as [4]

$$h_m = \frac{\lambda}{d} \left\{ 2\epsilon \left[1.10 + 0.555 \left(\Pr \mathrm{Re} \frac{d}{L} \frac{1}{4\epsilon^2} \right)^{0.55} \right] \right\} \quad (1)$$

where λ is the liquid thermal conductivity, d is the equivalent diameter of a channel defined as $2ZW/(Z + W)$ such that W is the channel width and Z is the channel height, ϵ is the aspect ratio defined as Z/W, Pr is the Prandtl number, Re is the Reynolds number, and L is the channel length. The expression is valid for a Reynolds number of less than 2300.

In the Re > 2300 region the flow is turbulent such that the following relationship is written [5]:

$$h_m = \frac{\lambda}{d} 1.11 \left[\frac{\mathrm{Re}^{0.2}}{(L/d)^{0.8}} \right]^{0.275} 0.023 \, \mathrm{Re}^{0.8} \Pr^{0.4}. \quad (2)$$

The unit channel surface area S is written as

$$S = Z(W + Z)L. \quad (3)$$

Using (1)–(3), and neglecting the fin efficiency, the final form of the thermal resistance per unit channel R_s is expressed as $1/(h_m S)$.

Fig. 2 shows the thermal resistance R_s as a function of the channel width W when the channel height Z changes. In this figure, taking conventional pump use into consideration, the pressure drop ΔP is chosen to be 2000 Kgf/m².

As can be seen, the channel width giving rise to the maximum thermal resistance exists. In the region where the channel width is less than the value which maximizes R_s, the thermal resistance decreases depending on the increase of the average heat transfer coefficient. This is caused by the thermal boundary layer becoming thinner. On the other hand, in the region where the channel width is greater than the maximum point of R_s, the thermal resistance decreases depending on the enlargement of the channel's surface area and on the increase of the heat transfer coefficient. This results from the increase in the liquid flow rate. Furthermore, the thermal resistance is lowered when the channel height increases because of the enlarging channel surface area.

Since reducing the channel height is desirable for ensuring a

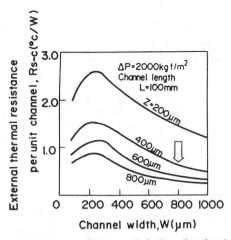

Fig. 2. External thermal resistance per unit channel as function of channel width W with channel height Z as parameter.

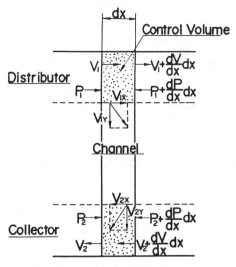

Fig. 3. Mathematical model of complete flow distribution system.

shorter via-hole length, we selected a channel height of 400 μm. In addition, although the thermal resistance can be lowered in regions where the channel width is small, it is fairly sensitive to channel width changes in such regions. Where the width exceeded 800 μm, the decrease in the thermal resistance becomes less and finally levels off. On the other hand, enlarging the channel width is undesirable in maintaining sufficient via-hole clearance.

Taking these thermal resistance and via-hole clearance relationships into consideration, we therefore decided upon a final channel width of 800 μm. Under this condition we achieved a thermal resistance of 0.5°C/W, which is almost equal to that obtained by conventional liquid cooling [1].

Coolant Distributor and Collector Structures

It is recognized that a uniform liquid flow distribution is essential to maintaining a nearly isothermal device temperature, which in turn achieves uniform electrical characteristics. Because the liquid flow distribution is sensitive to the coolant distributor and collector dimensions, we designed an optimal structure incorporating the continuous flow model as well as the results from a discrete flow distribution system. In our procedure, we dealt only with the counter flow type because the parallel flow type is generally undesirable for obtaining a uniform velocity profile.

A mathematical model which assumes that the branching flow can be replaced by a narrow rectangular duct flow is shown in Fig. 3. The individual branching channels at a distance ΔL are replaced by an equivalent rectangular slot having length $L = N\Delta L$ and width Nd/L, where N is the number of channels. Applying the continuity and momentum equations to a control volume element of the model and incorporating friction loss yields the following differential equations for velocities, V_1 and V_2, which are in the x-direction in the distributor and collector, respectively.

Assuming that V_x equals zero, the following equation can be written for the distributor:

$$A_1 \frac{dP_1}{dx} = -2\rho A_1 V_1 \frac{dV_1}{dx} - A_1 \xi \frac{1}{D_1} \frac{\rho V_1^2}{2} \qquad (4)$$

where A_1 is the cross-sectional area of the distributor, P_1 is the pressure, D_1 is the equivalent diameter, ρ is the liquid density, and ξ is the friction loss factor.

Using the velocity component V_x, which joins the branching flow, and the collector flows which are assumed to be $\alpha_c V$ where α_c generally takes the value of 0.4, the following equation can be written for the collector:

$$A_2 \frac{dP_2}{dx} = -\rho A_2 (2 - \alpha_c) V_2 \frac{dV_2}{dx} - A_2 \xi \frac{1}{D_2} \frac{\rho V_2^2}{2} \qquad (5)$$

where A_2 is the cross-sectional area, P_2 is the pressure, and D_2 is the equivalent diameter.

The flow rate q_d at discrete channels which relates (4) and (5) is written as

$$\frac{dV_1}{dx} = -\frac{q_d}{A_1} = \frac{Ca}{A_1 \Delta L} \sqrt{\frac{2(P_1 - P_2)}{\rho}}$$

$$= \frac{Ca}{A_2 \Delta L} \sqrt{\frac{2(P_1 - P_2)}{\rho}} = \frac{q_d}{A_2} = \frac{dV_2}{dx} \qquad (6)$$

where C is the flow coefficient and a is the cross-sectional area of the discrete channel. Since it is difficult to obtain a general solution for the simultaneous equations (4)–(6), the numerical calculation was performed using the Milne method.

The nonuniformity ratio, defined as the ratio of the maximum to minimum flow rate for the individual channel, is shown in Fig. 4 as a function of the equivalent diameter ratio, defined as the channel equivalent diameter d divided by the distributor equivalent diameter D. Here the collector equivalent diameter was chosen to be equal to that of the distributor. As is clear, a completely uniform flow has a nonuniformity ratio equal to unity. An increase in the equivalent diameter ratio or in the number of the branch channels results in a larger nonuniformity. When the number of branch channels is increased, the equivalent diameter ratio should especially be small to obtain a uniform flow. An equivalent diameter ratio of less than 0.1 is therefore chosen.

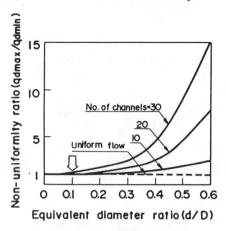

Fig. 4. Nonuniformity ratio versus equivalent diameter ratio with number of branching channels N as parameter.

Under a channel cross-section condition of 400 μm high $\times$ 800 μm wide as indicated in the prior section, the equivalent diameter d of each channel becomes about 533 μm. Assuming the distributor and collector widths to be 10 mm, for example, the coolant distributor and collector heights become 3.6 mm. Fabricating the coolant distributor and collector is easy using the same punching process as is employed in via-hole or channel formation. Therefore, the coolant distributor and collector can be formed into a single body within the substrate.

PACKAGE FABRICATION

A conventional and improved process flow outlines for package fabrication are compared in Fig. 5. To strengthen the metallic mold for channel punching, the mold pitch has been chosen to be 5.08 mm, which is twice that of the via-hole or I/O pin center spacing. This meant, of course, that channel punching had to be performed twice during the channel formation operation. Furthermore, the coolant channels were arranged into a staggered two-story construction so that a cross section of several channels would be equalized, with the punching action of the channels prevented from negatively influencing the location of the via-hole formation.

The process flow consists of five steps. First, green sheet 1 is channel-punched. Second, green sheet 2 is channel-punched shifting the position 2.54 mm compared with that for green sheet 1. Third, via-hole punching is done following green sheet 1 and 2 stacking to ensure accurate alignment. Fourth, the conductor layers and lid cover are stacked onto green sheets 1 and 2. Finally, the I/O pins are co-fired and brazed into the substrate.

The process shown in Fig. 5(b) represents an improvement over the process in Fig. 5(a). The cross section of several channels, in particular, the channel width, could be better equalized compared with the process of fabricating all channels into a single green sheet. This improvement is dramatically shown in Fig. 6. When the channels were punched into one identical green sheet, the width of the initially punched channels was enlarged, while that of the subsequently punched channels was narrowed by the shear stress resulting from the second punching.

By cutting open the center portion of a package, the flow rate distribution was measured using a collector tube to collect

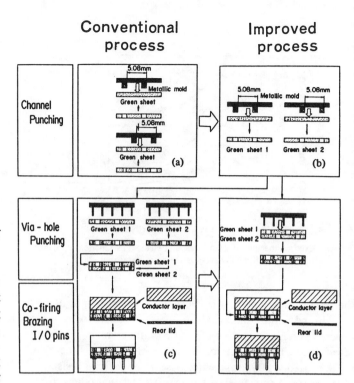

Fig. 5. Package fabrication process. (a) Punching channels into same green sheet. (b) Improved channel-punching process. (c) Via-hole punching process by which green sheets 1 and 2 are stacked after via-hole punching. (d) Improved via-hole punching process.

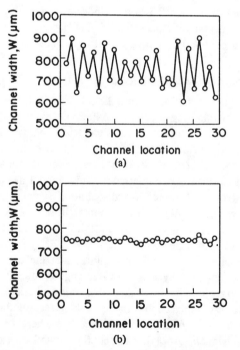

Fig. 6. Channel width versus channel location. (a) Using process shown in Fig. 5(a). (b) Using improved process shown in Fig. 5(b), with coolant channels arranged into staggered two-story construction.

the outflow from each channel. The flow distribution obtained for each channel is shown in Fig. 7. The flow rate ratio was determined by the flow rate of each channel divided by the average flow rate over all channels. The solid curve represents the analytical result by numerically solving simultaneous

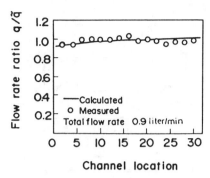

Fig. 7. Flow rate ratio versus channel location when total flow rate is 0.9 l/min.

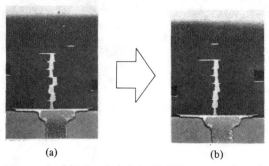

(a)   (b)

Fig. 8. Photograph of fabricated via-hole. (a) Using process shown in Fig. 5(c). (b) Using improved process shown in Fig. 5(d), in which excellent interconnection is actualized.

equations (4)–(6). As can be seen, the flow rate excursion was small, and an almost completely uniform flow distribution could be realized due to the equalization of most of the channel cross sections. The maximum excursion shows an approximate five-percent increase, and a maximum eight-percent decrease. Furthermore, the agreement between the experimental and the calculated results is excellent. This confirms that the continuous flow model is a valid means to designing the coolant distributor and collector structure.

Because the via-holes are fabricated after the channels are punched, it is possible to avoid the discrepancies often occurring between layers. In fact, excellent interconnection is actualized as shown in Fig. 8. Furthermore, we confirmed experimentally that no water permeability exists and that the fabrication of the coolant channels near the via-hole does not negatively influence electrical performance.

COOLING CAPABILITY

Using the aforementioned design procedure, an indirect liquid cooling package was constructed which could be used to assess the cooling capability. The fabricated package dimensions were 85 mm wide × 105 mm long on which a 5 × 5 array of chips (10 mm² maximum) was mounted. The substrate featured 29 very fine coolant channels, six conductor layers, and 900 I/O pins. Table I lists the specifications of the fabricated package. A photograph of the overall package is shown in Fig. 9, with the lids for the coolant distributor and collector removed. Fig. 10 is a closeup of the package demonstrating that a uniform cross section of several channels was realized.

The cooling capability was measured using this fabricated package and was evaluated by computer simulation. The simulation was accomplished by the three-dimensional finite difference method (FDM). The heat balance equation was written for each node established in the center of the control volume. The equation for each node is written as

$$\sum_{j=1}^{N} \frac{1}{R_{ij}} (T_i + T_j) + Q_i = 0 \tag{7}$$

where T is the temperature, R_{ij} is the thermal resistance, Q is the amount of heat absorbed, and subscripts i and j are the number of nodes. The nodal T_i temperature is unknown, and there are N unknowns for N nodes. These simultaneous

TABLE I
FABRICATED PACKAGE SPECIFICATIONS

Chip	chip size	10 mm square maximum
	mounting chips	5 × 5 array
	number of I/O pins per chip	124
Substrate	outside dimensions	85 mm wide × 105 mm long × 2 mm high
	outside dimensions of coolant distributor and collector	7 mm wide × 77 mm long × 1 mm high
	number of conductor layers	6
	number of I/O pins	900 (30 × 30)
	I/O pin center spacing	2.54 mm
	via-hole diameter	200 μm
Cooling section	number of channels	29
	channel pitch	2.54 mm
	channel length	86 mm
	channel width	800 μm
	channel height	400 μm

Fig. 9. Photograph of fabricated package.

Fig. 10. Close-up of fabricated package; coolant channels are 800 μm wide and 400 μm high.

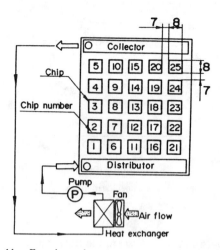

Fig. 11. Experimental setup to assess cooling capability.

equations for N are solved numerically. Equations (1) and (2) are used for calculating the heat transferred from the channel surface to the liquid flow.

Experimental Setup

The experimental setup shown in Fig. 11 was constructed to assess the cooling capability. A 5 × 5 array of chips measuring 8 mm^2 was bonded onto the package using high thermal conductivity epoxy resin. The heat generated in the package was transferred to the coolant, eliminated by an air-cooled heat exchanger having 1.5 kW heat removal capability. The coolant temperature deviation was maintained at less than 2°C. The cooling liquid was supplied to the package through a miniaturized pump.

Thermal Resistance

The excursion of thermal resistance from a chip junction to the inlet coolant is shown in Fig. 12 as a function of the chip location (see Fig. 11), which is perpendicular to the coolant flow. The solid curves represent the analytical results by FDM, assuming a completely uniform flow. It is clear that agreement between the experimental and analytical results is excellent. Because a completely uniform distribution of flow is achieved as shown in Fig. 7, it was possible to maintain the thermal resistance excursion at less than ten percent.

On the other hand, Fig. 13 shows the thermal resistance from a chip junction to the inlet coolant as a function of the chip location parallel to the coolant channel. It is clear that the thermal resistance increases along with the coolant flow because of the increase in the coolant temperature rise. When the total flow rate is increased, the thermal resistance is lowered, however, since the coolant temperature rise decreases due to the increase in the heat capacity and in the heat transfer coefficient as well.

Assuming that the allowable temperature rise is 50°C, the allowable heat dissipation per package at 0.3 l/min is 300 W, and 380 W at 0.8 l/min of flow rate. Neglecting the thermal diffusion, the temperature rise of the liquid coolant passing through the channels is almost 5°C at 0.8 l/min of flow rate. At 0.3 l/min, however, the coolant temperature rise increases to as much as 20°C. Therefore, the total flow rate must be 0.8 l/min or more to ensure a reduction in the coolant temperature rise.

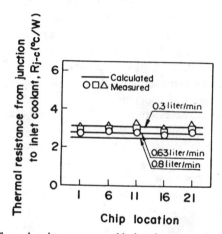

Fig. 12. Thermal resistance versus chip location perpendicular to coolant flow with total flow rate as parameter.

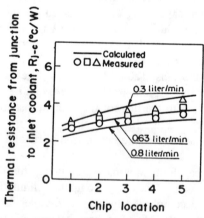

Fig. 13. Thermal resistance versus chip location along coolant channel.

Allowable Heat Dissipation

The allowable heat dissipation may be calculated to satisfy the upper limit of the chip junction temperature of 85°C. This chip junction temperature is obtained using the equation

$$T_j = T_{\text{in}} + \Delta T_{\text{allow}} + \Delta T_c = T_{\text{in}} + R_{j-c}P + \Delta T_c \leq 85°c \quad (8)$$

where T_j is the chip junction temperature, T_{in} is the inlet coolant temperature, ΔT_c is the temperature rise of the coolant, R_{j-c} is the overall thermal resistance per chip, and P is the amount of heat generated in a chip. In general, T_{in} is maintained at less than 25°C since the heat is drawn off through an externally provided heat exchanger, and T_c is kept below 10 or 15°C when three or four packages are connected in series [6]. Therefore, from (8) the allowable temperature rise ΔT_{allow} is 50 or 55°C.

The allowable heat dissipation for the fabricated package is shown in Fig. 14 as a function of the total flow rate. In this case the allowable temperature rise is assumed to be 50°C and the highest possible value of R_{j-c} is used. The allowable heat dissipation increases when the total flow rate increases because R_{j-c} decreases as shown in Fig. 13.

This cooling system allows the chip to dissipate up to 16 W and the package itself to dissipate up to 400 W at a flow rate of 1.0 l/min. This value is equal to or larger than that for conventional indirect water cooling.

Discussion

The allowable heat dissipation depends on the thickness of the conductor layer since the heat generated in the chips is conducted from the device junction to the channel surface by way of the multilayered alumina substrate. The allowable heat dissipation dependent on the conductor layer thickness was calculated by FDM with the calculated results shown in Fig. 15. The solid curves represent the results of a 5 × 5 array of 8-mm² chips mounted on the substrate, while the dotted curves represent the results for 10-mm² chips. The open circle represents the experimental results. In this case the total flow rate is chosen to be 1.0 l/min.

When the multilayered substrate thickness is increased, the allowable heat dissipation decreases because of the increase in the internal thermal resistance. At a 6-mm thickness (about 35 layers), however, the package can dissipate up to 12 W per chip (8 mm²) and 300 W per package using epoxy resin as a die bonding material. Furthermore, reducing the internal thermal resistance increases the allowable heat dissipation. Moreover, enhancing the die bonding area using a higher thermal conductivity medium lowers the internal thermal resistance as well.

Changing the bonding medium from epoxy resin to Sn/Pb solder, for example, increases the allowable heat dissipation to 14.5 W per chip and to 360 W per package. Thus the cooling capability limit can be extended by over 20 percent through a reasonable design change. Additionally, enlarging the chip size to 10 mm² increases the allowable heat dissipation to 450 W per package, owing to the increase in conduction area, and extends the cooling capability by over 50 percent.

To further clarify the advantages of this innovative cooling technique, we compared the currently obtainable volume power densities (the allowable heat dissipation for a package's outward volume) as shown in Fig. 16. The measured volume power density for the proposed technique exceeds 17 kW/l (300 W/(10.5 cm × 8.5 cm × 0.2 cm)) with dramatic improvement being actualized over the conventional cooling techniques. This density represents a tenfold increase over conventional indirect liquid cooling and equals that of immersion cooling [7].

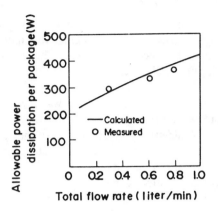

Fig. 14.   Allowable power dissipation per package as function of total flow rate, when allowable temperature rise is 50°C.

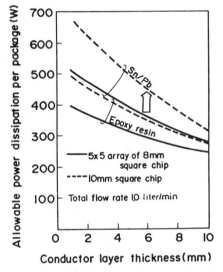

Fig. 15.   Allowable power dissipation per package as function of conductor layer thickness with chip size and die bonding material as parameters.

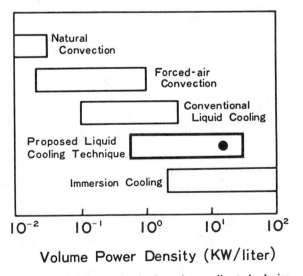

Fig. 16.   Volume power density for various cooling technologies.

CONCLUSION

A new technique for increasing the cooling capability and for effecting high-density stack packaging for multichip package application has been developed. The technique mounts VLSI chips on a multilayered substrate which incorporates very fine coolant channels.

The optimal structure for cooling sections was investigated by computer simulation and experiment, with the former being implemented. The obtained numerical solution of the coolant flow distribution ensured that the coolant distributor and collecter structure dimensions could be determined to meet the uniform velocity distribution condition. Additionally, the channel cross section was designed to be 800 μm wide $\times$ 400 μm high to achieve a lower thermal resistance.

Based on the results of these structures, an indirect liquid cooling package was fabricated which mounted a 5 $\times$ 5 array of VLSI chips on a substrate measuring 85 mm $\times$ 105 mm. An allowable heat dissipation higher than 400 W per package (mounting 8 mm^2 chips) was realized at a flow rate of 1.0 l/min. Furthermore, since the thickness of the cooling section is smaller than 1 mm, the volume power density was found to increase to 17 kW/l or more, a value tenfold greater than that obtained by conventional indirect water cooling and equal to that obtained by immersion cooling.

This new packaging is thus particularly suited to future needs, especially in terms of its excellent cooling capability, high-density packaging which reduces the overall wiring length, chip size enlargement, increased the number of conductor layers, and so on. Accordingly, this technique will find broad application to large-scale high-density stacked multichip packaging.

REFERENCES

[1] S. Oktay and H. C. Kammer, "A conduction-cooled module for high-performance LSI devices," *IBM J. Res. Develop.*, vol. 26, pp. 1–12, 1982.
[2] T. Watari and H. Murano, "Packaging technology for the NEC SX supercomputer," *ECC*, pp. 192–198, 1985.
[3] U. P. Hwang and K. P. Moran, "Boiling heat transfer of silicon integrated circuit chips mounted on a substrate," *ASME HTD*, vol. 20, pp. 53–59, 1981.
[4] K. Fujikake, "Study on corrugated finned heat exchanger," *J. JSME*, pp. 241–249, 1977.
[5] L. M. K. Boelter, NACA TN, pp. 1451, 1948.
[6] R. C. Chu, U. P. Hwang, and R. E. Simons, "Conduction cooling for an LSI package: A one-dimensional approach," *IBM J. Res. Develop.*, vol. 26, pp. 45–54, 1982.
[7] Cray Research, Inc., "The Cray-2 computer system," 1985.

Part 4

Thin Film Multilayer Modules on Silicon

THIN film multichip modules on silicon provide chip packaging with higher density, better resolution, and matched thermal coefficient of expansion. With multiple layers of metallization and dielectrics, silicon serves a circuit board where multiple bare die are interconnected. Because the earliest use of a lower dielectric constant interconnect insulation on silicon was the use of silicon dioxide, a paper from Mosaic has been selected to show this application. More recent advocates of silicon dioxide dielectric include Tuckerman of Livermore Laboratories and N-Chip, but the included article only partially describes the technology.

Perhaps the papers that had the highest visibility for military applications were those from Rockwell, Cedar Rapids; the Evans paper from SAMPE; and the Hagge paper from the ECC. It is the change to organic dielectric that has permitted deposition of more layers and the use of copper circuit metal. The RISH program in England provides the most recent papers (Early). The AT&T papers and the IBM Boblingen paper show other advocates of this approach.

ADVANCED PACKAGING CONCEPTS —
MICROELECTRONICS MULTIPLE CHIP MODULES
UTILIZING SILICON SUBSTRATES

Alva A. Evans

John K. Hagge

Rockwell International Corporation

Collins Avionics Divisions

Cedar Rapids, Iowa 52498

Abstract

Advancements in speed and circuit density of integrated circuits used in high performance systems are creating a need for materials and processes capable of providing high density multilayer interconnections with increased reliability. These needs can be met by a multiple chip packaging technique which utilizes existing technology found in integrated circuit (IC) and hybrid fabrication processes. Multiple chip packaging offers flexibility of combining various IC technologies, i.e., CMOS, GaAs, Bipolar... etc., within the same package and eliminates the necessity for redundancy required by wafer scale integration (WSI). Silicon substrates with thermal coefficients of expansion matching that of the chip devices and having good thermal dissipation are used to support a multilayer, passive interconnect system. Aluminum conductor layers, 25 micron wide, separated by polyimide dielectric are used to form the interconnect patterns. Chip devices are attached to the substrates with epoxy adhesives and wire bonded to the interconnect circuitry. Standard hermetic hybrid packages are used to the final closure. A twelve chip reprogrammable memory module which operates at 25 MHz has been fabricated to demonstrate process feasibility and electrical performance of the interconnections.

1. INTRODUCTION

1.1 Background

Multiple chip modules using silicon substrates is a technology which complements industry efforts to reduce size and weight and improve reliability.[1][2][3] Our objective is to develop an interconnect/packaging subsystem which is smaller, lighter and more reliable than Surface Mount Technology (SMT).

1.2 Approach

Normally, the choice of a new packaging concept results from looking at a multitude of parameters, i.e., cost, weight, electrical performance, reliability, etc., weighing the gains in some areas against the losses in other areas and selecting the optimum path to follow. Studies of various interconnect concepts[4] led to the conclusion that a concept consisting of multichip modules with silicon interconnect substrates (SIS's) was a viable approach — significant gains with little or no losses.

This approach provides direct connection of IC die to a substrate, thereby eliminating one level of packaging and interconnect as shown in Figure 1.

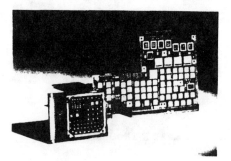

Figure 1. Significant Size and Weight Reductions are Possible By Using Multichip Modules Instead of SMT Circuit Assemblies. The Three-Inch by Four-Inch Multichip Module in the Foreground Contains All the Circuitry of the Larger SMT Circuit Assembly.

The approach includes the fabrication of the SIS using proven IC technology, but with the addition of power and ground planes to afford control over electrical characteristics

such as impedance and noise coupling. The features (conductor lines and spaces) will be at least an order of magnitude larger than for IC's, thus improving yield. When compared to SMT, in which leadless chip carriers (LCC's) are mounted on multilayer boards (MLB's) improvements are possible in all areas — smaller size and weight, better reliability, and better thermal performance. Although several types of substrates have potential benefits in a multiple chip module, SIS substrates appear to hold the highest potential gains in size, weight and reliability.[4]

This project has been structured to take advantage of new materials and to utilize mature multidiscipline technologies in clever new combinations. The approach is to combine the best of semiconductor, hybrid and printed circuit interconnection technologies, and to infuse into the program the latest packaging advances from VLSI, wafer scale integration, and VHSIC packaging programs.

2. SILICON SUBSTRATE DEVELOPMENT

2.1 Substrate Material

A thorough analysis of materials[5,6] resulted in the selection of silicon as the best choice for substrates. Silicon's thermal coefficients of expansion matches that of the IC chip devices, it has a relatively high thermal conductivity as shown in Figure 2, silicon wafers of various sizes and properties are readily available, IC processing technology has developed equipment and processes for handling silicon wafers, and finally, a goal was established to make SIS compatible with possible future expansion to include active devices in the wafer along with interconnect features.

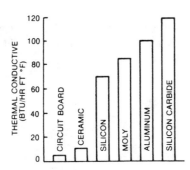

Figure 2. The Thermal Conductivity of Silicon is Greater Than Epoxy or Ceramic.

ELECTRICAL	
Dissipation Factor:	(1 kHz): 0.002
Dielectric Str:	4000 vts/mil
Vol. Resistivity:	10^{16} ohm-cm
Surface Resistivity:	10^{15} ohm
Dielectric Constant:	(1 kHz): 3.5

Figure 3. Cured Film Properties.

SIS processes were developed consisting of aluminum conductor material deposited on silicon wafers using conventional thermal evaporation techniques.[7,8,9] Conductor lines of 25 micron width and spacing and 2 micron thickness were patterned by a subtractive approach where a pattern is defined by photolithography using standard chrome/glass masks. Unwanted material is removed by wet etching.

A wet chemical process was developed for forming via holes (interconnections between layers) in a 2-4 micron polyimide film. 75 micron diameter via holes were found to be optimum for this process. Experiments with polyimide of several viscosities were conducted to characterize film thickness versus coater spin speed. The results are shown in Figure 4.

Metal layers are separated by polyimide dielectric. Polyimide resin has been evaluated extensively[10-16] as a dielectric material for IC fabrication. When fully cured polyimide films are heat resistant, mechanically tough, and chemically resistant, they provide good step coverage by planarizing surface topography and possess good electrical properties. The properties of the film are shown in Figure 3.

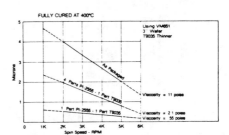

Figure 4. Thickness of Polyimide Films was Optimized Experimentally.

Pinholes in the dielectric layers were reduced by applying the polyimide through a filter. Polyimide diluted with thinner was selected since it provided the highest viscosity which could be easily applied through the filter. A spin speed was chosen which yielded a cured film thickness of 1.8 to 2.2 microns. Typical interlayer via connections are shown in Figure 5.

PHYSICAL	
Tensile Str (ultimate):	19,000 psi
Elongation:	10%
Density:	1.39 gm/cu cm
Refract. index (Becke Line):	1.70
Flexibility:	180° bend, no cracks

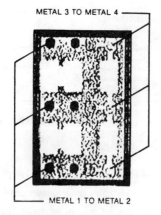

METAL 3 TO METAL 4

METAL 1 TO METAL 2

VIEW A. POWER AND GROUND
INTERLAYER CONNECTIONS

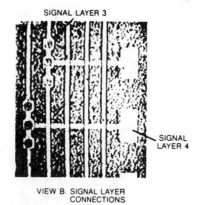

SIGNAL LAYER 3

SIGNAL
LAYER 4

VIEW B. SIGNAL LAYER
CONNECTIONS

Figure 5. Interconnections Among Two
Signal Layers and Power and Ground
Planes Are Made Using 75 Micron Vias
(X44 Magnification).

Electrical shorts between power and ground
planes were eliminated by a double dielectric
coating process. A first coat of polyimide was
applied, imaged using a via mask, and
removed where vias (and pin holes) were
defined. A second coat of polyimide was then
applied and imaged with a different mask
with the same via pattern, but with a different pattern of pin holes, since pin holes are

random. A high precision mask aligner was
used to achieve the critical alignment of via
holes of coats 1 and 2, shown in Figure 6.

X350 MAGNIFICATION

Figure 6. Double Coat Process Used to
Achieve Short Free Power and Ground
Planes.

Precise alignment of the two coats is shown
for these vias. Figure 7 shows the nine
masks needed to make a four layer SIS,
including 1A and 1B for separating ground
and power planes.

MASK	LAYER	MATERIAL
VIA-4	Diel.	Polyimide
Metal-4	Signal	Aluminum
VIA-3	Diel.	Polyimide
Metal-3	Signal	Aluminum
VIA-2	Diel.	Polyimide
Metal-2	Power	Aluminum
VIA-1B	Diel.	Polyimide
VIA-1A	Diel.	Polyimide
Metal-1	Ground	Aluminum
	Insul.	SiO_2
	Subs.	Silicon

Figure 7. Nine Masks Are Required to
Fabricate SIS With Two Signal Layers
and Power and Ground Planes.

Investigations were undertaken to determine
if any limitations existed as to the type of
profile that could be provided (rectangular,
circular, irregular) for silicon substrates, rec-

110

tangular shapes are provided with IC dicing technology. Other contour shapes were achieved with wet etching or laser cutting. Experimental results showed that both methods could provide any arbitrary profile or cut-out in silicon.[17] However, laser provided the best edge quality.

3. MULTIPLE CHIP MODULE DEVELOPMENT

3.1 Design

Fifteen designs of silicon wafer interconnect substrate have been completed. These designs have permitted evaluation of a number of design methods including: manual, Computervision CADDS-3 Printed Circuit CAD System, Daisy Chipmaster Integrated Circuit CAD system, Racal-Redac Printed Circuit CAD System.

Use of the Daisy CAD System provides the most direct route to tooling since it is set up for direct input to glass-mask making equipment. Two representative CAD output designs are shown in Figure 8.

Both digital and RF circuits have been adapted to SIS. Functional testing was completed on a 32-BIT Reprogrammable Memory Module, Figure 9, with excellent results.

(a) Frequency Synthesizer

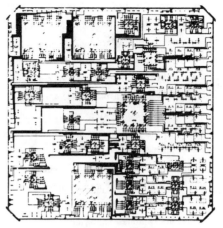

(b) Digital Processor

Figure 8. Two Typical CAD Designed SIS Outputs

Figure 9. Chip Memory Module Using SIS Interconnect System.

This 12-chip module operated successfully at 25 MHz, exceeding design requirements by 5 MHz. The interconnect system was totally transparent to the memory operation, causing no degradation of electrical signals.

3.2 Mechanical Design

The mechanical design issues of thermal expansion mismatch, vibration, and cyclic altitude pressure variations have been evaluated.[18] The principal mechanical design issues are related to providing large (2" x 2") silicon substrates with adequate structural support to survive vibration, shock and thermal cycling environments, while maintaining the significant SIS advantages of small size, low weight and high reliability. Although unsupported silicon wafers are quite fragile, after bonding to a support core they become quite rugged. Matched thermal expansion and high thermal conductivity of the core material is important. Molybdenum, aluminum nitride, silicon carbide, and graphite composites are candidate materials.

Several concept models were designed and built to demonstrate the significant packaging improvements available with SIS technology.

Figure 1 shows a concept model built to demonstrate a miniaturized version of a Digital Data Processor Interface (DDPI). The large Invar-core surface mount circuit board in the background is replaced by 61 IC's on a SIS. The IC's can be wire-bonded, TAB mounted, or flip-chip soldered and are encapsulated with a silicone gel encapsulant. The SIS module utilizes a high density connector with 180 I/O on 0.050 inch centers.

Figure 10 shows the same DDPI redesigned into an SIS unit in a large hermetic hybrid package. The wide range of packaging flexibility of the SIS approach is also demonstrated in Figure 11(b) which shows a large hermetic SIS unit along with numerous SMT devices mounted to each side of a 3/4-ATR (SEM-E) module. A rail-mounted thermal core provides high heat removal capacity and a new SEM-E connector provides 250 I/O capability.

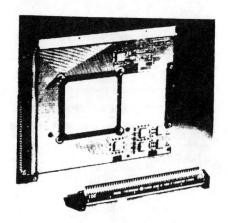

Figure 10. Many Options Exist for the Next Level of Packaging. This Example Shows Two SIS Modules (one on the rear) Mounted on a SEM-E Module

3.3 Thermal Design

Thermal design guidelines have been established[18] for several configurations. Finite element modeling has been used to define dissipation limits for several packaging approaches. A typical finite element model is shown in Figure 11(a).

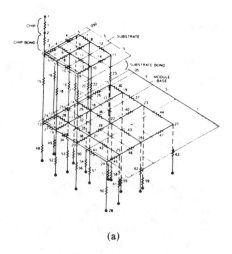

(a)

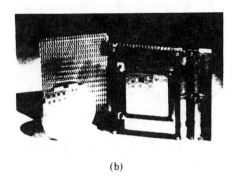

(b)

Figure 11. Finite Element Modeling Was
Used For Developing Module Concepts

This model was used to design the high
power air-cooled module shown in Figure
11(b). This concept model carries a 50-watt
SIS module on either side of an air cooled
core, for a total of 100 watts in a 4.6 x 6.0 x
8.5 inch module.

The finite element results for this high
power module are shown in Figure 12 for
several configurations. The importance of
using highly conductive materials to spread
the heat away from the IC is evident in this
figure.

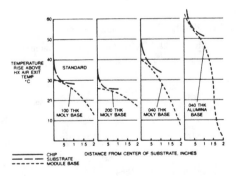

Figure 12. Finite Element Results for
Various Module Bases for the Model
Shown in Figure 11(a).

An additional thermal improvement has been
developed by eutectic bonding the IC to a
silicon chip carrier as shown in Figure 13(a).
The high conductivity silicon provides good
thermal spreading to decrease IC junction
temperatures as shown in Figure 13(b).

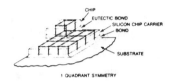

(a)

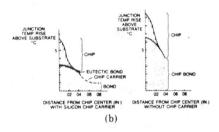

(b)

Figure 13. Finite Element Results For
Silicon Chip Carrier

3.4 Assembly

Assembly of the 12-chip module (Figure 9)
was accomplished with conventional hybrid
assembly techniques for substrate attach, die
attach, and wire bonding. The use of silicon
instead of a conventional ceramic substrate
presented no assembly problems. Available
adhesives were reviewed and MIL-approved
types were selected. Substrate and die attach
were accomplished with conventional high
conductivity epoxies. Monometallic wire
bonds were made between the aluminum
pads on both the silicon IC's and silicon substrate using 1 mil aluminum (1% silicon)
bonding wire.

4. CONCLUSION

Silicon substrates and multilayer interconnects using aluminum metallization and polyimide dielectric layers provide an advanced packaging concept which allows flexibility to meet many design criteria, size and weight reduction over surface mount technology, and increased functional reliability. Further work is underway to characterize all process and material parameters of this technology, extend the concept to include packaging of RF circuitry, and insertion into major electronic systems.

5. ACKNOWLEDGMENTS

The authors would like to thank K. Nelson for the design layout of the modules described in this paper, and H. Hopp for review and editing the paper.

6. REFERENCES

1. Lyman, J., "Military Moves Headlong Into Surface Mounting", Electronics, July 10, 1986, p 93-98

2. Staff, "Reliability: The Air Force Strikes Back", Aerospace America, July 1986, p 16-18

3. Letter from General Lawrence A. Schantze, Commander Air Force Systems Command, to Mr. Robert Anderson, Chairman of the Board and Chief Executive Officer, Rockwell International Corporation, 19 July 1986

4. Hagge, J.K., "Advanced Packaging Concepts For Multi-Chip Modules, Part I Literature Survey", Rockwell Internal Document, Janaury 1986

5. Hagge, J.K., "Advanced Packaging Concepts For Multi-Chip Modules, Part III. Properties of New Materials", Rockwell Internal Document, February 1986

6. Hagge, J.K., "Advanced Packaging Concepts For Multi-Chip Modules, Part IV Properties of Substrate Materials", Rockwell Internal Document, February 1986

7. Maissel, L.I. and Glang, R., "Handbook of Thin Film Technology", pp 1-4 thru 1-123, McGraw-Hill, 1983

8. Burggraaf, P., "A Review of Metal Deposition Equipment and Application Trends", Semiconductor International, November 1985

9. Rhodes, S.J., "Multilayer Metallization Techniques For VLSI High Speed Bipolar Circuit", Semiconductor International, March 1981

10. Zglar, S., "Dielectric Characterization of Polyimide Thin Films", Abstracts of First Technical Conference on Polyimide, 1982 pp 143-144.

11. Rothman, L.B., "Properties of Thin Polyimide Films", Journal of Electrochemical Soc., Vol. 127, October 1980

12. Mittal, K.L., "Polyimides - Synthesis, Characterization and Applications", (50 papers on Polyimide), Proceedings of the First Technical Conference, Vol. 1 and 2, Plenum Press, 1984

13. Samuelson, G., and Lytle, S., "Polarization Effects in Polyimides", Journal of Electrochemical Soc., Vol. 131, No. 11, Nov 1983

14. Bateman, J., "Recent Advances in Photoimagible Polyimides", SPIE, Vol 539 pp 157-180

15. Grengoritsch, A.J., "Polyimide Passivation Reliability Study", IBM Bulletin E-20950, IBM System Product Division, Essex Junction, VT

16. Senturia, S.D., "Polyimide as an Interlevel Insulator in Integrated Circuits", Proc. ANTEC, Soc. Plastics Eng. April 1985, pp 414-416

17. Monk, D., and Hagge, J.K., "Preliminary Study of Etching and Laser Methods to Profile Contours and Cutouts In Silicon Substrates", Rockwell-Collins Working Paper, Nov 1986

18. Hagge, J.K., "Advanced Packaging Concepts For Multi-Chip Modules, Part II MC² (Module, Core, Connector) Packaging Options", Rockwell Internal Document, April 1985

Multichip Packaging Design for VLSI-Based Systems

CHARLES J. BARTLETT, MEMBER, IEEE, JOHN M. SEGELKEN, AND NICHOLAS A. TENEKETGES

Abstract—The introduction of many VLSI devices into system designs is placing new requirements on the packaging technologies that are used to interconnect devices and assemble systems. These requirements include the assembly of high pin out (up to 500 I/O's) devices, the ability to sustain synchronous system operation at frequencies up to 100 MHz, and cooling at thermal loads greater than 1 W/cm². A new packaging technology is described that will overcome many of the limitations of conventional packaging. The new technology has the capability for assembling devices with more than 200 I/O's, a maximum signal lead length less than 20 cm, a power and ground lead inductance less than 0.1 nH, a signal lead capacitance less than 20 pF, and a cooling capability greater than 1 W/cm². The substrate for the proposed new packaging technology is a silicon wafer. The power and ground are distributed by means of copper planes on either side of the substrate. Two signal layers are positioned above the power plane using a polyimide dielectric material. The signal leads are a minimum of 10 μm wide. With 10-μm-thick polymer, the line capacitance is ~ 1 pF/cm. Copper metallization is used to achieve a resistance of 10 Ω/cm for the minimum width signal leads with 2-μm-thick conductors. Devices are attached to the interconnection substrate by means of solder. The solder technology was chosen because it is repairable, provides a low-inductance (< 1 nH) connection between chip and substrate, and enables I/O pads to be positioned over the area of the chip. The matched thermal expansion coefficient between the substrate and chip results in a mechanically stable solder attachment. This new packaging technology has been demonstrated by means of a prototype vehicle. Three chips from the WE® 32100 chip set: CPU, memory management unit (MMU), and math accelerator unit (MAU), have been flip-chip bonded and interconnected together on a 1.3 × 3.0-cm substrate. The chips have been spaced 500 μm apart. The assembly has been bonded to a metal heat sink using a compliant adhesive layer to minimize stresses in the silicon substrate. The chip-to-heat sink thermal resistance for the assembly is 5°C/W (junction to case). A 160 I/O multilayer printed wiring board package has been used to fan out from the edge of the substrate to pins on a 100-mil grid. Wire bonds electrically connect the substrate to the package and provide for chip backside grounding. A protective cover completes the package assembly. In summary, significant improvements in system performance can be achieved with the new multichip technology. Compared to conventional packaging, a factor of 3 improvement in system operating frequency is expected when IC's are designed specifically for the technology. Additionally, system size and power dissipation can be reduced by a factor of seven and 30 percent, respectively.

INTRODUCTION

A NEW packaging technology is being developed at AT&T Bell Laboratories. This technology, which is referred to

Manuscript received February 11, 1987; revised July 15, 1987. This paper was presented at the 37th Electronic Components Conference, Boston, MA, May 11–13, 1987.

The authors are with AT&T Bell Laboratories, 600 Mountain Avenue, Murray Hill, NJ 07974.

IEEE Log Number 8717173.

® Registered service mark of the American Telephone and Telegraph Company.

as advanced VLSI packaging (AVP), is designed to meet the needs of future VLSI-based systems. It is expected that these systems will incorporate the use of mixed device technologies, for instance, MOS and bipolar, silicon and GaAs. Multichip packaging allows IC's to be tested individually before attachment so that only functional devices are assembled.

The number of I/O's on large VLSI chips will increase well above 200 as design rules shrink below 1 μm. Increasing the number of I/O's above 200 on a 1 cm² chip with a 200-μm minimum center-to-center spacing on the I/O pads will require that the positions of the bond pads not be constrained to a single row around the perimeter. An area array of pads allows the maximum bond pad size for test probe access. It also has other advantages in that signals internal to the device no longer have to be routed to the outside edge with the associated penalties in routing space and electrical performance.

Frequencies for synchronous system operation are increasing. The WE 32200 chip set has an operating frequency of 24 MHz. The military VHSIC II program has a targeted system operating frequency of 50 MHz. Synchronous operation at these frequencies will require short interconnection lengths, low capacitive loads on IC output buffers, and low dielectric constant insulating materials.

Signal rise times are expected to decrease with increasing operating frequencies. For MOS devices the short rise times are achieved at the expense of large current spikes which give rise to voltage noise when coupled through the power and ground inductance. Very low inductance chip connections will be required to minimize the noise in VLSI-based systems with a large number of simultaneously switching output buffers [1].

Finally, power dissipation will increase above 1 W/device, even for CMOS devices, as operating frequencies increase. The combination of the increasing power dissipation per device and the increasing functional densities for systems means that the cooling capability of future packaging must be above 1 W/cm².

Current system packaging can be characterized by IC's packaged in individual packages which may be either through-hole or surface mounted to printed wiring boards (PWB's) [2]. The assembled circuit packs are generally interconnected by means of a backplane and separable connector [3]. This packaging technology is characterized by signal lead lengths that can be greater than 50 cm, inductance of the power and ground leads greater than 1 nH, capacitance of the signal leads greater than 100 pF, and a maximum cooling capability on the order of 0.2 W/cm². High-performance VLSI-based systems will require at least a factor of 5 improvement in these parameters.

Reprinted from *IEEE Trans. Components, Hybrids, Manuf. Technol.*, vol. CHMT-12, no. 4, pp. 647–653, December 1987.

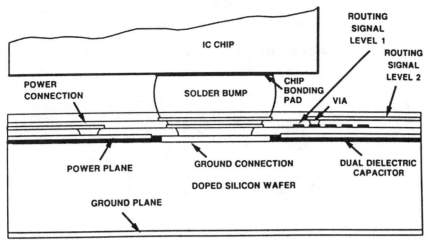

Fig. 1.   AVP cross section.

ELECTRICAL ENVIRONMENT

The AVP interconnection substrate is constructed to provide an excellent electrical environment in which to route signal paths and distribute voltage reference levels. The substrate provides a microstrip configuration for two signal levels over a continuous power plane which acts as a high-frequency ground, as shown in the cross section of Fig. 1. The metallization of both signal lines and voltage planes is plated copper. The use of copper assures flexibility in altering line thickness and hence line resistivity as designs require. Line resistivity of 10 Ω/cm is achievable with 2-μm-thick copper and 10-μm linewidths. Good electrical contact between interconnection levels is achieved with solid nickel vias which are also plated. Signal level microvias measure 10 μm in diameter, and power/ground vias measure 90 μm in diameter. The polyimide that forms the dielectric between metallized levels has a dielectric constant $E_R = 3.5$. Characteristic impedances of 50–70 Ω can be achieved with dielectric heights $h_1 = 10$ μm and $h_2 = 6$ μm as shown in Fig. 2 for linewidths and spaces of 10 μm. The dimensions also provide low capacitive loading on the signal paths. Fig. 3 shows capacitive loading C_W for variations in path width, space, and dielectric height as simulated with two-dimensional finite-element (FEM) analysis [4].

Noise with the new packaging technology is considerably reduced over that with current technologies. Crosstalk among adjacent signal paths is kept to a minimum through the use of the metal plane that is continuous under the signal lines. Effective coupling is further reduced by having very short net lengths and line spaces greater than minimum spacing for most designs. Induced power and ground noise that results from many IC output drivers switching simultaneously is reduced dramatically through the use of solder bump chip attachment and an integrated decoupling capacitor. The simultaneous switching noise,

$$V_{max} = NL \frac{di}{dt}$$

where N is the number of drivers, L is the inductance of the path from the IC pad to the voltage reference planes, and di/dt

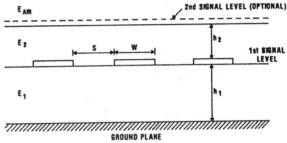

Fig. 2.   Capacitance modeling. Two-dimensional wire cap simulation. Quasi-three-dimensional simulation for crossover region.

is the single driver switching current rate, is best reduced by decreasing L. Table I shows how solder bump chip attachment compares with wirebonding and tape automated bonding (TAB) using inductance simulation tools [5]. Reductions up to a factor of 20 can be achieved in V_{max} with the flip-chip bump attach. To further ensure a low-inductance path, decoupling capacitance should be placed extremely close to the chip I/O. This is accomplished in the AVP technology by building a decoupling capacitor plane directly into the silicon substrate. The silicon wafer is highly doped to be conductive and is plated on the backside to act as a ground. A dual dielectric of SiO_2 and Si_3N_4 deposited on the top side of the silicon forms a defect-free capacitor that can be accessed directly under the solder bump. Capacitances of 25 nF/cm^2 have been achieved using this approach.

SYSTEM PERFORMANCE

Advantages at the system level come primarily from the ability to compact complex high I/O applications into a small physical space. AVP technology enables the extension of the high level of integration of VLSI devices to the system level, allowing as many as 100 complex VLSI chips to be interconnected on a single wafer. Fig. 4 shows the relative packing density of bare IC chips mounted on three interconnection substrates. The devices assumed are 1-cm^2 200-I/O 50 000-gate chips randomly interconnected. The Rent's exponent for peak wire congestion in both the x and y signal planes was assumed to be $b = 0.7$ [6]. The plot shows that with AVP technology no routing space beyond that for the IC's and an

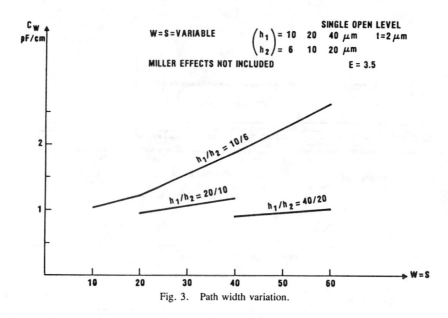

Fig. 3. Path width variation.

TABLE I
CHIP ATTACHMENT TECHNOLOGIES

	Cross Section (mil)	Length (mil)	Inductance[a] (nH)
Wire bond	1 (diameter)	50	0.9
TAB	1 × 3	50	0.7
Solder bump	4 × 4	2	<0.05

[a] 25 percent of I/O are power/ground.

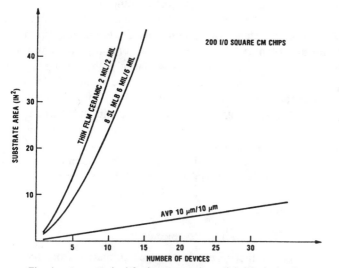

Fig. 4. Area required for interconnection—digital logic density.

associated guard band is needed for interconnection. More conventional technologies, such as two signal level ceramic and eight signal level multilayer board, exhibit quadratic growth in the required interconnection area as device counts increase. The ability to place large-scale systems or subsystems on AVP wafers leads to minimal electrical parasitics, which results in higher system operating frequencies and lower power dissipations.

Interconnection path delays for MOS VLSI devices can be greatly reduced by decreasing capacitive loading. This can translate into reduced cycle times and higher synchronous operating frequencies. Capacitive loading is kept to a minimum through the elimination of single-chip packages and the use of short interconnect lengths. Single-chip packages can add between 5 and 10 pF per package. Signal nets on multilayer boards (MLB's) may easily exceed 50 cm in total length, while nets on AVP substrates rarely exceed 10 cm total length. Output drivers that are used to drive board-level loads typically must drive more than 100 pF. Most interconnection paths for AVP require drivers designed only for 25 pF loading. In addition, solder bumping removes the requirement of having to route internally generated signals to the outside edge of the IC, resulting in minimal IC parasitics. Fig. 5 shows a simple circuit function packaged in conventional chip carrier and MLB technology and the same function packaged in AVP technology. By tailoring the size of the CMOS driver to the load, the total delay through the circuit is reduced by approximately 50 percent. A large part of this delay reduction is accounted for within the smaller driver circuit. With IC's designed specifically for AVP, synchronous clock frequencies above 50 MHz should be achievable with complexities well above 1 million logic gates on the wafer.

Power dissipation levels are also reduced by short interconnect lengths and small capacitive loads. Power dissipation of an output CMOS driver is approximately $P = 0.5 C_L V^2 f$, where C_L is the load to be driver, V is the voltage swing, and f is the frequency of voltage swings. For high I/O VLSI designs, output drivers can account for up to 40 percent of the total chip power dissipation. With a reduction of C_L by factors of four, total chip power can be reduced up to 30 percent.

WE 32100 MODULE SUBSTRATE DESIGN

The first implementation of the AVP technology is a small laboratory test vehicle consisting of the WE 32100 central processing unit (CPU), memory management unit (MMU), and math accelerator unit (MAU). The chip I/O's range from 81 to 120 I/O. The substrate is not much larger than the chips

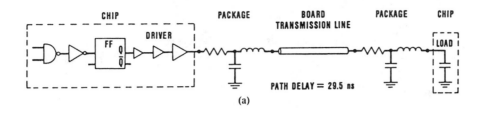

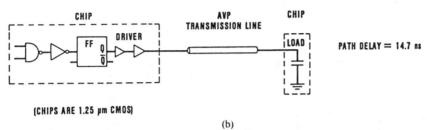

(a)

(b)

(CHIPS ARE 1.25 μm CMOS)

Fig. 5. Technology comparison.

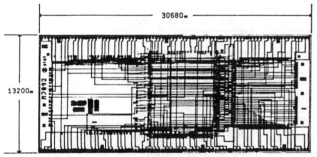

Fig. 6. Layout of the 32100-chip set substrate.

Fig. 7. 160 I/O pin grid array package for 30-chip assembly.

themselves, measuring 1.3 × 3.0 cm as shown in the layout of Fig. 6. The chips are spaced 500 μm apart with most of the routing taking place under the chips. Given abundant routing area, signal paths were widened to 20 μm, and all signal nets were brought to the periphery of the wafer. This allowed full test access at the periphery of the substrate since all three devices are tristatable. Numerous prototypes were fabricated, assembled, and packaged. These assemblies were demonstrated at clock frequencies of 14 MHz, the operational limit of the IC chip set itself.

WE 32100 MODULE PACKAGE DESIGN

A 160 I/O pin grid array package was designed for the three-chip assembly (see Fig. 7). The package consists of a four-layer PWB with minimum design dimensions of 4-mil lines/6-mil spaces. The PWB is fabricated using BT (bismaleimide-triazine epoxy blends) material to accommodate thermosonic wire bonding at 150°C. A rectangular hole is routed out of the center of the package to provide space for a metal heat sink/structural support member. The AVP interconnect substrate with the three chips flip-chip mounted is assembled to the metal heat sink (aluminum) using a thick layer of compliant adhesive (silicone gel material). The heat sink structure is assembled to the PWB package, and the assembly is then thermosonically wire bonded using 1-mil-diameter gold wire (ball, wedge) bonding. A protective plastic cover designed with alignment tabs provides mechanical protection for the wire-bonded assembly. Although this package is intended to be used as a laboratory test vehicle, extensions of this packaging technology are being investigated. For example, chip encapsulation materials are presently being developed and evaluated.

There are two particular aspects of this packaging approach that are unique and critical for future AVP package designs. These innovations are the attachment of the large silicon substrates (greater than 1 cm square) to a metal heat sink and a low thermal resistance heat conduction path providing effective cooling for the flip-chip assemblies.

HEAT SINK AS STRUCTURAL SUPPORT

The attachment of the silicon substrate to a metal heat sink has been modeled analytically [7], [8] to estimate the stresses acting on the bimaterial assembly during thermal cycling. Stresses arising in the assembly due to the thermal expansion mismatch of its constituents include: normal stresses $\sigma_1(x)$ and $\sigma_2(x)$ acting over the cross sections of the silicon (1) substrate and the metal heat sink (2), and shearing stresses $\tau(x)$ and normal (peeling) stresses $p(x)$ acting in the adhesive layer.

Fig. 8 presents some sample data and corresponding calculated stresses for the WE 32100 test vehicle. The results of these equations are summarized in the following.

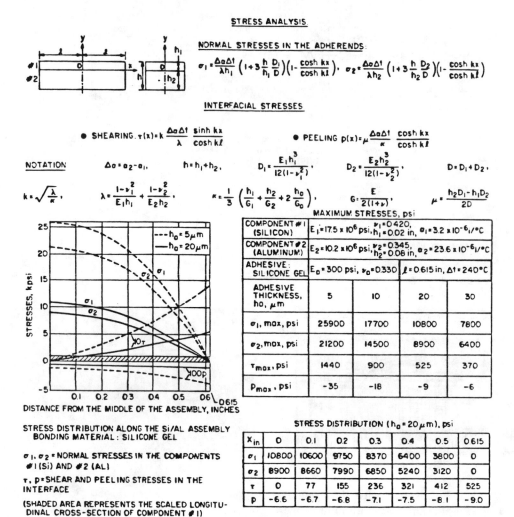

Fig. 8. Stress analysis.

The compliance of the adhesive is proportional to its thickness and inversely proportional to its shear modulus G, where

$$G = \frac{E}{2(1+\nu)}, \qquad \begin{array}{l} E = \text{Young's modulus} \\ \nu = \text{Poisson's ratio.} \end{array}$$

As compliance of the adhesive layer is increased, a redistribution of shearing stresses $\tau(x)$ and the peeling stresses $p(x)$ occurs in such a way that the maximum stresses at the edges of the assembly decrease and the stresses in the inner part of the assembly increase—tending to make the stress distribution more uniform. The increased compliance of the adhesive layer also results in a reduction of the maximum normal stresses acting over the cross sections of the silicon substrate and the metal heat sink. An effective design selects an appropriate adhesive material that provides sufficient compliance such that all stresses generated are less than critical stresses. Significant thermal shock and thermal cycling tests have been performed to evaluate the static and fatigue performance of the assembly. These critical stresses are estimated on the basis of experimental data and an appropriate factor of safety to avoid static and fatigue failures. The results provide a sound basis for a reliable attachment.

HEAT TRANSFER PERFORMANCE—FLIP-CHIP ASSEMBLY

A general heat transfer model was formulated to evaluate the thermal performance of the AVP packaging approach. Fig. 9 describes the physical configuration of the model. It consists of a chip (where the heat is generated) flip-chip attached using a solder bump technology to a silicon substrate attached to a metal heat sink. The following results were derived from the model.

• For the assumed condition of forced-air convection 400 ft/min in a Fastech™ (see [3]) environment, the thermal resistance of the heat sink to air was calculated and measured to be approximately 9°C/W.

• The thermal resistance of the chip to heat sink conduction path was calculated and measured to be 5°C/W. Approximately 40 percent of the conducted heat flows through the 2-mil air gap between the chip and the silicon substrate. The remaining 60 percent of the conducted heat flows through the solder bump attachments.

Based on these results, forced-air convection packages can provide effective cooling for 1-cm² chips that dissipate less than 3 or 4 W each. By changing from forced-air convection to water-cooled heat sink structures, the package can provide effective cooling for 1-cm² chips that dissipate up to 10 to 12 W each.

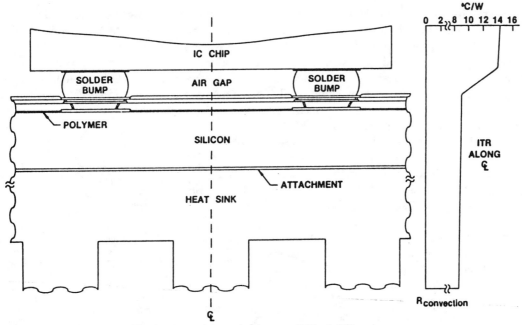

Fig. 9. Internal thermal resistances (ITR) of AVP package.

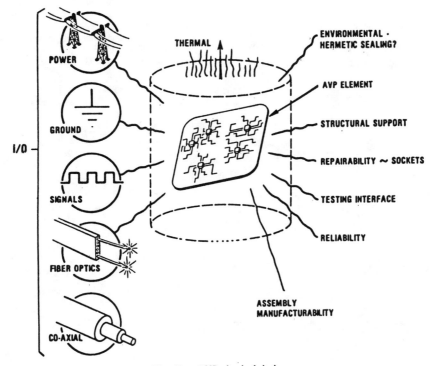

Fig. 10. AVP physical design.

FUTURE DESIGN CHALLENGES

The WE 32100 module design is but a modest beginning. The AVP technology has significant application to a host of designs, including large-scale switching networks, digital-signal processing networks, parallel processing systems, and certain military systems. Typically, these designs employ large numbers of high I/O chips running at high clock frequencies dissipating significant thermal energy. The system and electrical design challenges are considerable. Modules requiring interconnection of more than 10 000 chip I/O's will require more advanced layout systems than are presently available. Complex systems (greater than 1 million gates) operating a high clock frequencies (up to 100 MHz) will require advanced electrical and functional simulation techniques to ensure system operation without breadboarding. Complex circuitry in very small physical space also demands more advanced methods of providing test access than currently available for PWB's and ceramic hybrid integrated circuits (HIC's), as well as a means of reducing test access through design for testability approaches. By meeting these challenges, systems may be designed in ten percent of the physical space with three times the clock frequency, and between 20 to 70 percent of the power dissipation associated with current technologies.

The physical design challenges of developing packaging approaches for AVP are outlined graphically in Fig. 10. Input and output conductive paths may take various forms. Metallic conductors composed of power, ground, and signals could number between a few (<100) to thousands (2000 or 3000). Fiber optics and coaxial cables offer large bandwidth capability, and their interface to the AVP substrate must be considered. Significant power dissipation of 100 W to perhaps greater than a kilowatt (over a 2-in-square area) pose cooling challenges. Structural support for large silicon assemblies (greater than 2 in square) need to be carefully designed. Assembly of packaging structures and their repairability (failed devices, wire bonds, packages, etc.) have to be addressed. System reliability must be guaranteed through appropriate environment protection (hermetic or nonhermetic). These are but an outline of some of the major considerations and challenges for providing an integrated packaging approach such as AVP. In summary, the development of AVP technology presents numerous challenges for both system designer and physical designer; however, the future system benefits derived from AVP demand that these challenges be met.

ACKNOWLEDGMENT

The authors wish to acknowledge their AT&T Bell Laboratories colleagues of the Advanced VLSI Packaging Laboratory who contributed to the work discussed in this paper. Specific appreciation is expressed to Y.-C. Lee, H. T. Ghaffari, and David Kniep for the heat transfer modeling and experimental evaluations, M. F. Jukl and F. Remite for layout of the WE 32100 substrate, and L. H. Cong and M.-S. Lin for some of the circuit simulation and electrical modeling.

REFERENCES

[1] L. W. Schaper and D. I. Amey, "Improved electrical performance required for future MOS packaging," *IEEE Trans. Components, Hybrids, Manuf. Technol.,* vol. CHMT-6, pp. 283–289, Sept. 1983.

[2] C. J. Bartlett, "Advanced packaging for VLSI," *Solid State Technol.,* pp. 119–123, June 1986.

[3] W. L. Harrod and W. E. Hamilton, "The Fastech™ integrated packaging system," *Solid State Technol.,* pp. 107–114, June 1986.

[4] R. E. Bank, W. M. Coughran, Jr., W. Fichtner, D. G. Rose, and R. K. Smith, "Computational aspects of transient device simulation," in *Process and Device Simulation.* New York: North Holland, 1986, pp. 239–264.

[5] A. J. Rainal, "Computing inductive noise of chip packages," *Bell Syst. Tech. J.,* vol. 63, pp. 177–195, Jan. 1984.

[6] D. C. Schmidt, "A model of the impact of integrated circuits on printed wire routing," in *Proc. IEPS, 1st Ann. Conf.,* Cleveland, OH, Nov. 1981.

[7] E. Suhir, "Stresses in adhesively bonded bi-material assemblies used in electronic packaging," presented at the MRS 1986 Spring Meeting, Palo Alto, CA, Apr. 1986.

[8] E. Suhir, "Calculated thermally induced stresses in adhesively bonded and soldered assemblies," presented at the ISHM Int. Symp. Microelectronics, Atlanta, GA, October 6–8, 1986.

HIGH-DENSITY MULTICHIP INTERCONNECT (HDMI)
Kevin P. Shambrook, Ph.D., and Philip A. Trask
Building 700, Mail Station E2107
Microelectronic Circuits Division
Hughes Aircraft Company
Post Office Box H
Newport Beach, CA 92658-8903

Abstract

VHSIC II chips require a new packaging technology to take advantage of their high operating speed and packing density. Packaging can increase the packing density as much as the gate count in a new generation of signal processors. High density substrates create a need for new materials and for the processes, discipline, and clean room facilities used to manufacture semiconductors. Larger ICs and large substrates lead to either large packages or acceptance of alternate passivation techniques. The design and testing of these multichip packages require a system-level approach, forcing the integration of system, IC, and packaging design. The ability to integrate all of these disciplines in a timely manner may provide the key competitive edge in the next decade.

Introduction

The operating speeds of VHSIC II ICs will require a systems approach to packaging wherein the chip and package must be characterized as a single entity. The package interconnects must be correctly matched to the chip in terms of impedance. At operating frequencies of 100 MHz or greater, conventional types of IC packages will not permit the propagation of signals from the package to the chip without severe loss of signal or signal distortion [1,2]. New packages will be required that answer this need, thus insuring that signal propagation can occur without distortion, excessive propagation delays, or crosstalk between signal lines, caused by inter-line capacitance. The full benefits from improved IC performance will be realized only to the extent that packaging density can be increased and RF performance of the interconnections tailored to the circuit requirements.

At present, discrete decoupling capacitors are required in hybrid packages to minimize the effects of rise-time overshoot, which results in "ground-bounce" phenomenon. This adds an additional space requirement for the discrete capacitors, making the package substrate larger than is required for only the IC chips. As both the packing density and the complexity of the ICs are increased, the power density of the package substrate will rise, necessitating an improvement in power dissipation by the package. A new approach to packaging technology is needed to answer these critical requirements. It should have high packing density while dissipating high power and provide for impedance matching. A systems approach must be used that treats the problem of packaging the next generation of ICs as one of designing the package to suit the ICs instead of the ICs to suit existing packages.

Hughes is responding to this challenge with the High-Density Multichip Interconnect (HDMI) Program. The density inherent in silicon processing has always tantalized packaging engineers with the possibility of Wafer Scale Integration: Simply put, the chip is made as big as the wafer. The evident problems of yield and the desirability of mixing technologies have kept this an elusive dream [3]. The immediate needs [4] of military programs have dictated an approach that uses technologies developed in the last few years that can be pushed to production maturity with manageable risk. This paper will focus on the choice of substrate technology, but the total concept would not be viable without the developments in large ceramic packages, self-testing of ICs, and developments in large-scale multilayer routing programs. Production viability will also probably depend on the producibility of Tape-Automated Bonding (TAB) for chips with over 300 pads.

The objective of HDMI is to provide the capability of fabricating a military-qualified multichip substrate with high yield and full testability that will meet the needs of military customers, who require higher density, lighter weight, and hermetic packages designed to function properly with the VHSIC II series of integrated circuits.

This paper will give the rationale for using a polyimide dielectric, outlining some of the constraints, problems, and areas that need further characterization. The trade-offs enabled the production of useful circuits in 1988, with the potential for matching the VHSIC progress in the future.

Overview

The Package

The next generation of integration, which uses VHSIC II circuits, will require larger packages, such as 3" x 3" or 4" x 2". This requirement is dictated by the partitioning problems and the density. There is a large loss of density and speed at package boundaries; hence, the larger the package, the better. The constraints on size, then, come in physical form for both the package and substrate. Assuming a hermetically sealed package is required, it is difficult to design a large, thin package because of lid deflection, the difficulty of sealing a large package, the planarity, and the package strength required. Making the package thicker negates the purpose, since packaging density is a volume measure: the package height translates into module pitch. The current goal is a 2" x 4" x 150-mil package. (Two possible styles are shown in Figure 1.) Note that the package height and high I/O density (over 350 I/Os, under 50-mil pitch) precludes the use of leads through glass beads. It is likely that the lid will have to be supported by posts in the center.

The Substrate

There are three major issues in the development of HDMI substrate technology. The first issue is substrate fabrication yield. The yield is a function of defect density of the individual layers in the process. Control of defect density is paramount in the list of tasks required for a successful demonstration of substrate yield. The Hughes Microelectronics Center (HMC) in Newport Beach has been working on defect density reduction for over three years and has reduced the defect density on a 13-layer process used to make 2-micron HCMOS ICs from ten defects per square centimeter to less than one defect per square centimeter. During this period, the corresponding IC yield has increased by a factor of 10, in direct correlation to the reduction and control of defect density.

The second issue is the testability of the substrate. The necessity of having a 100-percent

Reprinted from *Proc. 39th Electron. Components Conf. (ECC)*, pp. 656–662, 1989.

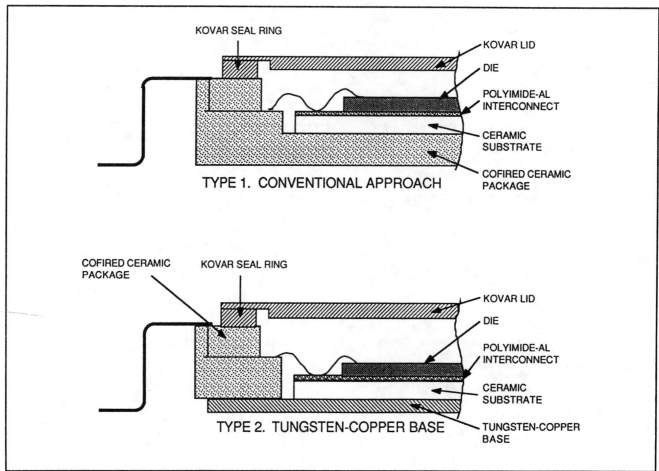

Figure 1. Two Package Styles: Type 1 -- Conventional Approach (Ceramic), Type 2 -- Tungsten-Copper Base

good substrate prior to committing to die attach means that all conductor paths on the substrate must be tested both for continuity between external pads and internal chip wire bond pads and for shorts between conductor lines.

The third issue of great importance is the ability to successfully assemble a large area multilayer substrate into a hermetic package without substrate cracking, attach the large ICs to the substrate without die cracking, wire bond (or TAB bond) the ICs to the substrate and the substrate outer lead pads to the package, and have the whole assembly pass temperature cycling and burn-in testing specified in MIL-STD-883C.

The substrate size is constrained by several pragmatic factors. One factor is yield, or, correspondingly, defect density. Another is the equipment and state of the art in semiconductor processing. Most semiconductor fabrication lines are set up for 4-inch diameter wafers today and are moving to 6 inches. A 6-inch wafer can give a square substrate of 3.8" x 3.8". The choice of 1.8" x 3.8" for the substrate is pragmatic. It is possible to get two substrates from a 6-inch wafer and one from a 5-inch wafer. There are not many 6-inch wafer lines in production for silicon, and there are only captive lines for polyimide substrates [5]. There are no known 8-inch wafer fabrication lines for polyimide substrates. However, 8-inch equipment is being developed for use in the next two-year period.

The Substrate

Benefits to be derived from HDMI technology in the building of military-qualified multichip packaging are:

a) Greatly increased substrate packing density, due to the reduction in width and spacing of the interconnects.

b) An improvement in electrical performance in terms of operating frequency, signal delay, and distortion.

c) A reduction in interconnect crosstalk, resulting from minimizing the length of the interconnections between IC chips.

d) A corresponding reduction in the size and weight of the packaged system, which arises from the use of a few larger packages instead of many smaller ones.

Figure 2 shows a typical high-density substrate for digital circuits, which consists of the base, ground, and power layers (ideally with a "decoupling" capacitance between them); two signal layers; and a pad (top surface) layer, separated by dielectric. The die are mounted on the top surface and either wire bonded or TABed. The separation between die is constrained by the pad size required for wire bond, including area for rework. The interconnect density is high since the die may have over 300 pads, requiring a 4-mil pitch to achieve the interconnection in two layers. The typical thick film multilayer interconnect used for hybrids could not be used for two reasons.

First, the minimum pitch of 20 mils (typically 40 mils) required for producibility in thick film technology would require a large number of interconnect layers. Second, the capacitance resulting from the large geometry interconnects would not allow the CMOS circuits to function at the maximum operating frequency. The next generation of bipolar ECL circuits

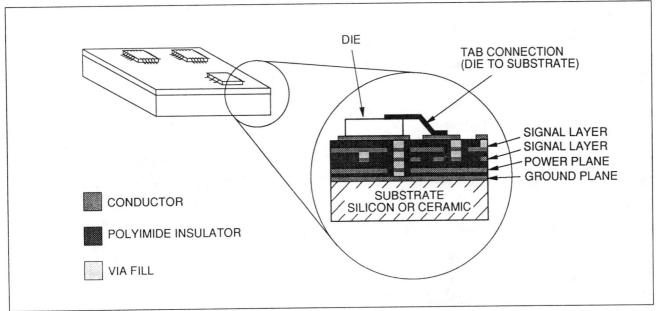

Figure 2. Typical High-Density Substrate for Digital Circuits

will require controlled line impedance (50 ohms or more), which is not achievable with "thick film" technology.

HDMI substrate technology offers multiple choices of base material, decoupling capacitor dielectric, signal dielectric, interconnect metallization, and via types. Analog circuits would benefit from further ground planes for signal isolation, which would mean additional layers, which can be built using HDMI technology.

The Signal Dielectric

Polyimide is the front runner for the signal dielectric today. The choice of dielectric is the key decision that influences both the choice of base and the design rules. Polyimide is chosen for its low dielectric constant (K = 3.6 typically, with K = 2.8 available) and its adaptability to fine line processing. It is normally applied by spinning and is self-planarizing, so polyimide does not suffer from the topology problem that can result in poor step coverage and plagues both the semiconductor and thick film processes.

It is possible to build substrates with 12-micron line width, but the typical sizing today is shown in Figure 3, where the width is 25 microns. The proportions in Figure 3 produce approximately a 50-ohm line (with K = 3.8).

The "4 times W" pitch results in reasonable crosstalk for digital signals (lines do not typically run parallel for more than 2 inches). As the trace width "W" is reduced, the dielectric thickness can be proportionately decreased while retaining the 50-ohm impedance. Thinner dielectric will reduce the substrate's tendency to warp; however, narrower traces will also result in higher trace resistance, which is generally undesirable, particularly for analog circuits.

Substrate warpage is probably the most significant producibility problem today. It is caused by the shrinkage of the polyimide as the curing takes place, by the large TCE (40) of many polyimides, and also by stress in the the metallization (e.g., copper can crystallize with temperature cycling). Warpage may be a problem not only for assembly, but also for automatic substrate handling and pattern lithography during substrate fabrication. The photolithography

must focus sharply over the total 6-inch dimension, so the wafer must be flattened by vacuum fixtures. For low defects, it is necessary to handle the wafer automatically in cassettes, and warpage causes sticking. Warpage can also cause delamination of layers, if the adhesion is not strong, or cracking of the base. The warpage problem increases with the size of the substrate and total number of substrate layers.

Various polymers are being evaluated for low dielectric constant, low TCE, and low moisture absorption. The outgassing of the polyimide in the hermetic package is another critical parameter, since this is additive to any epoxies used for die attach and substrate attach.

The Interconnect and Vias

Gold, aluminum, and copper have all been used successfully for interconnect metals. Gold is expensive. Aluminum has a higher resistivity than copper, but may be used in low-power circuits and has the advantage of being well understood for semiconductor-type processes. Copper has the advantage for higher currents because of lower resistivity and the ease of plating to get thicker lines.

The vias are often the weak link in any interconnect technology. Several techniques to form vias appear to be adequate at this time and fall into three main categories: unfilled, filled, and post. Generally, the vias need to be staggered in all

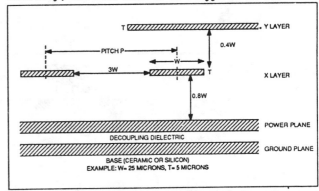

Figure 3. Typical Sizing of Conductors and Dielectric Thickness

categories, but filled vias can be stacked. For the unfilled and filled vias, a hole is made in the polyimide and then the metal layer, when deposited, coats the sides and bottom of the hole, creating the via (see Figure 4). The post process, on the other hand, first forms the via as a post on metal layer 1 and then spins on the dielectric (polyimide), etching or lapping it back to expose the post before the next metal layer is applied.

The Base and the Package

The three main contenders for base material are silicon, ceramic (alumina), and aluminum nitride. However, other bases are also being used: aluminum for high power and beryllium for high strength. Combinations are also possible; for example, silicon on aluminum nitride. An attractive base being developed is a high-temperature cofired ceramic multilayer base with a high-value integral decoupling capacitance between power and ground.

In more conventional bases, the trade-off is between ceramic and silicon. Ceramic is the short-term winner because the warpage problem will be less and because it is compatible with conventional hybrid package materials. The higher modulus of elasticity of ceramic gives one-third the warpage of silicon of the same thickness, and thickness is significant in the package height. A rule of thumb is that the base thickness for ceramic should be 20 times the polyimide thickness for a 4-inch substrate. For example, if the polyimide is 50 microns total buildup, then the ceramic would be 40 mils. Fifty microns is low (100 microns is better for low capacitance), and with silicon it would require 120 mils! These rules are very gross, but they illustrate the current framework. As lower TCE polyimides and lower temperature processes are developed, the warpage problems will be eliminated and silicon will be viable. Silicon has the long-term advantage of being able to incorporate MOS decoupling capacitors and active devices; for example, the bare substrate could have a "built-in" self test, similar to the die. Since the package may also be ceramic, it is natural to develop an inte-grated package where the substrate becomes the package base (see Figure 5).

CAD and Testing

Sophisticated CAD tools are required at all levels--system partitioning, chip design and simulation, substrate design, and back annotation--with an emphasis throughout on design for testability. Current CAD routing tools need further development. The semiconductor routers are very good for two layers of signal, but they cannot handle special ground plane designs, staggered vias, and the other design rules peculiar to each process. The hybrid routers are good at multiple layers, but not as efficient at two layers with over 3000 pin pairs in the netlist. Ultimately, the routing of the substrate should be integrated with the chip design.

A further step in density and power dissipation comes by placing the die in cavities (reducing the package height) and butting the die together (hence, no I/O on the butted side). It is not easily handled with current CAD tools.

Microcircuits containing a number of VHSIC die are going to be costly to throw away if there is a fault. It must be possible to locate the fault and rework the circuit--by removing and replacing a die, if necessary. However, the rework of such a dense circuit is difficult, and it is essential to reduce rework to a minimum by testing the components individually prior to assembly. Testability must be built into the die design. As much as 30 percent of the gates may be devoted to self-test, which also affects performance by adding time delays. There are several approaches to testing the die before attaching them to the substrate.

The first is "soft" probing (using probes or some contactor that does not damage the die), but this approach requires a lot of tooling. Since it is easier to test the die in a package, another approach is to wire bond the die to a package with "long" wires, so that the wires can be cut and rebonded to the substrate after testing. This approach does not lend itself well to production.

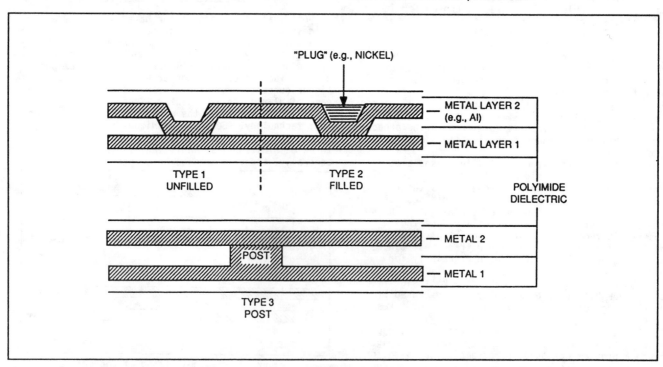

Figure 4. Via Categories: Type 1 -- Unfilled, Type 2 -- Filled, Type 3 -- Post

126

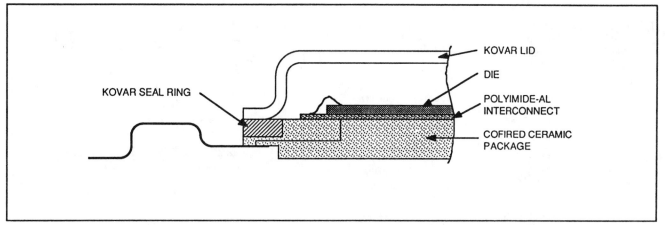

Figure 5. An Integrated Package

A preferred approach would be to use Tape-Automated bonding (TAB). TAB enables pretesting and burn-in, in theory. In practice, more development of tape is required for the structure to withstand the burn-in temperatures without warpage. TAB also requires die with "bumps," which are not normally available, and there are logistic problems in purchasing wafers from vendors, especially in the small quantities required for prototypes or military programs. Many semiconductor vendors are developing TAB in-house, and the best solution would be if they start selling die on tape for multichip circuits.

The substrate must also be fully tested before assembly, and this may require special design rules for testability. The current approach to testing is to use a sensitive probe to measure the capacitance at each pad. All pads in the same net should measure the same value. This checks for connectivity and will also show if nets are shorted together. This method may need to be supplemented by resistance testing.

Another approach being developed is to extend the E-beam technology currently used for scanning ICs to the wider field of view required for a substrate. Both of these approaches require post-processing of the CAD database to provide the test parameters. In short, CAD development is essential for improving the performance and producibility of future designs.

Packaging Density

How does packaging density relate to functionality and performance? A Hughes processor will be used as an example, since it can be compared directly in three different programs using different technologies. Table I shows performance history.

TABLE I. PERFORMANCE HISTORY

Program	Year	Clock Rate MHz$_3$	Relative Processing Power	Volume (Cu.Ft.)	Performance Unit Volume
A	1978	7	1	1	1
B	1986	10	2	1	2
C	1990	25	44	2	22

The increase in performance is a function of three factors: silicon density, packaging density, and clock rate. Programs B and C are compared in Table II, showing a packaging compression factor of 3.15 at the module level. Table II shows the potential functionality increase to be 18; the actual functionality increase is the performance/unit volume (22/2 = 11) plus the gates dedicated to "Built-in Test" and the

added Floating Point functions (approximately 35%): 11 x 1.35 = 14.85. The discrepancy between the potential and actual is 1.21 (18/14.85 = 1.21). Hence, 21% has been lost in relative packaging compression beyond the module; that is, connectors and backplane and thermal management. Note that the silicon area-to-substrate ratio is 29%, but the silicon-to-module area is 16%. Hence, there is a potential area gain of approximately two by going to Chip on Board. The volume gain is even greater by eliminating the package base and substrate thickness.

TABLE II. PERFORMANCE FACTORS

Module Density

Program	Gates/ Module	Sq. In./ Module	Gates/ Sq. In.	Ratio
B	75,000	45	1,667	1
C	300,000	25	12,000	7.2

Silicon Density

Program	Gates/ Chip	Sq. In./ Chip	Gates/ Sq. In.	Ratio
B	8,000	0.16	50,000	1
C	40,000	0.35	114,000	2.28

Packaging Compression: 7.2/2.28 = 3.15
Potential Functionality Increase: Silicon Factor (2.28) x Packaging Factor (3.15) x Clock Factor (2.5) = 18

Conclusion: The Density Gain

It has been shown that the volume gains from packaging are exceeding the gains from silicon, and there are further potential gains from packaging. The other performance factor, clock rate, is also going to be paced by packaging in the future. Having VHSIC II chips in the current packages will be like having a race car in freeway traffic. Packaging is going to be the key to overall performance and cost competitiveness.

Acknowledgments

This paper is based on substrate and package development supported by teams from Microelectronic Circuits Division, Radar Systems Group, and Hughes Microelectronics Center.

References

[1] W. E. Pence and J. P. Krusius, "The fundamental limits for electronic packaging and systems," *IEEE Trans. Components, Hybrids, Manufact. Technol.*, pp. 176-183, 1987.

[2] O. A. Palusinski, J. C. Liao, P. E. Teschan, J. L. Prince, and F. Quintero, "Electrical modeling of interconnections in multilayer packaging structures," *IEEE Trans. Components, Hybrids, Manufact. Technol.*, pp. 217-223, 1987.

[3] C. A. Neugebauer and R. O. Carlson, "Comparison of wafer scale integration with VLSI packaging approaches," *IEEE Trans. Components, Hybrids, Manufact. Technol.*, pp. 184-188, 1987.

[4] M. L. McConkey and H. B. Dussault, "Application of wafer scale integration to spaceborne signal processing," *Government Microcircuit Applications Conference (GOMAC) Digest of Papers*, pp. 509-513, 1987.

[5] G. Messner, "Cost-density analysis of interconnections," *IEEE Trans. Components, Hybrids, Manufact. Technol.*, pp. 143-189, 1987.

Appendix: Description of the HDMI Test Substrate

The High-Density Multichip Interconnect (HDMI) test substrate is designed to serve as a vehicle for substrate vendor evaluation. The design is based on five levels of metal interconnect with four interlayer dielectric layers of polyimide (see Figure 6). Staggered vias are used throughout, and there are no thermal vias. The design rules followed in the CAD layout are compatible with those of Raychem Corporation. This test substrate was designed to provide data on several potential problems in the fabrication and packaging of HDMI substrates. There are several groupings of test structures, each designed to answer different questions. A complete description of the test substrate, including test structures is given below.

Size

    Substrate: 1.8 by 3.8 inches
    Outer Lead Bonding Pads (Perimeter Pads):
        368 pads on 0.030 inch centers
    Pad Size (Perimeter Pads):
        0.010 wide by 0.020 inches long
    Layers, from the substrate up (see Figure 6):
        OPTIONAL POLYIMIDE LAYER ON SUBSTRATE FOR
            PLANARIZATION (NOT REQUIRED ON SILICON
            SUBSTRATE, REQUIRED ON ALUMINA)
        GROUND METAL LAYER
        VIA 1
        POWER METAL LAYER
        VIA 2
        SIGNAL 2 METAL LAYER
        VIA 3
        SIGNAL 1 METAL LAYER
        VIA 4
        PAD METAL LAYER -- DIE ATTACH AND
        INTERCONNECT
            BONDING PADS

Test Structure Description

The HDMI test substrate is comprised of the following separate test blocks.

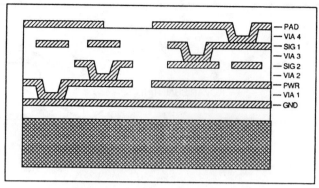

Figure 6.   HDMI Cross Section

Circuit Functional Block

A functional circuit consisting of four 32K gate array chips interconnected together and terminated on both an internal ring of probe pads (which make a pattern 1.5 by 1.5 inches square) and perimeter bonding pads (on the edges of the substrate). The Circuit Functional Block occupies approximately half of the HDMI test substrate. The gate array chips are HAC 1133 designs. The 1133 is a 32K base array custom metallized to function as eight 512-bit shift registers. The circuit is configured such that one, or all four, of the 1133 chips may be exercised. Data may be entered into the shift registers on one chip and retrieved, or clocked into the next chip and retrieved, through all four chips. The complete circuit thus has a maximum of 32 512-bit shift registers.

The purpose of this Circuit Functional Block is to characterize the electrical parameters of the test substrate in terms of circuit operating frequency limitations, power dissipation capability and thermal modeling, back annotation, and identification of pertinent physical and mechanical problems associated with both chip attachment and wire or TAB bonding from chip to the substrate outer lead bonding (OLB) pads. These OLB pads are sized to allow use of multiple wire bonds (in the case of repair of single defective wire bonds) or removal and replacement of an entire gate array chip. The 1133 is a HCMOS II gate array design with a 2-micron channel length capable of high operating frequency. The die size is 0.408 x 0.408 inch. The frequency may be continuously raised with a corresponding increase in power dissipation. There are no thermal vias under the chip attachment bonding pads. This represents a worst case for power dissipation. The use of thermal vias would thus be expected to result in much improved power dissipation properties.

Process Characterization Module

The Process Characterization Module consists of three test structures:

1. A serpentined metal pattern of Signal 1 metal over a serpentined metal pattern of Signal 2 metal. Both Signal 1 and Signal 2 serpentine patterns are composed of two individual conductors with a separate probe pad on each end of each conductor. This arrangement allows the determination of both intra-layer and inter-layer shorts and opens.

2. A via chain consisting of Signal 1 and Signal 2 metal with Via 3 vias between. This structure allows the measurement of continuity between signal levels and some measure of via

resistance. There are 76 Via 3 vias and two Via 4 vias to connect to the pad layer.

3. A capacitor consisting of two parallel plates of metal separated by one layer of polyimide dielectric. There are five versions of this capacitor structure, one for each of the four polyimide dielectric layers, and a fifth consisting of polyimide layers 2 and 3.

A single Process Characterization Module consists of one each of the three test structures described above. There are five versions, corresponding to the five capacitor dielectrics, with tests 1 and 2 the same for all. The Process Characterization Module tests terminate on 40 pads that are intended to be contacted using a probe card interfaced to a computer-driven parametric tester manufactured by Keithley.

There are both individual 40-pad modules and rows of 40-pad modules that are interconnected as long chains of the test structures. This is to allow comparison of small area structures with larger areas of the same structure. The intention is to use these structures to characterize defect densities useful in modeling yield. The size of these 40-pad Process Characterization Modules is 2400 by 2000 microns.

Frequency Characterization Structures

Microline and stripline conductors of minimum line width are included. A single long conductor spaced far from a set of four conductors of minimum pitch is provided to enable measurements to be made of characteristic impedance and crosstalk. These structures are on Signal 1 and Signal 2 metal layers, and there is a separate Signal 2 metal layer with Signal 1 metal layer in a covering sheet for the stripline set. There are microstrip conductors on the test substrate with internal pads, and on the PPTS-2 Circuit Functional Block with perimeter pads that can be bonded to the package.

Packaging Test Structures

A single long conductor of minimum width is provided that runs completely around the perimeter of the substrate, between the border (edge) and the perimeter wire-bond pads. This conductor is terminated on two of the perimeter pads and serves as a crack monitor for detecting substrate cracking at substrate-to-package bonding. This crack monitor works by measuring the continuity of the conductor before and after substrate attachment. An open conductor indicates that a crack has occurred that severed the conductor.

There is also a long serpentined single minimum-width conductor that runs halfway along the long axis of the substrate and terminates on two other perimeter pads. This serves as a stress monitor for the area of the substrate to be used for mounting a support post between package lid and substrate.

Other Structures

There is a set of lithography monitor lines of various pitch that are not terminated on perimeter pads, but have internal pads. There are two large die-attach pads with inner lead bonding pads spaced at minimum separation. These are sized to fit LLSI 10129 gate array chips and will be used to study the capabilities of attaching die and wire bonding at these dimensions. Rework of chip attach will also be investigated. The die size for these chip bonding pads is 0.587 by 0.587 inch.

Multichip Assembly With Flipped Integrated Circuits

K. Gail Heinen, Walter H. Schroen, Darvin R. Edwards, Arthur M. Wilson, Roger J. Stierman,
and Mike A. Lamson

Texas Instruments Inc., P.O. Box 655012, MS 477, Dallas, TX 75265

ABSTRACT

A multichip module process has been developed using flipped chip interconnections. The process employs plated copper bumps for superior thermal transport characteristics, active silicon as a substrate material for matched expansion properties, on-chip interconnection metallization that allows bumps to be placed over the active circuitry, and use of conventional wafer fabrication facilities for low cost production. Details of the fabrication process will be described and a case example given.

INTRODUCTION

Multichip assembled functions have become increasingly interesting to the system user to improve density, system performance, and/or reduce costs. To the semiconductor supplier, multichip assemblies may be of interest as a means to control or expand component market share through the ability to offer unique functions, or as a means of early market introduction of system components that cannot be readily integrated on silicon. Examples include blends of chips from various processing technologies into multichip systems whose monolithic integration might be years away. Multichip modules allow mixtures of analog, digital control, and digital signal processing technologies. In advanced systems, multichip modules can allow combinations of silicon and gallium arsenide circuits in the same application.

As the number of interconnections and chip I/O's increases, not only is board space consumed, but conventional printed wiring board (PWB) design and manufacturing technology is challenged. The need for very high density, efficient interconnection has resulted in the development of blends including circuit board, hybrid and integrated circuit technologies. Multichip modules can simplify these concerns by reducing board level complexity. Additional cost incentives for multichip packages may be realized through the decrease in system and board level design time made possible by forming many interconnections within a single package.

In an attempt to respond to customer needs while utilizing the technologies, resources, and available circuit designs available within Texas Instruments, the multichip packaging approach described here was developed. This interconnect technology employs flipped chips using plated copper bumps. The interconnect is formed by solder connection of the copper bumps to thin film metallized substrates having thermal expansion coefficients (TCE) matching that of silicon.

For successful design and fabrication of multichip assemblies, an organized methodology similar to that which has proved successful in design and assembly of single VLSI circuits has been used. This approach applies (a) computer aided modeling of the circuit and package for electrical, thermal, and mechanical simulation, (b) test chips for process development and failure mechanism testing, and (c) fabrication of actual demonstration circuits. Verification of function and reliability was then made through temperature cycle testing of (-65°C to 150°C), exposure to accelerated moisture environments, and measurement of heat dissipation properties. This approach and an example of its application to a multichip module that demonstrated successful performance on the first design pass will be discussed.

FLIP CHIP INTERCONNECTION FOR MULTICHIP MODULES

Comparative density and electrical performance analyses have been reported by a number of authors on the merits of interconnection technology for multichip packaging. They have agreed that for high performance systems, flip chip bonding has the greatest number of advantages [1]-[3].

However, the technology has not been widely accepted. The major concerns with flip chip interconnection are stress on the solder connection during temperature cycling, thermal dissipation of the assembly and lack of bumped chips. Several workers have proposed and are pursuing substrate materials with thermal expansions coefficients that match silicon to minimize stress on the solder joint [4], [5]. Silicon is an obvious choice of substrate materials; it also allows fabrication of active and passive components directly in the substrate. This may provide a solution to many multichip requirements, but costs must be considered. For small multichip modules, the active substrate and package cost may be readily justified. In considering larger substrates, yields and processing difficulty must be carefully analyzed for cost tradeoffs.

Features of the flipped chip interconnection technology to be described here are depicted in the cross sectional view of Fig. 1. From the top down, this includes the flipped chip, the barrier metal, the copper bumps, the solder connection to substrate, the signal lines, a layer of polyimide dielectric, a meshed ground plane, and the active silicon substrate. For final assembly, the module can be direct attached to a board as in chip-on-board technologies, mounted and wire bonded on a carrier, or assembled in a cavity-type package.

BUMP PROCESSING

Much has been published from work at IBM since the 1960's on flip chip assembly [6]. In the early work, solder bumps were formed on the chip surface and copper balls were embedded into the solder. This was replaced by solder bumps

Reprinted from *Proc. 39th Electron. Components Conf. (ECC)*, pp. 672–680, 1989.

where solder run out along metal traces was contained by the use of glass dams, thereby controlling the collapse of the solder bumps whose surface tension maintained proper offset between the chip and substrate. Other reports from Mitsubishi, AT&T and Hitachi suggest that solder is typically used for flip chip bonding. Ball formation may be by evaporation, sputter or plating [7]-[10].

An alternate approach, described here is electroplating copper bumps capped with sufficient solder to provide reflow bonding. Copper provides a superior thermal conduction path from the chip, as was shown by Totta [4], and thermal resistance can be greatly influenced by the geometry, number and location of bump interconnections. Other potential advantages of copper bumps are that the standoff between the chip and substrate is controlled by the bumps, and, since solder is not in direct contact with the barrier metallization, concern is eliminated for gold solubility in solder or copper / chromium intermixing requirements in the barrier metals as is needed with solder bumps.

A process flow diagram is given in Fig. 2. The steps of the process are as follows: 1) deposition and pattern of a polyimide stress buffer layer, 2) deposition of a barrier metal between the aluminum of the circuit bond pad and the copper of the bump, 3) resist deposition and patterning, 4) barrier metal etch, 5) copper flash of entire wafer, 6) thick resist application and pattern for bump plating, 7) plating of the copper bumps through the openings in the thick resist, 8) solder cap plating, 9) thick resist removal, and finally, 10) copper flash etch. Details of each step follow.

Chip passivations of silicon nitride, silicon dioxide or combinations of these can be sufficiently thick and defect free to tolerate the metallization and etching processes without yield loss. However, it was found that thermal stresses from the mass of copper bumps on the passivation caused passivation cracks at reflow temperatures. Therefore, the first step in the process was to apply a more compliant polyimide layer as a stress buffer if bumps were to be placed over the active circuits. This successfully prevented cracking of the brittle passivation.

A number of barrier metal combinations are suitable for covering the aluminum bond pads and other exposed metal areas on the chip prior to bump plating. Properties of concern are adhesion to the chip passivation and aluminum, stress, process latitude and selectivity, thickness and cost. Metals which provide adhesion are chromium, titanium, or titanium:tungsten (Ti:W). Some metals which act as barriers to the aluminum and which are inexpensive enough to apply thickly for good step coverage across the chip are nickel and copper. Typically, the exposed metallization surface is gold in an attempt to minimize corrosion and other reactions.

In this work, the metal scheme consisted of 1KÅ Cr, 5KÅ Cu and 5KÅ Au based on performance in accelerated moisture testing and on ease of processing. Metals were deposited by sputtering in a single vacuum pump down in order to avoid contamination and oxidation of the interfaces. Photo patterning with conventional negative photoresists and proximity printing was more than adequate. Chemical etching of the gold and copper was done with an ammonium iodide/iodine mixture and the chromium by caustic ferricyanide. The patterned multilayer metal overlap of the passivation was 13 microns (0.5 mils) on either side of the bond pad which acted to seal the aluminum against attack by the etches.

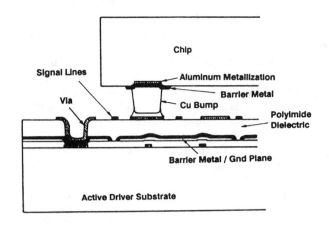

Fig. 1. Cross sectional depiction of the flipped chip assembly described in this paper.

Polyimide Deposition & Pattern
↓
Barrier Metal Deposition
↓
Barrier Metal Resist Deposition & Pattern
↓
Barrier Metal Etch
↓
Copper Flash
↓
Thick Resist Lamination & Pattern
↓
Copper Bump Plating
↓
Solder Cap Plating
↓
Thick Resist Strip
↓
Copper Flash Etch
↓
Cleanup

Fig. 2. Flow of bump plating process.

A plating electrode was formed by sputtering a thin film, 5KÅ to 10KÅ, of copper over the wafer. Vias for bump plating were provided by dry film photoresist. Dry film photoresist is typically available in 25 to 50 microns thickness, so, to provide the desired bump height of 85 to 100 microns, it was necessary to laminate two resist layers together.

Process development efforts showed that it was difficult to pattern and plate uniform bumps when the bump height was greater than the bump diameter. Spacing between bumps was also a concern; patterning accuracy suffered when the bump-to-bump space was less than the thickness of the resist. Therefore, a 1:1 rule was developed stating that the diameter of the bump was to be the same or greater than the height of the bump, and the bump-to-bump space was the same or greater than the bump diameter. 100 micron diameter bumps with 100 micron spaces were used when bump heights were on the order of 85 microns.

Bump planarity is defined as the height difference between the tallest and shortest bumps across a wafer. In order to achieve a planarity of +/- 0.5 mil (+/- 12.5 microns) across a 100 mm wafer, it was necessary to fabricate a plating fixture which allowed uniform wetting of the wafer and a uniform plating solution flow rate. An acid plating bath was used. The planarity of the copper bumps across a 100 mm wafer in a well controlled process is typically less than 10 microns, or +/- 5 microns, for 85 micron tall bumps.

The solder cap for reflow connection is plated immediately following the copper plating. Solder plating composition selected was a 10/90 tin/lead fluoborate bath because it is more readily controlled than lower lead ratio baths (such as 5/95 Sn/Pb) and yet allows for higher reflow temperatures than 60/40 tin/lead solder, an important property when considering later board assembly temperature profiles. While the fluoborate bath used does present significant waste disposal and automation concerns, the solder quality with respect to composition control and voids was superior to that of sulfamate baths. The solder is allowed to mushroom over the resist 13 microns. Cross sections of a bump before and after reflow are given in Fig. 3.

To complete the processing, the photoresist was removed, and the copper electrode was etched with dilute sulfuric acid/peroxide solutions. Oxidation, chemical attack and contamination of the solder requires very careful control at the etch and clean up processes.

DESIGN CONSIDERATIONS

Because existing chips designed for wire bonding with peripheral bond pads were used, it was necessary to relocate some bumps when the pads were too close to each other for proper patterning or too close to the scribe street for saw clearance. In addition, for optimum assembly yield, for reduced stress, and for improved thermal performance of the assembly, more than one bump per I/O was sometimes used. In such cases, the barrier metal layer was used to provide connection to the I/O's and to define bump pads at alternate locations closer to the chips's centers (Fig. 4). This resulted in bumps over active circuit areas. Using the barrier metallization as a signal layer with chips designed specifically for bump processing, it is possible to design chips around circuit constraints rather than pad constraints; that is, it is possible to place buffer circuits where they make the most sense from a circuit design standpoint rather than where they are required for periphery bond purposes.

A driving concern during development of a design rule set was to insure minimum yield loss due to the flip chip assembly process. This is an important criterion since yield loss due to bump processing and assembly occurs after most "value added" manufacturing has occurred. On chip design rules to maximize bump patterning and metallization interconnect yield were concerned with insuring good step coverage, designing for lowest interconnect resistance, maximizing processing yield, and, in the case of special driver buffers, insuring matched electrical characteristics from one lead to another lead. The first two concerns were addressed by the same

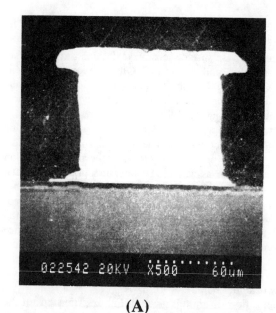

(A)

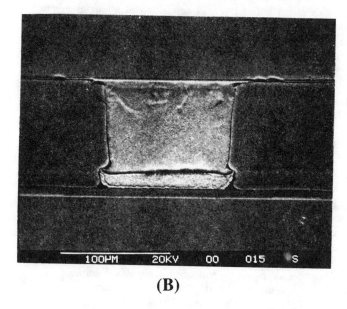

(B)

Fig. 3. Cross sections of a bump (A) before and (B) after solder reflow.

design technique; making the interconnect lines as wide as possible (>50 microns in width). Since bump interconnect density on the chip surface is typically low, the process tolerance and yield can be maximized by designing wide lead to lead spaces (>50 microns). The electrical criterion was addressed by matching the lengths, widths and spaces of the critical leads. Similar design rules were used for the interconnect of the substrates.

SUBSTRATE PROCESSING

Material selection for substrates was limited to those with thermal expansions coefficients approaching that of silicon, and with insulating or readily passivated surfaces to support conductor patterns. High thermal conductivity was desirable and other process related concerns such as flatness, smoothness, and strength were considered. Materials found suitable were silicon carbide, aluminum nitride, copper/invar/copper and silicon. Silicon was ideally suited for the process equipment available and all emphasis described here is limited to silicon. The silicon wafers were 100 mm diameter, polished, and oxidized. A minimum wafer thickness of 0.5 mm (20 mils) to avoid breakage during substrate processing was required.

Substrates fabricated with single level metal patterns used similar thin film processes to those described for chip barrier metals. The interconnect metal was, however, different from that used for the chips. Chromium was used as the adhesion metal on all substrates, copper of varying thickness provided high conductivity and the metal to be solder contacted was palladium. The number of metals compatible with solder reflow is somewhat limited; gold, copper, silver, nickel, platinum and palladium are readily identifiable. Copper and silver were eliminated on the basis of reactivity or migration potential, gold on cost and excess solubility in solder, platinum on cost, and nickel on being somewhat difficult to solder. Palladium is readily wet by solder, has limited solubility in solder, and is amenable to thin film processing.

For multilayer substrates, polyimide was the dielectric, and copper was the conductor sandwiched between chromium layers which provided adhesion. Aluminum was not considered in these applications because the module was not to be assembled in a hermetic package, and the reliability of thin aluminum conductors in polyimides may not be sufficient for non-hermetic applications [11].

To form the dielectric levels, PIQ Coupler-3 was spun on the substrates and cured. Polyimide was then spun on the substrates and soft baked. Photoresist was applied and soft baked, and the dielectric via pattern was exposed and developed. The vias were etched, and the substrates subjected to a partial imidization bake. Solvent cleanup after this step removed the resist, but not the partially-cured polyimide. After resist cleanup, the substrates completed polyimide curing at 400°C.

The substrate was designed with large alignment targets at the corners of each chip (Fig. 5). Targets consisted of "frames" formed by two wide metallization lines which fit the chip corners and allowed for both hand assembly and machine assisted assembly. Alignment is performed by matching the chip corners to the "frames". Accuracy of the alignment can be determined by visually sighting on the two lines. For accurate alignment, the lines should be of equal width. Offset alignment will result in one line appearing wider than the other at inspection. The positional accuracy of alignment using such a structure is a function of the sawing accuracy, typically on the order of +/- 13 microns. Therefore, the bump lands on the substrate were appropriately oversized to be 15-18 microns per side larger than the bump. Self aligning of the assembly during reflow was observed, showing these rules to be sufficient for good assembly yields.

ASSEMBLY PROCESS

It was necessary to flux the solder capped bumps before reflow assembly under all reflow conditions tested. A water white rosin flux in the form of a paste was found to improve process yield. The paste was spread onto a platen in a thin layer approximately 18 microns thick, and the solder bumped surface pressed into the layer. This ensured that only the

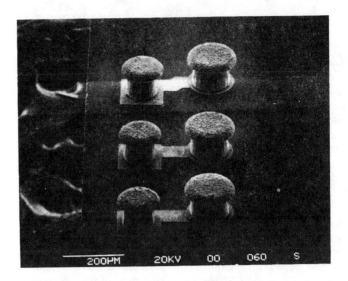

Fig. 4. S.E.M. photograph showing redundant bumps placed over active circuit area.

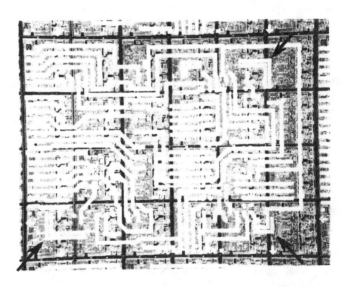

Fig. 5. Image of substrate interconnect pattern. Note the large split frame alignment marks to assist in manual chip alignment.

solder caps would be fluxed and prevented the flux from being transferred to the chip surface. The chip was then aligned on the substrate either manually or with the aid of split beam optics. The paste flux held the chips in place while they were hot chuck reflowed at 350°C in a nitrogen gas flow.

MODELING AND DESIGN VERIFICATION

Finite element modeling was used to simulate thermomechanical stresses resulting from process temperature excursions and to help define design rules. Macroscopic models demonstrated that since the substrate and chip are both rigid materials, stresses are localized in the bump connections and are concentrated near the interface areas. In order to provide the necessary resolution, mesh sizes were decreased in critical areas as shown in Fig. 6.

Simulations predicted that as the height of the bump was increased, stress at the silicon to copper interface was reduced. Further simulations studying solder thickness requirements showed that with thin solder caps (2.5 microns), stresses were particularly high near the silicon interface, but could be reduced by increasing the solder thickness (Fig. 7). These results were verified experimentally and a controlled solder thickness of 13 microns was selected.

Stress simulations predicted increasing levels of stress on bumps placed further from the geometric center of the chip. In design, corner locations were avoided or supported by redundant bump contacts.

A test chip and substrate combination was designed to test the proposed process and design rules (Fig. 8). Tests included continuity monitoring of daisy chains, leakage measurements between the chains, and four point probe of individual bumps [12]. The daisy chains were at differing distances from the chip center to monitor bump reliability under different stress conditions. Fig. 9 gives a brief schematic of the test configuration for this chip.

Table 1 shows temperature cycling results from this test chip using a variety of substrate materials. As expected, only substrates with expansion coefficients matched to that of silicon performed acceptably. Data from the four point probe measurements (Fig. 10) demonstrates that significant solder fatigue had not occurred. The data indicate that assemblies of flip chip devices on silicon may be subjected to temperature cycling over a range of -65°C to + 150°C.

Thermal modeling was also performed using Finite Element Analysis (FEA). The structure modeled is shown in Fig. 11. In this instance, the multichip module was mounted on a SIP carrier and various carrier materials were simulated. Two separate power dissipation conditions were investigated, one low power and one high power. The modeling assumptions were that junctions were uniform across the chips and that the thermal resistance through the chip metallization was minimal. Natural convection was assumed. In one instance, thermal vias (copper "plugs") were included in the carrier to increase heat transport to a heat sink on the back of the carrier. The heat sink was simulated by forcing the back of the carrier to 40C in all cases. The simulations demonstrated the effectiveness of bumps in heat transfer, but showed that the thermal properties of the carrier and of the material that attaches the module to the carrier control the system's heat dissipation (Table 2).

A thermal test chip used at Texas Instruments was modified to accept the bump process (Fig. 12). Two bump patterns were used to test the contribution of bumps to heat transport; one with 12 bumps, and one with 101 bumps. Heat is generated in this chip by passing current through low value (50 ohm) resistors. The temperature at the chip surface is then detected by passing a know current through a calibrated forward biased diode and measuring the voltage developed across the junction. Such a test can provide a very accurate measurement of junction temperature.

Table 3 shows the results for the 12 bump vs. 101 bump case on a silicon substrate. Note the small contribution to heat transport of the additional bumps when not strategically located. This data indicates the effectiveness of the copper bumps in transporting heat from the circuit.

The electrical modeling software used for this effort was a 2 dimensional capacitance/inductance calculator based on the numerical integration of the Green's function around the

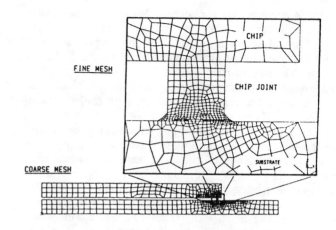

Fig. 6. Finite Element Analysis (FEA) mesh used to calculate stress on bumps. Notice grading of mesh size from coarse to fine.

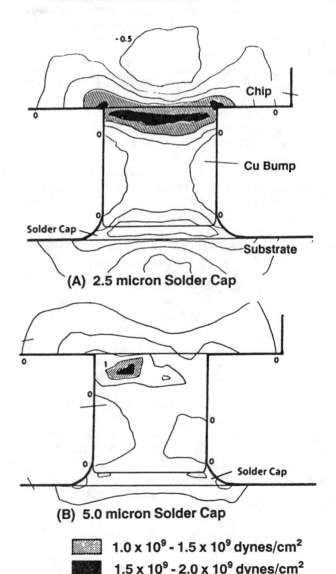

(A) 2.5 micron Solder Cap

(B) 5.0 micron Solder Cap

▨ 1.0 x 10⁹ - 1.5 x 10⁹ dynes/cm²
■ 1.5 x 10⁹ - 2.0 x 10⁹ dynes/cm²

Fig. 7. Modelled tensile stresses on reflow soldered Cu bumps. Two solder thicknesses were studied.

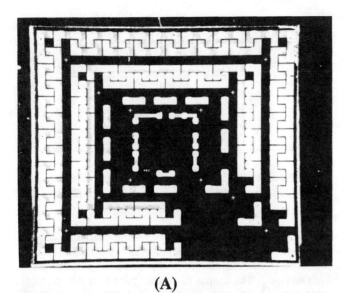

(A)

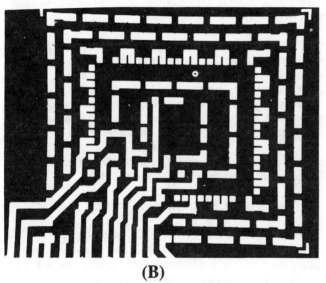

(B)

Fig. 8. Bump test chip (A) and substrate (B) used for bump process and substrate evaluation.

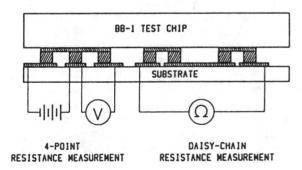

Fig. 9. Test schematics for daisy chain and 4-point probe of bumps on test chip of Fig. 8.

Substrate TCE Fails
-65°C to 150°C

Material	TCE (ppm/ C)	%Fail	@ cycles
Glass	10.0	100	@ 50
Alumina	6.5	100	@ 100
AlN	4.0	0	@ 1000
SiC	3.0	0	@ 600
TIKOTE	3.0	0	@ 1000
		9	@ 3000
Si	2.3	2	@ 1100
		3	@ 3000

Table 1.Temperature cycle results for bump test chip with a variety of substrate materials. (Note: TIKOTE is a SiC material with a graphite core.)

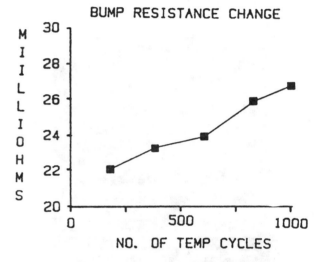

Fig. 10. Resistance increase of individual bumps as a function of temperature cycling (Four point probe measurement) (-65°C to 150°C).

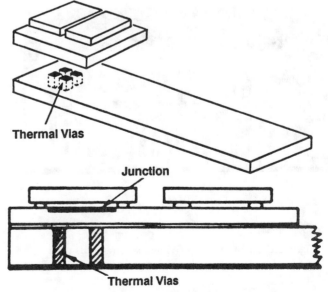

Fig 11. Outline of multichip module and module carrier used for thermal Finite Element Analysis (FEA) simulation. Note thermal vias.

135

FEA Thermal Simulations: Junction to Case Temps.
Multichip Module Carrier Material Matrix

Carrier Material	Module Power (W)	Max. Junct. Temp. (°C)	T_{jc} (°C)
Alumina	0.5	48.0	16.0
PC Board	0.5	85.0	90.0
PC Board (Cu Vias)	0.5	44.0	8.0
SiC	3.0	50.5	3.5
Alumina	3.0	79.3	13.1
PC Board	3.0	254.0	72.0
PC Board (Cu Vias)	3.0	53.0	6.5

Table 2. Thermal simulation results for multichip system. Two powers dissipations are used.

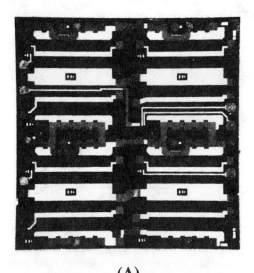

(A)

(B)

Fig. 12. Thermal test chip (A) and substrate (B) used for thermal analysis of bump process.

Thermal Test Chip Results (Silicon Substrate)

Airflow (LFPM)	Power (W)	T_{ja} (°C/W) 12	101 (# bumps)
0	2	30.6	27.3
0	6	30.2	26.0
0	10	28.2	24.9
250	6	18.9	14.3
500	6	16.0	11.5

Table 3. Junction to ambient (T_{ja}) temperatures measured with thermal test chip on silicon substrate.

periphery of each conductor to solve for the charge on neighboring leads. The potentials are preset in the problem, so the capacitance value is equivalent to the total charge on the conductor surfaces. The output of the program is in the form of capacitance and inductance matrices of the order of the lead count. The inductance is computed from the inverse of the capacitance matrix where all dielectric constants are set to unity. This is the common Transverse Electromagnetic mode approximation. The 2D modeling is repeated several times along the length of the conductor group of interest.

The capacitance, inductance and calculated resistance were transferred to a SPICE electrical simulation program where the electrical behavior such as cross talk and attenuation was determined based on typical input waveforms.

The dielectric thickness, metal thickness and width as determined by process capability were optimized for the application requirement based on electrical models.

DEMONSTRATION VEHICLE

In order to demonstrate the applicability of flip chip assembly to low cost multichip packaging, the integrated circuits required to fabricate a telecom subscriber line card system were selected. The requirements were 1) to provide customer system integration with cost reduction, 2) board space reduction, and 3) enhanced performance. Critical system level performance factors were gain, noise, longitudinal balance and transhybrid loss.

Texas Instruments chips selected to fabricate the transformer line card function included a DS3680 line driver, TCM 4209 SLCC performing the combined functions of supervisor control, hybrid and attenuation and a TCM 2914 COMBO circuit performing the functions of CODEC and FILTER. In the demonstration module, no discrete components were included on the module itself, but were instead placed on the system board in conventional fashion.

The driver circuit was used to form the silicon substrate. Since a single driver chip was too small to support the SLCC and COMBO chips to be flipped on its surface, a matrix of 4 x 6 driver chips was used, giving a substrate dimension of 8 x 10.5 mm (Fig. 5). To avoid mask redesign, a single new passivation mask was made which opened pads only on the one driver circuit to be used and which patterned a new scribe line.

Although the system clock rates (2 MHz) and the analog frequencies of this application were relatively low, a stringent cross talk specification of -40dB was required for certain conductors. Loading capacitance was also required to be less than 10pF for various output lines. Design techniques involving electrical modeling and simulation were therefore required in order to insure a workable layout.

Several features in the design were manipulated to meet the maximum cross talk criteria. A split ground plane for the analog and digital sections was added to the substrate to reduce the coupling between analog and digital signals. Grounded conductor strips were placed adjacent to sensitive lines on the substrate to increase shielding. Certain leads were positioned at maximum distances from noise sources on both the substrate and the external package lead structure. After each design change, electrical modeling and simulation programs were run to determine the improvement in operation.

Special features of this assembly were a meshed ground plane to minimize stresses in the metal and polyimide and a large number of redundant bumps on both flip chips to optimize strength, thermal dissipation and yield through double electrical contact.

Each circuit was multiprobe tested prior to assembly. The bumped wafer yields were similar to conventional non-bumped chips processed at that time. The substrates were tested for functionality of the driver circuit, continuity on major lines, and leakage between metal ground and interconnect layers. The first pass yield on the substrate wafers was 96%.

System functionality was tested twice. First, the reflowed multichip substrates were multiprobe tested using a system test program that combined functional, parametric and timing interactions of the three circuits after flip chip assembly. The substrates were then attached to a ceramic carrier using silver filled conductive epoxy, wire bonded, pins attached to form a SIP configuration (Fig. 13) and tested for system functionality a second time. System performance of the modules was compared to both DIP and SIOC plastic packaged components assembled on a subscriber line function board. System performance of the multichip module was found to equal or exceeded the performance of the system assembled with individually packaged units in all tests.

To better meet the needs of standard assembly handling equipment, the system was shrunk to allow the substrate to be placed in a 44 PLCC plastic cavity package. An earlier demonstration of single level metal substrate with three flipped chips had demonstrated reliability in excess of conventional IC requirements (Fig. 14).

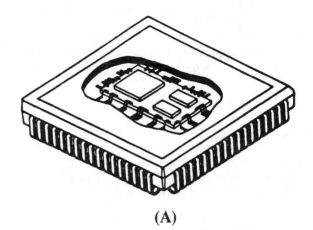

(A)

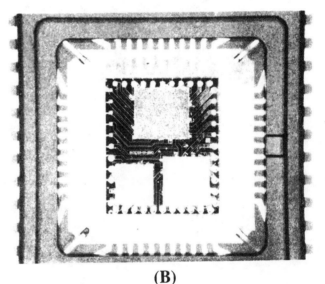

(B)

Fig. 14. Multichip system in 44 PLCC plastic package. The diagram (A) shows a cut-away overview of the final package. The photograph (B) depicts the multichip module in the cavity package after wire bond but before lid attach. The multichip module shown here is not the line card demonstration vehicle, but was an earlier module forming a telephone function.

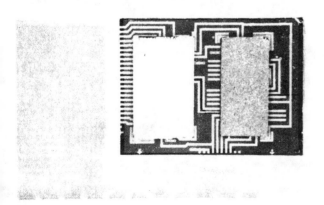

Fig. 13. Line card demonstration module on ceramic SIP.

CONCLUSIONS

A multichip fabrication process has been developed which allows assembly of multiple IC's in a single package component. The process offers three major advantages; 1) use of active silicon material as a substrate, 2) placement of bumps at any location on the chips to be flipped, and 3) high thermal dissipation through use of copper plated bumps. In addition, the process is amenable to standard IC fabrication techniques including masking, spin resist deposition and development, and wafer handling.

A demonstration vehicle has been described which used the developed process to fabricate a multichip assembly that is smaller in size and with better electrical performance than its counterpart assembled with conventionally packaged IC components.

137

The development of adequate fabrication equipment, such as precision pick and place machines, will be necessary for volume manufacturing and mass marketing of multichip modules. Characteristics of such equipment will include fast, high precision alignment of the bumped chips to the substrate and the ability to handle wafers diced on flex-frames.

ACKNOWLEDGEMENTS

The authors would like to gratefully acknowledge the contributions of Steve Groothuis and Jesus Martinez for the preparation and analysis of the FEA results. In addition, Brad McCarty was indispensable for his skills with processing of the material used for these evaluations.

BIBLIOGRAPHY

[1] C. A. Neugebauer and R. O. Carlson, NEPCON, 1987, p. 227.

[2] Charles A. Bartlett, John M. Segelken, Nicholas A. Tenketges, IEEE - 37th ECC, 1987, p. 518.

[3] Capt. B. J. Donlan, J. F. McDonald, R. H. Steinvorth, M. K. Dhodhi, G. F. Taylor, and A. S. Bergendahl, VLSI Systems Design, January 1986, p. 54.

[4] P. A. Totta, IEEE - 21st ECC, 1971, p. 275.

[5] Clinton C. Chao, Kenneth D. Scholz, Jaques Leibovitz, Maria L. Cobarruvias, and Cheng C. Chang, IEEE - 38th ECC, 1988, p. 276.

[6] L. S. Goldman and P. A. Totta, Solid State Technology, June 1983, p. 91.

[7] Masanobu Kohara, Shin Nakao, Kazuhito Tsutsumi, Hiroshi Shibata, and Hidefumi Nakata, IEEE - 35th ECC, May 1985, p. 180.

[8] Kazuhito Tsutsumi, Masanobu Kohara, Hiroshi Shibata, and Hidefumi Nakata, IJHM, December 1984, p. 38.

[9] Dr. T. S. Liu, W. R. Rodrigues de Miranda, P. R. Zipperlin, Solid State Technology, March 1980, p. 71.

[10] Takeo Yamada, Kanji Otsuka, Ken Okutani, and Kunizo Sahara, Proc. 5th Ann. IEPS, 1985, p. 551.

[11] A. M. Wilson, Proc. 3rd Intl. Conf. on Polyimides, SPE, Ellenville, NY, November 1988, to be published.

[12] R. A. Young, W. R. Rodrigues de Miranda, IEEE-IRPS, 1982, p. 194.

A SERIES OF DEMONSTRATORS TO ASSESS TECHNOLOGIES FOR SILICON HYBRID MULTICHIP MODULES

L C Early

Mars Electronics
on Secondment to the RISH Project
RSRE, St Andrews Road, Malvern Worcestershire, WR14 3PS, UK

Abstract

In order to exploit the full advantages of today's high density, high speed and high I/O count VLSI devices, an advanced interconnect technology is required. One such technology is that of silicon hybrid multi-chip modules (MCMs). This paper will illustrate the advantages of using silicon as a hybrid substrate through a series of process and functional demonstrators. These demonstrators also enable the various technologies available for chip attach and substrate interconnect fabrication to be investigated and these will be discussed.

Introduction

The concept of using silicon as a hybrid substrate is not new, indeed the earliest reported work[1] was in the late 1960s. However, the rapid advances in monolithic integration through the 1970s and 80s satisfied the existing demands for system integration. Now, as chip size increases and I/O counts in the range 200-300, are not unusual, single chip packaging is becoming prohibitively expensive and wasteful of circuit board area. Consequently, considerable effort is being directed toward the multi-chip module approach, using advanced substrates to interconnect bare dice. This approach reduces packaging overheads, reducing interconnect delay time and improves reliability by the removal of a level of interconnect hierarchy. Silicon, as a substrate, has some additional advantages for multi-chip module applications: no expansion mismatch between substrate and Si chip, resulting in improved reliability; a well established technology database utilising tried and tested processes for fine line interconnect; and the possibility of including active devices in the substrate. The degree of interest is illustrated by the increasingly large number of technical publications on this subject[2,3].

Technologies Available for Silicon Substrate Based MCMs

These split into two main areas, those concerned with interconnect fabrication and those concerned with chip attach.

The properties required for the interconnect include low dielectric constant dielectrics for high speed signal propagation, coupled with good planarising properties. Most interconnect systems require between 2-4 layers of relatively thick metallisation (approx 2-3 um) to reduce line resistance, and this has resulted in the substantial use of organic dielectrics, typically polyimide, with either aluminium/silicon metallisation (as used in conventional IC processing) or alternatively copper based systems. However, copper based systems can require additional processing, normally using adhesion and barrier metallisation layers, if used with polyimide dielectrics. Copper based interconnect has the advantage of lower resistivity enabling both finer linewidths and larger substrates.

Chip attachment and connection splits into three main areas, epoxy or similar die attach with conventional wirebonding, face-up or face-down tape automated bonding (TAB), and peripheral or area array solder reflow flip-chip. The choice of technologies to be employed for chip attach is somewhat application specific, with no one technology having overall superiority. Wirebonding is a very well established, flexible, technology and is suitable for a large number of applications where the area overhead is not so critical and the costs and time involved in either post-processing wafers for solder reflow flip-chip attach, or commissioning custom tapes for TAB are prohibitive. For high speed applications, solder reflow flip-chip may be the optimum choice due to the very much reduced lead parasitics. Alternatively, high pin count VLSI devices might lead to the use of fine pitch tape automated bonding or perhaps solder reflow flip-chip if packing density is also important.

Process Evaluation Vehicles

Interconnect Technologies

A test substrate has been designed which can be fabricated with up to four levels of metallisation (see Figure 1). There are various structures which enable a given technology to be evaluated. These include simple metal resistivity meanders and multiple level via chains and capacitors, along with more complex line coupling test structures in a variety of forms. Also, there are devices for assessing planarity and resolution achieved using a particular process.

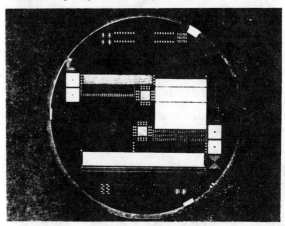

Figure (1)   Substrate Interconnect Evaluation

To date, these test substrates have been fabricated with both aluminium and copper metallisation systems, using polyimide dielectrics. Via chains have been successfully fabricated in the range 10 to 40 um, with successful resolution of tracks down to 5 um. Substantial work has been undertaken to optimise via formation[4] using a variety of techniques employing Reactive Ion Etching.

Reprinted from *Proc. 39th Electron. Components Conf. (ECC)*, pp. 557–561, 1989.

Work is currently under way to confirm electrical simulation results obtained for characteristic impedance and crosstalk.

Chip Connect and Attach Technologies

As previously mentioned, this program is evaluating three main methods of chip attach, conventional wirebond, tape automated bonding and solder reflow flip-chip. A number of test modules have been designed to assess the reliability and capabilities of these techniques.

Solder Reflow 'Flip-Chip'

This technology has several advantages over the alternatives. Area array bonding is possible, thus providing a method for interconnecting high I/O count devices, with improved interconnect performance due to reduced parasitics. There also exists the possibility of renewing and replacing a non-functional device. Technology input[5] has enabled the development of a process capable of post-processing 'off the shelf' and custom parts and the assembly of high density flip-chip modules (Figure 2).

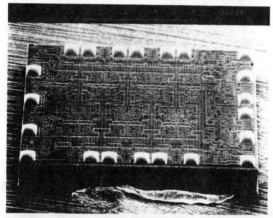

Figure (2) Post-Processing Capability

To assess reliability and investigate a variety of bond geometries, a series of continuity modules have been fabricated and evaluated. These range from 240 um bonds on a 300 um pitch to 40 um bonds on a 100 um pitch in both perimeter and array configuration. Continuous chains of >570 bonds using 90 um bonds on a 200 um pitch have been successfully fabricated. Figure 3 shows a typical module. A selection of these parts are currently undergoing reliability trials.

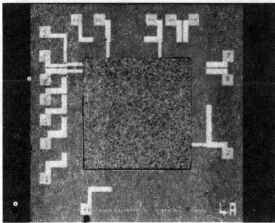

Figure (3) Flip-Chip Continuity Module

Design Environment

Advanced interconnect technologies such as have been described require a similarly comprehensive CAE system. Consequently, in parallel with process development, RISH is developing an integrated CAE system. This has been achieved by linking standard PCB layout software with a full custom IC layout suite via in-house written software. In this way, complex circuits can be autorouted on many layers and then translated into a full custom IC layout environment where layout parasitics can be extracted along with running standard design rule checks and mask pattern generation software. A methodology has been developed for creating substrate footprints for components to be attached by either flip-chip or wirebond connection techniques from a single source file. Routing problems can occur because the router is configured to handle parts with standard pinouts eg 0.1" pitch DIL packages, whereas in this application very few ICs have commonality of I/O positions, with varying chip size, pad size and pad pitch. Thus attention needs to be paid to the autorouter set-up to achieve optimum results. This system has been developed in conjunction with the design and fabrication of a number of functional demonstrator vehicles which are silicon hybrid implementations of existing designs.

A typical process cross-section is illustrated in Figure 4, showing in this case wirebond chip attach/connection. An integral decoupling capacitor is formed between the power and ground layers with automatic connections formed to the supply planes as autorouting is performed on the two signal layers.

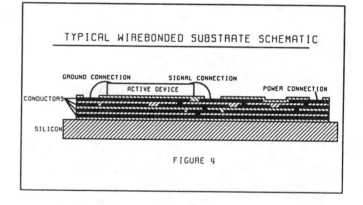

TYPICAL WIREBONDED SUBSTRATE SCHEMATIC

GROUND CONNECTION   SIGNAL CONNECTION

ACTIVE DEVICE   POWER CONNECTION

CONDUCTORS

SILICON

FIGURE 4

Typically, tracking is 25 um wide on a 75 um pitch, with line capacitance varying between 0.5 - 2 pf/cm depending on the routing layer, line resistance is approximately 6Ω/cm. A design route has also been established to provide a post-processing capability for 'off the shelf' semiconductor devices to be solder bonded. Accurate bond pad information is derived from measurement and processed to enable mask generation.

140

Functional Demonstrators

In order to understand the problems encountered with real designs a series of 'real' demonstrators have been designed and fabricated. The first RISH demonstrator vehicle implemented a MIL STD 1553 multiplexed databus remote terminal and bus controller on a two chip module. The device was fabricated with both wirebond and solder reflow flip-chip technologies to establish an in-house capability, although in this instance no effort was made to utilise the superior packing density of the flip-chip technology and fairly conservative design rules were used (Figure 5). The device was placed on a ceramic carrier to facilitate I/O connection - note this is not an optimised package. The ratio of active silicon area to overall package area gives a useful indication of substrate/packaging overhead. This circuit was originally fabricated with surface mounted devices on a co-fired ceramic substrate, and the active silicon area represented some 5% of the total device area. This may be compared with 12% for the silicon hybrid in the non-optimised ceramic carrier or 35% if one considers just the substrate.

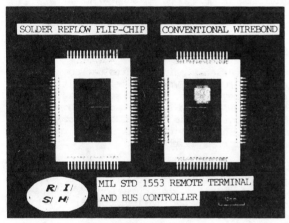

(Figure 5)

There is an ever-increasing requirement for complex microprocessors coupled with high speed, high density memory and the second RISH demonstrator has integrated the Inmos T800 32 bit floating point transputer with 32K bytes of DRAM. Also included on this hybrid are surface mount passive components for decoupling and for the phase locked loop on the transputer (Figure 6).

Figure (6)

The final substrate dimensions (23 x 21.5 mm) show between three and four times reduction in module size over a conventional PCB implementation. It is interesting to note that a single packaged T800 transputer is larger than the more complex multi-chip module.

The third RISH demonstrator vehicle was aimed at integrating a larger number of chips, thus providing an increase in complexity from the previous two devices. This module is a superset of the first RISH functional demonstrator namely the MIL STD 1553 remote terminal and bus controller with the addition of a VME sub-system interface. This increases the chip count to ten, with the addition of two SRAMs, two more ASICs and four line drivers.

This device has been fabricated with chips connected by conventional wirebonds and also by solder reflow in a flip-chip mode. Figure 7 illustrates the two substrate layouts placed adjacent to each other. The substrate for flip-chip bonding has peripheral fan out tracks to enable placement into the same package as the wirebond version.

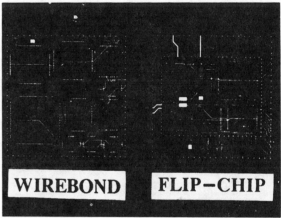

Figure (7)   Substrate Comparison

A standard commercially available package was chosen, a 92 pin 0.05" pitch Kovar flat pack, and the substrate for the wirebond version was routed to fit this rather than be as closely packed as the transputer module. However, the substrate for the flip-chip bonded version has been laid out to fully utilise the increased packaging density of flip-chip bonded devices.

A comparison can be made between the existing co-fired ceramic version of this circuit, and the silicon hybrid implementation. Figure 8 shows the wirebonded version and the original implementation. The surface mount co-fired ceramic version is split into two sections, but the chip count is not directly equivalent to the MCM version since there has been some integration of the smaller devices into VLSI circuits and gate arrays. Analysing the ratio of active silicon area to packaged substrate area we have 5% for the existing ceramic implementation, 25% for the wirebond silicon substrate MCM, and 53% for the flip-chip solder bonded MCM. (This assumes the flip-chip bonded substrate is packaged without the 'extender tracking' and in a suitable package). Additionally, considering the flip-chip bonded substrate alone, the maximum achievable ratio of active silicon area to substrate is 75% with the standard RISH chip spacing rules and this design achieves approximately 65%.

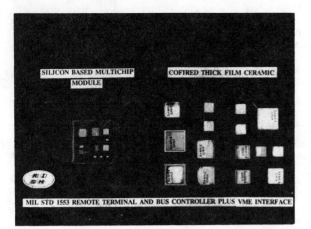

Figure (8)

As stated in the introduction, one unique advantage that silicon has as a hybrid substrate is the capability of integrating active devices in the substrate. To investigate this concept and establish compatability of the RISH multi-level interconnect with common IC processing technology, a test vehicle has been fabricated using a wafer of Dynamic RAM as the substrate, post processing with four layers of interconnect and then attaching a T800 transputer by solder reflow flip-chip bonding (Figure 9). This device has the same configuration as the previous transputer module but is fabricated on an array of eight DRAM chips

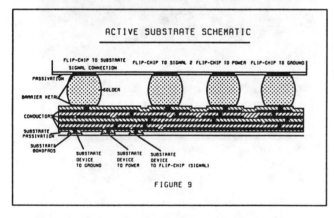

of which four are selected by wirebonding. This allows up to 50% redundancy of the active devices in the substrate. Figure 10 shows a post-processed wafer. This module occupies some 37% of the area taken up by the conventional wirebond version previously described.

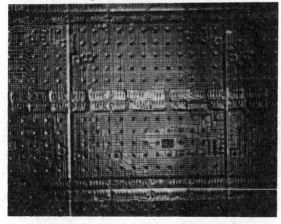

Figure (10)   Post-Processed DRAM Wafer

Future Work

Thermal Considerations

As MCMs reduce in size, their thermal performance requires careful consideration due to the increased power density. RISH are currently fabricating a test vehicle to investigate the thermal characteristics of Si based MCMs. A substrate with chips attached by solder reflow flip-chip and conventional wirebonding will be assessed and compared with the results from thermal modelling. Some chips will be able to dissipate up to 25 watts and there are 16 chips per substrate. Power densities of up to 25 W/cm^2 can be set up to evaluate the effects of power cycling on bond continuity. This module will be used to test a high performance package designed specifically for this application.

High Interconnect Density

None of the functional modules previously described have demonstrated the capability of silicon to handle very high density interconnect. A future demonstration vehicle, a digital switch module used in telephone exchanges, will illustrate the benefits of the fine line interconnect achievable on silicon-based MCMs.

Tape Automated Bonding (TAB)

TAB is increasingly being used to package fine pitch high I/O count VLSI devices. Additionally, it enables devices to be pre-tested after inner lead bonding (ILB) prior to outer lead bonding (OLB). Also the leads have better interconnect performance in terms of capacitance and inductance than conventional wirebonding.

TAB can be employed in a 'face-up' (active face of chip away from substrate) or alternatively a 'face-down' or 'flip' configuration. The face-down method requires a smaller footprint, and can use straight leads which do not require forming. RISH are currently investigating both techniques.

Figure 11 shows a 50 x 50 mm single layer substrate which will contain nine face-up TAB devices. These have varying OLB pitches from 0.007" to 0.01" and are connected in a serial continuity chain, shorts between adjacent bonds and open circuits will be detectable on any row of leads on any chip. All inner lead bonds will be pre-tested. Both solder reflow and gold thermocompression OLB will be used and evaluated.

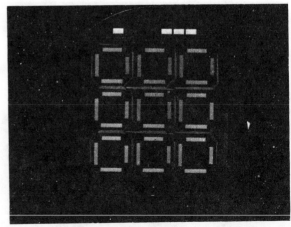

Figure (11)   Face-Up TAB Substrate

Figure 12 shows a second 50 x 50 mm substrate which will contain nine face-down TAB devices. These devices have straight leads on a 0.004" pitch, with 90 bonds/edge resulting in some 3240 bonds per substrate. Gold thermocompression will be employed for OLB.

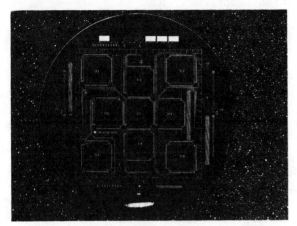

Figure (12)   Face-Down TAB Substrate

Both sets of substrates are currently being assembled and, after packaging, will be thermally cycled and reliability tested.

Conclusions

A series of MCM demonstrators fabricated on silicon substrates have been described. The advantages of the various interconnect methods and chip attach techniques have been discussed. However, straight translation of existing designs on to silicon hybrid MCMs, whilst reducing device dimensions does not fully utilise the technology. To gain maximum advantage, the system components require re-designing in view of the fact that the component output buffers are no longer required to drive the large capacitive loads normally associated with conventional packaging and assembly onto PCBs. This greatly reduces the power consumption of the individual chips and consequently of the module as a whole. There is also the option of distributing the component I/O over the whole of chip surface if using flip-chip solder attach thus reducing chip size if dictated by I/O count. Fully integrated systems could be designed with suitable components in the substrate (high power devices, passive components) and a mixture of VLSI devices and possibly high frequency GaAs parts mounted on the surface.

Acknowledgments

This paper reports on the work by the RISH team of which the author is a member.

References

[1]   P Kraynak and P Fletcher, Wafer Scale Assembly for LSI, IEEE Transactions on Electron Devices, Vol: ED15, No 9, Sept 1968.

[2]   C J Bartlett, J M Segelken and N A Teneketges, Multi-Chip Packaging Design for VLSI Based Systems, Proceedings 37th ECC, 1987.

[3]   J K Hagge, Ultra-Reliable Packaging for Silicon-on-Silicon WSI, Proceedings 38th ECC, 1988.

[4]   M J Rutter, Via Formation in Thick Polyimide Layers for Silicon Hybrid Multi-Chip Modules, Proceedings 1989 ASM Electronic Materials and Processing Congress.

[5]   A G Munns, Flip-Chip Solder Bonding for Microelectronic Applications, Metals and Materials, Jan 1989. (The Journal of the British Institute of Metals).

HIGH-DENSITY MULTILAYER INTERCONNECTION WITH PHOTO-SENSITIVE POLYIMIDE DIELECTRIC AND ELECTROPLATING CONDUCTOR

Kunio Moriya, Takaaki Ohsaki and Kohsuke Katsura

Musashino Electrical Communication Laboratory, NTT.

Musashino-shi, Tokyo 180, Japan

ABSTRACT

Advanced LSI devices will require increased performance from electronic packages. We have developed high density multilayer interconnection with a photo-sensitive polyimide dielectric and an electroplating copper conductor to meet these needs. The main accomplishment has been the development and application of a selective electroplating and photolithographic process to achieve a fine copper conductor and a small via hole through the relatively thick photo-sensitive polyimide for high density interconnection. In via hole formation, the problems of an over-hanging configuration and an increase in via contact resistance had been encountered. These problems have been solved by a simple chemical treatment and a thin titanium film coated on the lower copper layer. The minimum conductor and via hole achieved are 3μm wide and 15μm in diameter through the 10μm thick polyimide layer. The results of evaluation tests of a four-layer interconnection indicate that they are all reliable and stable.

INTRODUCTION

Advances currently being made in LSI circuit density and speed are necessitating the development of high-performance wiring for LSI chip interconnetion; greater wiring density and smaller signal propogation delay are required to support high speed logic chips.

Multilayer interconnection technology has been a key to success in microelectronic packaging. Recent developments in the fabrication of multilayer interconnection with a relatively thick polyimide dielectric layer have expanded into high speed LSI chip packaging (1)(2)(3). Although many fabrications have been attempted, no approach with a photo-sensitive polyimide dielectric layer made the formation of high-density multilayer interconnection possible.

This paper describes multilayer inter-connection technologies involved in the formation on a substrate of both fine patterns with electroplating copper and small via holes with photo-sensitive polyimide dielectric layers. Photo-sensitive prepolymer polyimide was employed because the required interlayer could be achieved by spin coating, and the via hole opening could be formed by merely applying exposure and development processes (4)(5). This technology thus results in low cost fabrication of high-density multilayer interconnections because of the short process turnaround. Also, signal propagation delay is minimized because of a low dielectric constant and low conductor resistance. This paper will cover the following :

1. Details of fabrication
   A. Fine line conductor formation
   B. Dielectric layer and via hole formation
   C. Electrical via connection
2. Application
   A. Multilayer construction
   B. Tests and evaluations
3. Conclusions

DETAILS OF FABRICATION

Polyimides as an interlayer dielectric have a low dielectric constant in cured form and possess leveling and planarizing characteristics. Furthermore, photo-sensitive polyimides have the advantage of shortening the interlayer patterning process. Copper as a conductor pattern is likely to be the most practical material in terms of cost, reliability and resistivity.

The main process of multilayer fabrication compared with conventional methods using polyimide dielectric and sputtered conductor is shown in Figure 1.

(1) A thin chromium/copper film is sputtered and a copper conductor is electroformed by additive plating with photoresist. Base metallization is then etched off.

(2) A photo-sensitive polyimide is spincoated and prebaked.

(3) Small holes are easily formed by the same process as the usual photoresist technologies.

(4) Thermal curing of the photo-sensitive polyimide turns it into a polyimide end product.

(5) Base metallization for both the second conductor layer and via connection is accomplished in a sputtering system, and the

Reprinted from *Proc. 34th Electron. Components Conf. (ECC)*, pp. 82–87, 1984.

copper conductor is electroplated selectively. After this, the base metallization layer is etched off.

Use of photo-sensitive polyimide as the interlayer thus results in a simplified fabrication process.

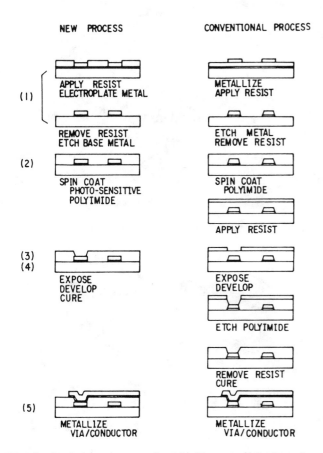

Fig.1. Main process of multilayer fabrication (in cross section).

For multilayer packaging, the required thickness of the interlayer is quite different from that when it is used for fabrication of multilevel devices such as LSI chips; it is needed to be on the order of 10μm or more. Therefore, small via hole fabrication through the thick photo-sensitive polyimides becomes extremely difficult. The fabrication process has been investigated accordingly. A high-density conductor is also required for multilayer packaging, then fine pattern formation of the copper conductor with a vertically square cross section has been accomplished.

This section describes the high-density multilayer interconnection process with photo-sensitive polyimide dielectric and electroplating copper conductor.

A. Fine Line Conductor Formation

Fine geometory conductor patterns corresponding to increases in wiring capability must be developed to densely interconnect chips. Attainment of minimum voltage drops and signal propagation delay calls for high aspect ratio (thickness / width). This can best be accomplished by using selective electroplating technology with high conductance copper. The metallization process used was a combination of sputtering and electroplating. Thin chromium followed by copper conductors were sputtered on a substrate (or a polyimide plane) for a plating base prior to the resist coating. The above several micron meters of copper can then be achieved by selective electroplating.

Since the photoresist patterns are applied for selective electroplating, the steepness of the edge of the patterns is quite important. We investigated thick photoresist pattern formation (several micron meters) with a vertical sidewall . To achieve fine line resolution and vertical sidewall definition, much attention to the developing and baking processes was given. Consequently, we found that AZ 4620 photoresist was suitable for fabricating thick selective electroplating patterns with a vertical side wall. Exposure energy of over 260 mJ/cm² was needed to obtain complete exposure of the thicker resist and vertical side wall profile. The postbake temperature of over 110 °C lessened the steepness of the wall edges as shown in Figure 2 (a). A 10μm thick photoresist pattern after postbaking at 90 °C is shown in Figure 2 (b). Very sharp edges and a high degree of uniformity were obtained.

(a) after postbake at 120°C.

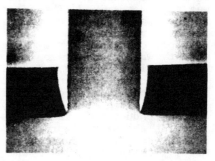

(b) after postbake at 90°C.

Fig.2. Influence of the postbake temperature on the photoresist vertical side wall profile.

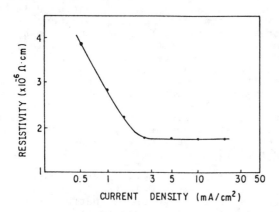

Fig.3. Relationship between current density and copper resistivity.

 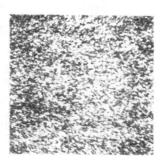

(a) electroplated at 5 mA/cm².

(b) electroplated at 35 mA/cm².

Fig.4. Influence of the current density of electroplating on the conductor surface structure.

To obtain low resistivity in an electroplated copper conductor, the effect of current density on the resistivity was determined using copper sulfate solution. Figure 3 shows that a current density above 30 mA/cm² results in excellent electrical resistivity (1.87 X 10Ω-cm) ; it is near bulk for copper. The effect of the current density on the structure of the electroplated deposit was also determined. The specimen plated at the lowest current density led to a large nodular structure as shown in Figure 4 (a). The sample obtained at the highest current density was found not to have a square cross section geometry. A density of 35 mA/cm² was selected to achieve smooth uniform surface and square cross section (see Figure 4-b). Utilizing this processing method, copper conductor formation was accomplished. As shown in Figure 5, conductor patterns with vertically square cross section were formed ; line width of 10μm and thickness of 5μm were attained. Conductor resistance of 3Ω/cm, conductor adhesion strength of 600 kg/cm² or more were realized.

Consequently, conductors with maximum conductor thickness of 10μm, and minimum conductor line width of 3μm shown in Figure 6 were able to be fabricated with this process.

Fig.5. SEM photo of 5μm thick, 10μm wide electroplating copper patterns.

(a) 10μm thick, 10μm wide.

(b) 3μm thick, 3μm wide.

Fig.6. Maximum thickness and minimum line width of copper conductors fabricated.

B. Dielectric Layer and Via Hole Formation

Polyimide dielectric layer thickness on the order of 5μm in cured form could be achieved with one spin coating by applying a higher viscosity polyimide and slowing spin speed to 2000 rpm for 20 seconds. A good leveling of the multilayer structure, which was the capability of covering the under layer irregularity, was obtained by applying two or more spin coats.

With regard to the via hole formation, the photo-sensitive polyimides allowed a direct production of patterns by conventional photolithographic techniques. The photo-sensitive polyimide used was negative working photopolymer. Via hole patterns were generated by ultraviolet (UV) exposure through a photomask, and then the unexposed parts could be chemically dissolved. Careful temperature control of the photo-sensitive polyimide before and after the photolithography process was required to obtain a good via hole configuration. In the case of the dielectric layer thickness on the order of 10μm , two spin coating, prebaking at 80 °C for 30 minutes with a hotplate and 780 mJ/cm² exposure energy were required to achieve this thickness. A cure process carried out on a hotplate is superior to that carried out in an oven. Hard curing was accomplished at 350°C in an N₂ atmosphere for 30 minutes.

Problems in patterning the via hole with a small tapered wall profile had been encountered by the experiments of thick photo-sensitive layer ; a somewhat overhanging wall profile had been observed as shown in Figure 7. Problems encountered were thought to be the result of UV absorption at the bottom being less than that at the top layer ; this in turn resulted in higher soluability at the bottom of the thick photo-sensitive layer.

Fig.7. Cross section of overhanging via wall profile.

Attempts were made at improving the overhanging by changing exposure, development and baking processing. Longer post baking resulted in a small improvement, but it was insufficient for a small tapered profile. A large amount of irradiation caused difficulty in fabricating a small via hole configuration. After many development solution concentrations, the best results occurred with a simple treatment using chemicals such as an alkaline solution. Figure 8 (a) shows the cross section of the via hole accomplished through 10μm thick photo-sensitive polyimide layer with the chemical treatment. Figure 8 (b) is an oblique close-up of the via hole. A small wall slope and round shoulder for the via hole can be achieved merely by applying exposure, developing and baking process with the simple chemical treatment.

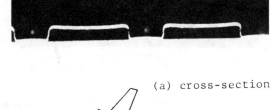

(a) cross-section

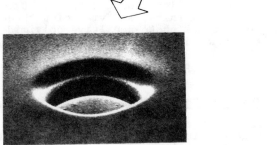

(b) oblique close-up

Fig.8. Photograph of the 25μm diameter via hole accomplished through 10μm thick photo-sensitive polyimide layer after chemical treatment.

By applying these technologies, experiments of small via hole fabrication were carried out. Consequently, we have been able to fabricate via holes with 15μm in diameter as shown in Figure 9. This technology thus results in short process turnaround and small via formation.

Fig.9. Possible minimum via holes, 15μm in diameter, through 10μm thick layer.

C. Electrical Via Connection

For multilayer packaging, it is very important to realize a low via contact resistance and reliable via connection. To obtain high density multilayer interconnection, we tested the fabrication and metallization of small wall slopes for via holes. A small wall slope can increase the conductor line density compared with conventional chemically etched via holes having a wall angle of near 45°, or less (7)(8). Once the via hole has been formed, the next step in the process is the adhesive/conduction sublayer deposition. As via holes have vertical walls, it becomes difficult to accomplish the uniform metallization sufficient for the electroplating sublayer. The sputtering is suitable for the sublayer formation. In case of chromium/copper metallization on the lower copper conductor layer, via contact resistance had increased even after the electroplating was completed. We were concerned about the possibility of some chemical interaction between the copper and the coated photo-sensitive polyimide. Therefore, Auger analysis was carried out on the contact surface. From the result of the analysis, it was found that a complex (or residue) between the copper and photo-sensitive polyimide had been formed as shown in Figure 10. This complex could not be dissolved away in the development process

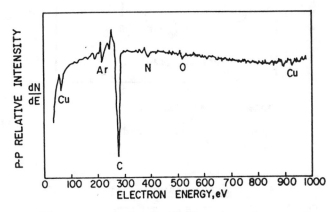

Fig.10. Auger analysis indicating the complex generation (note appearance of C signal in curve) between copper and photo-sensitive polyimide in case of Cr/Cu metallization.

147

of photo-sensitive polyimides. Consequently, the via contact resistance became as large as several hundred milliohms per via.

Many processes were attempted to avoid the complex generation, then the results indicated that thin titanium or chromium evaporated on the copper conductor prior to the photo-sensitive polyimide spin coating filled the role of a good protection against the complex generation. From the results of Auger analysis, no complex was observed as shown in Figure 11. Titanium was selected in view of the process facility. Titanium of a thickness of 500-1000 Å could be easily etched off in the via contact area after the via hole formation and before the next base metallization step. Selective electroplating was then achieved to the desired thickness in controlled plating baths and current density. Removing the photoresist and etching the base metal completed the conductor formation and via connection process. Via contact resistance of only a few milliohms was achieved with this process (see Figure 12).

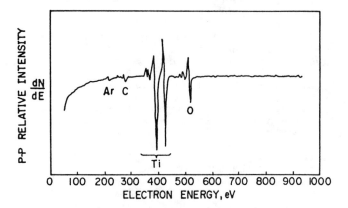

Fig.11. Auger analysis indicating no complex formed (note lack of C signal in curve) after evaporating titanium on the copper conductor.

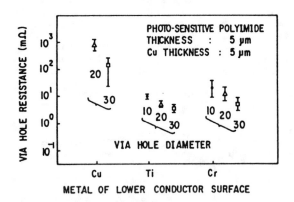

Fig.12. Effect of the evaporated thin film on via contact resistance.

APPLICATION

On the basis of these experiments, a four-layer interconnection was fabricated which can be used to assess potential processing technologies.

A. Multilayer Structure

The demonstration patterns fabricated with photo-sensitive polyimide dielectric and electroplating copper conductor on a substrate with 3 inch diameter are shown in Figure 13. It is a four-layer interconnection; ; it has four conductor layers, each 5μm thick, and three polyimide dielectric layers, each 10μm thick. Patterns had both a conductor width of 10μm and a via hole diameter of 25μm : a two via hole structure as shown in Figure 14 was included. The required interlayer thickness has been found to be above 10μm in considering such electrical conditions as characteristic impedance and crosstalk. This thickness was achieved with two coats applied by spinning. As a result, the capacitance and resistance of transmission lines were 1 pF/cm and 3Ω/cm respectively.

Fig.13. Photograph of the demonstration surface patterns.

(a)

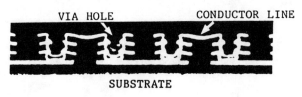

(b)

Fig.14. Cross section of the four-layer interconnection. Conductors are 10μm wide, 5μm thick, and via holes through 10μm thick photo-sensitive polyimide are 25μm in diameter. Two via hole structures are included to assess potential processing technologies.

B. Tests and Evaluations

Interconnection reliability was investigated with special attention paid to the insulator resistance and the small via having a vertical wall configuration. Contact resistance through vias is of major concern to multilayer construction. Test patterns contain about 70 vias that can be tested as one series. The test circuits were evaluated by performing thermal shock, humidity and high temperature tests.

The test results of thermal shocks from -65 to +125°C indicated that even after 1000 thermal shock cycles, either of via hole configuration survived without any recorded failures, as shown in Figure 15. Also, via/conductor resistance change was less than 10% after 1000 cycles. The humidity test was carried out at 80°C / 90 percent RH with 12 V DC bias. The result showed the insulater resistance of the photo-sensitive polyimide interlayer was more than $10^9\Omega$-cm after 1000 hours. A high temperature test was also performed at 85°C with 150 V DC bias, and no problems were observed in the multilayer interconnection.

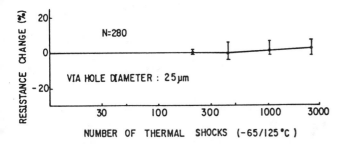

Fig.15. Via connect resistance change after thermal shocks.

The results of these experiments show that the interconnections are all practically reliable, stable and satisfactory in either via hole configuration shown in Figure 14 ; therefore, we can conclude that multilayer interconnections with photo-sensitive polyimide interlayers and electroplating copper conductors are suitable for the use of high-density microelectronic packaging.

CONCLUSIONS

High-density multilayer interconnection technology for high speed LSIs has been developed with photo-sensitive polyimide and electroplating copper. Our main accomlishment has been the development and application of selective electroplating and photolithographic processes to the fine geometry of multilayer configuration. In via hole formation using photo-sensitive polyimide, the short turnaround processes were developed merely by applying exposure, development and baking processes with a simple chemical treatment to obtain a small tapered via wall configuration. Conductors with a square cross section having low electrical resistivity were completed by determining thick resist patterning processes

and current density. Low via contact resistance was realized with a titanium protective thin film which was applied to avoid a complex (residue) on the lower conductor layer caused by the interaction between the copper and the photo-sensitive polyimide coated. Possible minimum conductor and via holes are 3μm wide and 15μm in diameter with a 10μm thick polyimide layer. As a result of the evaluations of a four-layer interconnection fabricated, this technology was found to be applicable for large scale LSI chip packaging.

REFFERENCES

(1) N.Goldberg, "Design of thin film multichip modules" 33rd Electronic Components Conference, pp.610-615 , 1983.

(2) R.P.Vidano, J.P.Cumming, R.J.Jensen, W.L.Walter and M.J.Helix, "Technology and design for high speed digital components in advanced application", ibid. pp.334-342.

(3) J.Shurboff, "Polyimide dielectric on hybrid multilayer circuits" ibid. pp.610-615.

(4) R.Rubner, W.Bartel and G.Bald, "Production of highly heat-resistant film patterns from photoreactive polyimide precursors", Siemens Forsch.-u. Entwickl.-Ber. Bd.5, Nr.4, 1976.

(5) F.Kataoka, F.Shoji, I.Takamoto, I.Obara, M.Kajima, H.Yokono and T.Isogi, "Characteristics of highly photoreactive polyimide precursor", First Technical Conference on Polyimides, Extended Abstracts, p 158, 1982.

(6) K.Sato, S.Harada, A.Saiki, T.Kimura, T.Okubo and K.Muraki, " A novel planer multilayer interconnection technology utilizing polyimide", IEEE Trns. PHP-9, No.3, pp.176-180 , Sep.1973.

(7) K.Mukai, A.Saiki, K.Yamanaka, S.Harada and S.Shoji, "Planer multilevel interconnection technology employing a polyimide", IEEE Journal of Solid-State Circuits, SC-13, No.4, pp.462-467, 1978.

(8) Y.Harada, F.Matsumoto and T.Nakakado, "A novel polyimide film preparation and its preferential-like chemical etching techniques for GaAs devices", Journal of Electrochemical Society, Vol.130, No.1, pp.129-134, 1983.

HIGH-DENSITY MULTICHIP MODULE BY CHIP-ON-WAFER TECHNOLOGY

By

Susumu KIMIJIMA, Takeshi MIYAGI, Toshio SUDO
R&D Center Toshiba Corp.
Kawasaki Japan

Osamu SHIMADA
Electronic Packaging & Assembly Center Toshiba Corp.
Yokohama Japan

ABSTRACT

A high-density module for image processing was developed by chip-on-wafer technology. A silicon wafer was used as the substrate and the LSI chips were flip-chip bonded to the silicon wafer by bumps in chip-on-wafer technology. A primary merit for using a silicon wafer is the little induced thermal stress which affects the bumps. The module contained a digital signal processor, SRAMs and other periphral LSIs. A total of sixteen chips were bonded on the wafer. The LSIs were connected to each other by copper/polyimide multilayer interconnections consisting of eight copper conductive layers and polyimide dielectric layers. The characteristic impedance for the signal lines was controlled to 50 ohms. The LSIs were connected to the wafer electrically and mechanically by solder bumps, which were formed on the LSI bonding pads. A 188 pin AlN ceramic package was used for the module in order to obtain high heat radiation and high reliability. The occupied area for the module was reduced to 20%, compared with the size for conventionally assembled DIPs on a PC board.

INTRODUCTION

Recently, electronic systems are becoming more speedy and compact. Many high speed and highly versatile LSIs which are fit for high performance systems have been developed. But conventional packaging or assembly technology does not fit such LSIs any longer because of: their large number of I/O pins, large sized packages, signal propagation delay caused by their signal line length, characteristic impedance mismatching between the signal lines and devices, and their high power consumption. Therefore, such LSIs require a new packaging technology which contribute to their capabilities and make the system more compact.

The chip-on-wafer technology has attracted great interest in order to meet these requirements. A silicon wafer, copper thin film signal lines, polyimide dielectric layers and flip-chip bonding are the characteristics of the chip-on-wafer technology. The chip-on-wafer technology consists of a silicon wafer as a substrate with copper/polyimide multilayer interconnections, and copper cored solder bumps which connect the LSIs with the wafer. Figure 1 shows a cross-sectional view of the chip-on-wafer technology.

The authors have developed a high-density image processing module, named a DSP-module (Digital Signal Processor module), by the chip-on-wafer technology. The LSIs used in the module were a 32-bit digital signal processor (T9506), twenty-four 64-Kbit SRAMs (TC55416) and forty-four 8-bit latches (TC74HCT374A). One memory chip included two 64-Kbit SRAMs and one latch chip included 16 latches in order to decrease the number of chips. As a result, sixteen chips of LSIs were mounted on the silicon wafer. The module was contained in an AlN ceramic package. An AlN ceramic package has several thermal merits. The chip-on-wafer technology is also suitable for assembling high speed devices, because of its short signal line and its controlled characteristic impedance for the signal lines.

SUBSTRATE

A 50x64 mm silicon wafer was used as the substrate for the module. A silicon substrate has three merits. 1) Since both the LSIs and the substrate are silicon, there is very little thermal stress, caused by the thermal expansion difference between the LSIs and the substrate, induced in the bumps. Table 1 shows the thermal expansion coefficients and thermal conductivities of some materials for substrates and packages. 2) The thermal conductivity of silicon is ten times as high as that of alumina ceramics which are used as substrates in ordinary hybrid ICs. The module obtains high heat radiation through the silicon substrate. 3) Active or passive devices are able to be formed in the silicon substrate by standard IC techniques.

Copper/polyimide multilayer interconnections were formed on the wafer. The conductive layers were evaporated Ti/Cu/Ti thin films. The eight conductive layers included three layers for signals, two for power supply, two for ground and one for bump connection. Polyimide dielectric layers have two merits. 1) The polyimide layer thickness and the signal line width and thickness control the characteristic impedance for

the signal lines. 2) The signal propagation velocity in the signal lines is high. This is because the dielectric constant of polyimide ($\varepsilon_r=3.2$), which is lower than that of alumina ceramics ($\varepsilon_r=9.5$), makes the signal-ground capacitance low.

Though the input frequency of the module was 10 MHz, the multilayer interconnections were designed for above 100 MHz devices to investigate higher speed module assembly. The copper signal lines were T=4 μm thick, W=50 μm wide and P=200 μm pitch. For a 50 ohm characteristic impedance, the polyimide layers between copper layers are required to be 40 μm thick in a normal plane ground. A forty micrometer thickness for one layer does not match the thin film process. Therefore, a bias mesh pattern was adopted for the ground and power supply layers. The polyimide dielectric layers were designed to be 10 μm thick in the case of this bias mesh pattern ground. The mesh pitch was 141 μm and line width was 20 μm. Figure 2 shows the bias mesh pattern ground. Figure 3 shows the characteristic impedance controlled by the aperture of the mesh pattern ground and the polyimide thickness. The aperture is defined by the mesh pitch P_G and line width W_G as $A=(P_G-W_G)^2/P_G^2$. The conductive layers were connected with each other through the 50 μm square viaholes formed in the polyimide layers. The characteristic impedance for the signal lines measured 48 ohms using a sampling oscilloscope (Model 7854, Tektronix) and a high frequency probe (Model 1002, CASCADE MICROWAVE). The signal line resistance was less than 1 ohm/cm and the contact resistance through the viaholes was less than 5 milliohms. The cross-talk for the two parallel signal lines was -35 dB at 100 MHz. The signal line length was 3 cm. Figure 4 shows the transmission characteristics, (a) waveform distortion at 100 MHz, (b) waveform distortion at 1 GHz, and (c) cross-talk, for the multilayer interconnections. The electrical performance for the multilayer interconnections was sufficient for devices operating at above 100 MHz. Figure 5 shows the layout of the DSP-module substrate.

BUMPS

The LSI chips were flip-chip bonded to the silicon wafer electrically and mechanically by the bumps. Flip-chip bonding has two merits. 1) An LSI chip with a large number of bonding pads is able to be bonded at one step. 2) The LSI bonding area is equal to the LSI chip area.

Copper cored solder bumps were formed on the LSI bonding pads by electroplating. First, copper cores were electroplated through the viaholes formed in the photoresist layer by photo-lithography on the bonding pads. Next, solder was electroplated on the copper cores. The distance between the wafer and the LSI chips was kept constant by the copper cores. The bump diameter was 150 μm and the minimum bump pitch was 180 μm. Figure 6 shows the photograph of a solder bump. Figure 7 shows the photograph of the bump connection.

AlN PACKAGE

The wafer was glued and wire bonded to a 67x83 mm AlN ceramic package, which was designed for the module. A total of 188 I/O pins were arranged in a flat lead type on the four sides of the package. The thermal expansion coefficient of AlN ceramics is close to that of silicon (see Table 1). Though the large silcon wafer was glued onto the AlN ceramic package, the warp of the wafer caused by the thermal expansion difference between the wafer and the package was small compared with the warp of a wafer glued onto an alumina ceramic package. The thermal conductivity of AlN ceramics is ten times as high as that of alumina ceramics, and therefore the heat radiation through the AlN ceramic package is higher than that through an alumina ceramic package. These are the reasons why an AlN ceramic package was used for the module. Figure 8 shows the photograph of the module mounted on the AlN ceramic package.

CONCLUSION

A module was developed by the chip-on-wafer technology. Sixteen silicon chips, which included 69 LSIs, were connected to the copper/polyimide multilayer interconnections formed on the silicon wafer. High heat radiation and high reliability will be enhanced by using an AlN ceramic package. The main merits of the DSP-module are: high-density, compactness and decrease in the number of I/O pins. The module size was three time as large as T9506, which was used in the module and ordinarily mounted in a 43x43 mm PGA package. The silicon chip area to substrate area ratio for the module was approximately 0.5. As a functional unit, consisting of a CPU, memories and their interfaces made into the module, the total of 188 I/O pins were less than the 208 I/O pins of the CPU(T9506). The number of I/O pins can be decreased by making a functional unit into a module. This is another merit of the chip-on-wafer technology.

In the future, many high speed and high-density modules should be developed by this chip-on-wafer technology.

ACKNOWLEDGMENTS

The authors wish to acknowledge K. Suzuki of the Medical Engineering Lab. for his advice and T. Yanazawa of the R&D Center for his cooperation.

REFERENCES

1) R. K. Spielberger, Charles D. Huang, William H. Nunne, H. Mones, Darrell L. Fett, Freddie L. Hampton, "Silicon-on-Silicon Packaging", IEEE Trans. on CHMT, Vol.CHMT-7, No. 2, June, 1984.

2) H. Reichl, "Silicon Substrates for Chip Interconnection", Hybrid Circuits, No. 11, September, 1986.

3) Masato Iwabuchi, Katsumi Ogiue, Kazuo Nakamura, Shuuichi Nakagami, Satoru Isomura, Shigeo Kuroda, Seiichi kawashima, "A 7ns 128K Multichip ECL RAM-with-Logic Module", Proceedings ISSCC, February, 1987.

4) Charles J. Bartlett, John M. Segelken, Nicolas A. Teneketges, "Multi-Chip Packaging Design For VLSI-Based Systems", Proceedings ECC, May, 1987.

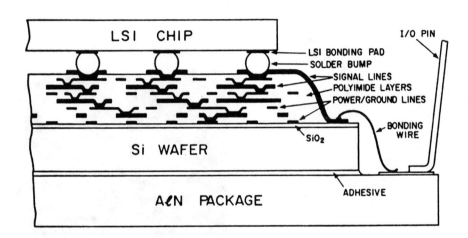

Fig. 1. CROSS-SECTIONAL CONSTRUCTION OF CHIP-ON-WAFER TECHNOLOGY

Table 1. THERMAL PROPERTY OF SOME MATERIALS FOR SUBSTRATES AND PACKAGES

Materials	Thermal Expansion Coefficient $\times 10^{-6}$ (Room Temp. ~400℃)	Thermal Conductivity W/m·K (Room Temp.)
Si	3.4	150
AlN	4.5	170
Al$_2$O$_3$	7.0	17

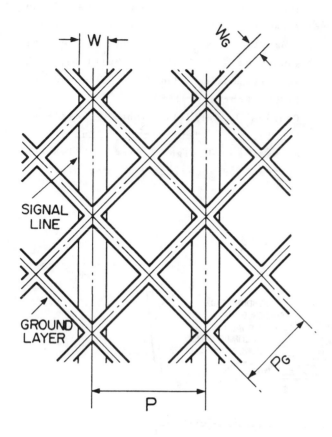

Fig. 2. BIAS MESH PATTERN GROUND AND SIGNAL LINES

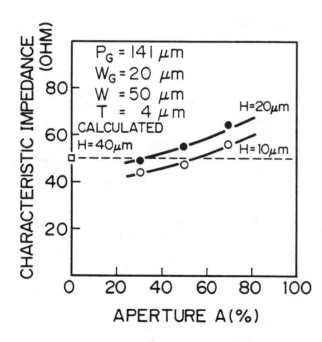

Fig. 3. DEPENDENCE OF CHARACTERISTIC IMPEDANCE ON APERTURE

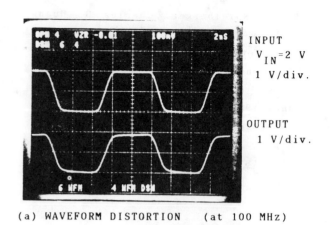

(a) WAVEFORM DISTORTION   (at 100 MHz)

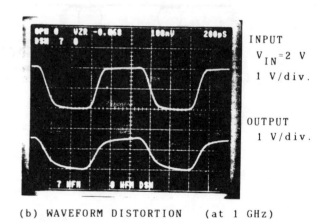

(b) WAVEFORM DISTORTION   (at 1 GHz)

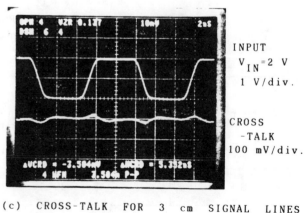

(c) CROSS-TALK FOR 3 cm SIGNAL LINES
(at 100 MHz)

Fig. 4. TRANSMISSION CHARACTERISTICS OF SIGNAL LINES

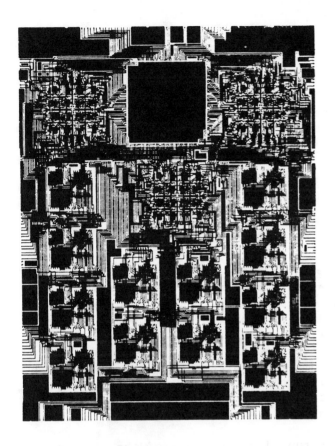

Fig. 5. LAYOUT OF SUBSTRATE FOR DSP-MODULE

Fig. 7. BUMP CONNECTION

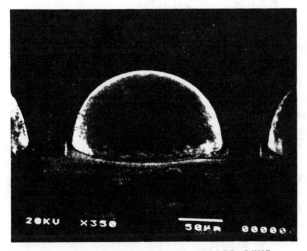

Fig. 6. SEM PHOTOGRAPH OF SOLDER BUMP

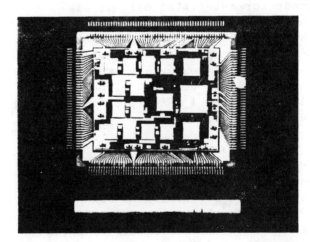

Fig. 8. DSP-MODULE WITH AlN CERAMIC PACKAGE

AN ELECTRICALLY PROGRAMMABLE SILICON CIRCUIT BOARD

Albert A. Bogdan
Mosaic Systems, Inc.
Troy, Michigan

ABSTRACT

The paper describes an electrically programmable, very high density, microelectronic die interconnect system built on a silicon substrate. It describes its unique features. The unusual features permit a design engineer to go from logic design to a programmed Silicon Circuit Board ready for die mounting within 1 to 2 days.

1. INTRODUCTION

Mosaic Systems, Inc. has developed an electrically programmable silicon circuit board which it calls UNIPRO SCB. The UNIPRO SCB is a high density, microelectronic die interconnect system, mounted on a silicon substrate containing a "sea of bonding pads" vertically connected to a matrix of conducting lines separated by an electrically programmable amorphous silicon "antifuse". It is the highest density interconnect system and the only one permitting engineers to go from logic design to a programmed board ready for integrated circuit die mounting in one day. Figure 1 shows two (2) UNIPRO SCBs tiled together to hold a 1.1 megabit SRAM. The entire package is approximately two (2) square inches. It is composed of seventeen (17) 64,000 by 1 SRAMs plus the six (6) buffers to drive the memory.

The original electrically programmable interconnect system's development was started by Dr. Herbert Stopper and Dr. Robert R. Johnson in 1983 and completed in early 1986. The system is a three (3) inch by three (3) inch patterned interconnect area on a four (4) inch wafer. See Figure 2. It contains almost 20,000 inches of wiring. By mid year, Mosaic Systems completed the development of a more typical hybrid size system. It is a nominal one (1) inch by one (1) inch Segment size.

2.0 DESCRIPTION

The Segment provides a greater degree of partitioning for larger systems and to meet the larger market demand. The Segment contains 720 inches of internal wiring. The wafer scale system can mount 240 SSI size die or a smaller number of larger die. The Segment can interconnect 36 SSI size (86 mils on a side) die or a smaller number of larger size die.

A functional cutaway of the interconnect system can be viewed in Figure 3. I will describe it piece by piece.

2.1 MOUNTED ON SILICON

Silicon was specifically selected as the substrate since the original conceivers of the concept wanted to be able to build a wafer scale integrated computer capable of operating in liquid nitrogen. In order to do so, they required a material that would have the same Temperature Coefficient of Expansion as the ICs to reduce thermal stresses. Silicon, of course, is a perfect match with the silicon IC. Silicon is, also, an excellent thermal conductor thus reducing hot spots, which in turn reduces the IC junction temperature and distributes the heat over a larger surface area for more even cooling.

Note: The author is now President of Strategies Plus, a private consulting firm in Huntington Woods, MI.

2.2 POWER DISTRIBUTION SYSTEM

A metal layer is deposited on the silicon substrate, followed by a silicon dioxide dielectric layer which in turn is followed by an additional metal layer. Each layer is vertically connected to 420 bonding pads distributed over the surface of the Segment.

The metal, dielectric, metal layers form a large distributed capacitor of approximately 40 nanofarads for the Segment size interconnect system and 400 nanofarads for the wafer size system. The distribution system has a very small, not easily measurable, inductance, thereby enhancing the high frequency bypassing effect of the distributed capacitance. Mosaic Systems has found that the system eliminates the need for adding discrete bypass capacitors for high frequency noise suppression. If additional low frequency filtering is required an electrolytic capacitor can be mounted in parallel across the two plane power system at the edge.

The power system can be represented by the equivalent circuit shown in Figure 4. R-Dis is the resistance of each layer from the power entry source to the center of each plane. R-Spread is the resistance from a voltage or ground pad to the same power entry point. A quick computation show that even for ten (10) amperes, the maximum voltage drop is only 140 millivolts. It is an excellent power distribution system.

2.3 SEPARATED BY AN AMORPHOUS SILICON "ANTIFUSE"

The matrix of lines in the Segment is separated at 101,061 intersections by a an amorphous silicon "antifuse". The wafer product has over 1.6 million "antifuses". "Antifuse" is a word Mosaic Systems, Inc. coined to represent a material that has a normally high resistance. However, when a voltage above a certain threshold is applied the material switches to a low

resistance. The opposite action of a fuse.

Figure 5 shows that the nominal resistance is approximately two hundred megohms in its normal state. When a voltage is applied above a threshold voltage the resistance switches to a nominal four ohms with the high end of the distribution being eight ohms. Presently it is set for a nominal 20 volts. Every intersection has a minimum 14 volts applied to stress test it, to assure that no intersections can be switched at voltages lower than that voltage during normal circuit operation. The threshold voltage is proportional to the thickness of the amorphous silicon material. If a higher threshold voltage is required, it can be designed into a new Segment.

The amorphous silicon, upon switching forms a 1 micron filament between the two layers. Physicists argue as to what occurs in that filament. Some argue that a crystalline structure is formed while others argue that a metal migration occurs to form a silicide. It doesn't matter. The material forms a permanent short circuit. Tests have shown that once switched, the material does not switch again. It is a permanent connection. Studies by Raytheon indicate the material, at normal operating conditions has a failure rate of 10^{32} years.

The "antifuse" feature makes several important contributions to the UNIPRO SCB design. The first is obvious, It permits us to establish a set of standard off the shelf parts that can be customized by electrical programming. The Segment and the Wafer are just two examples. Through the use of electrical programming we can manufacture the part prior to the determination of a custom design and thereby, take almost the entire substrate customization process out of the production loop. In a thick film ceramic hybrid the hybrid manufacturer after receiving a net list needs to perform a place and route process, design the screens, make them, and produce a customized substrate ready for IC assembly. With the UNIPRO SCB all of the steps are eliminated after place and route. Even the routing is simplified since all of the lines are in place. We then substitute a programming step which takes approximately 5 to 10 minutes depending on the number of connections. The programming program is an automatic output of the place and route program.

The second and almost as important feature is that the "antifuses" provide the addition of redundancy. Mosaic Systems has developed a product that is very defect tolerant. As you can imagine, producing a wafer scale interconnect system, on silicon, that is defect free is a difficult task. But with the "antifuses" the interconnect system can withstand defects since the "antifuses" provide significant redundancy. If defects are discovered during the testing and programming operation they are automatically programmed around. A new wiring path is established. This permits higher real yields and, thereby, the production of a fairly low cost interconnect system.

The programming feature, also, permits us to produce to connect lines with different electrical characteristics in a manner to optimize delays. We can tune the lines.

2.4   MATRIX OF LINES

The aluminum lines are deposited and etched into a complex architecture to provide the line structure necessary to optimize the transmission line characteristics to reduce delay times. The lines are of varying widths and lengths. The inductance, capacitance, and resistance for the Segment is shown in Figure 6. The transmission characteristics operate under what is normally called the "lossy line" principal. Mosaic Systems, Inc. has learned to use what is normally considered to be negative properties of a transmission line (high resistance) and convert them to positive benefits. The delay times are altered depending

on the order the connections are made. If timing is required below 2 Nanoseconds for the present design Segment or 350 picoseconds for the new high performance Segment then significant care has to given to both die placement and order of line connection.

Mosaic System, Inc. has, also, discovered that by cascading the lines in a proper proprietary manner, based on the geometries designed into the line structure, the lines can be tuned to optimize transmission delays. For the present Segment design, the signal delay will be equal to the highest possible speed (speed of light) for a silicon dioxide dielectric for about half the distance across a Segment. We call this the critical distance.[1] For the new high performance design that distance has been extended to almost three inches. The optimum delay for the present Segment and the new high performance Segment is shown in Figure 7. The delay across almost 2 inches of the present Segment is less than 2 nanoseconds, while for the new high performance system the measured delay is approximately 350 picoseconds.

Although literally hundreds of lines run parallel to each other for fairly long distances, there is no measurable crosscoupling. The amount of crosscoupling in a transmission line is proportional to line diameter, the distance the lines are from ground and inversely proportional to the distance separating the lines. In the UNIPRO SCB the cross-sectional area of the lines is very small while the distance between the lines is fairly large in comparison to their separation from ground. Due to the line geometries, the Segment becomes essentially noisefree.

Another benefit, due to the lossy line transmission characteristics of the UNIPRO SCB, is that neither source or load terminators are required to eliminate ringing. The proprietary design permits the line resistance to perform effectively as a distributed terminator to absorb the energy of the what would normally be a wave form reflecting off the unloaded termination. The SRAM developed by Mosaic Systems, Inc totally eliminates the undershoot that historically has created problems for an SRAM design engineer. The elimination of noise, ringing, and transients permits the design engineer to convert the noise margin in his circuit to improved performance systems.

The only significant problem with the lossy lines is that they cannot as easily handle DC. The line resistance act to attenuate the signal thereby eliminating noise margins. Therefore, the design engineer needs to use push pull drivers or other design structures to eliminate the DC component which would be needed to drive terminators.

2.5 "SEA OF BONDING PADS"

The lines are connected through vertical conductors to 2758 signal pads on the surface of the UNIPRO SCB Segment (Figure 7) and to 12,480 signal pads on the wafer interconnect system. The Segment is organized into 36 cells. The layout of a cell is shown in Figure 8. The line shown running through the Voltage and Ground pad strip indicates the boundary between adjacent cells.

The aluminum pads, at 4.7 mils on a side, are designed for ease of wire bonding and rework, if necessary. The 11.8 mil pitch permits a very high density interconnection of the die to the Segment. This is particularly useful for high padout ASIC or VHSIC devices and acts to reduce their total footprint. The pads are connected to the matrix of lines in a manner to provide the design engineer the greatest flexibility in the design of the interconnect system.

The small bond pitch permits the UNIPRO SCB to handle some relatively high padout die in a much smaller footprint. Standard single device packaging technology tends to connections at much larger pitches than the UNIPRO SCB's 11.8 mil centers. Due to the high fan out, the die may end

representing as little as 15 % of the total package area. Therefore, for multi-ASIC die the UNIPRO SCB can provide a footprint that is as much smaller than other interconnect technologies. The UNIPRO SCB can also interconnect a group of IC representing standard cells to form a fast turn-around ASIC substitute for prototyping or low volume requirements.

All internal connections are made to NET lines that run vertically in the Segment. There are 150 NET lines. All of the NET lines are connected to large pads on the north and south side of the Segment. This means that every internal node comes to the edge of the Segment, thereby, significantly improving the ease of troubleshooting. Every die mounted on the UNIPRO SCB can be tested in situ.

In the X direction there are three types of signal lines. The C line is only the length of a cell. Each C line has only one pad connection with the capability of connecting to 24 net lines. As a general rule, all internal die output signals are connected to a C pad. The S pads are connected to PAD lines that run the length of the Segment. Each PAD line is connected to three or four S pads distributed in an east-west direction.

In addition, the low resistance X pads are connected to I/O pads on the east and west side of the Segment. Most of the lines run across half the Segment. Forty eight of the X pads at the extreme east and west end of the Segment are not connected to any amorphous silicon "antifuses". They have been included to permit a higher voltage input to the Segment for either EPROM programming, programmable logic arrays, or for higher voltage analog signal inputs to an A/D converter or threshold detector or as a higher voltage power supply input.

The programming process is done on a prober containing a proprietary set of probe cards. An outer probe card establishes connection with all of the net lines. The inner probe card makes contact with all of the signal pads within a cell. All of the net lines are grounded except the one to be connected. 10 volts is applied to the net line. A minus 10 volts is applied to the probe contacting a pad we want to connect to the net line. The "antifuse" switches, as the resistance falls a current limiter automatically switches in to limit the current to 150 milliamperes for several milliseconds. The process is repeated, all under computer control, until the entire interconnect system is formed. On the average the process takes about five (5) minutes.

By the summer of this year, a complete UNIPRO SCB programming system and the place and route software will be available from Data I/O and FutureNet to permit the design engineer to design and program his own custom interconnect package within 1 to 2 days after finishing his logic design. The package will have the smallest footprint, profile, or volume of any packaging scheme available except for an all silicon ASIC. In order to make the software readily available it will be designed to operate on a PC/AT system.

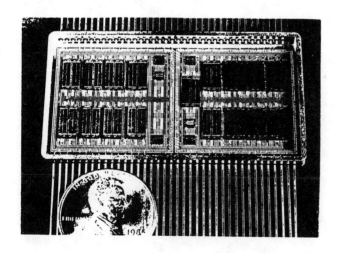

Figure 1   UNIPRO SCB based 1.1 megabit SRAM

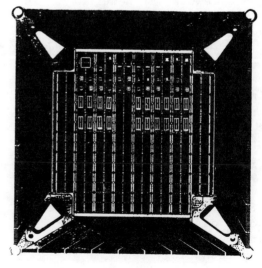

Figure 2   UNIPRO SCB Wafer Size Interconnect System

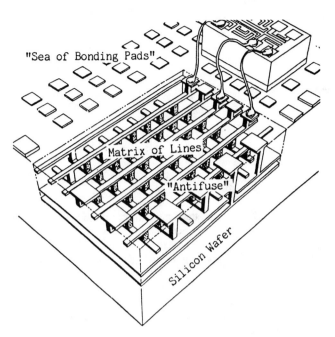

"Sea of Bonding Pads"

Matrix of Lines

"Antifuse"

Silicon Wafer

Figure 3 Functional Cutaway of a UNIPRO SCB

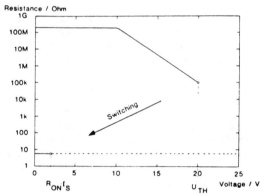

Figure 5 Switching Characteristics of Amorphous Silicon "Antifuse"

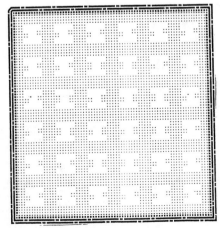

Figure 7 Pad Layout of UNIPRO SCB Segment

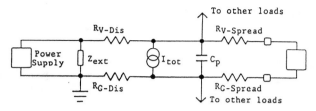

RG-Distr.   <1.5 milliohms
RV-Distr.   <3.0 milliohms
RG-Spread < 70 milliohms
RV-Spread <140 milliohms
Cp > 0.4 uf

Figure 4 Equivalent Circuit for UNIPRO SCB Power Distribution System

UNIPRO SCB TRANSMISSION LINE PARAMETERS

LINE TYPE		L	C	R	CRITICAL DISTANCE
		pH/mm	pF/MM	Ohms/mm	mm
SERIES	C	250	0.4	3.4	10
2000	N	90	1.2	2.1	5
	S	250	0.4	3.4	10
SERIES	C	170	0.5	0.5	48
3000	N	210	0.4	1	32
	S	530	0.2	2.5	29

Figure 6 Electrical Parameters of UNIPRO SCB Transmission Lines

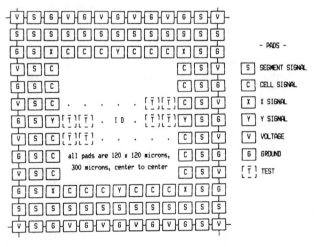

- PADS -

S	SEGMENT SIGNAL
C	CELL SIGNAL
X	X SIGNAL
Y	Y SIGNAL
V	VOLTAGE
G	GROUND
[T]	TEST

all pads are 120 x 120 microns, 300 microns, center to center

Figure 8 Layout of Typical UNIPRO SCB Cell

MULTICHIP PACKAGING TECHNOLOGY WITH LASER-PATTERNED INTERCONNECTS

Andrew T. Barfknecht, David B. Tuckerman, James L. Kaschmitter,
and Bruce M. McWilliams

Lawrence Livermore National Laboratory
P.O. Box 5503, M/S L-271
Livermore CA 94550

Abstract

A multichip silicon-on-silicon packaging technology has been developed which incorporates laser-patterned thin-film interconnects. This technology is particularly suited for application in high speed, high power, and high I/O systems where its unique characteristics provide many advantages over more traditional methods. The laser-patterned thin-film interconnects allow higher I/O densities and better electrical performance than wire bonds or TAB. The face-up, thin-film eutectic die attach technique used provides much lower thermal resistance between the substrate and the chips than solder bump die attach can achieve. In addition, laser-patterned interconnects demonstrate superior ruggedness and fatigue resistance under thermomechanical cycling and shock. This technology has been used to produce a 10-chip memory module, samples of which have been tested to relevant methods of MIL-STD 883C.

Introduction

Thin film multichip modules appear likely to supplant multilayer ceramic (MLC) modules as the packaging method of choice for future high performance VLSI systems. Thin film modules promise superior electrical and thermal characteristics on circuit modules that are much smaller and lighter than their MLC counterparts. [1-13]

Much of the work on thin film modules has relied on flip-chip solder bump techniques for chip-to-substrate interconnect.[14,15] Such a face-down die attach method can provide a dense area array of interconnects with good electrical performance, but the limited contact area provides poor heat conduction between the chip and the substrate. As in MLC modules which use flip-chip interconnect, elaborate methods will presumably be necessary to provide additional conduction paths for heat dissipation from high speed, high power chips.[16,17]

Alternatively, face-up die attach methods can provide the very low thermal resistance necessary between the die and the substrate for high power applications.[16,18] The standard techniques for chip-to-substrate electrical interconnect used with face-up die attach systems are wire bonding and tape automated bonding (TAB). Both of these techniques provide poorer electrical performance than solder-bumps in high speed applications.[15,19] Their major drawback, however, is the limited density of interconnects allowed compared with that possible with solder bumps.

Recently we described a method of producing thin-film chip-to-substrate interconnects by laser patterning.[1] Such interconnects incorporate the vastly superior power dissipation capabilities of face-up die attach with high I/O density and electrical performance close to that of solder bumps *(Table 1)*. These laser-patterned interconnects are also completely exposed for visual inspection, which is generally not possible with flip-chip attachment.

Several other groups have recently developed novel methods of producing thin film chip-to-substrate interconnects.[11,13,20-22] As opposed to our technique in which the chips are mounted on top of the substrate using a very thin (~5 μm) metallurgical bond, these other methods rely on embedding the chips in a cavity in the substrate and using a polymeric compound to bridge the relatively large gap (~100 μm) between the chip and the substrate.

As a further demonstration of our technology we have designed and built a complex silicon-on-silicon multichip module with laser-patterned interconnects *(Figure 1)*. A description of the ten-chip memory module follows.

Table 1. Performance of Chip-to-Substrate Interconnects

	Laser-patterned thin film hybrid	Wire bond	TAB	Flip-chip
I/O pitch	25 μm	200 μm	100 μm	400 μm
Pinout (1 cm² die)	1600	256	400	625+ (area)
Lead inductance	0.25 nH	1-2 nH	1 nH	0.05 nH
Heat transfer from IC	Very high	High	High	Low

Reprinted from *Proc. 39th Electron. Components Conf. (ECC)*, pp. 663–667, 1989.

Figure 1. A completed ten-chip memory module, shown without its hermetic package.

Design of the 10-chip memory module

The hybrid module incorporates ten static RAM chips on a 4.58 cm x 4.58 cm substrate. The SRAM's are National Semiconductor NVC63Q64 8k x 8 self-timed memories, developed under the VHSIC program as a radiation-hardened part. The chips are organized in the module to give an 8k deep x 80 bit memory array. Typical access times for these chips are 35 to 55 nsec, although the designed system access time for the module is several times slower. The chips are run in the fully asynchronous mode; the data-in and data-out lines for each bit are tied together at the chip surface. The design thus requires 30 thin film interconnects from each chip to the module substrate, for a total of 300 such interconnects. The module has 120 external I/O pads which include redundant ground and power leads distributed about the periphery.

The 10 chips are arranged in two rows of five. The chips are rotated such that their pads for address and control signals face a central bus on the module (*Fig. 2*). The data lines fan out above and below the two rows of chips. Pads were provided for four surface mount chip capacitors for on-module decoupling of the power supply if necessary.

The design was implemented in a two-level metallization scheme using gold conductors and plasma-enhanced chemical vapor deposited (PECVD) silicon dioxide interlayer dielectric. The ten die are attached to the substrate with a thin-film gold-silicon eutectic bond. The use of gold metallization reduced the number of process steps by allowing the first level interconnect and the pads for the die attach to be formed in the same deposition and patterning steps. Gold was also used for its low resistivity, a necessity for thin film interconnects, and for its corrosion resistance. PECVD oxide was used as the interlayer dielectric because of its thermal and mechanical stability and low dielectric constant, and because it allows conformal coverage of the gross topography of our modules.

In order to reduce electrical noise and to enhance the radiation tolerance of the parts, the substrate of each die is grounded by making electrical contact to the gold-silicon eutectic bond at the backside of each die. The contact was made by etching a via through the oxide at the base of each die and routing a ground lead to the via.

For the intended application of this module, the required operational speed was relatively slow; therefore, the lines on the module did not need to be optimized for the shortest possible signal propagation delay. The film thicknesses and line geometries were not chosen for high speed electrical performance, but rather were chosen for manufacturability with good yield in our modest facilities. Minimum geometries were 100 microns in first level metal, and 20 microns in second level metal. First metal was approximately 1 micron thick, second metal 3 microns, and the interlayer dielectric 3.5 microns.

Six patterning steps were used in the design. The three mask levels done before die attach were accomplished with conventional photolithography and wet chemical etches. After die attach, the three remaining patterning steps were done with laser etch processes in conjunction with "dry" etching techniques (reactive ion etching and/ or ion beam milling). The computer-controlled laser apparatus automatically compensates for any displacement of the chips from their optimal positions on the substrate, and allows the patterning of the thin-film chip-to-substrate interconnect by maintaining the beam focus while writing on the sides of the bevelled chips.

The address/control bus, power buses, data line fan out and pads for the die attach are implemented in the first level metal. The second level metal was used primarily for the chip-to-substrate interconnects. In other regions of the hybrid the second level metal served as a ground plane.

Since the module uses CMOS parts and is run fairly slowly, power dissipation is low. The thin film die bond provides very efficient heat

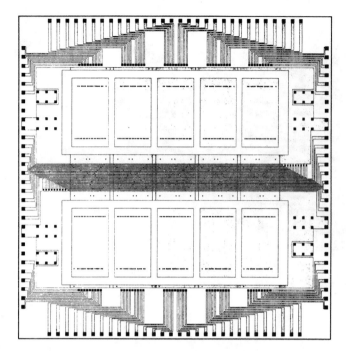

Figure 2. Enlarged view of the first level metal pattern of the multichip memory module. Each side is actually 4.58 cm long. The address/control bus is visible between the two rows of die attach pads. The design is very conservative and was not intended to demonstrate the ultimate packing densities achievable with this technology.

Table 2. Process Flow

Pattern mesas
Grow base dielectric
Sputter-deposit M1
Pattern M1
Deposit thin passivation layer
Pattern die attach pads

Test substrate

Attach die
Bevel die
Deposit interlayer dielectric
Pattern vias (*laser process*)
Sputter-deposit M2
Pattern M2 (*laser process*)
Deposit topside passivation
Pattern wire bond pads (*laser process*)

Test module

Attach module to Kovar package
Wire bond package leads to module pads
Seal lid

Final test

transfer from the chips to the substrate. Mounting the module inside a standard Kovar package with a compliant, thermally-conductive epoxy provides adequate heat sinking. For higher power applications, we will bond the silicon substrate to a liquid-cooled heat sink.[23] Gold wire bonds were used to connect the 120 external I/O pads of the substrate to the leads of the surface-mountable Kovar package. A lid is resistance-welded to the package to provide hermeticity.

Process Technology

Our multichip memory modules were fabricated using technology that is essentially a two-level metal extension of that used for the two-chip module we described earlier.[1] The process flow used for these ten-chip modules is outlined in Table 2 and described below.

Standard 100 mm silicon wafers with 1 micron of thermal oxide were used as starting material. Mesas for the die attach were formed by patterning the oxide, which then served as an inorganic mask for the anisotropic silicon etch. The surface of the wafer surrounding the mesas was removed to a depth of 75 microns.

The base dielectric of 0.5 micron thermal oxide was then grown by wet oxidation. The first level metallization (M1) of Ti:W/Au/Ti:W was sputter deposited, and patterned using conventional photolithography and wet etches. A thin (0.5 micron) layer of PECVD oxide was then deposited as a protective coating over the first level metal. The final step in the fabrication of the substrate is the patterning of openings to the die attach pads. At the same time, small vias are made at the extreme ends of all signal lines.

The substrate is then completely tested. All signal lines are checked for continuity from end to end and for shorts to other lines. This testability is one of the advantages of a hybrid technology over mono-lithic wafer-scale integration; fully tested chips are attached to the fully tested substrates, facilitating good yields for the completed modules.

Chips are attached by a thin film Au/Si eutectic bond formed *in vacuo*. We have developed this technique to consistently yield bonds which are completely free of voids. The quality of each die bond is verified by acoustic microscopy.

After die attach the sides of the chips are bevelled at 60 degrees to the horizontal and a portion of the mesa at the base of the bevel is removed (by micromachining) to insure a smooth surface on which to pattern the chip-to-substrate electrical interconnect. The thin (3-5 micron) metallurgical bond and the matched thermal expansion coefficients of the chips and the substrate allow the subsequent formation of thin-film metal signal lines over the joint. Die placement errors are typically 25-75 microns which is well within the automatic routing capabilities of the computer controlled laser-patterning apparatus.

Approximately 3.5 microns of PECVD oxide is then deposited over the entire module to serve as the interlayer dielectric. A thin coating of amorphous silicon is deposited over the oxide to serve as an inorganic mask. The via pattern (for contacts to both the substrate and the pads on the chips) is etched in the amorphous silicon mask by laser writing in a chlorine atmosphere.[24] The pattern is transferred to the silicon dioxide by reactive ion etching.

The second level of metal is then sputter-deposited. Again Ti:W/Au/Ti:W metallurgy is used. Reactive sputtering to form a TiN layer has been used to improve the barrier layer properties between the Al pads on the chips and the Au interconnect. The M2 layer is patterned by forming an inorganic mask of PECVD oxide. Layers of silicon dioxide and amorphous silicon are deposited over the entire module;

the silicon is etched with the laser apparatus (including forming the pattern on the bevelled sides of the chips); this pattern is then transferred into the underlying oxide by reactive ion etching. The oxide is then used to mask the gold during ion milling.

A thin layer of PECVD oxide is then deposited to act as a protective coating over the module. Vias through this topside coating of oxide for the wire bond pads are formed by the laser patterning process in a manner completely analogous to that used to form vias in the interlayer dielectric.

The finished module is then tested, assembled into the metal package, and tested again.

Results

Numerous functional 10-chip modules were produced. The very conservative design rules enabled us to obtain reasonably good yield. Samples have been subjected to, and passed, applicable tests from MIL-STD 883C (*Table 3*). As was found for the two-chip hybrids[1], these multichip modules are remarkably robust: ten-chip modules have survived repeated thermal cycling from 400 C to -196 C with no damage to the interconnect or die attach; in fact, working samples survived with no loss of functionality. Functioning samples have been vibration tested to 20 g at up to 20,000 Hz without failure. Samples within the Kovar package have survived constant acceleration of 5,000 g; unpackaged samples have survived testing to 30,000 g.

Four of these modules have been integrated into a solid-state recorder experiment being conducted by Fairchild Space Corporation for the Air Force's P87-2 mission, a satellite which will map the Van Allen Belts. Those modules received additional testing as part of the recorder, including 100 hours of thermal cycling over the mission temperature profile of -23 C to 55 C; a random vibration test at 11.3 g RMS; and an EMI test conforming to MIL STD 461A RS03, class A2.[25] The modules will be tested throughout the one year mission lifetime, so that radiation effects and module integrity can be monitored. This mission is scheduled to fly in the 3rd quarter, 1989, and will represent one of the first applications of VHSIC parts in space.

Conclusions

We have demonstrated our laser-patterned thin film interconnect technology on a complex multichip module. The unique capabilities of this technology will enable its use in future VLSI systems that require high speed, high power, high reliability, extremely low weight and volume, and large numbers of I/O's.

Acknowledgments

This work was performed under the auspices of the United States Department of Energy by Lawrence Livermore National Laboratory under contract W7405-Eng-48, with support from the VHSIC Program.

We thank D. Ashkenas, L. Kohnfelder, and P. Rasmussen for software and hardware development, and M. DeHaven, L. Evans, J. Feikert, G. Griggs, E. Schmitt, and J. Thompson for technical assistance.

We thank L. Capots, R. Chitty and S. Bryant of Fairchild Space Company for their assistance with the P87-2 mission and space qualification.

Table 3.

MIL-STD 883C Tests Completed on Functional 10-Chip Hybrids

Method 1015.4	Burn-in	240 hours @ 125 C
Method 2007.1 Condition A	Vibrations	20 g peak acceleration
Method 1010.5 Condition B	Temperature Cycle	-55 C to 125 C
Method 1011.4	Thermal Shock	-55 C to 125 C
Method 2001.2 Condition A	Constant Acceleration	5,000 g
Method 2002.3 Condition A	Particle Impact Noise Detection (PIND)	

References

[1] D. B. Tuckerman, "Laser-Patterned Interconnect for Thin-Film Hybrid Wafer-Scale Circuits," *IEEE Elect. Dev. Lett.*, EDL-8, 540-543, 1987.

[2] C.J. Bartlett, J.M. Segelken, N.A. Teneketges, "Multichip Packaging Design for VLSI-based Systems," *IEEE Trans. Components Hybrids Manufact. Technol.*, CHMT-12, 647-653, 1987.

[3] C.W. Ho, D. A. Chance, C. H. Bajorek, R.E. Acosta, "The thin-film module as a high-performance semiconductor package," *IBM J. Res. Develop.*, 26, 286-296, 1982.

[4] S.H. Wen, J. Kim, J. Hurst, N. Chou, "Thin Film Wiring for Integrated Electronic Packages," *Materials Research Society, Fall Meeting*, Paper I 2.3, 1987.

[5] C.C. Chao, K.D. Scholz, J. Leibovitz, M.L. Cobarruviaz, C.C. Chang, "Multilayer Thin Film Substrate for Multichip Packaging," *Proc. 38th Electronics Components Conference, IEEE*, 276-281, 1988.

[6] J.K. Hagge, "Ultra-reliable Packaging for Silicon-on-Silicon WSI," *Proc. 38th Electronics Components Conference, IEEE*, 282-292, 1988.

[7] R.K. Spielberger, C.D. Huang, W.H. Nunne, A.H. Mones, D.L. Fett, F.L. Hampton, "Silicon-on-Silicon Packaging," *IEEE Trans. Components Hybrids Manufact. Technol.*, CHMT-7, 193-196, 1984.

[8] R.J. Jensen, J.P. Cummings, H. Vora, "Copper/polyimide Materials System for High Performance Packaging," *IEEE Trans. Components Hybrids Manufact. Technol.*, CHMT-7, 384-393, 1984.

[9] T.A. Lane, F.J. Belcourt, R.J. Jensen, "Electrical Characteristics of Copper/Polyimide Thin Film Multilayer Interconnects," *IEEE Trans. Components Hybrids Manufact. Technol.*, CHMT-12, 577-585, 1987.

[10] M. Kohara et al, "Thermal Stress-Free Package for Flip-Chip Devices," *IEEE Trans. Components Hybrids Manufact. Technol.*, CHMT-7, 411-416, 1984.

[11] R.W. Johnson, J.L. Davidson, R.C. Jaeger, D.V. Kerns, "Silicon Hybrid Wafer-Scale Package Technology," *IEEE J. Solid State Circuits*, SC-21, 845-851, 1986.

[12] T. Yamada, K. Otsuka, K. Okutani, K. Sahara, "Low Stress Design of Flip-Chip Technology Si-on-Si Multichip Module," *Proc. 5th Int. Electronic Packaging Conf.*, 551-557, 1985.

[13] C.A. Neugebauer, R.O. Carlson, R.A. Fillion, T.R. Haller, "High Performance Interconnections between VLSI Chips," *Solid State Tech.*, 31 (6), 93-98, 1988.

[14] L.F. Miller, "Controlled Collapse Reflow Chip Joining," *IBM J. Res. Develop.*, 13, 239-250, 1969.

[15] D.J. Pedder, "Flip Chip Solder Bonding for Microelectronic Applications," *Hybrid Circuits*, 15, 4-7, 1988.

[16] A. Bar-Cohen, "Thermal Management of Air- and Liquid-Cooled Multichip Modules," *IEEE Trans. Components Hybrids Manufact. Technol.*, CHMT-10, 159-175, 1987.

[17] Y.C. Lee, H.T. Ghaffari, J.M. Segelken, "Internal Thermal Resistance of a Multi-Chip Packaging Design for VLSI-Based Systems," *Proc. 38th Electronics Components Conference, IEEE*, 293-301, 1988.

[18] M. Mahalingam, J.A. Andrews, "TAB vs. Wirebond — Relative Thermal Performance," *IEEE Trans. Components Hybrids Manufact. Technol.*, CHMT-8, 490-499, 1985.

[19] D. Herrell, D. Carey, "High-Frequency Performance of TAB," *IEEE Trans. Components Hybrids Manufact. Technol.*, CHMT-10, 199-203, 1987.

[20] R.W. Johnson, T.L. Phillips, R.C. Jaeger, S.F. Hahn, D.C. Burdeaux, "Thin-Film Silicon Multichip Technology," *Proc. 38th Electronics Components Conference, IEEE*, 267-275, 1988.

[21] J.G. Black, S.P. Doran, D.K. Astolfi, M. Rothschild, D.J. Ehrlich, "Laser-Direct-Written Interconnection Techniques for Multiple-Chip and Wafer-Scale Systems," *Materials Research Society, Fall Meeting*, Paper B 2.5, 1987.

[22] M. Ohuchi, A. Hongu, M. Sairo, A. Iida, K. Yoshida, H. Odaira, "A New LSI Interconnection Method for IC Cards," *IEEE Trans. Components Hybrids Manufact. Technol.*, CHMT-10, 310-313, 1987.

[23] D.B. Tuckerman, R.F.W. Pease, "High-Performance Heat Sinking for VLSI," *IEEE Electron Dev. Lett.*, EDL-2, 126-129, 1981.

[24] F. Mitlitsky, D.B. Tuckerman, B.M. McWilliams, "Process for rapid laser patterning of thin films," *Conf. Lasers Electro-Optics Tech. Dig. Series 1987*, 14, 274-275, 1987.

[25] Richard Chitty, Fairchild Space Corporation, private communication.

MULTI-CHIP MODULES WITHOUT THIN FILM WAFER PROCESSING

Dick Pommer and John Chiechi
UniStructure, Inc.
Irvine, California

Abstract

Integrated circuit (IC) component densities are rapidly increasing in electronic equipment, creating a new form of packaging: multi-chip modules (MCM). Today's systems require multi-chip modules with silicon-to-substrate densities in the 40% to 60% range.

It is the density of the component's outer lead bond pitch and the module's wiring requirements that determine the necessary substrate density and construction. The majority of industry has assumed that thin film copper or aluminum metalization, with polyimide dielectric, on a ceramic base, manufactured on a wafer processing line was the required substrate technology for multi-chip modules.

This paper will look at the component packages required to support 40% to 60% silicon area to substrate area. We will also review the electrical criteria needed to support current high speed digital designs. Power density (Watt/in^2) will be established as a parameter for substrate heat dissipation. This criteria will then be used in example circuit routes for a variety of multi-chip modules to establish conductor features and layer count. These module parameters have been used by USI to develop a substrate technology that bridges the gap between conventional printed circuit fabrication and thin film processing on ceramic.

Introduction

An electronic multi-chip module can be defined as follows:

> Logical functional blocks grouped to maximize
> interconnect within a compact assembly, yet
> minimize interconnect to the next level.

With this definition, two types of MCMs currently exist, hybrids and daughter cards. In the past, electronic modules (hybrids) had bare IC die wire bonded to ceramic substrates encased in a "hermetic can". As the number of die per module increased, die fallout at final test created unacceptable module yields. The solution to the assembly yield was adding small electroformed leads to the die that allowed burn-in testing and screening of the die prior to assembly. TAB (tape automated bonding), FLC (flexible leaded components)(ref. 1), and BEAM LEAD are typical names associated with the die mounted in very small leaded packages. Today, high density hybrids with thin film organic substrates are called multi-chip modules.

Daughter cards also fit the definition of a multi-chip module. In the past, daughter cards used DIPs (dual in-line packages) for component packages. As densities of cards needed to increase, pin grid arrays, surface mount packages, fine pitch quad flat packs, etc. were developed to increase component densities. Now chip-on-board/TAB-on-board is being used. The only difference in the components for daughter cards is whether the components are housed in environmentally protected, plastic packages or ceramic, hermetic packages.

Using the logic block definition for MCMs, USI (UniStructure, Inc.) developed its multi-chip module substrate technology in response to three system packaging issues:

1) Component packaging density obtainable
2) Electrical criteria required to support high density digital systems
3) Mechanical / thermal requirements

Reprinted with permission from *Proc. 9th Int. Electron. Packaging Conf.*, pp. 211–229, 1989. Copyright © 1989 by the International Electronics Packaging Society, Inc.

The substrate produced by this technology could be used in either "hermetic cans" or environmentally protected (organic encapsulated) applications.

Background : System Packaging Issue

Component Packages

To demonstrate the effect that changing the component package has on substrate requirements, USI compared a Motorola MC68030 microprocessor in various packaging forms (see Figure 1a). Assumptions included a fine pitch package with an outer lead bond pitch of .008 inch. Also, the microprocessor was operated at 2.6 watts on a substrate measuring 5 in. by 5 in. which was populated entirely with the same die. The effective die package area was .400 inch x .420 inch. The die itself measured .330 in. by .350 in. An evolution is visible which transforms the die from a relatively low density component into a more efficiently packaged component. The percentage ratio of silicon die area to substrate area occupied by the package is used as the measure of packaging efficiency. As seen in Figure 1b, the silicon area per square inch of substrate occupied by the 68030 package increases from 3% for a dual in-line package to 68% for a fine pitch package such as TAB/FLC. To support these higher density packages, substrate characteristics change drastically. Routing densities increase because of the magnitude change in the number of pins per unit area. Finer lithography is needed to transform the design into photographic tools. The surface feature pitches of USI-designed product are in the .008 inch to .012 inch range. This results in MCM designs with 40% to 60% silicon to substrate area. The penalty paid is that power density goes up, greatly increasing the need for efficient heat dissipation.

Electrical Criteria

The substrates to be designed had to operate with signal speeds in the 25 to 100 MHz range with areas the size of daughter cards. It was important that the dielectric chosen would support these signal speeds and areas. As discussed in past literature (ref. 2) for a fixed sized substrate, lowering the dielectric constant of the polymer could increase the signal speed significantly. Lowering the dielectric constant would save on power and reduce crosstalk. Because water has an adverse effect on the dielectric constant, the selected dielectric should have low moisture absorption. This led to a careful review of potential polymers (see Table 1).

TABLE 1   Polymer Candidates for Dielectric

Property	BCB*	FEP+	Cynate Ester	Pyralin LX #	Pyralin 2545 #	Silicone	RO2800&
Dielect. Const. (1 MHz)	2.6	2.1	2.9	2.9	3.5	2.7	2.8
Dissip. Fact. (1 Mhz)	.0005	.0002	.005	.002	.002	.001	.002
Vol. Reist. (Ohms-cm)	9×10^{19}	$>10^{18}$	1.1×10^{18}	$>10^{16}$	$>10^{16}$	4.6×10^{15}	1×10^{8}
Moisture Absorp. (%)	.2	<.05	1.6	.5	2-3	.08	<.1
CTE (ppm/°C)	50	90	44	3	20	262	19
Elast. Mod. (psi)	--	--	--	1.4×10^{6}	2.3×10^{5}	321	1.5×10^{5}
Flex. Mod. (psi)	475	8.1×10^{4}	450	--	--	--	--

* Benzocyclobutene
+ Fluorinated ethylene propylene
# Pyralin is a registered trademark of the DuPont Company.
& RO2800 is a registered trademark of the Rogers Corporation.

Signal line attenuation in the substrate was also a concern. Our designs had signal speeds low enough that attenuation was the sum of conductor series resistance, skin-effect loss, and dielectric loss tangent (ref. 3). This meant that the substrate technology must support copper conductors, be capable of making fine line features that were thick (up to .0014 in.) and be able to minimize the loss tangent of the dielectric. The high power demands of the new IC devices packaged close together resulted in the need for voltage and power planes as thick as .0028 in. We felt that the electrical requirement for the conductor and plane thickness precluded the use of normal thin film technology.

Mechanical Considerations

Size:

The substrate technology had to accommodate size ranges which included a typical hybrid (2 - 8 sq. in.), a military daughter card (30 sq. in.) or the next-generation high-speed commercial substrate (up to 100 sq. in.)

Thermal Dissipation:

For present designs with up to 60% silicon on the substrate, power densities range from 10 Watts per sq. in to 20 Watts per sq. in. At this level of density, heat dissipation is critical. For many commercial applications, through-the-lid cooling provides an adequate path for heat transfer. However, for rigorous commercial or military applications, heat must be transferred through the substrate and then to a cooled core.

A thermal analysis was done on the feasibility of using via posts to conduct heat from the substrate surface to the core (see Figure 2). The analysis used calculations, finite element models, and actual test measurements to determine the thermal vias needed. For a power density of 15.5 Watts/in^2, a reasonable thermal resistance of .72 $^\circ$C/Watt from chip to substrate core was achieved using an array of .010 inch square posts on .050 inch centers (400 post/in^2). By utilizing an array, adequate routing channels are still available.

Two issues determine the need for lower CTE in the substrates; fatigue of copper leads bridging the silicon die to the substrate surface and the stress between large surface-mounted die and the substrate surface. One option is to match the silicon's CTE of approximately 2.5 ppm/$^\circ$C. However, this results in a large difference between the CTE of the copper (17 ppm/$^\circ$C) and the substrate. Expensive substrate materials would be required or the leads of the TAB/FLC could be lengthened, thus increasing the component package size. Another option is to decouple the stress on the silicon die by using a low modulus, compliant adhesive (ref. 4) or a low modulus solder such as 98/2. In a survey of USI customers, the CTE range requested varied greatly as the applications. Most customers would accept substrates in the 7 - 10 ppm/$^\circ$C range if we used a compliant die attach adhesive, and this would allow a balance between short lead stress and die stress.

Since the substrate core material chosen must be able to provide adequate thermal conductivity and constrain the thermal expansion of the dielectric over a customer's specified range, as well as support the physical size of the substrate. Table 2 represents some of the materials that might be used as substrate cores.

TABLE 2   Candidates for Substrate Cores

Material	CTE 10^6/$^\circ$C	Thermal Conduct. W/m$^\circ$K
Aluminum	23.0	240
Aluminum Nitride	3.3	230
96% Alumina	6.6	20
Copper/Molybdenum (20% Copper)	7.2	197
Invar	1.5	11
Kovar	5.3	50
Molybdenum	5.0	146
Silicon Carbide	3.7	270
Tungsten	4.5	200

Design Case Studies

Initially, we did not have enough density data to assume whether thin or thick film photolithography was required. The decision was made to look at a range of designs and let the routing results drive the process methodology to be used (see Figure 3).

Nine different designs were evaluated and can be summarized as follows:

1) Microprocessor MCM with 44 active die and 32 passive chips.
   Size: 1.497 in. x 1.800 in.

2) Microprocessor MCM with 16 active die and 11 passive chips.
   Size: .804 in. x 1.386 in.

3) Miscellaneous random logic MCM with 14 active die and 9 passive chips.
   Size: .882 in. x 1.218 in.

4) Memory MCM with 10 active die and 4 passive chips.
   Size: .948 in. x 1.270 in.

5) SEM - E (Standard Electronics Module) memory card with 247 active die and 98 passive chips.
   Size: 5.340 in. x 5.850 in.

6) SEM - B microprocessor card with 4 284-lead active parts, 3 other active parts and 8 passive chips.
   Size: 1.030 in. x 5.030 in.

7) High speed DSP (Digital Signal Processing) MCM with 107 active die and 50 passive chips.
   Size: 2.000 in. x 4.200 in.

8) Analog-to-digital MCM with 42 active die and 92 passive chips.
   Size: 1.000 in. x 3.330 in.

9) High speed CMOS/ECL (Complementary Metal-Oxide Semiconductor/Emitter-Coupled Logic) computer card with 66 active die and 67 passive chips.
   Size: 5.2 in. x 5.4 in.

The component footprint features were varied from .012 inch centers down to .008 inch centers for the test routes yielding silicon to substrate ratios from 35% to 70%.

Results : Design Criteria

As a result of the case studies, USI established the following criteria:

Via/Footprint Relationship:
Because of customer requests, three methods of interconnecting the footprint to solid post vias were required : vias under the footprint, vias in the interior of the footprint, and vias on the exterior of the footprint. Vias located directly under the footprint pads make routing more difficult because routing channels through the footprint are reduced by the vias. Vias located on an area array in the interior of the footprint pads, using a transition pattern, allow for the optimization of routing channels. The vias can be located in strategic patterns so as to open routing channels in selected directions. Vias located outside the footprint pads allow for "cut and jump" repairability, but utilize more area on the surface of the MCM, i.e. reduce % silicon to substrate.

Via Type:

Three types of solid post vias allow for various interconnections within the MCM. Stacked or staggered through-vias are used to electrically connect one side of the MCM to the other. Stacked through-vias are used as a heat transfer path if not used electrically. Buried vias (vias that connect traces within a signal pair) and blind vias (vias that connect two different signal pairs) are also used to make electrical interconnections and could be stacked or staggered through the substrate.

Trace Widths:

Using .0012 to .002 inch wide conductors on .004 inch centers and "invisible" vias (vias that are contained with the width of the trace), all 9 of the case studies were routed in two circuit layers. Using .003 to .004 inch wide lines and invisible vias on .010 inch centers, some of the case studies required 4 signal layers. The worst case occurred with .004 inch wide traces and .008 inch square vias, on .012 inch centers, which required 6 circuit layers. However, the larger features would produce much higher yields on large area substrates and can be more cost competitive than finer features over large areas. A trade-off exists between the use of thermal posts for the conduction of heat to the core and the area removed from available routing channels. When .010 inch square thermal posts in .050 inch center arrays were used, all the case studies could be routed in two layers.

USI Substrate Capability

Based on the considerations outlined above, a MCM substrate line was installed at USI with the following capabilities:

TABLE 3   Substrate Core Characteristics

Size: Panel: 12 in. x 12 in.
Substrate image size up to 10 in. x 10 in.

Substrate Base Materials:
Aluminum
Brass
Copper
Copper/Invar
Copper/Molybdenum
Copper/Tungsten
Invar
Molybdenum
Tungsten

Metal bases were chosen over ceramic, silicon carbide,etc. because of metal's relatively low cost and robust handling characteristics for large sized MCM's. Metals also provide USI with the ability to tailor CTE and thermal conductivity. The ease of selectively removing the metal with photolithographic and wet chemical etching techniques allows additional design flexibility.

Liquid resins were necessary for the process to be additive and to encapsulate the fine features. USI has developed UV (ultraviolet light) curable liquid resins with low dielectric constants and reduced moisture content. Because the resins contain 100% solids, high layer counts can be attained without residual stresses (Shrinkage is reduced greatly because there are no solvents to drive off.) For higher performance applications, blends of polyimide resins are being used but have a higher substrate cost.

TABLE 4   Liquid Resin Properties

	Aliphatic Resin	Polyimide
Dielectric Constant (1 MHz)	2.5	2.9 - 3.5
Dissipation Factor (1 MHz)	.0014	.002
Volume Resistivity (Ohms-cm)	$>10^{17}$	$>10^{16}$
Operation Temperature (°C)	160	400+
% Moisture Absorption	1.2	.5 - 3
CTE (ppm/°C)	140	3 - 20
Modulus of Elasticity (PSI)	1×10^4	$2 \times 10^5 - 2.2 \times 10^6$

Table 5   Conductors and Plane Characteristics

	Avg inches	Avg microns	Minimum inches	Minimum microns
Line Width (+/- 5%)	.002/.004	50/100	.0008	20
Line Spacing	.002/.004	50/100	.0016	40
Vias (solid copper posts)	.002/.008	50/200	.0008	20
Line Thickness	.0007/.0012	18/ 30	.0006	15
Plane Thickness	.001/.003	25/ 75	.0006	15
Routing Density	640-320 in/in²*		1600 in/in²*	

* Based on 4 layer circuit at 80% theoretical route efficiency

TABLE 6   Surface Metalization

	Avg inches	Avg microns	Minimum inches	Minimum microns
Nickel	.0002	5	.00004	1
Gold (99.99% pure)	.0002	5	.00004	1
Fused Tin/Lead	.002	50	.0002	5

Substrate Processing

Referencing Figures 4 and 5:

A)   Prepare a metal panel that will serve as the substrate's base. The panel (annealed if required) is ground and polished (planarized) to provide a flat and parallel surface from which to begin.

B)   Liquid resin is flow coated over the panel in a thickness of approximately .001 inch. The dielectric is cured. The epoxy resin is initially UV cured followed by an oven cure at approximately 180 °C. For the polyimide resin, a stepped temperature vacuum cure is completed, followed by a post cure at 420 °C. The dielectric is then planarized. If the substrate contains thermal or feedthrough vias that connect to the core, the vias would be imaged and electroformed before the first layer of dielectric would be applied.

C)   An adhesion layer is sputtered over the panel approximately 1000 angstroms thick. Copper is sputtered approximately 4000 angstroms thick.

D)   Dry film photoresist is laminated on the panel. The thickness of the photoresist is approximately the thickness of the finished conductor.

E)   The conductor pattern of the first layer is imaged on the resist and the photoresist is stripped with aqueous solution. Currently, dry film photoresist is used for features down to .002 inch lines and spaces. For finer lithography, we use laser photoablation.

F)   Copper is electroplated on the panel with a thickness of .001 inch for lines and .001 inch to .003 inch for planes.

G)   Another layer of dry film resist is applied to form the vias. The resist is slightly thicker than the final via height.

H)   The via image is exposed.

I)   The solid copper vias are electroplated.

J)   The photoresist is then stripped. The initial sputtered adhesion layer of metal is flash etched. When using polyimide resins, the exposed copper conductors are coated with electroless nickel approximately 1000 angstroms thick.

K)   Liquid dielectric resin is flow coated over the panel to encapsulate the lines and vias.

L)   The dielectric is planarized to the correct thickness required by the electrical criteria. The tops of the vias are now exposed.

The process is then repeated for the subsequent layers until all the layers are added. (This fabrication sequence is similar to MCC's solid post additive process. ref 5) The surface features are completed last. Figure 6 shows a typical cross-section of a completed substrate. The locking pads give excellent mounting pad peel strengths. Figure 7 thru 9 show examples of substrate processing.

Substrate Acceptance Testing

Appropriate industry-standard acceptance specifications are not currently developed for MCM substrates. Since specifications do exist for printed circuit board substrates, USI chose to pattern in-house environmental testing around these specifications with any additional tests as needed. The summaries of the test results below are for product built with USI epoxy resin on both copper/molybdenum and brass bases.

Table 7   Temperature Cycling

    Per MIL-STD-202 Method 107
    -65 $^{\circ}$C for 1/2 hour
    Transfer Time = Seconds
    125 $^{\circ}$C for 1/2 hour
    Transfer Time = Seconds
    Boards cycled 100 times = Passed
    (Boards were continuity tested every 25 cycles)

    Equipment: Blue M Dual Thermal Shock Test Cabinet
               WSP-109DMP3

Table 8   Thermal Shock

    -70 $^{\circ}$C for 30 seconds (Using Dry Ice and Acetone)
    Transition Time = Seconds
    15 $^{\circ}$C for 30 seconds (Using Hot Fusing Oil)
    Transition Time = Seconds
    Boards cycled 100 times = Passed

Table 9 Moisture and Insulation Resistance Testing

   Per MIL-P-55110, IPC-TM-650, Method 2.6.3 Class 3
   25-65 ^{O}C at 90-98% Relative Humidity for 160 hours
   Polarization Voltage : 100 VDC
   Final I.R. Measurements at 50 VDC
   Pass = 500 megohms minimum

   Within Layer: Between Layers:
   4.0 to 10.0x10^{10} Ohms   2.0 to 8.0x10^{10} Ohms

   Sample Size, 6 parts = Passed

Table 10   Internal Water-Vapor Content (Inside a
            hermetically sealed device)

      Per MIL-STD-883 Method 1018, Procedure 1
      3000 ppm maximum water content at 100 ^{O}C for Class 5
      Sample Size, 6 parts = Passed

Table 11 Electrical Tests

   Continuity (on all substrates)
     Tested at 5 VDC
     100% of Circuits
     PASS : 2 Ohms maximum

   Insulation Resistance (on all substrates)
     Tested at 250 VDC, 25 megohm
     All nets to each other
     PASS : Current leakage not to exceed 100 microamps
     (Also can capacitance map nets)

   Impedance (per customer requirements)
     Sampling of Critical Nets
     Design Goal : 50 Ohms
     Actually measures :
        Microstrip = 53 Ohms +/- 10%
        Stripline = 47 Ohms +/- 10%
        Sample Size = 5

Summary

     By using both thin film and thick film processing techniques, USI is able
to produce multi-chip modules without the commonly used combination of ceramic
and polyimide. Metal cores allow for large substrates, tailorable CTEs ranging
from 5 to 17 ppm/^{O}C, thermal conductivities of 200 W/m^{O}K, as well as ease of
workability.   The photographic techniques of the process combined with
electroforming of copper conductors and solid vias will support 40% to 60%
silicon to substrate ratios with the capability of higher ratios in the future.
The low dielectric constant, liquid resins naturally support high-speed digital
electronics.   UniStructure's Very High Density Interconnects (VHDI) provide
system designers with flexible base technology for building multi-chip modules.

References

1. Tony Johnson, UniStructure, Inc., "The Flexible Leaded
    Component (FLC) : Reliable "Bare-Chip" Packaging," 9th
    Annual IEPS Conference, pp. TBD, September 1989

2. John Balde, IDC Corporation, "Low Dielectric Constant-
    The Substrate of the Future," 7th Annual IEPS
    Conference, Volume two, pp. 860-892, 1987

3. Evan E. Davidson and George A. Katopis, "Package
    Electrical Design," (Chapter 3), Microelectronics
    Packaging Handbook, pp. 128-130, Van Nostrand Reinhold,
    New York, 1989

4. Brian J. Sova, Lori A. DeOrio, and Tony K. Johnson,
    McDonnell Douglas Electronics Company, "Materials
    Engineering Considerations in Bonding Surface Mount
    Technology Printed Wiring Board to Aluminum Heat
    Sinks," SAMPE Conference, 1987

5. Tony Pan, S. Poon, and Brad Nelson, Microelectronics and
    Computer Technology Corporation, "A Planar Approach to
    High Density Copper Polyimide Interconnect," 8th Annual
    IEPS Conference, pp. 174-189, November 1988

	DIP	PGA	SMT	FINE PITCH
Silicon/Inch²	3%	6%	10%	68%
Pins/Inch²	31	69	110	762
Surface Feature Pitch	.100	.100	.025	.008
Avg. Watts/Inch²	.63	1.40	2.24	15.49
Substrate Area Reduction Comparison	1	.45	.28	.04

MC68030 Microprocessor At 2.6 Watts - 128 Pins
Effective Component Package Size: .400 inch x .420 inch
                    (Die Size): .330 inch x .350 inch

Substrate Size: 5 x 5

Figure 1a: Packaged Part Comparison

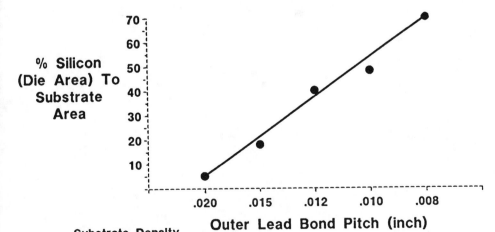

% Silicon (Die Area) To Substrate Area

Outer Lead Bond Pitch (inch)

Substrate Density Assumptions:

  Graph Based On 9 Current Designs
  Parts Effectively Use 80% Of Available Area
  I/O Occupies .100 inch Around Substrate Perimeter

Figure 1b: Silcon To Substrate Density

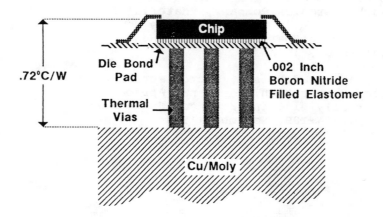

Assumptions: – 15.5 Watts/in²
            – Vias Occupy ≅ 4% of Area Under Die
            – Via Height = .012 in.

Result: – Thermal Resistance From Junction Temperature
          to Top of Core = .72°C/Watt

Figure 2: *Power Dissipation Through Substrate*

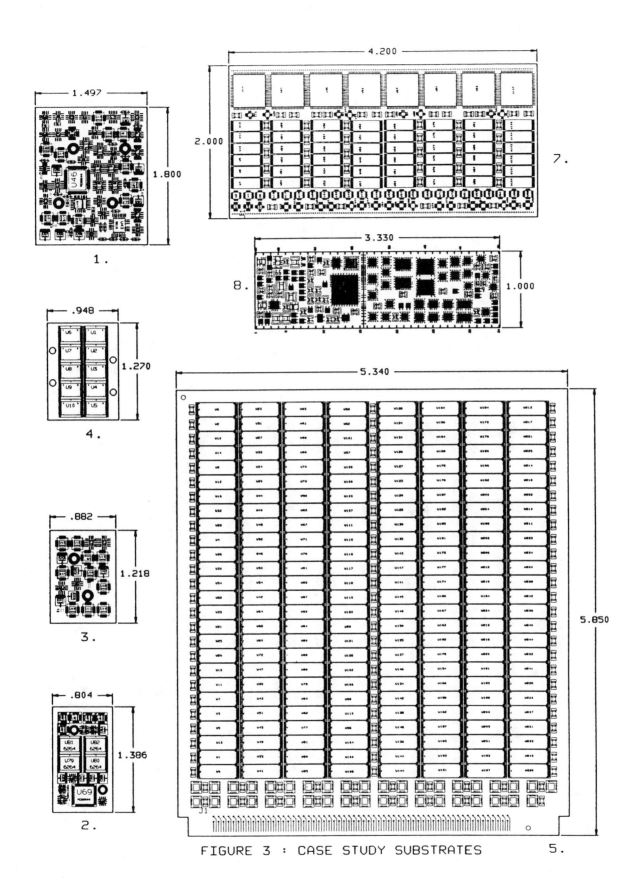

FIGURE 3 : CASE STUDY SUBSTRATES

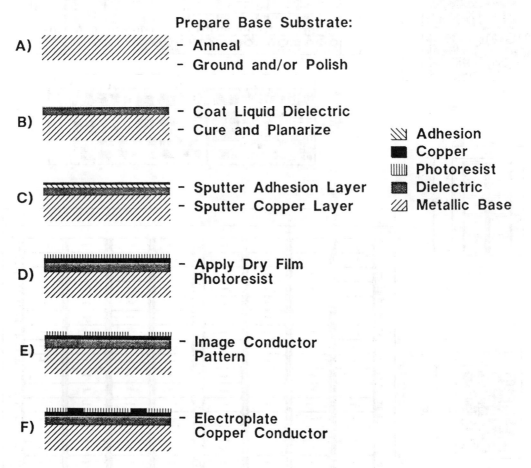

Figure 4: Overview of Manufacturing Process

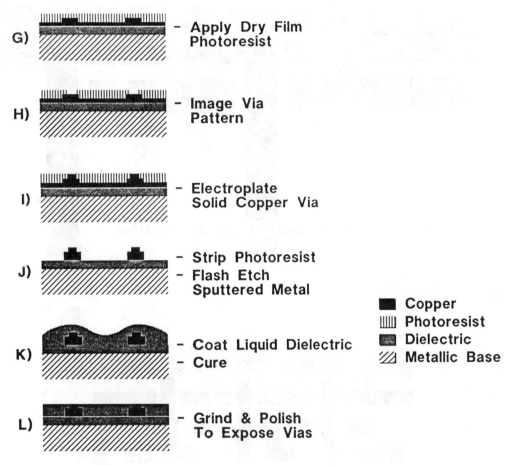

G) – Apply Dry Film
Photoresist

H) – Image Via
Pattern

I) – Electroplate
Solid Copper Via

J) – Strip Photoresist
– Flash Etch
Sputtered Metal

K) – Coat Liquid Dielectric
– Cure

L) – Grind & Polish
To Expose Vias

■ Copper
▥ Photoresist
▦ Dielectric
▨ Metallic Base

Figure 5: *Overview of Manufacturing Process*
(Continued)

LOCKING PADS/TRANSITION LAYER

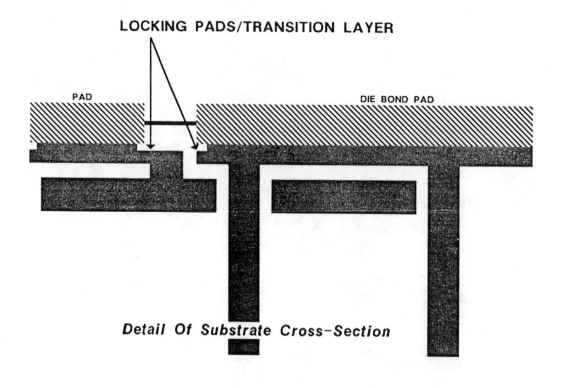

Detail Of Substrate Cross-Section

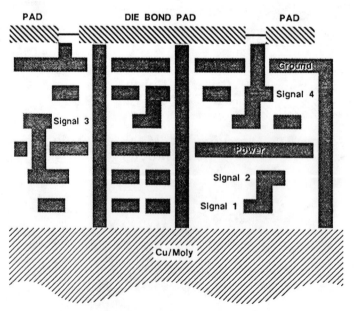

Figure 6: Typical Substrate Cross-Section

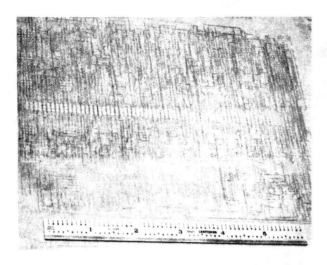

a.) .004″ line, .001″ thick electroplated conductor pattern single image on 12″ x 12″ panel.

b.) 20 micron line, 25 micron space in .002″ (50 micron) photoresist. High aspect, straight side walls done with laser photoablation.

c.) Via posts on lines after flash etching and prior to dielectric coating.

Figure 7: Examples of Substrate Processing

a.) Via formed in .0007"
(18 microns) photoresist.
Holes are left to right:
25, 15, 10 and 5 microns.
Holes formed by laser
photoablation.

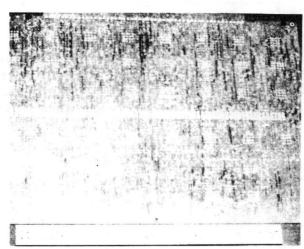

b.) Panel after dielectric
coating and planarizing
to expose vias. Large
square via patterns
are thermal posts.

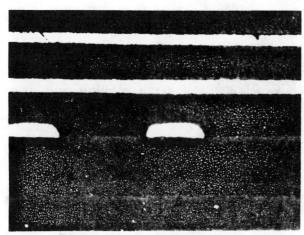

c.) Cross-section shows
the planarity of the
dielectric layer.
Conductor is .004" with
plane and conductor
thickness .001".

Figure 8: Examples of Substrate Processing (Cont.)

a.) Cross-section of typical stacked via structure. Via .004″ square.

b.) .002″ square buried via between signal pairs. Various conductor widths.

c.) Example of locking feature on die attachment pads.

Figure 9: Examples of Substrate Processing (Cont.)

PACKAGING DESIGN OF A SiC CERAMIC MULTI-CHIP RAM MODULE

by

Ken Okutani, Kanji Otsuka, Kunizo Sahara and Kazuyoshi Satoh

Device Development Center of Hitachi Ltd.
1450 Josuihon-cho, Kodaira shi, Tokyo, Japan 187

ABSTRACT

A new type of multi-chip silicon-carbide module technique is described. Face-down multiple LSI chips are jointed by controlled collapse bonding to a silicon substrate that provides interconnections. Then, the silicon substrate with LSI chips is bonded by gold-tin eutectic to a new type of SiC ceramic base. Finally, the package is completed by aluminum wire bonding and silicone gel encapsulation.

The advantages of this module are as follows.
(1) Excellent thermal matching.
(2) Low thermal resistance.
(3) High packaging density.
(4) The use of conventional IC process to fabricate silicon wiring substrate.
(5) High reliability.

INTRODUCTION

Fig.1 shows a photograph of a SiC ceramic multi-chip module including six chips of 1K bit ECL RAM , a silicon substrate and a SiC ceramic base. This multi-chip module is designed to realize following items.
(1) Thermal resistance; less than 6° C/W.
(2) High reliability of frip chip solder joints under ambient temperature cycles and LSI power cycles.
(3) High packaging density; six-chip RAM module in 27.4mmX27.4mm package size with 108 outer leads for surface mounting.
(4) Sealing; silicone gel encapsulation

The development has been undertaken to ascertain the feasibility of above design concept and has achieved a package with high reliability and high cost performance.

PACKAGING DESIGN AND ASSEMBLY TECHNIQUE

Fig. 2 shows the cross section of the SiC six-chip RAM module. Multiple RAM chips are faced down and jointed by 95Pb/5Sn solder ball bonding to silicon substrate to provide interconnection and heat dissipation. Each RAM chip has 25 bumps for I/O and additional 52 bumps for heat dissipation. Then, the assembled silicon substrate is tested completely and after the testing, it is bonded by gold-tin eutectic to a SiC ceramic base at 295°C in nitrogen atmosphere. The die bonding temperature 295°C is lower than the liquidus temperature of 95Pb/5Sn solder joint. The thermal resistance of gold-tin eutectic bonding layer is less than 0.001°C/W, and no thermal resistance problem will occur for die bonding. Then, ultrasonic aluminum wire bonding is performed to provide a connection between the silicon substrate and the lead frame that connects to the outside. Then, the assembled silicon substrate and aluminum wire are filled with dispensed silicone gel that has two functions in this module. One is the protection of the assembled silicon substrate and aluminum wire from the outside humidity. The other is the protection of memory cells from alpha particle bombardment, memory cells are separated from solder bumps by more than $100\mu m$ and filled with silicone gel in this narrow space. In general, silicone gel encapsulation is well known as the method for high reliability sealing which is based on the silicone gel characteristic such as lower interfacial stress between gel and packaging parts, and micro penetration force into the micro crevices after the curing. Then, the lid and aluminum fin for heat dissipation were attached to the module by adhesive silicone rubber that has the elastic chracteristics. Consequently, no stress problem induced by lid and aluminum fin attachment. Finally, the lead frame is cut and shaped into the style of gull wings to provide surface mounting on the PCB board, and surface mounting has the advantage of high packaging density.

This SiC multi-chip module is successfully designed about the hierarchy systems concerning to the assembly process temperature and yield consideration. And the functions of this module structure are clearly divided into three kinds according to the aspect of the module, the upper side aspect provides heat dissipation, the beneath aspect provides the assembly and maintenance area, and the peripheral side aspect provides the connection with the outside world.

THERMAL RESISTANCE

In order to reduce the thermal resistance of this module less than 6°C/W, packaging materials and structure are well designed. Fig.3 shows the thermal conductive path of the RAM module. The thermal resistance between six RAM chips and silicon substrate R_{RS} is

$$R_{RS}=(R_{RC}+R_{CP}+R_B+R_{MP})/6 \qquad (1)$$

All symbols in the equation are indicated in Fig.3. Firstly in order to reduce R_{RS}, each RAM chip was designed 77 bumps (Fig.4); 25 bumps for I/O and additional 52 bumps for heat dissipation. The additional 52 bumps reduce 75% of R_B and R_{RS} is reduced

to 1.44°C/W. Secondly, the use of gold-tin eutectic die bonding has the advantage of thermal conduction, the thermal resistance of this layer is less than 0.001°C/W and no heat resistance problem for die bonding will occur. Thirdly, the use of SiC ceramic base has the advantage of high thermal conduction. Table 1 summarizes the physical properties of the SiC ceramics. The thermal conductivity of SiC ceramic is 270 W/m.k and which is 13.5 times larger than that of Al_2O_3, and the thermal resistance of SiC base is only 0.025°C/W. Thermal resistance from silicon substrate to ambient air R_{SA} is

$$R_{SA}=R_S+R_{DB}+R_{SiC}+R_{SiA}+R_A$$

$$=0.08+0.001+0.025+0.14+4.1$$

$$=4.35°C/W \qquad (2)$$

The total thermal resistance of SiC RAM module is

$$R_T=R_{RS}+R_{SA}=1.44+4.35=5.79°C/W \qquad (3)$$

The experimentally evaluated thermal resistance is 5.5°C/W and it corresponds with the simulated value 5.79°C/W, This value is satisfactory for high speed bipolar RAM. (1W/chip and 4.5 ns Address access time)

MATCHING IN THERMAL COEFFICIENT

Matching in thermal coefficient is a must in packaging. In this SiC RAM module, matching in thermal expansion coefficient is excellent. The expansion coefficient of LSI chips and silicon substrate is the same, and the expansion coefficient of SiC ceramic base is also almost the same. Therefore in the ambient temperature cycle test, no thermal stress problem will occur. And this module was good for over 1000 cycles in the $-35°C \rightleftarrows 150°C$ temperature cycle test. However, as for the LSI power cycle test, the temperature difference between LSI chips and silicon substrate causes the slight thermal mismatch. But, the excellent reliability of solder joint was evaluated by coffin-Manthon formula calculation as follows.

The number of cycles to failure (N_f) is in the following form.

$$N_f = c \cdot f^{1/3} \cdot \gamma_{max}^{-2} \cdot \exp\left(\frac{\Delta E}{K \cdot T_{max}}\right) \qquad (4)$$

$$\gamma_{max} = \frac{1}{(D_{min}/2)^{2/\beta}} \left(\frac{V_j}{\pi \cdot h_j^{1+\beta}}\right)^{1/\beta} \cdot \alpha \cdot \Delta T \cdot d \qquad (5)$$

where

γ_{max}; maximum local strain on the joint

$c=0.05$; constant determined by solder composition
$E=0.123$ eV; activation energy
$\beta=0.6$; material constant
$K=8.6 \times 10^{-5}$ eV/ K; Boltzman constant
$f=1$ cycle/day; frequency of the temperature cycle
$T_{max}=338°K$; maximum Temperature
$d=2.21 \times 10^3 \mu m$; distance of each of the joints from the neutral point
$\alpha=3.5 \times 10^{-6}$; expansion coefficients of silicon
$T=R_{RS} \times W$
   $=1.44 \ (°C/W) \times 6(W)$
   $=8.64°C$; temperature difference between the RAM chip and silicon substrate
$D_{min}=90\mu m$; contact diameter of the joint
$h_j=70\mu m$; height of the joint
$V_j=6.25 \times 10^5 \mu m^3$; solder volume

the maximum local strain is

$$\gamma_{max}=0.0017 \qquad (6)$$

the number of cycles to failure (N_f) is

$$N_f=1.2 \times 10^6 \ (\text{correspond to 3300 year}) \qquad (7)$$

Consequently, according to all above the consideration, no thermal stress problem will occur under both ambient and power cycle conditions, and reliability is enhanced.

FUTURE FEASIBILITY

In order to develop this type of SiC multi-chip module in the near future, various new methods have been considered and undertaken to ascertain the feasibility of total high performance.

(1) Reduction of the thermal resistance

The degree of LSI integration becomes higher, reduction of the thermal resistance is inevitable in packging. In order to reduce thermal resistance, a heat conduction through hole structure under the solder joint is considered.(Fig.5) This structure provides the higher thermal conductivity with the SiC substrate.

(2) Reduction of electrical resistivity

As VLSI propagation delay time becomes shorter, interconnection delay in the substrate between VLSI's becomes a serious problem, and the electrical resistivity reduction is a must in high speed computer systems. In order to reduce the electrial resistance, copper interconnection on the substrate must be developed.

RESULT

A SiC ceramic multi-chip module has been successfully desiged, processed, assembled and tested from concept to finished module.

The thermal resistance of this SiC multi-chip module was simulated 5.79°C/W and experimentally evaluated 5.5°C/W with 3m/s forced air.

This module was good for over 1000 cycles in the ambient temperature cycle test and simulated no problem in LSI power cycle test.

High packaging density was ascertained by six RAM chips in 27.4mmX27.4mm package size with 108 outer leads for surface mounting.

Silicone gel sealing provides the high reliability in environmental acceleration test.(Table 2)

CONCLUSION

The SiC RAM module has performed satisfactorily in testing and high packaging density.

This SiC multi-chip module technique will be applicable not only to RAM module but also to very large scale VLSI logic and RAM hybrid module in the near future.

ACKNOWLEDGEMENTS

The authors would like to express their sincere thanks to S. Kuroda, A. Anzai, S. Hososaka; Hitachi Device Development Center, F. Kobayashi; Hitachi Kanagawa Works, for their constant encouragement and advices.

REFERENCES

1) C. Huang, W. Nunne, D. Spielberger, A. Mones, D. Fett and F. Hampton, "SILICON PACKAGING A NEW PACKAGING TECHNIQUE," IEEE Proceedings of CICC, 1983

2) Kanji Otsuka, Yuji Shirai and Ken Okutani, "Sealing Mechanism of Silicone Jelly Encapsulation with High Reliability," 34th Electronic Components Conference Proceedings, 1984, pp88-94

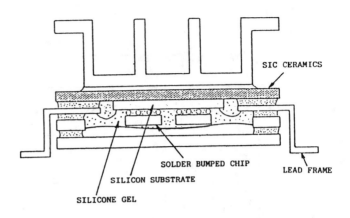

Fig.2   CROSS SECTION OF MULTI-CHIP MODULE

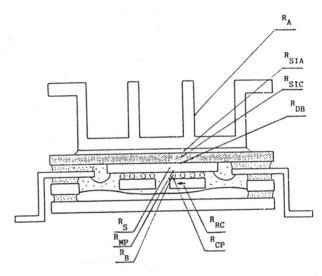

Fig.3   SCHEMATIC DIAGRAM OF THERMAL RESISTANCE

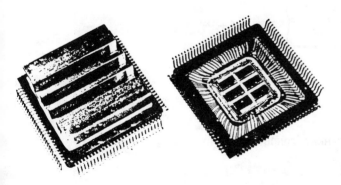

Fig.1   SiC CERAMIC MULTI-CHIP MODULE

Fig.4   MICROPHOTOGRAPHS OF THE RAM CHIP

TABLE 1. PROPERTIES OF NEW TYPE SiC CERAMICS AND VARIOUS MATERIALS

MATERIAL	THERMAL CONDUCTIVITY (W/m·K)	THRMAL EXPANSION COEFFICIENT (10^{-6}/°C)	ELECTRICAL RESISTIVITY (Ωcm)	DIELECTRIC CONSTANT (1MHz)
NEW TYPE SiC CERAMICS	270	3.7	$>4\times10^{13}$	42
Al_2O_3 CERAMICS	20	6.7~7.5	$>10^{14}$	8~10
BeO CERAMICS	240	8.0	$>10^{14}$	6~8
ALUMINUM	230	25.7	2.7×10^{-6}	——
SILICON SINGLE CRYSTAL	125	3.5~4.0	——	12

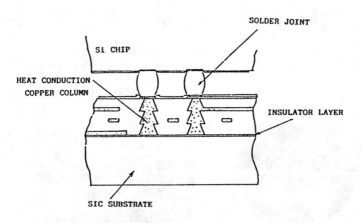

Fig.5 SCHEMATIC DIAGRAM OF HEAT CONDUCTION AND
INTERCONNECTION THROUGH HOLE

Table 2   Environmental acceleration test results of the devices

Test items	condition	results (No. of failures /No. of samples)				Failure mode
		192hr.	360hr.	500hr.	1000hr.	
High temperature operation	125°C	0/45	0/45	0/45	0/45	
High temperature storage	150°C	0/32	0/32	0/32	0/32	
Low temperature storage	-35°C	0/32	0/32	0/32	0/32	
High humidity/ operation cycle test	65°C/95%R.H.	0/32	0/32		0/32	
		320cycles	600cycles	1000cycles		
Temperature cycling	-35°C⇄150°C	0/45	0/45	1/45 (*1)		Leak current

*1 Leak current by a conductive foreign particle in wafer process.

ULTRA-RELIABLE PACKAGING FOR SILICON-ON-SILICON WSI

John K. Hagge
Advanced Technology & Engineering
Avionics Group
Rockwell International Corporation
400 Collins Road NE
Cedar Rapids, Iowa 52498

ABSTRACT

Silicon-On-Silicon WSI (Wafer Scale Integration) Packaging provides electronic equipments with significant reductions in size, weight, cost, and IC junction temperatures, together with significant increases in reliability and high speed electrical performance. The new packaging technique combines semiconductor lithography techniques, printed circuit multilayer techniques, and hybrid multichip module techniques. The silicon substrate has multiple layers of metallization and dielectric and serves as a "silicon circuit board". This paper discusses the advantages of using silicon instead of conventional ceramic as the substrate material, and reviews the published status of this technology at other organizations. While conventional hybrid packages are successfully being used in early implementations to reduce size and weight, there exists an untapped potential for significant reliability improvements by switching to packages specifically designed for silicon substrates. Several potential packaging approaches are reviewed and results are presented for the fatigue life and thermal performance of silicon substrates.

INTRODUCTION

Conventional circuit board packaging of electronics typically uses multilayer circuit boards to interconnect single-chip IC packages. The functions of the single-chip packages are to mount and protect the IC chips, to provide an interface for testing individual chips, and to provide an interface for assembling them onto the interconnecting circuit board. This approach has worked well for many years, but driven by a desire for miniaturization, single-chip packages have become a limitation to further miniaturization. The trend for reduction in package size in the evolution from Dual-In-Line to Chip Carrier to Small Outline packages, is now being counteracted as package sizes begin to grow to accommodate the high I/O count of VLSI and VHSIC chips. But in all cases the single-chip package occupies many times more space than the actual IC. The single-chip packages also add weight, package

cost, an impediment to heat removal, longer signal paths, and additional levels of interconnection which reduce the reliability of the electronic assembly. An alternate to single-chip packaging is needed which provides adequate mounting, protection, testability, and interconnection.

Hybrid circuit packaging has long used multiple-chip packaging to accomplish miniaturization. However, testability issues, low density interconnections on substrates, and high costs present limitations to this technology. A new multichip packaging approach comes from using semiconductor lithography techniques to build a high density, multilayer thin film interconnect on a silicon wafer substrate as shown in Figure 1. The IC chips can be mounted very close together thus avoiding single-chip packaging limitations. A four-to-one size reduction over surface-mount packaging can be achieved as shown in Figure 2. The silicon-on silicon packaging approach is extremely flexible. The chips attached to the silicon substrate can be LSI, VLSI, VHSIC chips or discrete chip components. Whereas other approaches are developing complex processes to grow both silicon and gallium arsenide devices on the same wafer [1-3], the ICs in this new approach can be pre-tested and any mixture of GaAs, MOS, or Bipolar devices. Similarly, the die-to substrate interconnect can be chosen from wire-bond, flip-chip, or TAB techniques. The package protecting the multichip substrate can be fully hermetic for harsh environments or a low cost package with the ICs protected by silicone gel encapsulants. The package configuration can range from flatpacks to pin-grid-arrays to integral package-connector modules.

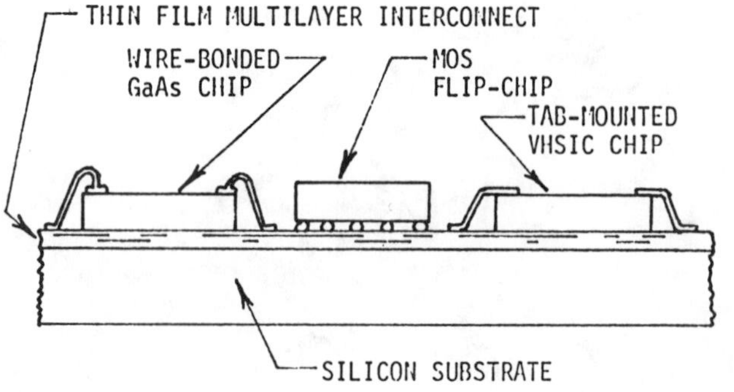

THIN FILM MULTILAYER INTERCONNECT
WIRE-BONDED GaAs CHIP
MOS FLIP-CHIP
TAB-MOUNTED VHSIC CHIP
SILICON SUBSTRATE

FIGURE 1.   CONSTRUCTION OF INTERCONNECT ON SILICON

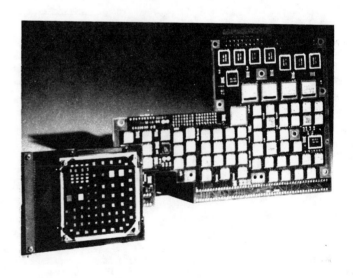

FIGURE 2.   MODULE IS 1/4 SIZE OF SMT PWB

Reprinted from *Proc. 38th Electron. Components Conf. (ECC)*, pp. 282–292, 1988.

ADVANTAGES OF SILICON-ON-SILICON PACKAGING

The choice of silicon for the substrate material instead of conventional alumina provides many advantages:

* The substrate coefficient of thermal expansion is an exact match with large silicon VLSI and VHSIC chips, providing higher reliability.
* Silicon substrates give lower IC junction temperatures since the silicon substrates have thermal conductivity 5 to 10 times higher than conventional ceramic substrates.
* They can have active devices (such as mass memory) built directly into the substrate, with the other chips mounted on top of the substrate.
* Wafers can use IC lithography techniques to make interconnect line/space features of 10 to 50 microns permitting higher IC densities than competing technologies as shown in Table A.
* The higher density packaging results in shorter line lengths, giving 2 to 3 times shorter delay times compared to surface mount packaging.
* The improved density creates room for redundancy in the circuit design for higher reliability.
* The semiconductor interconnection technology used is more reliable than solder joints or thick film techniques.
* Wafers are compatible with automated IC processing equipment for low cost, high volume production of substrates.
* The substrates have high yield since line/space sizes are ten times larger than the IC equipment was designed for.
* High purity, low cost polished silicon wafers are widely available, at less than $1 per sq. in.
* A selection of dielectrics for the multilayering is available with dielectric constants ranging from 2.6 to 14.
* Silicon wafers are extremely flat and dimensionally stable, eliminating dimensional control and process control problems associated with conventional thick film and co-fired substrates.

PACKAGING TECHNOLOGY	LINE/SPACE SIZE (micron)	PACKAGING EFFICIENCY*	TYPICAL ICs/IN²
INTEGRATED CIRCUIT	0.3 to 3	100%	–
SILICON-ON-SILICON	10 to 50	30–60	8–20
CHIP-ON-BOARD	100 to 200	15–30	5–9
HYBRID THICK FILM	125 to 250	10–30	3–8
SURFACE MOUNT PWB	125 to 250	6–14	2–4
VHSIC PACKAGES	100 to 200	5	.3–1
DIP PWB	200–300	1–3	1–2

* packaging efficiency=100xIC area/footprint area

TABLE A. COMPARISON OF SEVERAL PACKAGING TECHNOLOGIES

INDUSTRY STATUS OF SILICON-ON-SILICON WSI

The advantages and characteristics of various aspects of silicon-on-silicon packaging have been widely discussed and information on the development work of several dozen organizations has been published [4-114]. Some of the companies working on this technology are listed in Table B. The increasing number of recent articles is indicative of the growing interest in silicon-on-silicon packaging. The present status of this technology is essentially

that the potential for size and weight reduction, improved thermal performance, and improved high speed electrical performance have all been adequately demonstrated by several groups. The technology has moved out of the research lab, and some production implementations have been made in equipments being delivered to customers [e.g., 12,40,43,78,91]. Specific examples exist of successful packaging designs. However, widespread implementations which demonstrate the full potentials of improved reliability and low cost of silicon-on-silicon packaging are still to come. The remainder of this paper will focus on some significant advancements possible in electronics equipment reliability through development of packaging approaches specifically designed for silicon substrates, and on the key developments which must occur to achieve those improvements.

ORGANIZATION	REFERENCES			
	to 1982	1983-1984	1985-1986	1987-1988
AT&T			[4]	[5-16]
AUBURN			[17-21]	[22-27]
GE	[28]	[29]	[30-35]	[36-38]
GIGABIT LOGIC			[39-40]	
GOULD			[41]	
HITACHI		[42-43]	[44-45]	
HONEYWELL		[46-47]		[48]
IBM	[49-51]	[52]	[53]	[54-56]
INTEL			[61]	[62]
LAWRENCE LIVMR	[63]		[64]	
LETI LABS			[65]	
MCC			[66]	[67-69]
MIT				[70-71]
MITSUBISHI		[72]		
MOSAIC	[113]		[73-77]	[78-83]
NORTHERN TELCM			[84]	[85]
NTT			[86]	
PLESSEY			[87]	
PRIME COMPUTER			[88]	
RAYCHEM			[89-90]	[91-93]
RAYTHEON				[79]
RENSSELAER POLY			[94-95]	[96]
ROCKWELL				[97-100]
STANFORD	[101]	[102]	[103-107]	[108-110]
TRILOGY TECH			[111-112]	
UNISYS			[111,113]	
OTHER				[114]

TABLE B. SILICON AS AN INTERCONNECTING SUBSTRATE

NEW APPROACHES TO RELIABILITY AND PARTITIONING

Some interesting studies have been done on the relationship between field life costs and reliability of avionics equipments. On one hand there is an issue of what sort of black box MTBF is required in an aircraft loaded with very many black boxes, to achieve a desired MTBF for a fleet of aircraft. This is of critical importance whether one is concerned with the defense preparedness of a military fleet or on-time departures of a commercial fleet. The second issue is the trade-off between initial equipment costs and life cycle maintainability and repair costs. The total life cycle costs include not only the costs of repair labor and replacement parts, but also the costs of maintaining repair depots, test equipment, training of repair technicians, maintaining up-to-date repair manuals, inventory and transport, expenses caused by delayed flights, etc.

There would be a tremendous benefit to avionics customers if the reliability of electronic equipment could be improved by an order of magnitude. Commercial customers are intrigued by the possibilities of "life-of-the-airframe" or twenty year mean time between failure electronics. Military customers can see a tremendous impact if their equipment could be built to guarantee a 100,000 hour mean time between replacement. Both have been led to ask the obvious question: What would it cost to design the next generation systems for this order of magnitude improvement in reliability? Such designs could obviously be a good investment for customers even if they cost substantially more than present equipment.

Improvements in the reliability of electronics equipments could be achieved by a four-fold approach: 1) reducing parts count, 2) screening to higher levels, 3) designing in fault tolerance, and 4) addressing the dominant failure mechanisms. Recent trends are toward use of higher and higher levels of integration in ICs. VLSI and VHSIC chips combine the function of dozens of less sophisticated chips which they replace. Even though the new chips are more complex, the assemblies are more reliable, the interconnections on silicon being more reliable than the solder joints they replace in single-chip designs. Much of this type of reliability improvement can also be achieved with silicon-on-silicon packaging. Consider Figure 3. In the conventional single-chip package surface mount design each chip-to-chip signal must travel from chip to package, to lead, to solder joint, to circuit

board, to solder joint, to lead, to package, and finally to the neighboring chip. The leads, solder joints, and circuit boards are all exposed to the ravages of environmental exposure. However, in the silicon-on-silicon design the signal travels directly from chip to substrate to chip, all within the protection of the hermetic environment. Six parts of the interconnection path, each with its own failure rate, are eliminated per chip-to-chip signal: two package metallization paths, two leads, and two solder joints. Additionally, a more reliable silicon substrate replaces the circuit board. The number of wire-bonds, TAB leads, or flip-chip bumps remain the same in both approaches. It is also evident that many passive components and their interconnections will move from the exposed environment on the circuit board to within the protected hermetic multichip module. Because of the dramatic size reductions with silicon-on-silicon packaging, it is also possible to devote some of the space saved to redundant circuits, environmental sensor chips, or self-test circuits that further improve reliability.

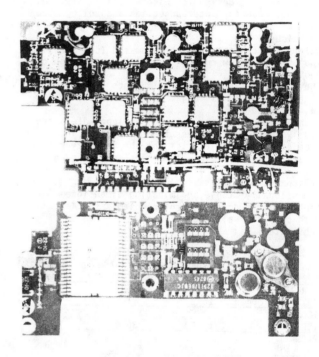

SURFACE MOUNT PWB DESIGN

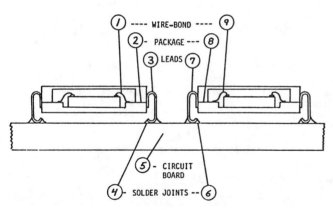

1987 SURFACE-MOUNT CIRCUIT BOARD ASSEMBLY

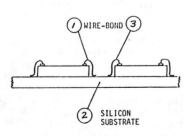

1989 SILICON-ON-SILICON
MULTIPLE CHIP MODULE
ASSEMBLY

FIGURE 3. MORE RELIABLE CHIP-TO-CHIP SIGNAL PATH

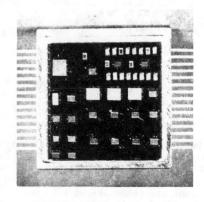

SILICON-ON-SILICON DESIGN

FIGURE 4. SYNTHESIZER DESIGN COMPARISON

Partitioning engineering is frequently neglected, yet a well engineered partitioning design has a strong effect on both equipment cost and reliability. Partitioning circuitry into multichip modules can greatly reduce the number of hermetic I/O required in an assembly, for the same reasons as demonstrated in Figure 3. This is also shown in the synthesizer module of Figure 4. To fit the space requirements of the previous generation of equipment, two surface mount circuit boards were designed to fit "piggy-back" together. The next generation of equipment is shown designed as a multichip module. Table C summarizes the improvements obtained when the two circuit boards are replaced by one multichip module. The use of silicon-on-silicon packaging not only provides a 3.5 times size reduction compared to surface mount packaging, it also allows elimination of 94 percent of the solder joints and 84 percent of the hermetic I/O. The connector contacts between the two circuit boards are obviously eliminated with the multichip module, but it is possible to eliminate additional connectors. If the total black box can truly achieve a twenty year MTBF, then it needs to be removable only at the box level, and mother-daughter card connectors can also be eliminated.

DESIGN	AREA (sq in)	QUANTITY solder joints	QUANTITY Bd-to-Bd I/O	QUANTITY hermetic I/O	QUANTITY connector contacts
two SMT PWBs	13.40	496	15	194	56
multichip module	3.84	30	0	30	0

TABLE C. COMPARISON OF SURFACE MOUNT AND SILICON-ON-SILICON PACKAGING

Electronic equipment field failures can be categorized in several ways. The failures may be related to unanticipated operating conditions, inadequate design, or latent defects resulting from marginal process control. Corrective design and manufacturing procedures can resolve these problems. Failures may also be categorized by the type of failed part, the most prevalent electronics failure areas being: semiconductor devices, solder joints, passive devices, circuit boards, connectors, and cabling. But in all of these categories, the bulk of all failures are really related to one of four fundamental failure mechanisms:
* mismatch of thermal expansion
* corrosion
* solid state diffusion
* loads exceeding material strength
Any packaging technique which strives to provide ultra-high reliability for electronics equipments must be designed meticulously to avoid these four failure mechanisms.

PACKAGING REQUIREMENTS FOR SILICON-ON-SILICON WSI

Silicon-on-silicon multichip modules can be seen to provide many of the basic elements needed to make significant improvements in the reliability of next generation electronics equipments:
* reduced count of solder joints, hermetic I/O, leads, exposed passive components, connector contacts, etc.
* available space for redundant, fault tolerant, and self-test circuitry
* matched die and substrate expansion, reducing stress in large ICs
* good heat conducting substrates to lower IC junction temperatures
* high yield manufacturing processes based on mature, automated semiconductor processes and equipment

To support the silicon substrate with the ultra-high level of reliability desired, two new generations of packages are required: a hermetic, high-rel version for military applications, and a low-cost non-hermetic version for commercial applications. These multichip modules are foreseen as taking the place of ceramic/plastic DIPs, chip carriers, and hybrid packages, and will need the usual requirements of those types of packages. Additionally, these new silicon-on-silicon module packages will have a clearly definable set of new requirements which must be met to provide the reliability improvements desired:
1. The package base must be closely matched in thermal expansion characteristics to silicon. (Candidate materials would include silicon carbide, aluminum nitride, molybdenum, graphite/epoxy, Kevlar/epoxy or combinations thereof. See Figure 5).
2. The package design needs to provide a family of large cavity sizes and high I/O counts. (See Table D.).
3. The package base needs to be a material of good thermal conductivity to help remove heat and to maintain low IC junction temperatures. (See Figure 6).
4. The package base must be of adequate stiffness to support the silicon substrate with bending deflections of less than .0002 inch per inch length of substrate.
5. The package lid must be of adequate stiffness and support to withstand the cyclic atmospheric pressure changes experienced between sea level and 100,000 feet altitude.

NUMBER OF I/O		PACKAGE SIZE inch(mm)		
2-SIDED	4-SIDED	LENGTH	WIDTH	HEIGHT
30	60	1.0 (25)	1.0 (25)	.16 (4)
74	110	2.0 (51)	1.0 (25)	.16 (4)
74	148	2.0 (51)	2.0 (51)	.16 (4)
90	180	2.4 (61)	2.4 (61)	.20 (5)
110	220	2.9 (74)	2.9 (74)	.24 (6)

TABLE D. DESIRED RANGE OF PACKAGE SIZES

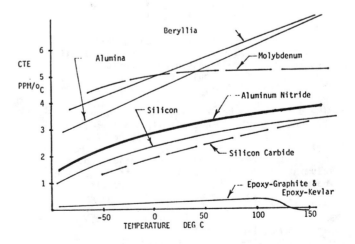

FIGURE 5. EXPANSION OF PACKAGING MATERIALS

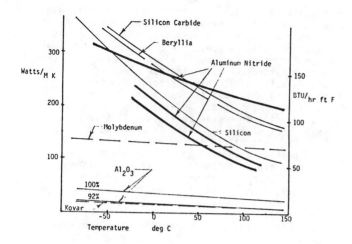

FIGURE 6.   THERMAL CONDUCTIVITY OF PACKAGING MATERIALS

6. The package must be compatible with a mechanical mounting method, such as corner tabs, bottom studs, or a constraint strap. It must not use leads for mounting; the mechanical and electrical connections must be separated to achieve the desired high reliability.

7. Package leads must be compliant enough for the low expansion, large dimension package to withstand mounting to conventional higher expansion glass/epoxy circuit boards. The leads must have sufficient compliancy such that a sideways X or Y force of .04 pound per lead will produce a lead deflection of at least .004 inch, giving an effective spring rate of 10 pounds/inch. Similar compliancy is required in the Z direction. It should be noted that this is considerably more compliant than the standard J-lead which has spring rates of 42, 88, and 178 pounds/inch in the X, Y, and Z directions, respectively [115]. Non-magnetic lead materials are preferred since they avoid the signal propagation problems experienced by leads made of iron-nickel alloys [116].

8. While recent improvements have been made in the lead-frame alloy [117], lead-frame finish [118], and glass-seal materials [119-120], co-fired I/O appear to have better potential for ultra-reliable packages than glass-sealed leads which can develop leakage or corrosion problems after lead flexure [121]. A multichip package is needed which has ultra-reliable hermetic seals at both the I/O and lid interfaces. The use of chip encapsulant materials is an acceptable option for enhancing or replacing the hermeticity requirement in some applications.

9. The packages need to have reasonable price and availability. They need to be competitive with presently available packaging technologies.

10. Some applications require that the package lids, or a section thereof, be transparent for EPROM devices.

11. The packages to be developed must be designed for optimum electrical, mechanical, and thermal interface to the next level packaging.

12. Internal package metallizations must be compatible with I/O to the substrate. For designs which use aluminum lines on the silicon substrate, aluminum pads on the silicon ICs, and aluminum wire-bonding, it would be advantageous to have aluminum pads inside the package wall so that a totally mono-metallic interconnection system could further enhance module reliability.

13. To assure ultra-high reliability, the new generation of packages will need to be screened beyond the requirements of MIL-STD-883. The

following tests have been estimated to be necessary to assure equipment lives of twenty years in military modules:

Series A: Tests on assembled modules:
  * High temperature aging: 4000 hrs @ 250C
  * Thermal shock: 1000 cycles, -65 to +150C
  * Pressure cycling: 20,000 cyc, 5 to 15 psia
  * Autoclave: 24 hrs @ 130C and 95% RH
Series B: Mounted to next level interconnect
  * Thermal cycling: 5000 cycles, -55 to +125C
  * Vibration: triple exposures to MIL-STD-883
  * Salt spray: triple exposures to MIL-STD-883

POSSIBLE PACKAGING APPROACHES FOR SILICON-ON-SILICON

Several recently available materials look attractive for ultra-reliable multichip modules. Both silicon carbide and aluminum nitride have the desired high thermal conductivity and expansion rates close to silicon as shown in Figures 5 and 6. Both have some prior success in microelectronic packaging [42-44, 122, 126-127]. Figure 7 shows a co-fired aluminum nitride package which was tooled to hold up to 1.8 x 1.8 inch silicon substrates and having up to 136 I/O. Compliant leads are attached to the sides of the package as shown in Figure 8.

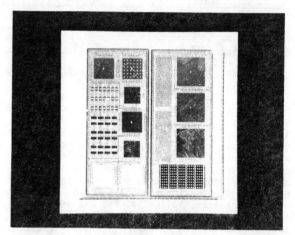

FIGURE 7.   CO-FIRED ALUMINUM NITRIDE PACKAGE

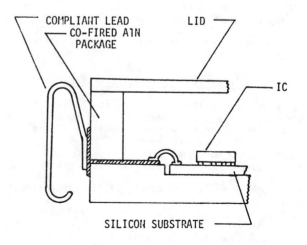

FIGURE 8.   COMPLIANT LEAD

FIGURE 9. LOW-COST NON-HERMETIC PACKAGE

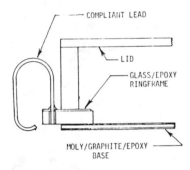

FIGURE 10. COMPLIANT LEAD

Reliability evaluations are currently in progress using the test substrate shown in Figure 12. This design features special ICs with junction temperature sensing elements, triple track corrosion monitors, several dozen passive chip components, over one thousand test wire-bonds, and several thousand test vias. These test substrates are mounted in both the conventional hybrid Kovar packages and in the new aluminum nitride co-fired packages. They are being sent through the rigorous environmental tests previously described to determine the base-line performance of each packaging technology.

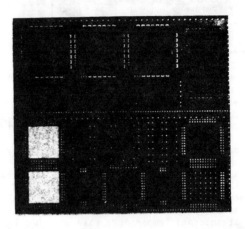

FIGURE 12. TEST SUBSTRATE FOR RELIABILITY EVALUATIONS

A non-hermetic, low-cost multichip package is shown in Figure 9. This construction uses a base made of graphite/epoxy laminated between two sheets of molybdenum foil. The ratio of moly-to-graphite/epoxy thicknesses is designed to provide a close match with the expansion of the silicon substrate. The lead I/O ring-frame is constructed of conventional circuit board material by standard circuit board processes. Compliant leads and a lid are attached as shown in Figure 10. Since this package is non-hermetic, the ICs are encapsulated with silicone gel for protection from the environment.

A further option is shown in Figure 11 where an integral module/connector housing carries both the silicon substrate and a hermetic connector. In this concept the housing would be made of a low expansion material like aluminum nitride, silicon carbide, or a graphite-aluminum matrix.

A separate series of tests were completed to determine the static and fatigue strengths of the silicon substrates. As expected, the static ultimate strength depends upon surface finish, with low defect wafers breaking around 100,000 psi and those imperfectly polished as low as 17.000 psi. Even with the lowest strength wafers a substrate can be bent to a .010 inch/inch bow without fracture. If static stresses are kept below 10,000 psi there is virtually no danger of silicon fracture.

Another series of tests exposed silicon wafers to cyclic bending to determine their fatigue properties. Figure 13 shows that silicon substrates are remarkably resistant to fatigue damage. Fatigue life for silicon substrates is into the hundreds of millions of cycles if the design strain range is kept below 0.0005.

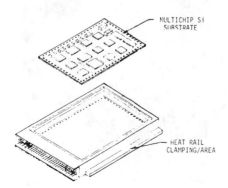

FIGURE 11. INTEGRAL MODULE/CONNECTOR

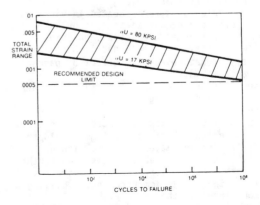

FIGURE 13. EXCELLENT FATIGUE LIFE OF SILICON

195

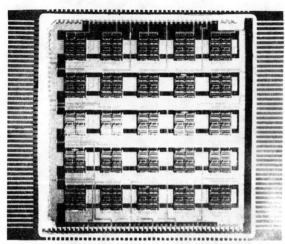

FIGURE 14. THERMAL TEST MODULE

A thermal test substrate was fabricated to put 25 test ICs into a 2.2 x 2.2 inch module as shown in Figure 14. A typical dissipation for real circuitry in this size module is around 10 watts. At this dissipation level the worst case junction-to-case temperature rise was less than 5 Centigrade degrees in the Kovar package, and is estimated to be less than 3 Centigrade degrees in the aluminum nitride package. This module has also been run at 50 watts dissipation with only a 26 Centigrade degree junction-to-case rise. It is apparent that internal temperature rise will not be an issue in properly designed multichip modules. The module's large flat bottom area provides an excellent, nearly isothermal area for heat removal. The only thermal issue with most multichip modules will be how well the thermal path through the next packaging level is designed.

Because of some industry concern about the environmental stability of aluminum nitride, some additional data was gathered to assure that aluminum nitride would be an acceptable material for ultra-reliable packages. The following data were reported by industry:

* Aluminum nitride samples passed MIL-STD-883 humidity tests with no degradation [128].
* Aluminum nitride samples exposed for 1300 hours to 90% RH at 80C showed no weight loss and no degradation in two separate series of tests [129].
* Aluminum nitride substrates placed in a pressure cooker for 100 hours showed no degradation in leakage current or bulk thermal properties. A thin, self-limiting layer of $Al(OH)_3$ did form on the exposed surfaces, but had no adverse effects on packaging performance [129].
* Aluminum nitride substrates heated in air above 1000C experienced no degradation of bulk electrical or thermal properties. A thin, self-limiting layer of Al_2O_3 did form on exposed surfaces, but had no adverse effect on packaging performance [129].
* Aluminum nitride substrates immersed in water for one month showed no weight change or change in surface condition [130].
* Aluminum nitride substrates placed in a pressure cooker for extended periods of time develop a thin, self-limiting layer of Al_2O_3 within the first hour on exposed surfaces. No further film growth occurs after that. The same results occur with the sample immersed in the boiling water or placed in the steam [130].
* Aluminum nitride samples experience a slight etch at grain boundaries when placed in alkaline solutions. Non-alkaline cleaners are therefore recommended [128].
* Aluminum nitride samples placed in a 80% RH humidity chamber for one month showed no weight loss and no effect on the surface structure as shown in Figure 15 [131].
* Aluminum nitride samples exposed to ten passes through an alkaline cleaning spray conveyor (pH=11) for a circuit board cleaning line showed a slight weight loss (.07 mg/cm^2/pass) and a slight etching of the surface at grain boundaries as shown in Figure 15 [131].

Further tests are in progress to determine if any of these results represent any significant risk in using aluminum nitride packages in ultra-reliable applications.

AS-RECEIVED

ONE MONTH IN HUMIDITY CHAMBER

TEN PASSES IN ALKALINE CLEANER
photos at 4000X

FIGURE 15. SURFACES OF ALUMINUM NITRIDE

SUMMARY

Silicon-on silicon packaging provides
well-demonstrated reductions in size and weight for
electronics equipments. With proper package design it
also holds the potential to greatly improve the
reliability of electronic systems. Utilization of new
high conductivity, low expansion materials will
considerably enhance these improvements. Development
of compliant leads is essential for a reliable
interface with the next level of packaging. With
these enhancements it is straightforward to repackage
what formerly occupied entire circuit boards into
multichip modules like those shown in Figure 16.

PROCESSOR MODULE WITH 54 COMPONENTS

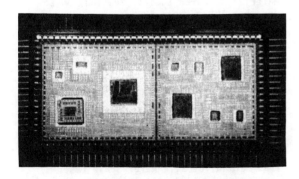

SYNTHESIZER MODULE

FIGURE 16. TYPICAL HERMETIC MULTICHIP MODULES

REFERENCES

[1]   K. Kaminishi, "GaAs on Si Technology", Sol St
Tech, Nov 1987, pp 91-97.

[2]   Staff, "Researchers Tilt Silicon To Grow Pure
GaAs On It", Electronics, Feb 10, 1986, pp 16-17

[3]   Staff, "Composite Wafers A Key For Future ICs",
Semiconduct Int, Dec 1986, pp 11-12.

[4]   C.J. Bartlett, "Advanced Packaging For
VLSI",Sol St Tech, Jun 1986, pp 119-123.

[5]   H.J. Levinstein, et al, "Multichip Packaging
Technology For VLSI-Based Systems", IEEE Int
Sol St Cir Conf, Feb 1987, pp 224-225.

[6]   S.K. Tewksbury, et al, "Chip Alignment
Templates For Multichip Module Assembly", IEEE
Trans CHMT, Vol CHMT-10, No 1, Mar 1987, pp
111-121.

[7]   C.J. Bartlett, et al, "Multichip Packaging
Design For VLSI-Based Systems", IEEE ECC,
Boston, May 1987, pp 518-525.

[8]   P. Singer, "Multichip Packaging on Silicon
Substrates", Semiconduct Int, Jun 1987, pp 34-36

[9]   L.A. Hornak and S. K. Tewksbury, "On The
Feasibility of Through-Wafer Optical
Interconnects For Hybrid Wafer-Scale-Integrated
Architectures", IEEE Trans Elect Dev, Vol
ED-34, No 7, July 1987, pp 1557-1563.

[10]   N.Teneketges, "Silicon Based VLSI
Interconnections", IEEE Comp Pack Workshop,
Brussels, Sept 1987.

[11]   E.N. Fuls, et al, "A High Density Interconnect
Technology For VLSI-Based Systems", IEEE/NBS
VLSI & GaAs Pack Workshop, Research Triangle
Park, Sept 1987.

[12]   C.J. Bartlett, et al, "Multichip Packaging
Design For VLSI-Based Systems", IEEE Trans
CHMT, Vol CHMT-10, No 4, Dec 1987, pp 647-653.

[13]   C.J. Bartlett, "Packaging For Wafer Scale
Integrated Systems", Mat Res Soc Mtg, Boston,
Dec 1987, Paper I6.5/L5.5.

[14]   C.P. Wong, "High Performance Silicon Gel As IC
Device Chip Protection", Mat Res Soc Mtg,
Boston Dec 1987, Paper I5.7/L4.7

[15]   K.L. Tai, "High Bit Rate, High I/O, High
Density and Low Power VLSI Chip Interconnection
On Si Wafer", Mat Res Soc Mtg, Boston, Dec
1987, Paper I2.4.

[16]   C.P. Wong, "High Performance Silicone Gel As IC
Device Chip Protection", EF Conf on Adv Mat &
Proc for High Dens Pack, Santa Barbara, Mar 1988

[17]   R.C. Jaeger, et al, "Active Silicon Hybrid
Wafer-Scale Packaging", SRC Project Summary,
June 1986.

[18]   J. Lyman, "The Latest Wafer-Scale Design Is A
Hybrid", Electronics, Mar 17, 1986, p28.

[19]   R.W. Johnson, et al, "Hybrid Wafer-Scale
Integration Packaging Technology", Southcon,
Mar 1986, Paper 19/3.

[20]   R.W. Johnson, et al, "Planar Hybrid
Interconnection Technology", ISHM, Atlanta,
Oct. 1986, pp 758-765.

[21]   R.W. Johnson, et al, "Silicon Hybrid
Wafer-Scale Package Technology", IEEE J Sol St
Cir, Vol SC-21, No 5, Oct 1986, pp 845-850.

[22]   R.W. Johnson, et al, "Planar Hybrid
Interconnection Technology", Int J Hybrid
Microelectronics, Vol 10, No 1, 1987, pp 28-36.

[23]   R.C. Jaeger, et al, "Hybrid Silicon Wafer Scale
Packaging", SRC Project Summary, Jun 1987.

[24]   R.W. Johnson, et al, "Hybrid Wafer Scale
Integration Utilizing Planar Interconnection
Technology", IEEE/NBS VLSA & GaAs Packaging
Workshop, Research Triangle Park, September
1987.

[25]   R.C. Jaeger, et al, "Test Section Design For
Evaluation of Cooling of Silicon Wafer-Scale
Packaging", IEPS, Boston, November 1987, pp
944-953.

[26]   J.S. Goodling, et al, "Wafer Scale Cooling
Using Jet Impingement Boiling Heat Transfer,
ASME Winter Meeting, December 1987.

[27]   R.W. Johnson and R.C. Jaeger, "Thin Film
Hybrids on Silicon", Engineering Foundation
Conference on Advanced Materials and Processes
For High Density Packaging, Santa Barbara,

March 1988.

[28] T.R. Anthony, "Forming Electrical Interconnections Through Semiconductor Wafers", J. Appl. Physics, Vol. 52, NO. 8, August 1981, pp 5340-5349.

[29] C.A. Neugebauer, "Approaching Wafer Scale Integration From The Packaging Point Of View", IEEE International Conference On Computer Design, Port Chester, NY, October 1984, pp 115-120.

[30] C.A. Neugebauer, "Comparison Of VLSI Packaging Approaches To Wafer Scale Integration", IEEE Custom Integrated Circuits Conference, 1985, pp 31-37.

[31] E.J. Lerner, "GE Method Overcomes Previous Wafer-Scale International Problems", Research & Development, October 1985, pp 51-52.

[32] A.J. Yerman, "High Density Digital Packaging," University of Wisconsin Microelectronics Symposium, Madison, January 8, 1986.

[33] C.A. Neugebauer and R.O. Carlson, "Comparison of VLSI Packaging Approaches", SRC Packaging Review, October 1986.

[34] R.O. Carlson and C.A. Neugebauer, "Future Trends in Wafer Scale Integration", Proc. IEEE, Vol. 74, No. 12, December 1986, pp 1741-1752.

[35] C.A. Neugebauer and R.O. Carlson, "Comparison of Wafer Scale Integration With VLSI Packaging Approaches", IEEE Computer Packaging Workshop, Split Rock, PA, May 1986.

[36] C.A. Neugebauer and R.O. Carlson, "Comparison of Wafer Scale Integration with VLSI Packaging Approaches", IEEE Trans. CHMT, Vol CHMT-10, No. 2, June 1987, pp 184-189.

[37] C.A. Neugebauer, "Material Limitations In The Higher Electronic Packaging Levels", Material Research Society Meeting, Boston, December 1987, Paper I 1.3.

[38] C.W. Eichelberger, et al, "High Density Interconnects Using Loser Adoptive Writing", Engineering Foundation Conference on Advanced Materials and Processes For High Density Packaging, Santa Barbara, March 1988.

[39] T.R. Gheewala, "Packages for Ultra-High Speed GaAs IC's," Microelectronics Journal, March/April, 1985, pp 30-37.

[40] R. Pound, "Packaging Links Fast GaAs Dice to High Speed Systems," Elect. Pack. Prod., August 1985, pp 70-74.

[41] P.L. Young, et al, "Thin Film Multi-Chip Module" IEEE/NBS VLSI Packaging Workshop, Washington, D.C., September 1985.

[42] K. Satoh, et al, "A High Speed Multi-Chip RAM Module with Thermal Stress Free Configuration," IEEE International Conference on Computer Design, Port Chester, New York, October 8-11, 1984 pp 569-572.

[43] K. Okutani, et al., "Packaging Design of a Silicon Carbide Ceramic Multi-Chip RAM Module, "Proc IEPS, October 1984, pp 299-304.

[44] T. Yamada, et al., "Low Stress Design of Flip Chip Technology For Silicon-On-Silicon Multi-Chip Modules", 1985 IEPS, Orlando, pp 551-557.

[45] A. Masaki, "Silicon-On-Silicon Packaging Technology," IEEE Computer Society Conf., Japan, January 27-28, 1986.

[46] C. Huang, et al, "Silicon Packaging - A New Packaging Technique," IEEE 1983 Custom Integrated Circuit Conference, Rochester, NY, May 1983, pp 142-146.

[47] C.D. Spielberger, et al, "Silicon-On-Silicon Packaging", IEEE Trans on CHMT, Vol CHMT-7, No. 2, June 1984, pp 193-186.

[48] R.J. Jensen, "Recent Advances in Copper/Polyimide Thin Film Multilayer Interconnect Technology For IC Packaging", Material Research Society Meeting, Boston, December 1987, Paper I 2.2.

[49] D.J. Bodendorf, et al, "Active Silicon Chip Carrier," IBM Tech. Disc. Bull., Vol. 15, No. 2, July 1972, p 656.

[50] M.J. Brady, "Fabrication Processes for a Silicon Substrate Package for Integrated Gallium Arsenide Laser Arrays," J. Electrochemical Soc., Vol. 125, No. 10, October 1978, pp 1672-1647.

[51] R.A. Laff, et al, "Thermal Performance and Limitations of Silicon-Substrate Packaged GaAs Laser Arrays," Appl. Opt., Vol 17, No. 5, 1 March, 1978, pp 778-784.

[52] Feinberg, I., et al, "Interposer for Chip-On-Chip Module Attachment," IBM Technical Disclosure Bulletin, Vol. 26, No. 9, February 1984, pp 4590-4591.

[53] Logue, J. and Ketchen, M., "Silicon-On-Silicon Packaging Possibilities," IEEE Computer Society Conf., Japan, January 27-28, 1986.

[54] S.H., Wen, et al, "Thin Film Wiring For Integrated Electronic Packages", Material Research Society Meeting, Boston, December 1987, Paper I 2.3.

[55] M.S. Goldberg, et al, "Chromium-Polyimide Interface Chemistry", Material Research Society Meeting, Boston, December 1987, Paper I 5.9/L 4.9.

[56] H.M. Tong and K.Saenger, "Diffusion of NMP in Polyimide and Its Impack On Polyimide/Polyimide Adhesion", Material Research Society Meeting, Boston, December 1987, Paper I 6.4/L 5.4.

[57] C.H. Yang, et al, "Studies on Interfacial Adhesion of Cu/Cr/Polyimide Interfaces", Material Research Society Meeting, Boston, December 1987, Paper I 6.6/L 5.6.

[58] D.C. Hofer, et al, "Internal Stress in Polyimides For Multilevel Interconnection", Material Research Society Meeting, Boston, December 1987, Paper I 5.4/L 4.4

[59] A. Gupta and R. Jagannathan, "Direct Writing of Copper Lines From Copper Format Film", Material Research Society Meeting, Boston, December 1987, Paper B2.8.

[60] D. Hofer, "Stresses In Polymers For Multilevel Interconnections", Engineering Foundation Conference On Advanced Materials and Processes For High Density Packaging, Santa Barbara, March 1988.

[61] E. Paper, "Assessing High Performance Packaging Technologies", SRC Packaging Review, October 1986.

[62] C.H. Ting, et al, "Applications of Thin Film Interconnection Technology for VLSI Multichip Packaging", ISHM, Minneapolis, September 1987, pp 151-156.

[63] I.P. Herman, et al, "Wafer-Scale Laser Pantography", 1982 Materials Research Society, Boston, November 1982.

[64] B.M. Williams and D.B. Tuckerman, "Wafer-Scale Integration", Lawrence Livermore National Laboratory Energy and Technology Review, December 1985, pp 1-9.

[65] Nicolas, G., "New Approaches for VLSI Interconnection", 5th NBS/IEEE VLSI Packaging Workshop, Paris, November 1986, p 69.

[66] M. A. Fischetti, "A Review of Progress at MCC", IEEE Spectrum, March 1986, pp 76-82.

[67] B.H. Whalen, "Materials Requirements for Advanced Packaging and Interconnect", Material Research Society Meeting, Boston, December 1987, Paper I 1.1.

[68] D. Herrell, "System Challenges: Power, Cooling and Test", Engineering Foundation Conference

on Advanced Materials and Processes for High Density Interconnect, Santa, Barbara, March 1988.

[69] B. Nelson and T. Pan, "Reliability Issues in Copper Polyimide Systems", Engineering Foundation Conference on Advanced Materials and Processes for High Density Interconnect, Santa Barbara, March 1988.

[70] S.D. Senturia and M.G. Allen, "The Role of Adhesion In The Microelectronic Applications of Polyimide", Material Research Society Meeting, Boston, December 1987, Paper I 6.1/L 5.1.

[71] J.G. Black, "Laser-Direct-Written Interconnection Techniques for Multiple-Chip and Wafer-Scale Systems", Material Research Society Meeting, Boston, December 1987, Paper B 2.5.

[72] M. Kohara, et al, "Thermal Stress-Free Package for Flip-Chip Devices," IEEE Trans. CHMT, Vol. CHMT-7, No. 4, December 1984, pp 411-416.

[73] H. Stopper, "A Wafer with Electrically Programmable Interconnections", IEEE International Solid State Circuits Conference, February 1985, pp 268-269.

[74] Staff, "Whip process Makes Small, Fast Computers", Elect. Pack. & Prod., June 1985, p 16.

[75] Staff, "Mosaic to Ship Wafer Packaging", Electronics, September 23, 1985, p 24.

[76] T. Costlow, "Switch To Silicon Gives Hybrids A Packaging Edge", Electronic Design, August 21, 1986, pp 33-34.

[77] J. Lyman, "Hybrid Industry Gets a Universal Substrate", Electronics, November 27, 1986, pp 38-42.

[78] Staff, "Mosaic Develops Silicon Circuit Board", Huybrid Circuit Technology, February 1987, p6.

[79] Staff, "Raytheon Buys 12% of Mosaic", Electronic News, May 4, 1987, p 35.

[80] Staff, "Tools For Silicon Circuit Board Debut", Electronic Engineering Times, April 13, 1987, pp 1, 13.

[81] A.A. Bogdan, "Electrically Programmable Silicon Circuit Board Expands Package Options", Hybrid Circuit Technology, April 1987, pp 9-14.

[82] R. Braun and S. Shart, "Silicon Circuit Boards", Machine Design, July 23, 1987, pp 95-98.

[83] R.Braun and S. Sharp, "How to Use Silicon Circuit Boards Effectively", Hybrid Circuit Technology, September 1987, pp 55-60.

[84] R. Pound, "Digital Systems and Mobile Phones Drive Telecom Packaging", Electronic Packaging & Production, april 1985, pp 112-120.

[85] Ad, "VLSI Packaging Process Development", Elect. Pack. & Prod., June 1987, p 132.

[86] H. Tsunetsugu, et al, "Multilayer Interconnections Using Polyimide Dielectrics and Aluminum Conductors, Int. J. for Hybrid Microelectronics, Vol. 8, No.2, June 1985, pp 21-26.

[87] R.D. McKirdy, et al, "Wafer Scale Integration: The Limits of Packaging", IEEE/NBS VLSI Packaging Workshop, Paris, November 1986, p 15.

[88] Ad, "Electronic Packaging and Interconnect Engineers", Electronic Packaging and Production, August 1986, p 140.

[89] C.A. Lofdahl, "Advanced Thin-Film Multi-Chip Packaging Product Development", Presentation to Rockwell, Cedar Rapids, IA, September 24, 1985.

[90] C.A. Lofdahl, "Advantages for Electronic Systems of High Density Interconnect (HDI) Packaging", Company Report, Raychem Corp.,

[91] C.A. Lofdahl, "Beyond Printed Wiring Board Densities: A New Commercial Multichip Packaging Technology", IEEE Computer Packaging Workshop, Brussels, September 1987.

[92] E. Bogatin, "HDI: An Enabling Technology for Multichip Packages", IEEE/NBS VLSI & GaAs Packaging Workshop, Research Triangle Park, September 1987.

[93] Raychem Corp., "HDI, High Density Interconnect Design Manual, Document HDI-S 7.4, November 24, 1987.

[94] B.J. Donlan, et al, "The Wafer Transmission Module", VLSI Systems Design, January 1986, pp 54-48, 88, 90.

[95] B.J. Donlan and J.F. McDonald, "A Placement and Routing System for Wafer Scale", IEEE Int. Conf. on CAD, 11 Nov. 86, pp 462-465.

[96] H.T. Lin, et al, "Electron and ION Beam Repair Strategies for the Wafer Transmission Module", IEEE/NBS VLSI & GaAs Packaging Workshop, Research Triangle Park, September 1987.

[97] A.A. Evans and J.K. Hagge, "Advanced Packaging Concepts - Microelectronics Multiple chip Modules Utilizing Silicon Substrates", SAMPE Electronic Materials and Process Conference, Santa Clara, June 1987, pp 37-45.

[98] Staff, "Chips-On-Silicon Could Replace SMT", Elect. Pack. & Prod., September 1987, p 14.

[99] A.A. Evans, "Multiple Chip Modules Utilizing Silicon Substrates", DuPont Symposium on High Density Interconnect, Wilmington, September 1987.

[100] J.K. Hagge, "Overview of High Density Packaging Configurations and Encapsulation Requirements", Engineering Foundation Conference on Advanced Materials and Processes for High Density Packaging, Santa Barbara, March 1988.

[101] D.B. Tuckerman and R.F.W. Pease, "High Performance Heat Sinking for VLSI", IEEE Electron Device Letters, Vol, EDL-2, No.5, May 1981, pp 126-129.

[102] D.B. Tuckerman and R.F.W. Pease, "Microcapillary Die Attachment for VLSI", U.S./Japan Symposium on VLSI, September 1983.

[103] C. Murray, "SRC-Stanford Effort Products ICs Cooling Technique", Semiconductor International, November 1985, p 28.

[104] O.K. Kwon and R.F.W. Pease, "Closely-Packed Transmission Lines as very High Speed Chip-to-Chip Interconnects", IEEE International Electronic Manufacturing Technology Symposium, San Francisco, September 1986.

[105] A. Paal and R.F.W. Pease, "Extending Microcapillary ATtachment:, IEEE International Electronic Manuafacturing Technology Symposium, San Francisco, September 1986.

[106] A. Paal and R.F.W. Pease, "Extending Microcapillary Attachment to Rough Surfaces", JEMT, September 1986.

[107] R.F.W. Pease, "Advances in Packaging for VLSI Systems", Invited Paper, IEEE International Electron Devices Meeting, Los Angeles, December 1986.

[108] R.F.W. Pease, et al, "High Performance Packaging for VLSI Systems", Advanced Research in VLSI "New Technologies for New Architectures", Stanford University, Stanford, CA, March 23-25, 1987.

[109] O.K. Kwon and R.F.W. Pease, "Discontinuities in Very High-Speed Lossy Transmission Lines for Chip-to-Chip Interconnection", U.S./Japan VLSI Symposium, September 1987.

[110] O.K. Kwon and R.F.W. Pease, "Closely-Packed Transmission Lines as Very High Speed

January 1986.

Chip-to-Chip Interconnects", IEEE Trans. CHMT, Vol. CHMT-10, No. 3, September 1987, pp 314-319.

[111] Staff, "Trilogy in $10M R&D Project", Electronic News, July 11, 1985.

[112] Staff, "Saratoga in Trilogy Facility/Financing Deal", Electronic News, December 1, 1985, p 40.

[113] J.F. McDonald, et al, "The Trials of Wafer-Scale Integration", IEEE Spectrum, October 1984, pp 32-39.

[114] J.W. Balde, "Small Dimensions or Low Dielectric Constant - The Competing Approaches to High Density Interconnect", Material Research Society Meeting, Boston, December 1987, Paper I 2.1.

[115] R.W. Kotlowitz, "Comparative Compliance of Generic Lead Designs for Surface Mounted Components", IEPS, Boston, November 1987, pp 965-984.

[116] C.C. Huang and L.LL. Wu, "Signal Degradation Through Module Pins in VLSI Packaging", IBM J. Res. Dev., Vol. 31, No. 4, July 1987, pp 489-498.

[117] A.J. Schmuck, "New Alloy for Glass-To-Metal Shals", NASA Tech Briefs, July/August 1986, p 100.

[118] W.B. Thomas, "Matched Glass-to-Metal Seal Improvements By Controlled Atmosphere Metal Oxidation", Solid State Technology, September 1986, pp 73-74.

[119] J. Ahearn, "New Glass Bead Developed", Hybrid Circuit Technology, September 1987, pp 73-75

[120] J.E. Ahearn, "A Duplex Glass Enhancement for Improving Hybrid Microelectronic Packages", ISHM, Minneapolis, September 1987, pp 275-279.

[121] K.A. Berry, "Corrosion Resistance of Military Microelectronic Packages at the Lead-Glass Interface", ASM Conference on Electronic Packaging, Minneapolis, April 1987, pp 55-61.

[122] N. Iwase, "AIN Packages with Matched TCE and Low Thermal Resistance", IEEE Computer Packaging Workshop, Japan, January 1986.

[123] F. Miyashiro and N. Iwase, "High Thermal Conductive AIN Ceramic Substrates and Packages", Engineering Foundation Conference on Advanced Materials and Processes for High Density Interconnect, Santa Barbara, March 1988.

[124] M. Terasawa, "Advanced Packaging Technology Using New Ceramics with High Thermal Conductivity and Low Thermal Expansion", Engineering Foundation Conference on Advanced Materials and Processes for High Density Interconnect, Santa Barbara, March 1988.

[125] A. Ikegami, et al, "Application of High Thermal Conductivity Silicon Carbide to VLSI Packaging, "IEEE/NBS VLSI Packaging Workshop, Washington, D.C., September 1985.

[126] S. Ogihara, et al, "Application to LSI Packages of Si C Ceramics with High Thermal Conductivity", Int. J. Hybrid Microelectronics, Vol. 8, No.2, June 1985, pp 16-20.

[127] N. Iwase, et al, "Aluminum Nitride Multi-Layer Pin Grid Array Packages", IEEE ECC, Boston, May 1987, pp 384-391.

[128] M. Hutfless, "Keramont Advanced Ceramic Products Corporation", Personal Communication, July 1987.

[129] K. Nakajima, "Tokuyama Soda Company", Personal Communication, July 1987.

[130] M. Hutfless, "Keramont Advanced Ceramic Products Corporation", Personal Communications, January 1988.

[131] J.K. Hagge, Rockwell Int., Unpublished Test Data, Nov. 1987.

ULTRA-RELIABLE HWSI WITH ALUMINUM NITRIDE PACKAGING

by

John K. Hagge
Rockwell International Corporation
Cedar Rapids, Iowa 52498

ABSTRACT

This paper discusses the further developments of ultra-reliable packaging approaches for Hybrid Wafer Scale Integration (HWSI) technology. A variety of approaches are shown using packaging fabricated of machineable, standard, and super conductivity grades of Aluminum Nitride. Applications include both digital and microwave circuits packaged in non-hermetic, hermetic/ compliant lead and plug-in multichip modules. Environmental stress testing in thermal cycling has shown silicon-on-silicon multichip modules to offer significant reliability advantages over conventional packaging approaches. Module sizes of several inch dimensions, several hundred I/O, and over 100 components are shown to be practical if adequate attention is paid to producibility and test strategies.

INTRODUCTION

Hybrid wafer scale integration has attracted considerable attention (see table I) because it offers significant potential improvements over Surface Mount Technology (1,2):

- 1/3 to 1/4 the size and weight
- significantly improved reliability
- reduced signal delay time
- Improved thermal performance
- Competitive costs

High density interconnecting substrates are fabricated on silicon wafers using semiconductor lithography techniques and multilayer thin film metallizations. Typically a power layer, a ground layer, and two signal layers are adquate to interconnect many dozens of VLSI ICs on a single wafer. Figure 1 shows a complex power/ground/2-signal layer silicon substrate which interconencts

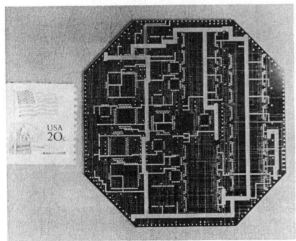

FIGURE 1.   COMPLEX 4 METAL LAYER SILICON SUBSTRATE INTERCONNECTING 134 COMPONENTS

FIGURE 2.   DUAL 1750A PROCESSORS PLUS ON-BOARD MEMORY AND SELF-TEST ON 3 INCH WAFER

134 components on a single wafer. Standard hybrid assembly techniques are used to wire-bond, flip-chip, or TAB connect ICs and passive chips on the silicon substrate. Figure 2 shows this substrate in an operational design which has packaged two 1750A processors plus on-board memory and self-test into a single multichip module.

INDUSTRY STATUS

Industry interest has grown rapidly in silicon-on-silicon multichip modules as reflected in the number of technical articles on silicon substrate technology. That data base has grown from a scant 15 articles in 1984 to over 200 articles by the end of 1988 as summarized in table I.

AT&T (4-16, 132-139, 244)
Boeing (145-146)
BPA (147-148, 233)
DEC (149)
DOW (239-240, 246)
DuPont (150-153)
GE (28-38, 154-159, 213, 231)
Gigabit Logic (39-40)
Gould (41)
Hewlett Packard (160)
Hitachi (42-45, 238)
Honeywell (46-48, 161-162)
Hughes (221-222)
IBM (49-56, 163-165, 213, 236)
Intel (61-62, 166)
Martin-Marietta (226)
MCC (66-69, 167-168)
Mitsubishi (72)
Mosaic (73-83, 113, 169-176)
Northern Telecom (84-85)
NTT (86)
Plessey (87, 177-179, 227)
Polycon (180-181, 230, 245)

Prime (88)
Raychem-APS (89-93, 182-183, 197, 219)
Raytheon (79)
RISH (187, 229, 243)
Rockwell (1, 97-100)
TI (184-185)
Toshiba (191)
Trilogy Tech (111-112)
Unisys (111,113)
Xerox (235)
Other (114,186-198, 219, 228,241,247-249)
AFIT (199-200)
NOSC (202)
RADC (201, 203-204)
Auburn (17-27, 140-144, 220, 232)
Lawrence Livermore (63-64, 205-206)
Leti Labs (65)
MIT (70-71, 207-208)
MCNC (209-210, 225, 234)
RPI (94-96, 211-216)
Stanford (101-110, 217)
U. Dayton (218)

Table I. Information Sources for Silicon-on-Silicon Packaging

EXAMPLES OF MINIATURIZATION POTENTIAL

Thin film metallization layers can be fabricated across full silicon wafers with high yeilds using 10 to 25 micron wide lines. Using such lines which are an order of magnitude finer than circuit board technology allows very dense packaging of IC chips. Figure 3 shows a silicon-on-silicon subassembly fabricated as part of a prototype DARPA miniature GPS receiver. The miniaturization which is achieved allows the entire GPS receiver to be packaged in a size smaller than a cigarette pack. Figure 4 shows the small size of the dual 1750A processor previously discussed. Figure 5 shows a single-board avionics computer which has been shrunk into a single multichip module. Figure 6 shows a redundancy manager unit (voter module) which monitors inputs from up to four of the computer modules of figure 5, and determines both correct operation and fault isolation for the four processors. Four of the multichip modules shown in figures 5 and 6 operate in the same space formerly occupied by the single-board computer. As evident in these examples, the technology is maturing sufficiently to allow fairly complex functions to be assembled in a single module. The last three modules shown contain 134, 70 and 90 components respectively.

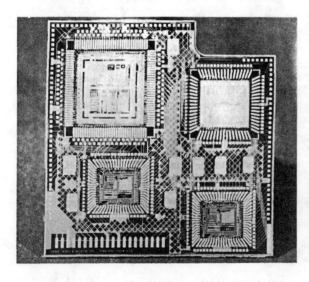 

FIGURE 3. 1 x 1" SILICON-ON-SILICON SUB-ASSEMBLY FOR DARPA MINIATURE GPS RECEIVER   FIGURE 4. SIGNIFICANT MINIATURIZATION OF SPACE-APPLICATION COMPUTER

FIGURE 5.   AVIONICS PROCESSOR/MEMORY IN SINGLE
2.4 x 2.4" MULTICHIP MODULE

FIGURE 6.   REDUNDANCY MANAGER UNIT FOR FOUR
PROCESSORS IN SINGLE MULTICHIP MODULE

PROPERTIES OF ALUMINUM NITRIDE

Packages for large silicon substrates can provide greater reliability when they are closely matched in thermal expansion characteristics to silicon. Aluminum nitride is one of the few packaging materials which provides that close match, as shown in figure 7. Its expansion rate ranges from 2.1 ppm/C at -55°C to 3.8 ppm/C at +125°C, approximately 0.5 ppm/C higher than silicon. The thermal conductivity of aluminum nitride has been shown to be strongly dependent on

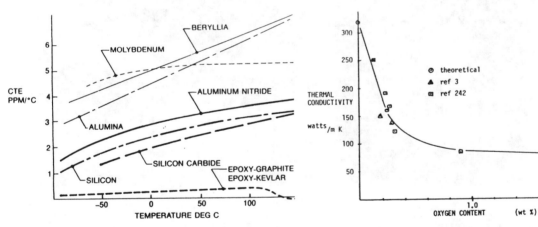

FIGURE 7.   EXPANSION OF PAKAGING MATERIALS

FIGURE 8.   AlN CONDUCTIVITY RISING WITH PURITY.

aluminum nitride with improved thermal conductivity as shown in figure 9. Thermal conductivity is a strong function of temperature, as shown in figure 10. The temperature range of most importance for packaging properties is the 70 to 90°C range (corresponding to semiconductor junction temperatures operating in the the 80-100°C range). In this range the machinable grades of aluminum nitride have thermal conductivities about 6 times higher than alumina or Kovar, the co-firing package grades of aluminum nitride have thermal conductivities equivalent to 6061-T6 aluminum, and the premium substrate grades of aluminum nitride are better than both beryllia and premium silicon carbide.

The electrical properties of aluminum nitride are similar to alumina with a dielectric constant around 8.9 and with a loss tangent of about .0008 at 1 MHz, both rising slightly by the 10 GHz range (3). The mechanical strength of aluminum nitride is better than alumina with an elastic modulus around 45 million psi and a flexural strength aroung 60 kpsi. Aluminum nitride has good environmental stability over typical military temperature ranges for all conditions except strongly alkaline exposures as previously reported (1).

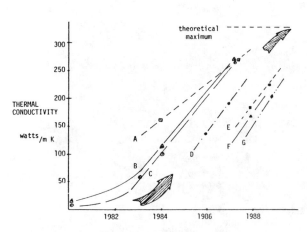

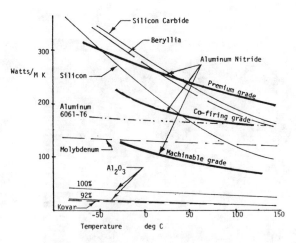

FIGURE 9. INCREASING CONDUCTIVITY FROM 7 VENDORS

FIGURE 10. CONDUCTIVITY OF PACKAGING MATERIALS

ALUMINUM NITRIDE PACKAGING APPLICATIONS

The machinable grades of aluminum nitride can be used to fabricate package cavities and lid cavities of fairly complex shapes. For example, the integral package/connector plug-in module shown in figure 11 can have its housing (figure 12) fabricated of the machineable aluminum nitride grade. This approach totally eliminates circuit board assemblies, replacing them with hermetic multichip modules which plug directly onto backplane connectors. Support and heat removal can be accomplished via edge rails and rack mounts such as is done with SEM-E modules.

FIGURE 11. PLUG-IN MODULE WITH HIGH DENSITY .050" CENTERS CONNECTOR

FIGURE 12. ASSEMBLY OF PLUG-IN MULTICHIP MODULE A. HOUSING (Al for non-hermetic, AlN for hermetic). B. DISTRIBUTION CIRCUIT.

The co-firing grades of aluminum nitride can be used for fabricating interconnect circuits for RF multichip modules, for distribution substrates such as shown in figure 12, or for hermetic packages as shown in figure 13. These co-fired packages provide significant improvements over conventional alumina or Kovar hybrid packages. The aluminum nitride provides a closely expansion-matched base to reliably support the large silicon substrates. The co-fired package walls give dependable hermetic I/O throughout extended field life, unlike glass-sealed leads of conventional packages, which are prone to leakage with any lead flexure. The excellent thermal conductivity of the aluminum nitride package provides efficient heat removal, and junction-to-case temperature differentials seldom exceed 10 degrees C in these packages. Some of the significant design features of these co-fired packages are as follows (see figure 13):
A.   The Si-on-Si packaging gives ultra dense circuitry, equal to or better than full Wafer Scale Integration (no wafer area need be sacrificed to redundant yield-raising elements). The closely spaced chips and short signal paths give faster electrical performance.
B.   The high purity aluminum nitride package construction gives expansion matched to silicon and high thermal conductivity. (No low conductivity alumina or Kovar).
C.   The co-fired I/O give ultra-high hermeticity over extended field life. (No glass-sealed leads to crack.)

204

D. The fine stranded copper leads give maximum flexibility to survive extreme thermal cycling, even with the packages mounted to ordinary circuit boards. (No stiff rectangular leads, and no constrained core circuit boards required.)

E. The thin silicone web maintains lead spacing during manufacturing or severe vibration, and maintains good flexibility over full temperature ranges.

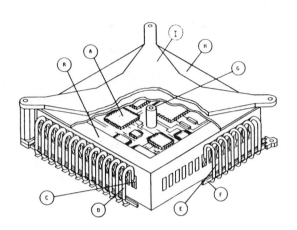

FIGURE 13. CO-FIRED ALUMINUM NITRIDE PACKAGE

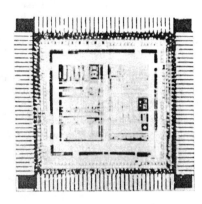

FIGURE 14. VLSI CHIP ON PREMIUM CONDUCTIVITY ALUMINUM NITRIDE CHIP-MOUNT

G. Support posts at any desired location to control lid flexure in pressure variations.

H. High strength metal lids of new materials closely matched in expansion to the aluminum nitride package.

I. High stiffness graphite/epoxy/moly retainers to hold the packages to the circuit board in severe vibration, minimizing vibration forces on leads or solder joints.

The premium grade of aluminum nitride, with its extremely high thermal conductivity, can be used for chip-mount carriers which provide good heat spreading away from high wattage ICs. Figure 14 shows a VLSI chip mounted to such a chip-mount carrier.

THERMAL CYCLING TESTS

Reliability evaluation modules were fabricated to assess the effects of various environmental stress exposures. The silicon substrates in these modules contained several thousand test vias and wire-bonds, corrosion-monitoring triple track chips, junction-temperature monitoring test ICs, passive test chips, and component and substrate mounting test features as shown in Figure 15. At

FIGURE 15. MULTICHIP MODULE DESIGNED FOR ENVIRONMENTAL STRESS TESTING CHARACTERIZATIONS

the start of these tests, the new aluminum nitride packages were not yet available, so performanc of silicon-on-silicon assemblies in conventional Kovar packages was evaluated. Full silicon wafers were also bonded to a variety of other base materials to develop fatigue prediction curves. The strain induced in the silicon from the expansion mismatch with the base can be calcuated to be:

$$\Delta e = \left(\frac{1-B}{1-v}\right) \Delta\alpha\Delta T$$

Where:
- Δe = The strain range experienced
- $\Delta\alpha$ = The difference in thermal expansion rates
- ΔT = The temperature range in thermal cycling
- v = Poisson's ratio, taken for silicon as 0.25
- B = A stiffness factor determined by the ratios of the modulus and thickness of the silicon substrate and base.

The stiffness factor B can be calculated from classical bimetal strip/plate equations and is plotted for convenience in figure 16. A plot of the calculated strain-range versus the experimentally observed number of cycles to failure gives a familiar fatigue-type curve as shown in figure 17. For the aluminum nitride base, the strain range in silicon can be calculated to be .0001 which projects to a fatigue life far in excess of 10,000 cycles. By contrast, silicon rigidly bonded (epoxy or solder) to Kovar has first failures around 600 cycles, and bonded to aluminum fails in the first cycle. Compliant silicones were found to provide adequate strain relief to allow 3 inch wafers bonded to aluminum to survive 1000 thermal cycles, but at a sacrifice of lower thermal conductivity in the adhesive joints. Other results of interest after thousands of thermal cycle tests include:

- There were zero die-mount failures for either the various silicon ICs or passive devices
- Out of thousands of test vias and thousands of monometallic aluminum wire-bonds, there were zero failures.

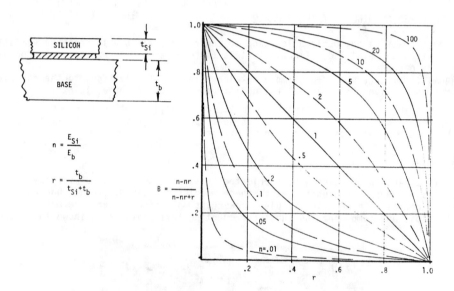

FIGURE 16. STIFFNESS FACTOR B FOR WAFERS BONDED TO BASES

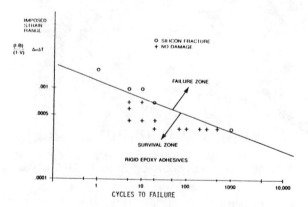

FIGURE 17. FATIGUE LIFE OF SILICON WAFERS
RIGIDLY BONDED TO MISMATCHED BASES

THERMAL PERFORMANCE

Finite element analyses were run for the aluminum nitride packages of the plug-in module (figure 11) designs. Each module has a typical module dissipation around 10 watts. Predicted maximum junction-to-case temperature rises were about 4C° for the plug-in module and 3C° for the flatpack module. These results agree with the outstanding low junction temperatures measured on previous designs of multichip modules.

PRESSURE CYCLING PERFORMANCE

Holes in silicon substrates can be easily laser cut to accommodate lid support posts as is evident in figures 2, 6, and 13. Such posts are an effective method of controlling lid flexure of these large packages caused by ambient pressure variations. (For reference, a plane flying three sorties per day would experience 20,000 such pressure cycles of 15 to 1 psia in its 20 year life.) Center posts on the designs shown were found to provide about a ten times reduction in both lid deflection and stress in the sealing weld, contributing considerably to high reliability in hermetic lid seals.

VIBRATION PERFORMANCE

Vibration tests are currently in progress and will be reported upon at a later date. For these tests a circuit board assembly containing a large memory function has been redesigned into a single small multichip module. Both the PWB and MCM versions are being run side-by-side in severe vibration and extended thermal cycling tests to quantify the improved MTBF provided by the new packaging technology. Since all resonant frequencies of the multichip module assemblies are considerably above the typical 0 to 2000 Hz environment of airborne equipment, vibration performance is expected to be considerably improved over circuit board assemblies.

COMPLIANT LEAD PERFORMANCE

Several studies have shown that the reliability of solder joints of leaded devices surface mounted to mis-matched expansion circuit boards, is a strong function of lead compliancy (224,225). Stiff leads impart significant strains into solder joints during thermal cycling, whereas compliant leads minimize such fatigue damage. As was evident in figure 13, the leads in the co-fired aluminum nitride package perform only the electrical interconnect function. The mechanical function of attachment to the circuit board has been transferred to a retaining strap arrangement. With this approach the leads can be made exremely compliant, using fine stranded copper wire as previously discussed. Thus, thermal expansion forces of the leads against the circuit board solder joints are reduced drastically compared to conventional flat leads as shown in table II.

	STANDARD FLAT LEAD .020 x .010 Kovar	"COMPLIANT" FLAT LEAD .020 x .004 Kovar	ULTRA-FLEX LEAD 60 x .001 dia Coper
X-Direction	.39 lb	.05 lb	.01 lb
Y-Direction	1.57 lb	.63 lb	.01 lb

TABLE II. Typical Lateral Forces Exerted Against Corner Solder Joints By J-Leads On 2 Inch Flatpack During Temperature Cyclng

SUMMARY

Hybrid wafer scale integration developments continue to show substantial advantages in size/weight/cost reductions and performance/reliability improvements. Aluminum Nitride and other newly developed packaging materials offer exciting packaging possibilities to support the emerging HWSI technology.

REFERENCES

1.   J.K. Hagge, "Ultra-Reliable Packaging For Silicon-on-Silicon WSI," Electronic Components Conference, Los Angeles, May 1988.
     Also IEEE-CHMT Transactions.
2.   G. Ginsburg, "Multichip Modules Gather ICs into A Small Area," Electronic Packaging & Production, October 1988, pp 48-50.
3.   N. Kuramoto, et al, "Translucent AlN Ceramic Substrate," IEEE Trans CHMT, December 1986, pp 386-390, vol. CHMT-9, no. 4.
4.-131. These references refer to identically numbered references previously listed in reference no. 1, above.
132. C.J. Bartlett, et al, "Multichip Packaging Design for VLSI-Based Systems," IEEE Trans CHMT, pp 647-453, vol. CHMT-10, no. 4.

133. S.K. Tewksbury, "FIR Digital Filters for High Sample Rate Applications," IEEE Communications Magazine, July 1987, pp 62-72, vol. 25, no. 7.
134. P. Franson, et al, "Fundamental Interconnection Limits," AT&T Tech Journal, 1987
135. A.C. Adams, et al, "High Density Interconnections for Advanced Packaging," Electrochemical Society Meeting, Philadelphia, PA, Spring 1987.
136. Y.C. Lee, et al, "Internal Thermal Resistance of a Multichip Packaging Design for VLSI Based System," EIA/IEEE ECC, May 1988.
137. L.A. Hornak, "Optical Interconnection Issues for Multi-Wafer Systems," Engineering Foundation Confrence on Advanced Materials and Processes for High Density Packaging, Santa Barbara, March 1988.
138. C.P. Wong, "Encapsulant Options for Chips Packaged in Multichip Modules," IEEE IEMT Workshop on Multichip Interconnection, Orlando, FL., October 1988.
139. J.J. Rubin, "Prototype Designs and Applications for Advanced VLSI Packaging," DuPont High Density Interconnect Symposium, October 1988.
140. R.W. Johnson, et al, "Hybrid Silicon Wafer-Scale Packaging Technology," IEEE ISSCC Technical Digest, 1986, pp 166-167.
141. R.W. Johnson, "Hybrid Silicon Wafer-Scale Packaging Technology," 1987, SRC Report no. T87113.
142. R.W. Johnson, "Thin Film Silicon Multi-Chip Technology," EIA/IEEE ECC, May 1988.
143. R.W. Johnson, "Thin Film Hybrids, VLSI & GaAs Packaging Workshop, September 1988.
144. R.W. Johnson, "Multilayer Thin Film Hybrids on Silicon," ISHM, October 1988.
145. C.C. Chen, et al, "Thermal Management of a High Performance Multi-Chip Module," EIA/IEEE ECC, May 1988.
146. P.L. Young, et al, "Impact of Fabrication of High Performance Multichip Packages on Future Manufacturing Technology," IEEE Int. Elect. Mfgr'g. Tech. Symposium, September 1986, San Francisco, p 2.
147. N.K. Pearne, "Packaging and Interconnection Developments Into the 90's," Printed Circuit World Convention IV, Tokyo, June 1987, Paper no. WCIV-83.
148. M. Sage, "Multichip Modules," ISHM, October 1988.
149. N.G. Sankar, "High Density Performance Substrates for Interconnecting Tape Automated Bond Packages," Printed Circuit World Convention IV, Tokyo, June 1987, Paper no. WCIV-26.
150. Staff, "Dupont to Manufacture Honeywell's THin-Film Multilayer Technology," Printed Circuit Assembly, April 1988, P.4.
151. M.J. Gurnick, "Copper Polyimide Multichip Package," DuPont High Density Interconnect Symposium, October 1988.
152. B. Merriman, "New Low Coefficient of Expansion Polyimide for Inorganic Substrates," DuPont High Density Interconnect Symposium, October 1988.
153. M.T. Pottiger, "New Developments in Photodefinable Polyimide," DuPont High Density Interconnect Symposium, October 1988.
154. C.A. Neugebauer, "High Performance Interconnection between VLSI chips," Solid State Technology, June 1988, pp 93-98.
155. J.E. Kohl, et al, "High Density Overlay for Bare-Chip Interconnect," GOMAC, Las Vegas, November 1988, Paper 15.2.
156. C.A. Neugebauer, et al, (GE), "Multichip Module Designs for High Performance Applications," IEPS, Dallas, TX., November 1988, pp. 163-173.
157. R. Carlson, et al, "A High Density Copper/Polyimide Overlay Interconnection," IEPS, Dallas, TX., November 1988, pp. 793-804.
158. R.O. Carlson and R.A. Fillion, "Update on GE High Density Overlay Interconnect", Dupont High Density Interconnect Symposium, October 1988.
159. W.M. Marcinkiewicz and F.J. Hale, "High Density Multilayer Thin Film Process for VHSIC Signal Processing," DuPont High Density Interconnect Symposium, October 1988.
160. C.C. Chao, et al (Hewlett-Packard), "Multilayer Thin Film Substrates for Multichip Packaging," EIA/IEEE ECC, May 1988.
161. T.J. Moravec, et al, "Manufacturing Multichip Packages for VHSIC: The Challenge of Thin-Film Techniques for Wafer-Size Substrates," November 1988, Paper 15.4, GOMAC, Las Vegas, NV.
162. R.J. Jensen, "Update on Honeywell Thin Film Interconnect Technology," DuPont High Density Interconnect Symposium, October 1988.
163. N.G. Koopman (IMB), "Chip-to-Package Interconnections," Engineering Foundation Conference on Advanced Materials and Processes for High Density Packaging, Santa Barbara, March 1988.
164. H. Sachdev and K. Sachdev, "Applications of Plasma Polymerized Organosilicon Films in Multi Level Metal Structures," DuPont High Density Interconnect Symposium, October 1988.
165. C. Feger, "Curing of Binary Mixtures of Polyimides," DuPont High Density Interconnect Symposium, October 1988.
166. L.E. Mosley, "Electrical Characterization of a Multichip Module," DuPont High Density Interconnect Symposium, October 1988.
167. T. Pan, "A Planar Approach To High Density Copper-Imide Interconnect Fabrication," IEPS, Dallas, November 1988, pp. 174-189.
168. J.T. Pan, "A Planar Approach to High Density Interconnect Technology," DuPont High Density Interconnect Symposium, October 1988.
169. R.R. Johnson, "Discrete Wafer-Scale Integration," in Wafer-Scale Integration, G. Saucier and J. Trilhe, editors, Elsevier Science, N Holland, 1986.
170. Staff, "Silicon Boards Offer Substrate Alternative," Circuits Manufacturing, February 1988, p. 16.
171. W.R. Iversen, "Amorphous Silicon May Be The Key," Electronics Week, April 8, 1985, pp. 25-26.
172. W.R. Iversen, "Amorphous Vias in Wafer Link Chips," Electronics, September 22, 1983, PP. 48-49.
173. D. Sylvester, "Wafer-Scale Integration: Down But Not Out," Electronic Business, November 1, 1984, pp. 50-52.
174. Staff, "Mosaic Talks to 4 Military Firms," Electronic News, October 27, 1986.

175. D.E. Meyer, "Wafer Scale Integration? Time to Get Serious," Semiconductor International, November 1986, p. 32.

176. L. Hyden, and J. Cofield, "Silicon Circuit Boards: Off The Shelf Convencience," Elect. Pack. Prod., April 1988, pp. 70-74.

177. D.J. Pedder, "Flip Chip Solder bonding for High Density Interconnections," Engineering Foundation Conference on Advanced Materials and Processes for High Density Interconnection, Santa Barbara, March 1988.

178. D.J. Pedder, "Flip Chip Solder Bonding for Microelectronic Applications," Hybrid Circuits (ISHM-UP), January 1988.

179. D.J. Pedder, "Future Trends in Surface Mount and High Density Interconnection Technology," 1st Int. Conf. SMT, IFS, UK, December 1987.

180. J. Reche, and R. Mace, "Implications of Properties of Thin Films in Wafer-Scale Multilevel Interconnects," SAMPE Electronic Materials and Process Conf., June 1988, Seattle.

181. J.J. Reche, "High Density Multichip Interconnect and Packaging Technology," IEEE IEMT Workshop on Multichip Interconnection, Orlando, FL, October 1988.

182. C.A. Lofdahl, "Beyond Printed Wiring Densities: A New Multi-chip Packaging Technology," IPC Spring Meeting, Holywood, FL, April 1988.

183. E. Bogatin, "Electrical Properties of Interconnect Materials and Structures For Multichip Modules," SAMPE Electronic Materials and Process Conf., June 1988, Seattle, WA.

184. P. Chatterjee, "Interconnection Bottlenecks Will Dominate Future VLSI Designs," Electronic Design, January 7, 1988.

185. W. Schroen, "Chip Packages Enter the 21st Century," Machine Design, February 11, 1988, pp. 137-143.

186. J.M. Perhot, et al, (Bull) "Wafer Scale Memory" in Wafer Scale Integration, G. Saucier Editor, Elsevier Science Publishers, 1986, pp. 99-114.

187. A.D. Trigg and J.W. Bailey, "Silicon Super Hybrids--An integrated Solution," Engineering Foundation Conference on Advanced Materials and Processes for High Density Packaging, Santa Barbara, CA., March 1988.

188. M. Reichl, "Silicon Substrates for Chip Interconnection," Hybrid Circuits-Europe, September 1986, pp. 5-7, no. 11.

189. M. Schneider, "Flip Chip Bonding Offers Packaging Alterntive," Hybrid Circuit Technology, March 1988, pp. 29-31.

190. G. Messner, "Price/Density Tradeoffs in Multichip Modules," ISHM, October 1988.

191. S. Kimijima, et al, "High Density Multichip Module by Chip-on-Wafer Technology", ISHM, October 1988.

192. To be published, "Hybrid Wafer Scale Integration," ASM Electronic Materials Handbook, Volume I, Packaging, Section 3F., C.A. Dostal, editor, ASM International, Metals Park, OH., 1989.

193. K. Umezawa, et al, "A High Performance GaAs Multichip Package," VLSI & GaAs Packaging Workshop, September 1988.

194. B. Beach, et al, "Pyrolin Dielectric for Multichip Module Applications," IEPS, Dallas, TX., November 1988, pp. 190-220.

195. D.J. Elliott, "Excimer Laser Ablation of Polyimide and Copper Films In IC Packaging Applications," DuPont High Density Interconnect Symposium, October 1988.

196. T. Miyashita, et al "Integrated Optical Devices Based on Silica Waveguide Technologies," SPIE Int. Symposium, Boston, MA., September 1988, Paper 993-46.

197. E. Bogatin, "Thin Film Multilayer Interconnect Technology for Advanced Packaging Systems," DuPont High Density Interconnect Symposium, October 11988.

198. W.C. O'Mara, "Substrates for Multichip Modules," DuPont High Density Interconnect Symposium, October 1988.

199. R.W. Mainger, "Investigation of a Hybrid Wafer Scale Integration Technique That Mounts Discrete Integrated Circuit Die In A Silicon Substrate," AFIT/GE/ENG/88M, March 1988, Thesis, Available NTIS, AD-A190-482.

200. G.L. Takahashi, "Fabrication and Electrical Characterization of Multilevel Aluminum Interconnects Used to Achieve Silicon Hybrid Wafer Scale Integration," AFIT/GE/ENG/87D-65, December 1987, NTIS AD-A190-519.

201. Deleted.

202. L. Martin, VLSIC Packaging Technology," RFP Solicitation no. N66001-88-R-0181, 23 February 1988.

203. R.S. Putnam, "Reliable Module for Wafer-Scale Integration," RFP Solicitation no. F30602-88-R-0009, 9 February 1988.

204. G. Blackburn, "Organic Interlevel Dielectrics: A Military Perspective," SAMPE Electronic Materials and Process Conf.," June 1988, Seattle, WA.

205. D.B. Tuckerman, et al, "Fabrication of Hybrid Wafer Scale Integrated Circuits Using Laser Patterned Interconnects," VLSI and GaAs Packaging Workshop, September 1988.

206. D.B. Tuckerman, "Laser-Patterned Interconnects for Thin-Film Hybrid Wafer-Scale Circuits," IEEE Electron Device Letters, November 1987, pp. 540-543, vol. EDL-8, no. 11.

207. D. Volfson, "Polyimide Planarization Method for a Plated Via-Post Multilayer Interconnect Technology," DuPont High Density Interconnect Symposium, October 1988.

208. S.D. Senturia, "Recent Results on the In-Situ Measurement of Mechanical Properties and Adhesion of Thin Films," DuPont High Density Interconnect Symposium, October 1988.

209. D. Nayak, et al, "Calculations of Capacitances of Chip-to-Chip Interconnects on a High Density Multi-chip Package," IEPS, Dallas, TX., November 1988, PP. 202-212.

210. T. Tessier, "Selecting Polymer Dielectrics for Thin Film Interconnects," DuPont High Density Interconnect Symposium, October 1988.

211. H.J. Greub, et al, "FRISC - A Fast Reduced Instruction Set Computer Using Advanced Bipolar Differential Logic, and Wafer Scale Hybrid Packaging," IEEE Int. Conf. Computer Design, 1987, pp. 478-484.

212. H.J. Greub, et al, "Architecture of a 32-Bit Fast Reduced Insertion Set Computer (FRISC) for Implementation With Advanced Bipolar Differential Logic and Wafer Scale Hybrid Packaging Technology," VLSI 87, Proc. IFIP TC 10/WG 10.5, Int. Conf. on VLSI, August 1987, Troy, NY, pp. 275-289.

213. J.F. McDonald, et al, "Multilevel Interconnections for Wafer Scale Integration," J. Vac. Sci. Technology, A-4(6), Nov/Dec 1986, pp. 3127-3138.

214. J.F. McDonald, et al, "New Systems for Wafer Scale Interconnection Fabrication for Packaging," EIA/IEEE ECC, May 1988.

215. J.F. McDonald, et al, "Optimized Focused Ion Beam Inspection and Repair of Wafer Scale Interconnections," Soc Phot-Optical Instrum Engrs, 1987, Proc vol 773, pp. 206-215.

216. JF. McDonald, et al, "Adaptive Discretionary Wiring for Wafer Scale Integration Using Electron Beam Lithography," SPIE , 1987, pp. 140-149, vol. 773.

217. O. Kwon and R.F.W. Pease, "Closely Packed Microstrip Lines as Very High-Speed Chip-to-Chip Interconnects," IEEE Trans CHMT, September 1987, pp. 314-320, vol. CHMT-10, no. 3.

218. J.P. Martino, "Technological Changes Affecting the PCB Industry," PC FAB, March 1988, pp. 68-77.

219. D. Grabbe, "High Density Multichip Packaging/Interconnection," IEEE Computer Packaging Meeting, Boynton Beach, FL., October 1988.

220. J.S. Goodling, et al, "Cooling Tomorrow's Computer Systems," Computers in Mechanical Engineering, May/June 1988, pp. 16-22.

221. Staff, "New Hybrid Packaging Technique," Solid State Technology, September 1988, p. 24.

222. R. Keeler, "Multichip Module Shaves Lines to 10 Microns," Electronic Packaging & Production, October 1988, p. 16.

223. R.W. Kotlowitz, "Comparative Compliance of Generic Lead Designs for Surface Mounted Components," IEPS Journal, October 1988, pp. 7-19.

224. W. Englemaier, et al, "Surface Mount Solder Attachment Reliability Figures-of-Merit Design for Reliability Tools," IEPS, Dallas, November 1988.

225. L.T. Hwang, et al, "The Effects of the Skin-Depth on the Design of a Thin-Film Package," IEPS, Dallas, TX., November 1988, pp. 213-232.

226. R.C. Landis, "Electronic Packaging For the Year 2000," IEPS, Dallas, TX., November 1988, pp. 535-543.

227. D.J. Pedder, et al, "Bond Design and Alignment in Flip Chip Solder Bonding," IEPS, Dallas, TX., November 1988, pp. 830-841.

228. G. Dehaine, et al, "Multichip Packages in the Espirit Program," NEPCON, Anaheim, CA., March 1989.

229. J. Bailey, "Multichip Development in the R.I.S.H. Program," NEPCON, Anaheim, CA., March 1989.

230. J. Reche, "High Density Interconnect for Advanced Packaging," NEPCON, Anaheim, March 1989.

231. L. Levinson, "High Density Interconnects Using Laser Lithography," NEPCON, Anaheim, March 1989.

232. R.W. Johnson, "Thin Film Multichip Hybrids: An Overview," NEPCON, Anaheim, CA., March 1989.

233. M. Sage, "The Future of Multichip Modules," NEPCON, Anaheim, CA., March 1989.

234. I. Turlik, et al, "Multichip Packaging for Supercomputers," NEPCON, Anaheim, CA., March 1989.

235. R. Bruce, "Using Multichip Modules for a High Bandwidth Bus," NEPCON, Anaheim, March 1989.

236. F. Andros and Pl Flynn, "CMP Multichip Packaging," NEPCON, Anaheim, CA., March 1989.

237. J. Craig and B.T. Merriman, "New Polyimides for High Density Interconnect," NEPCON, Anaheim, CA., March 1989.

238. M. Satou, et al, "PIQ Polyimide for Multichip Module Applications," NEPCON, Anaheim, CA., March 1989.

239. J. Carr, et al, "Benzocyclobutenes (BCB's) for High Density Interconnect," NEPCON, Anaheim, CA., March 1989.

240. M. Paquette, "Aluminum Nitride for the New Multichip Packages," NEPCON, Anaheim, CA., March 1989.

241. D. shanefield, "Comparison of Ceramics for Multichip Hybrids," NEPCON, Anaheim, CA., March 1989.

242. Y. Kurokawa, et al, "Highly Thermal Conductive Aluminum Nitride Substrates," ISHM, Minneapolis, MN., September 1987, pp. 654-661.

243. J.W. Bailey, "Silicon Super Hybrids - An Integrated Solution," IEEE Computer Packaging Workshop, Brussels, January 1989.

244. M.S. Dhanaliwala, "Multiple Channel Optoelectronic Packaging Technology for Optical Data Links," IEEE Computer Packaging Workshop, Brussels, January 1989.

245. J. Reche, "High Density Multichip Interconnect and Packaging Technology," IEEE Computer Packaging Workshop, Brussels, January 1989.

246. P. Garrou, "Dielectric Materials for MCMs," IEEE Computer Packaging Workshop, Brussels, January 1989.

247. R.J. Perko, "MMIC vs Hybrid: Glass Microwave ICs Rewrite the Rules," Microwave Journal, Novemer 1988, pp. 67-78.

248. D. Grabbe, "IC Packaging and Sockets for High Speed Devices and WSI," IEEE Computer Packaging Workshop, Brussels, January 1989.

249. I. Turlik, "Design Issues in Multichip Packaging," IEEE Computer Packaging Workshop, Brussels, January 1989.

Passive-Silicon-Carrier Design and Characteristics

H. SCHETTLER
IBM Laboratories
Boeblingen,Germany

Abstract

A silicon-on-silicon package for 9 VLSI chips is presented. The carrier is especially designed to support the high frequency current surges of complementary logic like CMOS or BICMOS. It contains integrated decoupling capacitors and uses 3 layers of metal. The chips are mounted via 'controlled collapse chip connection'. A /370 processor containing 9 CMOS chips has been assembled. Functionality and manufacturability have successfully been demonstrated.

Introduction

The progress of VLSI technology requires a common design optimization of chip and package. The presented paper shows such a design of a silicon-on-silicon package that is based on experiences that have been made with VLSI Processor Chips (1,2).

These VLSI chips showed clearly that the delta I problem of switching CMOS gates needs special attention. The referenced chips are designed in a 1.0 μm technology.

The Processor Chip 1 (2) generates a current surge of 1100 mA/ns at its voltage and ground terminals. The VLSI chips are packaged on 36 mm ceramic PGA's. With an effective ground and voltage line inductance of circa 0.5 nH each, the current surge causes power and ground line noise spikes of ca. 600 mV.

The current demand of a CMOS or BICMOS circuit depends on circuit rise and fall time (Tr,f), voltage swing (Vs) and load (Cl):

$$dI/dT = k * Cl * Vs / (Tr,f)^2$$

Table 1 indicates an extrapolation of the noise problem up to 0.8 μm (3;6) and 0.5 μm technology.

Technology	1.0 μ (2)	0.8 μ (3;6)	0.5 μ (estim.)
Delta T (ns)	0.8	0.4	0.3
Delta V (Volt)	5.0	5.0	3.4
dV/dT² (normal.)	1	4	5
dI/dT (ma/ns) chip pad	1,100 MEASURED	5,000	10,000

Table 1: Expected current surges for future VLSI's

Further packaging requirements for VLSI-based systems (4) are high connectivity (more than 500 Signal-IO's per chip),high wiring density high performance (low wire capacitances), good cooling capabilities (> 1 W/cm²) and low cost.

The Carrier Design

The stated requirements lead us directly to a Thin Film Package. Silicon as a substrate for TF package is well suited. All the process technology for thin film on silicon is available and well understood.

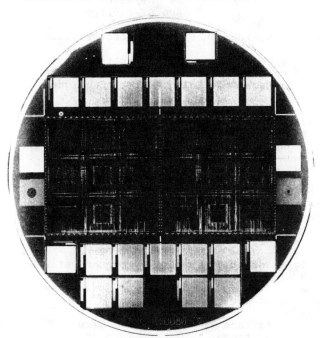

Fig. 1 : Passive Silicon Carrier on 125 mm Wafer

The proposed package employs 3 layers of metal and an integrated capacitor that is used for power supply decoupling.

Reprinted from *Proc. 40th Electron. Components & Technol. Conf. (ECTC)*, pp. 559–561, 1990.

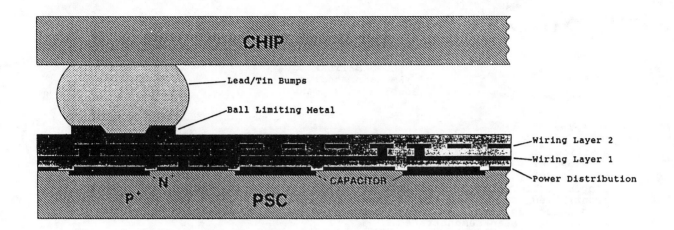

Fig. 2: Silicon carrier cross section (schematic)

The decoupling capacitor consists of n+ implanted stripes within the p+ doped substrate. The n+ region is connected to Vdd and the substrate is connected to Ground. The depletion layer(n+/p+) capacitance is 0.6 fF/μm² at 5 Volts at a breakdown voltage of > 8 Volts.

Substrate doping : Boron 3E17 1/cm**3

Implant doping : Arsenic 6E15 1/cm**2

Capacitor : 0.6 fF/μm²
64 nF/Chip
600 nF/Carrier

Table 2:Silicon Substrate Characteristics

The presented processor application is based on chips previously designed for a PGA package that allowed for 164 Signal IO's per chip only.

The shown silicon carrier has 320 Signal-IO's, 80 Vdd and 80 GND pads. Its size is 47*47 mm².The carrier size determines the maximum wire length which is ca.90 mm.The actual wires within time critical path is less than 50 mm. These specific wires are

electrically short compared to CMOS Rise and Fall times of ca. 2 ns :

T (50mm) = 300 ps < Tr/2

This means that control of transmission line impedance is less important. The RC time costant of the line is ca.0.02 ps/mm² and almost negligible.

Wire Resistance : 0.13 Ω/mm
Wire Capacitance : 0.15 pF/mm
Wire Inductivity : 0.2 nH/mm
Propagation delay : 6 ps/mm
Decoupling capacitor : 64 nF/Chip
Ground-Power Loop Ind. : < 1 pH/Chip
Decoupl.Series Res. : 0.7 mΩ/Chip

Table 3 : Wire characteristics and
Power Line Decoupling

Process

The silicon substrate employs a 5 micron process, using seven masks as shown in table 4. The integrated capacitor was fabricated in the boron doped silicon substrate by an implantation of arsenic. All metallisation layers are based on Lift-Off processes using image reversal of diazoquinone Novolak Resists.

Process step	layer thickn. μm	size/ width μm	wire pitch μm
Capacitor def.	0.5	50	
Metal 1	2.5	30	50
Via 1/2	2.2	5	
Metal 2	2.0	15	25
Via 2/3	2.5	5	
Metal 3	3.0	15	25
Terminal via	4.0	50	
Barrier layer	1.7	125	
Bump	120	125	

Table 4 : Process Flow and Feature Sizes

The 1st metal layer is directly evaporated on to silicon. All GND metal contacts the grounded substrate and all Vdd metal contacts the n+ implant forming the capacitor. With the exception of the M1 to M2 insulation layer, where Si-Nitride and polyimide were used, the dielectrics were formed with polyimides. The Vias were reactive ion etched, whereby flat slopes could be achieved with a short resist reflow before the RIE process.

Slope control was not a concern, because the design was restricted to one via size per layer only.

The terminal vias were structured in a direct develop and etch process of the alkali developer, which can dissolve the partially cured polyimide in the vias. As the Pb/Sn solder bumps cannot be reflowed directly to the aluminum, a barrier layer consisting of Cr/Cu/Au was deposited on the terminal Al pads.

This interface layer and the following Pb/Sn bumps were defined by a molybdenum shadow mask.

RESULTS

The first process problem arises with the size of the silicon substrate. No physical damage or process irregularities are allowed over a 25 cm² silicon area.

Compared to chip processing, where large defects destroy only a few chips out of the hundreds available, the manufacturing yield of silicon substrates is significantly impacted if a single carrier is deffective as there are only a few sites possible on a wafer.

We found that process engineers must pay much more attention to area defects than on small ones in the 1 μm range (5).

Special attention was given to the following items:

1. Implantation : Particles of > 3 μm diameter can cause leakage currents due to implantation shadowing under these particles.

2. Metallization : Fortunately M1 needs no defect sensitive insulation underneath. The huge metal area of Vdd and GND is directly attached to the silicon.
The relatively small critical areas of the wiring layers M2 and M3 are not too sensitive to particles.
Only the high pointed particles and metal residues from Lift-Off cause interlevel shorts.

3. Via Layers : Vias are no problem with respect to particles because of the extremely small critical area. However resist bubbles in the thick viscous resist can cause interlevel shorts.

A prototype silicon carrier has been built, tested and the 9 chips have been mounted. The first lot showed a good yield, which demonstrates the manufacturability of the process.

Fig.3 : Passive Silicon Carrier
cross section (μ-photograph)

Fig. 4: Complete processor

References:

(1) H.Schettler, G.Koetzle, U.Schulz, O.Wagner:'A Highly Flexible 60k-Gates CMOS Master Image' ESSCIRC86, Sept.86

(2) K.Klein,G.Koetzle,E.Miersch,H.Schettler, U.Schulz,O.Wagner:'Characteristics of a Set of 12.7 mm Processor Chips' IEEE Journal of Solid-State circuits, Vol.SC-22,No.5, Oct.87

(3) T.Takahashi, M.Kawashima, M.Fujita, I.Kobayashi,K.Arai,T.Okabe:'A 1.4 M-Transistor CMOS Gate Array with 4 ns RAM' ISSCC89 Digest, Feb.1989

(4) H.Levinstein, C.Bartlett, W.Bertram : 'Multi-chip Packaging Technology for VLSI-based Systems' ISSCC87 Digest, Feb.1987

(5) V.Kreuter, G. Wenz : 'Reinheitsanforderungen in einer Pilotlinie zur Herstellung hoechstintegrierter Schaltkreise', VDI-Berichte Nr.693,88

(6) H.Schettler, J.Hajdu, K.Getzlaff, W.D.Loehlein,C.W.Starke:'A Mainframe Processor in CMOS Technology with 0.5 μm Channel Length.' ISSCC90 Digest

Part 5
Thin Film Multilayer Modules on Ceramic or Metal

THE NEXT SHIFT in technology was to deposit the organic insulation directly on the ceramic substrate. In an early paper, a mainframe application is from NEC (Watari and Dohya) and, of course, the Jensen *et al.* paper of the first chapter. Lebow also demonstrated the advantages of this technology in applications for ITT. Other papers follow, including descriptions of the work at TI, Hughes, and MCC. The NTT paper by Ohsaki and others shows a higher density application, as does the MCNC papers showing the utilization of aluminum nitride and silicon carbide.

Lastly, there is the laminate approach applied to ceramic substrates. Polylithics fabricates the laminate first, then bonds on the ceramic substrate, while GE lays the laminate over the chips and forms the circuit. Both are different designs that seem to fill application niches at their companies, but have not been accepted for use elsewhere.

The mainstream approach seems to be deposited dielectric of polyimide or equivalent with aluminum or copper metal interconnect circuits. Leaders in this technology are Raychem with their HDI technology, and Polycon with either polyimide or BCB. The Rockwell papers show the possible applications. At the writing of this book, DEC had not yet produced an article on the technology of their recently an-nounced multichip module of polyimide on a metal substrate, a similar construction.

Tektronix has talked of their variation—deposition on copper—and Toshiba of their multilayer copper–polyimide. This is the NTT, NEC, and Honeywell approach, and that adopted by most of the interconnect substrate vendors. There are so many articles in this category we needed to be very selective, not because articles from others than the ones chosen here are of no interest, but because a whole book could be devoted to this approach.

One older technology applied to polymer processing—lift-off technology. Lift-off was once only applied to silicon dioxide processing, but now it is being used for multilayer polyimide and other dielectrics. André Schiltz of CNET, France Telecom Research, has produced a very good review article of his work and of all the comparative lift-off processes of the major involved companies. It is included as a recent overview of this technology.

Comparative analysis is always valuable. A recent review of the major multichip technologies has been made by Gustafsson of Ericsson. Perhaps this will put things together for many of the readers.

C. W. Ho
D. A. Chance
C. H. Bajorek
R. E. Acosta

The Thin-Film Module as a High-Performance Semiconductor Package

This paper discusses a multichip module for future VLSI computer packages on which an array of silicon chips is directly attached and interconnected by high-density thin-film lossy transmission lines. Since the high-performance VLSI chips contain a large number of off-chip driver circuits which are allowed to switch simultaneously in operation, low-inductance on-module capacitors are found to be essential for stabilizing the on-module power supply. Novel on-module capacitor structures are therefore proposed, discussed, and evaluated. Material systems and processing techniques for both the thin-film interconnection lines and the capacitor structures are also briefly discussed in the paper. Development of novel defect detection and repair techniques has been identified as essential for fabricating the Thin-Film Module with practical yields.

Introduction

Very Large Scale Integration (VLSI) in semiconductor technology is expected to result in significant improvements in device cost, performance, reliability, and function, and concomitant improvements in systems based on such devices. However, many of the benefits of VLSI will be lost without significant improvements in device packaging techniques. This paper first summarizes key VLSI trends in chip technology and their implications for device packaging, the limitations of current packages, and how these have been overcome by IBM's latest multichip modules based on Multilayer Ceramic Technology (MLC). The main body of the paper discusses a packaging approach being investigated in this laboratory, the Thin-Film Module, which features thin-film transmission lines and on-module decoupling capacitors integrated into the body of the module, improvements which are likely to be required to package high-performance VLSI devices in the late 1980s.

VLSI trends in chip technology

Projections for semiconductor chips [1] suggest that the number of circuits per chip for field effect and bipolar transistor logic chips and memory bits per chip will continue to double approximately every two years during the 1980s.

As the number of circuits on a logic chip increases, the ratio of input/output connections to the number of circuits decreases. This improvement, however, is not sufficient to avoid the need for a substantial increase in the total number of input/output connections to a group of circuits on a chip. As image size decreases, power per circuit decreases, but the countervailing trend of increasing integration results in a significant increase in total chip power to levels as high as 20 W/chip in high-performance applications. Similarly, although current per circuit also decreases, the total current that needs to be delivered to a chip also significantly increases, and could approach the 10-ampere level in high-performance bipolar devices. The need to simultaneously switch an ever-growing number of circuits on a chip at shorter pulse rise times and higher total currents exacerbates problems related to maintaining adequate signal/noise ratios due to chip and package inductance.

Chip and package inductance can give rise to what is called simultaneous switching noise. This inductance is particularly critical in the power supply distribution to off-chip driver circuits typically used in high-performance digital systems (Fig. 1). As depicted in this figure, the

Reprinted with permission from *IBM J. Res. Develop.*, vol. 26, no. 3, pp. 286–296, May 1982. Copyright © 1982 by International Business Machines Corporation.

simultaneous switching of many such drivers can cause fluctuation of the voltage level at the individual driver transistors by an amount ΔV:

$$\Delta V = LN\frac{\Delta I}{\Delta t}, \tag{1}$$

where L is the effective chip and package inductance, N the number of drivers switched simultaneously, ΔI the current switched by each driver, and Δt the current rise time. Fluctuation of the power supply voltage level can cause delay in the rise time of interchip signals in active nets and, if excessive, can cause noise in quiet nets which can lead to false switching of receivers connected to the quiet nets. The latter effect can cause errors in computation.

Future VLSI drivers are expected to have ever-increasing switching speeds (shorter rise times) and increasing numbers of drivers switched simultaneously per unit area of chip or package, while maintaining approximately constant driver switching current. Referring to Eq. (1) and Fig. 1, containment of simultaneous switching noise to acceptable levels will therefore require significant reductions in chip and package inductance.

Implications for device packaging

The trends discussed in the previous section have direct implications for packaging of the devices. Future packages must be capable of providing corresponding increases in the number of input/output connections and in the wiring capacity to interconnect chips. Sophisticated cooling techniques and reduced package inductance will also be essential. Attainment of maximum performance for a given system design and circuit family calls for minimizing signal propagation delay by minimizing interchip distances and the dielectric constant of interchip and intermodule transmission lines. Improving the reliability and reducing the cost of future systems requires minimizing the number of physical interconnections between chips.

Continuing the present practice of containing the function of chip packages to that of a space transformer, packaging one chip to a module, exclusive use of additional package levels, cards and cables to provide intermodule (interchip) wiring, and reliance on peripheral input/output connections to the chip and module will not permit the necessary improvements in chip and package inductance and input/output connections and will result in a loss of many of the benefits of VLSI.

A very attractive packaging approach which can provide many of the improvements needed to capitalize on VLSI is IBM's recently announced multichip MLC-based module approach (Fig. 2) [2, 3]. This technology makes it possible to package as many as one hundred logic chips on a single

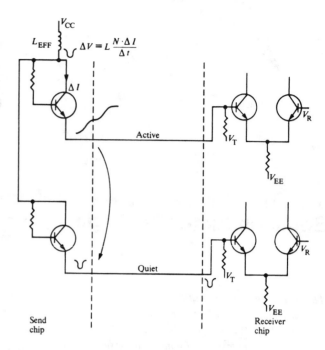

Figure 1 Simultaneous switching noise.

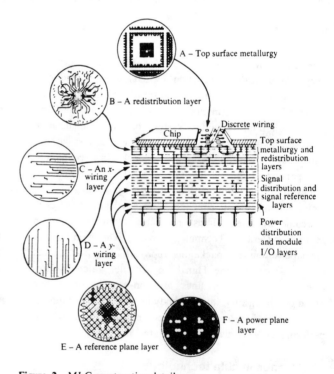

Figure 2 MLC construction detail.

ceramic substrate. At present levels of device integration, a single substrate can support as many as 60 000 logic circuits. This substrate no longer serves the simple role of a space expander; it also provides for permanent chip interconnec-

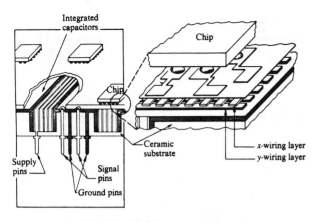

Figure 3 The Thin-Film Module.

Table 1 Impact of VLSI trends on package.

	Present	*1985–1990*
Semiconductor		
Lithography ground rules (μm)	2.0	1.0
Chip size (mm)	4.6	4.0
No. of circuits	600	2500–5000
No. of signal contacts	96	289–400
Voltage swing (V)	1.1	0.5–1.0
Rise time (ns)	1.1	0.2–0.4
Package (normalized multipliers)		
Package inductance		
Rise time effect	1.0	3.5
Delta I (switching current)	1.0	2.0
Noise intolerance	1.0	2.0
	1.0	14
Wiring density	1.0	3.0–4.0
Cooling	1.0	2.0–3.0

tions throughout the entire chip area which until now had to be made at the card level. It reduces the number of input/output connections at this level, the overall wire lengths (resulting in shorter time of flight between groups of circuits), and the power, while improving performance at the system level. The reduced number of interconnections between dissimilar packaging levels results in higher reliability and also makes it possible to improve system performance and cost by savings at the card, board, cabling, power supply, and frame levels. It also features means of performing engineering changes via discrete wires bonded to metal pads surrounding each chip site.

The extension of such multichip ceramic modules to accommodate more chips with larger numbers of input/output connections to the chip and to the module, and the adaptation of this technology to semiconductor and magnetic bubble memory devices, are expected to make it the leading device package candidate of the 80s. However, further improvements will be required to serve the needs of VLSI devices expected beyond this time frame.

The level of improvement likely to be required in critical package functions is summarized in Table 1. The first part of Table 1 compares bipolar chip parameters of present devices with those expected toward the end of this decade. The second part compares the corresponding level of improvement that must be attained in packages to support such devices. Specifically, the rise times of driver circuits are likely to decrease by a factor of 3 to 4. The total number of drivers switching simultaneously is expected to increase by a factor of 4 as the current level per driver decreases to approximately one half of present devices, resulting in a factor of 2 increase in the total current switched simultaneously per chip. The reduced voltage level will decrease noise margins by a corresponding amount. These changes will require reducing future package inductance by more than an order of magnitude below that of present packages. The 4- to 10-fold increase in the number of circuits per chip is expected to require a 3- to 4-fold improvement in wiring capacity, and improvement in cooling capability by a factor of 2 to 3 beyond that of present packages.

The Thin-Film Module
One approach being investigated in this laboratory to achieve the aforementioned package improvements for high-speed digital systems is the Thin-Film Module (Fig. 3). It features the use of multilayer photolithographically defined interchip wiring and the use of decoupling capacitors integrated into the body of the substrate to control simultaneous switching noise. The combination of thin-film lines and integrated decoupling capacitors offers a very good potential to reduce the effective package inductance by more than an order of magnitude below that of present packages.

The following sections describe our efforts to date to understand the electrical characteristics of thin-film transmission lines and of power distribution networks which incorporate decoupling capacitors in the body of the substrate, as well as the materials and processes required to achieve the necessary structures.

• *Thin-film lines*
The majority of wires for chip interconnections in modules, cards, and boards used in modern digital systems are strip transmission lines. IBM's latest MLC multichip modules also use such strip lines. Figure 4(a) depicts the cross section of the strip lines in the MLC modules. The dimensions shown are also representative of strip lines in the cards and boards which carry such modules. The cross-sectional area

of strip lines used to date, even with use of relatively resistive metals such as Mo in the ceramic modules, has resulted in transmission lines with small resistive loss, thereby enabling interchip and intermodule signal propagation with negligible signal distortion. As discussed in the previous sections, future VLSI trends demand that future packages provide significant improvement in interchip wiring capacity. This objective can be attained via a combination of miniaturization of strip transmission lines to improve line density and use of additional wiring planes to improve total line capacity.

Figure 4(b) depicts the cross section of thin-film strip lines which could be fabricated with modification of state-of-the-art photolithographic processes used in chip fabrication. However, consideration of practical dielectric and metal material properties, dimensions, and processes used for such strip lines readily suggests that thin-film strip lines with characteristic impedance of 50 Ω will have significant resistive loss. For example, a 50-Ω strip line of 5 × 9 μm of Cu will have a resistance per unit length of 4 Ω/cm. On the other hand, the potential for a 20-fold improvement in wiring density relative to the lossless case is very attractive vis-à-vis achieving significant wiring capacity while using as few as two wiring planes in future VLSI packages. The latter potential motivated us to attempt to understand the extent to which thin-film lossy strip lines could be used to provide the chip interconnection function in future packages.

Pulse propagation, coupled noise, and dispersion of high-speed pulses in various configurations of lossy thin-film strip transmission lines have been modeled and experimentally investigated in our laboratory. The details of these studies are reported in [4]; the main conclusions are summarized below.

● *Pulse propagation on lossy transmission lines*
High-speed pulse propagation on uniform lossy transmission lines is characterized by the general transmission line equations [5–8]

$$\frac{\partial v}{\partial x} = Ri + L\frac{\partial i}{\partial t},$$

$$\frac{\partial i}{\partial x} = C\frac{\partial v}{\partial t}, \tag{2}$$

where R, L, and C are resistance, inductance, and capacitance of the line per unit distance. No assumption on the magnitude of R is made; however, the conductance G through the dielectric material is known to be small for most practical dielectric materials and is therefore neglected here.

The solution for an infinitely long lossy line characterized by Eq. (2) in the frequency domain is

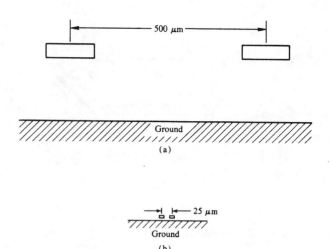

Figure 4 Strip transmission line dimensions for (a) lossless and (b) thin-film lines.

$$V(\ell, s) = e^{-\gamma(s)\ell}V_{in}(s), \tag{3}$$

where

$$\gamma(s) = \sqrt{(R + sL)sC} = s\sqrt{LC}\sqrt{1 + R/sL} \tag{4}$$

is the propagation constant, $V_{in}(s)$ is the Laplace transform of the input voltage, and ℓ is the distance from the source $V_{in}(s)$ to the point of interest on the line. It can be shown [9] that the inverse transformation of (3), which is the time domain solution of the line for a step input $V_{in}(s) = 1/s$, is

$$v(\ell, t) = e^{-R\ell/2Z_0}U(t - \ell\sqrt{LC}) + g(\ell, t)U(t - \ell\sqrt{LC}), \tag{5}$$

where the first term is an attenuated step function and the second term, represented by $g(\ell, t)$, is a rather complicated mathematical expression which is slow-rising like an RC circuit. The function $U(t)$ is a unit step function. The lossy line therefore behaves like an LC line and an RC line combined. A typical solution is plotted in Fig. 5. In this paper, we are primarily interested in the high-speed pulse propagation, and hence we concentrate on the fast-rising exponential term in Eq. (5).

When a pulse propagates on a lossy transmission line, it is attenuated exponentially, as described previously. When the attenuated pulse hits an open circuit, however, the current drops to zero. A reflection starts to propagate in the opposite direction, which also causes the voltage to double at the receiving end. If the line length, the line resistance, and its characteristic impedance Z_0 are within a certain range, the reflection can restore the lost pulse amplitude and the pulse rise time to a large extent, and yet not cause excessive ringing and distortion at the receiving end. Information can

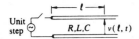

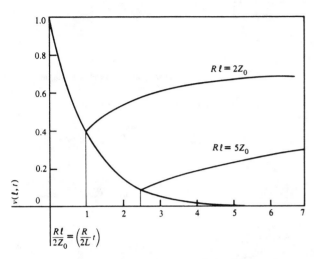

$$\frac{R\ell}{2Z_0} = \left(\frac{R}{2L}\,t\right)$$

Figure 5 Solutions to general lossy transmission line equations:

$$\frac{\partial v}{\partial x} = L\frac{\partial i}{\partial t} + Ri,$$

$$\frac{\partial i}{\partial x} = C\frac{\partial v}{\partial t},$$

$$v(\ell, t) = \left[e^{-R\ell/2Z_0} + \frac{R\ell}{2Z_0} \int_{t=\ell\sqrt{LC}}^{t} \frac{e^{-Rt/2L}}{\sqrt{t^2 - (x\sqrt{LC})^2}} \right.$$

$$\left. \times\, I_1\left[\frac{R}{2L}\sqrt{t^2 - (x\sqrt{LC})^2} \right] dt \right] u(t - \ell\sqrt{LC}).$$

therefore be transmitted on the lossy line for receivers which have high input impedance compared to Z_0 attached to the open end of the line.

It is easily seen that if the product $R\ell$ is too large, the exponential term in the solution of Eq. (5) becomes insignificant and the line behaves practically as an RC line. However, if $R\ell$ is very small, such that it approaches the lossless case, assuming that the pulse rise time is still small compared with the total line delay $\ell\sqrt{LC}$, the line behaves like an LC line, and sustained reflections make meaningful transmission of information very difficult unless a clamp circuit is used to suppress the reflections. It can be shown that if we choose the total line resistance in the range

$$\frac{2Z_0}{3} \le R\ell \le 2Z_0, \tag{6}$$

the pulse shape received at the open end is very similar to the transmitted pulse. For a uniform line, $R\ell = 2Z_0$ establishes

the maximum line length $\ell_{max} = 2Z_0/R$ and $R\ell = 2Z_0/3$ establishes the minimum line length $\ell_{min} = 2Z_0/3R$. In practical situations, it is often the case that the pulse rise time is not so small compared with the delay of the minimum length lines; i.e.,

$$t_{rise} \ge 2\ell_{min}\sqrt{LC}. \tag{7}$$

In such cases, the reflections on a line whose length is shorter than ℓ_{min} do not cause too much pulse distortion, since before the pulse can have a chance to double itself, the second reflection from the low-impedance voltage source starts to come in and pulls the pulse in the other direction. Hence the minimum line length condition is not critical, because the pulse shape is in general maintained at the receiving end. In those cases, it is important to see that the open-circuit lossy line can be used to transmit high-speed pulses with virtually the same speed as the lossless line. The attenuated pulse amplitude is restored due to the doubling effect, subject only to the condition that a maximum line length is not exceeded. For example, in a 50-Ω line system, with a line resistance per unit length of 4 Ω/cm, which can be readily achieved with 5 × 9-μm copper lines, the maximum line length for pulse propagation with small distortion is 20 cm.

The preceding analysis thus shows that lossy thin-film transmission lines can, within certain limits, be used to propagate high-speed pulses with low distortion for distances of up to 20 cm, provided that such lines are terminated with an impedance much higher than the characteristic line impedance in combination with a clamp circuit to suppress reflections in short lines. Two important attributes of such thin-film lines in semiconductor packages are the attainment of high wiring density and the reduction of power dissipation due to elimination of the power dissipated in the terminating resistor of matched lossless transmission line designs.

A complete design of a transmission line structure also requires assessing coupled noise effects. The coupled noise characteristics of lossy transmission lines for both the near-end noise (NEN), which propagates toward the driver side of the quiet line, and the far-end noise (FEN), which propagates toward the receiver side of the quiet line, are to a first order identical to those of lossless designs, with the exception that loss exacerbates far-end noise since line resistivity increases the degree of inhomogeneity of lines above and beyond the inhomogeneities caused by crossover lines and by the vias which are needed to interconnect transmission line layers in lossy or lossless designs [4]. Coupled noise is also expected to depend strongly on specific line-to-line and line-to-ground-plane dimensions. We have evaluated several alternatives involving the use of one or more ground planes to meet the following requirements:

$Z_0 = 50 \ \Omega$,

$\ell_{max} = 20$ cm,

line pitch = 25.4 μm (1000 lines per inch),

coupled noise $\leq 10\%$,   (8)

where the line pitch is the center-to-center distance of two adjacent lines, and the coupled noise indicates the worst-case coupled voltage into a quiet line between two adjacent active lines switched simultaneously.

Figure 6 depicts the cross section of strip lines which meet the aforementioned requirements. All characteristic dimensions are shown in micrometers. The dielectric constant of the insulator is 3.5. The closed nature of this triplate design makes it ideally suited for use in a structure like the Thin-Film Module because it minimizes any electrical interaction between the fan-out layer shown in Fig. 3, needed to interconnect transmission lines with semiconductor chip terminals, and the x-y transmission lines. Connection between the fan-out layer and the lines and between lines requires vias of approximately 6.0 μm diameter. Representative signal responses and coupled noise for the indicated input voltage for various line lengths of this triplate design are shown in Fig. 7. The worst-case transient current amplitude is 11.2 mA for the input voltage amplitude of 600 mV. The propagation delay at the 50% signal level is 66.2 ps/cm and the worst-case coupled noise is 8.8%. The previously mentioned voltage oscillations at the high-impedance receiver can be seen in all cases, especially for the shortest line lengths. These reflections can be suppressed by means of a clamp network using Schottky barrier diodes at the receiver. The 20-cm line length case exhibits the beginning of the RC-type slowdown on the voltage incident at the receiver. Avoidance of this effect, which can cause undesirable delay in long lines, requires restricting the maximum line length to approximately 20 cm. This restriction is not expected to be serious since lines 20 cm long should enable wiring of large arrays of VLSI chips on multichip modules using thin-film transmission lines.

These results have been verified experimentally with a high-speed pulse sampling oscilloscope in conjunction with coaxial probes and thin-film line test samples specially designed to minimize discontinuities at the probe-to-sample interfaces [4]. The input pulse, the measured pulse, and a simulated pulse are shown in Fig. 8 for a 50-Ω strip line with a conductor 12 μm wide and 5 μm thick. Figure 8 shows good agreement between the predicted and measured response.

Simultaneous switching noise

As mentioned previously, another key requirement in future VLSI packages is significant reduction of package inductance to minimize simultaneous switching noise. This noise

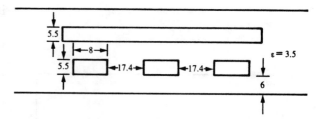

Figure 6 Triplate structure design for thin-film transmission lines. All dimensions are shown in micrometers. $\rho = 1.8 \ \mu\Omega$-cm, $R_0 = 4.1 \ \Omega$/cm, $L_{max} = 2Z_0/R_0 = 20$ cm, and $Z_0 = \sqrt{L_{22}/C_{22}} = 41 \ \Omega$.

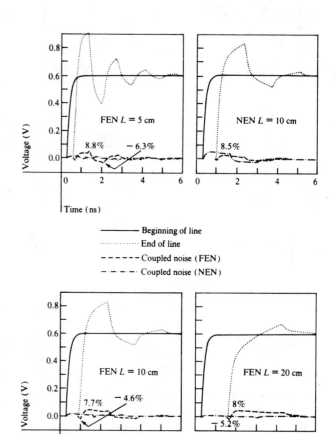

Figure 7 Signal and coupled noise responses for strip transmission lines of Fig. 6 for four line lengths.

arises when many input/output terminals in one VLSI chip attempt to communicate with other chips simultaneously during a specific instant of a machine cycle, and a large amount of current has to be supplied instantly to the chip through its power supply terminals by the package. The current is divided and comes out of the chip as signals from its drivers, through signal I/O terminals to the transmission lines on the package. If the package is incapable of supplying the current quickly, extra delays or even malfunction may

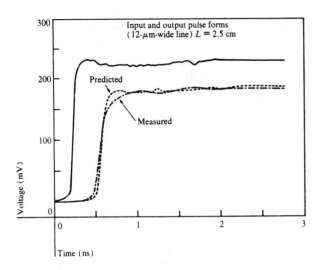

Figure 8 Comparison of theory and experiment for voltage response of 2.5-cm-long thin-film transmission line.

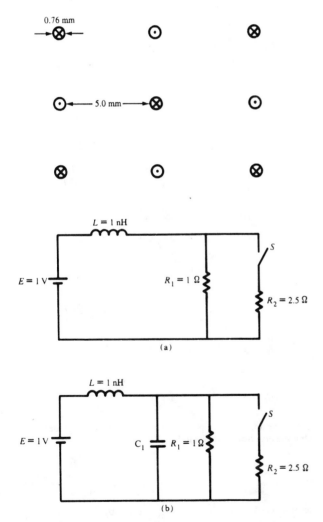

Figure 9 Package and chip power distribution equivalent circuits for modeling effect of (a) package inductance and (b) decoupling capacitor on simultaneous switching noise.

happen as a consequence. In this section a simultaneous switching example is used to illustrate the nature of the problem. The possible sources for supplying the driver currents and how an integrated-on-module capacitor design can be used as one possible solution to solve this important problem for the VLSI package are discussed.

Let us consider a set of on-module power supply pins, which bring power from the printed circuit board to the module, arranged in an array as shown in Fig. 9. If the pin-to-pin spacing and the length of the pin are 5.0 mm (200 mil), the total pin matrix inductance is calculated to be 1 nH. In the following discussions, 1 nH is used as the package inductance seen by the chip.

A typical bipolar chip draws current from the power supply for its internal circuits and external drivers. The power supply system of such a chip can be modeled with the equivalent circuit shown in Fig. 9(a), where R_1 represents the load due to the internal circuits, R_2 is the load of the drivers, and L is the package inductance. This example corresponds to a chip with internal circuit current of 1.0 A and with 20 drivers switching simultaneously into 50-Ω transmission lines which draw an additional 0.4 A at 1 V. When the switch S closes and the drivers start to draw current, the time constant of this simple R-L circuit is L/R_{eq}, where R_{eq} is the equivalent parallel resistance of R_1 and R_2. Hence the time constant is L/R_{eq} = 1.4 ns. Since the circuit needs about 2.2 times the time constant to reach 90% of the final value, the corresponding time is 3.08 ns, which is too large compared with the 0.3-ns rise time of such drivers. The inductance within the power supply of the package is therefore excessive for the chip and there is a simultaneous switching noise problem.

The classical solution to this problem is to add a decoupling capacitor C to the chip as shown in Fig. 9(b), to supply the current needed by the chip and decouple it from the package inductance. Assuming that the power supply seen by the chip can tolerate only a 5% variation during simultaneous switching of driver circuits as modeled by resistor R_2, the switching current in R_2 can only be supplied from two sources: the resistor R_1 and the capacitor C. A 5% drop of voltage at resistor R_1 of 1 Ω diverts 50 mA of its current to R_2. Since R_2 needs a total current of 0.95 V/2.5 Ω = 0.38 A to reach 0.95 V, the remaining current of 0.38 A − 0.05 A = 0.33 A has to be supplied by the capacitor C. The capacitor current is determined by $I = C(dV/dt)$ where, if dV = 0.05 V, dt = 1 ns and I = 0.33 A, C = 6.6 nF. A key requirement is minimizing the inductance of the capacitor and chip to the capacitor interface, since an inherent assumption made in Fig. 9(b) is that the inductance between capacitor C and the resistors R_1 and R_2 is much smaller than L and is ignored.

As previously mentioned, the use of thin-film wiring structures on the surface of a ceramic substrate for interconnection allows the decoupling capacitor to be built into the ceramic substrate. The VLSI chips can be mounted on the thin-film surface on top of the substrate and are separated from the capacitor substrate only by the thickness of the thin-film layers. The chip-to-capacitor substrate distance is at most 0.1 mm. If an array of solder balls is used for power supply interconnection, the lead inductance from capacitor to chip may be reduced to several picohenrys. Two types of structures for the integrated capacitor substrate are shown and discussed below. Each structure has its advantages and disadvantages, but the basic tradeoff appears to be between simplicity of capacitor substrate fabrication and inductance of the interconnection from chip to capacitor.

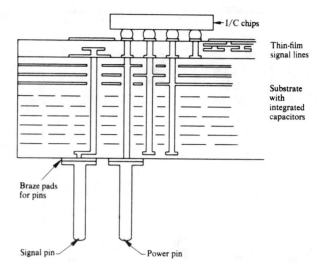

Figure 10 Cross section of substrate with integrated decoupling capacitance based on horizontally laminated power planes.

● *Integrated capacitor structures*
The structure shown in Fig. 10 is a multilayer ceramic (MLC) substrate supporting thin-film lines and a VLSI chip. The integrated capacitor substrate is the typical MLC structure with horizontal ceramic layers interspersed with metal layers. Clusters of vias are provided under each chip as shown in Fig. 11. Vias for three levels of voltage are shown arranged in rows. Each row of vias for the different voltage levels is connected to appropriate capacitor planes. For vias passing through a capacitor plane, donut areas (ceramic areas) isolate each via from the capacitor plane. Also shown in Fig. 11 are signal vias and reference voltage vias, which are on a larger grid spacing than the power vias passing through the capacitor plane. The desired capacitance is provided by varying the number of layers, thickness, and dielectric constant of the ceramic substrate with the following limitation. Signals must also travel from chips through vias in the substrate to other modules and input/output devices. The signal flight time t_D is given by the relationship

$$t_D = \frac{\sqrt{\varepsilon_r}}{c}\,\ell\,, \qquad (9)$$

where ℓ is the substrate thickness, ε_r is the dielectric constant of substrate material, and c is the speed of light. If one limits the flight time for a high-speed module to less than 100 ps, for a substrate dielectric constant of 50 the substrate thickness should be less than 4.2 mm for this horizontal capacitor structure design.

An optimum inductance path from an integrated capacitor substrate to chips would be obtained if the chips were mounted directly on the edges of the capacitor plates. A close approximation to this optimum design is the structure with vertical capacitor plates shown in exploded perspective in Fig. 12. The structure consists of three elements, the capacitor inserts A, the main body of the substrate with signal and reference vias B, and the redistribution layers C.

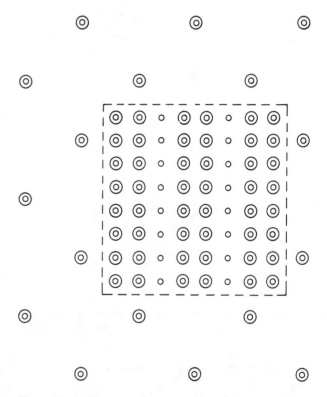

Figure 11 Arrangement of power supply and signal vias under each chip site of substrate shown in Fig. 10.

The inserts A are fabricated separately by laminating together a stack of ceramic and metallization layers of the appropriate pattern. In dicing the laminate to the desired size, tabs from the capacitor plates are exposed on the top for

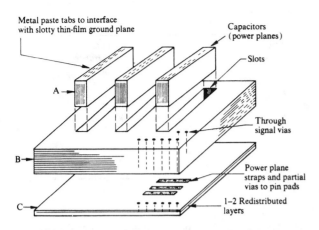

Figure 12 Assembly of substrate with integrated decoupling capacitance based on vertically laminated power plane.

connection to the thin-film structure, and on the bottom for connection to power plane straps. There is no via in these capacitor inserts. The main body of the structure, B, consists of ceramic layers with slots punched in the individual sheets which contain through vias for reference and signal. There is no metallization pattern on these layers. The redistribution layers C consist of at least two ceramic layers. The power plane straps are deposited on the top layer and, in addition, there are signal vias passing through all redistribution layers. On the lower layer or layers, the power and signal vias are redistributed by short lines of metallization to interconnect to pin pads on the bottom of the substrate, not shown.

The entire substrate is fabricated by stacking the redistribution layers C and the layers with punched slots B, then inserting presized laminates A in the slots. A final lamination is then performed, joining together the various elements in the "green" state. Cutting the module to size and sintering to burn off the organic binder to fuse the ceramic parts and metal together complete the fabrication process.

An integrated capacitor substrate may carry from 10 to 100 chips. If each chip requires a minimum capacitance of 20 to 40 nF, a total capacitance of 2 to 4 μF in a substrate up to 10 × 10 cm in a size is needed. The sheet thicknesses and numbers of layers for this capacitance and substrate size for several dielectric materials are given in Table 2. For the horizontal and vertical structures the numbers of layers range from 20 to 150 for materials with dielectric constants in the 50 to 10 range.

Depending on the number of layers and the dielectric constant desired, different dielectric and metal sets can be used. For example, alumina with a dielectric constant of 9 can be used with molybdenum or tungsten metallurgy

because of the high sintering temperature needed. On the other hand, glass ceramic material can be developed to obtain a dielectric constant between 10 and 100. Its sintering temperature is lower, and hence metals such as AgPd and Ni can be used.

Detailed analysis of the effective inductances of several horizontal and vertical integrated substrate designs indicates that the vertical design (Fig. 12) has two to three times lower inductance than the horizontal design [10]. However, both types of structures are expected to be able to support large arrays of chips switching simultaneously as many as 72 drivers per chip with better than 10% stability in the power supply at the chip terminals.

We are presently investigating materials and processes needed to implement such integrated capacitor substrates and we have partially succeeded in building satisfactory experimental structures. These results are beyond the scope of this paper and will be presented later.

Thin-film fabrication
Effort has also been devoted in this laboratory to developing material and processes for the thin-film structure. This section summarizes the essence of our preferred approaches and highlights the need to develop defect detection and repair techniques to achieve acceptable yield for a Thin-Film Module.

The wiring structure of the Thin-Film Module must possess the following characteristics: a minimum of five metal layers, two ground planes sandwiching two orthogonal wiring layers with dimensions exemplified in Fig. 6, and a topmost layer, the fan-out layer shown in Fig. 3, to connect the wiring layers to the semiconductor chip contacts. This fan-out layer can also be used to provide engineering change capability in a manner similar to that practiced on IBM's latest MLC multichip modules. Vias must be provided to vertically connect the various layers to the substrate vias and to make connections within the thin-film structure. The fan-out to chip contact and thin-film via to substrate via interfaces may require special metallurgical barriers to provide a reliable connection to the semiconductor chips and the substrate. Transmission line impedance control dictates the need to achieve better than 10% leveling at each film layer. A high degree of leveling and planarity is also required to achieve good image tolerance with photolithographic processes over large substrate areas and to minimize complications with subsequent assembly of devices with several hundred contacts per chip. The planarity requirements are comparable to those imposed on semiconductor device wafers and will most likely require planarizing of ceramic substrates prior to thin-film deposition. Minimizing resistive loss calls for wiring layers and vias with aspect ratios (height

Table 2 Geometrical requirements for integrated capacitors.

Substrate structure	Area of plates (cm × cm)	Dielectric thickness (μm)	Dielectric constant	No. of layers	Total capacitance (μF)
Horizontal	10 × 10	75	50	60	3.5
"	"	25	10	100	3.5
"	"	25	50	20	3.5
Vertical	0.5 × 10 (× 10 stacks)	25	10	150	0.26 (2.6)
"	"	25	50	30	0.26 (2.6)

to width ratio) and cross sections significantly larger than those typically required in semiconductor devices. Minimizing propagation delay calls for insulator films with low dielectric constant and low loss up to 1 GHz. The composite structure must be capable of withstanding repeated exposure to temperatures of several hundred degrees centigrade during processing and component attachment. This combination of properties cannot be achieved with state-of-the-art processes and will require significant materials and process innovations.

Cost, reliability, and resistivity considerations suggest that Cu is likely to be the most practical material for the metal layers. Polymer materials such as polyimide are likely to be best for the insulator. Such materials are known to have low dielectric constant in cured form ($\varepsilon \le 3.5$) and possess leveling and planarizing characteristics superior to those of inorganic films since they can be deposited as viscous liquid films which can flow during the curing process.

Both Cu and polymer films can be patterned by a variety of techniques. Subtractive techniques using liquid etchants are not likely to be adequate because these lead to undercutting under resist stencils, an undesirable effect vis-à-vis maximizing metal line and via aspect ratios and cross sections. This effect can be suppressed by use of dry etching techniques using reactive plasmas or by use of additive electroplating, electroless plating, or evaporation followed by a lift-off through a photoresist stencil. We have therefore emphasized using the latter techniques with partial success in key areas. The details of this work will also be presented later.

In manufacturing thin-film structures for both chips and packages, the resultant yield is usually gated by the size and the number of defects introduced into the structures by less than ideal manufacturing environments and processes. There are in general three major kinds of defects encountered in fabricating multilevel thin-film structures: intralevel metal line opens and shorts, interlevel metal line shorts, and

defective interlevel vias. Within a given level, the line opens and shorts are due to missing or excess of a portion of the metal lines. These can be the result of a mask defect, incomplete or excessive removal of photoresist in the additive approach, and particulate contamination. The defects are also process-dependent. For example, in electroplating the process starts at the bottom surface of the photoresist stencil. It is therefore important that the surface be clean and that there be no wetting problem or air bubbles between the electrolyte and the metal surface. In the lift-off process the metal is deposited through a photoresist stencil in a line-of-sight projection from the metal source in the evaporator. It is therefore important that there be no dust particles on the substrate surface that may block metal deposition.

An intralevel defect may or may not be "fatal" (affecting the electrical performance of the chip or package) depending on its size. For example, a metal line 8 μm wide may very well be able to tolerate a dust particle of 3 μm in size. On the other hand, a dust particle equal to or more than 8 μm in size which settles right on the line would cause a line open, a fatal defect. Similar considerations apply to defects located between metal lines which may or may not cause a short depending on their size. Experience has indicated that in a controlled clean-room environment, the density of defects occurring on a substrate is a strong function of the size of the defects [11]:

$$f(x) = k \frac{1}{x^3}, \qquad (10)$$

where x is the defect size, $f(x)$ is the probability density function of defects of size x, and k is a constant. Once a critical defect size x_0 is determined, beyond which the defect can become a fatal one, Eq. (10) can be integrated from $x = x_0$ to $x = $ infinity to calculate the total number of defects D_0 with a size equal to or larger than x_0 occurring in a given area:

$$D_0 = k_1 \frac{1}{x_0^2}. \qquad (11)$$

If we now define the critical area A_c to be an area on the substrate where a defect with a size x_0 or larger will cause a fatal defect, a Poisson distribution for a uniform distribution of defects over the whole substrate is given by

$$Y = e^{-A_c D_0}. \tag{12}$$

Note that Eq. (11) shows that the yield is really determined by the total number of defects that occur on the total critical area in a multilevel thin-film device. Another observation can be made with Eq. (12) if we substitute Eq. (11) into (12), using the fact that for a given structure on a square substrate of size y, if all horizontal dimensions of structures of this substrate are scaled proportionately to y, the critical area A_c is proportional to y^2. The expected yield can therefore be expressed as

$$Y = e^{-k_2 (y/x_0)^2}, \tag{13}$$

where k_2 is another constant. Equation (13) therefore states that the yield is only a function of the *ratio* of the substrate size *versus* the critical defect size. If we scale down the substrate horizontal dimensions and all its internal horizontal features linearly, the critical defect size also scales down with the substrate size. Therefore, the ratio of y and x_0 remains constant and so does the yield. This means that given the above assumptions on defect distribution, the yield for fabricating a given device does not change when all its horizontal dimensions are scaled up or down simultaneously. Hence, achieving high yield in fabricating a Thin-Film Module with a substrate size of 10×10 cm and a minimum feature of 10 μm would be comparable to achieving high yield in fabricating a chip with the same structure but with a chip size of 0.5 cm and a minimum feature of 0.5 μm, which surely exceeds the capability of present integrated circuit processes practiced in stringent clean-room environments.

The foregoing analysis suggests that achieving practical yields in building a Thin-Film Module with the complexity described in this paper will require development of novel defect detection and repair techniques for each layer in the thin-film structure. Electron-beam microscopy is likely to be best for defect detection since it offers the potential for contactless testing of what are likely to be delicate electrical structures [12]. Jet or laser-enhanced plating and etching are examples of potential defect repair techniques being investigated in this laboratory [13]. In our judgment, achieving practical yield in thin-film structures over large substrate areas through the use of novel defect detection and repair techniques is the principal obstacle to fabrication of the Thin-Film Module.

Summary and conclusions

Key trends in future VLSI devices and their expected impact on future semiconductor packages for high-speed digital systems have been summarized. Two key requirements for future semiconductor packages are significantly higher wiring capacity and containment of simultaneous switching noise to acceptable levels. A novel package concept, the Thin-Film Module, featuring thin-film strip transmission lines for interchip wiring and power supply decoupling capacitors integrated into the body of the module, has been proposed as an alternative to meet these two requirements. Thin-film transmission lines of dimensions required to provide the necessary wiring density and capacity will have significant resistive loss. However, analysis of high-speed pulse propagation and coupled noise has established that with certain restrictions lossy lines can indeed be used to wire future VLSI devices in multichip packages. Novel MLC approaches for integrated decoupling capacitor substrates and material and process considerations for the required multilayer thin-film wiring structures have also been discussed. Novel defect detection and repair techniques for thin-film structures will be essential for fabricating the Thin-Film Module with practical yields.

References

1. E. Bloch, "VLSI and Computers—Challenge and Promise," keynote address, IEEE Computer Society Conference, San Francisco, February 1980.
2. A. J. Blodgett, Jr., "A Multilayer Ceramic Multichip Module," *IEEE Trans. Components, Hybrids, Manuf. Technol.* **CHMT-3**, 634–637 (1980).
3. B. T. Clark and Y. M. Hill, "IBM Multichip, Multilayer Ceramic Modules for LSI Chips—Design for Performance and Density," *IEEE Trans. Components, Hybrids, Manuf. Technol.* **CHMT-3**, 89–93 (1980).
4. A. Deutsch and C. W. Ho, "Triplate Structure Design for Thin Film Lossy Unterminated Transmission Lines," presented at the 1981 International Symposium on Circuits and Systems, Chicago, April 27–29, 1981.
5. E. Weber, *Linear Transient Analysis*, Vol. II, John Wiley & Sons, Inc., New York, 1956, p. 383.
6. N. Arvanitakis, IBM General Technology Division, Endicott, NY, private communication.
7. C. W. Ho, "Theory and Computer-aided Analysis of Lossless Transmission Lines," *IBM J. Res. Develop.* **17**, 249–255 (1973).
8. C. W. Ho, "Thin Film Lossy Line Package," U.S. Patent No., 4,210,885, July 1, 1980.
9. W. C. Johnson, *Transmission Lines and Networks*, John Wiley & Sons, Inc., New York, 1950.
10. G. V. Kopcsay, IBM Thomas J. Watson Research Center, Yorktown Heights, NY, private communication.
11. C. H. Stapper, A. N. McLaren, and M. Dreckmann, "Yield Model for Productivity Optimization of VLSI Memory Chips with Redundancy and Partially Good Product," *IBM J. Res. Develop.* **24**, 398–409 (1980).
12. T. P. Chang, F. J. Hohn, P. J. Cohane, D. P. Kern, and W. H. Bruenger, "Electron Beam Testing of Packaging Models for VLSI Chip Arrays," *Proceedings of the 16th Symposium on Electron Ion Photon Beam Technology*, May 1981.
13. J.-Cl. Puippe, R. E. Acosta, and R. J. von Gutfeld, "Investigation of Laser Enhanced Electroplating Mechanisms," *J. Electrochem. Soc.* **128**, 25–39 (1981).

Received July 13, 1981

C. H. Bajorek is located at the IBM Research laboratory, 5600 Cottle Road, San Jose, California 95193 and the other authors are located at the IBM Thomas J. Watson Research Center, Yorktown Heights, New York 10598.

Copper/Polyimide Materials System for High Performance Packaging

RONALD J. JENSEN, JOHN P. CUMMINGS, AND HARSHADRAI VORA

Abstract—Advancements in the speed and input/output (I/O) density of integrated circuits (IC's) used in computers and other high performance systems are creating a need for new designs, materials, and processes capable of providing high density multilayer interconnections with controlled electrical characteristics. To meet these needs, we selected an interconnect materials system consisting of multiple layers of thin film copper conductor patterns separated by polyimide (PI) dielectric layers, fabricated on a ceramic substrate. We developed processes for depositing films and patterning high resolution features in both conductor and dielectric films, and established a process sequence for filling vias and building up multilayer structures. We investigated the stability of the Cu/PI materials system by determining the effects of cure conditions and humidity aging on the dielectric properties, internal stress, mechanical properties, and adhesion of PI. We then fabricated multilayer test structures, including a functional multichip ring oscillator circuit, to demonstrate process feasibility and the electrical performance of Cu/PI interconnections.

INTRODUCTION

Background

CONTINUAL ADVANCES in the speed and integration of integrated circuits (IC's) used in high performance systems have created a demand for higher density interconnections to accommodate large numbers of input/output (I/O) and to achieve short interconnect lengths for improved electrical performance. The lower limit on interconnect linewidth and spacing is dictated by a combination of processing constraints and electrical considerations such as resistive losses and crosstalk; further demands for high interconnect density necessitate multilayer structures. Multilayer interconnections with interspersed ground planes also afford better control over electrical characteristics such as characteristic impedance and noise coupling, which are becoming increasingly important at high operating speeds.

A number of material systems and process technologies for fabricating multilayer IC interconnections have been proposed. These include high density printed wiring boards (PWB's) [1], co-fired multilayer ceramic [2], [3], thick film systems with screened conductor pastes and glass/refractory dielectrics [4], thin film metallization with polymer interlayer dielectrics [5]–[10], and wafer scale integration using Si IC processing [11]. We believe that the most cost effective, extendable, and versatile interconnect technology is one using thin film multilayer structures with sputtered copper (Cu) conductor and polyimide (PI) dielectric.

Approach

Our approach for using multilayer Cu/PI interconnection in the packaging of single or multiple high performance IC's is to customize a co-fired ceramic substrate. A conceptual drawing of such a package is shown in Fig. 1. The ceramic substrate is a standardized part which provides power and ground wiring to the IC's and contains pins or pads for connection to the next level of packaging, e.g., a printed wiring board. Customization occurs when multiple Cu/PI layers are patterned on the ceramic to form ground planes and signal wiring for chip interconnection and I/O redistribution. The IC's are bonded to the Cu/PI interconnects by wire bonding, tape automated bonding (TAB), or area TAB. Potential applications of this technology currently under investigation include multichip packaging for high performance computers, the packaging of very high speed integrated circuit (VHSIC) chips for high speed signal processors, and the packaging of highly integrated GaAs IC's for gigabit/second digital systems.

The design of thin film multilayer interconnections is determined by a number of competing requirements relating to the wireability, electrical performance, and thermal management of the system, balanced against the processing capabilities of the interconnect technology. A number of authors have reviewed the electrical, thermal, and mechanical requirements for packaging high-density high-speed circuits [3], [6], [9], [10], [12]. The effect of these system requirements on interconnect geometries and material properties can be summarized as follows: 1) interconnect line lengths should be short to minimize media delay, resistive losses, and signal coupling, but the close spacing of IC's to achieve short interconnect lengths increases the power density and thus requires more efficient thermal dissipation from the IC's, 2) conductor lines must have a high aspect ratio (large thickness/width) in order to achieve high wiring density while maintaining a large conductor cross section for low resistive losses, 3) the spacing between lines should be large to reduce signal coupling, however, a large spacing reduces wiring density and thus requires more wiring layers, 4) a large dielectric thickness between signal lines and ground planes is required for high characteristic impedance (Zo) interconnections, but has the undesirable effect of increasing signal coupling between adjacent lines, 5) a high conductivity conductor is required for low resistive losses, and 6) a low dielectric constant (ϵ_r) dielectric material is desirable without exception because it will result in high Zo, low media propagation delay

Manuscript received March 1984; revised July 5, 1985. This paper was presented at the 34th Electronic Components Conference, New Orleans, LA, May 14–16, 1984.

R. J. Jensen is with Honeywell Physical Sciences Center, 10701 Lyndale Avenue South, Bloomington, MN 55420.

J. P. Cummings is with Honeywell Solid State Electronics Division, Plymouth, MN.

H. Vora was with Honeywell. He is now with Sperry Corporation, Eagan, MN.

Reprinted from *IEEE Trans. Components, Hybrids, Manuf. Technol.*, vol. CHMT-7, no. 4, pp. 384–393, December 1984.

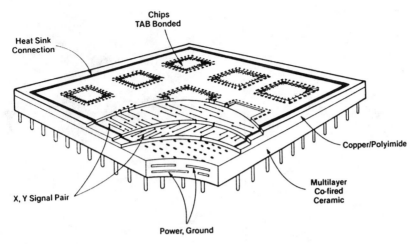

Fig. 1. Conceptual drawing of multichip package for computer applications which features multilayer Cu/PI customization of co-fired ceramic substrate.

and low coupling. In general, then, high package performance will be achieved by spacing IC's as closely as the cooling technique allows, by interconnecting the IC's through high aspect ratio lines patterned in a high conductivity material, by spacing the conductor lines as far apart as the signal wiring layers allow, and by separating signal lines from ground planes with a relatively thick layer (>10 μm) of a low ϵ_r dielectric material.

Copper is a desirable conductor material because of its high conductivity, low cost, and solderability. Furthermore sputtered thin film Cu can be deposited and patterned with a high degree of dimensional control using photolithographic techniques. Polyimide is an advantageous dielectric for a number of reasons. The polymer precursor, polyamic acid (PAA), is a viscous liquid which can be applied at the desired thicknesses of 10–50 μm by spinning or spraying, providing partial planarization of conductor lines and accurate thickness control. The fully cured PI is chemically inert, mechanically tough and flexible, and thermally stable to 450°C, which allows it to withstand fabrication processes such as metal deposition, photolithography, wet etching, and soldering. Polyimide can be patterned with high resolution by plasma etching. Finally PI has excellent electrical properties for high speed applications, i.e., a low dielectric constant ($\epsilon_r = 3.0$–4.0), a low dissipation factor over a wide frequency range, and a high breakdown voltage.

The purpose of this research has been to develop the basic processes required for fabricating thin film multilayer Cu/PI interconnections and to determine the material stability and electrical performance of these interconnections. In the first section of this paper we describe our characterization of the individual processes involved in depositing and patterning Cu conductor and PI dielectric, and then describe the overall process sequence used to fabricate multilayer Cu/PI structures. In the next section we present results from measurements of the stability of PI, specifically, the effects of humidity and process conditions on the dielectric properties, internal stress, mechanical properties, and adhesion of PI to various materials. In the final section we describe a ring oscillator circuit which has been fabricated to test the electrical performance of Cu/PI interconnections.

THIN FILM MULTILAYER PROCESSING

The fabrication of multilayer Cu/PI structures on a rigid substrate involves a repetitive sequence of thin film processes for depositing and patterning conductor and dielectric layers. Our basic approach is to deposit Cu conductor layers by sputtering and to pattern Cu by wet etching, ion milling, or in some cases selective plating. Polyimide dielectric layers are deposited by spin coating and patterned by reactive ion etching. We will discuss the characterization of each of these processes in more detail below.

Conductor Deposition and Patterning

Processes for depositing and patterning thin film Cu conductor were reported by us at an Electronic Components Conference [9]. Conductor lines of 10 μm width, thickness, and spacing were patterned by both a subtractive approach and an additive approach. In the subtractive approach a thin Cr adhesion layer and a Cu layer of the desired thickness (5–10 μm) are deposited by radio frequency sputtering (for Cr) and dc magnetron sputtering (for Cu). A pattern is defined by photolithography and the unwanted material is removed by wet etching or ion milling. In the additive approach, a thin preplate layer of Cr/Cu is first sputtered, and a negative image of the conductor pattern is defined in a thick photoresist layer. Copper is then electroplated into the spaces in the photoresist, and the resist and preplate layers are stripped away. Due to process simplicity and control over conductor dimensions, the subtractive approaches are currently preferred. Wet etching is satisfactory down to conductor linewidths of 25 μm and 1:3 aspect ratios, while ion milling is preferred for higher aspect ratios.

Polyimide Deposition

Polyimide films are deposited by either spinning or spraying a solution of polyamic acid onto the substrate, soft curing at 120°C to evaporate the solvent, and hard curing at 250–420°C to convert polyamic acid to PI. We investigated the spin coating process to determine the effects of material and process parameters on film thickness, uniformity, and planarity. Fig. 2 shows the effects of spin speed, spin time, cure

229

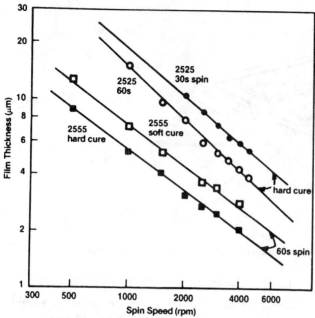

Fig. 2. Effect of spin speed on polyimide films thickness for different spin times, cure conditions, and types of polyimide.

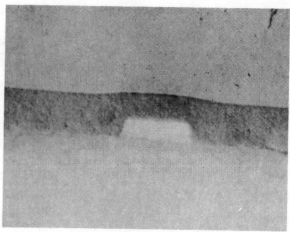

Fig. 3. Cross section of 10 μm thick conductor line planarized with five coatings (25 μm) of polyimide.

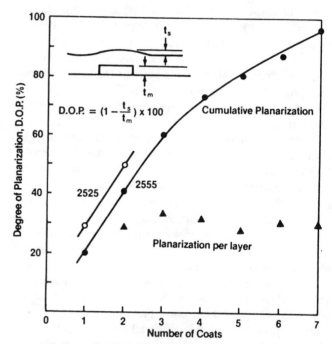

Fig. 4. Degree of planarization achieved with single and multiple polyimide coatings.

temperature, and the type of PI on the thickness of a single spin coating. Film thicknesses vary with spin speed according to a power law relationship: $h = k\omega^{\alpha}$ where h is the film thickness in micrometers, and ω is the spin speed in revolutions per minute. For soft cured (120°C) DuPont 2555 PI spun for 60 s, $k = 90$ and $\alpha = -0.7$. Hard curing reduces film thickness by an average of 25 percent. The higher viscosity DuPont 2525 PI (5–7 Pa·s) produces a film 2.5 to 3 times thicker than 2555 (1.2–1.6 Pa·s). Finally reduced spin times result in thicker films, as shown in Fig. 2 for 2525 PI spun for 30 and 60 s.

The largest practical thickness achievable by a single spin coating is on the order of 15 μm. Thicker films require multiple spin coatings, with a cure temperature of at least 150°C between coats to prevent solvent attack of underlying layers. The thickness of multiple coatings was found to be additive, i.e., thickness increased linearly with the number of coats. PI film thicknesses as great as 75 μm have been achieved using multiple spin coatings.

Due to hydrodynamic forces and the ability of polyamic acid to flow, a partial planarization of underlying conductor patterns is obtained using spin coating; however, the eventual solidification and subsequent shrinkage of the PI during cure results in some conformal topography. Fig. 3 shows a cross section of a 10 μm thick conductor line planarized by five coatings (25 μm) of polyimide. The degree of planarization (DOP), defined as the percent reduction in step height over a conductor line, is shown in Fig. 4 for multiple coatings of two different PI's. The incremental DOP achieved with each coating is approximately 30 percent for PI 2555. Except for a consistently low DOP on the first coating, the DOP is independent of the number of coats or the initial step height. The cummulative DOP after n coatings, which can be described as

$$(DOP)_n = 1 - (1 - DOP)^n$$

reaches 80 percent after five coats (19 μm) of PI 2555 or 50 percent after two coats (19 μm) of higher viscosity PI 2525. Thus, even though a high viscosity PI produces thick films with fewer coatings, the DOP for a given thickness will be improved if one uses a lower viscosity PI requiring more coats.

Polyimide Patterning

The patterning of features such as vias in PI is accomplished by reactive ion etching (RIE) in an O_2/CF_4 plasma. In order to etch high aspect ratio features in relatively thick (>2 μm) films, the pattern must first be defined in a masking material which has a low etch rate in O_2 plasmas and can be deposited in a thin layer on top of the PI. We have investigated a number of masking materials including evaporated Al, sputtered Ti, reactively sputtered SiN_x, evaporated SiO, and chemical vapor deposited SiO_x. All of these materials except Al can be etched in a CF_4-rich plasma, so that the mask etch, PI etch, and mask removal can be done in the same plasma system. The currently favored masking materials are the Si oxides.

The etching of high resolution features in relatively thick (10–50 μm) PI films places several unique requirements on the etch process which must achieve: 1) a high degree of selectivity (large PI etch rate/mask etch rate), 2) highly anisotropic etching (large vertical etch rate/lateral etch rate), 3) a high etch rate (>5 nm/s) for reasonable process times, 4) uniform etch rates over a large (25–200 cm^2) substrate, and 5) minimal formation of residue from involatile plasma products. We have achieved these etch characteristics by 1) modifying the electrode materials and electrode configuration in the plasma reactor, 2) accurately detecting end points using a laser interferometer, 3) wet etching Cr layers at the bottom of vias to remove residue, and 4) varying the controllable process parameters.

A number of trade-offs are involved in the selection of RIE process parameters. In general low pressure and high power density enhance anisotropy but reduce selectivity and etch uniformity and increase the likelihood of residue formation. Etch rates increase with increasing power and pressure. Increased CF_4/O_2 ratios enhance etch rates but reduce selectivity and therefore necessitate a thicker masking layer.

Desirable results are obtained by operating at power densities in the range 0.2–0.5 W/cm^2, pressures of 50–150 mtorr, and etch gas mixtures of 2–10 percent CF_4 in O_2. Under these conditions, structures such as the 37 μm wide line shown in Fig. 5 can be etched in a PI film 25 μm thick with a vertical/lateral etch rate ratio of 2:1 producing 14 μm of mask undercut. Vertical/lateral etch rates as high as 7:1 have been achieved at higher powers and lower pressures, but etch rates are slower and metallization step coverage is more difficult to achieve. The ability to control anisotropy and sidewall angles by varying process parameters is one of the great advantages of the RIE process for patterning polyimide.

Multilayer Process Sequence

The deposition and patterning processes described above have been combined in various sequences to fabricate multilayer structures. A crucial step in the buildup of these structures is the formation of vias for connection between layers. Several techniques have been investigated for forming vias, including those in which the via hole is first etched in PI and then filled by sputtering or plating, and other methods in which the via post is formed first and then coated with PI. The via fill approach which has proven most successful is also the simplest; the via holes are etched, a conformal conductor layer is sputtered over the entire surface including the via holes, and the vias and next layer metallization are patterned in one photolithography and wet etch step. Fig. 6 shows a via 100 μm square $\times$ 25 μm deep patterned by this technique.

The overall multilayer process sequence is described in Fig. 7. First a conductor layer is sputtered and patterned by photolithography and wet etching. Multiple coatings of PI are spin coated over the conductor pattern, providing partial planarization. A plasma etch mask is then deposited and

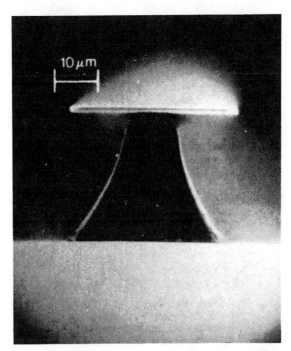

Fig. 5. Cross section of 37 μm wide line reactive ion etched in 27 μm thick polyimide film.

Fig. 6. Via 100 μm square $\times$ 25 μm deep filled by conformal sputtering.

patterned by photolithography. A CF_4 plasma is used to transfer the pattern into the mask, the PI is etched in an O_2 plasma which also strips away the photoresist, and the mask is removed by CF_4 etching. Vias and next layer metallization are then patterned by the conformal sputtering technique described above or by a selective plating process. The sequence of steps is repeated for additional conductor layers. For vias connecting through multiple dielectric layers, the conformal via must be stair-stepped in opposite directions at each level.

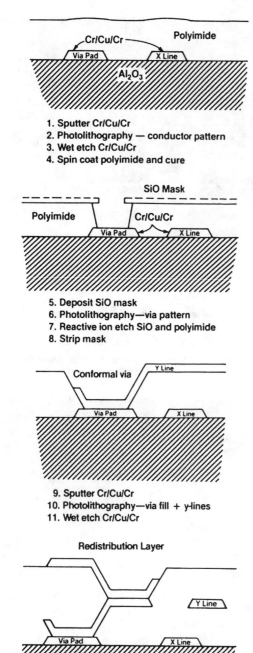

1. Sputter Cr/Cu/Cr
2. Photolithography — conductor pattern
3. Wet etch Cr/Cu/Cr
4. Spin coat polyimide and cure

5. Deposit SiO mask
6. Photolithography—via pattern
7. Reactive ion etch SiO and polyimide
8. Strip mask

9. Sputter Cr/Cu/Cr
10. Photolithography—via fill + y-lines
11. Wet etch Cr/Cu/Cr

12. Repeat steps 4-11 for additional layers

Fig. 7.   Process sequence for fabricating multilayer Cu/PI patterns.

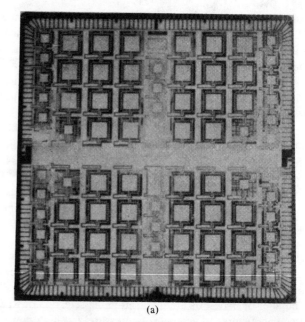

(a)

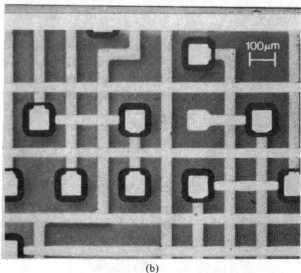

(b)

Fig. 8.   (a) Multichip pattern on 2 × 2 inch ceramic substrate containing three conductor layers separated by PI dielectric. (b) Magnified view of multichip pattern after second layer metallization, with 50 μm wide lines on 125 μm pitch and 100 μm square vias.

Multilayer Test Pattern Demonstration

The feasibility of using the multilayer processes described above was tested by fabricating the multichip pattern shown in Fig. 8(a). This structure is a half-scale reduction of a test pattern developed for the Micropackage in Honeywell's DPS 88 mainframe computer [13]. It is patterned on a 2 × 2 inch ceramic substrate and contains three conductor layers each 5 μm thick, which are separated by PI layers 25 μm thick. The first two conductor layers are 50 μm wide signal lines on a 125 μm pitch with 100 μm square vias, and the top conductor layer

contains chip attach and bonding pads. Fig. 8(b) shows a magnified view of the structure after second layer metallization.

A multichip pattern identical to the one shown in Fig. 8 but with dimensions twice as large was fabricated to test line continuity and via resistance for a three-layer structure. DC resistance measurements were made between outer bonding pads and inner chip bonding pads for 45 signal lines containing 2–13 vias per line. No lines were open, and line resistance ranged from 0.40 to 3.2 Ω depending on line length and the number of vias.

A variety of other test structures have been fabricated in order to evaluate the electrical and mechanical stability of PI and the electrical performance of Cu/PI interconnects. The results of these experiments are discussed in the next two sections.

STABILITY OF POLYIMIDE

The stability of the organic dielectric material in the Cu/PI material system is a primary concern for applications demanding high reliability. Based on our experience with PI and the results of other investigators [14]–[17] we identified four crucial material stability issues requiring further investigation: 1) effects of temperature, humidity, and cure conditions on the dielectric properties of PI, 2) internal stress in PI films resulting from thermal expansion mismatches between PI and substrate or metallization layers, 3) the mechanical properties of PI cured at different conditions, and 4) the effects of temperature and humidity aging on PI adhesion to substrate and conductor materials.

Dielectric Properties

Changes in humidity are known to cause changes in the dielectric properties of polyimide [14]; this will result in changes in the electrical characteristics such as the propagation delay and characteristic impedance of Cu/PI interconnections. In order to quantify these phenomena, we investigated the effects of humidity and cure conditions on the dielectric constant, dissipation factor, and breakdown voltage of thin films (5 μm) of DuPont 2555 PI. Parallel plate capacitor structures with 0.1×0.1 inch (2.5×2.5 mm) pads were fabricated on ceramic substrates, using PI cured in N_2 at different temperatures (250, 300, and 350°C) and different pressures (1.0 and 0.5 atm). Dielectric properties were measured with a capacitance bridge operating at 1.6 kHz. Measurements were taken immediately after processing and after aging the samples at room temperature in controlled humidity environments maintained with saturated salt solutions. Aging times of up to 500 h were required for some of the capacitor structures to reach equilibrium, due to the small diffusion cross section beneath the capacitor pad.

The dielectric constant increases linearly with increasing relative humidity, as shown in Fig. 9, with an average value of $\epsilon_r = 3.1$ at 0 percent relative humidity (RH) and a slope of about 0.01/percent RH. This data agrees quite well with data reported by DuPont for Kapton H, a flexible PI film [18]. Cure conditions have a very small affect on ϵ_r, however, a least squares fit of the data does show a consistent trend in which decreasing cure temperature and pressure cause a decrease in ϵ_r. Cure conditions have a much stronger effect on the dissipation factor, which roughly doubles in going from 250°C to 350°C cure temperature. The dissipation factor also increases by about a factor of three in going from 0 to 100 percent RH exposure. The dissipation factors measured in this work ranged from a tan δ of 0.7×10^{-3} to 3×10^{-3}, which in all cases is small enough (by at least a factor of ten) to ensure negligible resistive losses in the dielectric media at a frequency of several gigaHertz. Breakdown voltages decrease from 3.3 MV/cm at 0 percent RH to 1.6 MV/cm at 100 percent RH, and cure conditions have an unmeasurable effect on breakdown voltage. In all cases the breakdown voltage is high enough to ensure no dielectric breakdown.

The effects of humidity on the dielectric properties of PI, particularly ϵ_r, indicate a need to control the package moisture content in order to maintain constant electrical characteristics.

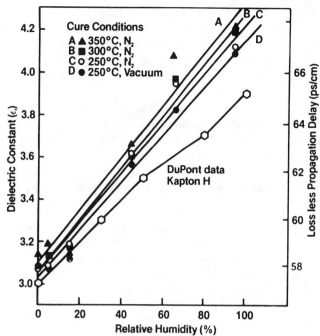

Fig. 9. Effect of humidity on dielectric constant of polyimide cured at different conditions.

However, since electrical characteristics such as Zo and propagation delay have a 1/2 power dependence on ϵ_r, they vary less than does ϵ_r with changes in humidity. For example, Fig. 9 shows that a 35 percent variation in ϵ_r (from 3.1 to 4.2) over the full humidity range results in a 16 percent variation in propagation delay (from 58 to 68 ps/cm). In most cases these variations in electrical characteristics will be within system design tolerances.

Internal Stress

If any mismatch occurs between the coefficients of thermal expansion (CTE) of PI and Al_2O_3 or Si substrates, the process of cooling from the hard cure temperature to room temperature will introduce internal stresses in the PI film. These stresses will be tensile if the CTE of PI is greater than the CTE of the substrate. If the tensile internal stress exceeds the fracture strength of the PI film, the film can crack and cause defects in the metallization patterns. Thus the magnitude of internal stress is an important reliability concern.

Using an X-ray double crystal lattice curvature technique, Goldsmith *et al.* [17] measured internal stresses in PI films deposited on Si substrates. The internal stresses were found to be tensile; the magnitude of the stress was observed to increase linearly with an increase in cure temperature and, for a given cure temperature, was independent of the thickness of the PI film for thicknesses of up to 24 μm. Goldsmith *et al.* have quoted a value of 50×10^{-6}/°C for the CTE of PI used in their work, as compared to typical CTE values of 3×10^{-6}/°C for Si and 7×10^{-6}/°C for alumina. Since the CTE of PI is much greater than the CTE of Si or Al_2O_3, PI films deposited on Al_2O_3 substrates will have tensile internal stresses of comparable magnitude to those developed on Si substrates.

In our study we measured internal stresses in DuPont 2525 and 2555 PI deposited on Si wafers, using an interferometric technique to measure wafer deformation [19]. The PI was soft cured at 120°C for one hour and hard cured at 250°C for one hour. The internal stress measurements are summarized in Table I, which shows that treating the Si wafer with the silane adhesion promoter causes only a small increase in internal stress.

Table I also shows that the measured internal stresses in DuPont 2525 and 2555 are independent of the PI thicknesses and are nearly equal, indicating no significant differences in the CTE of these PI's. Measurements of the CTE of PI 2525 done in our laboratory over a temperature range of 25–285°C have yielded an average value of $35 \times 10^{-6}/°C$.

Mechanical Properties

As discussed in the preceding section, internal stresses become an important reliability issue if they exceed the fracture strength of the PI. Therefore we measured the mechanical properties of PI films, including their Young's modulus, fracture strength, and fracture strain, and compared fracture strength to internal stress.

DuPont 2525 PI was spin coated onto Al_2O_3 substrates and cured at 120°C for one hour and at 285°C for one hour. The PI film was then peeled off the Al_2O_3 substrate. Tensile test specimens 5 mm wide and 25 mm in gauge length were cut from the peeled films and also from a 5 mil thick sheet of Kapton PI film. The specimens were then tested using an Instron testing machine at a crosshead speed of 1.0 mm/min. From the uniaxial tensile stress–strain curves of these films, the values of Young's modulus, fracture strength and fracture strain were calculated. These values are listed in Table II. Since the measured fracture strength of DuPont 2525 PI (200 MPa) is approximately a factor of five higher than the measured internal stress (40 MPa), it is concluded that the internal stresses alone are not sufficient to cause failures in the PI films.

Adhesion

In an effort to measure the adhesion of PI's to various materials, we have used peel tests extensively in our work. To obtain these data, several Al_2O_3 (MRC Superstrate) and Si substrates were coated with DuPont 2525 PI. Some of these substrates were first treated with a silane adhesion promoter (γ-aminopropyl triethoxy silane); others had a layer of Cr/Cu/Cr sputtered on the substrate prior to PI deposition. After hard curing the PI at 285°C, the substrates were scribed and broken and several 6 mm wide peel test specimens were peeled off the substrates at an angle of 90° using an Instron testing machine with a crosshead speed of 0.25 mm/min. The peel force (in grams of force) was divided by the width of the PI films peeled, and the ratio was taken as a measure of adhesion. This ratio was found to be independent of the width of PI film for widths of up to 15 mm, in agreement with the earlier observations of Rothman [16]. Some of the specimens were tested immediately after curing to obtain the "initial" values of peel strength. The remaining peel test specimens were allowed to age at room temperature and humidity for a period of approximately two months before testing peel strength; some of these specimens were then further aged in an 85°C/80 percent RH environment for a period of 16 h and peel tested.

TABLE I
INTERNAL STRESSES IN DuPont POLYIMIDE FILMS

Type	Film Thickness, μm	Stress MPa (10³ psi)	Adhesion Promoter
2525	8.5	30 (4.4)	None
	17	30 (4.4)	
2525	8.5	40 (5.8)	Silane
	17	40 (5.8)	
2555	3.5	50 (7.2)	Silane
	7.0	40 (5.8)	

TABLE II
MECHANICAL PROPERTIES OF POLYIMIDE FILMS

Type	Thickness, μm	Orientation*	Young's Modulus (GPa (10⁵psi))	Fracture Stress MPa (10³psi)	Fracture Strain, %
Du Pont 2525	25		4.8 (7)	200 (29)	25
Du Pont Kapton	125	11	4.8 (7)	180 (26)	40
	125	1	3.5 (5)	165 (24)	42

*With respect to rolling direction

The results of the peel tests are summarized in Table III. The data show that the silane adhesion promoter improves the initial adhesion of PI to both Si and Al_2O_3 and reduces, but does not completely eliminate, the adhesion loss due to aging in room temperature and in 85°C/80 percent RH environments. In general peel strengths of less than 5 gf/mm (50 N/m) are unacceptable for reliable packaging structures; therefore, an adhesion promoter is definitely required for the deposition of PI on Al_2O_3 substrates. The observed loss in adhesion of PI to Al_2O_3 after humidity aging also re-emphasizes the need to control the package moisture content. The adhesion of PI to Cr/Cu/Cr metallization after room-temperature aging is excellent (75 gf/mm) and does not degrade on subsequent aging in an 85°C/80 percent RH environment. Thus Cr/Cu/Cr ground planes can serve as effective adhesion layers and do not require the use of an adhesion promoter for PI adhesion.

ELECTRICAL PERFORMANCE

Several multilayer Cu/PI structures have been fabricated to test the electrical performance of Cu/PI interconnections. In a previous paper [10], we reported results from computer simulations and experimental measurements of a simple 30 cm long microstrip transmission line (one ground plane and one signal layer) transmitting 2 ns risetime pulses at 50 MHz. Based on the encouraging results of this test (66 ps/cm propagation delay, 5.2 mV/cm attenuation), we designed and fabricated additional vehicles to test the electrical characteristics of striplines (one signal layer between two ground planes) and multiple layer structures incorporating vias.

To demonstrate a functioning multichip circuit using Cu/PI interconnections, we designed a ring oscillator circuit containing two Fairchild 100K ECL IC's (F100102). The superimposed masks for the multilayer multichip package are shown in Fig. 10. A 2 × 2 inch ceramic substrate contains three independent ring oscillator circuits. Each circuit contains five gates arranged alternately on two chips; after each gate the signal must travel between chips through Cu/PI interconnect lines. A 40 cm long serpentine delay line may be inserted into the circuit for the measurement of propagation delay. Quiet

TABLE III
EFFECT OF AGING IN VARIOUS ENVIRONMENTS ON ADHESION OF POLYIMIDE

Substrate	Adhesion Promoter	Polyimide* Thickness, μm	Peel Strength, gf/mm (N/m)		
			Initial	After Aging At Room Temperature for 2 Months	After Subsequent Aging At 85°C/80% r.h. for 16 hrs
Silicon	None	25	44** (430)	33 (320)	0
	Silane	25	57** (560)	45 (440)	23 (230)
Al₂O₃	None	8.5	10 (100)	4 (40)	3 (30)
	Silane	8.5	22 (220)	19 (190)	17 (170)
		25	28 (270)	22 (220)	17 (170)
	Cr/Cu/Cr	25	75 (740)	75 (740)	75 (740)

*Du Pont 2525
**Film did not peel over the entire width of the sample. Had the film peeled over the entire width, the measured peel strength would have been greater than that quoted.

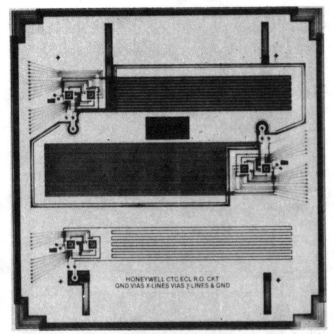

Fig. 10. Superimposed masks for three-layer Cu/PI test package containing three two-chip ring oscillator circuits.

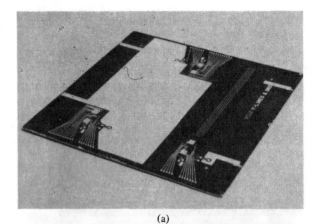

(a)

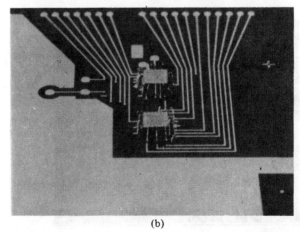

(b)

Fig. 11. (a) Completed ring oscillator test circuit on 2 × 2 inch substrate. (b) Close-up of two wire-bonded IC's in ring oscillator circuit.

lines run parallel to the driven delay lines to enable measurement of signal coupling. The delay line in the upper circuit in Fig. 10 is a stripline with 25 μm wide lines on a 125 pitch; the delay line in the middle circuit is a stripline with 50 μm wide lines on a 250 μm pitch and the lower circuit contains a 50 μm wide delay line running alternately on two different metallization levels, with via connections between levels.

The multilayer Cu/PI substrate was fabricated using the processes described in this paper. It contains three conductor layers: 1) a bottom ground plane, 2) a middle signal layer with serpentine delay lines and chip-to-chip interconnects, and 3) a top layer consisting of bonding pads, power supply lines, signal I/O lines, a ground plane for the striplines, and serpentine lines for the two-level delay line. To test the effect of dielectric thickness on electrical characteristics, the PI thickness between conductor layers was varied from 10 to 50 μm on different substrates. Chips were attached to the package with silver-filled epoxy and bonded with aluminum wire.

Fig. 11(a) shows a completed substrate populated with four IC's, and Fig. 11(b) shows a close-up of two wire-bonded IC's. The multichip substrate is mounted in a test fixture which contains termination resistors, decoupling capacitors, and 50 Ω microstrip lines terminated in coax connectors for interfacing with test equipment.

The ring oscillator circuit was first tested by packaging the chips in conventional flatpacks and interconnecting them on a printed wiring board. In this hardwired configuration the circuit oscillated at 95 MHz with a pulse risetime of 2 ns; the waveform at the output of the fifth gate is shown in Fig. 12(a). When the chips were mounted on the multilayer substrate with the Cu/PI interconnections, but without the delay line inserted, the oscillation frequency increased to 135 MHz and the rise time dropped to 1.5 ns, as shown by the waveform in Fig. 12(b). With the 40 cm delay line inserted into the circuit, the

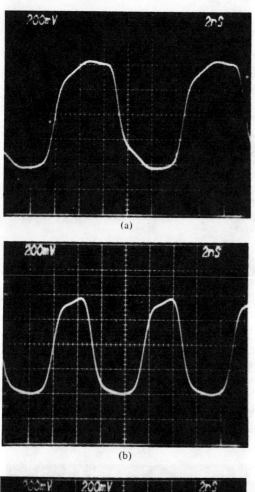

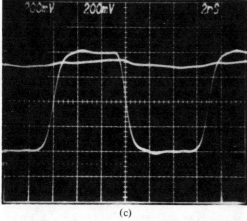

Fig. 12. Voltage waveforms at the output of the fifth gate in the ring oscillator circuit. (a) Conventionally packaged IC's connected on PWB. (b) IC's mounted on multichip substrate with Cu/PI interconnections between chips. (c) IC's on multichip substrate with 40 cm delay line inserted into Cu/PI interconnect circuit.

frequency decreased to 77 MHz, giving a value of 69 ps/cm for the propagation delay of the 50 μm wide stripline, which is close to a theoretical value of 62 ps/cm for the propagation delay of a lossless transmission line. The waveform in Fig. 12(c) shows that the delay line causes no attenuation and no change in the risetime of the output signal. Fig. 12(c) also shows the signal coupled onto a quiet line running parallel to and coplanar with the driven delay line. The amplitude of this coupled signal is 0.1 that of the driven signal (−20 dB

coupling), which is an acceptable level for coupling for most high performance system designs.

CONCLUSION

The thin film Cu/PI materials system has the properties and processing capabilities required for high density multilayer interconnections in a variety of high performance packaging applications. The geometries required for multilayer interconnects is dictated by trade-offs between system performance requirements and processing capabilities. We have demonstrated the capability of fabricating three-conductor-layer structures with 25 μm wide lines on a 125 μm pitch, 50 μm vias, and 25 μm thick dielectric layers. We have also shown the extendability of the individual thin film processes to finer features and higher aspect ratios. Our analysis of the stability of PI has shown that 1) humidity affects the dielectric constant of PI and thus the electrical characteristics of Cu/PI interconnects, but the variations are within most design tolerances, 2) the adhesion of PI to Cr is very high and is unaffected by humidity, while adhesion to Al_2O_3 requires an adhesion promoter, and 3) internal stresses are developed in PI films due to thermal expansion mismatches between PI and the substrate, but these stresses are well below the fracture strength of the PI. Finally we have demonstrated the electrical performance of multilayer Cu/PI interconnections in a two chip ring oscillator circuit operating at 135 MHz. The Cu/PI interconnections achieve nearly lossless propagation delays (69 ps/cm) and low signal coupling (−20 dB), and significantly improve the oscillation frequency and signal risetime over that obtained by conventional flat pack/printed wiring board connections.

ACKNOWLEDGMENT

The authors would like to thank D. Saathoff, T. Gruchow, D. Walt, J. Lee, and C. Knudson for their valuable contributions in processing and testing the various structures described in this paper, and F. Belcourt and A. Rau for the design and testing of the ring oscillator circuit.

REFERENCES

[1] J. R. Bupp, L. N. Challis, R. E. Ruane, and J. P. Wiley, "High-density board fabrication techniques," *IBM J. Res. Dev.*, vol. 10, pp. 306–317, May 1982.

[2] A. J. Bodgett, Jr., "A multilayer ceramic, multichip module," in *Proc. Electronic Components Conf.*, 1980, pp. 283–285.

[3] C. W. Ho, D. A. Chance, C. H. Bajorek, and R. E. Acosta, "The thin film module as a high-performance semiconductor package," *IBM J. Res. Dev.*, vol. 26, pp. 286–296, May 1982.

[4] D. E. Pitkanen, J. P. Cummings, and C. J. Speerschneider, "Status of copper thick film hybrids," *Solid State Technology*, pp. 141–146, Oct. 1980.

[5] S. LeBow, "A method of manufacturing high density fine line printed circuit multilayer substrates which can be thermally conductive," in *Proc. Electronic Components Conf.*, 1980, pp. 307–309.

[6] N. Goldberg, "Design of thin film multichip modules," in *Proc. Int. Soc. Hybrid Microelec.*, vol. 4, 1981, pp. 289–295.

[7] M. E. Ecker and L. T. Olson, "Chip carriers for high density semiconductor dies," in *Proc. Int. Soc. Hybrid Microelec.*, vol. 4, 1981, pp. 251–257.

[8] J. Shurboff, "Polyimide dielectric on hybrid multilayer substrates," *Proc. Electronic Components Conf.*, 1983, pp. 610–615.

[9] J. P. Cummings, R. J. Jensen, D. J. Kompelien, and T. J. Moravec,

"Technology base for high performance packaging," in *Proc. Electronic Components Conf.,* 1982, pp. 465–478.

[10] R. P. Vidano, J. P. Cummings, R. J. Jensen, W. L. Walters, and M. J. Helix, "Technology and design for high speed digital components in advanced applications," in *Proc. Electronic Components Conf.,* 1983, pp. 334–343.

[11] D. Pelzer, "Wafer scale integration: The limits of VLSI?," *VLSI Design,* pp. 43–47, Sept. 1983.

[12] E. E. Davidson, "Electrical design of a high speed computer package," *IBM J. Res. Dev.,* vol. 26, pp. 349–361, May 1982.

[13] C. H. McIver, "Flip TAB, copper thick film create the micropackage," *Electronics,* vol. 55, pp. 90–99, Nov. 1982.

[14] S. MN. Zalar, "Dielectric characterization of polyimide thin films," in

Extended Abstracts of First Technical Conf. Polyimides, 1982, pp. 143–144.

[15] A. M. Wilson, "Polyimide insulators for multilevel interconnections," *Thin Solid Films,* vol. 83, pp. 145–163, 1981.

[16] L. B. Rothman, "Properties of thin polyimide films," *J. Electrochem. Soc.,* vol. 127, pp. 2216–2220, Oct. 1980.

[17] C. Goldsmith, P. Geldermans, F. Bedetti, and G. A. Walker, "Measurements of stress generated in cured polyimide films," *J. Vac. Sci. Technol.,* vol. A1, pp. 407–409, Apr.–June 1983.

[18] DuPont Technical Information Bulletin, "Kapton-Summary of Properties," No. E-50553, Aug. 1982.

[19] L. I. Maissel and R. Glang, *Handbook of Thin Film Technology.* New York: McGraw Hill, 1970, ch. 12, pp. 22–23.

Packaging Technology for the NEC SX Supercomputer

TOSHIHIKO WATARI, MEMBER, IEEE, AND HIROSHI MURANO

Abstract—Technological considerations in realizing high-speed super-computers are presented, focusing on large-scale integrated (LSI) chips, new circuit packaging technology, and a liquid cooling system. The Model SX-1 and SX-2 supercomputers employ a new circuit packaging technology achieving up to 1300 megaflops processing speeds with a 6-ns machine cycle. This new technology features a 1000-gate current mode logic (CML) LSI with 250 ps gate delay as a logic element, a 1 kbit bipolar memory with 3.5 ns access time for cache memory and vector registers, a 10 cm² multilayer ceramic substrate with thin film fine lines (25-μm width, 75-μm center-to-center), and a multichip package which contains up to 36 000 logic gates. A liquid cooling module is implemented for high-density high-efficiency heat-conductive packaging for the arithmetic processor. In addition, high-density high-speed packaging of 64 kbit static metal–oxide semiconductor (MOS) RAM's are used to implement large-capacity fast main memory.

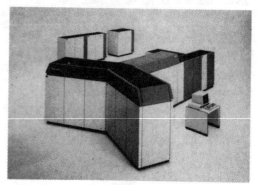

Fig. 1. Supercomputer SX-2.

INTRODUCTION

HIGH OPERATIONAL speed has always been a major objective in designing computers. In trying to attain maximum speed, using state-of-art technology, an increase in cost is inevitable. The cost/performance ratio and flexibility for all kinds of applications are major design considerations in general purpose computers. In the supercomputer, speed is the primary objective and the cost/performance ratio must be optimized for scientific applications. Packaging technologies for the SX computer system have been developed to achieve the highest class performance; 1300 megaflops (million floating point operations per second) [1]. An overview of the SX-2 computer is shown in Fig. 1. Its configuration is illustrated in Fig. 2. The scientific processing unit (SPU) is divided into two independent processors, called the arithmetic processor (AP) and the control processor (CP). The AP executes both scalar and vector operations in the user's job programs.

The main memory unit (MMU) has fast and large capacity storage. The AP determines the operational speed in conjunction with large and fast memories thus the packaging system of supercomputers becomes an important consideration in attaining high-speed processing and large capacity/high-speed memories. In the following sections we introduce large-scale integrated (LSI) packaging technologies and cooling system for the arithmetic processor and main memory.

Manuscript received March 30, 1985; revised June 21, 1985. This paper was presented at the 35th Electronic Components Conference, Washington, DC, May 20–22, 1985.

The authors are with the Computer Engineering Division, NEC Corporation, 1-10 Nisshin-cho, Fuchu City, Tokyo 183, Japan.

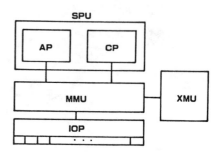

Fig. 2. SX system configuration. SPU: Scientific processing unit. AP: Arithmetic processor. CP: Control processor. MMU: Main memory unit. IOP: Input output processor. XMU: Extended memory unit.

LSI CHIPS

Supercomputer system performance is mostly determined by clock cycle time. The reduction of signal propagation delays from one logic element to another is, therefore, necessary to achieve a faster clock cycle. The use of high-speed logic chips and random access memories (RAM's) plays an important part in system performance, even though the signal propagation delays are approximately divided equally between chip delays and packaging delays in up-to-date high-density packaging technologies. A high-speed logic LSI and RAM were developed for the supercomputer. Table I shows the characteristics of the LSI logic chip, bipolar high-speed RAM, and large capacity MOS RAM utilized for the SPU and the MMU. Fig. 3 shows a tape automated bonding (TAB) LSI chip with a maximum of 1000 gates.

PROCESSOR PACKAGING

The SX system is a complete LSI computer. The use of high-speed LSI's and high-density packaging leads to the high speed switching of circuits and reduction in propagation delays

Reprinted from *IEEE Trans. Components, Hybrids, Manuf. Technol.*, vol. CHMT-8, no. 4, pp. 462–467, December 1985.

TABLE I
LSI CHIP TECHNOLOGIES FOR THE SX SYSTEM

CHIP	TECHNOLOGY	CAPACITY	SPEED	POWER	I/O	PACKAGING
LOGIC LSI	CML	1000 gates	250 ps/gate	5.7W TYP.	176	TAB
BIPOLAR RAM	CML	1K bits	3.5 NS ACCESS	1.35W TYP.	52	TAB
MOS RAM	STATIC-CMOS	64K bits	40 NS ACCESS	0.6W TYP.	32	LCC

Fig. 3.   LSI logic chip. TAB chip containing up to 1000 gates and 176 leads.

for signal. Both of these factors reduce machine cycle time. Furthermore high-density packaging allows the use of architectural approaches to obtain high performance, one of which is to enhance concurrency, without increasing the packaging volume, although concurrency (multiprocessors, pipeline processing, and parallel processing) usually results in more hardware volume. This new packaging technology features the following three levels of packaging and a liquid cooling system.

Level 1: A high-density high-pincount chip-carrier called FTC (flipped TAB carrier).
Level 2: A high-density multichip package (MCP) that consists of a multilayer substrate (MLS) with fine lines, higher speed connections, and high number of input/output (I/O) terminals.
Level 3: A high-density multilayer board mounting the above MCP's and an interconnection system including new zero insertion force (ZIF) connectors and high-speed coaxial cablings.

Liquid Cooling System: A cold plate and heat transfer block (HTB) that contains heat transfer studs form the liquid cooling module (LCM). Water circulating from the cooling unit (CLU) to the LCM can effectively cool the LSI and RAM chips.

LSI Chip Carrier

Logic LSI: A 1000 gate LSI device with 176 TAB leads is packaged in a 12-mm square chip carrier called the FTC (flipped TAB carrier). 128 leads of the LSI chip are for signal and 48 leads are for power and ground. The FTC consists of a multilayer ceramic substrate and a cap. The materials used, such as copper–tungsten alloys, have low thermal resistances and well-matched thermal expansion coefficients with silicon. The multilayer ceramic substrate of the FTC has 169 I/O

bumps arranged in a 13 × 13 matrix on its bottom surface and 176 TAB outer lead bonding pads on its top surface, respectively. The LSI chip is fixed in a face-down shape and 176 TAB leads are gang bonded simultaneously on the top surface of the FTC substrate. The LSI chip is die bonded in the FTC cap. Fig. 4 shows the overview of the logic FTC.

RAM: Four 1 kbit bipolar RAM chips are packed in a 14-mm square chip carrier (RAM FTC). The profile of the RAM FTC is almost the same as the logic FTC without the multichip structure. That is, four bipolar RAM chips are fixed individually in a face-down configuration and 52 TAB leads of each chip are gang bonded on the top surface of the substrate. 169 I/O bumps are arranged in the same way and also their foot prints are the same as that of the logic FTC. The RAM and logic FTC can be freely placed in any position on the MCP.

Multilayer Substrate

The key factors in reducing machine cycle time are reduction of signal propagation delays and the use of high speed LSI's. The MLS design therefore plays an important part in high-speed packaging technology. The requirements for the MLS to attain high-speed and high-density packaging are as follows.

High-Speed Wiring: Signal propagation delay t_d is given by the relationship:

$$t_d = \sqrt{\epsilon_r}/c$$

where ϵ_r is the dielectric constant of insulator material and c is the speed of light. The relationship shows the necessity for the use of lower dielectric constant material for the insulating layers.

High-Density Wiring: In order to reduce propagation delays for signals between LSI's, it is necessary to place LSI chips as close as possible on the MLS. The MLS should, therefore, use high-density wiring as well as supporting a large number of LSI's.

High Pincounts: Pincount P required for the MLS are given by Rent's rule [2]:

$$P = KG^r$$

where K and r are the experimental constants, and G is number of gates on the MLS. This relationship shows the MLS should have a capability of higher I/O pincount with increasing number of gates on the MLS. Fig. 5 shows a cross-sectional view of the MLS. The basic substrate of the MLS is a 100-mm square, 2.75-mm thick alumina ceramic substrate with tungsten metallized inner layers for power and ground and 2177 I/O pins brazed on its bottom surface. Signal layers consist of four polyimide insulative layers and five thin-film conductor layers, including two signal layers with two ground layers sandwiched around them, and a top metal layer with FTC attachment pads. The two ground layers reduce cross-talk noise in the signal layers (Fig. 6), and produce a uniform characteristic impedance in the signal lines.

By far the most innovative new development is the use of polyimide insulative layers for the multichip package substrate by combination with the alumina fired ceramic with I/O pins. By utilizing this new technology, a MLS with both high-speed

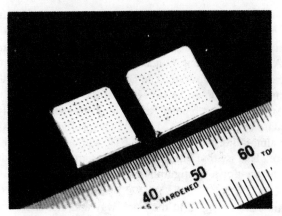

Fig. 4. Logic FTC and RAM FTC. The body of LSI chip is directly attached to 12 × 12 mm size copper–tungsten cap forming good thermal conduction path from chip to cap. Four RAM chips are packed in RAM FTC (14 × 14 mm size) in the same way as logic FTC.

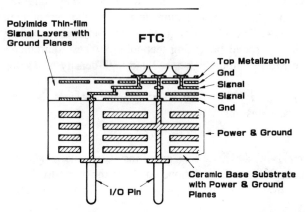

Fig. 5. MLS cross section. Tungsten metallized alumina ceramic base-substrate containing power and ground inner layers. Polyimide is utilized as insulative layers to realize high-speed high-density signal wirings. Thin-film fine-pitch signal lines are sandwiched between two ground mesh planes to reduce cross-talk noise.

high-density wiring and high pincounts can be achieved. The low dielectric constant of polyimide, $\epsilon_r = 3.5$, provides a signal delay time that is 60 percent of the conventional thick-film dielectric ceramic layers where $\epsilon_r = 9$. Moreover the distinctive features of polyimide such as a smooth surface, fine via hole formability, and low temperature curability (400°C max) make possible finer signal wiring on the ceramic substrate. Wiring with 25-μm width, 75-μm center-to-center grid, and 50-μm square via hole signal layers are formed. The signal lines utilize the thin-film line formation technique. The total maximum length of the signal line on the 100-mm square substrate is 110 m. Fig. 7 shows the fine signal lines on the substrate.

The low curing temperature of polyimide is very useful for the combination of fine line signal layers with high pincounts, and high power distribution ceramic substrates. That is, 400°C, nitrogen curing of polyimide does not cause oxidation of the tungsten–nickel–gold metallization on the ceramic substrate and the nickel–gold plated brazed pins. The fired ceramic base substrate has 2177 I/O pins brazed on a 2.54-mm (0.1 inch) staggered grid. Fig. 8 shows the overview of the MLS.

Multichip Package

The lower thermal resistance of the chip to the cooling mechanism, engineering change capability, and testability on the FTC mounted surface are matters of great importance for the design of the MCP shown in Fig. 9. The MCP consists of the MLS, with a maximum of 36 FTC's, and a flange. 36 FTC's are attached simultaneously on the MLS by tin–lead soldering techniques. A flange rings the entire bottom perimeter of the substrate. The flange thermally matches the perimeter of the substrate and is attached with fine measurement accuracy between I/O pins and flange registration holes. The flange that provides rough and fine pin guidance for connection to a base plate on the board allows for connection to a ZIF connector attached to the board. In addition the flange provides the base connection of the substrate to the liquid cooling module.

The engineering-change pad units are prepared for changes and probing for the MCP tests. The P-pad is connected to the internal signal "P"attern and the L-pad is connected to the chip "L"ead by way of the FTC. Each FTC is placed in the center of an FTC pad matrix, where 169 bump connections are made to 169 FTC pads on the top surface of the MLS. Each FTC pad is arranged on an 0.8-mm grid.

Liquid Cooling Module

The cooling design of the LCM requires meticulous care because the MCP generates 250 W maximum power. The LCM consists of a heat transfer block (HTB), a cold plate, and 36 studs which transfer the heat generated in the FTC (Fig. 10). Attachment of the LCM, HTB-to-MCP, is accomplished by use of the flange. The HTB removes heat by fine gap contact of stud-to-FTC. The stud is placed in machined holes in the inner surface of the HTB and has a unique shape to ensure the fine gap contact to the FTC. To make contact with the FTC, the HTB is loosely attached to the flange. Then, each of the 36 studs is adjusted individually and fixed in position in its hole. Every stud is fixed so as to keep a narrow gap between the face of the stud and the face of the FTC on the MCP. After all the studs are fixed, the HTB assembly is detached, a thermal compound is coated on the FTC, and then the HTB is re-attached to the flange to accomplish the LCM module assembly. The cold plate is fastened to the HTB and allows highly efficient cooling with water that circulates in the channel in the plate. Fig. 11 shows the overview of the LCM.

Multichip Package Connector

To achieve high speed and high reliability in a low-energy electrical connection, a new zero insertion force (ZIF) connector has been developed. As previously mentioned, the MLS has an array of 2177 connector pins. The pins mate with a corresponding array of bifurcated spring contacts whose tails are press fitted into plated through holes in the multilayer polyimide–fiberglass board. Fig. 12 shows the connector assembly which provides the zero insertion force structure. The connector assembly consists of seven connector housings, the base plate with two positioning pins, seven slide covers, and a slider. Seven connector housings cover all (2177) bifurcated spring contacts that are press fitted into the board.

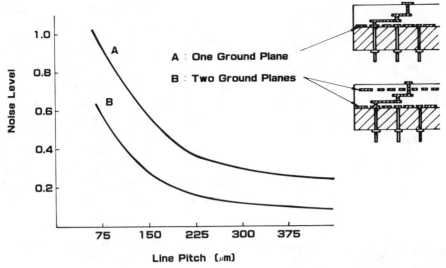

Fig. 6. Cross-talk noise level. Cross-talk noise level is compared in two cases. A: First ground mesh plane only formed directly on top surface of ceramic base substrate. B: Second ground mesh plane inserted to sandwich signal layers with first ground plane as shown in Fig. 5.

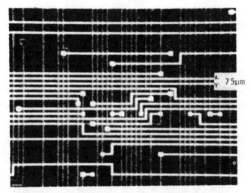

Fig. 7. Fine signal lines. A close-up of polymide signal plane showing 25-μm width 75-μm center-to-center fine lines and 50-μm via hole formed on polyimide film.

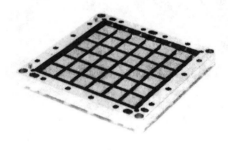

Fig. 9. Multichip package. Up to 36 FTC's mounted on MLS. Flange providing pin guidance for connection to board as well as base connection of substrate to cooling module.

Fig. 8. MLS overview. Ceramic base-substrate with 100×100 mm size and 2177 I/O pins brazed on its bottom surface effectively combining with high speed polyimide wiring layers.

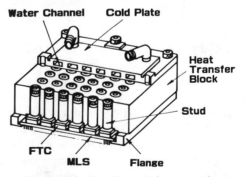

Fig. 10. Liquid cooling module cross section.

The slider and slide covers hold all (2177) pins of the LCM module and slides them into the bifurcated contacts without pressing the flange or MLS; consequently no distortion occurs between the pins, and MLS substrate. The two types of positioning pins, two long pins and three short pins, are attached on the base plate providing rough and fine pin guidance for the connection of the LCM modules. The LCM module connection and disconnection are easily achieved by the ZIF MCP connector.

The 545×460 mm (21.5×18.1 inch), 4.9-mm thick printed wiring board is designed to accomodate 125×125 mm LCM modules. Twelve LCM modules are mounted and

Fig. 11. LCM overview.

Fig. 12. MCP connector. 2177 MCP I/O pins can be reliably connected with zero insertion force connector. Slider surrounding seven slide covers and long positioning pins can be seen.

supplied power from the board. The board is comprised of 16 internal circuit planes for signal and power distribution. Eight planes of the board are for power and ground, six planes are for signal and two are for top and bottom surface patterns. The board also accommodates new coaxial signal cabling, terminating resistors, decoupling capacitors, and MCP mounting hardware in one unit with the capability for a maximum of 432 000 logic gates.

Since every LCM requires 2177 plated through holes in the board, the board is designed to accommodate a maximum of 26 124 through holes for the 12 LCM modules. Interconnections between the LCM's use six signal planes interlayed in the board and high-speed coaxial (HSC) cables which can be directly connected to each ZIF connector pin at the rear surface of the board. The board also has blind vias between the signal pair planes that are available to change the wiring direction. Since the ZIF connector pins are press fitted into the plated through holes in the board, rather than soldered, they provide high reliability and low manufacturing costs.

The press fitted connector pins also provide a direct connection to the coaxial cable on the rear surface of the board; i.e., the coaxial cable connector can be directly inserted to the pins. To interconnect between boards, a special board-to-board connector area is unnecessary, since the coaxial cable can directly connect from one signal pin to another on a LCM module located on another printed wiring board.

To achieve a 6-ns clock cycle, the signal delay from one LCM module to another must be reduced as low as possible, especially for board-to-board interconnections. A high-speed coaxial cable has been developed to meet this requirement. Since the HSC can be connected directly to the LCM module connector pins, the signals can travel the nearest distance from one LCM module to another and not through lines in the printed wiring board and board-to-board connectors. Furthermore the HSC cable has high-speed characteristics due to low-dielectric constant insulative material and a connector designed for low inductance. The signal propagation delay in the HSC cable is 3.8 ns/m. 1.08-mm diameter HSC cable provides high-density interconnection between boards.

WATER COOLING SYSTEM

The LSI chips (36) dissipate approximately 250 W maximum power in one LCM. The power density in the LCM module is calculated to be nearly 2.5 W/cm². This power density is approximately three times as high as NEC's conventional large general-purpose computer with an air cooling system. A water cooling system has been developed to remove the heat generated by the LCM's. A water cooling system can also provide compact packaging of the LCM's and boards, resulting in reduction of the signal wiring lengths. A 6-ns machine cycle has been achieved through use of this water cooling system. Fig. 13 shows the heat path from the LSI chips to the water. Total thermal resistance from junction-to-water is approximately 4.5°C/W. Fig. 14 shows the LCM modules and their water sleeve connections on the board. The water flowing in the cold plate on the LCM module is supplied by closed loop circulation from the cooling unit (CLU), which provides for heat exchange.

MAIN MEMORY PACKAGING

The performance of supercomputers is also characterized by large and fast memories. Even if the arithmetic speed is fast enough, high performance will not be attained without large and fast memories. The main memory system of the SX computer has 256 Mbytes maximum capacity, 6-ns machine cycle, and 512 way interleaves. To provide for the great number of RAM chips, three-dimensional packaging with an air cooling system is employed.

Memory Chip and Chip Carriers: 64 kbit static MOS RAM's with 40-ns access time, packed in leadless chip carriers (LCC) are used.

Single In-Line Package: To attain high-density mounting of the RAM and level converter chips, SIP's were designed. The SIP with 91.44 × 16.1 mm (3.6 × 0.63 inch) size and 36 I/O's accommodates five RAM LCC's, while a 71.12 × 16.1 mm (2.8 × 0.63 inch) size with 56 I/O's allows five level converter TAB's as shown in Fig. 15. A memory board with up to 100 SIP's (64 RAM SIP's and 36 level converter SIP's) provide two Mbytes of memory (Fig. 16).

The main memory system with 256 Mbytes and a 6-ns machine cycle consists of two memory cabinets; each memory cabinet has 128 Mbytes of memory. The memory cabinet accommodates two memory page frames. Up to 32 memory boards and 28 control boards in the page frame allow 64

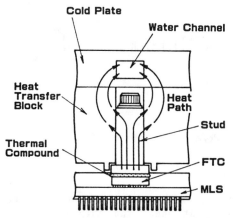

Fig. 13. Heat path in the LCM module. 4.5°C/W total thermal resistance from chip to water is realized.

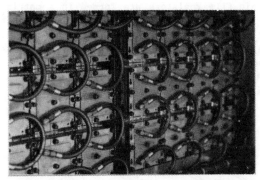

Fig. 14. LCM module water sleeve connections in the arithmetic processor unit.

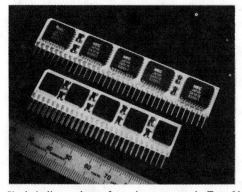

Fig. 15. Single in-line packages for main memory unit. Top: 64 kbit static MOS RAM chip carriers mounted on 91.44 × 16.1 mm (3.6 × 0.63 inch) SIP substrate. Bottom: Bipolar level converter TAB IC's mounted on 71.12 × 10.1 mm (2.8 × 0.63 inch) substrate.

Mbytes of memory. The memory and control boards are three dimensionally stacked in the page frames and cooled effectively by air. The air cooling system of the main memory allows high-density packaging of the RAM chips and easy repairability of the memory boards which contain 320 RAM chips.

Fig. 16. Two mega-bytes memory boards. 424.18 × 327.66 mm (167 × 129 inches) memory board mounting up to 100 SIP's (64 RAM SIP's and 36 level converter SIP's).

SUMMARY

The SX supercomputers introduce advanced LSI technologies, high-speed high-density multichip packaging technologies, an effective liquid cooling system, and high-density air-cooled main memory packaging. A machine cycle of 6-ns, 1300 mega-flops operation is realized by these advanced technologies. Compared with the NEC's highest class general purpose computer, S-1000(3), the new technology features include:

1) improved gate speed of the logic LSI: 500 ps/gate to 250 ps/gate;
2) number of gates on a logic chip increased by 5×: 200 gates to 1000 gates;
3) signal propagation time on the multilayer substrate reduced: 10 ns/m to 6 ns/m;
4) wiring density of the multilayer substrate increased by nearly 3×: total length = 5 cm/cm² to 14 cm/cm²;
5) I/O pin density on the multichip package increased over 4×: 5 pins/cm² to 22 pins/cm²;
6) cooling capability on the multichip package increased by 3×: 0.86 W/cm² to 2.5 W/cm².

ACKNOWLEDGMENT

The authors are indebted to many people who have contributed to this project, especially to those members of the Packaging Engineering Department and Circuit Engineering Department in the Computer Engineering Division with whom the details of the technologies were discussed, and the Production Engineering Department in the Computer Division who cooperated in the manufacture of the system. We would also like to especially thank Dr. Hisao Kanai for his suggestions and encouragements.

REFERENCES

[1] T. Watanabe, "Architecture of supercomputers—NEC supercomputer SX system," *NEC Res. & Develop.*, no. 73, pp. 1-6, Apr. 1984.
[2] B. S. Landman and R. L. Russo, "On a pin versus block relationship for partitions of logic graphs," *IEEE Trans. Computers*, vol. C-20, no. 12, pp. 1469-1479, Dec. 1971.
[3] H. Kanai, "Low energy LSI and packaging for system performance," *IEEE Trans. Components, Hybrids, Manuf. Technol.*, vol. CHMT-4, no. 2, pp. 173-180, June 1981.

A NEW MULTI-CHIP MODULE USING A COPPER POLYIMIDE MULTI-LAYER SUBSTRATE

Shinichi Sasaki, Taichi Kon, Takaaki Ohsaki

NTT Applied Electronics Laboratories
3-9-11, Midoricho, Musashino-shi, Tokyo 180, Japan

ABSTRACT

A new multi-chip module using a multi-layer substrate with a polyimide dielectric and a fine pattern of copper conductors is developed. This substrate contains small copper columnar vias, for reducing the thermal resistance of polyimide layers without causing a channel accommodation drop, and thin film resistors for terminated transmission. This new module can densely mount high-speed LSI chips that produce twice as much heat as those in conventional modules. Furthermore, this module can transmit high-speed pulses at over 2 Gbit/s.

INTRODUCTION

Advances in the speed and density of integrated circuits used in high-performance systems have created a greater demand for higher density packaging to improve electrical performance by reducing interconnection delay. To meet this demand, multi-layer substrates with a polyimide dielectric and a fine pattern of copper conductors have been developed.[1-8] These substrates have the advantage of high-speed signal transmission due to the low dielectric constant of a polyimide insulator. However, because the thermal conductance of a polyimide insulator is less than that of alumina, a multi-chip polyimide module needs a new cooling structure to reduce the thermal resistance of the polyimide insulator layer. Furthermore, terminated transmission is necessary for high-speed signal transmission without waveform distortion.

This paper presents a new multi-chip module that uses a copper polyimide multi-layer substrate containing small thermal vias for reducing the thermal resistance, and thin film resistors for terminating transmission.

STRUCTURE and FEATURES

Three-dimensional and side views of the new multi-chip module are shown in Fig. 1. Chips are bonded face-up to a substrate. The heat generated in the chips is conducted by the thermal vias to the alumina substrate and cooling fins, and from there dissipated into the air. Features of this new module are listed below.

(1) Many copper columnar vias in the polyimide multi-layer just under the chip reduce thermal resistance from the chip to the cooling fin.

(2) The copper columnar via diameter of less than 100 μm allows for many signal lines under the chip.

(3) Thin-film resistors are formed in a matrix on the alumina ceramic substrate. Matching resistors are interconnected only by random via-holes in the polyimide layer. These matching resistors enable high-speed signal transmission.

(4) Because I/O pins are attached to the alumina substrate using a low-melting-point material, the polyimide multi-layer is not damaged.

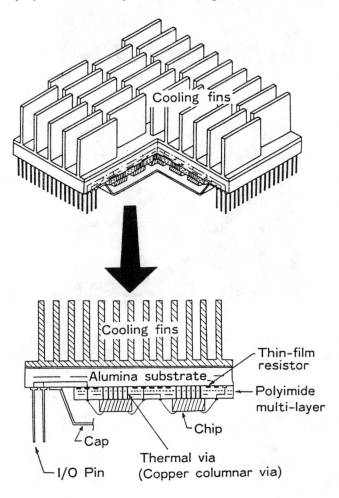

Fig. 1 New multi-chip module.

Reprinted from *Proc. 39th Electron. Components Conf. (ECC)*, pp. 629–635, 1989.

In this new module, if the thermal copper columnar via diameter is larger, the channel accommodation is lower, because the line area is reduced. On the other hand, if the via diameter is smaller to maintain the channel accommodation, then its cooling capability decreases.

The design concept of this module aims at increasing cooling capability without a channel accommodation drop and at achieving high-speed signal transmission without a reduction in LSI chip mounting density.

In this report, a small copper columnar via for cooling is formed on the same grid used by the interconnection vias. Furthermore, terminated transmission is necessary for high-speed signal transmission without waveform distortion. Therefore, thin film resistors for terminated transmission are formed on the substrate near the LSI chips to keep the interconnections short. Figure 2 shows a flow-chart for determining the structure and dimensions of a module.

First, parameters for power dissipation, interconnection channels (that is, circuit scale), line-width (line pitch), insulator layer thickness, and substrate dimensions are given. Then, the pitch and diameter of the copper columnar vias, and the number of conductance layers are set provisionally, and the channel number, from dividing line-space to line-pitch, is calculated. If the channel number is less than the specified value, the diameter of the copper columnar via is reduced, and the channel number is recalculated. If this recalculated channel number complies with the specified value, a thermal conductance of the polyimide substrate is calculated using an approximate formula, and is then compared with that of a ceramic substrate. If the polyimide substrate thermal conductance is smaller than than that of the ceramic substrate, the pitch and diameter of the vias and the layer's number are set again. This procedure should be repeated until all values are acceptable.

Next, the cooling capability of all parts of a module are estimated using the 3-dimensional node analysis method. In this method, the copper polyimide multi-layer substrate is modeled to a material which has the composite thermal conductance of a copper columnar via and a polyimide insulator.[13] This thermal conductance is written as

$$\lambda = (1-x)\lambda_p + x\lambda_c \qquad (1)$$

where

$$S_c = x \cdot S_{chip}$$

λ_p: Polyimide thermal conductance
λ_c: Copper thermal conductance
S_c: Copper columnar via area per chip
S_{chip}: Chip area

After simulating the maximum chip junction temperature, the optimum substrate structure of the thin-film resistors and the stripline are determined using a computer simulation program, for example SPICE, to examine their high-frequency electrical properties.

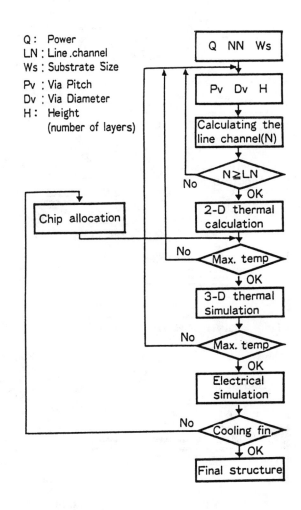

Q : Power
LN : Line channel
Ws : Substrate Size
Pv : Via Pitch
Dv : Via Diameter
H : Height (number of layers)

Fig. 2 Design flow chart of the new module.

In producing multi-chip modules using copper polyimide multi-layer substrates, the processes following polyimide multi-layer substrate fabrication must be carried out at temperatures below the temperature limit of polyimide. Each process must be carried out at a lower temperature than the previous ones. In addition, these thermal vias in this new module must be filled with plating copper to reduce the thermal resistance of the polyimide layer. Figure 3 outlines the fabrication processes.

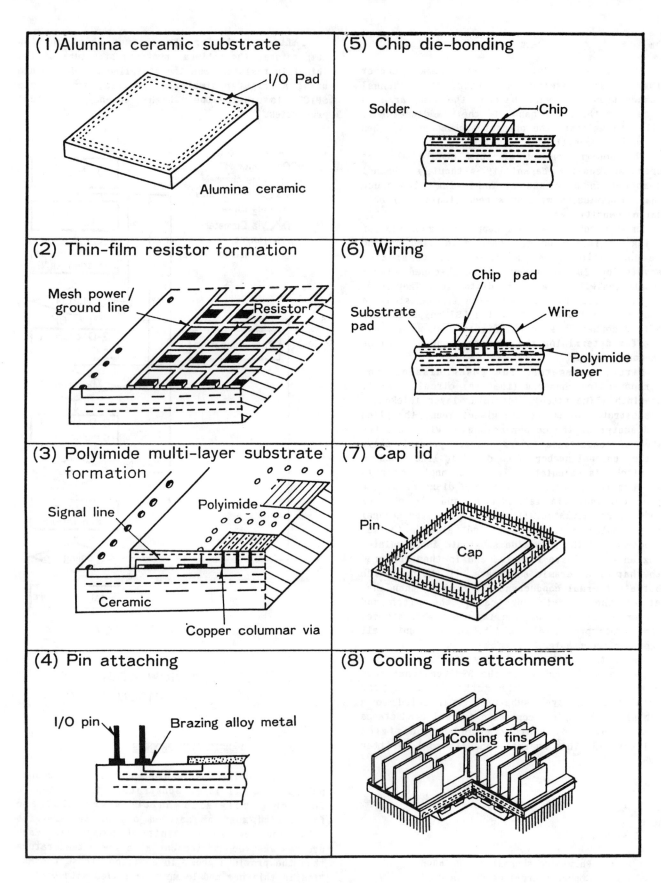

(1) Alumina ceramic substrate

I/O Pad

Alumina ceramic

(2) Thin-film resistor formation

Mesh power/ ground line

Resistor

(3) Polyimide multi-layer substrate formation

Signal line

Polyimide

Ceramic

Copper columnar via

(4) Pin attaching

I/O pin

Brazing alloy metal

(5) Chip die-bonding

Solder

Chip

(6) Wiring

Chip pad

Substrate pad

Wire

Polyimide layer

(7) Cap lid

Pin

Cap

(8) Cooling fins attachment

Cooling fins

Fig. 3 Fabrication process.

(1) CERAMIC SUBSTRATE

This ceramic substrate, which is made by a multi-layer co-firing process, has many I/O signal lines and a mesh layer of power/ground lines.

(2) THIN-FILM RESISTOR FORMATION

First, thin film of high-resistance metal is formed on the surface of the ceramic substrate, using thin-film processing technology. After this film is patterned, termination resistors are formed.

(3) POLYIMIDE MULTI-LAYER SUBSTRATE FORMATION

Using photo-sensitive polyimide, the conductor layer and insulator layer are formed by electroplating and electroless plating technology.[1] Then, thermal vias are filled with copper during the same signal vias process. This module has a stripline structure in which the signal line is sandwiched between the mesh power/ground layer, and its characteristic impedance is 50 Ω.

(4) PIN ATTACHING

In the bonding process for pins, chips, and the cap, it is necessary to keep the temperature lower than 400℃, which is the limiting temperature for polyimide. Pins are attached to the substrate using a brazing alloy with a lower melting point and sufficient strength.

(5) CHIP DIE-BONDING

In a nitrogen atmosphere furnace, chips are mounted to the multi-layer substrate using a solder.

(6) WIRING

The chip pads are connected by wire to the pads on the polyimide substrate.

(7) ATTACHING THE CAN LID

Next, a metal cap is attached to the module, using a brazing alloy.

(8) COOLING FINS ATTACHMENT

Finally, cooling fins are attached to the opposite side of the ceramic substrate.[14-15]

The thin-film resistors fabricated on the ceramic substrate are shown in Fig. 4. There is a resistor in each wiring grid. Figure 5 shows a photograph of the fabricated multi-chip module. This module has a 6-layer polyimide substrate to which 36 chips are mounted. The module is about 70 mm square and has 288 I/O pins. The signal line width is 50 μm and thickness is 5 μm. The diameter of the thermal copper columnar vias is 100 μmφ.

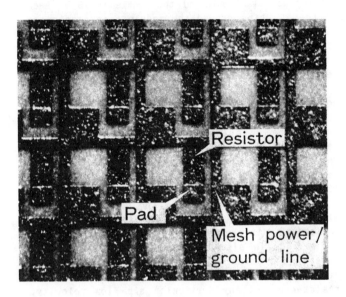

Fig. 4 Thin-film termination resistor.

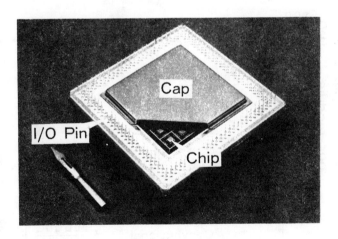

(a) Top

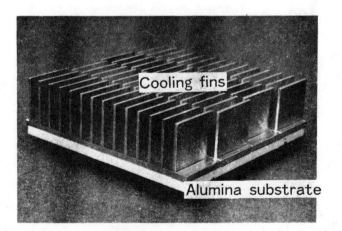

(b) Reverse side

Fig. 5 Fabricated with 288 I/O pins.

Module is 70 mm square with

36 chips (3 mm square) mounted.

THERMAL CHARACTERISTICS

The cooling capability was measured using this fabricated module and was evaluated by computer simulation. The simulation was accomplished by the three-dimensional node analysis method.

(1) TEMPERATURE DISTRIBUTION

Figure 6 shows a temperature distribution chart for a module surface of forced air convection. The total power dissipation of the module is 26 W. Each of the 20 chips around the module area dissipates 0.5 W, and each of the 16 chips in the central area dissipates 1.0 W. The air flow velocity at the inlet of the cooling fins is 6 m/s, and the temperature is 25℃.

The maximum temperature of a conventional module that has no thermal copper columnar via reached 74℃, with temperature near the chip increasing steeply. In contrast, the maximum temperature of the new module containing the thermal copper columnar vias is only 53℃, with temperature near the chips increasing gradually.

(2) THERMAL RESISTANCE

Thermal resistance from the chip junction to air coolant versus air flow velocity were measured, and the results are shown in Figure 7. Fin height is 15 mm and fin pitch is 4 mm. Chip size is 3 mm square. For each module, the lower the air flow velocity, the greater the drop in thermal resistance. At any particular air flow velocity, thermal resistance of the new module is about half that of a conventional module. As can be seen, the agreement between the experimental and calculated results is excellent.

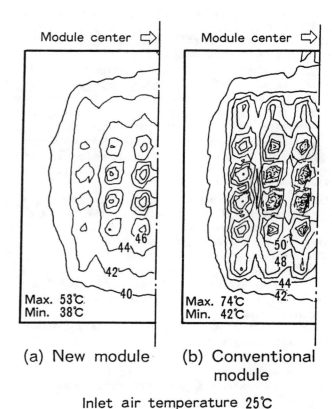

(a) New module   (b) Conventional module

Inlet air temperature 25℃
Inlet air velocity   6 m/s

Fig. 6 Temperature distribution.

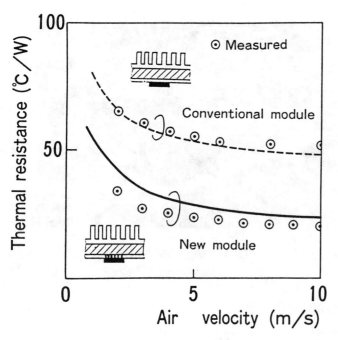

Fig. 7 Thermal resistance versus air flow velocity.

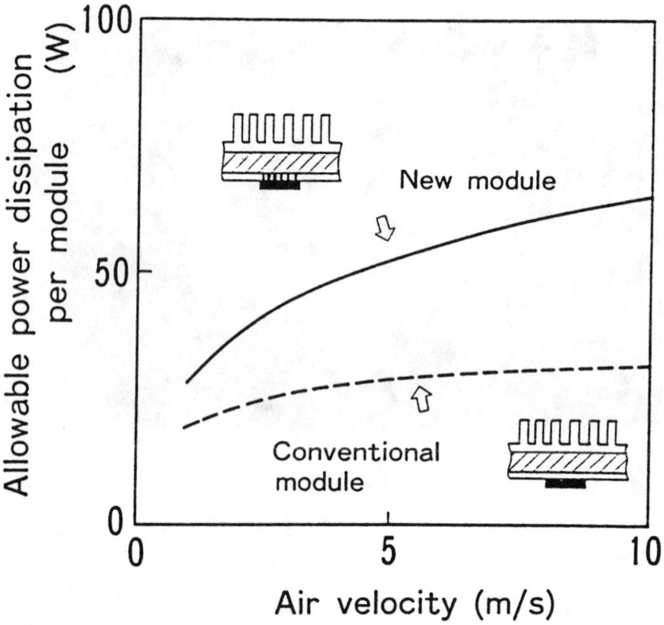

Fig. 8 Allowable power dissipation per module.

In the new module, the thermal resistance of the polyimide multi-layer can be reduced less than 5% of the total module resistance as compared with the conventional polyimide multi-layer, which forms 90% of the total resistance. The new module's resistance is also lower than that of a ceramic substrate.

(3) ALLOWABLE POWER DISSIPATION

Figure 8 shows allowable power dissipation per module as a function of coolant air flow velocity, when the allowable chip junction temperature is 60℃. Allowable power dissipation per module rises to 60 W, which is about twice that of a conventional polyimide module that has no copper columnar via.

ELECTRICAL CHARACTERISTICS

(1) REFLECTION AND DISTORTION

The electrical reflection and distortion characteristic of this module containing thin-film termination resistors as compared with a conventional substrate using chip-resistors is shown in Fig. 9. The electrical characteristic is measured using a Time Domain Reflectometer. There is no wave distortion because the connecting length between the LSI chip pad and the termination resistor can be shortened to about 100 μm, the same size as the polyimide substrate thickness.

Humidity, thermal-shock, and the high-temperature test caused little change in resistance, demonstrating that this resistor has a high enough reliability for a termination resistor in a multi-chip module.

(2) PULSE TRANSMISSION CHARACTER

The propagation delay time of 25-μm-wide, 5-μm-thick signal lines is 60 ps/cm, which is 40% less than for the alumina ceramic substrate. This is because the signal line has a low dielectric constant. Figure 10 shows the eye-pattern diagram for the module driven by a 2-Gbit/s NRZ pseudo-random pulse train. The openings in this pattern are sufficient to indicate that the fabricated module can transmit high-speed pulses at over 2 Gbit/s.[1][10-12]

CONCLUSION

To improve LSI chip mounting density, electrical transmission capability, and cooling capability without a channel accommodation drop, a new multi-chip module that uses a copper polyimide multi-layer substrate containing both thermal small copper-columnar vias and thin-film termination resistors was proposed.

This new module can densely mount high-speed LSI chips, which produce twice as much heat as those in conventional modules. Furthermore, it can transmit high-speed pulses at over 2 Gbit/s.

This new module will find many applications as a high-speed, high-power multi-chip compact module.

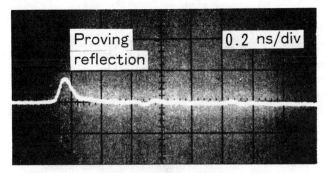

(a) New module with thin-film resistor.

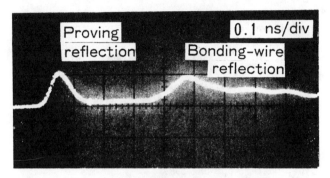

(b) Conventional module with chip resistor.

Fig. 9 Output waveform of Time Domain Reflectometer

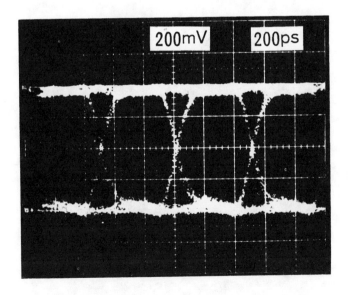

Fig.10 Eye-pattern diagram.

ACKNOWLEDGMENT

The authors wish to thank T. Yasuda, S. Yamaguchi, and T. Kishimoto for their valuable contributions in calculating and testing the modules described in this paper.

REFERENCES

[1] T. Ohsaki et al., "A fine-Line Multilayer Substrate With Photo-Sensitive Polyimide Dielectric And Electroless Copper Conductions," Proceedings of IEMT, pp. 187, 1987.

[2] N. Goldberg, "Design of thin film multichip modules," Proc. ISHM, vol. 4, 1981, pp. 289-295.

[3] J. Shurboff, "Polyimide dielectric on hybrid multilayer circuits," Proc. 33th ECC, 1983, pp. 610-615.

[4] K. Moriya, T. Ohsaki, and K. Katsura, "High-density polyimide multilayer interconnection with photo-sensitive polyimide dielectric and electroplating conductor," Proc. 34th ECC, 1984, pp. 82-87.

[5] R. J. Jensen, J. P. Cummings, and H. Vora, "Copper/polyimide material system for high performance packaging," IEE Trans. CHMT, vol. CHMT-7, pp. 384-393.

[6] T. Watari and H. Murano, "Packaging technology for the NEC SX supercomputer," IEEE Trans. CHMT, vol. CHMT-8, pp. 462-667, Dec. 1985.

[7] H. J. Levinstein, C. J. Bartlett, and W. J. Bertram, Jr., "Multi-chip packaging technology for VLSI-based systems," ISSCC Digest Tech. Papers, pp. 224-225, Feb. 1987.

[8] P. G. Rickerl, J. G. Stephanie, and P. Slota, Jr., "Evaluation of photosensitive polyimides for packaging applications," Proc. 37th ECC, 1987, pp 220-225.

[9] T. A. Lone, F. J. Belcourt, and R. J. Jensen, "Electrical characteristics of copper/polyimide thin film multilayer interconnects," Proc. 37th ECC, 1987, pp. 614-622.

[10] B. T. Clark and Y. M. Hill, "IBM multichip multilayer ceramic modules for LSI chip-design for performance and density," IEEE Trans. CHMT, vol. CHMT-3, pp. 89-93, March 1980.

[11] H. Ichino, M. Suzuki, K. Hagimoto, and A. Konaka, "Si bipolar multi-Gbit/s logic family using super self-aligned process technology," Ext. Abst. 16th Conf. Solid State Devices and Mater., 1984, pp. 217-220.

[12] M. Suzuki, K. Hagimoto, H. Ichino, and S. Konaka, "A 9-GHz frequency divider using Si bipolar super self-aligned process technology," IEEE Electron Device Lett., vol. EDL-6, pp. 181-183, April 1985.

[13] Y. C. Lee, H. T. Ghaffari, and J. M. Segeken, "Internal thermal resistance of a multi-chip packaging design for VLSI-based systems," 38th ECC Proc., 1988, pp. 293-301.

[14] T. Kishimoto, E. Sasaki, and K. Moriya, "Gas Cooling Enhancement Technology for Interggrated Circuit Chips," IEEE CHMT, vol. CHMT-7, NO. 3, Sep. 1984.

[15] T. Kishimoto and S. Sasaki, "Cooling characteristics of diamond-shaped interrupted cooling fin for high-power LSI devices," IEE Electronics Letters, vol. 23, No. 9, 23rd April, 1987, pp. 456-457.

HIGH-PERFORMANCE VLSI THROUGH PACKAGE-LEVEL INTERCONNECTS

Louis Liang, J.D. Wilson, N. Brathwaite, L.E. Mosley, D. Love

Intel Corp.

3065 Bowers Ave.
Santa Clara, CA 95051

145 S. 79th Street
Chandler, AZ 85226

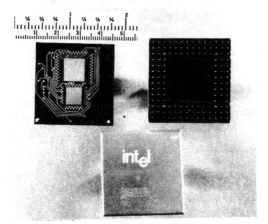

Figure 1. The C4 multichip module.

ABSTRACT

In this paper a packaging technique utilizing Controlled Collapse Chip Connection (C4) to integrate two existing high performance VLSI devices into a single pin grid array (PGA) package is described. The design and layout of the multilayer ceramic package, electrical and thermal performances, and future enhancements are discussed. The performances of the new package are also compared to a companion two-cavity wirebonded evaluation package. Finally, the potential use of this technique as a substitute for "wafer level integration" is examined.

I. INTRODUCTION

With the rapid advances in wafer fabrication process technology, IC designers are always tempted to increase chip level integration at an ever faster pace. Although computerized design and process modelling have significantly reduced both cost and time required to bring highly complex ICs, such as microprocessors (MPU), to the market place; the development costs could still run into millions of dollars and take years to accomplish. A packaging technique utilizing flip chip bonding (C4) has been developed to manufacture the next generation VLSI quickly and cost effectively. Conformance to existing package outline and pin out is achieved through multichip package level interconnects of existing components into a single package.

In this particular work, an advanced MPU is combined with a coprocessor in a multilayer ceramic PGA package using flip chip bonding as shown in figure 1. It is then a plug-in replacement for the existing MPU. This work differs significantly from prior work on multichip flip chip bonding [1] in that it uses components originally designed for wire bonding and the new package conforms fully to the original discrete package outline and with comparable manufacturing cost. A wirebonded companion evaluation package, as shown in figure 2, with two cavities, 21 layers, co-fired ceramic PGA [2] is compared to the new package to study electrical, thermal, design, and manufacturing attributes of C4 bonding.

Figure 2. The wirebonded multichip module.

II. PACKAGE DESIGN AND CONSTRUCTION

II.a. Package Design and Circuitry

There are several critical performance and form factor requirements imposed on this 132-pin multichip PGA package by both customer and time/cost objectives. The package design must support the operation of both devices at a frequency of 25-MHz or more while allowing the devices to maintain reliable junction temperatures at power dissipation values in excess of 3 watts. Four major constraints were imposed upon the design of this package: Dice, designed for wirebond interconnection, must be bonded to the package utilizing controlled collapse chip connection technology. The pin count, pin placement and package style were fixed by the need for plug compatibility with the wirebonded package design of the MPU [2]. The overall package

Reprinted from *Proc. 39th Electron. Components Conf. (ECC)*, pp. 518-523, 1989.

body thickness was constrained by the existing C4 package cap design. And finally the packaging cost must be comparable to the combined packaging cost of the discrete components being replaced.

In figure 3 a schematic view of the ceramic package substrate cross section is shown. The lay-up shows 8 dielectric layers and 9 conductor patterns including the surface layers for chip bonding and package pin brazing. The package was constructed by cofired ceramic technology except for the surface Thin Film Bond layer where, because of the fine lines and spacings required, thin film technology was used. It was necessary to allocate 2 layers for chip-to-chip and chip-to-pin interconnections. Further, it was desirable to shield each of these signal layers from each other to minimize layer-to-layer crosstalk. Each signal interconnect layer is thus surrounded by either ground or power planes. The remaining available conductor layers were allocated to either power supply or ground return planes.

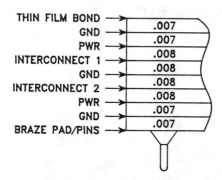

Figure 3. Ceramic package substrate cross section identifying conductor patterns and dielectric layer thicknesses (inches).

The interconnection circuitry for bus, address and control signals was designed for performance and interconnectability. Internal traces for the majority of the signal nets were designed for 35-ohm impedance by using 6-mil line widths on 20-mil centers with 8-mil ceramic tape thickness. Signal traces were also routed externally on the Thin Film Bond layer: (1) to provide fan-out from the fine pitch of the substrate solder bond pads and (2) to alleviate internal route crowding. From bond pad pitches as fine as 7-mils on the Thin Film Bond layer, signals were fanned out with 3-mil thin film traces to the larger pitch of the cofired ceramic vias. The thin film signal routes, which alleviate internal crowding, were designed with the same design rules as the internal routes. A photograph of the Thin Film Bond layer is shown in figure 4. Note the oversized via capture pads at the terminal points of the fan-out near the chip bonding sites. These oversized pads are necessary to guarantee connection between the cofired ceramic vias and the surface thin film circuitry.

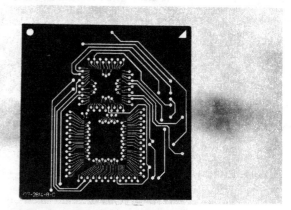

Figure 4. The Thin Film Bond layer showing solder pads, fanout traces, and some signal routing.

II.b. Manufacturing Highlight

Because of the tight pitch bond pads of the dice (which are designed for wire bonding), the standard circular shape C4 bumps cannot be used. Instead oval shape C4 bond pads were used, with the minor axis parallel to the die edge, in order to maximize bump volume without causing shorts between bond pads [3]. The other advantage of oval shape bumps is greater bond pad placement flexibility as shown in figure 5. Fabrication of a reusable metal mask with a minimum hole dimension lower than the ultimate pad height is usually difficult, and it requires special processing to minimize potential short related yield loss. Due to the tight pitch (7 mils) involved, the wafer bumping and chip join processes also required modifications. Reduction in evaporation rate of Lead/Tin for the solder bumps coupled with chip join reflow process changes resulted in significant yield improvements: to 100%.

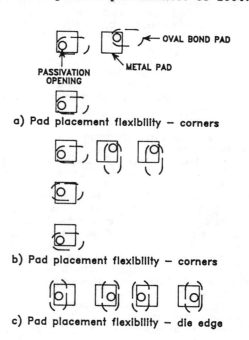

Figure 5. Terminal metal design flexibility.

The need to select a material and process readily available from the substrate manufacturer, good solderability and the requirement for conduction lines with low ohmic resistance, made copper an appropriate choice of metallization material. The thin film circuitry consisted of a "tri-metal" system of a base metal or adhesion promotor, the conducting metal and a covering, non-solderable, metal. A Cr-Cu-Cr system was used in this package.

Two methods of thin film deposition, E-beam evaporation and sputtered film, were evaluated. Experimental results showed sputtered films to have at least 50% better adhesion than E-beam evaporated films and confirmed previously published observations [4,5]. Sputtering was therefore chosen as the method of thin film deposition for this package.

The use of solder interconnects for attaching semiconductor devices to ceramic substrates have been practiced for over 25 years since first announced as Solid Logic Technology (SLT) in 1963. Advances in this technology have resulted in C4 interconnects, a technique well described in the literature [6,7,8]. In the C4 assembly method, solder bumps on the device and corresponding solder pads on the substrate are placed in a juxtaposition and reflowed. The collapse of the system is controlled by a combination of design and process conditions, resulting in a single solder column or joint.

Following the chip joining operation, the module is encapsulated using a crimped metal (aluminum) cap and a flexibilized epoxy backseal as shown in figure 6. Since the package pins used in this package are previously gold plated, no lead-finish operation is required.

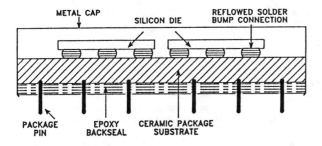

Figure 6. Schematic cross-section of a standard-assembled C4 package with bonded chips, metal cap, and epoxy backseal.

TABLE 1

Capacitance and inductance data for the C4 and wirebond multichip packages.

SIGNAL TYPE	Capacitance, pF		Inductance, nH	
	C4	WB	C4	WB
Control/Address	9.24	6.59	11.40	13.40
Data Bus	13.30	17.50	10.30	13.30
Power – Gnd	1740	3500	0.17	0.45

III. PERFORMANCE AND CHARACTERIZATION

III.a. Electrical Performance

Parametric characterization of each signal net and the power-ground paths in this package was performed in the laboratory. Capacitance values were obtained with a Hewlett-Packard 4192A impedance analyzer. The measurements were made at 1-MHz and 1 Volt peak-to-peak with respect to power and ground shorted together. This technique represents measurement of signal line load capacitance. The accuracy of this technique is 0.05pF which is basically the background capacitance of the probes. Inductance measurements were made using a test board specially designed for measuring PGA packages. The unique feature of this method is that the signal line inductance can be measured with respect to all grounds simultaneously. Complete details on the design and modelling of the test board will be described in a coming publication [9]. The inductance values were obtained using a Hewlett-Packard 4191A impedance analyzer at 25-MHz with an accuracy of 10pH for power and ground measurements and 200 pH for signal pin measurements.

Table 1 shows representative capacitance and inductance data for both the C4 and the wirebonded packages for each of the three signal types: control/address, data bus and power supply. For the C4 package, control and address signals are single lines (2-pin nets) connecting an external package pin to a solder bump at either chip. Data bus signals are routed between the two chips as well as to an external package pin (3-pin nets). The power supply and return paths are distributed on contiguous planes. The data in table 1 illustrates some interesting differences between the two designs. The control line capacitance difference reflects the fact that the wirebonded package utilizes the MPU in the same cavity as the single chip package with optimized pin-to-chip interconnect. The data bus capacitance in the wirebond package is higher, however, and indicates that the C4 package routing is accomplished more efficiently with less net length. The difference between power and ground capacitances simply reflects the difference in number of layers allocated to power and ground. The wirebonded package has a total of 13 layers dedicated to either power or ground while the C4 package has only 5 layers.

The difference in inductances for the C4 and wirebonded packages for all signal types is clearly the result of the reduction of interconnect inductance. While the control line capacitance of the C4 package was 2.5pF higher than the wirebond package, the inductance was 2nH lower in spite of the longer average net length. The data bus lines, as would be expected from the capacitance data, had on the average 3nH less inductance. The power and ground inductances demonstrated the most dramatic difference between the two packages. The C4 package has less than one-half the loop inductance of the wirebonded package.

It has been determined by regressing total net capacitances with total net lengths that the calculated capacitance per unit length for each package type is nearly identical. A value of 7.98 pF/inch was obtained for the case of the wirebond package while a value of 8.3 pF/inch was obtained for the C4 package. The 4% variation in capacitance per unit length could easily be due to the variation in dielectric constant of the ceramic tape material [10]. Similarly when inductances and line lengths are regressed for the two packages, the values 10.1 and 10.0 nH/inch are obtained for the wirebonded and C4 packages respectively.

Table 2 shows a comparison of ground and power path loop inductances for the two package types. The inductance data is presented as a function of increasing number of power-ground pin-pairs. The table data shows clearly the significant reduction in inductance due to the difference in the inductances between a wirebond and a solder bump. From the inductance data in table 2 one may compute the value of an individual interconnection [2] provided the interconnection inductance dominates the plane and via inductances. The inductance value obtained for the C4 interconnection (0.38nH) indicates the plane inductance is dominant. The self inductance using a round wire model [11] of a single solder bump is approximately 8pH. The inductance of a single wirebond (2.6nH) from the experimental method is in good agreement (8%) with the wire model calculation.

III.b. Device Characterization

Performance characterization of the C4 multichip package comprised voltage-frequency plots and I/O timing analysis for each of the VLSI devices in the module. Characteristic voltage-frequency response data for the C4 package type is shown in figure 7. The pass-fail plot of figure 7 represents five (5) packages whose frequency at failure was averaged for each power supply voltage tested. The power supply was varied from 4 to 6 volts in increments of 0.1 volt. Each package was functionally tested at clock period increments of 1 nanosecond. The corresponding operating frequencies ranged from 10 to 40 MHz. From this data it may be seen the C4 package provides ample margin for 25-MHz operation at both low and high voltage tolerance levels. The voltage margin at higher frequencies is largely attributable to the low power and ground path inductance in the C4 package (see Table 2). The reduced power-ground inductance also has an especially beneficial effect for ground bounce noise. Lower power and ground inductances mean that higher di/dt values from simultaneous switching outputs can be tolerated without causing excessive noise in the ground or false output switching. Although pin reduction is not possible for this particular package, it is clear from the foregoing that a reduction in the number of pins required for the power supply would be feasible. Pin reduction, in general, will result in cost reduction due to construction simplification and reduced precious metal plating area.

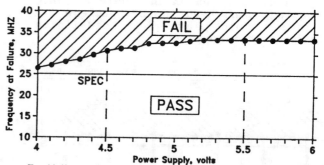

Figure 7. Voltage-frequency response of C4 multichip module.

TABLE 2.

Power—ground loop inductance as function of the number of pin pairs.

PIN PAIR (#)	PACKAGE TYPE:	
	C4 (nH)	WIREBOND (nH)
1	4.70	6.71
2	2.34	3.11
3	1.57	2.15
4	1.00	1.94
5	0.77	1.48
6	0.64	1.22
7	0.55	1.04
8	0.46	0.92
9	0.41	0.85
10	0.37	0.76
11	0.34	0.71
12	0.31	0.66
13	0.28	0.61
14	0.26	0.59
15	0.23	0.54
16	0.21	0.51
17	0.21	0.50
18	0.19	0.47
19	0.18	0.46
20	0.17	0.45
21	0.17	0.45

TABLE 3.

Thermal data for C4 and wirebond packages. All values listed are °C/watt units.

	WIREBOND		C4 WITH THERMAL CAP		C4 WITHOUT THERMAL CAP	
	θja	θjc	θja	θjc	θja	θjc
WITHOUT HEAT SINK						
AF=0	18.5	4	18.0	3	23.0	8.0
AF=200	12.5	4	12.5	3	15.5	8.0
WITH HEAT SINK						
AF=0	13.5	4	13.0	3	19.5	8.5
AF=200	7.0	4	6.5	3	11.5	8.5

III.c. Thermal Performance

The standard C4 package, as shown in figure 6, compares unfavorably to the wirebonded package because of high junction-to-case resistance. This deficiency can be overcome by enhancement methods [12,13] and as demonstrated in figure 8. Enhancement is, however, not required for this particular package for 25 MHz operation. Table 3 contrasts the thermal performance attribute of the C4 package, a thermally enhanced version as well as the wirebonded package. Data is presented for conditions 0 and 200 linear feet per minute airflow (AF) and for the packages with and without external heat sink attachment. It is apparent from the junction-to-case resistance of the enhanced C4 package that the thermal grease permits a heat removal path equivalent to the ceramic package. The heat sink used for this evaluation was designed for applications requiring less than 25°C rise junction-to-air. Normal applications do not require the maintenance of such low junction temperatures. In the worst power case the standard C4 package will allow the junction temperature to rise only 40°C above ambient.

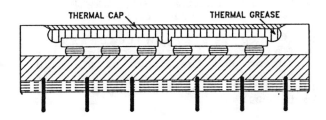

Figure 8. Schematic cross-section of a thermally enhanced C4 package.

III.d. Mechanical Reliability

The mechanical reliability of C4 on ceramic substrates has been well studied [14,15]. While the use of oval pads in C4 packages had been published previously, no data is available on the relative reliability of oval as contrasted to round pads. Initial modelling predicted that if the same volume of solder is used with oval pads (as for round pads), the mechanical reliability of C4 joints would be at least equivalent. This was indeed supported by the test results. The experimental results are 40-60% better than the theoretical prediction using modified Goldman's equation for a given distance to neutral point (DNP).

IV. DISCUSSIONS

Although the rapid advances of silicon processing and design technology have been a formidable competitor to multichip efforts, the flexibility, low cost, reliability and short design cycle offered by this packaging technique has many advantages.

IV.a. Design Cycle

Design of a simple, multichip module with existing VLSI devices requires about 3-5 weeks to complete the schematic, transfer chip geometry database information to a CAD program, route the devices, and check the design for manufacturability. The package vendor will require additional time dependent upon the complexity of the package for completion of manufacturing plans and print checks. The actual manufacturing turnaround time will be on the order of 6-12 weeks. The total design to manufacture cycle time may require 12-19 weeks until both design and manufacturing interests are satisfied. Electrical, mechanical and thermal modelling and simulation times are not factored into this time frame. These times may well increase and become significant for the design cycle as the requirements for higher performance and lower cost manufacturing drive beyond today's sophistication.

IV.b. Design Flexibility

In addition to the short design cycle, this new packaging technique also permits mixed technologies, for instance, Bi-CMOS, Analog-Digital, and special customization such as addition of ASIC peripheral chips to standard processors. More importantly, it allows the customer to use the existing PC board because of plug-in compatibility. This represents a major cost saving at the systems level. Future applications may include socket and pin reduction in a PC board, prolonging product life cycle (by expanding the scope of applications), and stop gap implementation of next level integration before the next generation silicon products become available.

IV.c. Low Cost VLSI with Known Reliability and Performance

The package cost of this device is about the same as the combined cost of two discrete packages being replaced. Existing test programs are used in manufacturing as the dice are tested independently as discrete components. All production test fixtures, sockets, shipping trays, etc. are used with little to no modification. Since the existing discrete components are well characterized and field proven, the manufacturing yield, performance and reliability are predictable. While C4 interconnection provides significantly lower inductance, C4 on ceramic substrate does have temp cycle limitation. Enhancement, however, can be achieved by either selecting a thermal coefficient of expansion (TCE) compatible substrate material or other mechanical means. Similarly, thermal performance can be enhanced with the application of thermal cap and thermal grease. The cost, mechanical reliability enhancement and thermal improvement trade-off offer the customer a choice, which can be tailored to meet his specific operating environment without the need to replace the socket or modify the PC board.

V. CONCLUSION

This work demonstrates an effective packaging technology to achieve high performance VLSI through multichip package level interconnects within the confines of the same package outline and pin out of the existing discrete components being replaced. Combination of multilayer co-fired ceramic and thin film top layer provides a simplified approach to complex packaging at low cost.

The ability to integrate proven products of known performance, yield and products with dissimilar wafer fabrication processes, significantly reduces risks and engineering resources required for new product development. In some applications, direct "plug" compatibility of the multichip package with the existing package substantially eases customer's design constraints and frees up board space for additional system level integration. The flexibility and short development cycle, provided by this packaging technique, open new windows of opportunity for marketing. They also extend product life cycle while allowing design and product engineers an expedient way to preview next generation product performance inexpensively.

Potential advantages provided by multichip package level large scale integration not only improve time to market, but also offer the possibility of overcoming the initial high cost of new wafer fabrication process technology and lower yields associated with the next generation larger scale designs. Flip chip bonding further offers density, cost and performance advantages which, when combined with multichip package level interconnects, may offer a viable alternative to traditional silicon level "wafer scale integration" approach.

ACKNOWLEDGEMENT

The authors would like to express their sincere appreciation to S. Sen for doing the complex device characterization, to K. Hose who first designed the electrical circuits, to M. Aghazadeh for thermal measurements and analysis, and to A. Saddler and N. Holden for their contributions to illustrations and graphics. We would also like to thank R. Blish, B. Jahsman, C. Steidel, and R. Sundahl for many valuable discussions and critiques of this work, to L. Regis who had tirelessly championed for this program and finally to the entire C4 Technology team, without whose dedication and innovation, this work never would have been possible.

REFERENCES

[1] C.J. Bartlett et al, "Multi-Chip Packaging Design for VLSI-Based Systems", Proceeding, 37th ECC, May, 1987.

[2] J.D. Wilson, L.E. Mosley, M.A. Schmitt, and B.K. Bhattacharyya, "Electrical Characterization of a Multichip Module", Proceeding of Semicon/East Conference, 1988

[3] D.L. Brownewell, "Solder Bump Flip Chip Fabrication Using Standard Chip and Wire Integrated Circuit Layout", Part A, Proeeding, 1974 ISHM Conf., p.26.

[4] M. Coulton et al, "Adherence of Cr and Ti films deposit on Alumina, Quartz and Glass --- Testing and Improvement Of E-Beam Deposition Technique" RCA Review, June 1979.

[5] J. P. Cummings and L.S. Weinman, "Preparation and Characterization of Patterned Chromium/Copper Films Deposit on Alumina", Int. J. Hybrid Microelectronics, 4, 1, 1981.

[6] L.F. Miller, "Controlled Collapse Reflow Chip Joining", IBM J. Res. Develop., 13, 239 (1969).

[7] E.M. Davis et al, "Solid Logic Technology: Versatile, High Performance Microelectronics", IBM J. Res. Develop., 8, 102 (1964)

[8] L.F. Miller, "Joining Semiconductor Devices with Ductile Pads", ISHM Hybrid Microelec. Symp., Chicago, 1968

[9] L. Mosley, B.K. Bhattacharyya, D. Mallik, " A New Technique for Measuring the Inductance of Pin Grid Array Packages", to be presented at Japan IEMT Symposium, Nara, Japan, April 1989.

[10] D.Stys, B.K. Bhattacharyya, " A Technique for Monitoring the Consistency of Package and Dielectric Constant in the Manufacturing Process", Proceeding, 38th ECC, Los Angeles, May 1988.

[11] L.J. Giacoletto (Ed.), Electronics Designer's Handbook. New York, McGraw-hill, 1977, Ch 3, pp. 42.

[12] Y.C. Lee et al, "Internal Thermal Resistance of a Multi-chip Package Design for VLSI-BASED Systems", Proceeding, 38th ECC, Los Angeles, May, 1988.

[13] A. Bar-Cohen, "Thermal Management of Air and Liquid-Cooled Multi-Chip Modules," ASME National Heat Transfer Conference, Denver 1985.

[14] L.S. Goldman, "Geometric Optimization of Controlled Collapse Interconnections", IBM J. Res. Develop., 13, 251 (1969)

[15] P.A. Tobias, N.A. Sinclair and A.S. Van, "The Reliability of Controlled-Collapse Solder LSI Interconnections", IBM J. Res. Develop. 13, 260 (1969).

HIGH DENSITY MULTILEVEL COPPER-POLYIMIDE INTERCONNECTS

by

Stephen Poon, J. Tony Pan, Tsing-Chow Wang, and Brad Nelson
Microelectronics and Computer Technology Corporation
Austin, Texas

Abstract

A versatile, high performance, high density, five-layer-metal, planar interconnect technology has been developed with copper conductors and polyimide dielectric. Reliability is designed into this technology with emphasis on high thermal stability and low stress. Fifty micron pitch, controlled impedance interconnects suitable for high speed signal transmission with sub-100 picosecond rise time pulses have been fabricated. Features of the technology include pattern electroplating of copper with nickel overcoat, and polish planarization of conductor and via layer pairs. With surface planarity achieved by this approach, stacked vias can be utilized as anchors for bond pads and as thermal pillars for heat dissipation. The fabrication process developed for this technology and signal transmission properties of the interconnects will be presented in this paper.

Introduction

In order to fully expolit the improvements in speed and level of integration of advanced VLSI circuits in high performance systems, technology for high density interconnection of these devices must be developed. As circuit speeds are improved, propagation delay due to long signal transmission paths in conventional packaging has become a limiting constraint on system speed. The increased number of input/output connections as the level of integration is increased on VLSI devices have created a demand for higher interconnect density. Since control of simultaneous switching noise is necessary for maintaining adequate signal to noise ratio for proper device operation as high input/output circuits are used at high operating frequency, there is a greater need for interconnects with reduced lead inductance and line capacitance. In aerospace and military electronics, a high density, reliable, and light weight package can significantly enhance system functionality and performance. Hence, a compact, reliable, high density package with input/output connections in close proxmity is required to satisfy the demands in high speed, high performance electronic systems.

A number of technologies for fabrication of interconnects that can fulfill these requirements have been proposed. They include high density printed wiring boards [1], multi-layer ceramics [2], thick film hybrids [3,4], thin film hybrids [5], and wafer scale integration [6]. Comparisons of these competing VLSI packaging approaches have been made [7,8]. The thin film hybrid approach with multi-layer metal interconnect has been predicted to become the dominant technology for the first level of packaging in high performance systems [6-9]. This technology can achieve density and performance levels similar to wafer scale integration at much reduced risk because only good circuits and substrates are interconnected. It also allows flexibility for state of the art VLSI with mixed device technologies such as CMOS, bipolar, and GaAs to be used in the same system. Commercial application of this technology has been successfully demonstrated in one form on a high speed computer with a machine cycle time of six nanoseconds [10].

In this paper, the approach used in selecting the materials system employed in the development of a high density multi-chip interconnect is reviewed. The process approach and fabrication technologies chosen for developement in fabrication of the interconnects are then described with an emphasis on reliability. Finally, an evaluation of the electrical performance of the interconnects are presented.

Materials Consideration

The design of an interconnect system requires tradeoffs between competing requirements relating to density, delay, crosstalk, loss, noise, and heat dissipation. High density and short interconnect length must be carefully balanced against performance requirements limited by crosstalk, attenuation due to resistive loss, and efficient heat transfer to the next level of packaging. In order for a substrate to be cost effective, the technologies developed must not sacrifice yield in meeting these objectives. Materials and processes which can effectively control variables that affect conductor and interlayer dielectric thicknesses and uniformities are desired since they determine process yield and the critical transmission line properties. The performance goal chosen in the development of this technology are 50 microns pitch interconnect lines with 50 ohms characteristic impedance. In the following section, motivations which affected the choice of materials used in this work are discussed.

Three basic components are required in fabrication of an interconnect system. These are materials for the substrate, the interlayer dielectric, and the conductor. Cost, ease of processing, thermal and chemical stability are considered in the selection of each material. To ensure system integrity, the materials must also be compatible with each other as well as the VLSI devices. In general, closely matched TCE materials should be used where possible to minimize thermal stress for enhanced reliability. High thermal conductivity materials are useful for transferring heat to the next level of packaging. Low density materials are preferred in order to reduce package weight in aerospace applications. In the case of the substrate, additional areas that must be considered are ease of packaging, mechanical strength, and stability. A comparison between major substrate materials in use are shown in Table 1.

Materials with low dielectric constant are desirable for interlayer dielectric since propagation delay is proportional to the sqaure root of dielectric constant. Lower dielectric constant also results in reduced line capacitance. For a given characteristic impedance, dielectric thickness can be reduced to allow higher line density while keeping aspect ratio constant. All these contribute to improve both speed and performance of the interconnects. Chemical and thermal stability are also important in the evaluation of interlayer dielectric due to the chemical exposure and the high temperatures that the interconnects might be subjected to in subsequent processing steps and in the operating environment. Other material dependent properties that must also be considered include adhesion, stress, and loss tangent. A comparison between alumina, epoxy, polyimide, teflon, and silicon dioxide are shown in Table 2.

Conductor metal used in a high density interconnect must have both excellent electrical and thermal conductivity. Low resistivity is critical in high density systems since it minimizes the metal's cross sectional area required for a given design rule. High thermal conductivity is useful if heat dissipated from the devices is to be conducted away through the interconnect and its substrate. Interlayer contact resistance, electromigration, and resistance to thermal and chemical degradation must also be considered in the selection of conductor materials. Table 3 compares the bulk resistivity, thermal conductivity, and TCE of metals in use for multilayer circuits.

In view of the above considerations, Cu and polyimide are selected as the conductor and interlayer dielectric materials for the fabrication of interconnects. Copper is selected due to its low resistance, high thermal conductivity, availability, and low cost. Polyimide is a suitable interlayer dielectric because of its low dielectric constant, high thermal and chemical stability, good planarization characteristics, and ease of processing for achieving the dielectric thickness range for controlled impedance requirements. It is noted Teflon has a lower dielectric constant compared to polyimide and with otherwise similar properties. The principle constraint which prevented its application is due to the poor adhesion it has with most materials. The copper-polyimide interconnects have been fabricated on silicon, ceramic, and metal substrates. Silicon is used due to its availability at low cost and good surface finish. Ceramic is used where hermetic seal and better TCE match in the package between the substrate, interlayer dielectric, and conductor materials are needed to minimize substrate warpage and stress. Metal substrates are required for high power circuits where heat dissipation is the limiting constraint.

Process Approach

A variety of approaches for fabrication of high density interconnects have been reported [5,9,11-20]. A summary of these different approaches are shown in Table 4. In most of these approaches, polyimide is chosen as the interlayer dielectric. Copper and aluminum are the most common conductors, on either a ceramic or silicon substrate. It is also interesting to note

258

that photo-sensitive polyimides are being pursued strongly in Japan while standard polyimides are more commonly accepted in the US. The benefits and limitations of these approaches are evaluated, then a new process approach which can effectively integrate the advantages of these technologies is developed to fabricate a high reliability, high density, controlled impedance interconnect.

One of the potential advantages of multi-chip interconnect technology can be reliability enhancement due to the reduced number of connections compared to conventional packaging. However, as an increased number of components are mounted on a multi-chip module, interconnect reliability becomes more important because of the value of components added to the substrate. Some of the more critical areas related to reliability that need to be studied are adhesion, stress, and planarization.

Interaction and adhesion of copper on polyimide has been studied extensively [21-23] by IBM. Poor macroscopic adhesion at the copper-polyimide interface was generally reported and attributed to weak interaction between copper and the underlying polyimide. Incomplete interface formation and islanding of copper on polyimide was observed by x-ray spectroscopy [22]. Migration of copper-rich precipitates into polyimide which can potentially change the dielectric properties of polyimide was also demonstrated by transmission electron microscopy for a polyimide on copper interface [23]. Hence, an adhesion/diffusion barrier layer must be placed between copper and polyimide for long term reliability of copper-polyimide interconnects. A variety of metal-polyimide systems have been under investigation with particular focus on Cr [24], Ti [25], Ni [26], and Al [27]. All these materials have better metal-polyimide adhesion compared to copper, with Cr and Ti showing the best adhesion. Recently, a study in the degradation of adhesion between Cr- and Ti-polyimide interfaces after thermal cycling was reported [28]. The Ti-polyimide interface was found to degrade more readily than the Cr-polyimide interface, mainly due to interaction between the metal and moisture absorbed by the polyimide. Similar results were also obtained at MCC when Cr and Ti adhesion to polyimide was compared using a 90 degree peel test. Thus, Cr is the preferred adhesion layer for copper-polyimide interconnects. To prevent delamination between copper and polyimide on the sidewalls of the conductor features, a nickel overcoat can be used as a barrier between polyimide and copper surfaces. The differences in adhesion between a copper feature overcoated with brown oxide and a similar feature overcoated with Cr/Ni after environmental test is shown in Figure 1.

Film stress should be minimized since it causes substrate warpage and can result in delamination and reliability problems. Generally, film stress at room temperature is composed of intrinsic stress and thermal stress. Intrinsic stresses of plated metals and polyimides have been measured and were found to be insignificant compared to the thermal stress. Thermal stress due to TCE mismatch between the substrate, metals, and polyimide contribute almost exclusively to the total stress and must be controlled since repeated thermal cycling over a range of more than 300 degrees C is required in the fabrication of a multi-layer interconnect with polyimide dielectric. The delamination force here for the same level of stress is much greater than that encountered in an IC process due to the greater thickness of films required for the interconnect. In-situ measurements of stresses generated in polyimide films have been studied and were reported to be in the range of 60 MPa at room temperature [29]. Recently, low TCE polyimides which can better match the TCE of silicon and ceramic substrates were introduced by Hitachi [30] and DuPont [31]. The stress-temperature relationship between a conventional polyimide material and a low TCE polyimide material on a silicon substrate have been measured and is shown in Figure 2. Significant reduction in stress is observed for the low TCE material. It should be noted that the relaxation temperature of the low TCE material is higher than the cure temperature of 350 C. This is also more desirable since TCE usually increases dramatically above the glass transition temperature for many materials. A detailed analysis of stress formation in polyimides and metals used in the fabrication of high density interconnects will be presented in a separate paper [32]. Other properties that must be considered in the selection of polyimide materials should also include self-adhesion, modulus, elongation, moisture absorption, and its effect on the dielectric properties. Both conventional and low TCE polyimides are used in the fabrication of the substrates. An oxygen plasma treatment of the cured low TCE polyimide surface is generally recommended by the vendors to achieve good adhesion between polyimide films.

Planarization is one of the most critical considerations in fabrication of multi-layer interconnects. As the aspect ratio (height to linewidth) increases due to the requirement of high density, planarization has become an increasingly important issue that needs to be resolved. There are generally two approaches used for fabrication of multilayer interconnects. They are the staggered via approach and the stacked via approach. A schematic comparison between these two approaches is shown in Figure 3. In general, the latter approach is more desirable for density and reliability reasons since the vias can be stacked and filled with metal. However, the staggered via approach is more commonly used due to ease of processing. A brief overview of planarization techniques suitable for multi-level metallization has been presented [33]. These techniques include thermal flow and etchback after spin-on for dielectric planarization; and lift-off, bias sputtering, and selective deposition for metal planarization. Among these, selective deposition of metal offers the shortest route for true planarization of an interconnect structure. However, selective deposition has been pursued since the early 1980's but has not yet been successfully demonstrated in a high volume, commercial application.

An alternative approach to obtain the planarization necessary for stacked vias was demonstrated [4] by electroplating of pillars in conjunction with mechnical polishing. In multi-chip interconnects, stacked vias can also be utilized as thermal pillars and as anchors for bonding pads. Since polish planarization offers the most direct approach for planarization of interconnects, it is developed to meet the requirements for interconnect fabrication. An additional advantage with the polishing approach stems from the small area ratio of pillar to field in most pratical applications. Since the pillar area is relatively small compared to the field area, the polishing rate is slowed considerably as planarization is achieved. This property of polishing is extremely useful in the control of dielectric thickness. A SEM micrograph which demonstrates the planarity that can be achieved with the polishing technique is shown in Figure 4.

Fabrication

The fabrication process developed for high density copper-polyimide interconnects at MCC have been presented [34]. The features of this approach include additive electroplating of conductor features, complete isolation between copper-polyimide interfaces, use of matched TCE materials, and stacked via planarization. It is believed that these features can contribute significantly to the reliability of the interconnects.

A schematic of the process flow is illustrated in Figure 5. Polyimide is first spin coated on a ceramic substrate and polished. This polishing step is done to planarize and expose the metal contacts on a co-fired ceramic when it is used as a substrate for fabrication of the high density interconnects. Otherwise, it is done as a dummy step. Adhesion layer and plating interconnect are sputtered on the polyimide surface. In this case, a thin layer of Cr/Cu/Ti is used. Photoresist is then spin coated and exposed to define the pattern for conductor plating. Electrolytic plating is used to obtain 5 micron thick copper conductors. Electroplating was selected because the technology is mature and potentially low cost. Electrolytic bath chemistries are in general better understood and can be better controlled compared to electroless baths. The concern with this approach is plating uniformity over a large area since electrolytic bath deposition can be limited and controlled by the distribution of current density over the substrate area and may be influenced strongly by pattern distribution. However, metal uniformity can be improved by other techniques which include using additives in the electrolyte and pluse plating. After conductor plating and stripping of the photoresist, a thicker layer of photoresist is spin coated and patterned on the conductor features to plate up pillars. Again, electrolytic copper plating is used to obtain the appropriate pillar thickness. A cleaning step is done prior to pillar plating to remove the copper oxide on the surface to ensure low contact resistance and good adhesion. After photoresist strip, a 1 micron thick Ni overcoat can be applied over the copper features to prevent corrosion and delamination problems. The sputtered layer is stripped by wet etch. A SEM micrograph for part of a via chain at this step is shown in Figure 6.

To complete fabrication for one layer of interconnect, polyimide dielectric is spin coated on the metal features to partially planarize the surface. Both low TCE and conventional polyimide materials are used. Since planarization is not entirely dependent on the coating process in this approach, high dielectric planarity is not critical as long as all the plated features are covered and spaces between features are filled with polyimide. Multiple spin and hot plate bakes are required to achieve the dielectric thickness and thickness control desired for this application. The degree of planarization after polyimide deposition is strongly dependent on both aspect ratio and feature proximity effects. An example of the planarization obtained after polyimide deposition is shown in Figure 7. Finally, the features are polished to expose the pillars and planarize the surface for contact to the next level of interconnect. The Ni overcoat on top of pillars is also polished off to minimize contact resistance.

Included in the test vehicle for the development of this technology are high density transmission lines with 50 micron pitch intended to achieve 50 ohm impedance. A schematic of the stripline structure is shown in Figure 8. Five layers of metal, one each for power and ground, two for signal, and a top bonding layer were designed and have been fabricated. A SEM micrograph of a completed interconnect structure is shown in Figure 9. Interdigit, serpentine, and via chain structures were also designed and placed in the test vehicle for use as a yield monitor. Photographs of these structures fabrication on a mesh ground plane are shown in Figure 10.

Electrical Performance

The multilayer interconnect lines fabricated with the MCC medium film process technology described above include both microstrip and stripline with one bottom or both top and bottom reference planes. The typical cross sections of the copper lines are designed to be 15 by 5 microns with a pitch of 50 micron and a dielectric thickness of 15 micron. This gives a DC resistance of 2.3 ohms/cm and a characteristic impedance of 51.6 ohms and a saturated near end crosstalk (NEXT) of 2.8% for coupled strip lines.

The impedance and crosstalk were derived from the calculated capacitance matrix using a dielectric constant of 3.5 for polyimide through finite element solutions based on the general method for modeling lossy tramsmission lines developed at MCC [35]. MCC has also developed methods of characterizing lossy transmission lines [36]. This includes measurments of the high frequency impedance, crosstalk, delay, and losses.

Figures 11 and 12 show the high frequency impedance, time domain transmission and crosstalk measurements of microstrip lines with one bottom reference plane. The impedances measured at the beginning of the line (point A) of figure 11 is about 55 ohms for the specific substrate. Figure 12 shows the transmission signal, far end (FEXT) and near end (NEXT) crosstalk for microstrip lines on both solid and meshed ground planes. The FEXT and NEXT are 1.5% and 5% respectively for this specific case. Crosstalk is slightly higher for the meshed checkerboard patterned ground plane. The propagation delay is about 62.4 ps/cm and attenuation is about 2.8%/cm, which are consistent with a dielectric constant of 3.5 and an attenuation constant of $R/(2*Z)$. Where R is the line DC resistance and Z is the impedance.

Similar results for strip lines with two reference planes have also been measured, and the agreements of impedance and crosstalk between experimental results and theoretical calculations are generally within 10%.

The high speed test results demonstrated that the medium film technology will be able to support sub-100 ps rise time high performance systems.

Conclusions

Technology developed for fabrication of a high density multi-chip interconnect has been demonstrated. Controlled impedance lines with 50 micron pitch suitable for high speed signal transmission have been obtained. The materials considerations, process approach, and fabrication sequence developed are presented along with the electrical parameters measured on the interconnects. Initial reliability testing data of the interconnects have been obtained with no significant degradation observed.

Acknowledgements

The authors wish to acknowledge the efforts of all their colleagues in the fabrication team who contributed to the development of process technologies described in this paper. We also would like to thank members of the design, testing, and materials analysis group who provided valuable contributions to this project. One of the authors (TCW) would particularly like to acknowledge L. Smith and U. Ghoshal for valuable discussion and S. Sommerfelt for measurement of the high speed test results.

References

1. J.R. Bupp, L.N. Challis, R.E. Ruane, and J.P. Wiley, "High-Density Board Fabrication Techniques", IBM J. Res. Dev., vol. 10, pp. 306-317, May 1982.

2. A.J. Blodgett and D.R. Barbour, "Thermal Conduction Module: A High-Performance Multilayer Ceramic Package", IBM J. Res. Dev., vol. 26, pp. 30-36, January 1982.

3. D.E. Pitkanen, J.P. Cummings, and R.E. Acosta, "Status of Copper Thick Film Hybrids", Solid State Technology, vol. 23, pp. 141-146, October 1980.

4. S. Lebow, "High Density/High Speed Multi-Chip Packaging", Proceedings of the 6th International Electronic Packaging Conference, pp. 417-423, November 1986.

5. R.J. Jensen, J.P. Cummings, and H. Vora, "Copper/Polyimide Materials System for High Performance Packaging", Proceedings of the 34th Electronic Component Conference, pp. 73-81, 1984.

6. S. Westbrook and W. Lukaszek, "A Multilevel Interconnect Test Vehicle for Wafer Scale Integration", Proceedings of the 3rd VLSI Multilevel Interconnect Conference, pp. 373-379, June 1985.

7. M. Terasawa, S. Minami, and J. Rubin, "A Comparison of Thin Film, Thick Film, and Co-Fired High Density Ceramic Multilayer with the Combined Technology: T&T HDCM (Thin Film and Thick Film High Density Ceramic Module), Proceedings of ISHM Symposium on Microelectronics, pp. 607-615, October 1983.

8. C.A. Neugebauer and R.O. Carlson, "Comparsion of Wafer Scale Integration with VLSI Packaging Approaches", IEEE Transactions on Components, Hybrids, and Manufacturing Technology, vol. 10, no. 2, pp. 184-189, June 1987.

9. C.W. Ho, VLSI Electronics: Microstructure Science, vol. 5, pp. 103-143, Academic Press, New York, 1982.

10. H. Murano, Y. Narita, T. Watari, H. Matsuo, and M. Okano, "Water Cooling Multichip Packages", Nikkei Electronics, pp. 242-266, June 1985.

11. S.H. Wen and J.I. Kim, "Thin Film Wiring for Intergated Electronic Packages", Materials Research Society Symposium Proceedings, vol. 108, pp. 81-87, 1988.

12. T.A. Lane, F.J. Belcourt, and R.J. Jensen, "Electrical Characteristics of Copper/Polyimide Thin Film Multilayer Interconnects", Proceedings of the 37th Electronic Components Conference, pp. 614-622, 1987.

13. E. Bogatin, "Beyond Printed Wiring Board Densities: A New Commercial Multichip Packaging Technology", Proceedings of Nepcon East '87, pp. 208-226, June 1987.

14. C.J. Bartlett, J.M. Segelken, and N.A. Teneketges, "Multi-Chip Packaging Design for VLSI-Based Systems", Proceedings of the 37th Electronic Components Conference, pp. 518-525, 1987.

15. H. Stopper, "A Wafer with Electrically Programmable Interconnections", Proceedings of IEEE Intl. Solid-State Circuits Conference, pp. 268-269, 1985.

16. C.C. Chao, K.D. Scholz, J. Leibovitz, M.L. Cobarruviaz, and C.C. Chang, "Multilayer Thin-Film Substrate for Multichip Packaging", Proceedings of the 38th Electronics Components Conference, pp. 276-281, 1988.

17. S. Lebow, "High Density/High Speed Multi-Chip Packaging", Proceedings of the 6th International Electronics Packaging Conference, pp. 417-423, 1986.

18. T. Ohsaki, T. Yasuda, S. Yamaguchi, and Taichi Kon, "A Fine-Line Multilayer Substrate with Photo-Sensitive Polyimide Dielectric and Electroless Copper Plated Conductors", Proceedings of the 3rd IEEE/CHMT International Electronics Manufacturing Technology Symposium, pp. 178-183, October 1987.

19. Technical Bulletin, NTK Advanced Product Group, "NTK High Density and High Precision Wiring Ceramics", November 1988.

20. H. Takasago, M. Takada, K. Adachi, A. Endo, K. Yamaha, T. Makita, E. Gofuku, and Y. Onishi, "Advanced Copper/Polyimide Hybrid Technology", Proceedings of the 36th Electronic Components Conference, pp. 481-487, 1986.

21. P.O. Hahn, G.W. Rubloff, J.W. Bartha, F. Legoues, R. Tromp, and P.S. Ho, "Chemical Interactions at Metal-Polymer Interfaces", Materials Research Society Symposia Proceedings, vol. 40, pp. 251-263, 1985.

22. R.C. White, R. Haight, B.D. Silverman, and P.S. Ho, "Cr- and Cu-Polyimide Interface: Chemistry and Structure", Appl. Phys. Lett., vol. 51, no. 7, pp. 481-483, August 1987.

23. S.P. Kowalczyk, Y.H. Kim, G.F. Walker, and J. Kim, "Polyimide on Copper: The Role of Solvent in the Formation of Copper Precipitates", Appl. Phys. Lett., vol. 52, no. 5, pp. 375-376, February 1988.

24. N.J. Chou, D.W. Dong, J. Kim, and A.C. Liu, "An XPS and TEM Study of Intrinsic Adhesion Between Polyimide and Cr Films", J. Electrochem. Soc., vol. 131, no.10, pp. 2335-2340, October, 1984.

25. F.S. Ohuchi and S.C. Freilich, "Metal Polyimide Interface: A Titanium Reaction Mechanism", J. Vac. Sci. Technol. A, vol. 4, no. 3, pp. 1039-1045, 1986.

26. N.J. Chou and C.H. Tang, "Interfacial Reaction During Metallization of Cured Polyimide: An XPS Study", J. Vac. Sci. Technol. A, vol. 2, no. 2, pp. 751-755, 1984.

27. L. Atanasoska, S.G. Anderson, H.M. Meyer, Z. Lin and J.H. Weaver, "Aluminum-Polyimide Interface Formation: An X-Ray Photoelectron Spectroscopy Study of Selective Chemical Bonding", J. Vac. Sci. Technol. A, vol. 5, no. 6, pp. 3325-3333, 1987.

28. K. Seshan, S.N.S. Reddy, and S.K. Ray, "Adhesion of Thin Metal Films to Polyimide", Proceedings of the 37th Electronic Components Conference, pp. 604-608, 1987.

29. C. Goldsmith, P. Geldermans, F. Bedetti, and G.A. Walker, "Measurements of Stresses Generated in Cured Polyimide Films", J. Vac. Sci. Technol. A, vol. 1, no. 2, pp. 407-409, 1983.

30. S. Numata, T. Miwa, Y. Misawa, and D. Makino, "Low Thermal Expansion Polyimides and Their Applications", Materials Research Society Symposium Proceedings, vol. 108, pp. 113-124, 1988.

31. B.T. Merriman and J.D. Craig, "New Low Coefficient of Expansion Polyimide for Inorganic Substrates", presented in the 2nd DuPont Symposium on High Density Interconnect Technology, October 1988.

32. J.T. Pan and S. Poon, "Film Stress in High Density Thin Film Interconnect", to be presented at Materials Research Society Symposia, April 1989.

33. A.N. Saxena and D. Pramanik, "Planarization Techniques for Multilevel Metallization", Solid State Technology, vol. 29, no. 10, pp. 95-100, 1986.

34. J.T. Pan, S. Poon, and B. Nelson, "A Planar Approach to High Density Copper-Polyimide Interconnect Fabrication", Proceedings of the 8th International Electronics Packaging Conference, pp. 174-185, November 1988.

35. U. Ghoshal, T.C. Wang, and L.N. Smith, "Modeling of Advanced Multichip/Wafer Scale Interconnects", Proceedings of the 5th VLSI Multilevel Interconnect Conference, pp. 265-272, 1987.

36. T.C. Wang, S. Sommerfeldt, and L. Smith, "High-Density Controlled-Impedance Copper/Polyimide Interconnects", Proceedings of the 6th VLSI Multilevel Interconnect Conference, pp. 252-260, 1988.

37. S.M. Sze, Physics of Semiconductor Devices, John Wiley, 1981, pp. 851-852.

38. W.C. O'Mara, "Substrates for Multichip Modulus", presented in the 2nd DuPont Symposium on High Density Interconnect Technology, October 1988.

39. E.A. Brandes, Editor, Smithells Metals Reference Book, Butterworths, 1983, Chapter 14, pp. 1-2.

40. D.W. Wang, "Advanced Materials for Printed Circuit Boards", Materials Research Society Symposium Proceedings, vol. 108, p. 130, 1988.

41. D.A. Doane, "Materials and Fabrication Techniques for Packages and Circuit Boards at Microwave Frequencies: An Overview", Proceedings of the 8th International Electronic Packaging Conference, p. 651, November 1988.

TABLE 1

Properties of Substrate Materials

	TCE (ppm/K)	Thermal Conductivity (W/cm K)	Bending Strength (Kg/mm2)
GaAs[37]	6.86	0.46	
Silicon[37]	2.6	1.5	
Alumina(96%)[19]	7.7	0.2	33
Silicon Carbide[38]	2.8	2.7	45
Beryllia[38]	4.7	2.36	20
Aluminum Nitride[19]	3.8	1.4	50
Aluminum[39]	23.5	2.38	
Copper[39]	17.0	3.97	

TABLE 2

Properties of Dielectric Materials

	Dielectric Constant	Dissipation Factor	Tg (C)	TCE (ppm/K)
Polyimide[31]	3.3	.002	>320	40
Epoxy[40]	4.0	.02	130	60
Alumina(96%)[19]	9.3	.0003	>1500	7.7
Teflon[41]	2.1	.00045		
Silicon dioxide[37]	3.9			0.5

TABLE 3

Properties of Conductor Materials [39]

	Resistivity (micro ohm cm)	Thermal Conductivity (W/cm K)	TCE (ppm/K)
Aluminum	2.67	2.38	23.5
Copper	1.69	3.97	17.0
Silver	1.63	4.25	19.1
Gold	2.20	3.16	14.1
Tungsten	5.4	1.74	4.5
Molybdenum	5.7	1.37	5.1
Nickel	6.9	0.89	13.3

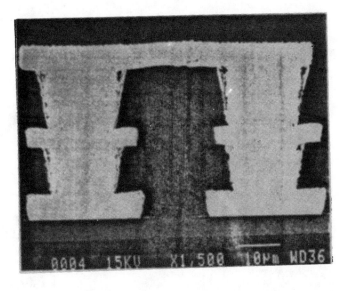

(a)

(b)

Figure 1. Comparison of polyimide adhesion to copper with (a) brown oxide and (b) nickel overcoats

TABLE 4

Comparison of Fabrication Approaches
for High Density Interconnects

	Substrate	Metal(W,T,P)	Dielectric(T, Via Pitch)	Adhesion Layer	Planarization
IBM[9,11]	Ceramic	Copper(8,5.5,25)	Polyimide(6,>6)	Cr, bare sidewall	Stagger vias
Honeywell[5,12]	Ceramic	Copper(25,5,125)	Polyimide(25,>25)	Cr, bare sidewall	Stagger vias
Raychem[13]	Silicon	Aluminum(25,5,75)	Polyimide(10,100)	Not require	Stagger vias
ATT[14]	Silicon	Copper(10,2,20)	Polyimide(10,>10)	Cr/Ti, Ni sidewall	Ni via fill
Mosaic[15]	Silicon	Aluminum(10,1,22)	Oxide(1,22)	Not require	Stagger vias
HP[16]	Ceramic	Copper(15,6,35)	Polyimide(15,20)	Cr, bare sidewall	Stagger vias
Augat[17]	Multiple	Copper(100,25,250)	Polyimide(50,>75)	None, 260 C	Polishing
NTT[18]	Ceramic	Copper(25,6,50)	Photo PI(23,>50)	Cr, bare sidewall	Stagger vias
NTK[19]	Ceramic	Multiple(25,5,65)	Photo PI(25,90)	Include in metals	Metals via fill
Mitsubishi[20]	Ceramic	Copper(50,5,100)	Photo PI(10,>50)	None	Stagger vias

265

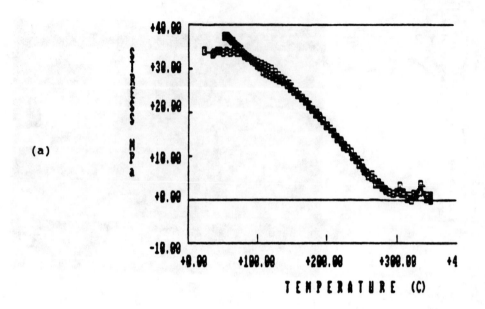

(a)

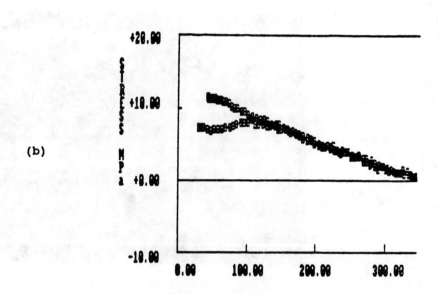

(b)

Figure 2. Stress-temperature relationships between (a) conventional
polyimide and (b) low TCE polyimide on a silicon substrate

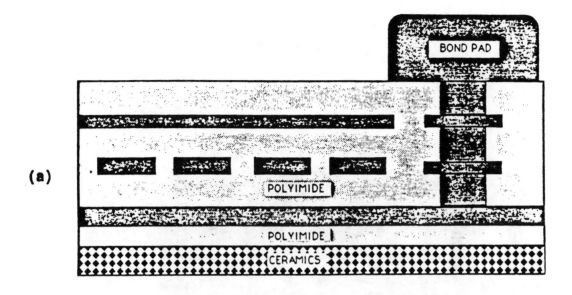

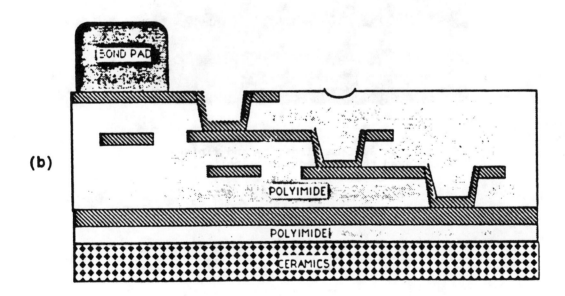

Figure 3. Comparison between (a) stacked pillar and
(a) staggered via approaches for multilayer interconnect

Figure 4. Surface planarity and finish
achieved through mechanical polishing

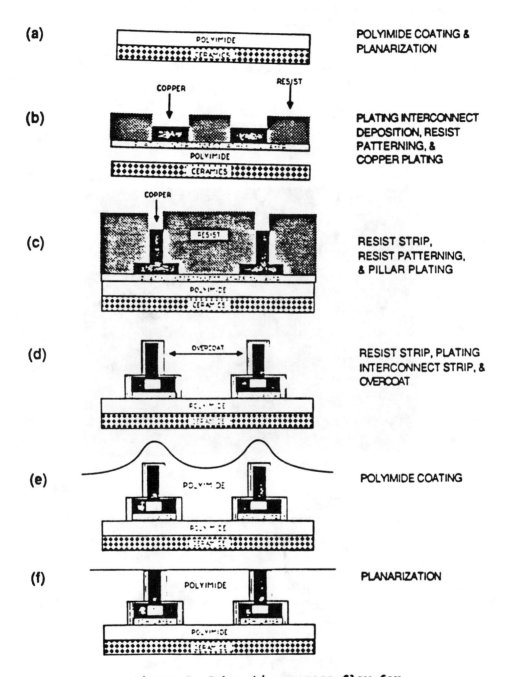

(a)   POLYIMIDE   CERAMICS   POLYIMIDE COATING & PLANARIZATION

(b)   COPPER   RESIST   POLYIMIDE   CERAMICS   PLATING INTERCONNECT DEPOSITION, RESIST PATTERNING, & COPPER PLATING

(c)   COPPER   RESIST   POLYIMIDE   CERAMICS   RESIST STRIP, RESIST PATTERNING, & PILLAR PLATING

(d)   OVERCOAT   POLYIMIDE   CERAMICS   RESIST STRIP, PLATING INTERCONNECT STRIP, & OVERCOAT

(e)   POLYIMIDE   POLYIMIDE   CERAMICS   POLYIMIDE COATING

(f)   POLYIMIDE   POLYIMIDE   CERAMICS   PLANARIZATION

Figure 5. Schematic process flow for
multilayer interconnect fabrication

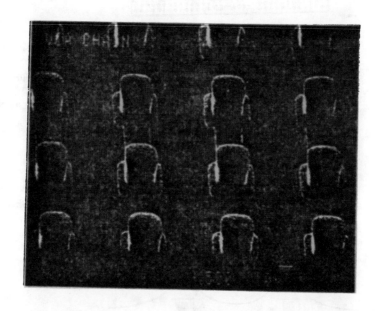

Figure 6. Electroplated pillar and conductor on substrate

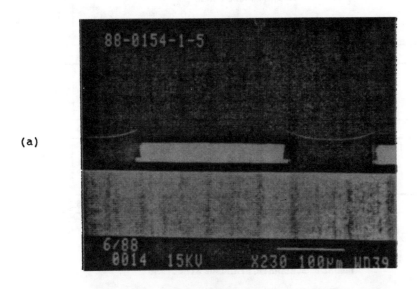

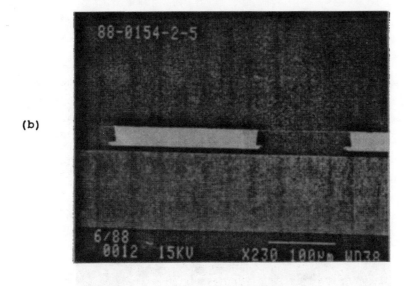

Figure 7. Planarization of conductor features after
(a) polyimide deposition and (b) mechanical polishing

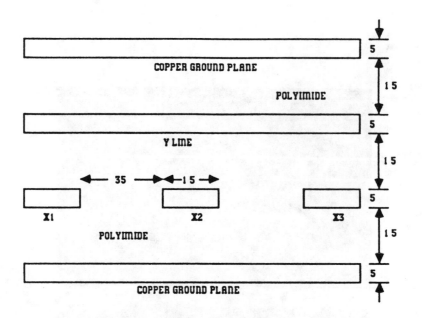

FIGURE 8. SCHEMATIC STRIPLINE STRUCTURE

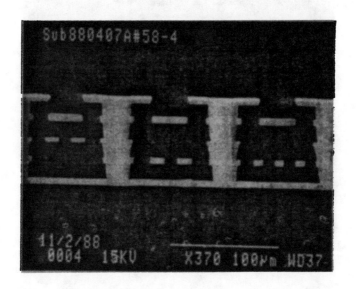

Figure 9. Five-layer metal interconnect

(a)

(b)

(c)

Figure 10. (a) Interdigit, (b)serpentine,
and (c) via chain structure

273

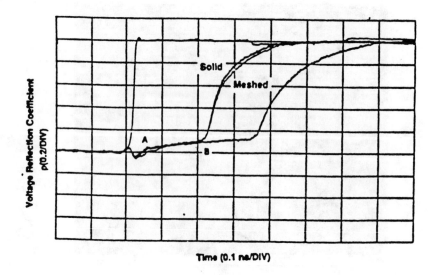

Figure 11. TDR of two different length lines measured with Hypres PSP-1000. Points A and B correspond to the beginning and end of an 1.425 cm transmission line

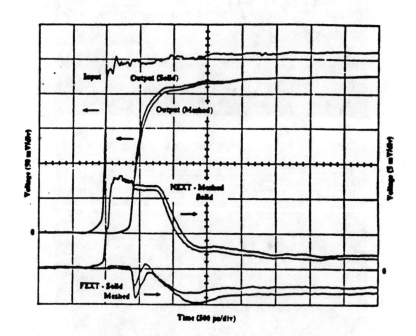

Figure 12. A comparison of the crosstalk between two 6 cm long 50 ohms terminated microstrip lines on meshed and solid ground planes. Crosstalks are in 10X scale.

274

High Density Hybrid IC Module by MCPH Technology

Mitsuyuki TAKADA, Tetsuro MAKITA, Eishi GOFUKU
Kurumi MIYAKE, Atsushi ENDO, Kohei ADACHI
Yoshiyuki MORIHIRO, and Hayato TAKASAGO

Materials & Electronic Devices Laboratory
Mitsubishi Electric Corporation

1-1, Tsukaguchi Honmachi 8-Chome
Amagasaki, Hyogo, 661 Japan

ABSTRACT

A unique copper/polyimide technology, MCPH (Mitsubishi Copper Polyimide Hybrid), which consists of wet-metallized copper conductors, air-fired RuO_2 (ruthenium-oxide) resistors and polyimide insulators, was developed and reported by us. [1][2] The MCPH technology was further improved and advanced to boost the packaging density of hybrid IC modules.

Recently, miniaturization of hybrid IC modules is desired for application to various electronic equipment, such as VTRs, CD players, etc. In addition, miniaturization of the hybrid IC module results in an increase in the number of hybrid IC modules per substrate to be processed, and can lower the production cost effectively. Therefore, our major effort of the recent MCPH development has been focused on miniaturizing hybrid IC modules by realizing fine-line copper conductors and/or fine-sized thick film resistors.

For further studies of the MCPH technology, a TEG (Test Element Group) was designed for a 106 mm⁻ substrate. Various process-parameters were examined and determined through the studies using this TEG. Consequently, both 20 μm line & space copper conductors and 300 μm-length or –width RuO_2 resistors were made possible. Moreover, an advanced laser trimming technology for the fine-sized RuO_2 thick film resistors was studied and developed.

Also, an effort was made to stabilize each process and confirm its reliability.

Using the advanced MCPH technology, an MCPH with the same circuit as a conventional hybrid IC was designed and completed in order to estimate the MCPH effectiveness. As a result, the packaging density is approximately 10 times more than the conventional one. Moreover, in the case of this newly-developed MCPH, 48 pieces of the MCPHs can be simultaneously processed using a 106 mm⁻ substrate. Therefore, the advanced MCPH technology is very suitable for mass production of high density hybrid IC modules.

INTRODUCTION

The MCPH was fundamentally constructed of wet-metallized copper conductors, polyimide insulators and RuO_2 thick film resistors. Compared to conventional thick film hybrid ICs, the main features are listed as follows;

- Fine-lining by photolithography
- Low resistivity conductors
- Low dielectric constant insulators

In addition, the reliable, conventional RuO_2 thick film resistors can be fully utilized in combination with copper conductors, so that the wide resistivity range and excellent characteristics of resistors can be realized as well as the conventional ones.

The object of our studies was to improve and advance the MCPH processes, and consequently to boost the packaging density of hybrid ICs more and more. By determinating each process parameter precisely, much finer patterns could be realized. Applying these improved and advanced processes to an actual circuit hybrid IC, the effectiveness of the MCPH was estimated.

The MCPH technology has realized remarkable improvements of hybrid ICs, and further progress is expected as shown in Fig. 1.

STUDIES

The cross-sectional structure of the MCPH is shown in Fig. 2. Studies for the improved and advanced MCPH were conducted mainly regarding the following items.

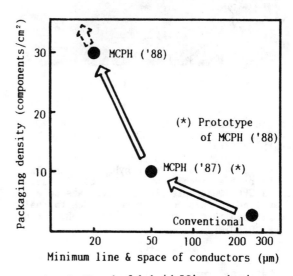

Fig. 1. Trend of hybrid IC's packaging.

1. Fine conductor line formation down to 20 μm line & space.
2. Miniaturization of thick film resistors.
3. Advanced laser trimming technology for the miniaturized thick film resistors.

For these studies, a TEG was designed for a 106 mm⬚ substrate as shown in Fig. 3, and by using this TEG, the process parameters were closely

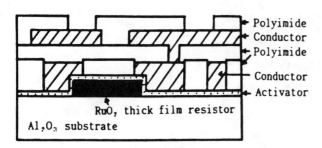

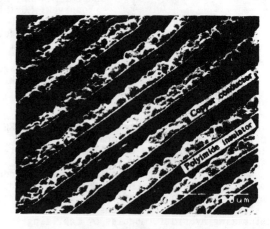

Fig. 2. Cross-sectional structure of MCPH.

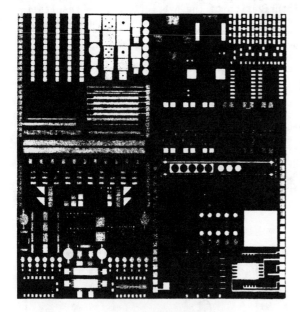

Fig. 3. MCPH TEG.

examined and determined. This TEG contains 20 μm line & space conductors, 300 μm minimum-width or -length RuO$_2$ thick film resistors and etc.

The improved and advanced processes were applied to an actual circuit hybrid IC for VTRs and their effectiveness was evaluated. In addition, a QFP (Quadruple Flat Package) was selected and applied to this actual circuit MCPH to realize fine pitched interconnections with a PWB. (Printed Wiring Board)

PROCESSES

Fine Pattern Formation

The first metallization layer is full-additively wet-deposited by using an activator paste containing glass-frits, which is screen-printed and fired at 630 °C. A photoactive polyimide layer is coated onto the fired activator paste and patterned by photolithography. To optimize the glass-frits particle size of the activator paste and the firing condition, the surface of the fired activator paste layer have become rather suitable for patterning the photoactive polyimide. In addition, the exposure- and/or developing-conditions of the photoactive polyimide were exactly determined and controlled.

As a result, 20 μm line & space conductors could be realized as shown in Fig. 4.

Fig. 4. First metallization layer. (20 μm line & space).

On the other hand, the second metallization layer is formed by a semi-additive copper deposit method. In this process, especially the electroplating conditions were closely investigated to attain a good uniformity of deposited copper thickness. To optimize the electroplating conditions as well as others, such as the patterning conditions of a photoresistant material, 20 μm line & space conductors with a good thickness uniformity have also become available as shown in Fig. 5.

Fig. 5. Second metallization layer. (20 μm line & space).

Miniaturization of Thick Film Resistors

In general, hybrid ICs contain many resistors as components, therefore, to boost the packaging density, it is very effective to reduce the total footprint area of the thick film resistors. The MCPH technology realizes a termination between a RuO$_2$ thick film resistor fired at 850°C and a wet-metallized copper conductor formed by using photolithography, therefore, the distance between terminations of both sides can be shortened campared with conventional thick film systems. Moreover, in the MCPH processes, thick film resistors are printed on a relatively flat surface of a substrate before forming conductors, so that a high resolution printing of RuO_2 resistive pastes is easily achieved.

By using the newly designed TEG, various sized resistors and/or terminations were formed as shown in Fig. 6, and their characteristics were closely investigated. Fig. 7 shows the proportionality between the resistor length and the resistance value, which represents the excellent ohmic contact which is formed between the RuO_2 thick film resistor and the copper conductor. Also, the other characteristics, such as voltage-current characteristics, were evaluated and excellent results were obtained.

Fig. 6. Various sized RuO_2 resistors in MCPH TEG.

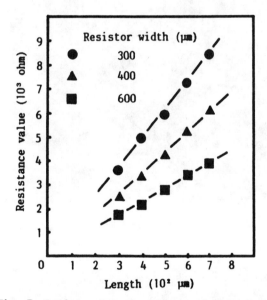

Fig. 7. Resistor length versus resistance value.

New Laser Trimming

Thick film resistors must be trimmed for adjusting the resistance value to the designed value. The conventional laser trimming method employs laser cutting, so that its application to miniaturized MCPH thick film resistors is very difficult for the following reasons.

1) Ordinary beam diameter of YAG laser is approximately 60 μm, which is not enough to cut the miniaturized resistors of the MCPH precisely.
2) The remaining width of a resistor becomes very small, so that the power consumption level is extremely lowered.

To solve the above problems, a new laser trimming technology was developed and applied to the MCPH thick film resistors. The newly developed laser trimming technology is called LARCS (Laser Activated Resitivity Control System), which adjusts the resistance value without cutting. LARCS employs a surface modification by pulsed YAG laser. Through LARCS, the resistance value is adjusted by decreasing the resistivity of a part of the RuO_2 thick film resistors. The correlation between the pulse number and the resistance value is shown in Fig. 8.

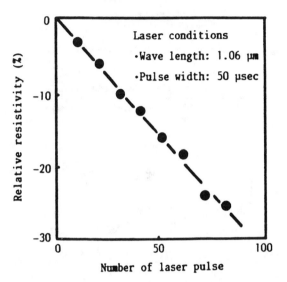

Fig. 8. Change of resistivity by LARCS.

LARCS does not employ laser cutting, so that it is very advantageous for applications to the miniaturized MCPH thick film resistors. Moreover, the resistance value is easily controlled by employing a pulse number as a controllable parameter.

APPLICATION

An MCPH with the same electrical circuit as a conventional type hybrid IC was designed and fabricated to estimate the effectiveness of the improved and advanced MCPH technology mentioned before. As an example of the conventional type hybrid IC, a typical thick film hybrid IC, which is currently manufactured and used for VTR sets, was selected.

Fig. 9 shows the comparison of the conventional hybrid IC and the MCPH. Both of these are fabricated on a single side of the substrate, so that the miniaturizing effect of the MCPH can be seen easily from Fig. 9. The specifications are listed and compared with each other in Table 1.

In this MCPH, by utilizing many fine conductors down to 20 μm width, the total footprint area of the conductor wirings can be remarkably reduced compared with the conventional one, where 200 μm or wider conductors are typical.

Also, thirty thick film resistors of 300 μm-length or -width were employed, and, in addition, each resistivity of three resistive pastes was optimized by using a personal computer, the designing program of which had been developed and reported by the authors.[3] Consequently, the total footprint area of the thick film resistors

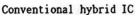

Conventional hybrid IC

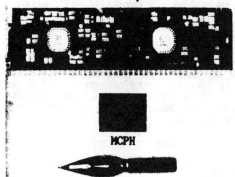

MCPH

Fig. 9. Conventional hybrid IC versus MCPH.

Table 1. Comparison of specifications

	Conventional	MCPH
Substrate		
Size (mm x mm)	80.6 x 23.0	16.6 x 12.5
Area (cm²)	18.5	2.1
Conductors		
Minimum width (μm)	250	20
Minimum space (μm)	250	20
Package style	SIP	QFP
I/O terminal pads		
Pitch (mm)	1.78	0.8
Size (mm x mm)	1.3 x 2.0	0.5 x 0.4
Packaging density (components/cm²)	3.4	30.3

became very small compared with the conventional one.

In addition, to boost the packaging density of this MCPH, assembled components were carefully selected. Two flip-chip ICs were prepared and small sized (1.6 mm length x 0.8 mm width) chip capacitors and/or resistors were fully employed as shown in Table 2.

Table 2. Components list of MCPH

Component	Dimensions (mm x mm)	pcs.	Sum of footprint area (mm²)
Chip Capacitor	1.6 x 0.8	26	33.3
Chip Resistor	1.6 x 0.8	3	3.8
Transistor	2.1 x 2.0	2	8.4
IC	3.3 x 3.2	1	10.6
	2.8 x 2.9	1	8.1
TFR (*)	---------	30	7.6
Total	---------	63	71.8

(*) Thick Film Resistor

Moreover, to achieve many interconnections between I/O (Input/Output) terminal pads on the MCPH and I/O leads, a QIL (Quadruple-In-Line) type package, in which the I/O terminal pads are formed in quadranglar line near the periphery of the square substrate, was selected. In general, the QIL can decrease the mean distance between the assembled components and the I/O terminal pads compared with SIL (Single-In-Line) and/or DIL (Dual-In-Line) types, so that the high facility for the high density packaging design was obtained. The I/O leads were connected with the terminal pads and then transformed like a flat package of LSIs. Therefore, this MCPH could be assembled on another PWB by using SMT (Surface Mount Technology) and also could contribute to the high density assemblies of PWBs.

As a result, the packaging density of this MCPH reached 30.4 components/cm², which is approximately 10 times more than the conventional one.

Fig. 10. MCPH substrate containing 48 pieces.

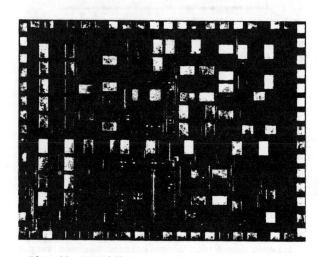

Fig. 11. Magnified view of a piece of MCPH.

Moreover, this MCPH was designed for a 106 mm☐ Al₂O₃ substrate and 48 pieces can be simultaneously fabricated, therefore, the production cost can be remarkably lowered and excellent productivity was obtained. Fig. 10 shows the completed substrate which contains 48 pieces of hybrid ICs. Also, Fig. 11 shows the magnified view of a piece of the MCPH shown in Fig. 10.

EVALUATIONS

For the evaluation of electrical properties, a computer controlled test system was specially designed and applied. This test system, the probing portion of which is shown in Fig. 12, can probe and test all the pieces on a substrate step by step automatically. By using this test system, the evaluation time can be shortened and the measured data can be easily arranged by the computer, therefore, the effective and efficient evaluation has become available.

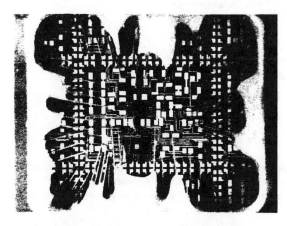

Fig. 12. Test system for MCPH.

Various electrical and/or mechanical properties of this advanced MCPH were evaluated. Especially, the insulation characteristics of the narrow spaced conductors were closely investigated. In the case of 20 μm line & space conductors, the initial insulation resistance is more than 10^{13} ohms, and the high voltage insulation up to 1000 volts can be obtained for both the first- and second-metallization layer.

Also, some chamber tests, such as a high-temperature storage test, a pressure cooker test, thermal cycling test and etc., were employed. Through these tests, the MCPH shows excellent stability of electrical and/or mechanical properties.

CONCLUSIONS

The following conclusions have been reached regarding the high density MCPH described above.

1) Processing technology was advanced to boost the packaging density.

    a) 20 μm line & space copper conductors have become available.
    b) Small sized thick film resistors, the minimum width or length of which is 300 μm have become available.
    c) New laser trimming technology for small-sized thick film resistors was developed.

2) High density MCPH with the same circuit as the conventional hybrid IC was fabricated and the effectiveness of the MCPH was proved clearly.

    a) The packaging density is approximately 10 times the conventional one.
    b) An automatic test method was established.

3) The reliabilities were evaluated and established sufficiently.

The MCPH technology can boost the packaging density of hybrid ICs remarkably, and, in addition, can supply a suitable package style for various board assemblies. Therefore, the MCPH will be widely used in many fields.

ACKNOWLEDGEMENT

The authors wish to acknowledge Dr. Shoji Hirabayashi and Dr. Teruhiko Yamazaki for their many helpful discussions.

REFERENCES

[1] Hayato Takasago, Mitsuyuki Takada, Kohei Adachi, Atsushi Endo, Kurumi Yamada, Yoichiro Onishi, and Yoshiyuki Morihiro, "Fine-line, multilayer hybrids with wet-processed conductors and thick-film resistors," in 34th Electronic Components Conf., pp. 324–329, 1984.

[2] Hayato Takasago, Mitsuyuki Takada, Kohei Adachi, Atsushi Endo, Kurumi Yamada, Tetsuro Makita, Eishi Gofuku, and Yoichiro Onishi, "Advanced copper/polyimide hybrid technology," in IEEE Trans. Components, Hybrids, Manuf. Technol., vol. CHMT-10, No. 3, pp. 425–432, Sept. 1987.

[3] Mitsuyuki Takada, Hayato Takasago, Hideaki Otsuki, and Yoichiro Onishi, "High-density multi-stacked module," in International Microelectronics Conf., pp. 447–452, 1984

MULTICHIP PACKAGING FOR SUPERCOMPUTERS

By

I. Turlik[1], A. Reisman[2], R. Darveaux[3], L-T. Hwang

Microelectronics Center of North Carolina
P.O. Box 12889, Research Triangle Park, NC 27709

[1]MCNC Resident Professional
Bell Northern Research
P.O. Box 13478, Research Triangle Park, NC 27709

[2] also with the Department of Electrical and Computer Engineering
North Carolina State University
P.O. Box 7911, Raleigh, NC 27650

[3] also with the Department of Materials Science and Engineering
North Carolina State University
P.O. Box 7907, Raleigh, NC 27695

Abstract

A thin film multi-chip packaging design was chosen as the basis for a high performance, high power dissipation vehicle suitable for VLSI/ULSI applications. The package supports 25 chips (1x1cm), each capable of dissipating as much as 40 watts. With 100,000 gates, (about 400,000 devices at a 1μm groundrule level), and 530 I/O's per chip, approximately 260 m of wiring length will be required to interconnect these 25 chips.

A new equation for predicting pin counts on actual chips in a system environment is presented. This equation is based on a Rent's type relationship, $I/O's = C (\#gates)^X$. Also, an existing empirical relationship was employed to determine the average wire length as a function of gate count. We discuss the implications of thin-film interconnections on signal parameters, such as characteristic impedance, voltage attenuation, signal delay, coupling noise, and skin effect.

A thermal module was designed to transfer heat efficiently from high power dissipation chips to a liquid coolant via forced convection. The temperature distribution of the thermal module was obtained by solving 2-D heat conduction equations for isolated heat sources (chips), and heat sinks (the water channels) using finite difference techniques. The maximum chip temperature rise relative to inlet water temperature at a 40 watt/chip dissipation level is 12°C, and the maximum temperature variation between chips assuming all are powered except one is also 12°C for a "translator chip" design and 18°C for a "deformable heat sink" design. Clearly, if the power dissipation is less than 40 watts/chip, the temperature rise will be proportionally less than 12°C for the first design and less than 18°C for the second design. The thermal performance of the package was also analyzed as a function of thickness and thermal conductivity of the deformable die bonding material used in the second configuration.

*A major part of this paper was presented at NEPCON West, Anaheim, CA, March 1989

Introduction

Each new development in technology makes possible enhancement of system performance. Such enhancement involves packing devices together more closely and improving their performance, with special provisions to assure required reliability and low cost/function. As a consequence, today's packaging challenges arise because of the continuing ability of the technologist to integrate more components in a single integrated circuit chip. The maximum number of devices is at present about 8 million per state of the art DRAM chip, (each bit contains two devices), and about a million per logic chip. The device density has, on average, doubled annually for almost two decades [1]. It is anticipated that such advances will continue to be made during the 1990's.

A total system technology strategy requires making engineering tradeoffs among device, circuit, chip interconnection, and packaging delays to maximize system performance. Perhaps the most important thing is to identify potential "show-stoppers" and address these early on. The factors that need to be addressed with respect to technology include: system size, speed, power, heat dissipation, wiring capacity, and input/output capability. In such a complicated situation, it is quite possible that alternative strategies to balance chip delay against interconnection and package delays may produce similar results at the system level.

As can be seen in Table I, which tabulates the performance and technology used in modern high performance computers and supercomputers, Bipolar, CMOS, and GaAs technologies have each found an application. The technology choice for each application represents a tradeoff between performance and cost. An important trend in reducing the hierarchy in packaging is the use of multichip modules, some of which are listed in Table II. These modules, designed to accommodate 4-200 chips, combine the function of the single-chip package and the printed circuit board. Multichip modules containing VLSI logic chips (10^{16}- 10^{21} devices/chip [1]) can, together with associated DRAM memory, often be packaged on just a few mother boards to build an entire system. It should be noted, that even the most advanced multichip package technology relies on relatively primitive fabrication technology when compared to the most aggressive chip technology. Because monolithic integrated-circuit technology has been the driving force behind electronic growth, it would appear logical to look to integrated-circuit techniques for system level packaging. This has however been the exception rather than the rule.

The concept of using silicon as a substrate for a semiconductor package is known in the art [2-7]. Major problems associated with these packages are that the power distribution systems are limited, due to the material properties of silicon. Silicon substrates, in general, do not make a good packaging medium unless a metallic reference plane is used on the surface. This is due to the fact that silicon itself constitutes a lossy dielectric, as well as creates slow wave propagation due to the substrate resistance and the position of the signal-image current in the substrate. Consequently, in most cases its application is currently limited to low performance systems.

As chip density penetrates into the ULSI regime (10^{21}-10^{26} devices/chip) with more than 2 million logic transistors per chip, and as devices operate at higher and higher speeds, increasingly dense interconnection and packaging systems become necessary. Sooner or later, the ability to remove heat may become a fundamental, technological limitation. At the GHz level, CMOS power dissipation begins to approximate NMOS levels because power dissipation/circuit increases with switching frequency (it is CV^2f, where C is the circuit capacitance, and f is the switching frequency). One solution being explored at MCNC is liquid cooling, using microchannels [8]. The high density modules being explored are capable of dissipating 1000 watts from a 4"x4" substrate with only a 12-18°C steady state worst case chip temperature rise [8,9,10].

In the present paper, two new design concepts for a high performance multichip package for the VLSI/ULSI era are presented. We also discuss new approaches to alleviating some electrical and thermal design problems, such as skin effect, crosstalk, signal reflections and power dissipation. Relative to power dissipation, we discuss thermal stress issues and material requirements anticipated for future packaging strategies.

Package Description

A multichip, thin film interconnection package is currently in the development stages at the MCNC. It was chosen as the basis for a high performance, high power dissipation vehicle suitable for VLSI/ULSI applications, at or below room temperature, if necessary. The package supports a 5x5 array of 1 cm x 1 cm chips on 1.8 cm pitch, each capable of dissipating as much as 40 watts with a very small temperature rise. Cut-away views of two multichip package designs are shown in Figure 1. The "deformable heat sink" design is shown in Figure 1a, and the "translator chip" design is shown in Figure 1b. The heat sink and thin film interconnection processing would be identical for the two designs, but the die attachment process and structural configuration are different.

The electrical requirements of high performance packages have dictated the design rules for the thin film interconnections. The interconnection structure consists of up to 7 layers of copper conductors in a polyimide dielectric with ~10 μm wide signal lines and 15-30 μm spaces, depending upon design (the spacing is greater for a microstrip configuration than for a stripline configuration). To maximize density, straight walled non-nested vias are used with selective electroless nickel via filling. Manufacturing processes for such structures are now being developed at MCNC.

For either design, the thin film interconnection structure is built on an AlN substrate. This substrate material was selected for its superior thermal, electrical, and mechanical properties, as well as its compatibility with the thermal expansion coefficient of silicon (TCE_{AlN} = 4.1E-06/°C, and TCE_{Si} = 2.6E-06/°C).

The heat sink is densified, reaction-sintered silicon carbide loaded with 15% silicon to fill interstices. Silicon carbide was chosen for its high elastic modulus, low TCE, and high thermal conductivity. Aluminum nitride can also be used. The heat sink has 15, 1 mm x 5 mm water channels designed to accommodate a total flow rate of 500cc/s (7.9gpm). In the prototype module, the channels were made by cutting grooves in a silicon carbide block and epoxy bonding or brazing a cover plate to the top. Ultimately, the channel structure will be made in a single firing step, or an improved bonding technique, which is currently under development, will be employed.

Deformable Heat Sink Design

In this design, the thin film interconnection structure is built on a 10.2 cm x 10.2 cm (4in x 4 in) substrate which has molybdenum conductors between the thin film interconnections and the I/O pins. Solder bumps are used to attach the integrated circuits using a process different then the C-4 (control collapse chip connection) process introduced by IBM. The nominal dimensions of the solder bumps are 125 μm dia x 75 μm high on a 250 μm pitch. Each chip can accommodate a 37 x 37 bump array for a total of 1356 connections (3 bumps removed from each corner and one from the center for registration).

The large number of solder connections improves thermal conduction [11] and mechanical reinforcement [7] between the chip and substrate. Thermal conduction through the solder bumps is important when a significant amount of heat is generated in the chip-to-chip interconnections and terminating resistors. An analysis (given in [12]) shows the heat generation on the substrate becomes significant as line density and switching frequency increase. Mechanical reinforcement between the chip and substrate is important in the deformable heat sink design because the solder bump array will be stressed as the die bond material deforms to accommodate thermally induced displacements.

Pure indium is being investigated as the bonding medium between each die and the heat sink. Indium has a high thermal conductivity, deforms at low stress, has excellent fatigue characteristics, and it can be bonded within the desired process temperature hierarchy relative to die attachment. The indium bond is made by a reflow process to metallization present on both the heat sink and on the chip backside. The support ring around the package periphery prevents deflections due to mechanical loading, but all thermally induced displacements must be absorbed by the indium die bond.

The MCNC deformable heat sink design is similar to the Mitsubishi multichip module shown schematically in Figure 2. However, the Mitsubishi package has at least a 10x greater device-to-ambient thermal resistance than the MCNC design due to Mitsibishi's air cooling, low thermal conductivity adhesive between the ceramic cap and heat sink, and 30μm gas gap between each copper plate and the ceramic cap [13]. The gas gap provides mechanical compliance for thermally induced displacements in the structure. In the MCNC deformable heat sink design (Fig. 1a), all thermally induced displacements must be absorbed by the indium die bonds, so each solder bump array will be subjected to stress as the indium deforms. An analysis to determine the maximum stress on solder bumps during a power up is given in reference [14].

Translator Chip Design

In this design, a translator chip is used to connect each die to thin film interconnections on the substrate by means of an areal solder bump array. The translator chip functions like a 3-level, area array TAB connection [15]. The process sequence for the package would be as follows 1) fabricate a 12.7cm x 12.7cm (5" x 5") AlN substrate with 25 square holes, 2) insert the IC chips and planarize using polyimide, 3) fabricate a thin film interconnection structure on the substrate and translator chips, 4) solder bump the translator chips to pads on each die and the substrate, 5) braze I/O pins to the interconnection metal on the package periphery, and 6) attach heat sink. One advantage of this design is that each die can be reworked individually.

The translator chip design is similar to the NTT multichip module shown schematically in Figure 3 [16]. However, the MCNC design has a much greater I/O capability due to the area array translator chips, and it has at least a 10x smaller device-to-ambient thermal resistance.

Electrical Issues

Package Wirability

The problems in interconnecting large numbers of logic circuits, e.g., the number of I/O's/chip, availability of wiring channels etc., are widely recognized. Simple wirability and I/O algorithms are given to enable estimation of a 90% probability of wiring success during automatic placement and wiring chips. These algorithms are based on Rent's Rule [17] and a wirability model proposed by Heller, Donath, and Mikhail [18,19]. These relationships define: 1) the number of I/O pads necessary for a logic block, based on extensive literature analysis, and 2) the average wire length as a function of gate count.

Signal pin counts for real chips reported in the literature are plotted in Figure 4 as a function of the number of gates/chip in CMOS Logic Gate Array technology. In Figure 4, the least squares fit to this data is given by the empirical relationship

$$I/O's = 3.201G^{0.434} \qquad (1)$$

I/O's are the number of I/O's per chip, and G is the number of gates/chip. The coefficient and exponent are significantly different from the original values, which were predicated on bipolar technology data. Based on this relationship, it is predicted that VLSI CMOS at 1 μm groundrules, on a chip 1 cm on an edge, which accommodates 100,000 logic circuits [20], requires approximately 474 signal pins (the total number of pins should be increased by at least 10% for power and ground connections). Indeed, based on these findings it might be expected that future chips will be unable to accommodate the required number of I/O's using peripheral attachment techniques.

The I/O requirement is one of the key driving parameters for all levels of packaging: chip, chip carriers etc. It is to be recalled from geometrical scaling theory, for example, that in a scaled chip, the total current that has to be brought to a chip increases by the scaling factor even though each circuit requires a decrease in current by this same scaling factor. This places an additional burden on I/O count if chip interconnection current densities which are already degraded by the scaling factor are not to be degraded further due to the increased current requirements to the chip as a whole. These considerations are probably the strongest ones for changing to areal contact technologies.

In the design of high performance systems, the interconnection capacity of the entire packaging hierarchy, from chips on upward, should be known from the beginning. Too little available package wiring capacity discovered at a late stage in the system design process could mean severe degradation of the system performance, or unpredictable extension of schedule. A predictive empirical wiring algorithm suitable for random logic cells was developed by Heller et. al. [18]

$$R = K(1 + 0.1ln(M))M^{1/6} \qquad (2)$$

where R is the average wire length in entity (cell) pitch units, K is found to lie between 0.5 and 1, and depends upon the efficiency of placement, M is the number of entities (when M is large, the efficiency of placement decreases, therefore for a large value of M, a correspondingly lower random value of K results.

The total capacity (line length available) for a regular pattern of wire distributed among n layers of metal patterns is given by:

$$TFC = u \cdot l \cdot w \cdot (ch/mm) \cdot n \qquad (3)$$

where u = 0.5-0.6, is the utilization factor of the wire (at best there is 50% efficiency in actual wiring [21]), w and l are the substrate dimensions, ch/mm are the number of channels per unit length.

Based on equations 1, 2, and 3 one can estimate the total wiring length necessary to interconnect entities (in our case logic chips), and the number of layers needed. For example, to package 25 CMOS chips with 100,000 gates/chip, 474 signal I/O's are needed per chip, and over 257 m of interconnection wire is required (assuming a fanout of two). For 8μm wide lines on a 32μm pitch, no more than two pairs of x-y signal layers will be needed. Table 3 shows a comparison between the published wiring lengths for IBM's TCM module in a 3033 system technology, and the calculated values based on our analysis. The results are in excellent agreement. In addition, Table 3 shows calculated values of two MCNC module designs, one for 25 chips and the other for 16 chips.

If the interconnection capacity of the module is adequate, then electrical performance or noise barrier considerations become the gating parameters. By employing thin-film interconnections (i.e., line width ~10μ, and line pitch 20-40μ) interconnection capacity can be increased significantly, and the number of signal layers can be reduced substantially. For example, advanced thick film module designs, such as the IBM TCM, require 33 or more layers of wiring and ceramic each to interconnect 100 chips. On the other hand, because of the increased wiring density possible, micrometer rather than mil dimensions, thin film modules need only 5-10 layers, depending on the chosen interconnection structure, ie. microstip, stripline or coplanar configuration.

Signal Propagation

Shrinkage of line cross-section in thin film modules leads to a concomitant increase of line resistance. The lines become lossy, and behave differently than the comparatively lossless lines on a PC board. Thus, design guidelines for the thin-film interconnections have to be established. A detailed analysis of lossy thin-film interconnections with microstrip and stripline structures in terms of ringing, crosstalk, power dissipation, and loading characteristics is given by Nayak et al. [12].

As the clock frequency of digital circuits increases, shorter device rise times are required. When rise time is short compared to interconnection propagation time, the interconnections must be designed as properly terminated transmission lines to minimize ringing. In addition, the skin effect associated with current redistribution at high frequencies must be addressed, because it results in signal amplitude degradation.

With an alternating current (e.g., sinusoidal signal) in a conductor, the current density over the cross section of the conductor is not uniform, tending to be redistributed toward the surface of the conductor as frequency is increased. This phenomenon is called the "skin effect," and it can result in attenuation of the signal amplitude. Using Fourier analysis, a pulse can be expressed as a summation of many sinusoidal signals with different amplitudes and phases. Since pulse width is analogous to wavelength, the shorter the pulse width is, the larger is its effective frequency, and the broader the range of frequency it can encompass. For input signals with short rise times, the skin effect in the interconnection must be considered in order to determine propagation characteristics of the line, such as characteristic impedance, propagation time, and attenuation coefficient.

In our analysis, copper lines 8 μm wide, 4 μm thick, and isolated 12 μm above the ground plane were employed. The dielectric constant of the insulator was 3.5 (this is a typical dielectric constant of polyimide). Figure 5 shows the calculations of voltage attenuation (per unit length) for the thin-film copper line as a function of frequency. The curves a and b in Fig. 5 were obtained by using Assadourian's [22], and Pucel's [23] formulas, respectively, in which skin and ground proximity effects were considered. The formulas were applied over a limited frequency range. The lower bound of the range was at a frequency where the skin depth is one third of the conductor thickness, and the upper bound was at 10 GHz.

Weeks' algorithm [24] was also applied to obtain the resistance and inductance of the line as a function of frequency (curve c). It was found that the attenuation behavior predicted using Weeks' algorithm are in a close agreement with that obtained using the Assadourian and Pucel equations at high frequencies. Thus, the skin effect of the conductors, and the proximity effect of the ground are properly, and directly accounted for. The solid curve in Figure 5 (curve d) represents attenuation without the skin effect, with the assumption that the inductance is fixed at the value of a low frequency (10 MHz), and the resistance is fixed at the dc value. From curves c and d, the skin and proximity effects at high frequencies are clearly depicted.

The performance of long lines (15 cm) can be impaired by the skin effect if the input pulse width is equal to or less than 0.5 ns and rise time of 0.1 ns (for 4μm x 8μm copper lines). Therefore, the skin effect can not be neglected in circuit simulations of long lines when the input pulse width is less than 0.5 ns. However, for short lines (1 cm), the width of the input pulse can be as short as 0.2 ns without noticeable degradation in performance.

Because of the short rise time assumed in our analysis (i.e., 100 ps), the interconnections on our package were treated as distributed circuits. It is known that reflection occurs at either end of a transmission line if it is not properly terminated. For lossless transmission lines, parallel and series termination are commonly applied, and a termination resistance equal to the line characteristic impedance is used. In parallel termination, a resistor is added in parallel with the load. In series termination, a resistor is inserted between the driver and the line.

For a *lossy* transmission line (such as the thin-film interconnections on a multichip package), the termination issue becomes more complicated since the line attenuation is a function of length, and the characteristic impedance is frequency dependent. For long lossy lines $(8cm < \ell < 18cm)$, the high resistance of the line is able to damp the reflection noise, so termination is unnecessary (*i.e.*, the line should be open). Open termination for long lines provides adequate signal output, and minimum propagation delay. For very short lines, $(\ell < 1cm)$ where signal attenuation due to the resistance of the line is negligible, termination is required. The high frequency limit characteristic impedance, R_o is recommended for termination of very short lines [25]. If R_o is employed, parallel termination is preferred over series termination because it results in less reflection, less crosstalk, and slightly better speed. For short to intermediate length lines $(1cm < \ell < 8cm)$, the optimum terminating resistance depends on line length and the method of termination (series or parallel). More details on the design and termination of thin film interconnections for high speed systems are provided by Nayak et al. [12].

Power Dissipation

Semiconductor devices are being scaled to submicron dimensions to achieve the best performance. If one applies precise geometrical scaling of the device, current and voltage decrease by the scaling factor α, the density of circuits/unit-chip-area increases by a factor of α^2, while the power dissipation/device decreases by a factor of $1/\alpha^2$. For a constant size chip, this would result in constant power dissipation/unit-chip-area as dimensions are scaled. In practice, however, the power supply voltage, V is not scaled to V/α, but is either kept constant (5V for NMOS and CMOS), or is scaled by an electrical scaling factor k, where $\alpha/k > 1$. This introduces an undesirable increase in power dissipation/unit-chip-area if device density increases by α^2.

A high performance microchannel water cooled heat sink has been designed and tested at MCNC [8,9]. Computer simulations were used to optimize the design, then a prototype module was built. The prototype module was similar to the translator chip configuration (Figure 1b) and it had a thin layer of gallium for die attachment. The worst case chip-to-chip temperature variation was 12°C when 40W of power was dissipated in all chips except one [8,9]. The initial thermal analysis was extended to the deformable heat sink design (Figure 1a) and the calculated worst case chip-to-chip temperature variation was 18°C [10].

Heat generated by the high density interconnection network between chips compounds the thermal management challenge of a thin film, multichip module. High density interconnections are necessary in VLSI applications which have several hundred signal I/Os per chip. In such a package, heating due to resistive losses in the thin film lines and terminating resistors must be accounted for. The heat flux from chip-to-chip interconnections will be depend on driver characteristics, switching frequency, method of termination, transmission line configuration, and interconnection density. An analysis of the heat generation in lossy interconnections and terminating resistors for the MCNC package is given in [12], and the impact on package thermal performance is discussed in [10].

A 2-D finite difference approach was employed in the thermal analysis computer simulations [8,10]. The cross section used in the analysis of the deformable heat sink design is shown in Figure 6. Heat was assumed to be uniformly generated in the device layer of the silicon chip. Heat generation on the substrate was neglected in the present analysis. From symmetry considerations, the chip centerline and the centerline between chips were modeled as insulating boundaries; i.e. no net heat flow was allowed across either boundary. Also, the air gap between chips was assumed to be insulating (natural convection was neglected).

Heat was assumed to be removed both by forced convection to water in the channels and by free convection to air contacting the substrate surface. The ambient water and air temperatures were assumed to be 25°C. The heat transfer coefficients at the surfaces were calculated using the relations in reference [26]. At the channel surfaces, $h=3.57W/cm^{2}{}^{\circ}C$, and on the substrate surface, $h=0.0012W/cm^{2}{}^{\circ}C$. Radiant heat transfer and lateral conduction to the package support structure, piping, etc., were neglected. Consequently, the analysis represents a worst case situation.

An isothermal plot for power dissipation of 40 W/cm² on each chip is shown for the deformable heat sink design in Figure 7. In the simulation, the indium die bond thickness was assumed to be 800μm and the water temperature was 25°C. The maximum package temperature is 41.9°C (16.9°C above water temperature). This temperature occurs at the chip centerline in the device layer. The temperature variation across the chip is 1.8°C, and substrate temperature is very uniform. Since the water experiences a 0.5°C temperature rise as it travels through the module, the worst case chip-to-chip temperature variation is 17.4°C (assuming one chip at the water inlet is unpowered and all other chips are fully powered).

This 17.4°C chip temperature rise for the deformable heat sink design is somewhat greater than the 11.4°C rise for the translator chip design [8] because of the following design changes: 1) the die bond thickness increased by 700%, 2) the die bond thermal conductivity increased by 190%, 3) the die bond width decreased by 44%, 4) a thermal contact resistance was assumed at the soldered interfaces, and 5) the SiC heat sink thermal conductivity is 48% less than was originally modeled. The combination of these factors increases the device-to-water thermal resistance by about 55% over the translator chip design. Nonetheless, the temperature rises for both designs are very small, considering the power dissipation of 40 W/chip. Obviously, at more conventional dissipation levels, $\leq$ 5 W/chip, the temperature elevations would be proportionately less.

Structural Considerations

The structural design requirements for a package must address manufacturing, yield, and long term reliability issues. Manufacturing and yield issues encompass the process temperature hierarchy for bonding, curing, and soldering operations, the mechanical tolerances required for assembly, and the ability to rework defective parts. Long term reliability issues depend on the compatibility of various materials in the structure, and the severity of the use environment. Material failure in use can result from such phenomena as corrosion, electromigration, brittle fracture, creep fatigue, etc.

Material Compatibility

As an example of how material choice can affect structural design, we will discuss thermally induced stress in flip-chip solder bumps. As indicated previously in Figure 4, the number of I/O's required on a chip increases with the number of logic gates. For ULSI applications, several hundred chip I/O's will be required. In a flip chip package, solder fatigue limits the maximum number of connections which can be made to the chip. In a typical flip chip configuration, where the attachment is to an alumina substrate, fatigue is due primarily to stress from the in-plane thermal expansion mismatch between the chip and the substrate. However, a large number of I/Os *can* be achieved in a flip chip package by using a substrate with a thermal coefficient of expansion (TCE) close to that of the silicon chip.

In Figure 8, the worst case thermal expansion mismatch between the chip and substrate is approximately the same for a 140 I/O chip bumped to an Al_2O_3 substrate or a 1356 I/O chip bumped to an AlN substrate. The in-plane thermal expansion mismatch is calculated from

$$\delta = (\Delta T_{substrate}\, \alpha_{substrate} - \Delta T_{chip}\, \alpha_{chip})DNP \qquad (4)$$

Where δ = thermal expansion mismatch, ΔT = temperature change, α = thermal coefficient of expansion, and DNP = distance from neutral point of the solder bump array.

For example, the worst case thermal expansion mismatch can be calculated for the chips in Figure 8 during a system power up. Assuming the chip heats up by 30°C, the substrate heats up by 30°C, and the bumps are on a 250μm pitch, the worst case expansion mismatch is 0.23μm for the 140 I/O chip and 0.27μm for the 1356 I/O chip (see Table IV).

To estimate the stress level in the solder as a function of time, we must know the dimensions of the components, the rate at which the package heats up, and the constitutive relation for the chip, substrate, and solder materials (stress = f(strain, strain rate, temperature)). For a power up transient followed by a constant temperature hold time, the stress level will reach a peak quite rapidly, then decay exponentially as the solder undergoes creep deformation.

Since a package and its devices are composed of many materials with different thermo-mechanical properties, there is always a potential for thermally induced stress failure. In addition to the in-plane shear stresses discussed above, there will also be tensile and compressive stresses. Also, components will warp due to temperature gradients within layered structures of dissimilar materials. To estimate thermally induced stresses in a complex package configuration, finite element analysis can be used. By judiciously using simulation tools and choosing the proper materials, one can substantially improve package performance.

Material Design

As die size increases, a compliant bonding medium is needed to prevent fracture of the die during thermal or power cycling of the package [27,28]. Similarly, in a flip-chip package where heat is removed from the backside of the die, a compliant contact medium is needed to prevent failure of the solder bump connections during thermal transients. Compliance can be achieved through mechanical means (e.g. spring loaded contacts to the heat sink), or through material design (i.e. using a soft (deformable), thick bonding layer). In general, increasing the compliance of the backside contact will result in a thermal "penalty."

Computer simulations were run to determine the effect of die bond material thickness and thermal conductivity on the performance of the MCNC deformable heat sink package design (Figure 1a). The maximum chip temperature as a function of die bond thickness is shown for a CMOS package dissipating 2W/chip in Figure 9a and for an ECL package dissipating 40W/chip in Figure 9b. In the simulations, three different material thermal conductivities were assumed, and heating due to the chip-to-chip interconnections was neglected.

For the CMOS package (Figure 9a) it is seen that die bond thermal conductivity is not critical because adequate performance can be achieved with a relatively poor thermal conductor between each chip and the heat sink. For a thermal conductivity of 0.01W/cm°C and thickness of 400μm, the maximum chip temperature is only 33.6°C. Therefore, a thermal paste would probably be an adequate bonding medium for a low power CMOS package. On the other hand, for an ECL package (Figure 9b) the die bond thermal conductivity is very important. Even for a void free indium die bond (approximately the k = 0.80W/cm°C curve in Fig. 9b), the maximum chip temperature is 40.7°C at a thickness of 400μm. It is seen that the temperature rise is not a strong function of thickness for the k = 0.80W/cm°C curve. Therefore, a thick indium layer can be used to improve mechanical compliance in the structure, while sacrificing very little thermal performance. An analysis to determine the maximum stress on the solder bumps for the deformable heat sink design is given in [14].

Summary

Major problems associated with the design of a high performance package were discussed: I/O count for VLSI/ULSI devices, wiring capacity of interconnections on the substrate, signal propagation, and thermal performance. We employed a simple algorithm for estimating the wiring length of interconnections and provided a modified Rent's rule to estimate the number of I/O's for a package suitable for VLSI/ULSI applications. Two designs were discussed, and both indicate good potential performance. For the "translator chip" design, if the 25 chips are all powered at 40W/chip, the maximum temperature rise is 12°C. For the "deformable heat sink" design, in the all powered situation, the maximum temperature rise is 18°C. The thermal performance of the deformable heat sink package was also analyzed as a function of the thickness and thermal conductivity of the die bonding material. It is clear that a package has to be treated as whole rather than as discrete electrical, thermal, and material issues. To achieve an optimum design, a series of iterations are needed to balance the key performance parameters.

TABLE I. TECHNOLOGIES FOR HIGH PERFORMANCE COMPUTERS

Computer	Speed (GF)	Cycle Time (ns)	Logic Size (Gates/Chip)	Delay (ns)	Tech	Memory	Reference
Hitachi S-810(20)	0.63	14	550	0.35	ECL	4.5ns-1K,40ns-16K	32
Hitachi M-68X			2000	0.2	ECL	4.5ns-4K,256K	33
NEC SX-2	1.3	6	1000	0.25	ECL	3.5ns-1K,40ns-64K	33
Fujitsu VP-200	0.5	14	1300	0.35	ECL	5.5ns-1k,55ns-64K	33
Cyber 205	0.4	20	250	0.7	ECL	45ns	34,35,36
Cray 1	0.16	12.5	4	0.75	ECL	1K	35,36
Cray 2	1.6	4.1	16		ECL		37
Cray 3	~20	2	500		GaAs	1K GaAs	38
ETA GF10	10		20K		CMOS	64K	35,39
Convex C	0.06	100/50	8K		CMOS	256K	40
IBM 308		26	704	1.2	TTL		41
IBM 309		18.5					

TABLE II. MULTI-CHIP MODULES

MFG	Chips	Substrates	Wiring Levels	Pins	Power (Watts)	Cooling	Reference
IBM (TCM)	100-133	Al_2O_3	33	1800	300	Pistons/He/Water	41
NTT	25	Al_2O_3	25	900	400	Water	42
Hitachi S-810 M-68X	9	Si/SiC	9	96	11	Air	43
Honeywell	4	Si				Air	2
Mosaic (EX)	40-200	Si	2 signal	840		Water/LN_2	3
Trilogy	32	Si/Mo	6	2244	950-1200	Water	4
NEC SX	36	AlN	8	2177	500?	Water	33

TABLE III. COMPARISON OF TYPICAL LOGIC WIRING LENGTHS AT MODULE/CARD LEVEL

	3033 Technology equivalent	TCM	MCNC (16)	MCNC (25)
Actual (Ref [44])	935 m	130 m		
Calculated		130 m	148 m	257 m

TABLE IV. FLIP CHIP / SUBSTRATE THERMAL EXPANSION MISMATCH

Assumptions: $TCE_{Si} = 2.6E\text{-}06/°C$, $TCE_{Al_2O_3} = 7.0E\text{-}06/°C$, $TCE_{AlN} = 4.1E\text{-}06/°C$

$\Delta T_{substrate} = 30°C$, $\Delta T_{chip} = 30°C$, Solder Bump Pitch = 250μm

Substrate Material	Solder Bump Array Size (I/O)	Worst Case DNP (μm)	Thermal Expansion Mismatch (μm)
Al_2O_3	140	1768	0.23
AlN	1356	6021	0.27

REFERENCES

1) A. Reisman, Proc. IEEE, 71, pp 550-565, May (1983).

2) C. Huang, W. Nunne, D. Spielberger, A. Mones, D. Fett, and F. Hampton, CICC, 142 (1983).

3) R.R. Johnson, IEEE Int. Cong. Comp. Design, 101 (1984).

4) D.L. Peltzer, VLSI Design, 43, September (1983).

5) C.F. Taylor, et.al. Proceedings 1985 Custom Integr. Circuit Conf, Portland, Oregon, May (1985).

6) R.W. Johnson, J.L. Davidson, R.C. Jaeger, D.V. Kerns, IEEE Journal of Solid State Circuits, SC-21, 5, pp 845-851, October (1986).

7) K. Okutani, T. Yamana, O. Kanji, and K. Sahani, Proc. 5th Annual Int. Electronic Packaging Conf, October (1985), pp 551-557.

8) L-T. Hwang, I.Turlik, A. Reisman, J Electronic Mat, 16, pp 347-355, (1987).

9) D. Nayak, L-T. Hwang, I. Turlik, A. Reisman, J Electronic Mat, 16, 357, (1987).

10) R. Darveaux, L-T. Hwang, A. Reisman, I. Turlik, Submitted to J. Electronic Mat.

11) A. Reisman, M. Berdenblit, C.J. Marz, and A.K. Ray, J Electronic Mat, Vol 11, No 2, (1982) pp 391-411

12) D. Nayak, L-T. Hwang, I. Turlik, Submitted to IEEE Trans CHMT.

13) M. Kohara, S. Nakao, K. Tsutsumi, H. Shibata, H. Nakata, IEEE Trans CHMT, Vol 6, No 3, September 1983.

14) R. Darveaux, L-T. Hwang, A. Reisman, I. Turlik, Proc 39th ECC (1989).

15) A. Reisman, C.M. Osburn, L-T. Hwang, J. Narayan, US Patent 4,774,630.

16) T. Kishimoto, T. Ohasaki, Proc 36th ECC (1986).

17) W.E. Donath, IEEE Trans Circ Syst, CAS-26 (4), pp 272-277, (1974).

18) W.R. Heller, W.E. Donath, W.F. Mikhail, Proc 14th Design Automation Conf, New Orleans, LA, (1977), pp 32-43.

19) W.R. Heller, C.C. Hsi, W.F. Mikhail, Design and Test, August (1984), p43.

20) C.M. Osburn, A. Reisman, J Electronic Mat, 16, p223 (1987).

21) G. Messener, W. Knausenbergh, IEEE Trans on CHMT, Special Packaging Issue, (1988).

22) F. Assadourian, E. Rimai, Proc IRE, pp 1651-1657, December (1952).

23) R.A. Pucel, D.J. Thesse, C.P. Herting, IEEE Trans Microwave Theory Techniques, Vol MTT-16, No 6.pp 342-350, June (1968).

24) W.T. Weeks, L.L. Wu, M.F. McAllistor, A. Singh, IBM J Res Devel, Vol 23, No 6, pp 652-660, November (1970).

25) L-T. Hwang, D.Nayak, I. Turlik, A. Reisman, Submitted to IEEE Trans Circuits and Systems.

26) W.M. Kays, M.E. Crawford, *Convective Heat and Mass Transfer*, McGraw-Hill, New York, 1980.

27) R.K. Shukla, N.P. Mencinger, Solid State Technology, July, (1985), pp 67-74.

28) E. Suhir, Proc IEEE 37th Elec Comp Comf, (1987), pp 508-517.

29) A.J. Blodgett, Jr., Scientific American, July, (1983).

30) Semicustom Design Guide, Summer 1986, and 1987 IEEE International Solid State Circuits Conference Digest of Technical Papers.

31) D.L. Ferry, IEEE Circuits and Devices, July (1985), p39.

32) S. Nagashima, Y. Inagami, T. Odaka, and S. Kawabe, ICCD, 238 (1984).

33) H. Horikoshi and Y. Inagami, ICCD, 227, (1984).

34) D. Barkai and K.J.M. Moriarty, "Applications and Development on the CDC Cyber 205."

35) P.J. Klass, Aviation Week & Space Technology, June 4, (1984).

36) D.B. Davis, High Technology, May (1984).

37) "The CRAY-2 Computer System," Cray Research, Inc.

38) "Supercomputers Hit Their Stride," Electronics, 44, March 10 (1986).

39) I/O ETA Systems, ETA Systems Publication, 2 (2), (1986).

40) "Convex C-1 Computer System," Publication 570-000180-000, Convex Computer Corporation, Richardson, TX (1985).

41) E.E. Davidson, ICCD, 573 (1984).

42) T. Kishimoto and T. Ohsake, IEEE ECC, 595 (1986).

43) F. Kobayashi, K. Ogiue, G. Toda, and M. Wajima, Proc IEEE ECC, 379 (1984).

44) A.J. Blodgett, Jr., IBM J Res Develop, Vol 26, No 1, January (1982).

45) A. Bar-Cohen, IEEE Trans CHMT, Vol 10, No 2, June 1987.

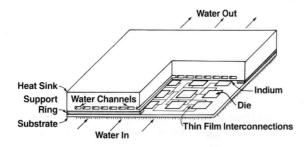

a) Deformable heat sink design.

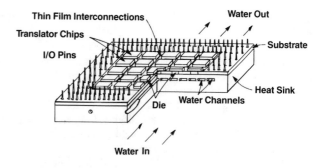

b) Translator chip design.

Figure 1. MCNC multichip package designs.

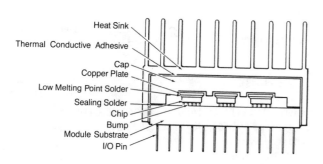

Figure 2. Mitsubishi multichip package [13,45].

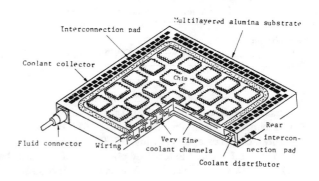

Figure 3. NTT multichip package [16].

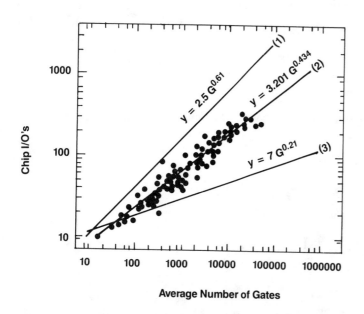

Figure 4. Chip I/O count for various technologies: (1) primarily bipolars - Ref [29], (2) CMOS gate arrays - Ref [30], (3) functionally partitioned chips - Ref [31].

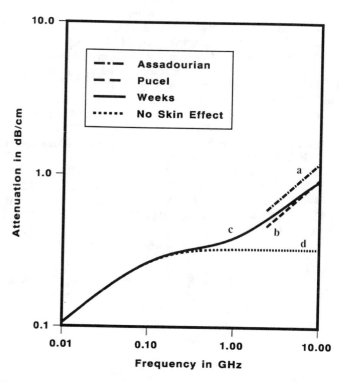

Figure 5. Voltage attenuation per unit length for lossy interconnections.

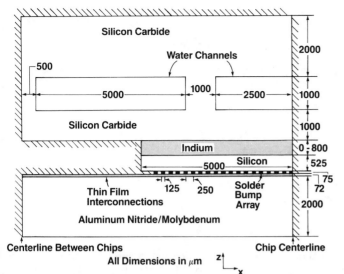

Figure 6. 2-D cross section for finite difference analysis.

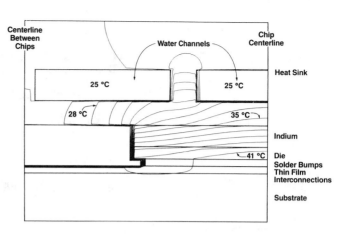

Figure 7. Isothermal plot at 1˚C intervals for deformable heat sink design. Power dissipation = 40 W/cm^2 on chip, indium thickness = 800 μm, and water temperature = 25˚C [10].

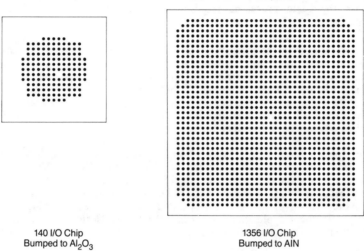

140 I/O Chip
Bumped to Al$_2$O$_3$

1356 I/O Chip
Bumped to AIN

Figure 8. Equivalent configurations for flip chip / substrate thermal expansion mismatch.

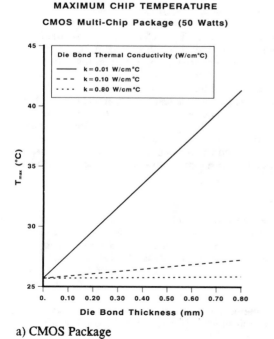

a) CMOS Package

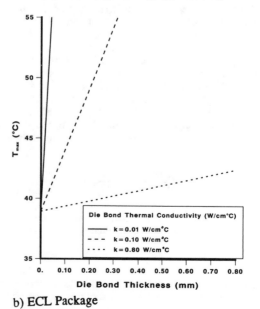

b) ECL Package

Figure 9. Thermal performance of MCNC deformable heat sink design.

289

Process Considerations in Fabricating Thin Film Multichip Modules

T. G. Tessier[1], I. Turlik[1], G. M. Adema[2],
D. Sivan[2], E. K. Yung[2] and M. J. Berry[2]

[1] MCNC Resident Professional
Bell Northern Research
P. O. Box 13478, Research Triangle Park, NC 27709
(919)-248-1854

[2] Microelectronics Center of North Carolina (MCNC)
P. O. Box 12889, Research Triangle Park, NC 27709
(919)-248-1484

Introduction

As is the case with most advanced technologies, the drive towards high performance multichip packaging has given rise to a number of engineering challenges. These challenges can be grouped into one of three categories: (i) design and fabrication of the thin film interconnections, (ii) integrated circuit/substrate compatibility and (iii) module assembly and reliability. MCNC's thin film multichip packaging program has been structured to address the major issues in each of these areas. This paper will discuss the basic processes required for fabricating thin film multilayer copper/polyimide interconnections.

A countless number of issues related to the processing of thin film interconnections are currently being explored to satisfy the demand for high density interconnections, to accommodate large numbers of inputs and outputs (I/O's), and to achieve short interconnection lengths for improved electrical performance. Substrate material choices to date have essentially been limited to silicon, Al_2O_3, and to a much lesser extent AlN, SiC or glass/ceramics. Though the ceramic materials listed generally possess superior thermal (TCE and/or thermal conductivity), electrical, and mechanical properties for these packaging applications, extensive use of silicon has been reported in the literature. This preference for silicon has been based on the much greater availability of equipment and tooling for the processing of silicon wafers. As the design complexity of thin film interconnections fuels a demand for larger substrate sizes, conventional thin film equipment and processes may no longer be appropriate. Various options for the processing of these larger substrates include plating instead of vacuum deposition for conductor fabrication, and spray coating instead of spin coating for the deposition of polymer dielectrics. Development of polymer dielectrics with enhanced properties for thin film applications remains an area of ongoing activity. In order to yield working interconnections a host of approaches are being investigated. One alternative mentioned repeatedly in the literature is built-in redundancy. However, the redundancy should be introduced in a manner that does not substantially reduce the interconnection density. For those applications, including the MCNC design, where maximum density is required, the ability to inspect for defects either by visual inspection or by pattern recognition followed by repair of these defects at each level is essential.

The design of MCNC's thin film multichip module dictates the use of a 4" x 4" co-fired AlN or SiC substrate with an area array of >2000 external I/O's. A high density (10 μm lines, 20-30 μm spaces) copper/polymer dielectric thin film structure used to interconnect 25 high performance logic chips each with an area of 1 cm² will be fabricated on the substrate. Integrated circuits (IC's) are attached to the substrate by solder bumps. The back side of each IC is bonded to a water-cooled heat sink via a compliant die attach. In order to achieve these objectives a number of technical hurdles need to be overcome. Our approach is to evolve the necessary components of technology through experiences in fabricating test structures and prototypes with gradually increasing degrees of complexity. This paper will provide an overview of the technical issues currently being addressed at MCNC related to the fabrication of such a thin film structure on Al_2O_3 and AlN and the development of a flip chip (C-4) process for IC attachment. Most of the experimental results reported here are based on observations made during the fabrication of a 4" x 4", 128 chip memory module as well as with test structures used to optimize individual process steps. Details of MCNC's thin film process, including metallization choices, patterning techniques, and polymer dielectric selection as well as process implications will be discussed. The C-4 process for chip bumping and joining will be presented. The progress towards integrating these technologies into a process for fabricating a working module will be reported. Approaches for module inspection and repair as a means of enhancing the overall yield will be discussed.

Processing of Thin Film Interconnections

The fabrication of multilevel interconnections involves a repetitive sequence of thin film processes for depositing and patterning conductors and dielectric layers. Our basic approach is to deposit Cr/Cu/Cr conductor layers by e-beam evaporation and to pattern conductors by using a lift-off technique. Polyimide, used as the interlevel dielectric, is deposited by spin coating. Vias are formed by reactive ion etching of the polyimide. The electroless plating process is used for via filling.

Patterning

Conductor lines can be defined by either "subtractive" or "additive" methods. With subtractive metal processes, the maximum linewidths are defined by the footprint of the photoresist if the etch process is anisotropic. However, after wet chemical etching, large etch biases are encountered since this process is isotropic. For example, a 6 μm composite structure of Cr/Cu/Cr will require a bias of 6 μm per side. So a mask pattern of 37 μm is needed to achieve a linewidth of 25 μm. This results in a larger pitch than can be achieved with additive methods. Since the metal linewidths are defined by the top of the photoresist in the additive method combined with the fact that there is no etch bias, higher density designs are attainable. As the need for higher density continues to grow, this attribute stands out as a key advantage of the additive method.

Electro- and electroless plating, evaporation and sputtering are metallization techniques currently being used to deposit composite metal structures for thin film applications [1-3]. Discussion here will be limited to the considerations of composite metal depositions relative to the additive vs. subtractive methods. Composite metal structures such as Ti/Pd/Au, Cr/Pd/Cu/Cr or Cr/Ni/Au each have a different set of requirements which must be met by the etchants used. The etchant compatibility between metals in the structure as well as with dielectric materials, photoresists, and substrates needs to be determined for each composite structure when using the subtractive method. The additive method eliminates these concerns; another key advantage.

Evaporation was chosen for ground, power, and signal layers because it provides the flexibility to deposit materials in a composite structure, using the additive method. There are limitations associated with additive evaporation of composite structures which are related to the properties of the photoresist stencil used. For instance, the thickness of the metal deposited through the stencil is governed by the thickness of the photoresist. Also, the total deposition time could be increased dramatically if intermittent cooling periods are required to keep the substrate temperature safely below the reflow temperature of the photoresist. To prepare the liftoff stencils required for additive evaporation, all photolithographic processing utilized an amine vapor-based image reversal process. Depending on the thickness of the metal to be evaporated, either 2 μm thick Shipley Microposit 1400-31 films or 5 μm thick Shipley Microposit 1650 films were applied. Coatings were manually applied and subsequently baked in a nitrogen purged oven at 100°C. The pattern exposure for the image reversal process was performed on a Karl Suss MA 56 mask aligner in the proximity mode. The amine vapor-based treatment followed by a flood exposure reversed the positive image in the photoresist. The flood exposure dose of 3500 mJ/cm² was performed on a Canon PLA-501F mask aligner. Substrates were immersion developed using Shipley MF312. Subsequent overdevelopment produced the negatively sloped sidewall profiles required for a liftoff stencil.

Yield loss results from contact printing when photoresist is pulled from the substrate onto the mask. While the mask can be cleaned so the defect pattern is not transferred during subsequent use, the substrate must either be reworked or repaired. In the subtractive case, the missing material results in opens. In the additive case, shorts are formed. Since the removal of excess material is more expedient than adding missing material, the additive method better facilitates repair. Reducing this type of defect could be accomplished by shifting from contact to proximity mode printing. This may not eliminate the problem however, if the proximity system in use requires each substrate to contact the mask during the initial plane/gap setting.

A previous study [4] determined the conditions which yield the largest process window using the contact mode for the pattern definition exposure. Since the proximity mode has a different aerial exposure profile than the contact mode, it was necessary to determine the exposure gap and dose required to obtain the desired sidewall slope. A matrix of experiments was performed where the exposure gap varied as 10, 20, 30 and 40 μm and the dose varied as 170, 255 and 340 mJ/cm². Held constant for these experiments were the flood exposure dose and the development conditions. All the points in the matrix showed a slope greater than the minimum 97° requirement. The resulting linewidths, however, were all less than the mask design. They ranged from 0.04% to 4.9% below the designed values.

Metallization

The choice of metal in a multilevel interconnection structure is based on its resistivity, cost, adhesion properties to the dielectric layer, deposition technique, and resistance to long-term corrosion. The three metals of choice are gold, aluminum and copper for their superior conductivity and ease of deposition. Cr, Ti, Mo, and Ti/W are commonly used adhesion and diffusion barrier layers between polyimide and Cu. The metallurgy used in our fabricated thin film modules consisted of evaporated chromium/copper/chromium ground, power, and signal layers with electroless nickel interlevel connections. This metallization scheme provides adequate conductivity while remaining within the size constraints which are necessary to achieve high interconnection density.

Fabrication of Signal Lines and Ground Planes

The fabrication of ground, power, and signal layers was accomplished using a Balzers BAK 760 electron-beam evaporator to deposit a 1000Å chromium layer followed by a 40000Å copper layer and a 1500Å chromium layer. The metal was patterned using the liftoff process. The chromium layers serve as adhesion layers as well as diffusion barriers between the copper and the polyimide. Copper was chosen because of its high conductivity.

The system geometry and the chamber temperature (and its effect on the substrate temperature) are key parameters when utilizing a liftoff metallization process. If the substrate temperature reaches 120°C the photoresist templates on the wafers will reflow which will result in poor or unsuccessful liftoff. However, an elevated deposition temperature is desirable because the adhesion between the photoresist and the underlying polyimide is enhanced. Because of these considerations, the chamber is heated to 50°C before the deposition of metal is initiated. It was found that this elevated chamber temperature provides adequate photoresist adhesion, yet the substrate temperature remains low enough so that the entire metallization sequence can be performed without an interruption of the deposition cycle.

As for the system geometry, it is well known that the angle of incidence of the vapor with respect to the substrate is a critical parameter in liftoff processing. Variations in the angle of incidence of metal vapor impinging on the substrate surfaces are both a function of the substrate size and its location within the chamber. A 90° angle of incidence of the vapor with the substrate is desirable, however any angle of incidence which is greater than the slopes of the photoresist walls is acceptable as subsequent removal of the photoresist is still viable. It is typical for evaporators which are designed for use with liftoff processing to have substrate carriers which position the substrate so that vapor leaving the center of the crucible arrives at the center of the substrate at a normal angle. With such a design, the critical geometries are the substrate size and the throw distance (the distance between the substrate and the crucible). In order to maintain an acceptable variation in the angle of incidence ($90°±10°$), larger substrate sizes require longer throw distances. For instance, the throw distance required for a 4" square substrate is 20 inches while the throw distance required for an 8" square substrate is 40 inches.

The thickness uniformity across a substrate is also a function of the substrate size and the throw distance. As expected, uniformity decreases with increasing substrate size and increases with increasing throw distance. Deposition uniformity is also dependent upon the angle between the surface of the source and the surface of the substrate, the optimum substrate position being located directly over the source. The thickness uniformity obtained in our system for a 4" square substrate is ±5%.

The reaction of copper with polyimide is a concern, in terms of both long term reliability of the interconnections and the degradation of the dielectric properties of the polyimide. It has been shown that polyamic acid reacts with copper eventually leading to the formation of cuprous oxide, which can diffuse and aggregate with the aid of the solvent carrier in the polyamic acid solution [5]. Although the top and bottom surfaces of our copper interconnections are protected by the chromium layers, sidewall coverage is inadequate and therefore a passivation scheme is necessary. Two different methods to accomplish this objective are presently being pursued. The first method involves the deposition of chromium directly onto the metal features via a physical vapor deposition technique. The second method uses the deposition of electroless nickel to passivate the sidewalls.

Via Fill Process

The requirement of straight-walled vias compounded with the temperature limits imposed by the dielectric layers narrows the choices of via filling processes considerably. Electroless (catalytic) deposition of metal into vias was found to be the most suitable process currently available [6]. The electroless plating process is a selective deposition process which is normally performed at temperatures below 100°C. Metal is deposited onto catalytic surfaces by the chemical reduction of metallic ions in an aqueous solution.

Many different electroless plating solutions are available today. Several factors need to be examined when choosing an electroless solution for this particular application. The deposited metal must provide adequate conductivity. The metal must be stable at a temperature of 400°C since subsequent processing of the module necessitates this temperature excursion. The plating solution must be compatible with the dielectric present so solutions with a high pH must be avoided if polyimide is being used as the dielectric. Uniformity of plating and a moderate plating rate are critical since typical via heights are 8-16 μm. A nickel/1%-boron process (Allied-Kelite solutions) was selected because the low pH (pH = 6) of the plating solution is compatible with polyimide. In addition, the resistivity and plating rate (> 6 μm/hour) of the metal is acceptable.

This selective via filling process is composed of three steps. The first step entails removing the chromium in the exposed areas. This is accomplished by etching the chromium in an ultrasonic bath of 1:1, $HCl:H_2O$ solution. Care must be taken in assuring that the etch is complete and the exposed copper is free of contamination before proceeding to the next step.

Since the exposed copper surfaces are not catalytic to the subsequent deposition of nickel-boron, a palladium chloride aqueous solution must be used to produce a catalytic surface. This change in the surface occurs through a substitution reaction between the copper and the palladium ions. The immersion time of the substrates in this solution is only long enough to allow a few atomic layers of palladium to form on the copper surfaces. Once the copper surfaces have been 'activated' by using the palladium chloride solution they are catalytic to nickel-boron deposition. This deposition is accomplished by using an aqueous solution which contains nickel chloride and sodium borohydride. Optimum deposition parameters include a bath temperature of 60°C and continuous agitation of the solution. Vias that were filled using the process outlined above are shown in Figure 1.

Further enhancements to the via filling process are being evaluated. One process variation being examined is the use of an evaporated palladium layer between the copper and chromium layers as a substitution for the palladium activation step. Alternative plating solutions that provide enhanced conductivity are also being investigated.

Interlevel Dielectrics

Polymeric materials are the most common dielectrics used in thin film interconnection technologies, due to their low dielectric constants and good planarization properties. Although polyimides have traditionally been the most widely used interlayer dielectric, considerable effort to develop new materials with enhanced properties for this application is currently underway. Most of the commercially available options are analogs of the conventional polyimide chemistry, however new non-polyimide based materials including bis-benzocyclobutane (BCB) and polyphenylquinoxaline (PPQ) coatings [7], have also been receiving serious consideration. These materials possess a host of dielectric material properties that impact the processability and the long term reliability of a thin film substrate. Some of the major criteria are listed here. Prospective dielectrics must be thermally stable enough to withstand the numerous thermal excursions associated with fabrication of the thin film substrate itself, as well as the subsequent final module assembly processes, in our case to temperatures as high as 400°C. The dielectric must be compatible with all processing steps associated with the evaporation/lift-off metal process for deposition and definition of the Cr/Cu/Cr conductor layers, and our electroless nickel via filling process. For example, as mentioned above, a clean residue-free via etch of 8 to 16 μm thick layers of dielectric is required to facilitate activation of the underlying metal layer and the selective deposition of nickel. Additionally, the dielectric must adhere well to the underlying substrate, the selected metallization (Cr/Cu/Cr), and to itself to avoid short and long term delamination failures of the interconnection structure. Good chemical and moisture resistance of the dielectric material are also essential requirements to ensure manufacturability and long term reliability, especially in the case of applications involving non-hermetic assemblies. A detailed overview of commercially available dielectric materials was carried out as part of our thin film packaging effort and has recently appeared in the literature [8]. A summary of these evaluation results and those for PPQ are shown in Table 1.

Based on the above results, the low stress and acetylene-terminated polyimides as well as the BCB and the PPQ offerings showed greatest potential for this application. Of these materials, three were chosen for use in fabrication of a memory module. Dupont's PI2525, a popular industry standard and the dielectric used in MCNC's baseline thin film process was selected as well as Dupont's rigid, low stress polyimide (PI2611D) and Dow Chemical's BCB-1. Reactive ion etching as well as moisture absorption and release experiences with these 3 materials during the fabrication of this prototype will be outlined in the remainder of this section.

Reactive Ion Etching

In order to achieve maximum density in thin film designs, it is necessary to minimize the area required for the vias interconnecting the conductor planes. This is achieved using straight-walled non-nested vias. Because relatively thick layers of dielectric (8-16 μm) were required for the memory module thin film interconnection, an anisotropic, high rate (> 1 μm/min), reactive ion etching process is needed. A split cathode magnetron RIE system designed and built at MCNC [9] was used to achieve these process requirements. It was found that the minimal amount of etch residue obtained with this process was not detrimental to the subsequent selective electroless nickel via filling process outlined above.

In Table 2 and Figure 2, the differences in via dimensions obtained with our anisotropic via etch process are compared to that expected for isotropic via etching processes. The range of sidewall angles shown in Table 2 are representative of those obtained with conventional wet and dry etching processes. As shown in this example, isotropic etch processes result in a widening of the upper via dimensions to 20 or 30 μm for a nominal 10 μm diameter via in a 10 μm thick dielectric layer. As would be expected, this difference between top and bottom via dimensions increases with increasing dielectric thickness. Though isotropic via technologies are currently being used successfully elsewhere to satisfy less demanding design guidelines, high density designs necessitate straight-walled vias.

The masking layer initially used for etching through the layers of dielectric was a 2000 to 5000Å thick plasma enhanced chemical vapor deposited (PECVD) SiO_2 or Si_3N_4 layer. This layer was deposited directly onto the dielectric at temperatures between 300 and 310°C. Though past experience had shown that good etch selectivity was achievable with these masking materials in O_2/N_2 plasmas, our study of commercially available dielectric materials [8] demonstrated that the thermal stability of a number of otherwise potentially useful dielectrics including BCB-1, were incompatible with these high temperature mask deposition processes. Additionally, mask deposition at temperatures at or below the T_g of the dielectrics was observed to result in considerable cracking of the masking layer. Propagation of these cracks completely through underlying, thick dielectric layers was even observed in extreme cases. Lower temperature etch mask alternatives to the above PECVD masks were studied and included SiO_2 and Si_3N_4 films deposited in a downstream microwave plasma reactor, evaporated aluminum and a phosphorus doped positive acting photoresist [10]. In all of these cases, the mask deposition temperatures were held below 100°C. Since the evaporated aluminum masking process remains the most readily available of the low temperature alternatives available to us and this metal mask is non-erodible in both silicon-containing (BCB) and non-silicon containing (PI2525, PI2611D) dielectric etching conditions, it was selected for use in the fabrication of the memory module. Typical via etch results obtained with the aluminum mask through thick layers of PI2525 and PI2611D are shown in Figure 3.

The silicon component of BCB-1 incorporated in the chemical structure to enhance the adhesion of this material dictates a change in etching conditions from the straight O_2/N_2 plasma based process that was used for PI2525 and PI2611D. Under these etching conditions, a passivating layer of SiO_2 quickly forms on the surface of the BCB-1 thereby preventing subsequent etching. Incorporation of SF_6 or CF_4 into the etch gas mixture allowed removal of the passivating SiO_2. Use of fluorine containing gases to accelerate the etch rate of non-silicon containing polyimides in conventional RIE etching systems has also been reported in the literature [11], with a resulting loss of etch anisotropy. It was much more difficult to establish a satisfactory end point for the BCB-1 using the requisite fluorine-containing etch gas mixture, since the underlying Cr, if exposed, will be etched by this etch gas mixture. If the Cr layer was inadvertently etched enough to expose the underlying copper, the exposed copper conductors and adjacent regions were removed during efforts to remove the aluminum masking layer in standard commercial $H_3PO_4/HNO_3/CH_3COOH$ based etchants. This same end-point problem has also been one of the major obstacles for others involved in making the transition from aluminum to copper conductors. In response to this problem, techniques to more selectively remove the aluminum masking layer have been developed [10]. This situation was not encountered in the etching of PI2525 or PI2611D, since the underlying metallurgy remained unaffected by the O_2/N_2 plasma process.

Effects of Moisture Absorption

The moisture absorption results for the 3 dielectric materials selected for use in the fabrication of the memory module varied considerably when measured using a standardized testing procedure involving immersion of coated silicon wafers in boiling water [8]. Under these conditions, the PI2525 polyimide absorbed 1.7% moisture by weight, whereas the low stress PI2611D polyimide and the BCB absorbed 0.5 and 0.3% respectively. The significant variation in moisture absorption observed in this testing procedure for these 3 materials accurately predicted differences in the relative ease of controlling moisture release and fabrication of these substrates, and subsequent assembly of the complete module. Uncontrolled outgassing of moisture from underlying layers of dielectric during subsequent high temperature processing can result in irreversible damage to the thin film structure as highlighted in the following two examples encountered during the fabrication of the module. Figure 4(a) shows damage to an aluminum masking layer in an non-thermally bonded sample, as a result of moisture outgassing from the underlying polyimide layer (PI2525) during etching in the magnetron RIE system. Under normal etching conditions, the substrate is thermally bonded to the water-cooled cathode of the reactor using glycerin, resulting in an etch temperature of 25-30°C. Improperly bonded substrates can heat up quite rapidly to 250-300°C, resulting in violent outgassing as mentioned above. Damage to the underlying polyimide layer can also be quite extensive as shown in Figure 4(b). Presumably since PI2611D and BCB-1 films tend to absorb and retain much less moisture than PI2525, no damage to the aluminum masking layers or to the dielectric layers themselves was observed with these materials whether the sample was kept at room temperature or was allowed to heat up. A more rigorous pre-metallization polyimide bake-out cycle was found to reduce the sensitivity of the PI2525 fabricated thin film structure to moisture outgassing damage. Preliminary results from flip-chip assembly of integrated circuits onto completed substrates suggest similar damage may occur during the solder reflow part of module assembly. Work is continuing to more clearly understand and control moisture uptake and release in our thin film structure during fabrication and in the field.

293

Integration of Technologies

The processes for fabricating the individual components of a multi-layer thin-film substrate, described above, were integrated to fabricate a multi-level thin film structure on an alumina substrate. A polyimide coating was applied to the alumina surface for electrical isolation and for reduced surface roughness. Then the deposition of the first level metal of Cr/Cu/Cr was done by the liftoff technique. A second coating of polyimide was applied and vias were opened by reactive ion etching. Via filling by electroless Ni onto Cu was performed after removal of the covering Cr. The cycle between metallization and via filling was repeated to build the second metal level and to expose I/O pads for C-4 connection. A SEM micrograph of the cross section of the structure is shown in Figure 5.

Defect Characterization/Repair Technology

For a wafer scale integrated (WSI) process with a 10 μm linewidth on a 25 μm pitch the predicted yields were 97% for 0.1 defects/cm² and 88% for 0.5 defects/cm² [12]. This high yield, according to the authors [12], for wafer scale integration reflects the sophisticated tooling and low-particulate environment in which the wafers are processed. For these geometries, the wires are immune to the faults caused by small defects, and those defects which have large size occur less often in a controlled environment. However, this prediction is overly optimistic. The low yield experienced with thin film modules cannot be properly explained by conventional particulate contamination statistics. A more realistic yield prediction can be obtained by using a simple random defect model, derived from Poisson statistics:

$$Y = e^{-(DA)},$$

where
Y = yield,
D = defect density,
A = active area.

Using this model and a defect density of .1/cm², one finds that the predicted yield is 90% for an area of 1 cm² and is effectively 0% for an area of 100 cm². This predicted zero yield illustrates the necessity of a repair step at each interconnection level.

The multichip modules under consideration involve complex circuitry over a large area, typically 10 cm x 10 cm. The pattern is integral and not confined to discrete chips and therefore defective sections cannot be discarded. Interconnection lines can be as long as 8 cm, with 10 μm linewidths and 20 μm spaces. Process experience indicates a need for a circuit repair technology to achieve maximum yield. After each processed layer, the modules will have to be scanned for defects and they will have to undergo a repair step when appropriate. The repair work will be aimed at filling opens and gaps and removing shorting defects.

Defects occurring in circuitry can be mask related, material related, or process/machine related. Mask related defects occur in repetitive and identical patterns, and are therefore easier to control. In hybrid scale lithography they can be drastically reduced by using proximity exposure instead of contact exposure, as outlined in the patterning section. Material and process related defects tend to be random in nature. However defects occurring within a certain lot can often be related to process/machine problems, whereas material related defects tend to appear with changes of processing baths, of process gas cylinders, and of other processing materials.

The defects that are considered repairable under this program are visible voids in interconnect lines or vias, and visible protrusions or residues between lines. Defects are considered acceptable, i.e. not causing rejection if the size of the defect is smaller than 50 percent of the line/via width for voids and smaller than 50 percent of the gap width between lines for protrusions and residues, as outlined in MIL-STD-883C [13].

Optical scanning is essential for locating sites needing repair. Only visible defects are considered repairable. Defective sections where no visible problem can be detected would be practically non-repairable . Consequently, electrical probing, while very useful as a screening tool, would not suffice for inspection/repair applications since the exact defect location cannot be determined.

To improve the yields of thin film interconnection a new process approach has to be developed. For example, redundant metallization deposition and patterning utilizes a plurality of metallization depositions to replace a cost effective single metallization to reduce the chances of voids or opens in a metallization pattern. This repetition of a simple metal and liftoff patterning will substantially extend the process steps, however, it will also drastically increase the the processing yields of the thin film module. The probability of random voids or opens that are formed in double patterned layers is minimal.

In order to repair gaps in lines and in via fillings we are presently developing a process to do localized selective metallization using a Research Devices 5010 Pattern Generator to delineate the repair pattern. The generator can be used as a 'defect locator' to store the location and shape data of defects in a CAD-like file. This data can be later used to expose a repair pattern on the device, using the alignment capability of the system. In Figure 6 we show a localized exposure of photoresist obtained with the 5010 Pattern Generator. Shorts caused by metallic bridging between lines have been removed successfully in our laboratory by laser ablation. For that end we have used a Xenon laser (The Florod Corporation Xenon Laser Cutter, Model LFA). In the cutting process, the laser pulse causes local melting of the metal line. The metal flows away from the cut and solidifies, resulting in a balling up of metal at the conductor ends, see Figure 7. If the repair is done after polyimide deposition, it is necessary to provide windows in the polyimide for any metal cutting because exposure to a laser beam causes charring of the polyimide. In the absence of a window, this creates a conducting carbon residue typically providing a leakage path of a few thousand ohms across the cut in the conductor. Cutting through oxide has been demonstrated successfully.[14,15] Alternatively, metallic shorts could be removed by selective etching, using the same pattern generator. Also, implementation of built-in redundancy might increase the cost effectiveness of multichip modules. Placing redundancy lines along with engineering change pads at the top surface of the module allows the re-routing of lines through this reconfigured grid.

Chip Bumping and Joining Processes

The prevalent technologies for chip-to-substrate interconnections include wire bonding (WB), tape automated bonding (TAB), and controlled collapse chip connection (C-4). The C-4 technology, a two-dimensional interconnection by nature, was selected because of the smaller inductance and capacitance of the solder joints when compared to bonding wires or TAB leads. The key feature of the C-4 technology is the deposition of a thin film structure and solder on the I/O pads. The thin film structure, usually termed ball-limiting metallurgy (BLM) or under-bump metallurgy (UBM), serves the purpose of an adhesion layer, a diffusion barrier, and a solderable base of controlled wetting area. The solder pads are aligned and tacked onto the matching footprint on the substrate, and connection to the substrate is made during the reflow of the deposited solder. Surface tension of the molten solder prevents the chip from collapsing during reflow.

Evaporation through metal (e.g. Mo) masks and electroplating are known methods of solder deposition. In solder evaporation processes, mechanical fixtures are necessary to align and clamp the metal mask to the wafer or to the substrate. In designing such mechanical fixtures one needs to take into account factors such as the evaporation conditions, mechanical properties of mask/fixture and silicon, wafer size, layout of I/O's, dome curvature, etc. As the wafer size increases and as the I/O size decreases for higher I/O density, achieving the goal of accurate pattern transfer by mechanical means as well as uniform deposition of solder on all I/O's becomes more difficult. In electroplating, pattern transfer is done by photolithography, and no mechanical parts are needed. Photo processing has inherently higher accuracy in pattern registration as compared to the mechanical alignment scheme for the solder evaporation process. Electroplating was selected for solder deposition in our process.

The process sequence for our electroplating based C-4 technology was as follows:

- depositing a UBM layer as the plating base
- laminating Riston dry-film photoresist
- UV-exposure for pattern transfer
- developing the photoresist
- plating solder into cavities formed in the developed photoresist
- stripping the photoresist
- etching the UBM using plated solder as the etch mask
- reflowing solder pads, if desired
- aligning and tacking chip to substrate
- reflowing for chip joining.

A detailed description of the individual steps is presented below.

A multi-layer, thin film structure was evaporated onto the wafer surface to serve as the plating base. This film had the composition of 1000Å Cr, 1000Å 50/50 graded Cr/Cu, 10000Å Cu, and 1000Å Pd. The evaporation was done in a multi-source e-beam system (Balzers BAK 760). The Cr layer served as the adhesion layer as well as a barrier for solder diffusion into the Al metallurgy underneath. Since Cr is not wettable by solder, a Cu layer was incorporated into the structure to be wetted by and alloyed with solder. The graded Cr/Cu layer was incorporated to prevent complete leaching of Cu by solder. It has been reported in literature [16] that solder balls would spall off the I/O pads in the absence of the graded Cr/Cu phase. The Pd layer served as an oxidation barrier. Since the I/O sites on the chip surface could be delineated through this film during photo imaging of the dry-film photoresist, patterning the deposited film for alignment purposes was unnecessary. At the end of the C-4 process flow, the part of this thin film structure left on the chip served as the UBM.

Thick (2-3 mils) dry-film photoresist was selected for patterning the sites for bump plating in order to provide the desired thickness (3-5 mils) of the solder bumps for fine-pitched I/O's. Thin (~ a few microns) liquid photoresists could also be used, but lateral growth of the electrodeposits, commonly known as "mushrooming", takes place. Mushroom-shaped deposits limit the minimum pitch of I/O's. Nonetheless, processing with liquid photoresist should be considered when the I/O pitch allows its use, since processing with liquid photoresists is compatible with IC fabrication, and is also superior in pattern transfer than the dry-film photoresists.

A hot-roller laminator (DuPont HRL-24) was used to laminate DuPont's Riston dry-film photoresists onto the wafers. Vacuum lamination was also tested and proved successful. No cracking of either the silicon wafer or the passivation dielectric was detected. Contamination from the Riston film was of no concern since the active devices on the wafer have been isolated by the deposited UBM layer.

UV-exposure was done with a Canon mask aligner (PLA-501F) to register the pattern of solder bumps onto the photoresist. A 500 W Xe/Hg vapor lamp was the light source, and typical dosage was ~70 mJ/cm². Soft contact mode was selected. The exposed dry-film photoresist was developed by spraying a 1% soda ash solution onto the wafers. To assure the long term reliability of the solder joints, the floor of the cavities should be large enough to seal the Al I/O metallurgy underneath and should also be free of residues of photoresist.

The plating bath was a methanesulfonic acid bath (Lea-Ronal, SolderOn acid), and was formulated by mixing Pb-concentrate, Sn-concentrate, and a catechol-based additive (MHS-L) with the acid. This additive was for plating matte lead. The composition of the plated solder was not significantly affected by either the current density (5 - 20 mA/cm²) or the mass transport rate, but rather was sensitive to the ratio of Pb- and Sn-components in the bath. Bath composition was selected to deposit solder in the 90/10-95/5 (wt %) Pb/Sn range.

The plating cell was designed to be an open-tank, continuous-flow unit. The wafer was the cathode. A lead sheet served as the anode. The Sn content in the bath decreased gradually with time, which was no concern in the laboratory operation, but in production mode the use of anodes having the same composition as the solder deposits and monitoring of bath activities should be practiced. During the plating process, the wafer surface was continuously swept by electrolyte, but the blind-via geometry limited the supply of metal ions to the recessed area where plating took place. Consequently, most of the plating was done at a current density below 15 mA/cm² to avoid adverse effects due to deficiency of mass transport, such as hydrogen evolution and loose, powdery deposits. After solder plating, the wafers were stripped of the dry-film photoresist in 3% KOH solution.

SEM micrographs for as-plated solder pads from two batches of processed wafers are shown in Figure 8. The plated solder columns replicated the cavity shape exactly, as is evidenced by the ridge-like features on the side wall of both solder columns. The difference in the appearance of the two plated solder columns points to the presence of a wide window of operation in photo-processing. The "mushrooming" phenomenon is illustrated in Figure 8(a). For this sample, the solder column had a vertical sidewall and a foot around its base. In Figure 7(b) a crevice can be seen at the outer edge of the base of the solder column, indicating the presence of a lip of photoresist residues around the circumference of the cavity floor, and limits the reliability of the C-4 joints. The presence of such a crevice facilitates undercutting during wet etching of the UBM layer, and would lead to detachment of pads if I/O metallurgy is exposed to the etchants. Solder deposition could be achieved via the liftoff process due to the slightly inverted (overhang) side wall of the developed photoresist, giving the potential of a simpler C-4 process using only photolithography and evaporated solder.

In etching the UBM, one desires selective etching of the Pd, Cu, and Cr layers, and the etching should incur minimal undercutting. An etchant was selected for etching Pd and Cu altogether, while leaving the Cr layer intact so that devices on the wafer could be protected during Pd/Cu etching. The HCl-based Cr etchant incurred gas evolution and we have observed, in at least one case, shattering of the passivation dielectric during Cr etching. The use of potassium ferricyanide-based alkaline etchant alleviated this problem. In some test samples including the one shown in Figure 8(a), the UBM structure did not contain the Pd layer and selective etching of UBM was achieved by using ammonium persulfate as Cu etchant and potassium ferricyanide or HCl as Cr etchant.

Reflow of deposited solder pads in N_2 was successful, with the application of various R and RMA fluxes. The peak temperature in the reflow process was ~360°C, which was beyond the recommended range of use for all fluxes tested. Charred flux residues were left on the wafer, and could not be effectively removed by common fluoro-chloro hydrocarbons. The charred debris could prevent successful chip joining during the second reflow. Fluxless reflow in H_2 was also investigated and proved successful. It is advisable not to reflow the bumped dies before they are tacked for joining, and fluxless reflow in H_2 is recommended if reflow prior to joining is necessary.

A custom-built flip-chip bonder (Laurier Inc.) was used for tacking and aligning chips to substrate. The machine used a dual-path optics to acquire images of both the chip I/O's and the substrate pads. The operator moved the substrate stage until the images overlapped. The chip collet and the substrate stage were typically held at 40°C, and the die was held in place by flux before a permanent bond was to be formed during the reflow process.

As a test vehicle, we fabricated solder bumps in an area array format on Si wafer, beginning with deposition of Al I/O pads. The pitch of the I/O pads was 10 mils, and diameter of the base for solder bumps varied from 4.5 mils to 6 mils. A SEM micrograph for the reflowed solder bumps is shown in Figure 9(a), and shown in Figure 9(b) is the cross section of such a chip joined to another chip, which serves as a substrate with an identical footprint.

The C-4 process was also tested on a commercial memory chip, which had square I/O pads on the periphery, with one I/O pad three times larger than the others. A square pad of a solder-nonwettable metal was deposited at the center of this large I/O pad to create two pads of a size equal to the other I/O's. This step was necessary to avoid the stretching of solder joints at I/O sites with low ratio of solder volume to wetting area. Shown in Figure 10(a) are the reflowed solder bumps on one edge of the memory die. The solder bumps in the middle of the chip were deliberately added to improve the mechanical integrity of the overall bonding structure. Shown in Figure 10(b) is the cross section of a memory die after being joined to a multi-layer thin-film substrate. In order to seal the exposed I/O pad, the diameter of the UBM layer underneath the solder needs to be theoretically at least 1.42 times the side of the square pad or bigger if tolerance of photo-processing and wet etching are granted. The via opening through the passivation dielectric is made so that conformal coverage and electrical connection can be made to the I/O's by the deposited UBM. No practical advantage is to be gained by exposing the entire I/O pad. Misalignment and UBM undercutting tolerances are better when the via opening size is smaller relative to the base of the solder pads. A 1-mil opening would be large enough for a micron-thick dielectric, and also circular via openings are more adaptable for C-4 processing.

It is imperative that flip-chip bonded modules be repairable. The repairing process involves detachment of the defective chips and the removal of solder residues. Design of a hot gas dressing station has been published in literature [17], and the issue of space requirement for repairing flip chips with this machine was also addressed, which the designer of a multi-chip package should bear in mind. For closely bonded dies with peripheral I/O's, a potentially viable scheme for removing solder residues is to place a heated precision-machined solder-wettable plate inside the cavity after the detachment of the defective chip, and to remove solder residues by wicking.

The C-4 process outlined above can be applied to commercial chips. Implementation of this process could be simpler if the chips are specifically designed for the C-4 process, namely having small circular via openings of uniform size in an area array format located near the center of the die.

Summary

The key process steps; patterning, conductor deposition, reactive ion etching, dielectric deposition, via filling process, and chip to substrate bonding, utilized in the fabrication of high performance thin film multichip modules were discussed. The integration of the developed process steps for the fabrication of a 128-chip memory module which served as a demonstration vehicle was described. The importance of repair techniques for yield enhancement was addressed, and solutions for defect repairs were proposed. Thin film multichip packaging technology offers high density interconnection capability and design flexibility. To make this technology economically viable, a number of issues still need to be resolved.

Acknowledgment

The authors are grateful to personnel in MCNC's central laboratories for their contributions to the progress of this project. Special thanks go to Mr. D. A. King for his efforts in processing and sample preparation. The authors would also like to thank DuPont Electronics in Research Triangle Park, NC, for allowing us to use their facilities and for assisting us with photoprocessing.

References

[1] R. W. Johnson, "Thin Film Multichip Hybrids: An Overview," proceedings NEPCON West, March, 1989, pp. 655-682.

[2] J. J. H. Reche, "High Density Multichip Interconnect for Advanced Packaging," proceedings NEPCON West, March, 1989, pp. 1308-1318.

[3] L. M. Levenson, C. W. Eichelberger, R. J. Wojnarowski, and R. O. Carlson, "High-Density Interconnects Using Laser Lithography," Proceedings NEPCON West, March, 1989, pp. 1319-1327.

[4] S. K. Jones, R. C. Chapman, G. Dishon, and E. K. Pavelchek, "Image Reversal and Lift-off: A Versatile Combination for ULSI Processing," Proc. Eighth International Technical Conference on Photopolymers (SPIE), Oct. 1988, pp. 279-291

[5] S. P. Kowalczyk, Y. H. Kim, G. F. Walker, and J. Kim, "Polyimide on copper: The role of solvent in the formation of copper precipitates," Applied Physics Letters, Feb. 1988, vol. 52, no. 5, pp. 375-376.

[6] W. J. Bertram, Jr., "High-Density, Large Scale Interconnection for Improved VLSI System Performance," IEDM Technical Digest, 1987, pp. 100-103.

[7] J. J. H. Reche, "Material Selection For VLSI Multichip Modules," 3rd Int. SAMPE Electronics Conference, 1989.

[8] T. G. Tessier, G. M. Adema and I. Turlik, "Polymer Dielectric Options for Thin Film Packaging Applications," 39th Electronic Comp. Conference, 1989, pp. 127-134.

[9] G. J. Dishon, S. M. Bobbio, T. G. Tessier, Y.-S. Ho and R. Jewett, "High Rate Magnetron RIE of Thick Polyimide Films for Advanced Computer Packaging Applications," Journal of Electronic Materials vol. 18, no. 2, 1989.

[10] T. G. Tessier, G. M. Adema, S. M. Bobbio and I. Turlik, "Low Temperature Etch Masks for High Rate Magnetron RIE of Polyimide Dielectrics in Thin Film Packaging Applications," 3rd Int. SAMPE Electronics Conference, 1989, pp. 85-93.

[11] R. J. Jensen, J. P. Cummings and H. Vora, "Copper/Polyimide Materials System for High Performance Packaging," 34th Electronics Comp. Conference, 1984, pp. 73-81.

[12] J. F. McDonald, A. J. Steckl, C. A. Neugebauer, R. O. Carlson, and A. S. Bergendahl, "Multilevel Interconnections for Wafer Scale Integration", J. Vac. Sci. Technol. A, vol. 4, no. 6, Nov/Dec 1986, pp. 3127-3138.

[13] MIL-STD-883C, Test Methods and Proceedings for Microelectronics, Method 2017.3, 8/1983.

[14] M. Mitani et. al., "Laser Cutting of Aluminum Thin Film with No Damage to Under-Layers," 29th General Assembly of CIRP, Sept. 1979, pp. 113-116.

[15] J. C. Logue, et. al., "Techniques for Improving Engineering Productivity of VLSI Designs," IBM J. Res. Develop., vol. 25, no. 3, May 1981.

[16] B. S. Berry and I. Ames, "Studies of the SLT Chip Terminal Metallurgy", IBM J. Res. Develop., 286-96, May, 1969.

[17] K. J. Puttlitz, "Flip-Chip Replacement Within the Constraints Imposed by Multilayer Ceramic (MLC) Modules," J. Electron. Mat., 13(1), 29-46, 1984.

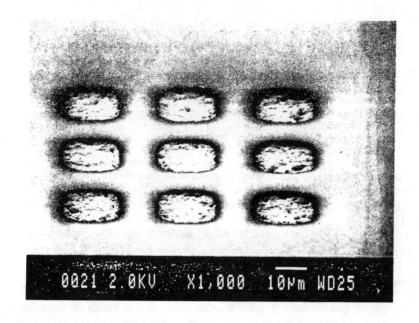

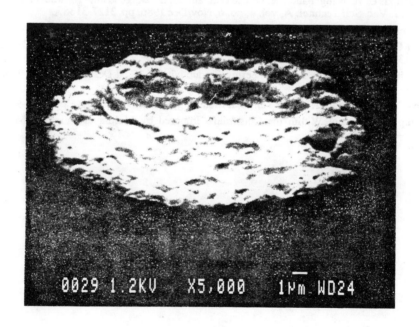

Figure 1. Electroless Ni-B filled vias

Material Type	Dielectric Constant	Water Absorption (wt. %)	T_g (°C)	Planarization (%)
Polyimide	3.4-3.8	1.7	300-320	24-34
Fluorinated PI	2.7	0.7	<320	25-30
Silicone PI	2.9	0.9	<300	22
Acetylene PI	2.9-3.4	0.7-1.4	<320	91-93
BCB	2.6	0.3	<360	91
PPQ	2.7	0.8	365	25
Low Stress PI	2.9	0.4	>400	25-28

Table 1. Dielectric properties summary

Isotropic via etch

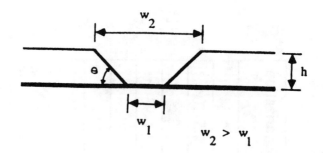

Anisotropic via etch

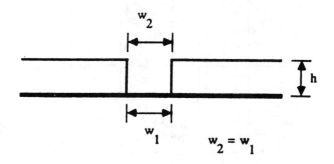

Figure 2. Isotropic and anisotropic etch profiles in thick dielectrics

Sidewall	Thickness of Dielectric Layer		
Angle	10μm	15μm	20μm
45°	30.0	45.0	60.0
60°	21.6	27.1	33.1
70°	17.3	20.9	24.6
90°	10.0	10.0	10.0

Table 2. Upper via dimensions (w_2) for isotropic and anisotropic etch profiles where $w_1 = 10$ μm

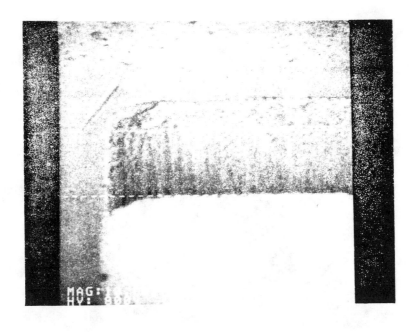

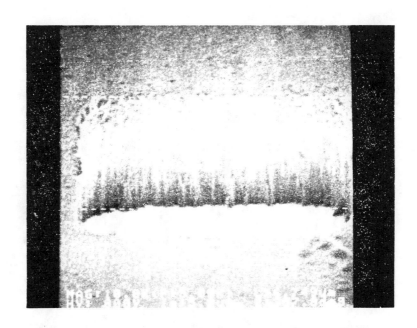

Figure 3. Typical via etch results obtained with PI2525 and an evaporated aluminum mask

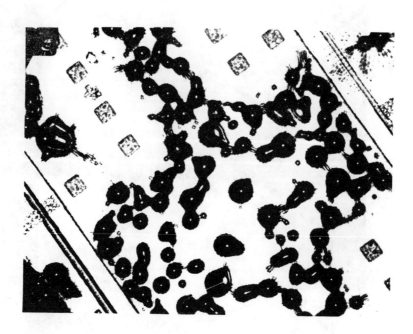

Figure 4(a). Heating induced moisture outgassing damage in a thin film structure fabricated with PI2525 (Non-bonded)

Figure 4(b). Heating induced moisture outgassing damage in PI2525 layer viewed after aluminum mask removal

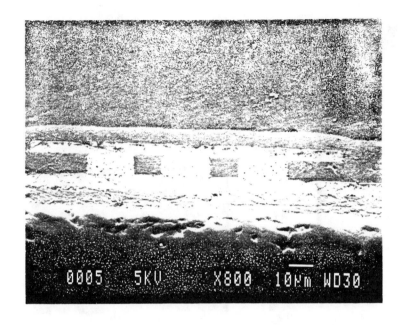

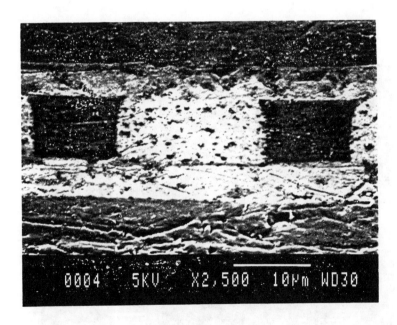

Figure 5. Cross sections of a multilayer thin film subtrate

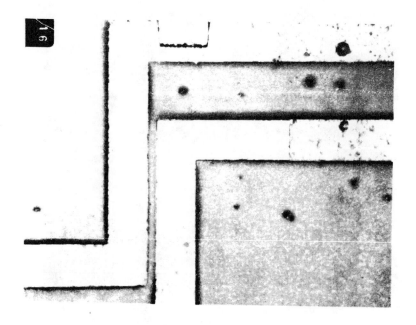

Figure 6. Localized exposure with a pattern generator

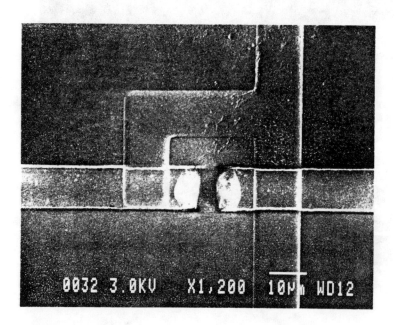

Figure 7. Laser cutting of conductor line

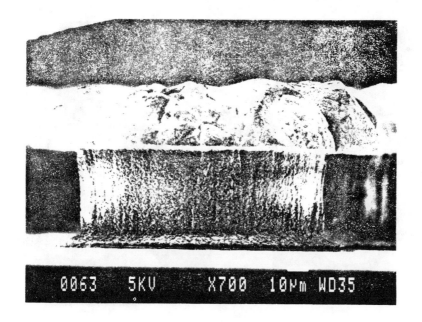

8(a).

8(b).

Figure 8. As-plated solder pads
8(a). Nominal pad diameter: 4.5 mils
8(b). Nominal pad diameter: 5.5 mils

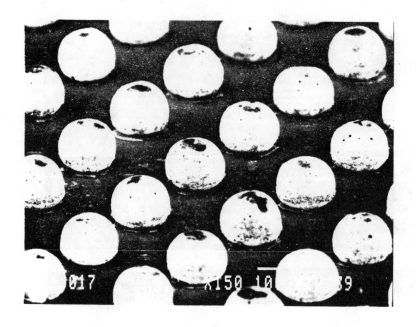

9(a).

9(b).

Figure 9. Solder bumps and joints
9(a). Area array of solder bumps after reflow in H_2
9(b). Cross section of solder joints

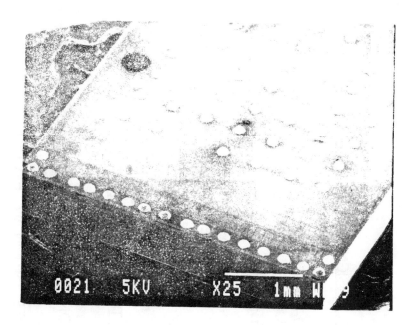

10(a).

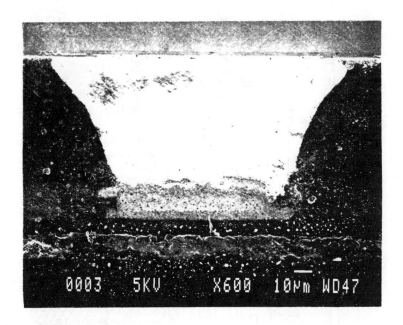

10(b).

Figure 10. Solder joints between a memory chip and a substrate
10(a). Solder bumps on a memory chip
10(b). Cross section of a memory chip joined to a substrate

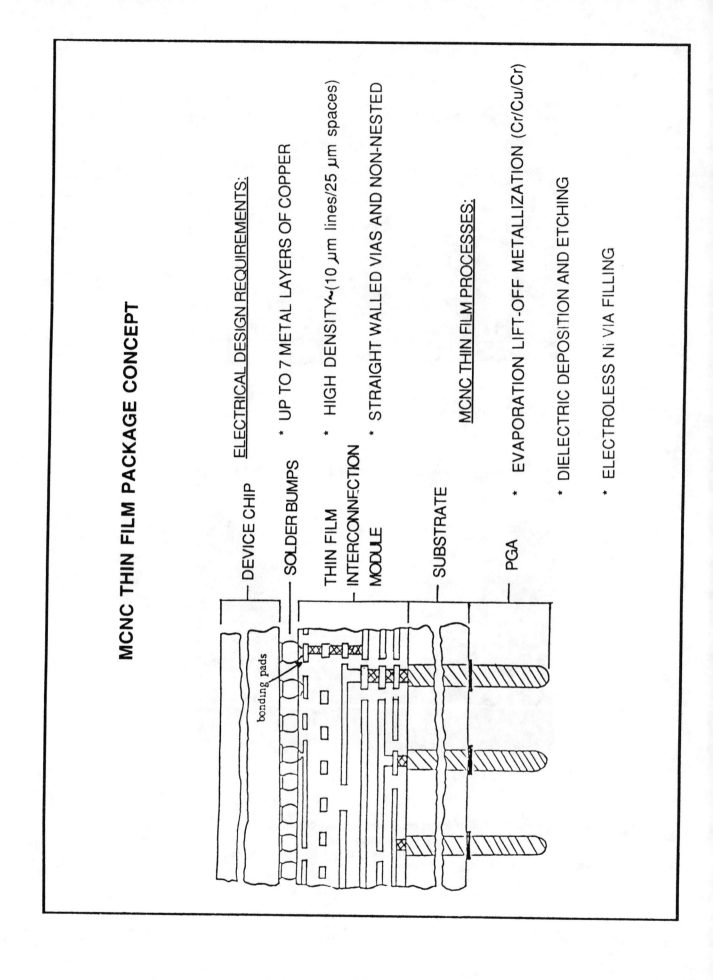

MCNC THIN FILM PACKAGE CONCEPT

ELECTRICAL DESIGN REQUIREMENTS:

* UP TO 7 METAL LAYERS OF COPPER

* HIGH DENSITY ~(10 μm lines/25 μm spaces)

* STRAIGHT WALLED VIAS AND NON-NESTED

MCNC THIN FILM PROCESSES:

* EVAPORATION LIFT-OFF METALLIZATION (Cr/Cu/Cr)

* DIELECTRIC DEPOSITION AND ETCHING

* ELECTROLESS Ni VIA FILLING

DEVICE CHIP

SOLDER BUMPS

THIN FILM
INTERCONNECTION
MODULE

SUBSTRATE

PGA

bonding pads

EVAPORATION / LIFT-OFF

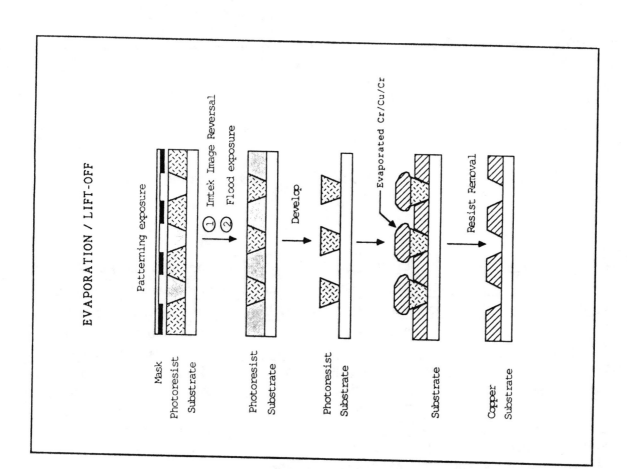

ELECTROLESS NICKEL VIA-FILLING
PROCESS FLOW

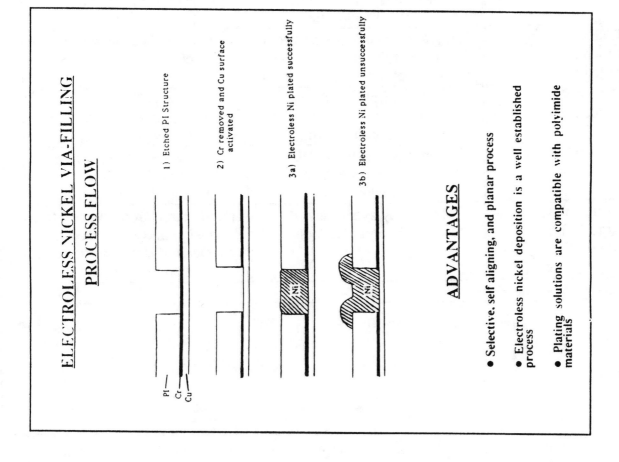

1) Etched PI Structure

2) Cr removed and Cu surface activated

3a) Electroless Ni plated successfully

3b) Electroless Ni plated unsuccessfully

ADVANTAGES

• Selective, self aligning, and planar process

• Electroless nickel deposition is a well established process

• Plating solutions are compatible with polyimide materials

309

COMPARISON OF PROCESSES FOR SOLDER DEPOSITION

Factor/Process	Electroplating	Evaporation through Mo mask
Solder purity	lower, higher chance of voids	higher, lower chance of voids
Contamination	possible, during etching of UBM	much less a concern
Control of solder composition	varying bath composition	varying source temperature or source composition
Complexity of process steps	generally straightforward except for etching of UBM and developing Riston	mask/wafer thermal mismatch an issue in mask design; design of mask clamping mechanism tricky
Photolithography cycles needed	2(1), one (none) for UBM and one for bump site openings	none
Solder wetback	generally, no solder base is defined by photolithography	yes using smaller mask openings for UBM deposition
Difficulty in process automation	manual steps in cutting wafers off photoresist film, alignment during photolithography cycles	manual steps in aligning and clamping the Mo mask
Process speed (throughput)	~2 hrs for plating 3 mils; can increase throughput via multi-wafer plating process	??? pumping down, evaporation, and cooling; mask clamping
Concerns	controlled etching of UBM resolution of Riston corrosion of Al	accuracy of mask patterns; maintaining alignment throughout processing
Extension to large (8") wafers	limited by availability of photolithographic tools	need much larger evaporator; design and fabrication of masks and clamping devices extremely difficult
Capital investment	low	high

A FINE-LINE MULTILAYER SUBSTRATE WITH PHOTO-SENSITIVE POLYIMIDE DIELECTRIC AND ELECTROLESS COPPER PLATED CONDUCTORS

Takaaki Ohsaki, Toyoshi Yasuda, Satoru Yamaguchi, and Taichi Kon

NTT Electrical Communications Laboratories
3-9-11, Midoricho, Musashino-shi, Tokyo 180, Japan

ABSTRACT

A new fine-line multilayer substrate with a photo-sensitive polyimide dielectric and electroless copper plated conductors is developed. To realize the characteristic impedance and resistance of the signal lines necessary for high-speed transmission, a thick polyimide dielectric layer and via-holes 20 μm or more in depth are required. Careful temperature control of the thick photo-sensitive polyimide both before and after the photolithography process is required to obtain a good via-hole configuration, and electroless plating is employed in forming such deep via-holes. A conductor adhesion strength of 500 kg/cm^2 is realized by innovative chemical treatment and sintering. Repeated thermal shock tests indicate that the 50 μm square via-holes through the 23 μm thick polyimide layer are highly stable and reliable. The propagation delay of the fabricated 25 μm wide, 6 μm thick strip lines is 60 ps/cm, which is 40 % less than for a co-fired multilayer ceramic substrate. The fabricated substrate can transmit high-speed pulses at over 2 Gbit/s.

INTRODUCTION

Continual advances in the speed and integration scale of the integrated circuits used in high performance systems have created ever greater demands for higher-density packaging to ensure reduced interconnection delays for improved electrical performance. Multichip packaging on a co-fired multilayer ceramic substrate has already been used in high-performance computer systems because it provides high packaging density [1-2]. However, conventional co-fired multilayer ceramic technologies using a screen printing method can only produce conductors wider than 70 μm and via-holes larger than 100 μm. Furthermore, since alumina ceramic has a high dielectric constant, signal propagation delays become large [3].

Recent developments in the fabrication of multilayer interconnection with a polyimide dielectric layer have expanded into high-density LSI multichip packaging [4-11]. Polyimide is chemically inert, mechanically tough, and thermally stable up to 400°C. Furthermore, since polyimide has a low dielectric constant, signal propagation delays are shorter. As another advantage, it is possible to form fine conductors on it, because the conductor lines are fabricated by sputtering and photo-etching processes. However, for higher-speed transmission, thick polyimide dielectric layers 20 μm or more in depth are required to realize adequate characteristic impedance and resistance in the signal lines. Also small via-hole formation through the thick polyimide dielectric is extremely difficult by the conventional sputtering and photo-etching processes.

This paper describes a new fine-line multilayer substrate formed on a ceramic substrate. It features small via-holes in photo-sensitive polyimide dielectric layers and fine patterns produced by electroless copper plating. Photo-sensitive polyimide is employed because the required polyimide layer can be achieved by spin coating, and the via-hole openings can be formed by applying only exposure and

development processes [6][10]. Electroless copper plating is employed to form such deep via-holes, because the electroless plating technology possess uniform metallization characteristics even on vertical surfaces and copper is a desirable conductor material because of its high conductivity, low cost, and good solderability. This technology thus results in low cost fabrication of high-density multilayer substrates.

In the first section of this paper the individual processes involved in conductor patterning, polyimide patterning, and via connection are described. The last section presents an evaluation of the electrical properties of a substrate fabricated using these processes.

MULTILAYER PROCESSING

The fabrication process for a multilayer substrate compared with the conventional method that uses a non photo-sensitive polyimide dielectric and a sputtered conductor is shown in Fig. 1. (1) A thick copper conductor is formed by electroplating with a photoresist after a thin copper film is deposited by electroless plating. The electroless plated film is

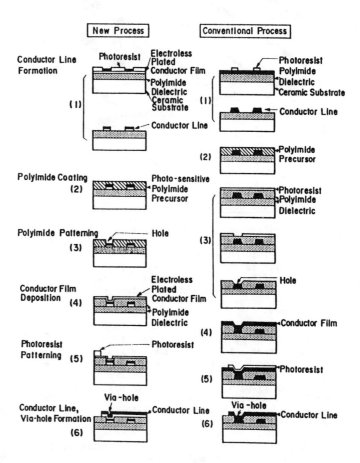

Fig .1. Fabrication processes for multilayer substrates.

Reprinted from *Proc. IEEE Int. Electron. Manuf. Technol. Symp.*, pp. 178-183, 1987.

then etched off. (2) A photo-sensitive polyimide precursor is spincoated and prebaked. (3) Small holes are easily formed by appling only exposure and development processes. Thermal curing of the photo-sensitive polyimide precursor turns it into a polyimide dielectric. (4) A thin copper film for both the second conductor layer and the via connection is then deposited by electroless plating. (5) Photoresist patterns are formed for selective electroplating. (6) A copper conductor is electroplated selectively. After this, the electroless plated film is etched off. (7) Processes (1)-(6) are repeated, and a multilayer substrate is formed. Thus, the use of a photo-sensitive polyimide dielectric and electroless copper plated conductors result in a simpified fabrication process.

CONDUCTOR PATTERNING

A fine conductor is required for high density packaging. Consequently, formation of copper conductors with fine patterns and cross-sections with vertical edges has been investigated. This can best be acomplished by using selective electroplating technology. Since relatively thick photoresist patterns are needed for selective electroplating, the steepness of the edge of the patterns is quite important. The optimum conditions for photoresist pattern formation were determined by experiments [6].

Fig. 2 shows a conductor pattern that has vertical cross-section edges. The line width is 25 μm and the thickness is 6 μm. The resistance of the line was 1.3 Ω/cm.

Fig. 2. Conductor pattern with 25 μm wide, 6 μm thick lines.

POLYIMIDE COATING and PATTERNING

For multilayer packaging, it is very important to realize low via contact resistances and reliable via connections. Thin chromium plating on the copper conductor prior to the photo-sensitive polyimide spin coating was adopted to prevent chemical interaction between the copper and the coated photo-sensitive polyimide, and to decrease the via contact resistance [6].

The thickness of the polyimide dielectric was designed to be 23 μm in its cured form to realize a 50 Ω characteristic impedance on the signal lines. This is very thick compared to conventional polyimide technology. Polyimide coating was achieved with spin coating by applying a higher viscosity polyimide precursor and slowing the spin speed. Photo-sensitive polyimide, a negative working photopolymer, allows direct production of patterns by conventional photolithographic techniques. After coating the polyimide precursor onto a substrate and prebaking it, the photo-sensitive polyimide was exposed and developed. Developing was done by batch immersion with ultrasonic application. Ultrasonic developing provided the most versatile means of uniform

developing of a thick films. As development time and via-hole configuration depended on developer temperature, developer temperature was controled to within ±0.5 °C. Careful temperature control of the photo-sensitive polyimide both before and after the photolithgraphy process was required to obtain a good via-hole configuration. Rinsing was done by batch immersion with ultrasonics, using isopropyl alcohol as the rinse solvent. Finally, the substrate was postbaked and final cured at 400 °C for 30 minutes, and the protective chromium film on the copper conductor in the via contact holes was etched off.

By applying these techniques, experiments on small via-hole fabrication were carried out, and it was established that 50 μm square via-holes could be successfully fabricated through a 23 μm thick photo-sensitive polyimide layer (see Fig. 3). This technology thus results in a short process turnaround time and small via-hole formation.

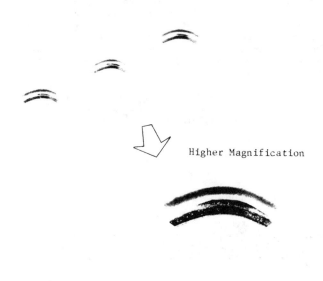

Higher Magnification

Fig. 3. 50 μm square via-holes through 23 μm thick photo-sensitive polyimide layer.

ELECTRICAL VIA CONNECTION

Once the via-holes have been formed, the next step is metallization. However, since the via-holes are 23 μm deep and have vertical walls, it is difficult to accomplish uniform metallization by conventional sputtering. As a result, electroless copper plating is employed in forming these deep via-holes because the electroless plating provides uniform metallization even on vertical surfaces. Then, after a thin electroless plating, a thick copper conductor is deposited by electroplating. The detailed metallization process is shown in Fig. 4.

From the aspect of improving the adhesion force of the electroless plated film, an appropriate surface treatment is required. Various chemical roughening and wetability promoting processes were examined. Out of this, an innovative chemical treatment using hydrazine hydrate, which acts to open an imide ring, was developed that promoted uniform metallization of the polyimide surface. A conductor adhesion strength of 500 kg/cm² or more was realized by sintering after the electroless plating (see Fig. 5).

Fig. 6 shows a cross-section of via-holes fabricated using the new technology compared with those using conventional technology. The via-holes are 50 μm square with a 23 μm thick polyimide layer. Uniform metallization was formed on the deep hole

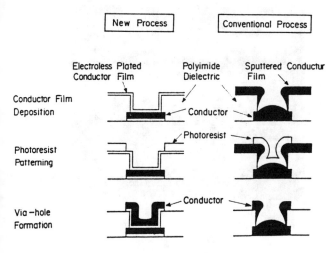

Fig. 4. Metallization processes for the via-holes.

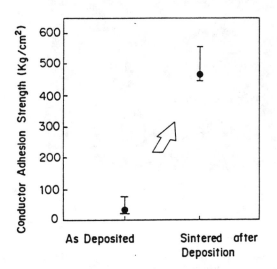

Fig. 5. Effect of sintering on conductor adhesion strength.

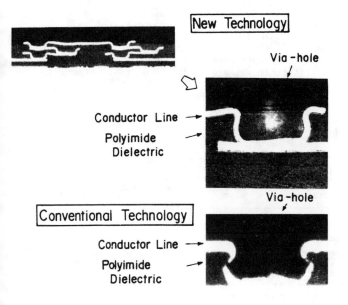

Fig. 6. Cross-section of via-holes. 50 μm square via-holes through 23 μm thick polyimide layer.

walls and a via contact resistance of 10 mΩ was achieved. After 3000 cycles of a thermal shock test from −65°C to +125°C, the fabricated substrate was found to be highly stable and to operate reliably (see Fig. 7).

ELECTRICAL PROPERTIES

Based on the results of these investigations, we designed and fabricated a test substrate with a photo-sensitive polyimide dielectric and electroless copper plated conductors to evaluate its electrical properties. The substrate, measuring 90 mm by 90 mm, consists of four metal layers : two layers of orthogonal signal lines sandwiched between power and ground planes in an offset stripline configuration; and a top layer for the attachment of chips and other components. The striplines were designed to have a characteristic impedance of 50 Ω in order to be compatible with other circuits.

The fabricated substrate is shown in Fig. 8 and a cross section of the substrate is shown in Fig. 9. The signal lines are 25 μm wide by 6 μm thick and the polyimide dielectric is 23 μm thick. The size of the via-holes is 50 μm square.

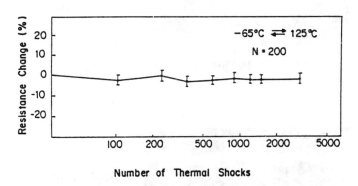

Fig. 7. Via-connection resistance change after thermal shocks.

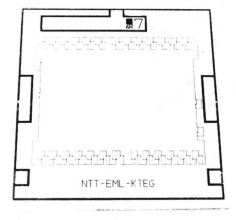

Fig. 8. Fabricated 90 mm by 90 mm substrate compusing four metal layers.

313

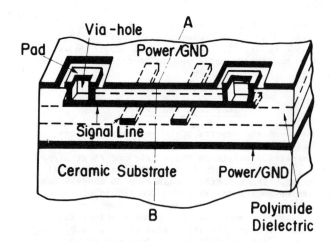

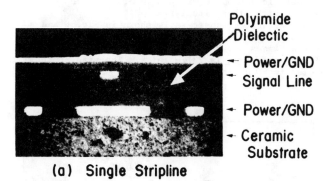

(a) Single Stripline

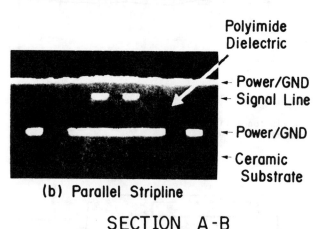

(b) Parallel Stripline

SECTION A-B

Fig. 9. Cross section of the substrate. (a) Single
stripline. (b) Parallel stripline. Signal
lines are 25 μm wide by 6 μm thick and polyimide
dielectric is 23 μm thick.

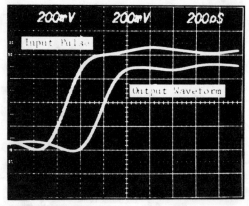

Fig. 10. An input pulse and its output waveform
propagated through a 5-cm long stripline.

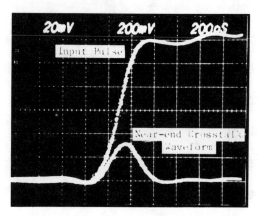

Fig. 11. Near-end crosstalk waveform of a 1-cm long
parallel stripline on a 50 μm pitch. (Input pulse
200 mV/div., Crosstalk Waveform 20 mV/div.)

FUNDAMENTAL TRANSMISSION PROPERTIES

Fig. 10 shows a sampling scope display of an
input pulse and its output waveform propagated through
a 5-cm long stripline. The input pulse is a
square-wave pulse with a rise time of 300 ps. There
is no reflection or ringing. The propagation delay
was approximately 60 ps/cm and the attenuation was
found to be 20 mV/cm.

The crosstalk ratio was determined by measuring
the coupled voltage at the near-end of an adjacent
line terminated at both ends with 50 Ω. The coupled
voltage of a 1-cm long parallel stripline on a 50 μm
pitch was 32 mV (see Fig. 11). The amplitude of this

coupled signal is 0.03 that of the driven signal,
which is an acceptable coupling level for most high
performance system designs.

The measured values are summarized and compared
with those of a conventional co-fired multilayer
ceramic substrate in Table I [3][12]. Very good
agreement was obtained between the measured and
calculated values. The propagation delay of the
signal lines (60 ps/cm) is 40 % less than for the
multilayer ceramic substrate. This is because the
signal line has a low dielectric constant and low
conductor resistance.

TABLE I   CHARACTERISTICS OF FABRICATED SUBSTRATE

	New Substrate	Co-fired Multilayer Ceramic Substrate
Characteristic Impedance	50 Ω	50 Ω
Dielectric Constant	3.3	9.5
Line width	25 μm	100 μm
Via-hole Size	50 μm □	100 μm φ
Line Resistance	1.3 Ω/cm (1.3 Ω/cm)	0.8 Ω/cm
Line Capacitance	1.2 pF/cm (1.3 pF/cm)	2.3 pF/cm
Propagation Delay	60 ps/cm (61 ps/cm)	100 ps/cm

( ) are calculated values.

314

APPLICATION

To demonstrate a functioning multichip circuit using this substrate, test circuits were made up that contained two high speed Si bipolar chips fabricated using super self-aligned process technology [13][14]. The chips consist of 2-input OR/NOR gates (see Fig. 12). Terminating resistors of 50 Ω were fabricated on the same chip for high-speed signal transmission. The signal swings are 800 mV, with transition rise and fall times of approximately 150 ps. The chips were attached to the substrate with epoxy and bonded with gold wire. Fig. 13 shows a close-up of the two wire-bonded chips. The multichip substrate was mounted in a test fixture containing decoupling capacitors and 50 Ω semi-rigid co-axial cable for interfacing with the test equipment.

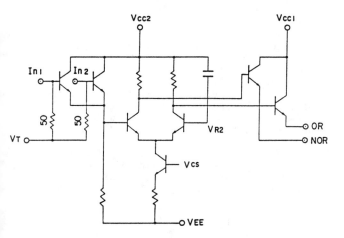

Fig. 12. Configuration of 2-input OR/NOR gates. Terminating resistors of 50 Ω are fabricated on the same chip.

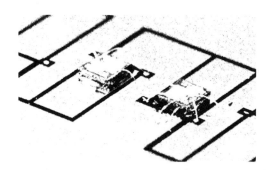

Fig. 13. Close-up of two wire-bonded chips on the substrate.

The chip to chip interconnection consisted of a serial net of two circuits. A square-wave pulse was inputed. Fig. 14 shows a sampling scope display of an output waveform transmitted by the two chips through a 5-cm long stripline. Fig. 15 shows the eye-pattern diagram for the circuit driven by a 2 Gbit/s NRZ pseudorandom pulse train. It shows good eye-opening. This waveform indicates that the fabricated substrate can transmit pulses at speeds higher than 2 Gbit/s.

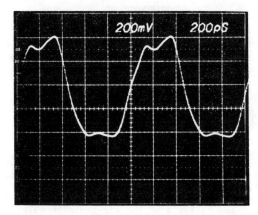

Fig. 14. An output waveform transmitted by two chips through a 5-cm long stripline.

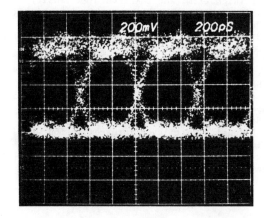

Fig. 15. Eye-pattern diagram for the circuit driven by a 2 Gbit/s NRZ pseudorandom pulse train.

CONCLUSION

A new fine-line multilayer substrate formed on a ceramic substrate has been developed. This new development features both small via-holes in photo-sensitive polyimide dielectric layers and fine patterns produced by electroless copper plating. To relize the characteristic impedance and resistance of the signal lines necessary for high-speed transmission, a thick polyimide dielectric layer and via-holes 20 μm or more in depth were required. Careful temperature control of the thick photo-sensitive polyimide both before and after the photolithography process was founded necessary to obtain a good via-hole configuration, and electroless copper plating was employed to form such deep via-holes. A conductor adhesion strength of 500 kg/cm² was realized by innovative chemical treatment and sintering, and a repetitive thermal shock test indicated that the 50 μm square via-holes through the 23 μm thick polyimide layer are highly stable and reliable. Based on the results of these investigations, a four layer substrate, measuring 90 mm by 90 mm, was fabricated. The propagation delay of the 25 μm wide, 6 μm thick striplines was 60 ps/cm, which is 40 % less than for a co-fired multilayer ceramic substrate. The fabricated substrate can transmit high-speed pulses at over 2 Gbit/s. This new substrate technology should be highly applicable to the large-scale multilchip packaging that will be developed in the future, including that for high-speed VLSI chips.

ACKNOWLEDGMENT

The authors would like to thank T. Suzuki, K. Katsura, H. Tsunetsugu, and N. Yamanaka for their valuable contributions in processing and testing the various substrates described in this paper, and M. Suzuki for the design of the high speed Si bipolar chips.

REFERENCE

[1] A. J. Blodgett and D. R. Barbour, "Thermal conduction module : a high-performance multilayer ceramic package," IBM J. Res. Develop , vol.26, no.1, pp.30-36, Jan. 1982.

[2] S. Cherensky, D. Genin, and I. Modi, "Electrical design and analysis of the air-cooled module (ACM) in IBM System/4381," in Announcement Booklet, IEEE ICCD, Nov.1983.

[3] M. Terasawa and S. Minami, "A comparison of thin film, thick film, and co-fired high density ceramic multilayer with the combined technology," in Proc. Int. Soc. Hybrid Microelec., vol.6, 1983, pp.607-615.

[4] N. Goldberg, "Design of thin film multichip modules," in Proc. Int. Soc. Hybrid Microelec., vol.4, 1981, pp.289-295.

[5] J. Shurboff, "Polyimide dielectric on hybrid multilayer circuits," in Proc. 33rd Electronic Components Conf., 1983, pp.610-615.

[6] K. Moriya, T. Ohsaki, and K. Katsura, "High-density multilayer interconnection with photo-sensitive polyimide dielectric and electroplating conductor," in Proc. 34th Electronic Components Conf., 1984, pp.82-87.

[7] R. J. Jensen, J. P. Cummings, and H. Vora, "Copper/polyimide material system for high performance packaging," IEEE Trans. Components, Hybrids, Manf. Technol., vol.CHMT-7, pp.384-393, Dec. 1984.

[8] T. Watari, and H. Murano, "Packaging technology for the NEC SX supercomputer," IEEE Trans. Components, Hybrids, Manf. technol., vol.CHMT-8, pp.462-667, Dec. 1985.

[9] H. J. Levinstein, C. J. Bartlett, and W. J. Bertram, Jr., "Multi-chip packaging technology for VLSI-based systems," ISSCC Digest Tech. Papers, pp.224-225, Feb. 1987.

[10] P. G. Rickerl, J. G. Stephanie, and P. Slota, Jr., "Evaluation of photosensitive polyimides for packaging applications," in Proc. 37th Electronic Components Conf., 1987, pp.220-225.

[11] T. A. Lone, F. J. Belcourt, and R. J. Jensen, "Electrical characteristics of copper/polyimide thin film multilayer interconnects," in Proc. 37th Electronic Components Conf., 1987, pp.614-622.

[12] B. T. Clark and Y. M. Hill, "IBM multichip multilayer ceramic modules for LSI chips-design for performance and density," IEEE Trans. Components, Hybrids, Manf. Technol., vol.CHMT-3, pp.89-93, March 1980.

[13] H. Ichino, M. Suzuki, K. Hagimoto, and S. Konaka, "Si bipolar multi-Gbit/s logic family using super self-aligned process technology," in Ext. Abst. 16th Conf. Solid State Devices and Mater., 1984, pp.217-220.

[14] M. Suzuki, K. Hagimoto, H. Ichino, and S. Konaka, "A 9-GHz frequency divider using Si bipolar super self-aligned process technology," IEEE Electron Device Lett., vol.EDL-6, pp.181-183, April 1985.

HIGH-DENSITY INTERCONNECTS USING LASER LITHOGRAPHY*

L.M. Levinson, C.W. Eichelberger, R.J. Wojnarowski, and R.O. Carlson

GE Corporate Research and Development
Schenectady, New York 12301

High-Density Interconnects (HDI) is a novel hybrid approach that uses polymer layer overlays laminated over bare chips mounted on a substrate. The overlays are laser-patterned with copper to connect the chips and I/O. The advantages of the HDI approach can be summarized as follows: the overlay layer makes the entire chip area available for interconnect lines; the interconnect has very high density and 2-mil pitch has been demonstrated; via and line formation are under computer control and no patterning mask is used. The HDI process is ideal for prototype or moderate volume production of high-performance circuits.

CONCEPT OVERVIEW

High-Density Interconnects (HDI) is a unique, novel, hybrid approach for the interconnection of high-performance integrated circuit (IC) chips in a dense configuration. This approach is currently in the active development stages at the General Electric Research and Development Center in Schenectady, New York. The original concept and preliminary process steps were developed during 1985-1986. The essence of the concept can be summarized as follows: bare IC chips are mounted face up on a substrate; a thin, flexible polymer film is laminated over the top of the IC chips; laser drilling is used to form vias in the polymer over the chip bonding pads; copper interconnects are defined on the polymer film to run connections from chip-to-chip and to the hybrid substrate input/output pads.

The HDI process has a number of advantages over more conventional high-density hybrid approaches. Some of these are as follows:

- The overlay layer makes the entire chip area available for interconnect lines.

- The interconnect has very high density; 2-mil pitch has been demonstrated.

- The bonds between the chip pads and interconnect are by thin film metallization; thus there are no wire bonds or solder joints on the hybrid.

- Via and line formation are under computer control, and no patterning masks are used. Chip misalignment is accommodated by computer-adaptive writing, i.e., the computer modifies the interconnect pattern to adjust for variations in chip placement.

- Copper is the conductor metallization.

- The chips can be placed to be almost touching, thereby giving maximum packing density. Slight gaps between chips provide strain relief.

- The interconnect technology accommodates any chip size and can accomodate mixed chip technologies.

- The process is ideal for prototype or moderate volume production because it is flexible and quickly implemented.

- The overlay can be removed and replaced repeatedly without chip damage.

- The overlay can be applied to a single chip in a socketable, single chip package for testing and burn-in, and then removed.

- The bare chips are mounted directly on the substrate. This facilitates heat removal.

FABRICATION

Figures 1, 2, and 3 show the basic assembly of the HDI structure, which consists of three layers: substrate, frame, and patterned overlay. To date, most of the frames and substrates have been alumina, because of its ready availability. We have, however, demonstrated the compatibility of our process with other substrate and frame materials such as silicon and Kovar.

The HDI fabrication process is as follows: Holes are cut by laser in a thin alumina frame, whose thickness is chosen to be a little less than the thinnest chips. The frame is bonded to a thicker alumina substrate with a glass layer. Power/ground lines and I/O pads can be formed on the frame by aluminum deposition and patterning. Tested bare chips are bonded into the recesses in the frame/substrate. An image scan locates the chips and their pad positions. A polymer glue layer (such as Ultem, a polyetherimide) is sprayed over the frame and chips and a polyimide sheet (Kapton) is laminated to the frame and chips under heat and pressure. Vias are then drilled by the laser over the chip pad positions. Any polymer debris and residual film can be removed by a plasma etch. The via formation step is followed by interconnect metallization. Chromium followed by copper is sputtered over the whole hybrid surface (including the drilled via holes) and the metallization built up

* Partially supported by the U.S. Air Force Weapons Laboratory

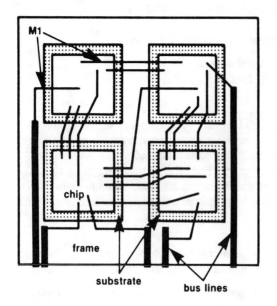

- **Assemble Frame/Substrate**

- **Pattern Bus Lines**
  Al deposit
  Pattern/wet etch

- **Die Attach**
  Ultem adhesive

- **Kapton Laminate**
  Ultem adhesive

- **Laser Drill Vias**
  Adaptive lithography

- **Metallize**
  Cr/Cu sputter
  Cu/Cr electroplate

- **Pattern Metal 1**
  Negative resist
  Adaptive laser lithography
  Wet etch metals

- **Second Dielectric**
  Spin coat polyimide

- **Repeat vias, metal**

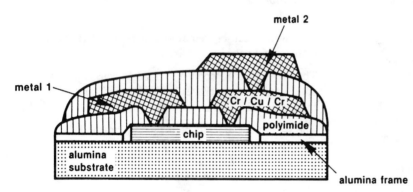

Fig. 1.   HDI process.

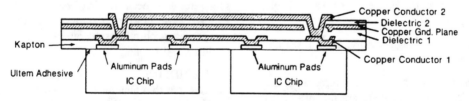

Fig. 2.   Schematic cross section of a two-signal-layer HDI structure with
a ground plane separating the two layers.

by copper plating followed by a chromium overcoat. Figure 4 shows two vias of different size and taper over an aluminum bonding pad. Via taper can be adjusted by varying the laser scan conditions. Via bottoms down to 10 μm are achieveable with high reliability. The metallization exhibits excellent step coverage and continuity in the via hole.

The signal metallization is defined by exposing a negative resist using the laser beam. Metal etching is used to remove the unwanted Cu/Cr. Chip misalignment is accommodated by computer-adaptive writing. The ideal chip pad locations and interconnect network are resident in software. Identification of the actual positions of two chip pads in a slightly misaligned chip serves to define the real position of all pads in a particular chip. The interconnect configuration is then slightly modified (shifted) to adapt to the actual chip pad positions.

The next signal layer is built up following a process sequence similar to that of the first layer. The polyimide dielectric is spun over the first metal layer. Via holes are opened with the laser, and metal deposition and patterning are carried out as before. Power and ground lines (if not already on the frame, or if more are desired) can be incorporated into the signal layers, or, for better impedance control, separate power and ground planes can be formed between, on top, or beneath the signal layers.

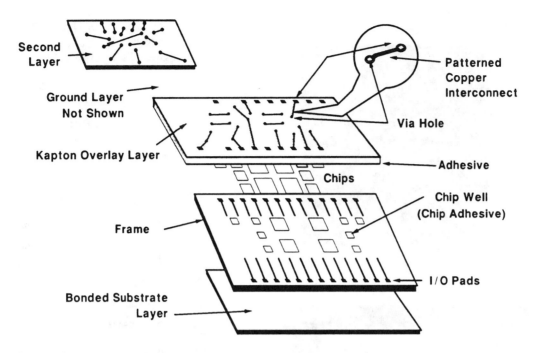

Fig. 3. Exploded view of HDI hybrid assembly.

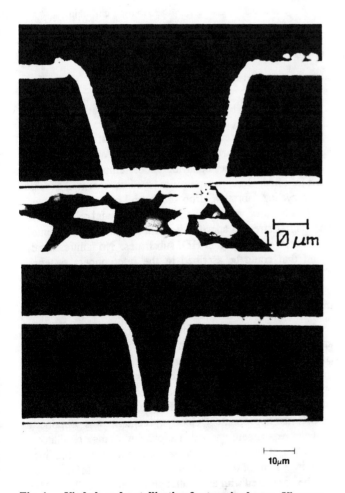

Fig. 4. Via hole and metallization for two via shapes. Vias are drilled in 25-μm Kapton film (overlay layer). Metallization is plated Cu on sputtered Cu/Cr.

RESULTS

Test substrates involving discrete chip functions, ring oscillators, and long chain interconnects through vias, have been fabricated to demonstrate line density, speed performance, signal fidelity, overlay removal and replacement, chip removal and replacement, durability under MIL STD 883 environmental and mechanical tests, and survivability under total and transient radiation dosage. As an example of one of these test substrates, Fig. 5 shows two views of the Pad Array Test Substrate. The four chips shown in the top view are 0.678 × 0.678 cm in size and fit into a frame cavity of 1.389 × 1.389 cm. Thus the average spacings between chips and chips to frame wall are only about 100 μm, dramatic proof of the close spacing of the chips possible with this overlay interconnect technique. The closeup in Fig. 5 shows 100-μm-wide by 5-μm-thick copper lines interconnecting short 23-μm by 1-μm-thick aluminum lines formed on an oxidized silicon chip. The copper lines are on two signal layers. A close look shows there are two sets of vias contacting the vertical copper lines, one set to connect from aluminum to first-level copper and a second set to connect first-level copper to second-level copper. A cross section of a two-level staggered via interconnect is shown in Fig. 6.

The chain resistance of the daisy weave pattern is very useful in demonstrating consistent near-100% yield in via connections and as a tool for proving interconnect durability through various stress tests. This and other test substrates were formed with copper lines of 25-μm width by 5-μm thickness, with 25- to 50-μm spacing, leading to a pitch of 50 to 75 μm. Still finer lines have been fabricated, down to 10 μm line and spacing, over limited lengths. Vias are typi-

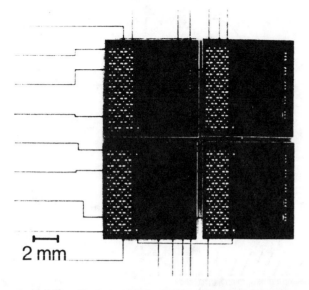

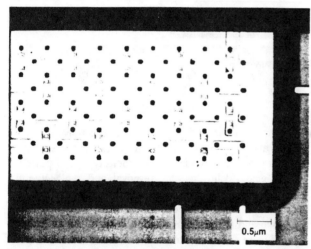

Fig. 5. Photo of pad array test circuit at two different magnifications. Note the high quality of the metallization on the curved polyimide film bridges between the chips. Laser adaptive lithography (see text) accommodates the evident misalignment of chips.

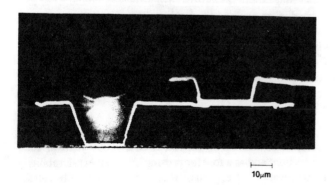

Fig. 6. Staggered vias showing interconnect between chip pad and first and second signal interconnect layers.

cally 25 by 25 μm (they can be laser-drilled to be square, rectangular, or circular), but can be made as small as 8 by 8 μm. Table 1 gives a listing of typical HDI parameters.

Table 1

HDI Process Parameters

Property	Value
Dielectric thickness	
Layer 1	25-30 μm
Layers 2, 3, 4,	10-30 μm
Line width	25 μm
Line spacing	25 μm
Via bottom	10-30 μm
Conductor thickness	2-5 μm
Chip-to-chip spacing	$\geq$ 100 μm
Thermal conductance through chip. substrate, and chip-to-substrate bonding layer	~1 °C/W
Die shear	>30 lb for 1 cm^2 chip
Dielectric properties	
Dielectric withstand	>1 kV
Dielectric constant	about 3
Interconnect capacitance	0.7-1 pF/cm

Any viable interconnect scheme must survive an appropriate set of environmental stresses. We have examined HDI interconnect behavior under temperature bakes, burn-in under bias, temperature cycling, thermal shock, power cycling, vibration, drop shock, and centrifuge. Populated substrates were visually examined and electrically tested after stress as appropriate. Table 2 summarizes some of the tests undertaken in HDI substrates. No failures were found that could be ascribed to the interconnect system. The few failures observed invariably derived from extraneous sources, such as IC chip failure or mechanical weakness in the Al$_2$O$_3$ substrate.

In order to evaluate the speed capability of HDI approach we have also fabricated CMOS ring oscillator circuits with various length of interconnect between the oscillator stages. A ring oscillator circuit is shown in Fig. 7. The operating delay per stage was measured and compared with calculated values using both RC and transmission line models as appropriate. Good agreement was obtained between experiment and calculations. A 7-stage oscillator, with each stage purposely separated from the preceding stage by 3.7 cm of interconnect, oscillated at 40 MHz. This is to be compared with an oscillation frequency of 220 MHz for a similar ring oscillator with close to zero stage-to-stage spacing. The delay per stage of 1.8 ns corresponds well with the calculated value.

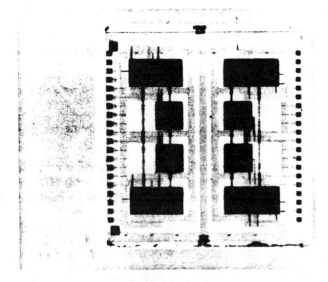

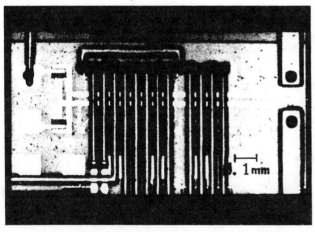

Fig. 7.   Photograph of ring oscillator circuit. The closeup shows interconnect to inverters using 25-μm pads.

Table 2

Stress Tests for HDI Substrates

Test Type	Description	Comments
Stabilization bake	150 °C, 48 h	No failures/changes
Temperature cycling	−55 °C to +150 °C	No failures/changes to 1000 cycles
Burn-in	125°C, 168 h with bias	No failures/changes from HDI process
Thermal shock	−65 °C to +150 °C liquid to liquid	No failures/changes to 1000 cycles
Thermal shock	Liquid nitrogen to hot plate at 150 °C	125 cycles No failures/changes
Power cycling	Self heating, 20 °C to 150 °C	1000 cycles No failures/changes
Variable frequency vibration	20 to 2000 Hz, 20 G level or 0.06 in. peak-to-peak	No failures/changes
Constant acceleration	7000 G, centrifuge	No failures/changes
Mechanical shock	5 drops to 1500 G level	No failures/changes if Al_2O_3 substrate reinforced
Polymer outgassing	Baked at 100 °C for 24 h after sealing in hermetic package	<1500 ppm moisture worst case, 500 ppm typical

In addition to various test circuits, a four-chip hybrid module was fabricated. This module is comprised of four identical 256-point Fast Fourier Transform Chips. These VLSI chips have 148 I/Os each; 352 I/Os are connected internally and 240 are brought out to the module periphery. Fig. 8 is a photograph of the completed module.

A major issue that must be addressed for any workable hybrid interconnect scheme relates to repair and rework. The HDI approach addresses rework by the demonstrated ability to remove the overlay layer and any faulty chip after test of the completed module. If the fault detected is minor, such as shorts or opens in the interconnect, or if a change in the interconnect pattern is desired, the overlay can be removed and a new overlay formed over the existing bare chips. If, in addition, one or more of the chips have been determined to be faulty, the overlay can be removed, followed by removal of the individual faulty chips. New chips are then bonded to the substrate and a new overlay formed. A set of experiments has confirmed that the HDI approach

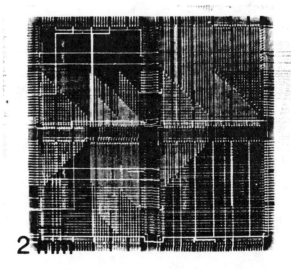

Fig. 8.   Photo of HDI interconnected 1024-point FFT signal processor. The four VLSI chips being interconnected have 148 I/O each. The circuit is 1024-point FFT signal processor composed of four 256-point FFT chips. 352 I/Os are connected internally; another 240 are brought out to the module periphery.

has extensive rework and repair capability. Details will be presented in a forthcoming publication.

SUMMARY

High Density Interconnect (HDI) is a unique, novel, hybrid approach currently in the development stages at General Electric. The approach uses polymer layer overlays laminated over bare chips mounted on a substrate. The overlays are laser-patterned with copper to connect the chips and I/O. The HDI approach has a number of advantages including the following:

- The overlay layer makes the entire chip area available for interconnect lines;

- The interconnect has very high density: 2-mil pitch has been demonstrated;

- Via and line formation are under computer control and no masks are required;

- The overlay can be removed and replaced without chip damage;

- Chips are mounted directly on the substrate for good heat dissipation.

Initial tests have indicated that the process is manufacturable, reliable, and repairable. We believe that HDI will prove ideal for prototype or moderate volume production of high performance circuits.

HIGH DENSITY/HIGH SPEED MULTI-CHIP PACKAGING

BY

SANFORD LEBOW

AUGAT MICROTEC
NEWBURY PARK, CA.

ABSTRACT

High lead count "Multi-Chip Packages" (MCP) have become a necessary crucial element in the packaging of Gallium Arsenide (GaAs), ECL and high speed CMOS devices, which provide from 200 megahertz to gigahertz performance. This paper describes the development of the Augat Microtec "additive" Copper plating and multilayer Polyimide lamination into a process which lends itself to quantity production of these high performance MCP packages. (Figure 1)

INTRODUCTION

The initial design goals for the multilayer structure were ambitious: - Standard signal line impedance of 50 ohms, - Incorporation of coaxial shielding in the structure, where required, - Development of acceptable levels of isolation (26 db), using rectangular coax construction, - Use of composite Polyimide dielectric, in combination with "additive" plating process and ultra fine line photographic printing, to produce three-dimensional multilayer structures.

Ceramic suppliers have struggled for years to satisfy these high speed MCP interconnect requirements, but cannot overcome the limitations of the physics of their material. The answer to this complex problem is being addressed and solved by the Augat Microtec "additive" three-dimensional process.

With engineering problems solved and the design goals met, the major effort at Microtec turned to the development of a process to manufacture these MCP packages at a high quantity, repeatable rate. During this period, a wide range of material processing problems were addressed: Namely: (1) Laser cutting of the polyimide for die cavities (Fig. 2) as an alternative to thermal clusters. (2) Incorporate rectangular coax construction for the transmission lines with typical 50 ohm characteristic impedance. (3) The design goal was not to exceed two signal layers, plus ground and voltage planes in most multi-chip designs. Our standard production process, 4 mil lines with 6 mil spaces, accomplished this objective. However, 3 mil lines with 5 mil spaces are in development. Ultimately 2 mil lines with 3 mil spaces will be required (5 mil pitch). (4) High temperature I/O lead attachment (Fig. 3)

The initial design of our MCP leaded Polyimide carrier was a 5-layer structure. It was fabricated utilizing the "additive" plating process, employing 2 or 4 mil glass filled Polyimide material for the dielectric between the conductor layers. Layer-to-layer connections were accomplished with plated Copper vias; all conductor and via patterning was done by photographic techniques. The signal line conductor width was 4 mils, with a 5-mil thick Polyimide dielectric, making the total ground-to-ground spacing 10 mils. The dielectric constant of the glass filled Polyimide is 3.8 -- which provides a characteristic impedance of 50 ohms.

The multilayer structure was physically and electrically attached to either a Copper or Ceramic base, with cavities so that semiconductor dies could be epoxy die bonded to the core. The die could be automatically wire bonded or TAB bonded, depending upon user capability. Input/output leads around the edge of the package were attached by high temperature solder flow.

FABRICATION PROCESS

The Microtec process is a combination of precision lithography, "additive" Copper plating and multilayer Polyimide lamination. Figure 4 describes details of the process from the initial photo-mask operation to the complete substrate, including a rectangular coax structure, a stripline construction and thermal vias.

In the first step, a high resolution photomask is applied to a flat panel and exposed in the patterns required for the first level signal and via locations. After cleaning, Copper is plated up through the windows of the photoresist to the required height of the Polyimide layer thickness.

A sheet of Polyimide prepreg is laminated over the plated vias; the prepreg flows flat against the base material and around the protruding solid Copper vias. The Polyimide material covering the vias is removed in a surface polishing operation, leaving the exposed vias and the cured Polyimide ready for the next fabrication layer.

Repetition of this process sequence yields a multilayer structure which may be constructed in a variety of geometries and configurations. Signal lines and ground elements may be structured as rectangular coax or stripline with thermal vias for heat dissipation.

The result of the process is a substrate which meets the rigorous requirements of GaAs packaging. Features as small as 0.002 inch lines and 0.003 inch spaces, vias as small as 0.003-in. by 0.003-in. and multilayer routing of signal and power lines allow I.C. devices to be very close together.

Alternatives in signal line and ground element relationships allow layout flexibility while attaining the exact characteristic impedance desired. Impedance is typically 50 ohms, but 25 and 60 ohm lines have been produced.

Precise manufacturing control and positioning of the large ground planes yield a controlled impedance line with minimal crosstalk. By using very large power and ground planes in the multi-layer structure, and keeping their separation minimal, the characteristic impedance of the power lines is typically 2 to 3 ohms and resistivity is negligible.

Thermal characteristics are also quite favorable. Clusters of solid Copper vias dissipate heat away from the device to larger heat sinks. In addition, to attain more direct thermal transfer in the multilayer structure, we have in development laser removal of the Polyimide for direct die attachment to the supporting metal base. This is accomplished by using a 50 Watt CO_2 laser for ashing away the Polyimide in these cavity locations.

Custom layer-to-layer circuitry can be designed to meet individual requirements for high performance, high density, controlled impedance, thermal management, or all four if needed. The fine line multilayer construction results in extremely short interconnections between devices, minimizing propagation delays and reflections. In addition, this fine line, additive Polyimide technology has the capability to produce solid vertical walls on each side of critical signal lines which can interconnect to top and bottom ground planes. These rectangular coaxial lines can provide the necessary isolation between signal lines where required, and furthermore, maintain the characteristic impedance of the line to match that of the device.

All metal lead frames or testable TAB leads may be added to the peripheral surface pads of the finished product by one-shot TC soldering techniques. A recent development at Microtec allows the leads to be TC bonded at 485°C with a dwell time of 4 seconds. This method for lead attachment provides a metallurgical interface which is superior to that of the brazed lead attachment provided on ceramic packages. Alternative base materials may also be chosen for their individual character-istic advantages. Materials now include Copper, Copper Tungsten, Alumina, BeO, Molybdenum and Polyimide.

PROPERTIES

Temperature: The composite structures can be operated continously up to 260°C. For short durations the structure can survive 300°C - up to one minute - without deterioration, for die attach or other high temperature processing. Wire bond can be accomplished by ultra sonic or thermal sonic technique, TC bonding should be avoided.

<u>Moisture</u>: Typically, less than .3% moisture absorption takes place when submitted to moisture testing, but more importantly, no voltage breakdown occurs after minimal drying time.

<u>ELECTRICAL</u> - <u>Capacitance</u>: Typical capacitance values of less than 2 picofarads per lineal inch are achieved using the 4 mil (.004:) buried line. <u>Resistance</u>: Typical values of 30 milliohms per square are inherent in the pure Copper conductor. <u>Inductance</u>: Less than 1.5 nanohenries of inductance is evident in the same 1 " length of .004" buried line. <u>Losses</u>: The losses exhibited by the composite are relatively high, approaching 1 db per lineal inch. However, it should be understood this technology allows for extremely close location to the chip, or from chip to chip, so minimum distances are attained as a natural function of the technology.

<u>CONCLUSIONS</u>

We have developed a multi-chip package (MCP) technology which has a number of advantages in material properties and processing capabilities for fabricating high density, high performance multilayer substrates. The MCP package has several unique features including fully shielded signal lines, cluster thermal columns or laser cut die cavities and high temperature I/O lead attachment techniques. With the introduction of high speed plating baths and vacuum laminating equipment we have now made this additive multilayer polyimide system cost effective. We believe that this packaging approach will become increasingly important for constructing various multichip packages, such as cache memories, micro processors and signal processors to name a few.

1. Lebow, S. "A Method of Manufacturing High Density Fine Line Printed Circuit Multilayer Substrates Which Can Be Thermally Conductive," Proceedings of the 30th Electronic Components Conference, p. 307 (1980)

2. Chalman, R. and E. Nogavich, "Multiple Functions of Blind Copper Vias in Polyimide Multilayer Structures," Proceedings of the 4th International Electronics Packaging Conference, (Oct. 1984)

3. Landis, R., "High-Speed Packaging for GaAs Interconnection," ECC/IEE (May 1985)

4. Nelson, L., "Success of GaAs Semiconductors Hinges on Packaging", PC Technology magazine (1985)

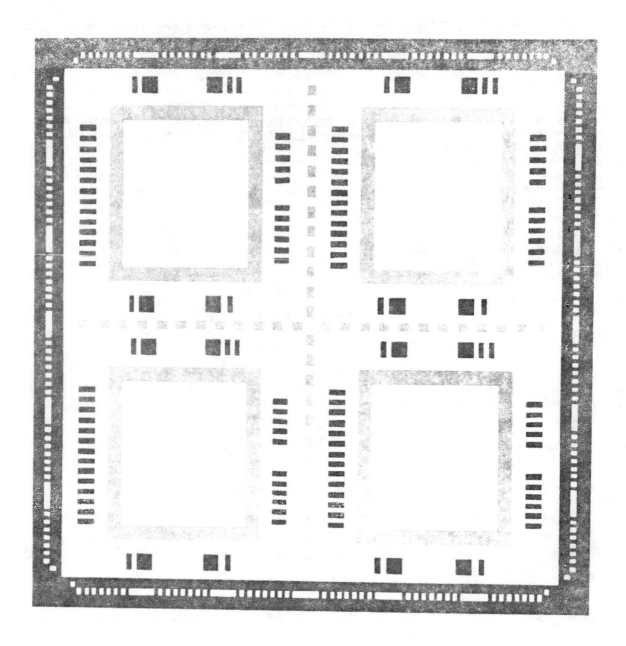

Fig. 1 Fine Line Polyimide Chip Carrier
for LSI Devices.

Fig. 2 Laser Cutting of the Polyimide
for Die Cavities

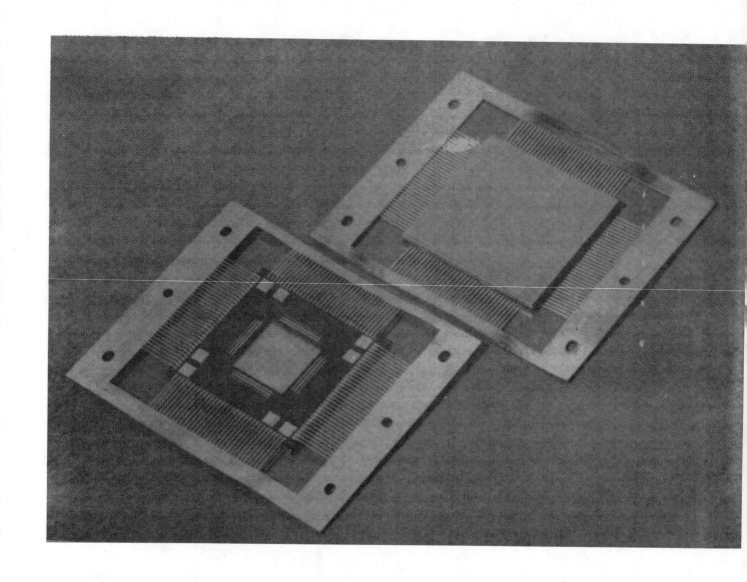

Fig. 3. High Temperature Lead Attachment

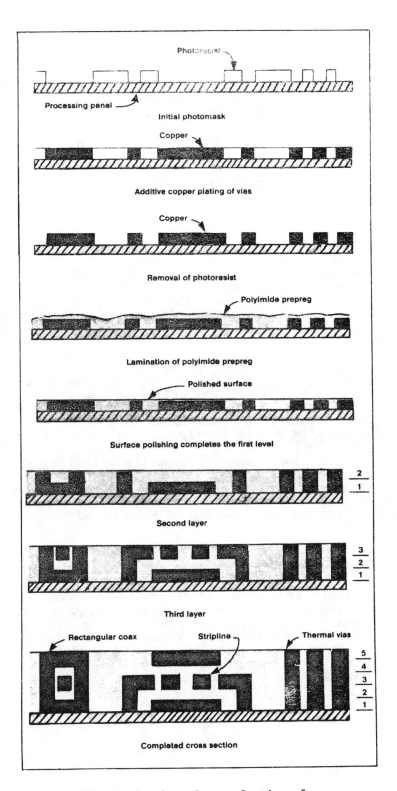

Photoresist

Processing panel

Initial photomask

Copper

Additive copper plating of vias

Copper

Removal of photoresist

Polyimide prepreg

Lamination of polyimide prepreg

Polished surface

Surface polishing completes the first level

2
1

Second layer

3
2
1

Third layer

Rectangular coax   Stripline   Thermal vias

5
4
3
2
1

Completed cross section

Fig. 4   Complete Cross- Section of
the Polymide Process

FLUOROPOLYMER COMPOSITE MULTICHIP MODULES

BY

Richard T. Traskos
Rogers Corporation
Rogers, Connecticut

and

Steven C. Lockard
Rogers Corporation
Chandler, Arizona

ABSTRACT

The need for improved material systems in multichip modules (MCM) is well understood. These improvements include lower dielectric constant, lower moisture absorption and lower modulus to reduce stress. Other issues critical in many MCM applications include improved impedance control, CTE matching to the die, thermo-mechanical reliability of the substrate and thermal management.

With the acquisition of Microtec, Rogers gained an additive copper/polyimide process for manufacture of multichip modules and other advanced circuitry. This process allows efficient routing of most MCM designs, low resistance copper conductors and solid copper thermal vias for heat removal. It also allows large substrates (4 x 4 inches and greater) to be manufactured, multiple up per panel.

Rogers has developed fluoropolymer composite materials with electrical and mechanical properties which address the needed MCM material improvements. Combining these materials with the Microtec sequential plating and lamination process yields multichip modules which offer unique solutions to the performance requirements of high speed, high density, high reliability MCM's.

INTRODUCTION: THE NEED FOR MCM IMPROVEMENTS

Many of the multichip modules being manufactured today are based on polyimide material systems. Some processes apply a liquid form polyimide precursor which is then cured; others use sheet form reinforced polyimide prepregs. These polyimide materials have known deficiencies which have been described in detail in past papers[1,2]. Material system improvements which are needed include both electrical and thermo-mechanical properties. The most important material improvements are summarized in Table 1.

Reprinted with permission from *Proc. 9th Int. Electron. Packaging Conf.*, vol. I, pp. 221–230, 1989.

Table 1   <u>Multichip Module Required Material Improvements</u>

    1. Lower dielectric constant and dissipation factor
    2. Lower moisture absorption
    3. Lower modulus to reduce stress
    4. CTE properties compatible with the MCM assembly
    5. More reliable copper/polymer interfaces

On the electrical side, lower dielectric constant is necessary to reduce crosstalk and propagation delay in a constant impedance environment. Lower dissipation factor is needed when signal risetimes are very short to reduce risetime degradation. Also, the stability of these electrical properties with variations in temperature, frequency and humidity is important in many applications.

Thermo-mechanical reliability is of critical importance in multichip modules. Current thermosetting polyimide materials used in MCM systems build up very high stress levels during the MCM fabrication process due to the shrink on cure and high modulus of these thermosetting systems and to the CTE mismatch between the polymer and metal in resulting MCM constructions. As a result, thermal cycling, typical of many end applications, often causes stress related failures in MCM substrates and assemblies. Lower modulus materials will reduce stress and when combined with CTE matching substrates will afford reliable MCM substrates and assemblies.

The reliability problems of copper-polyimide interfaces have been well documented[3,4]. Generally, chemical bonding between polyimide and copper is weak and various reactions, in many cases water-related, will further degrade the strength and reliability of the interfacial bond.

Water is very difficult to keep out of MCM structures made with polyimides. Water is usually generated during polyimide cure and the polyimide equilibrium water absorption is substantial, often greater than 1%, and the rate of water uptake can be rapid. In addition, since water diffusion rates in polyimide are usually high, any water in polyimide MCM structures can diffuse quickly to the problematic polyimide-conductor interfaces and cause corrosion.

To combat these interface problems, chromium is often used successfully as a barrier metal between copper and polyimide at interfaces. This, however, adds significant processing steps and costs to MCM structures which have many polyimide-copper interfaces. New material systems which are very low in water absorption and in water transmission rate and which form reliable bonds without many additional processing steps are desired.

In addition to material improvements, another area which needs attention is thermal management. Especially in high power applications, such as high speed ECL, the MCM substrate can play a critical role in removing heat from the ICs and in conducting to a cold plate or other system level thermal management scheme.

THE MICROTEC PROCESS

In 1988, Rogers acquired the Microtec Division of Augat and transferred the equipment and key people to Rogers facility in Chandler, Arizona. With this acquisition, Rogers gained a fully additive copper plating/sequential lamination process for the manufacture of multichip modules and other advanced circuitry. This process has been described in detail[5].

By electroplating copper into a pattern defined by a photoresist, selective patterns of copper traces and solid copper vias are produced. After stripping the photoresist, a sheet form dielectric material is laminated into place and a mechanical polishing operation is used to planarize the surface and expose the tops of vias. This process is repeated in a sequential manner to build multiple signal, power, ground, via and pad layers. The process, as acquired, uses a glass reinforced polyimide prepreg which has worked well to fill and flow in between fine features but has the limitations of polyimide described earlier.

Inherent to this process are the electrical and thermal property advantages of electro-plated copper. Signal conductors, typically 1 mil in thickness, exhibit low DC resistance, on the order of 200-300 milliohms per linear inch. Thus, transmission lines made using this process can be treated as lossless lines as opposed to most thin film processes which yield lossy lines. Solid copper thermal vias, as shown in Figure 1, conduct heat through the substrate and directly to a metal base plate with very low thermal resistance (on the order of 0.2 C/watt through the vias assuming 50 percent area covered under the die) while still allowing area under the die for signal routing.

Due to the sequential nature of this process and the fact that vias are nearly equal in size to the line widths, highly efficient signal routing results. Adjacent routing channels are not blocked by oversized vias and z-direction routing is quite flexible due to the resulting stacked vias which can begin and end at any layer with complete freedom as shown in Figure 2.

Another degree of flexibility is the substrate size. This process uses a standard panel of 15 x 18 inches which allows large substrates (4 x 4 inches and larger) to be made multiple up per panel. This capability is allowing system designers to consider partitioning their system in an optimal manner.

FLUOROPOLYMER COMPOSITE MATERIALS

Recent advances in fluoropolymer composite technology by Rogers have resulted in materials with electrical and mechanical properties needed for high speed, high density, high reliability circuit applications. By combining PTFE resins with ceramic particle fillers, the results are composites with electrical properties nearing that of microwave grade dielectrics but with significant improvements in mechanical properties.

Table 2 summarizes selected electrical and mechanical properties of Rogers RO2800 fluoropolymer composites. The electrical properties have been proven to be very stable with changes in temperature, frequency and humidity. Thermo-mechanical reliability has been proven through various surface mount and plated through hole reliability test programs[6,7,8]. In fact, compatibility with copper is best exemplified by 12 layer circuit boards passing 2800 thermal cycles with zero PTH failures and no loss of copper to polymer bond. RO2800 circuits have even survived high temperature flip chip processing conditions, due to the materials high temperature endurance and CTE/modulus properties. These results prove that CTE properties matched closely to copper combined with low modulus provide thermo-mechanically reliable substrates. In addition, X-Y CTE of the final substrate can be matched to components such as ceramic or silicon by bonding the substrate to a constraining core.

Table 2   RO2800 Selected Electrical and Mechanical Properties

Effective Dielectric Constant	2.8
Dissipation Factor	0.003 max
Low Water Absorption	0.13 % max after D48/50
Low Z-Direction CTE	24 ppm/C
CTE X-Y Plane	16 ppm/C
Tensile Modulus	120 KPSI

COMBINING FLUOROPOLYMER COMPOSITES AND THE ADDITIVE PROCESS

Initial test vehicles combining a version of RO2800 with the additive copper plating/sequential lamination process have been fabricated as a demonstration of the complete process. Two steps of the additive process which were suspect with RO2800 were fill and flow during lamination and polishing to a smooth surface without smear. In trials to date, the material filled effectively between vias and conductors after high temperature fusion bonding. It polished without smear due to a high concentration of ceramic filler. A surface view of one such vehicle is shown in Figure 3, with 1.1 mil line widths and 0.7 mil spaces.

Building RO2800 into the additive copper process combines the electrical, mechanical and thermo-mechanical attributes of each as described above to result in an advanced MCM technology with significant advantages compared to thin film approaches.

In addition, several processing advantages are expected from this material/process combination. The thermoplastic nature of these composite materials will allow multiple laminations without building additional stress into the substrate, thus overcoming one of the largest manufacturing problems inherent to polyimide. Also, adhesion and reliability of the copper/fluoropolymer interface will be greatly improved and will not likely require barrier metals for adhesion promotion or corrosion resistance.

CONCLUSION

The combination of the additive plating/sequential lamination MCM process and Rogers recent advances in fluoropolymer composites offers a viable alternative to thin film multichip modules. In fact, the Rogers' system offers several advantages including: electrical advantages, such as shorter delay, lower crosstalk, less variation with humidity and much lower DC resistance, and thermo-mechanical advantages, such as CTE properties well matched to copper and low modulus for substrate reliability and easy constraint to match silicon for reliable MCM assemblies. This advanced process/material combination will yield multichip modules with unique solutions to the performance requirements of high speed, high density, high reliability applications.

REFERENCES

(1) John D. Craig, William J. Lautenburger, "New Polyimides for High Density Interconnect Packaging", Proceedings of the National Electronic Packaging and Production Conference - West 1989, Anaheim, CA, Vol. 1, pp 925-926.

(2) Hidetaka Satou, Daisuke Makino, Shunichi Numata, Noriyuki Kinjo, "PIQ Polyimide for Multichip Module Applications", Proceedings of the National Electronic Packaging and Production Conference - West 1989, Anaheim, CA, Vol. 1, pp. 921-924.

(3) Y. - H. Kim, J. Kim, G. F. Walker, C. Feger, S. P. Kowalczyk, "Adhesion and Interface Investigation of Polyimide on Metals", J. Adhesion Sci. Technol., Vol. 2, No. 2, 1988, pp. 95-105.

(4) S. A. Chambers, K. K. Chakravorty, "Oxidation at Polyimide/Cu Interface". J. Vac. Sci. Technol., A6(5), Sep/Oct. 1988, p.3008.

(5) S. Lebow, "High Density/High Speed Multi-Chip Packaging", Proceedings of the 6th International Electronic Packaging Conference, November 1986, pp. 417-423.

(6) Arthur, D.J., "Electrical and Mechanical Characteristics of Low Dielectric Constant Printed Wiring Boards", IPC Technical Review, Vol. 27, No. 5, 1987, p. 10.

(7) C.S. Jackson, M.B. Elkins, "A New Approach to Reliable SMT Boards for Military Computers", Proceedings of the Surface Mounting and Reflow Technology Conference, January 1987.

(8) David J. Arthur, Elizabeth L. Kozij, "PTH Reliability of High Performance PWB Material", Proceedings of the 8th International Electronic Packaging Conference, November 1988, Dallas, TX, pp. 74-86.

MCM Cross-Section

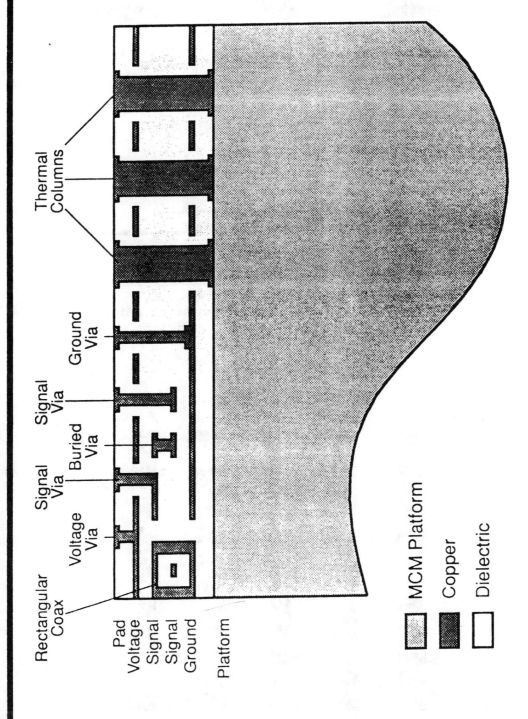

Rectangular Coax

Voltage Via

Signal Via

Signal Via

Buried Via

Signal Via

Ground Via

Thermal Columns

Pad
Voltage
Signal
Signal
Ground

Platform

MCM Platform

Copper

Dielectric

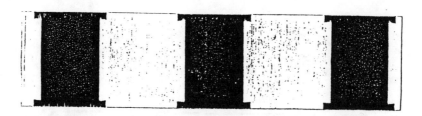

Figure 1. Solid copper thermal vias.

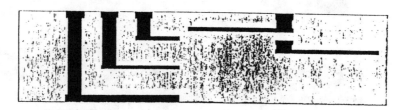

Figure 2. Solid vias which can start and stop at any layer allow efficient signal routing.

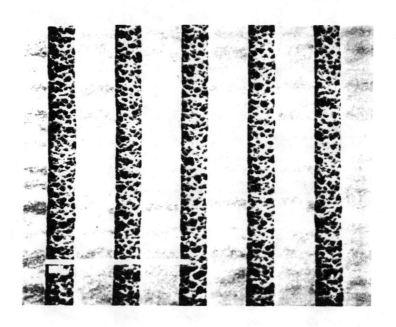

Figure 3. Surface view of 1.1 mil wide Copper Traces on RO2800 Fluoropolymer Composite

Reliability Testing

- Solder Float: 288°C, 10 Sec.
  (MIL-P-55110D)

- Thermal Cycling: -55°C to +125°C
  (MIL-P-55110D
   & MIL-STD-883C)

- High Temperature Exposure: 330°C for 90 Sec.

- Humidity Exposure: 85°C, 80% R.H.

Processing Advantages

- Sequential Lamination Without Build-Up
  of Stress

- Eliminates Need for Barrier Metals

- Improved Adhesion

- Large Substrates, Multiple Up Per Panel

Design Rule Objectives

	Near Term	Longer Term
Line Width (mils)	2.0	1.0
Line Spacing (mils)	3.0	1.0
Line Thickness (mils)	1.0	1.0
Via Size (mils)	3.0	1.0

Conclusions:

Low Dielectric Constant
Low Dissipation Factor   =   *Improved Electrical Performance*
Low Water Absorption

Low CTE
Low Modulus   =   *Improved Thermomechanical Performance, Improved Reliability*
Low Water Absorption
Good Adhesion

A NEW MULTICHIP INTERCONNECT TECHNOLOGY

Scott L. Jacobs and William E. Guthrie
Polylithics, Inc.
Sunnyvale, California

Abstract

A new multichip interconnect technology with significant improvements in manufacturability has been developed. This copper-polyimide technology supports circuit densities commensurate with integrated circuits. Small line widths and contact pads allow a "superchip" approach to multichip circuit design by sharply increasing the number of available signal ports. This approach lowers the cost of high density, high performance circuits by using smaller chips with higher yields. Standard integrated circuit die are used in multichip module construction without special processing, allowing designs with multiple sources and mixed technologies. The technology incorporates major advances in interconnect testing, chip contact size reduction, and rework processing. This paper describes the processes required for this new technology.

Introduction

Various approaches have been explored for combining several chips into one multichip module [1-3]. These approaches may be segmented into attempts to construct small circuit boards and attempts to construct larger integrated circuits. Both approaches improve circuit performance by reducing interconnect length and driver loading, especially through elimination of single chip packages. Our new multichip technology (see Fig. 1) focuses on using a copper-polyimide structure to construct larger integrated circuits. Module circuit densities are ten times larger than current VLSI chip designs, up to 15 cm^2. Nominal design rules for the interconnect structure are: 10 micron linewidth, 15 micron spaces, 2 micron conductor thickness, and 2 micron interlayer dielectric thickness. These rules were chosen to provide balanced electrical parameters, however lines as narrow as 2 microns can be fabricated. Excellent routability is achieved with only two signal interconnect levels and one power/ground interconnect level. Since this technology combines IC line densities, small contact pads, and controlled impedance transmission lines it is an ideal means of bridging the interconnection density gap between circuit boards and integrated circuits.

Dominant chip connection methods such as bond wires, solder bumps, and TAB tape require a chip pad pitch greater than 4 mils. This constraint often results in "pad limited" chip designs that employ serial data paths rather than parallel data paths. A multichip module technology must accommodate high internal I/O demand in order to realize the advantages of higher integration [4]. Employing a solderless flip-chip die attach method, this new process allows 12 micron square contacts at any location on a chip. The minimum via size is determined by the thickness of the interconnect system. For 12 micron contacts, a via aspect ratio of 1:1 is required. This powerful feature has important ramifications for chip design and thermal management.

1) Power and ground signals can be distributed by low inductance module traces rather than on-chip networks.

2) Engineering changes or metal options can be implemented in the module interconnect, simplifying chip manufacture.

3) As with other approaches, buffer size and power can be reduced.

4) Interconnect design rules are independent of chip power levels, because the thermal path is through the back side of the chips. No thermal posts are required [5].

Various copper-polyimide processes have been developed in recent years [5-8]. In addition to performance advantages derived from the flip-chip structure, our technology has manufacturing advantages over past methods [9]. The manufacturing advantages are:

1)   A flip–chip IC mounting technique uses standard semiconductor manufacturing methods and equipment.

2)   A "reach through via" process step creates uniform, high-yield electrical connections between chips and interconnect wiring.

3)   Use of glass substrates permits an optical comparator technique for interconnect inspection.

4)   The glass substrates are used as thin film carriers during manufacture, and are not consumed in the process.

5)   Salvage and rework operations are possible prior to final package assembly.

These manufacturing advantages result in a low cost interconnect technology with high reliability.

Process Overview

Figure 1 shows a test module halfway through the Polylithics process cycle.   A sample module with four connection test die is shown in the foreground.   A glass plate carrying the copper-polyimide interconnect structure is shown at the left.   A silicon wafer with the test die is at the right.

An overview of the process flow is shown in Figure 2.   The process begins with a transparent glass substrate. Layers of polyimide dielectric and chromium-clad copper (Cr/Cu/Cr) conductors are then deposited on the glass substrate.   The die are then optically aligned and mechanically attached to the interconnect with a thermally stable adhesive.   A ceramic ring is attached to the interconnect to form a support frame for each module.   Figure 1 shows a module at this stage of development.

The next phase of the process separates the modules from the glass plate.   With the top of the interconnect exposed, vias are etched through the polyimide to the contact pads on the chips, exposing pad metal and the buried copper traces.   The vias are then filled with aluminum, making the electrical connection.   Contacts at the edge of the module are ready for bonding to the next level of packaging.

Fabrication of Thin Film Signal Planes

Thin film fabrication requires wafer-shaped low expansion glass or quartz substrates as starting material. These substrates are commercially available in a variety of sizes.   A buffer layer of polyimide is spin-coated to a 2 micron thickness.   Using image reversal lift-off techniques, the first signal layer is fabricated.   A multi-source electron gun evaporator is used to deposit Cr/Cu/Cr conductors, chrome providing good dielectric adhesion and copper offering low resistivity.   This methodology was selected because complex etch chemistry for multi-metal systems is avoided. Additionally, high aspect ratios are achieved with narrow lines (fig. 3).   A polyimide dielectric layer 2 microns thick is applied after the lift-off step, creating a partially planarized surface.   The second signal layer is created according to the same procedure.

Substrate Testing

After the second layer is fabricated, an automatic optical inspection of the substrate is performed.   This test is done on a commercially available mask inspection tool, and identifies the precise location of all defects that create co-planar electrical opens and shorts.   Process improvement is facilitated by easy identification and cataloging of the defect mechanisms.   Mask repair systems can be used to repair traces in some cases.   This method fails to detect interlayer shorts and single layer opens at crossovers, but crossover area is minimized through orthogonal signal layer design.   Also, testing can be performed after producing the first layer.   The method can exhaustively test for shorts, unlike probe techniques which must perform N! tests to achieve the same coverage (N is the number of nets). Mechanical probing would not work for 12 micron pads, which is the lower limit provided by this interconnect technology.

Power/Ground Plane Fabrication

Subsequent to the non-destructive optical test, a final metal layer is fabricated on the substrates.   This layer is used for power and ground signal distribution.   This layer blocks transmitted light and obscures most of the signal layers for inspection purposes. Loose ground rules are employed on this layer, so the fatal defect density is extremely low. After a final coating of polyimide, substrate fabrication is complete.

Chip and Ring Attachment

Figure 4 illustrates the process sequence for attaching and connecting integrated circuits to the thin films. Like the substrate manufacturing process, this process employs existing, low-cost semiconductor manufacturing equipment.

Chip attachment is achieved using any tool capable of providing X, Y, Z and Ø motions with full wafer travel and one micron adjustment accuracy. Individual die are mounted on a vacuum chuck after being coated with an organic dielectric die attach material. An operator looks through the back side of the transparent substrate and aligns the chip under the correct module site. Pad windows in the power/ground layer permit easy and accurate alignment. Using Z motion control, the operator tacks the die to the thin films and releases the vacuum (fig. 4a). Each substrate is fully populated in this manner, one chip at a time. No electrical contact has been made to the chip at this point in the process.

Thin layers of free-floating polyimide are flexible, so a support structure is required for handling the multichip module. A ceramic ring is aligned and attached to the interconnect using the same method as for chip attachment. This ceramic ring also provides support for the bonding pads of the multichip module. Once the ceramic ring has been attached, each module becomes a mechanically stable unit that can be removed from the glass wafer (fig. 4b). A chemical process is used to remove the modules from the reusable glass wafer. The modules are then prepared for electrical interconnection.

Electrical Connection - The "Reach Through Via"

The "reach through via" is one of the major advances introduced in this process. The thin film surface formerly in contact with the substrate is coated with photoresist, exposed and developed (fig. 4c). Then, a 12 micron deep via is dry etched through the entire thin film structure, reaching the semiconductor chip contact pad (fig. 4d). Vias patterned over copper traces are etched only 2 microns deep. Figure 5 shows a top view of these two vias on an experimental sample. The chip pads and interconnect traces are then chemically stripped of surface oxides. Chrome and aluminum are deposited in the vias to form an electrical connection between the chip and the interconnect structure. This layer is then patterned and etched to electrically isolate the connections (fig. 4e), followed by a 400 °C sintering step to control via resistance.

Electrical Testing

The module is fully connected and testable at this point in the process. All thin film signals for external connection are routed to the edge of the module, where pads were exposed and coated with aluminum during the reach through via steps. Using special test jigs and fixtures, electrical testing is performed at maximum cycling rates. Since each chip pad is contacted with a reach through via, all internal nets can be probed to facilitate debug and failure diagnosis.

Rework and Repair

Modules failing electrical testing can be reworked or repaired. Good chips can be recovered from defective modules by etching away the reach through via aluminum. The chrome layer acts as an etch stop and protects the chip pads. Chips are pulled from the thin films and any residual organic materials left on the die are plasma stripped. For large modules with single chip failures, individual die removal and replacement steps may be performed. Aluminum is first removed as in the salvage operation. Bad die are isolated from their neighbors by mechanical barriers on both sides of the module. Special chemistry breaks the adhesion between die surfaces and the polymer layer, allowing removal of the bad chips. Subsequently, the die attach and reach through via steps are repeated with replacement die.

Final Packaging

Figure 6 shows several views of a nine chip module with TAB connections. In this configuration, a module is handled like a "superchip." Active burn-in and additional speed testing are facilitated with this approach. Simple encapsulated "chip on board" (COB) assembly techniques complete the packaging requirements if chip power dissipation levels are manageable.

Figure 7 illustrates an alternative packaging approach which provides hermeticity, surface mount capability and a thermal path suitable for heat fin cooling. The critical step of simultaneously attaching a base plate to all die and the ring is possible because the interconnect film is free-floating and flexible.

Wire bonding to the module is the most straightforward method, either using a COB approach or die attaching in a standard package such as a pin grid array. More exotic packaging schemes involve building metallized ceramic rings which provide film support and electrical connection. Reliability is enhanced in sealed cavity packages through vacuum bake operations during lid seal, driving off moisture absorbed by the polyimide.

Conclusions

A new multichip interconnect technology is being developed for multichip module applications. This technology promises improvements in manufacturability and module performance. Module densities approach wafer scale integration levels while keeping the chip yields of very large scale integration. This results in substantial cost/density improvements over monolithic approaches. The similarity to monolithic design, however, suggests that this new technology is truly a "superchip" approach.

The most important feature of this new technology is the "reach through via." This 12 micron square contact is a semiconductor-like structure which allows a 100x potential reduction in chip contact area. The reduction of contact pad sizes and chip-to-chip spacings eliminates the conventional design constraints posed by chip boundaries. The technology promises immediate improvements in integration and performance using off-the-shelf chips, but its full power will be exploited when IC design methodology treats the interconnect as additional metal layers in multichip circuits. Products created in this way will have a permanent role in bridging the gap between chips and circuit boards.

Acknowledgements

The authors wish to thank Dr. Alfred N. Riddle of Macallan Consulting, Milpitas, CA for his invaluable assistance in the preparation of this paper.

We greatly appreciate the fabrication support provided by Mark Walters of the Microelectronics Center of North Carolina.

References

[1] C.A. Neugebauer, and R.O. Carlson, "Comparison of Wafer Scale Integration with VLSI Packaging Approaches," IEEE Trans CHMT, pp. 184 - 189, June 1987.

[2] C.J. Bartlett, J.M. Segelken, and N.A. Teneketges, "Multichip Packaging Design for VLSI-Based Systems," IEEE Trans CHMT, pp. 647 - 653, Dec 1987.

[3] R.W. Johnson, J.L. Davidson, R.C. Jaeger, and D.V. Kerns, Jr., "Silicon Hybrid Wafer-Scale Package Technology," IEEE JSSC, pp. 845 - 850, Oct 1986.

[4] P.A. Venkatachalam, "High Performance Memory Chip Packaging Implications: Multichip Module vs. Single Chip Module," Proc 35th Elect Comp Conf, pp. 379 - 383, 1985.

[5] J.T. Pan, S. Poon, and B. Nelson, "A Planar Approach to High Density Copper-Polyimide Interconnect Fabrication," 8th Int'l Elect Pack Conf, pp. 174 - 189, 1988.

[6] R.J. Jensen, J.P. Cummings, and H. Vora, "Copper/Polyimide Materials System for High Performance Packaging," IEEE Trans CHMT, pp. 384 - 393, Dec. 1984.

[7] S. Lebow, "High Density/High Speed Multi-Chip Packaging," 6th Int'l Elect Pack Conf, pp. 417 - 423, 1986.

[8] R.O. Carlson, C.W. Eichelberger, R.J. Wojnarowski, L.M. Levinson, and J.E. Kohl, "A High Density Copper/Polyimide Overlay Interconnection," 8th Int'l Elect Pack Conf, pp. 793 - 804, 1988.

[9] Patent pending.

Figure 1. This picture shows several phases of the Polylithics interconnect process.

POLYLITHICS PROCESS FLOW

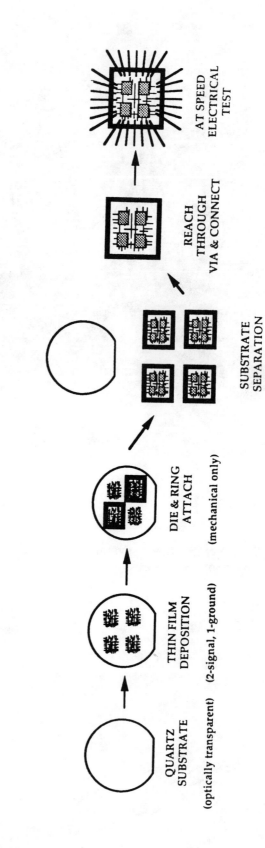

Figure 2. Overview of the new interconnect process.

Figure 3. SEM photograph of minimum geometry Cr/Cu/Cr
interconnect trace fabricated using lift-off
techniques.

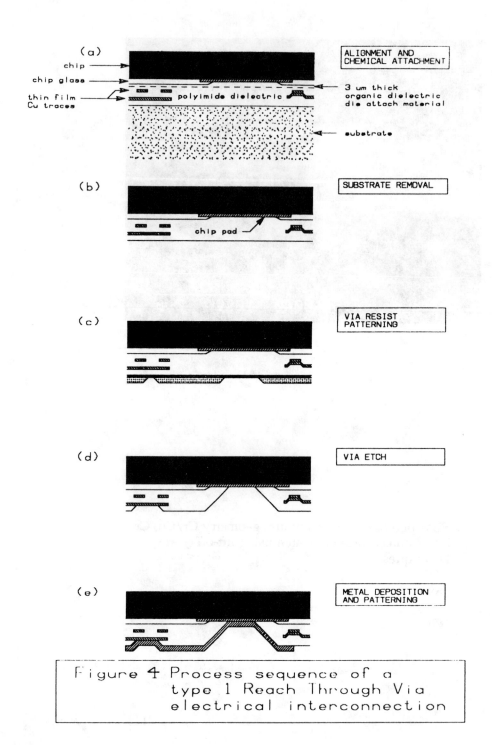

(a) ALIGNMENT AND CHEMICAL ATTACHMENT

chip

chip glass

thin film Cu traces

polyimide dielectric

3 um thick organic dielectric die attach material

substrate

(b) SUBSTRATE REMOVAL

chip pad

(c) VIA RESIST PATTERNING

(d) VIA ETCH

(e) METAL DEPOSITION AND PATTERNING

Figure 4 Process sequence of a type 1 Reach Through Via electrical interconnection

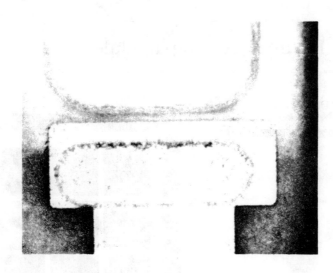

Figure 5. Photograph of "reach through vias." Top via reaches through 12μ of dielectric to aluminum chip pad. Lower via penetrates 2μ to copper interconnect trace.

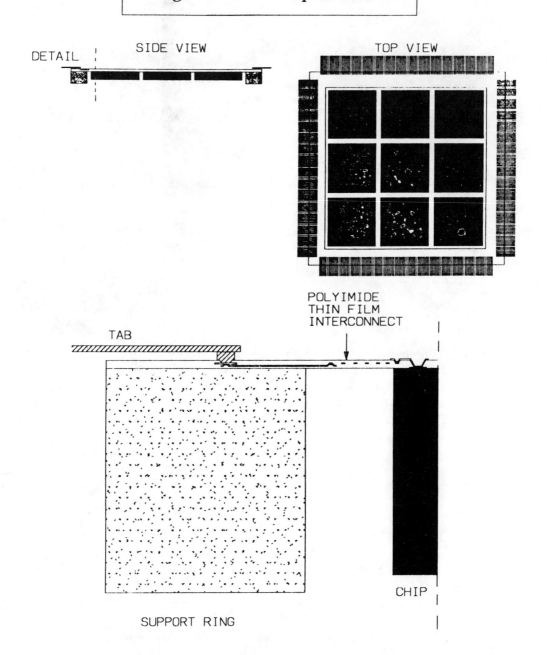

Figure 6. A 9 chip module.

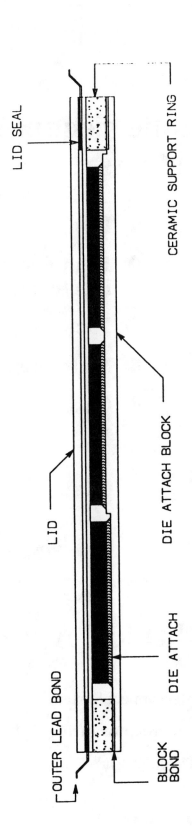

LID SEAL

CERAMIC SUPPORT RING

LID

DIE ATTACH BLOCK

OUTER LEAD BOND

DIE ATTACH

BLOCK BOND

Figure 7. Small protective carrier.

A Multichip Module Technology Must Address...

- **Thin Films**
- **Chip Bonding**
- **Heat Transfer**
- **Testing**
- **Reliability**
- **Repair**

- **Design**
- **High Frequency Electronics**
- **Packaging**
- **Materials**
- **Cost**

Highlights of Polylithics' Approach

- Copper / Polyimide Interconnect System

- Glass Substrates Used in Manufacturing Only

- Proprietary "Reach Through Via" Chip Connection

- Solderless Flip-Chip Structure

  - Area Bonding

  - Back Side Thermal Path

- No Special IC Die Processing Required

Polylithics, Inc

Principal Advantages

- ## Low Cost
  - No specialty substrates
  - No bump processing steps required
  - No TAB tooling needed
  - Chip salvage & module repair capability
  - Process uses past generation IC equipment

- ## High Performance
  - Short line lengths
  - Low loss chip connection
  - High I/O count

Polylithics, Inc

Process Overview

STEP 1.   Thin Film Manufacture

STEP 2.   Chip & Ring Attachment

STEP 3.   Reach Through Via Etch

STEP 4.   Electrical Chip Connection

STEP 5.   Packaging

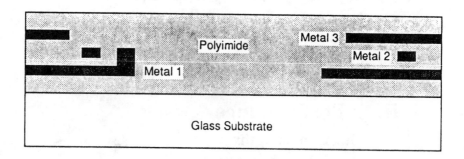

STEP 1. Thin Films Manufacture

Metal 3

Metal 2

Polyimide

Metal 1

Glass Substrate

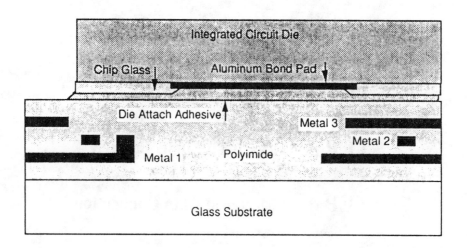

STEP 2. Chip Attachment

Integrated Circuit Die

Chip Glass

Aluminum Bond Pad

Die Attach Adhesive

Metal 3

Metal 2

Metal 1

Polyimide

Glass Substrate

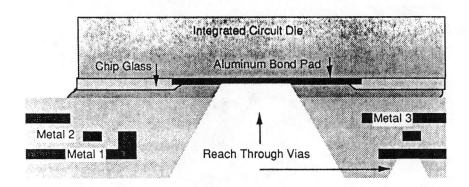

STEP 3. Reach Through Via Etch

STEP 4. Electrical Chip Connection

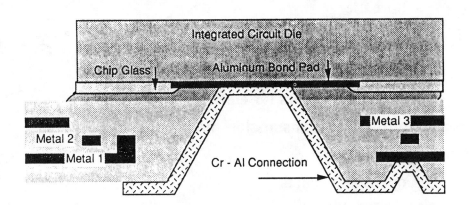

STEP 5. Packaging

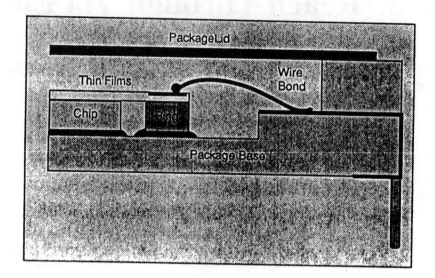

Testing

- ## INTEGRATED CIRCUITS
    - bare die
    - at speed
- ## THIN FILM SUBSTRATES
    - process monitors for electrical tests
    - automated transmission-mode optical inspection
    - 2 signal layers
    - opens and shorts
- ## MODULES
    - all nodes are observable
    - full AC testing before and after final packaging

A Copper/Polyimide Metal-Base Packaging Technology

HAYATO TAKASAGO, KOHEI ADACHI and MITSUYUKI TAKADA

Materials & Electronic Devices Laboratory
Mitsubishi Electric Corporation
1-1, Tsukaguchi-Honmachi 8-Chome
Amagasaki, Hyogo, 661 Japan

A unique substrate MCPM (Mitsubishi Copper Polyimide Metal-base) technology has been developed by applying our basic copper/polyimide technology.[1] This new substrate technology MCPM is suited for a high-density, multi-layer, multi-chip, high-power module/package, such as used for a computer. The new MCPM was processed using a copper metal base (110 × 110 mm), full copper system (all layers) with 50-μm fine lines. As for pad metallizations for the IC assembly, we evaluated both Ni/Au for chip and wire ICs and solder for TAB ICs. The total number of assembled ICs is 25. To improve the thermal dispersion, copper thermal vias are simultaneously formed by electro-plating. This thermal via is located between the IC chip and copper metal base, and promotes heat dispersion. We employed one large thermal via (4.5 mm$^\phi$) and four small vias (1.0 mm$^\phi$) for each IC pad. The effect of thermal vias and/or base metal is simulated by a computer analysis and compared with an alumina base substrate. The results show that the thermal vias are effective at lowering the temperature difference between the IC and base substrate, and also lowering the temperature rise of the IC chip. We also evaluated the substrate's reliability by adhesion test, pressure cooker test, etc.

Key words: Cu/polyimide composites, high performance packages, microelectronics packaging

INTRODUCTION

A high-density, functional circuit substrate/module has been earnestly demanded. In order to increase circuit density fine lines, greater speed and more efficient thermal heat dissipation is required. For computer applications, these are very important technologies. Various requirements are demanded in this field, such as;

1. Large amount of wiring accomodations
2. Improved or reduced signal propagation delays
3. High thermal dissipation capabilities
4. Reliable and mass-producible package.

Recently various metal-base substrates have been developed and released onto the market, however these metal-base substrates are mainly used in high-power applications, such as power modules. Materials employed in these modules are insufficient for high-speeds or a large amount of wiring accomodations.

Multi-layer ceramic substrates are also available for those computer package area, such as cofired low TCE substrates, etc. However, these substrates are expensive due to tooling costs, and, in addition, they have lower thermal capabilities in heat dispersions.

We considered these technological demands, and developed the high-density, metal-base circuit substrate technology MCPM (Mitsubishi Copper Polyimide Metal-base) by applying our basic copper/polyimide technology.[1,2,3]

STUDIES

Studies for the advanced computer substrate have been conducted with regards to the following technical items;

1. Copper metal base for high heat dissipation capabilities
2. Copper conductors because of low resistance
3. Photoactive polyimide because of low dielectric constant
4. Pad metallizing for various multi-chip LSIs.

We have made a major effort to develop a high quality monometal (copper metal base and copper conductor) system combined with a polyimide process enabling finer lines, smaller via sizes, multi-layering, high-power dissipation, and larger substrate use. This technology fully satisfies the requirements shown above.

FUNDAMENTAL STRUCTURES/PROCESSES

An MCPM schematic structure are shown in Fig. 1. The structure we developed has a copper metal-base (110 × 110 mm), full copper system (all layers) with 50-μm fine lines, and total three conductor layers. To improve the thermal dissipation, copper thermal vias are simultaneously formed by wet-metallizing. For use as the interlayer insulation, we selected a photoactive polyimide. By using this photoactive polyimide, the formation of small diameter vias and minute pad openings, required for the accommodation of the LSI chip, became easy.

(Received June 15, 1988; revised September 27, 1988)

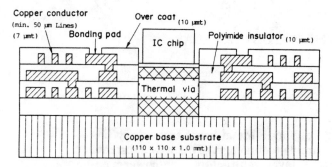

Fig. 1 — An MCPM schematic structure.

In other words, by combining copper metal-base, copper thermal vias, copper conductor and polyimide, and establishing the various related materials and process factors, unique package MCPM have been developed.

The fundamental process flow studied and developed consists mainly of the following four steps;

1. Thermal vias process
2. Conductor process
3. Interlayer insulation process
4. Assembly process.

These processes are shown in Table I with the related materials.

DETAILED MATERIALS/PROCESSES

Thermal Vias Process

Surface treating for the copper metal base is done by a proper chemical acid cleaning method in order to remove various contaminants.

A photoactive polyimide is then coated on the copper metal base and patterned openings for the thermal vias by photolithography. Then, the openings are electro-plated using the copper base metal as cathode and deposited copper.

To ensure the insulation between the base metal and upper conductor, the occurrence of pinholes and scratches in the polyimide insulation layer should be prevented. These pinholes and scratches are formed in the polyimide patterning and handling processes.

In the polyimide coating process, the insufficient wettability of the base materials (copper base metal or copper conductor) and micro air-voids in the polyimide resin cause pinholes. Also, in the polyimide patterning process, dust, stains and scratches on the photo mask are responsible for these defects. Especially, in the case of film mask, the dotted traces of binder which enhance adhesion between the base film and emulsion, remain on the transparent area as shown in Fig. 2. These traces become a serious problem concerning pinholes because of insufficient photopolymerization in the case of negative type photoactive polyimide use.

The resin residue which arises from the chemical interaction between the polyimide precursor and copper is retained on the copper. The residue was removed easily by hydrazine solution after post-baking. If the coated polyimide becomes thin in spots because of micro-voids and base-metal roughness, pinholes arise from penetration of hydrazine solution at the polyimide thinned points. This phenomena depends on the immersion time of hydrazine solution and the coated polyimide thickness. The polyimide coated substrate should be handled carefully to prevent scratches, cracks and other defects.

The thermal vias are formed by electro-plated copper deposition into openings in the polyimide layer

Table I. Fundamental Process Flow

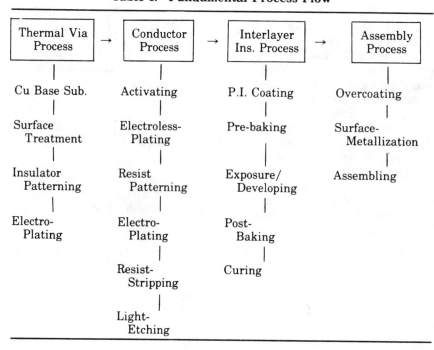

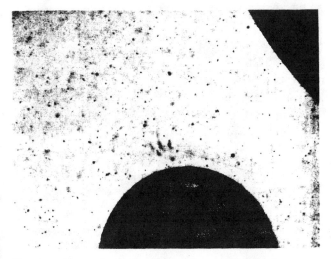

Fig. 2 — Dots and scratches observed on a Film Mask.

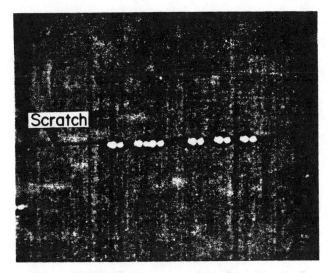

Fig. 4 — Spheres plated on a scratched line.

as mentioned above. At the same time, this procedure can check satisfactorily the unfavorable defects, such as sphere deposits formed by plating at polyimide pinholes. Figures 3 and 4 show the spheres plated at polyimide pinholes and on a scratch line respectively.

Conductor Process

Copper conductor is semi-additively formed onto the polyimide layer, the polyimide surface is conditioned, and then activated by a catalyst. Next, electroless-plating is done entirely covering the substrate surface. For a subsequent plating-resist process, we developed a new dry film process enabling 50-μm line and space. After this process, electroplating is applied and completed.

The dry film resist employed is an alkaline developing type and it is characterized by very thin

photo-resist layer and cover-coat film to increase the pattern resolution.

Electro-plating is done at room temperature by using sulfuric copper solution. By using a conventional plating tank, the plated thickness distribution was about ±30% of the average value without a shield board which can even current distribution effectively and about ±20% with a shield board. Furthermore, we improved the plating method by streaming the plating solution onto the substrate. By using our developed plating method, the plated thickness distribution was brought to within ±15%. Figure 5 shows minute pads with a 100-μm diameter on polyimide insulator and via-hole connections for interlayer electrical connection.

Interlayer Insulation Process

Photoactive polyimide is employed in order to shorten the related processes. Also, we employed a belt conveyor furnace for the curing. Typical via-hole for the interlayer connection is 80 μm in diameter. Also, we studied the polyimide thickness variation during the curing step, and designed the process specifications.

The decreasing curve of the polyimide thickness upon heat treatment temperature is shown in Fig. 6. As the treatment temperature becomes higher, the polyimide thickness become thinner because of evaporation volatile compounds and changing polyimide precursor to polyimide. In the case of a 15-

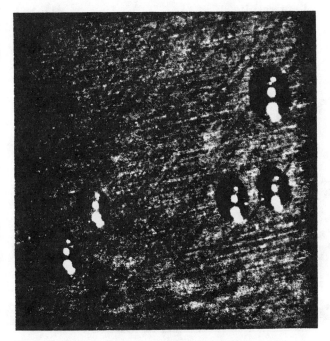

Fig. 3 — Spheres plated at polyimide's pinholes.

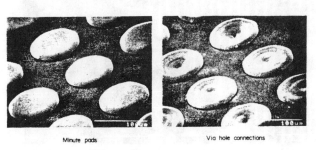

Minute pads   Via hole connections

Fig. 5 — Minute pads & via-hole connections.

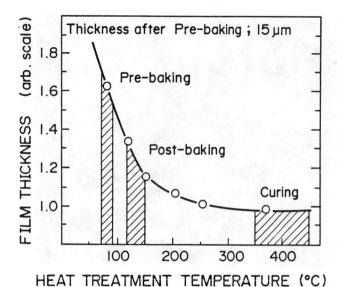

Fig. 6 — Obtained polyimide thickness vs heat treatment temperatures.

Fig. 8 — Ni/Au pads plated onto copper (100 μm^ϕ).

µm thickness after prebaking, the postbaked thickness and the cured thickness decrease respectively by factors of 0.76, and 0.63 compared to the prebaked thickness. These decreasing ratios change with the prebaking and/or postbaking temperatures.

Photoactive polyimide we employed tends to entirely absorb ultra-violet energy within a thin layer beneath the surface in the exposed area. Therefore, only a thin layer beneath the surface is unsoluble to the developing solution, and the underlying layer of polyimide can be attacked easily. Accordingly, the obtained polyimide pattern shape depends strongly on exposure and/or developing conditions.

Assembly Process

We evaluated two types of over-coating resins which are photoactive. One is a photoactive polyimide which is the same as the interlayer insulation, the other is an epoxy-acrylate dry film material which has a property for alkaline use. Figure 7 shows

the dry film resist pattern having 100-µm lines and spaces which are formed on the polyimide layer.

We also evaluated two types of pad metallization onto the copper. One is a Ni/Au metallization for wire bonding ICs. The other is solder pads for surface mounting LSIs, such as TABs, Flip-chip ICs, etc. Regarding the Ni/Au pad metallization, a suitable selection of the over-coating resins is required, because electroless Au solution ordinarily has a high alkaline value and is used at high temperatures. The two over-coating resins shown above have excellent chemical resistance.

Figure 8 shows minute Ni/Au pads which are plated onto the copper at 100-µm diameters. Figure 9 shows an example of reflow soldered TAB IC.

EVALUATIONS

We evaluated the following items and obtained excellent results.

1. Adhesion evaluations
2. Thermal dissipation capabilities
3. High-voltage insulation.

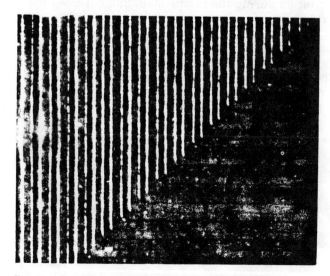

Fig. 7 — Dry film resist pattern having 100-µm L/S.

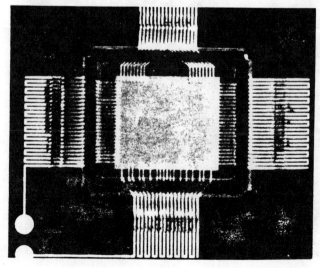

Fig. 9 — TAB IC on MCPM (reflow-soldered).

Adhesion Evaluations

Bond strength between polyimide and copper metal base was evaluated by employing various tests, such as pulling test, solder-float test, and solder-immersion test.

Figure 10 shows the pulling test method for the evaluation between polyimide and copper metal base. Aluminum stud (2.7 mm$^\phi$) is bonded onto the polyimide surface by epoxy resin, and is pulled vertically at a constant velocity (50 mm/min). We examined the changing surface roughness, Rz, of the copper metal base. And the results are: the average adhesion force was 5.3 kg/mm^2 (Rz = 1 μm) and 3.5 kg/mm^2 (Rz = 10 μm). All fractures were caused in the interface area of the epoxy adhesive and polyimide. From these results, the strength between the polyimide and copper base metal will exceed these values, and is sufficiently large.

In addition, solder float test and immersion tests were employed for the adhesion evaluation. Figure 11 shows a sample after solder float tested (250° C, 30 sec) having 50-μm patterned polyimide. And, Fig. 12 shows the immersion test results. This sample was solder dipped after the immersion test. The immersion test was done by employing high alkaline solution (pH = 12.5), and dipping the total substrate into this solution for 200 hr at room temperature.

Through these tests no failures were observed, therefore it can be concluded that polyimide has an excellent adhesion strength to the copper base substrate.

Thermal Dissipation Capabilities

Thermal dissipation capabilities are evaluated and compared to substrates having no thermal vias, and/or the copper/polyimide substrate formed onto an alumina substrate.

The heat dissipation properties were calculated by a computer analysis based on thermal resistance networks. The test module is 110 × 110 mm in size, made up of a total of 25 IC chips (IC size is 8 × 8 mm). We calculated the temperature excursions and distributions by using the differential method under the conditions of total 75 W thermal heat dissipations (3.5 W/chip). We assumed that each IC and/or each element was at the same temperature. And, a heatsink was attached to the backside of the copper metal base, and forced air cooling was applied.

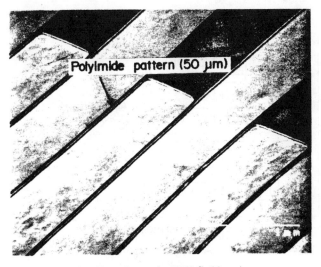

Fig. 11 — Solder floated sample (250° C, 30 sec).

Also, the area of the thermal vias is the same as IC size, and thermal resistance between ICs and each bonding pad (copper pad) is neglected. Detailed conditions employed for the calculations were as follows: the surface area of the heatsink in each element is 25 mm^2, the average velocity of the forced air cooling is 2 m/sec, and the ambient temperature is set at 25° C.

Figure 13 shows the thermal analyzed area and the conditions. Here, we calculated three cases shown below;

(a) Base substrate is copper metal base and thermal vias are formed
(b) Base substrate is copper metal base and no thermal vias
(c) Base substrate is alumina 96% and thermal vias are formed.

The calculation results of each temperature excursions and distributions are shown in Fig. 14. In this figure, shaded portions indicate the thermal vias.

When (a) and (b) were compared, the effect of the thermal vias was made clear. Results show that case (b) indicates a large temperature difference be-

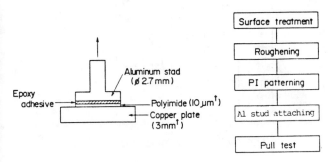

Fig. 10 — Pulling test method.

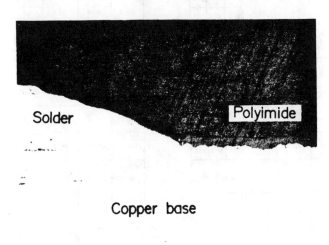

Fig. 12 — Solder dipped sample.

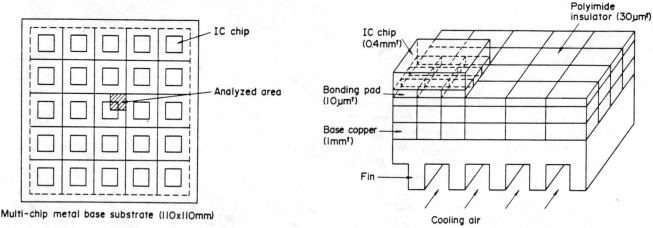

Fig. 13 — Thermal analyzed area and conditions.

tween the IC chip and copper metal base (max 33.6° C), however, case (a) shows a lower temperature difference (max 11.6° C). And, the temperature rise of the IC chip was reduced by about 20.5° C. When (a) and (c) were compared, the effect of the base substrate material was made clear. The calculation results in case (c), IC temperature was 159.7° C, and temperature difference between IC and base substrate became 122.4° C. Thermal conductivity of alumina is very low, and is approximately 1/27 of copper metal. Therefore, thermal vias formed on the alumina substrate were not effective.

High-voltage Insulation

High-voltage Insulation and reliabilities are evaluated. Figure 15 shows dc break-down voltage characteristics changing with the cured polyimide insulation thickness. In this figure, the straight line shows an initial break-down voltage, and the dotted line shows a break-down voltage after 300 hr pressure cooker test. Each dc break-down voltage was measured; a voltage sweep (rated 50 V/sec.) was applied by a probing contact (3 mm$^\phi$) to the polyimide surface formed on the copper metal base.

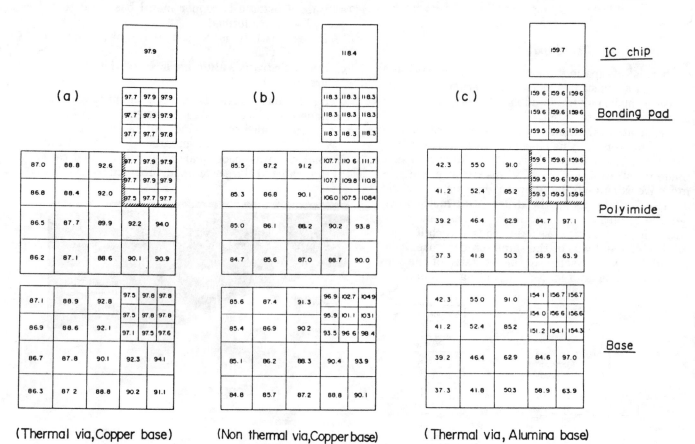

Fig. 14 — Calculation results of temperature distribution.

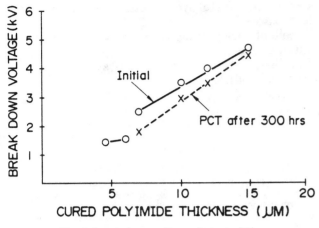

Fig. 15 — The dc break-down voltage characteristics.

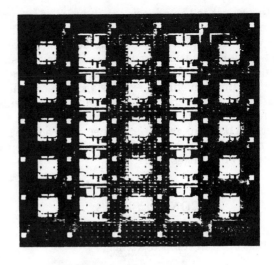

(a) MCPM (110x110 mm)

Results show that the initial break-down voltage is above 3.4 kV and remains 3 kV after the pressure cooker test (121° C, 2 atm).

From these evaluation results, we could observe the discontinuity of the initial break-down voltage at around 6 μm of polyimide thickness. We assume that this is caused by polyimide pinholes because of the insufficient polyimide thickness.

At our standard polyimide thickness, 10 μm, the initial break-down voltage was 3.4 kV, and even after the pressure cooker test this value became 3.0 kV.

CHARACTERISTICS

Table II shows the summarized characteristics of the copper metal base substrate. And, Fig. 16 shows the completed MCPM.

Regarding the conductor characteristics, we attained excellent low sheet resistivity by the adoption of copper conductors (3.0 m Ohm/ in 6 μm thickness). And insulation resistance related to both the interlayer and the surface exceeded 10^{13} Ohm, and even after the 200 hr pressure cooker test it still maintained 10^{11} Ohm.

Dielectric constant was 3.5 at 1 kHz. This value was enough for the low cross-talk and/or high speed signal propagation.

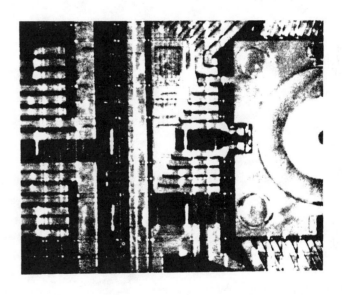

(b) Magnified view

Fig. 16 — The completed MCPM.

Table II. Summarized Characteristics of the MCPM

Properties	Conditions	Results
Conductor		
Sheet resistivity	6 μm^t	3.0 m Ohm/□
Adhesion strength	1.5 mm$^\phi$ pattern pull speed 5 mm/min.	2.5 Kg/mm2
Insulator		
Surface resistance	dc 100 V, 50 μm gap	10^{13} Ohm
	PCT 2 atm, 200 hr	10^{11} Ohm
Interlayer resistance	dc 100 V, 10 μm^t	10^{13} Ohm
	PCT 2 atm, 200 hr	10^{11} Ohm
Break down voltage	dc, 10 μm^t	3.4 kV
	PCT 2 atm, 300 hr	3.0 kV
Dielectric constant	Room temp., 1 MHz	3.5
Dissipation factor	Room temp., 1 MHz	0.003

With respect to the adhesive properties of the polyimide-overcoat or polyimide-polyimide interface, there were no negative symptoms because no blistering or delamination occurred in either the surface metallizing process or the assembly process.

Heat dissipation characteristics are also excellent because of the copper metal base and thermal vias.

Surface mounting LSIs are widely applicable for such as TABs, Flip-chip ICs, and so on. Pad metallization for these various LSIs has been developed and their reliabilities confirmed.

CONCLUSIONS

A high-density multilayer substrate for a computer has been developed with;

1. Copper monometal system with 50-μm fine lines and spaces
2. Excellent heat dissipation properties by the thermal vias
3. Excellent electrical characteristics
4. High reliability.

This technology is suited for high-density, high-power multi-chip packages.

REFERENCES

1. H. Takasago, *et al.* IEEE CHMT, Vol. CHMT-10, 425 (1987).
2. H. Takasago, *et al.* "Fine-line, multilayer hybrids with wet-processed conductors and thick-film resistors," Proc. 34th Electronic Components Conf., 1984, pp. 324–329.
3. Adachi, *et al.* "High Density Metal Base Substrate Technology for LSIs," Proc. Printed Circuit World Convention, June, 1987.

Very High Speed Multilayer Interconnect Using Photosensitive Polyimides

Robert L. Hubbard and Gail Lehman-Lamer

Tektronix Inc., Beaverton, Oregon

ABSTRACT

Recently, photosensitive polyimides have become commercially available that can be patterned directly. These materials are "negative-acting" and can be coated by spinning or spraying in layers 0.5μ to 60μ thick. After exposure the layers are developed and then cured. We describe a process for producing a polyimide interconnect package consisting of four layers of 25μ thick photo-patterned polyimide on a copper substrate using gold vias and lines.

Introduction

High density interconnects are defined here as multi-chip modules that allow for finer geometries than printed circuit boards but more density than thin-film hybrids. The predominant interest in these interconnects comes from the need for higher density packaging and therefore lower costs [1] [2]. Another need for such a technology arises from a need for packaging very high speed bipolar devices (gigabits per second). These devices require controlled impedance stripline structures with high I/O count.

These requirements dictate a minimum 4 mil spacing between shielding layers using a dielectric material with a dielectric constant less than 4. Ceramics have a dielectric constant of 9. Polyimide has a dielectric constant of 3.5 but is very difficult to layer into low stress films more than 50μ thick. If the lines are made thinner and narrower the polyimide could be thinned proportionately. For high speed circuits it is also necessary to keep runs less than a few centimeters long between devices. It would be advantageous to minimize crosstalk and power supply bounce by running all fast signals as pairs of balanced lines. The combination of all these electrical and physical restrictions result in specifications that include dielectric thickness of 25μ, dielectric constant of 3.5, gold or copper conductors 2μ thick and conductor widths of 25μ.

Multilayer polyimide interconnects (**MLP**) have been reported to successfully achieve this z-axis size reduction [3] [4] [5]. Very fine geometries are available by using photolithographic techniques to pattern deposited polyimides. In most cases however the objective was to reduce size and cost (increase density) and the dielectric layer thicknesses were 5-10μ. The conductors were typically copper and only two or three layers were necessary. The thin layers of polyimide were easily etched by RIE using typical semiconductor techniques. The increasing use of polyimides in the microelectronics industry seems to be concentrated in thin layers for passivation or simple interconnect systems just described.

Within the industry there is a general trend towards more components per chip, smaller gate delays and lower gate energies and gate power. As the junction area decreases and power increases the watts per centimeter rises to serious levels. Practical cooling schemes must be incorporated into the module design from the beginning. For the MLP modules described above the chips were typically mounted face down using "flip-chip" [6] [7] [8] or "Area Array" [9] bonding techniques. In the case where power will be high and some access to the IC after bonding is desired these techniques will not be satisfactory. Direct bonding of the IC to a metal substrate would be more desirable for heat dissipation.

For all of these reasons it was necessary to develop a package that had somewhat different characteristics than what was reported in the literature.

MLP Structure

Two ICs were designed to evaluate our MLP interconnect structure. A four layer interconnect scheme was drawn up with 1 mil polyimide thickness for each layer. Gold was chosen for the 1 mil wide by 2μ thick lines since gold was known to be more inert to polyimide. These dimensions were predicted to result in 50 ohm controlled impedance striplines.

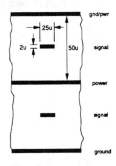

Figure 1

A metallic (copper) substrate was chosen for maximum thermal conductivity. Modeling has shown that the use of a simpler stainless steel substrate would produce a 26°C higher temperature seen by the junction. A thin layer of nickel was evaporated onto the copper for enhanced polyimide adhesion. Large vias were opened up through each layer of polyimide to allow the chips to be attached directly to the substrate. Vias that extended through more than one layer were "stair-stepped". This may not have been necessary but it eliminated any potential planarization problems. See **Figure 2** for a generalized diagram. Single Point Tape Automated Bonding (SPTAB) was chosen for connections from the IC for the finer pitch it allows. SPTAB also conducts more heat away from the IC than wire bonding.

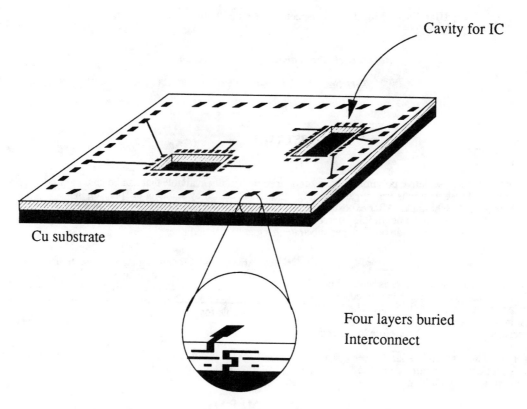

Cavity for IC

Cu substrate

Four layers buried
Interconnect

Figure 2

Process

A generalized scheme for the process flow is shown in **Figure 3**. The steps for the first layer are detailed and the next three layers are applied in a similar fashion in their turn.

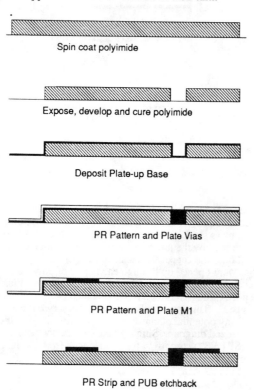

Spin coat polyimide

Expose, develop and cure polyimide

Deposit Plate-up Base

PR Pattern and Plate Vias

PR Pattern and Plate M1

PR Strip and PUB etchback

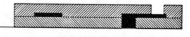

Spin and Pattern P2

Pattern and Plate Vias and M2

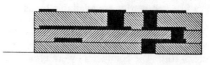

Layer 3

Layer 4

Figure 3

Since we were restricted to using a thick layer of polyimide (25μ) we decided to evaluate the newer photosensitive polyimides [10] [11] [12] [13]. Since these materials have similar properties as photoresists there might be a saving of as many as one-third of the processing steps for each layer. The disadvantages include higher cost and unknown processing characteristics. Of the materials available to us for evaluation we selected Ciba-Geigy Probimide 348. This material could be spun on to produce crack-free 25μ thick films consistently. The processing of this material can be done using standard photolithographic equipment and handling techniques. After imaging and developing the via wall angles were found to be approximately vertical. Final cure to 400°C completes imidization and solvent removal but also shrinks the film by about 40% in thickness. As the film shrinks around vias the wall angles become positively sloped (**Figure 4**). This positive slope is preferable for the next metallizations steps. As with other organic photo-steps, an oxygen RIE plasma "descum" is necessary to insure clean via openings prior to metallization.

Image Deformation at Cure

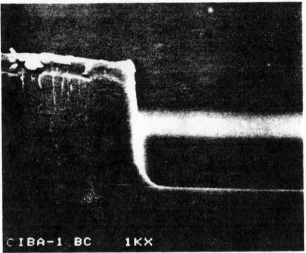

C I BA-1 BC   1KX

Before Cure

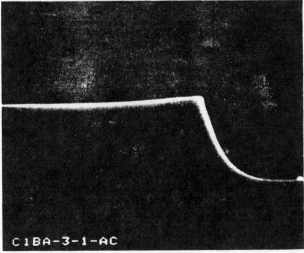

C1BA-3-1-AC

After Cure

Figure 4

A plate-up base is applied over the entire surface and patterned with standard photoresist for the plating step. After plating the vias to the level of the polyimide layer the photoresist is stripped. More photoresist is applied and patterned for the plating of signal lines. After plating the lines the remaining photoresist and plate-up base are removed again. This process is then repeated for three more layers. The polyimide is a very good planarizing material so the subsequent layers of 25μ completely cover the 2μ steps. The second metal layer is predominantly ground plane and covers most of the second polyimide layer. The final metal layer includes pads and test points.

After the IC has inner-lead bonds to the SPTAB the die is placed and bonded into the hole left in the MLP. Outer-lead bonds are then made to the MLP and the system is tested at speed. The module can then be encapsulated; the heat sink attached and the assembly connected with low pressure thin-film hybrid techniques to a printed circuit board.

Discussion

The electrical performance requirements for communications between very high speed devices placed restrictions on the materials and processes used to produce a multilayer multi-chip module. We found that photosensitive polyimides were able to be used in much thicker layers than is commonly reported and that multiple layers could be stacked. A combination of MLP and SPTAB technologies enabled the construction of a high density **and** high speed package suitable for complex analog and digital systems. Thermal management is accomplished by the use of direct attachment to a thermally conductive substrate and the use of SPTAB.

End Notes

1. Balde, J., *Electronic Packaging Materials Science III*, MRS Fall Meeting, 1987 (in press).

2. Whalen, B., *Electronic Packaging Materials Science III*, MRS Fall Meeting, 1987 (in press).

3. Jensen, R.J., Cummings, J.P., Vora, H., *IEEE Trans. Comp. Hyb. Manuf. Tech.*, **CHMT-7(4),** 384, 1984.

4. Kompelien, D., Moravec, T.J., DeFlumete, M., *Proceedings of the 1986 International Symposium on Microelectronics,* 749, ISHM, 1986.

5. Nguyen, P.H., and Russo, F.R., *Proceedings of the 1986 International Symposium on Microelectronics,* 702, ISHM, 1986.

6. E.M. Davis, W.E. Harding, R.S. Schwartz, J.J. Corning, **IBM J. Res. Develop., 8,** 102 (1964).

7. J.A. Perri, H.S. Lehman, W.A. Pliskin, J. Riseman, **Electrochemical Society Meeting**, October, 1961.

8. S. Bhattacharya and E.J. Sullivan, **IBM Tech. Discl. Bull., 23(2),** 575 (1980).

9. L.S. Goldmann and P.A. Totta, **Solid State Technology,** June 1983.

10. Saroyan, M. **Microelectronic Manufacturing and Testing,** August/September, 1986.

11. Araps, C.J., Czornyj, G., Kandetzke, S.M., Takacs, M.A., **USP 4,568,601,** 1986.

12. Pfeifer, J., and Rohde, O., **Second International Conference on Polyimides,** Mid-Hudson Section SPE, 1985.

13. Reche, J.H., **Semiconductor International,** pg. 116, September, 1986.

LIFT – OFF TECHNIQUES FOR MULTICHIP MODULES

by

A. SCHILTZ, M.J. BOUZID, D. JOURDAIN, J. TORRES, J. PALLEAU
J.C. OBERLIN, D. LAVIALE

CNET–FRANCE TELECOM
Chemin du vieux chêne
BP98 ZIRST–38243 MEYLAN – FRANCE

and

J. NECHTSCHEIN

CEMOTA BP3 69390–VERNAISON – FRANCE

ABSTRACT:

For multichip packaging, multilayer substrate with excellent characteristics have been required increasingly for miniaturizing and achieving high level functions.
Recent advances in HDMS (High Density Multilayer Systems) fabrication concerns mainly the use of the couple "Copper-Polyimide" to build Multichip modules for high speed computers [1-5].
In this paper, we report a newly developed technique for Multilevel interconnects fabrication using a Lift-off pillar and line fabrication technique. As metallic layer we used sputtered copper and as dielectric layer we used a new type of heat resistant polymer, a PolyPhenylQuinoxaline (PPQ) polymer.

I–Introduction:

In the fabrication of Integrated Circuits (CI), the trend today is to replace the conventional CVD-type dielectric materials by thermostable spin-on dielectric polymers such as Spin-On-Glasses [6,7] or Polyimides [8-11] because of their simple spin-coat processing, better planarization properties, low temperature cure and patterning versatility.
In the fabrication of Multilayer Thin Films for HDMS, this trend is stronger since metal/dielectric multilayers have to be fabricated with higher layer thickness, larger linewidths and higher number of levels.
In this paper we have developed a new multilevel interconnection fabrication process using a Lift-off pillar and line fabrication technique.

II–The HDMS test structure:

To demonstrate the multilevel Lift–off fabrication technique, a HDMS test structure is fabricated with 10 µm wide, 5 µm thick and 15 µm space lines interconnected by 10 µm diameter pillars.

The test structure will interconnect 4 chips and 2 capacitors with 256 internal I/O and 150 external I/O. The high integration density of lines (400 lines/cm) leads to the design of a 15*16 mm HDMS.

As shown in *Figure 1*, eight HDMS are fabricated at the same time on a 4" silicon wafer using conventional machines of the microelectronics industry.

III–The PPQ PolyPhenylQuinoxaline Polymer

For the fabrication of the HDMS test structure we used the IP200 PolyPhenylQuinoxaline polymer provided by the CEMOTA .

The IP200 polyphenylquinoxaline is synthesized by the condensation reaction of an aromatic tetra–amine with an aromatic tetra–ketone according to the chemical process represented in *Figure 2*. The reaction, performed at 20°C in a mixture of meta–cresol and xylene, yields a fully cyclised high molecular weight polymer having a storage stability of several years at room temperature.

The PPQ polymer is shown to satisfy all operating and processing criteria :

(1) Good dielectric properties:
Good dielectric properties are the major operating criteria in the use of dielectric polymers as intermetallic layers.
The dielectric constant (ϵ) and loss dissipation factor (δ) of PPQ were measured up to 350°C using a Hewlett Packard 4192 LCR meter.
As shown in *Figure 3*, the polymer dielectric constant ($\epsilon < 3$) is better than that of SiO2 ($\epsilon = 4$) and its loss dissipation factor is very low (10^{-3}– 10^{-4}).
As previously reported [12], the PPQ polymer has a lower dielectric constant than usual Polyimides ($\epsilon = 3$–4) and a lower loss dissipation factor.
Therefore it is well suited to HDMS fabrication.

(2) High thermal stability:
Thermal resistance is one of the major operating criteria in the use of dielectric polymers as intermetallic layers since temperatures as high as 450°C are reached during HDMS fabrication.
Dynamic thermogravimetric measurements were carried out using a Dupont 1090 Thermoanalyzer. *Figure 4* compares the thermal behaviour of the PPQ polymer and the Hitachi PIQ13 Polyimide up to 700°C in the air. The PPQ polymer exibits a better thermal resistance than PIQ13 since its thermal decomposition starts at 550°C whereas the PIQ13 decomposition starts at 500°C.

(3) Good adhesion:
Peel–Test measurements were performed using an Instron machine on 4 µm thick polymer layers spun on Silicon wafers.
Table 1 compares the results obtained for three Polyimides : IP300 from CEMOTA, PIQ13 from Hitachi and PI2555 from Du Pont de Nemours and the PPQ polymer.

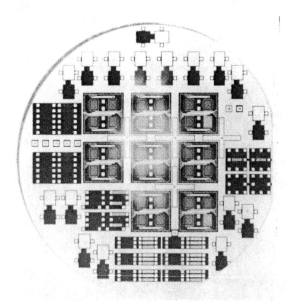

Fig.1–The HDMS Test Structure

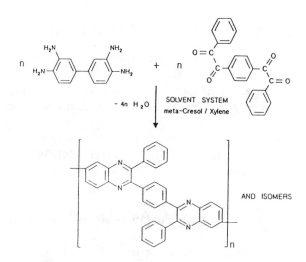

Fig.2–Synthesis of IP200 Polyphenylquinoxaline

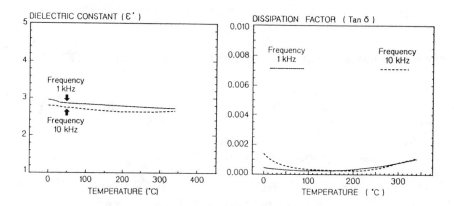

Fig.3–Dielectric properties of the IP200 PPQ polymer

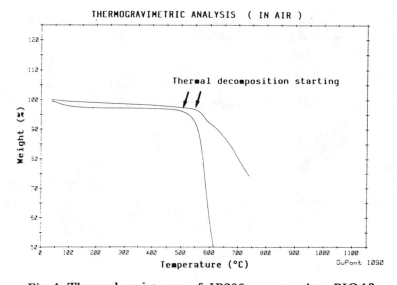

Fig.4–Thermal resistance of IP200 compared to PIQ13

After deposition the polymer layers were cured using a four–step thermal
cure procedure (100, 200, 350 and 450°C) [13]. To improve adhesion we used
the Hitachi adhesion promoter for the PIQ13 and the VM651 adhesion promoter from
Du Pont de Nemours for the other polymers.
As observed in Table 1, the PPQ polymer has adhesion results comparable to those
of PIQ13 Polyimide, whereas PI2555 and IP300 have better adhesion results.

However, after a 15 hours accelerated ageing treatment (85°C–80 % RH) the PPQ
polymer exhibits better adhesion results than Polyimides.

**Table 1– Comparison of Peel–Test adhesion measurements for three
Polyimides and the PPQ polymer**

Polymer	Type	Peel Force (g/cm)	Peel Force after 15 H/ 85°C–80% HR
IP 300	Polyisoindolo– quinazoline dione	318	147
PIQ 13	idem	207	145
PI 2555	PMDA–ODA	234	–
IP200	Polyphenylquinoxaline	207	177

(4) Good planarization properties:

Like polyimides, the PPQ polymer, as spin-on dielectric layer, fulfils the
planarization requirements:
Figure 5 compares the planarization properties vs isolated linewidth
for three polymers: PIQ13, PI2555 and PPQ.
Curves a,b and c trace the planarization properties for a 1.5 μm thick layer
deposited over 0.5 μm step height SiO2 features. Curve d traces the same
properties for a 5 μm thick PPQ layer over a 5 μm step height copper feature.
The degree of planarization is measured as the Relative Step Height (RSH), ie
the ratio of the final step height to the initial step height before
planarization.
As previously reported [7–11], the degree of planarization is observed to
increase with the line feature width. The PPQ polymer has planarizing
capabilities comparable but lower than that of Polyimides.

Figure 6 presents the SEM micrograph of a 10 μm wide, 5 μm high copper
feature planarized by a 5 μm thick PPQ polymer layer: after planarization the
final thickness over the feature is about 2 μm. Thus, it is obviously necessary
to spin-on a higher layer thickness (around 10 μm, [14]) in order to obtain a
5 μm thick dielectric layer between the metal lines.

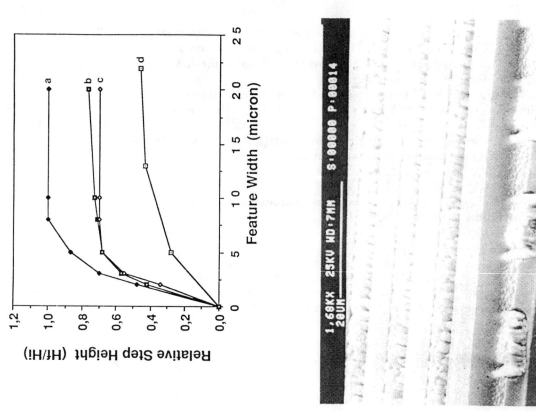

Fig.5–Planarization properties vs isolated linewidths:

a) 1.5 μm thick PPQ layer on 0.5 μm step height SiO2 features
b) 1.5 μm thick PIQ13 layer on 0.5 μm step height SiO2 features
c) 1.5 μm thick PI2555 layer on 0.5 μm step height SiO2 features
d) 5 μm thick PPQ layer on 5 μm step height copper features

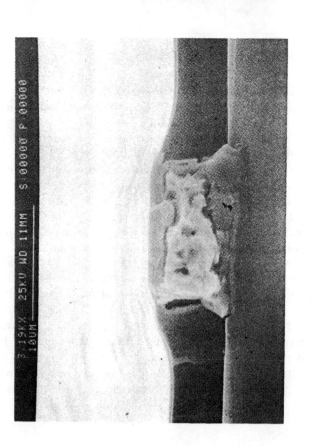

Fig.7–SEM micrograph of 10 μm wide, 5 μm high copper crossed lines fabricated using Lift–off process

Fig.6–SEM micrograph of a 10 μm wide, 5 μm high copper feature planarized by a 5 μm thick PPQ polymer

Figure 7 presents the SEM micrograph of the same feature type fabricated with the Lift-off process (see IV.B). In such a process, planarization properties are not important and thinner dielectric layers may be deposited since the Lift-off pillar fabrication technique ensures planar structure fabrication.

(5) Simple processing:

    -Since it is fully cyclised, the PPQ polymer solution may be stored at room temperature and it doesn't contain any precursor acid which could react with copper layer [15,16].
-The 5 μm thick dielectric layers are obtained by the coating of two 2.5 μm thick layers to avoid pinholes [10].
-After deposition, the PPQ layers can be cured using a simplified 3-step thermal cure (100,250 and 450°C) to eliminate all solvents and volatile product traces.
-The PPQ polymer is etched in a conventional Reactive Ion Etching plasma etcher using a mixture of O2 and SF6 gases to allow the growth of small grain copper structure. Indeed, RIE plasma etching in pure O2 was observed to lead to the growth of large columnar grain structure due to etched polymer surface roughness [16,17].
At high pressure (160 mTorr) an etch rate as high as 0.7 μm/min is obtained.

IV—The multilevel interconnect Lift-off pillar and line fabrication:

    Let us compare the Lift-off technique with the conventional Via-interconnect technique:

IV.A—The conventional Via-interconnect process [1-3,14]:

    As described in *Figure 8*, the conventional process consists in :
1-fabricating the first Metal line, 2- depositing the dielectric and etching holes to form the connection vias, 3- sputtering or evaporating a thin metal layer, electroless plating a thick metal layer and then wet etching the lines and via contours. The next steps are performed repeating the operations 8.1 to 8.3.
As an example, *Figure 8.4* represents a three levels structure. Such a process leads to planarization problems (see *Figure 5*) and requires high polymer thickness deposition to smooth the stacked multilayer structure [14].

IV.B—The Lift-off process fabrication:

    As observed by comparing *Figure 6* and *Figure 7*, the Lift-off technique obviously offers a superior linewidth definition and ensures a perfect planar multilevel structure fabrication [1,11,12].

Figure 9 describes the Lift-off line and pillar fabrication technique:

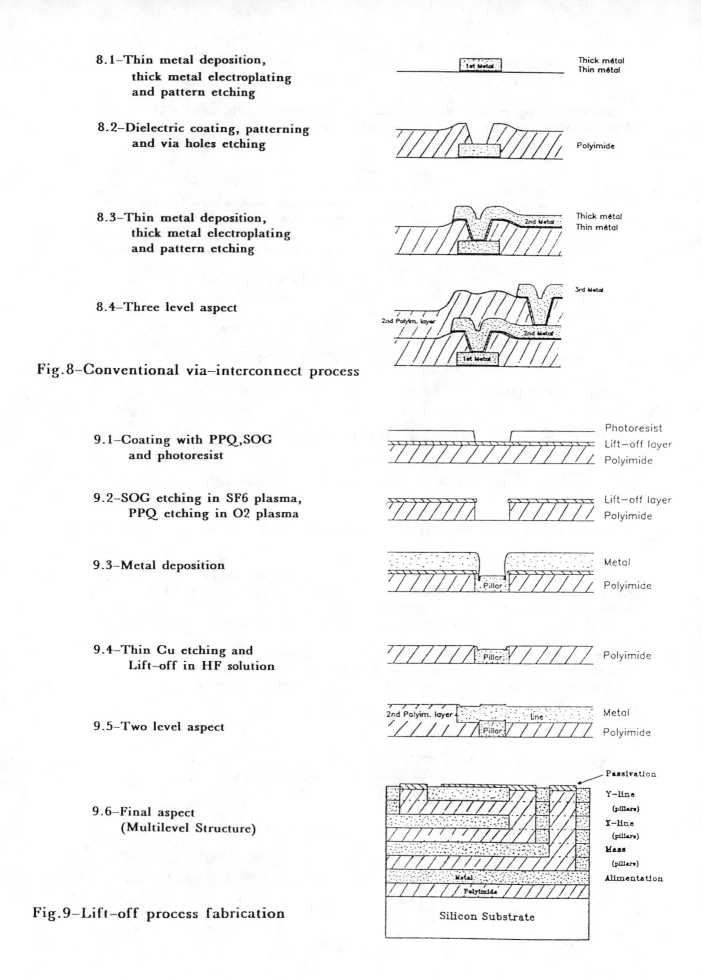

8.1–Thin metal deposition,
thick metal electroplating
and pattern etching

8.2–Dielectric coating, patterning
and via holes etching

8.3–Thin metal deposition,
thick metal electroplating
and pattern etching

8.4–Three level aspect

Fig.8–Conventional via–interconnect process

9.1–Coating with PPQ,SOG
and photoresist

9.2–SOG etching in SF6 plasma,
PPQ etching in O2 plasma

9.3–Metal deposition

9.4–Thin Cu etching and
Lift–off in HF solution

9.5–Two level aspect

9.6–Final aspect
(Multilevel Structure)

Fig.9–Lift–off process fabrication

First level fabrication (*Figure 9.1 to 9.4*):

(1)–After a 5 µm thick PPQ polymer 2–stage coating and cure, a 0.3 µm Polysiloxane (Accuglass 204 provided by Allied) SOG layer was spun–on and cured at 250°C. A HPR204 (Hunt positive resist) photoresist layer was then coated, exposed and developed to define the line hole pattern as shown in *Figure 9.1*.

(2)–The SOG layer was etched in SF_6 plasma in a RIE machine in anisotropic conditions (30 W–5 mTorr). The PPQ was etched in a mixture of O_2 and SF_6 plasma in isotropic conditions and since the SOG layer is non–erodible in O_2 plasmas [7], a controlled overhang was obtained as shown in *Figure 9.2*

(3)–As shown in *Figure 9.3*, a 5 µm thick copper layer was sputtered using a Nordiko machine. Since the deposition is not perfectly anisotropic, a thin metal layer has to be etched on the walls before the Lift–off step.

(4)–After etching the thin metal layer covering the walls in a $FeCl_3$ solution, the undesired copper blocks were removed ("lifted off") in a fully selective Fluorhydric acid solution which rapidly dissolves the SOG layer as shown in *Figure 9.4*. At this stage the remaining lines (or pillars) are fully embedded in the dielectric layer (see *Figure 7*).
   The result is planar.

Subsequent levels fabrication (*Figure 9.5 to 9.6*):

(5)–The next step is performed repeating operations 9.1 to 9.4 with the subsequent pattern to obtain the result shown in *Figure* 9.5. For demonstrative purposes, the SEM micrograph of *Figure 7* shows the superposition of two line levels 10 µm wide and 10 µm space crossed in perpendicular directions. The result is planar.

(6)–The following steps are performed repeating operations 9.1 to 9.4 so as to obtain a result such as the structure shown in *Figure* 9.6 (4 levels of lines and 3 levels of interconnect pillars).

V–Conclusion:

   The PPQ polymer is shown to satisfy all operating and processing criteria in the fabrication of multilevel interconnections.
The Lift–off technique allows fabrication of perfect planar multilevel interconnection structures and thus ensures the fabrication of HDMS.

References:

[1] J.F. McDonald,A.J. Steckl,C.A. Neugebauer,R.O. Carlson and A.S. Bergendahl, J.Vac.Sci.Technol.,A,(6),Nov/Dec,3127 (1986)

[2] R.J. Jensen,J.P. Cummings and Harshadrai Vora,IEEE CHMT, vol.7,Nb 4, Dec. (1984)

[3] M. Kohara,S. Nakao,K. Tsutsumi,H. Shibata and H. Nakata,Proc. IEEE,El. Comp. Conf.,180 (1985)

[4] T. Watari and H. Murano,Proc. IEEE,El.Comp.Conf.,192 (1985)

[5] T. Kishimoto and T. Ohsaki,Proc. IEEE,El.Comp.Conf.,595 (1986)

[6] S.K. Gupta and R.I. Chin, ACS Symposium,295,22,350 (1986)

[7] A. Schiltz, Microcircuit Eng. proc.,5,413 (1986)

[8] L.B. Rothman,J. Electrochem. Soc.,130,Nb 5,1131 (1983)

[9] H. Fritzsche,IEEE V–MIC Conf.,253 (1985)

[10] E.R. Sirkin and I.A. Blech,J.Electrochem.Soc.,1,Nb 1,123 (1984)

[11] P. Sanseau,A. Schiltz and J. Nechtschein,Makrom.Chem.,Makrom. Symp.,24, 349(1989)

[12] P. Sanseau,A. Schiltz,G. Rabilloud and L. Verdet,proc. Electrochem. Soc. Symp.,Vol.88–17,170 (1988)

[13] J. Nechtschein and A. Schiltz, Le Vide les couches minces,supl. S.M.T., Oct.,17 (1988)

[14] M.L. Zorilla, IEEE Eur. IEMT–EMT Symp.,22 (1988)

[15] D.Y Shih and al., J.Vac.Sci.Technol. A7(3),May/June (1989)

[16] A. Schiltz, J. Torres, J. Palleau, M.J. Bouzid and D. Laviale, VLSI Packaging Workshop–Garmisch 18–21 Sept. (1989)

[17] J.F. McDonald, H.T. Lin,H.J. Greub,R.A. Philhower and S. Dabral, IEEE Trans. on CHMT,Vol.12,N32,June (1989)

MULTI CHIP MODULES FOR TELECOM APPLICATIONS

KÅRE GUSTAFSSON AND GÖRAN FLODMAN

Ericsson Telecom

S-126 25 Stockholm

Sweden

ABSTRACT

The system performance can be improved by the use of MCM (Multichip modules) technology. Better signal transmission properties, lower power dissipation (if C-MOS chips are used) and decreased physical volume will be obtained. Today MCM is an expensive technique, but our cost analysis shows that it has the potential of becoming more cost effective than the standard technique of using singel chip packages. It can also become more cost effective to use MCM technique rather than to increase the number of gates on the silicon chip over a certain level. In order to study the properties of the MCM technology, a project has been started at Ericsson Telecom in which a processor application is built up on three different types of MCM substrates.

1. INTRODUCTION

One of the driving forces for the increased use of MCM technology has been the need for short distance between high speed chips (mainly GaAs and ECL). A high packaging density will lower the inductance, decrease signal delay as well as decrease the electromagnetic emission. If the higher packaging density allows for a design of a whole subsystem on one module, the gain in performance will be even higher, since the signal does not need to pass connectors and backplanes. The result is that the impact of the building practice on the total delay time will decrease singificantly.

The multichip technology can improve the performance even at lower signal speed. The power dissipation can be drastically reduced if C-MOS chips are used. This is due to the fact that the power dissipation for C-MOS chips are directly proportional to the capacitive load. Table 1 (ref 1) compares the capacitive load for a multichip application using TAB circuits with a "standard" solution using PGA packages. The increased packaging density for the MCM solution gives a lower capacitive load, which means that the power dissipation decreases from 16 W to 10 W. If the output drivers on the C-MOS chip had been designed to be used on a MCM solution, the power dissipation would have been reduced even more.

Reprinted from *Proc. 8th Int. Electron. Manuf. Technol. Symp.*, 10 pages, 1990.

Table 1 Comparison between a processor application using MCM technology and a traditional solution with PGAs on a PCB (ref 1).

	PGA + surface mount + FR4	TAB + wirebond + thinfilm board
Layout size	200 cm^2	50 cm^2
Critical conductor prop. delay	2.6 ns	0.8 ns
Critical conductor capacitive load 1)	67 pF	42 pF
Estimated power dissipation at 33 MHz	16 W	10 W
Package + board contrib. to cycle time at 33 MHz	8%	2.5%

1) On- chip capacitance 30 pF

If C-MOS chips are used, one has to be aware of the fact that the rise and fall time will decrease when the load capacitance decreases. This means that the design has to be treated as a high signal speed design, even if the clock frequency is moderate. Switching noise and electromagentic radiation might become a problem.

A determining factor for the future of the MCM technology will be the testability since it has strong influence on the final yield. MCMs has to be designed for testability (i.e. by using J-TAG). The availability of chips on tape has been improved during the last few years. This makes it easier to perform AC test on the chips before assembly on the substrates.

2. A MODEL FOR COST OPTIMIZATION OF MULTI CHIP MODULES
The cost for a PCB with attached MCMs can be approximated with the use of Knausenbergers assumption that the cost per unit length of a conductor is independent of the type of substrate (ref 2). This assumption has been modified by increasing the cost of the MCM level in order to adaept the model to an assumed cost level of commercially available MCM substrates. The test cost has been excluded in this calculation.

The MCM cost model assumes random logic. Four levels are used (figure 1). The basic level is the gate, then follows the chip, the MCM and the PCB level.

Figure 1, The different levels of the cost model

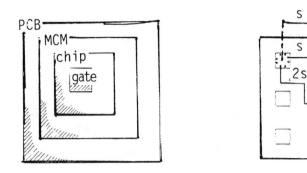

Each level can be regarded as a component which includes a number of subcomponents. The main questions is: How many gates should be used on each level if the total number of gates on the PCB level is given in order to minimze the total cost. All components on a given level are assumed to be equal.

The optimal number of chips on a multi chip module

The total cost for a system with a given number of gates has been calculated:

Total cost = CC*MC*PC*MAC*CAC

CC =   chip cost for 5000 gates on a chip

MC =   Cost for the MCM substrate assuming a relative cost of 0.07/cm. The necessary conductor length and the number of terminations are calculated using the formulas described below.

PC =   Cost for the PCB assuming a relative cost of 0.04/cm conductor length.

MAC = Cost for assembly of the MCMs on the PCB. Consist of a socket cost of 0.5 for each pin (No socket cost for the singel chip case) and an assembly cost of 0.2 for each pin.

CAC = The cost for assembly of the chip to the MCM = 0.1 for each pad on the chip.

A modification of Rents law has been used for the calculations of the number of terminations for the "components" (chips and MCMs) as a function of the number of gates. We have assumed that the exponent in Rents Law will decrease as a function of the number of gates according to the formulae below:

$$N_c = c*N^{a/(1+N/N_1)}$$

N_c = Number of connections

c = constant

a = constant in Rents law, = 0.4-0.6

N = Number of gates

$$N_1 = N_m*(ln(N_m) - 1)$$

N_m = The number of gates where N_c has a maximum

A modified version of Serafims formulae has been used to calculate the connector length.

Seraphins formulae is based on the following assumptions:

1 The components are evenly distributed with the pitch S (figure 1)
2 Each node has in average 1.5 connections
3 Each conductor goes between two nodes.

The average conductor length will be 1.5*S and the average length of conductors starting from each node will be 1.125*S

The modification of Seraphins formulae assumes connections between more distant nodes. p(n) is a probability factor that two nodes with a distance of n steps are connected. The average conductor length will then be:

$$X_m = \sum_{1}^{N} x(n)*p(n) \quad \text{with the condition} \quad \sum_{1}^{N} p(n) = 1$$

We have assumed that the probability of connection decreases in a geometrical serie as a function of the number of steps:

$$p(n) = (1-k)/(1-k^N)*k^{n-1}$$

which gives the following average length

$$x_m = (1-k)/(1-x^N)* \sum_{1}^{N} n*k^{n-1}$$

For large substrates where N approches ∞ the formulae will be reduced to:

$$x_m = 1/(1-k)$$

Assuming that Seraphims formulae is valid for large substrates gives k=1/3
The total length for each node will then be:

$$L = 0.75*S*x_m \quad \text{where}$$

$$x_m = 2.3^N/(3^N - 1) \sum_1^N n/3^N$$

Table 2 shows the result of the cost calculation for a system with 250 000 gates. The MCM cost in the case "1 chip per module" is the same as the IC package cost. The total cost according to this model has a maximum at two chips per module. The multichip solution will be cheaper that the singel chip solution if there are more than 5 chips per module. This is illustrated in figure 2.

Table 2. Calculation of relative cost for a system with 250 000 gates. The chip cost is 667. The cost for assembly of the chips on the modules is 438.

Chips per module	Modules per PCB	MCM size mm	MCM cost	PCB cost	Assembly cost	total cost
1	50	14	1320	1273	875	4573
2	25	18	1498	1317	1935	5855
5	10	25	1079	1284	1227	4695
10	5	32	886	1069	797	3856
25	2	48	647	729	351	2832

Figure 2, The total cost of a PCB with a given number of gates as a function of the number of chips on the MCM

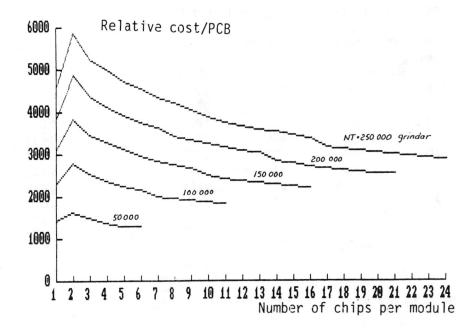

379

<u>Cost of MCMs compared to increased integration on silicon</u> In some cases the use of multi chip modules can be more cost effective than an incresed integration on Si, since the yield will decrease drastically when the number of gates increases above a certain level. If one assumes a cost per gate according to figure 3, the conclusion is that it is more cost effective to build a MCM using chips with 5000 gates per chip than to increase the integration on silicon using chips with 20 000 gates (table 3).

Figure 3, Cost per gate for a given IC technology as a function of the number of gates on the chip

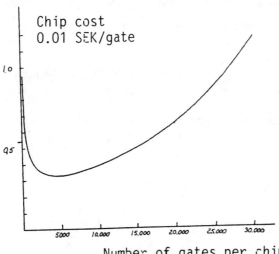

Number of gates per chip

Table 3. The relative cost for MCM and singel chip solutions using ICs with different numbers of gates for a system with 200 000 gates.

Gates per chip	chip per module	modules per PCB	chip cost	MCM cost	PCB cost	total cost
2000	2	50	394	775	1215	2384
2000	50	2	394	395	704	1493
5000	1	40	427	387	1234	2048
5000	20	2	427	360	704	1491
10000	1	20	565	275	1282	2122
10000	10	2	565	332	704	1491
20000	1	10	1076	196	1256	2528
20000	5	2	1076	277	704	2057

3. MCM SUBSTRATES

There is a large number of technologies that can be used in order to achive a high packaging density on the substrate. The material used as mechanical support can be ceramic, silicon, PCBs, metal, AlN or SiC, the conductors can be copper, aluminum or thick film conductors, the dielectric can be SiO_2, epoxy, polyimide or the newly devloped BCB. The chip can also be assembled to the MCM substrates by different methods, wire bonding, TAB bonding, flip-chip assembly or direct attach to copper conductors.

One way of characterizing the different types of MCM technologies are the following abbreviations proposed by Jack Balde.

MCM-D Multichip module - deposited technology. A base is used as mechanical support. The dielectric and conductors are deposited on the base using technology similar to the semiconductor production. (i.e. spin on polyimide and sputtered copper).

MCM-L Multichip module - laminated. Upgraded version of the standard PCB manufacturing technology.

MCM-C Multichip modules - multilayer ceramic. The "classical" multichip technology used by IBM for instance in the TCM modules. An expensive technology where the packaging density can be achived by the use of a high number of layers.

In order to evaluate the pros and cons of these three technologies, Ericsson Telecom is running a project where three different substrates are used for a processor application.

4. PROCESSOR APPLICATION

The application consists of three processor chips that are TAB bonded with 8 mil OLB pitch, 21 standard logic chips used for "glue logic" that are wire bonded and 16 S-RAMs that also will be wirebonded. There are totally more than five million gates on the module. The pad layer is shown in figure 4.

Figure 4, Padlayer of the processor application

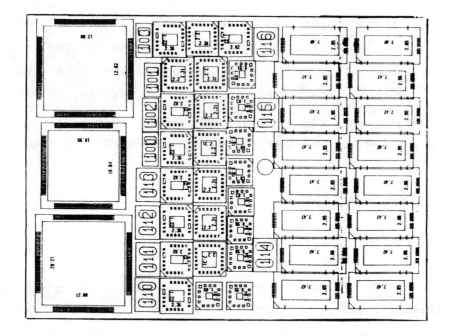

MCM substrates

Three different types of substrates are evaluated:

a. A commercially purchased HDI substrate with 25 um conductor width and 100 um
   pitch. The substrate is built on a metal base with copper conductors and
   polyimide dielectricum. Five metal layers are used. The size of the
   substrate is approximately 5*9 cm. (MCM-D)

b. Substrates based on an upgraded PCB technology are made inhouse at Ericsson
   Telecom. The conductor width is 50 um and the pitch 200 um. The system is
   designed for 50 ohm characteristic impedance. (MCM-L)

c. Substrates using multilayer ceramic technology are also made inhouse by
   Ericsson Components. The conductor width is 100 um and the pitch is 400 um.
   The number of metal layers is 7. (MCM-C)

The price level for commercially available HDI (High Density Interconnects) is
much higher than the level assumed in litteraure and in survey studies. Many
companies also seem to have difficulties in achieving in reality what they
promise in their prospects concerning conductor width, tolerances and number of
metal layers. Of course, multichip sustrates are not a mature product yet!

382

Some of the information that we have obtained in answers to our request for offers for a 5*9 cm MCM with 25 um lines and five metal layers is shown in table 4.

Table 4, Offered cost (1000 $) for a MCM 5*9 substrate with 100 μm conductor pitch.

Vendor	Delivery time weeks	Engin. cost	Cost 10 pcs	Total cost per MCM 10 pcs	Total cost per MCM 1000 pcs
A	12-14	21	38	5.9	1.5
B	13-15	24	36	5.9	0.6
C	14	26	8.4	3.4	0.7

5. MCMs FOR TELECOM APPLICATIONS

Hithertho, the telecom industry has not had the same need for very high speed and high power dissipation random logic systems as the main frame high speed computor industry. However, the first need for MCMs in telecom applications will be in high performance system where the better properties are needed. Most of the telecom applicatios will use C-MOS and BICMOS chips. A large gain will be the lowered power dissipation when C-MOS circuits are used. On a longer term, we belive that the use of MCM will increase also for high volume products.

For the next few years we do not need the very high packaging density that can be a hived through flip-chip assembly. Wire bonding is more flexible than TAB assembly, but the use of TAB simplifies the testing of the chips before assembly and clearifies the responsiblity if expensive chips have to be rejected. Therefore we prefer to have complex chips delivered on tape.

6. <u>CONCLUSIONS</u>

* Use TAB assembly of complex chips.

* The use of MCM will improve the signal transmission properties and lower the power dissipation for C-MOS circuits

* Today, the price is high for commercially available HDI substrates. This implies that MCMs will be used only in high perfomance products for the next few years

* Our price model shows that the MCM technology has the potential of becoming more cost effective than the use of singel chip packages also for high volume products, when the cost of the substrates reaches the cost level of 0.07/cm.

* The cost maximum for MCMs is reached for two chips per module. With more than 5-7 chips per module a lower price is reached for the MCM than for the singel chip solution

* The use of MCM technology can become more cost effective than the increase of the integration of silicon over a certain number of gates per chip

<u>REFERENCES</u>

1 Per Hedemalm, "Build your own multi chip module", presesentation at Computer packaging workshop 1989

2 W H Knausenberger, "Interconnection Costs of Various Substrates - The Myth of Cheap Wire", IEEE transactions CHMT, september 1984

<u>ACKNOWLEDGEMENTS</u>
The authors would like to thank all the people that are involved in this project, especially Krysztof Kaminsky, Per Hedemalm, Josef Bakszt, Göran Widlund, Mika Panttila and Monica Bakszt.

ALL THE EMPHASIS on multilayer deposition of organic dielectrics focused attention on polyimide materials, which leave much to be desired as interconnection layer dielectrics. Their comparative performance is presented in papers by Tessier *et al.* of MCNC and Lin of AT&T. Burdeaux *et al.* make the case for benzocyclobutane, and Craig and others from DuPont present a low thermal expansion polyimide.

Photosensitive polyimides can be useful, so a paper has been chosen in this area, and the debonding problem with copper conductors on polyimides is covered in the paper by IBM. Other dielectrics are possible: filled PTFE Teflon, as in the work by Rogers, or polyimide–siloxanes, as in the paper by Lee. Generally, the dielectric most used is a variety of polyimide, and this will be so until second sources make the alternative materials more readily available.

Polymer Dielectric Options for Thin Film Packaging Applications

T. G. Tessier[1], G. M. Adema[2] and I. Turlik[1]

[1] MCNC Resident Professional
Bell Northern Research
P. O. Box 13478, Research Triangle Park, NC 27709
(919)-248-1854

[2] Microelectronics Center of North Carolina (MCNC)
P. O. Box 12889, Research Triangle Park, NC 27709
(919)-248-1484

Abstract

Several classes of available polyimide based and non-polyimide based dielectrics have been recommended for use in thin film substrates. This paper provides a comprehensive survey of the commercially available types of materials considered for use in our advanced packaging program. Critical physical properties of these candidate dielectrics are compared, and related to their compatibility with established thin film processes.

Introduction

In recent years, the escalating circuit densities and switching speeds of VLSI circuits has fueled a surge of interest in thin film interconnection packaging technologies [1-4]. Over the past 4 years, the Advanced Packaging Group at the Microelectronics Center of North Carolina (MCNC) have been agressively pursuing the development of a thin film based high performance packaging technology to address this trend. A conceptual sketch of the MCNC multichip package is shown in Figure 1. In this packaging scheme, a metal/polymer dielectric thin film structure is built on a thick film aluminum nitride or silicon carbide substrate. These substrate materials were selected for superior thermal and mechanical properties, as well as their compatibility wutg thermal expansion coefficient (TCE) of silicon (TCE_{AlN}=4.1 x 10^{-6}/°C, TCE_{SiC}=3.7 x 10^{-6}/°C, TCE_{Si}=2.6 x 10^{-6}/°C). Solder bumps are used to attach the integrated circuits to the substrate using a process analogous to the C4 process first developed by IBM [5]. A water cooled microchannel heat sink is attached to the back of the chips. This heat sink is capable of dissipating at least 1000 W of power from a 4" x 4" module [6,7].

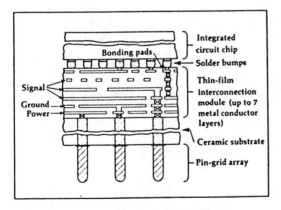

Signal

Ground

Power

Bonding pads

Integrated circuit chip

Solder bumps

Thin-film Interconnection module (up to 7 metal conductor layers)

Ceramic substrate

Pin-grid array

Figure 1: MCNC High Performance Package

The electrical requirements of this high performance package dictate the design rules for the thin film module. This substrate consists of up to 7 layers of copper conductors (4 μm thick) with 10 μm signal lines and 20 to 30 μm spaces. Interlevel isolation of metal layers is accomplished using 8 to 10 μm thick low dielectric constant polymer films. To maximize density, straight walled non-nested vias are used. Via filling is achieved using a selective electroless nickel plating technique. The processes required to manufacture such structures are being developed and various aspects of the MCNC thin film process have recently appeared in the literature [8-10].

Over the past two years, the MCNC Advanced Packaging Group has extensively evaluated commercially available candidate dielectric materials for use in our thin film structure. Though this work is an ongoing effort, considerable data has been accumulated comparing the key material properties of these dielectrics. A total of 20 materials from 6 different suppliers were included in this evaluation. Particular attention was given to material properties that affect the processability of these candidate dielectrics. This paper is intended to provide an overview of the various classes of polymeric materials that were considered for application in high density interconnection. Data comparing the chemical and thermal stability, moisture uptake and release, stress, adhesion, planarization and dry etchability of these materials are presented, and their resultant ease of processing are compared.

Dielectric Options

Polymer dielectrics used for thin film interconnection applications are expected to meet a number of general material requirements. The dielectric should show minimal moisture uptake and controllable release if the multilayer structure is to be manufacturable and have an acceptable level of reliability. The dielectric layer should planarize the rough topographies encountered in these thin film structures. The dielectric material should also be thermally stable in order to avoid any damage to the thin film structure due to uncontrolled outgassing during subsequent processing at elevated temperatures. Further, the dielectric should exhibit good adhesion to the a) substrate material whether it be silicon, alumina, aluminum nitride or silicon carbide, b) conductor metallization, and c) fully cured layers of the same material. Conversely, the proposed metallization must exhibit satisfactory adhesion to the dielectric. Since during the processing of these substrates, the dielectric could potentially be exposed to a host of organic and aqueous solutions required for subsequent processing steps, chemical resistance is also a critical requirement.

Polyimides typified by Dupont's PI2525 (Type III chemistry) and Hitachi's PIQ13 line are currently the most commonly used class of polymer dielectrics for thin film applications. These materials are generally supplied in the polyamic acid precursor form in an appropriate casting solvent. More recently, new materials have been suggested for use in multilayer structures. Fluorinated and silicone polyimide analogs of these "old standards" available in both polyamic acid and preimidized forms have been introduced to address some of the undesirable properties of polyimides (eg. substantial moisture absorption at ambient conditions - up to 3% by weight). Low stress polyimides with coefficients of thermal expansion that can be matched to that of the substrate and/or metallization are available [11]. This class of polyimides possesses material properties that make them worthy of consideration for this application. Acetylene terminated polyimides are of considerable interest for thin film applications as

Reprinted from *Proc. 39th Electron. Components Conf. (ECC)*, pp. 127–134, 1989.

demonstrated by a recent IBM Process Patent [12]. These materials are supplied in either a preimidized or isoimide form. The chemistry of Dow Chemical's Benzocyclobutene (BCB) based materials available in both silicon and very recently in non-silicon containing versions has been reported in the literature [13 14,15]. All of the materials evaluated in this study fall into one of the above classes.

Two other classes of potential dielectric materials were not included in this evaluation. Although paralene has been reportedly used elsewhere in the fabrication of thin film structures [16], it was not included in this study due to the unavailability of the necessary reactor to do this type of in-situ vapor phase polymerization. Photosensitive polyimides currently commercially available cannot give the degree of resolution called for in the MCNC design rules of <10 μm diameter vias in the 16 to 25 μm soft-baked film thickness required to obtain a final fully cured film thickness of 8 to 12 μm. Additionally, the sloped via sidewalls (usually in the 60 to 75° range) obtained after fully curing these thick photosensitive polyimide films with their effective doubling of upper via dimensions are incompatible with MCNC's anisotropic via requirements [17]. The use of photosensitive polyimides to accomodate less stringent design guidelines has been reported in the literature [18]. It is conceivable that photosensitive polyimides could be used as the passivation layer in most thin film processes thereby simplifying the processing of this last layer of dielectric.

The electrical properties of the conventional, fluorinated, low stress and silicone polyimides and BCB classes of dielectrics are compared in Table 1 for a frequency of 1 KHz. The rigid rod-like backbone structure of low stress polyimides results in a lowering of the dielectric constant relative to that observed for the helical, and loosely packed conventional polyimide chains. As might be expected, the more nonpolar silicone and fluorinated polyimide analogs have dielectric constants that are lower than those reported for the standard polyimides. Since the BCB family of materials are inherently much less polar polymers, they provide the lowest dielectric constant (2.6) of all of the materials evaluated in this study.

MATERIAL TYPE	DIELECTRIC CONSTANT *	DISSIPATION FACTOR *
STANDARD PI	3.4-3.8	0.002
FLUORINATED PI	2.7-3.0	0.002
SILICONE/PI	3.0-3.5	...
LOW STRESS PI	2.9	0.002
ACETYLENE TERMINATED PI	2.9-3.4	0.002
BCB RESIN	2.6	0.00045

* RANGE OF VALUES OBTAINED FROM SUPPLIER'S DATA SHEETS FOR A FREQUENCY OF 1 KHz.

Table 1: Electrical Properties of Dielectrics Evaluated

Evaluation Results

Solvent Compatibility

During the processing of multilayer thin film structures, the cured dielectric is exposed to a wide range of aqueous and organic based solutions. Acetone is used for the stripping of photoresist and isopropanol rinses are also useful. NMP is commonly used to strip improperly cast, wet polyimide films or to remove edge beads and is one of the most aggressive organic solvents used. This solvent therefore gives a good general indication of a dielectric's degree of stability in most

other organic solutions. Aqueous, ammonium hydroxide based solutions, such as Shipley MF-312 are commonly used for the developing of exposed positive photoresist layers. This is particularly important to the current MCNC process since the metallization lift-off process used to deposit the Cr/Cu/Cr conductors requires development of the pattern defining photoresist layer which is in direct contact with the interlayer dielectric [8]. As a multilayer thin film structure is being fabricated, the fully cured dielectric must also be resistant to the casting solvents of subsequent coating solutions. Photoresist films, though only temporarily in contact with the underlying layer are often cast from solutions that are quite aggressive. If multiple coats of the dielectric are to be applied either in building a multilayer structure or for achieving a given single layer thickness, the fully cured film must be resistant to its own casting solvent. In Table 2, the chemical stability of fully cured films of the candidate dielectrics in these processing solutions is compared.

	IPA	ACETONE	NMP	MF-312 DEVELOPER	SHIPLEY 1400 +VE RESIST	CASTING SOLVENT
STANDARD PI	+	+	+	+	+	+
FLUORINATED PI	+	+	-/+	+	+	-/+
ACETYLENE TERMINATED PI						
- FLUORINATED	+	+	-/+	+	+	-/+
- NON-FLUORINATED	+	+	+	+	+	+
SILICONE-POLYIMIDE	+	+	+	+	+	+
LOW STRESS	+	+	+	+	+	+
BENZOCYLCOBUTENE RESIN	+	+	+	+	+	+

Table 2: Dielectric/Solvent Compatibility

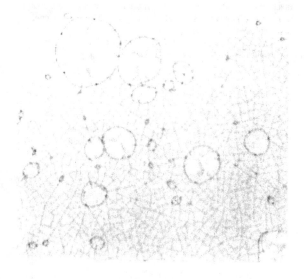

Figure 2: Casting Solvent Induced Cracking of a Fully Cured Polyimide During Recoating

Of the classes of materials included in this study, only the fluorinated polyimides and the fluorinated acetylene terminated polyimides displayed any signs of solvent incompatibility. Films of the other candidate materials, including the non-fluorinated acetylene terminated polyimides remained unaffected by prolonged exposure to the test solutions. A total of 5 commercially available fluorinated polyimides were evaluated including 2 acetylene terminated analogs. All of these dielectrics are supplied in proprietary NMP based solvent systems. Of the 5 materials tested, 2 of 3 fluorinated polyimides and 1 of 2 fluorinated acetylene terminated polyimides crazed when immersed in NMP or when the next coat of the same dielectric solution was applied onto a fully cured layer. An example of a particularly bad case of casting solvent induced crazing is shown in Figure 2. From these observations, it is clear that great care must be taken in selecting fluorinated polyimides that are suitable for thin film applications requiring multiple coats of dielectric.

Thermal Stability

The thermal properties of a particular dielectric material determine the range of processing temperatures and dwell times allowed during the thin film fabrication process and subsequent multichip module assembly. The decomposition and glass transition temperatures (T_g's) for the classes of dielectrics in this study are compared in Table 3.

	DECOMPOSITION TEMPERATURE [1] (in °C)	GLASS TRANSITION TEMPERATURE [2] (in °C)
LOW STRESS	620-650	>400
STANDARD PI	520-550	300-320
ACETYLENE TERMINATED PI	500-520	<320
FLUORINATED PI	>470	<300
SILICONE/POLYIMIDE	450	<300
BENZOCYCLOBUTENE (BCB-1)	430	<360

[1] TGA RESULTS OBTAINED FOR SAMPLES RUN AT 10° C /MIN. IN N$_2$

[2] T_g DATA OBTAINED FROM VENDOR SUPPLIED LITERATURE

Table 3: Summary of Dielectric Thermal Properties

The low stress polyimides were found to be the most thermally stable, having a thermal stability in the 620 to 650°C range, approximately 100°C higher than the traditional polyimides. The incorporation of fluorine or silicone containing moities into the basic polyimide structure appears to result in a 100°C lowering of the decomposition temperature of these analogs. The fluorinated and non-fluorinated acetylene terminated polyimides had decomposition temperatures that were comparable to those observed for their conventional polyimide counterparts. From the thermogravimetric analysis results, the silicon containing (XU-13005) BCB-1 exhibited the lowest thermal stability of all of the materials tested. In addition, films of BCB-1 were found to be sensitive to thermal oxidation during curing or other high temperature treatments (>350°C) if trace amounts of oxygen are present in a nitrogen purged oven. Considerable losses in film thickness, weight and electrical properties resulted. A new non-silicon containing version not currently available or included in this study, referred to as BCB-2 is reported to be less susceptible to this thermal oxidation [15]. This thermal sensitivity has also been reported for paralene [16]. Films of the other classes of dielectrics demonstrated good thermal stability and minimal property changes under similar processing conditions.

The low stress polyimides have the highest reported T_g's of the classes of dielectrics studied with values typically >400°C and the fluorinated and silicone polyimides had the lowest. Those materials

with T_g's below 300°C showed definite signs of incompatibility with MCNC's baseline PECVD Si_3N_4 etch mask process. When >1500A° thick films of PECVD Si_3N_4 were deposited onto fully cured fluorinated polyimide films at 300°C extensive cracking of the nitride masking layer was observed immediately upon removal from the reaction chamber. Closer investigation also indicated that the cracking in the nitride mask had propagated down through the 8 microns of dielectric below. For standard polyimides, like PI2525 which has a T_g of 308°C and using the same processing conditions, >3000 A° of nitride could be deposited without any sign of cracking. Based on these observations, the polyimide etch mask deposition temperature is now kept below the T_g of the dielectric used. Low temperature etch mask alternatives are currently under investigation to address this and other requirements and will appear in an upcoming publication [19].

Moisture Absorption and Release

Conventional polyimides are notoriously hydroscopic, absorbing as much as 3% moisture by weight. This ease of moisture uptake in turn impacts both the electrical properties [20,21] and the processability of these materials [22]. In Table 4, moisture absorption for the other classes of dielectrics investigated are compared to that of the standard polyimides. Test samples used in this evaluation, were prepared by coating both sides of a 4" silicon wafer with an 8 to 10 μm thick layer of dielectric. Weight gain was determined by weighing the test samples after an initial 3 hour dehydration bake in a vacuum oven at 100°C/1 mm Hg and after a 2 hour immersion of the samples in boiling water.

STANDARD PI (DUPONT PI2525)	1.7%
STANDARD PI (HITACHI PIQ13)	1.1%
ACETYLENE TERMINATED PI (NATIONAL STARCH)	1.4% (0.7%)
SILICONE/POLYIMIDE (GE SPI 2000)	0.9%
FLUORINATED PI (XHP3119)	0.7%
LOW STRESS PI (HITACHI PIQ L100)	0.5%
LOW STRESS PI (DUPONT PI2611D)	0.4%
BENZOCYCLOBUTENE RESIN (DOW XU13005.02)	0.3%

* SAMPLES SOAKED IN BOILING WATER FOR 2 HOURS

Table 4: Comparison of Dielectric Moisture Absorption

The standard polyimides absorbed the most water under the above test conditions (1.2-1.7%). The fluorinated and silicone polyimides showed typically 50% less water uptake than the standard polyimides. The fluorinated acetylene terminated polyimides absorbed 0.7% moisture by weight, about the same amount as the other fluorinated polyimides. The rigid chemical structures of the low stress polyimides promote much denser packing of their polymer chains than is achievable with the helical shaped chains of the other classes of polyimides. As a result of this dense packing, low stress materials absorb less than 0.5% moisture by weight. Since the BCB resin is the least polar dielectric in this study, it absorbed the least amount of water under these experimental conditions.

In the processing of thin film structures, it is critical that the moisture levels in the dielectric be kept as low as possible either by using materials that absorb a minimal amount of water or by carefully controlling the process so that exposures to moisture are minimized. For those thin film processes in which the standard polyimides have been successfully used, the controlled release of absorbed moisture is critical [1-4]. To maintain the integrity of these multilayer structures during their fabrication, moisture absorbed by the dielectric must be removed by dehydration bakes at temperatures below 150°C prior to deposition of the next level. This is especially crucial during subsequent high temperature processes such as dielectric curing or metal annealing. As shown in Figure 3, extensive damage to relatively thick (4 μm) metal layers can result from moisture outgassing. A dehydration bake in a vacuum oven at 100°C/1 mm Hg for 3 hours prior to the high temperature treatment was found to eliminate this metal blistering problem.

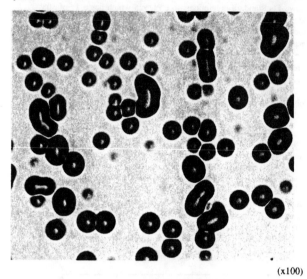

(x100)

Figure 3a: Metal Blistering Due to Moisture Outgassing From Underlying Polyimide Layer

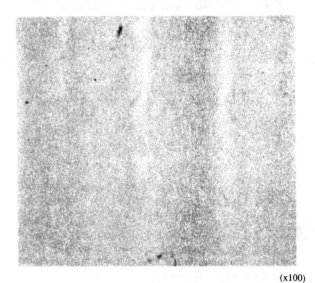

(x100)

Figure 3b: Effect of Pre-metallization Vacuum Bake of Polyimide Layer

Contrary to the general trend in the integrated circuit industry towards dry processing, thin film interconnection technologies are becoming increasingly dependent on wet processing. Additive metallization processes such as electroless nickel and copper plating, electroplated copper and Pb/Sn solder are gaining in popularity. These processes typically require immersion of the substrates in aqueous based plating solutions for lengthy periods at temperatures as high as 80°C. Subtractive metallization processes on the other hand, involve aqueous based etching solutions. Low moisture absorbing dielectrics

are expected to be much more compatible with these wet thin film processes, thereby simplifying thin film substrate fabrication, and more importantly, slowing the movement of water-borne ionic contaminants into the dielectric (eg. transport of aluminum ions during wet etching [19]). Work is continuing with the newer classes of dielectrics to determine the impact of their lower moisture uptake on wet processing.

Stress

The effect of the tensile stresses in cured dielectric films on the long term reliability of a thin film interconnection remains a topic of considerable debate [23,24]. Most of the dielectrics evaluated in this study have a TCE in the 25 to 65 ppm range compared to values of 2.3 and 9.0 ppm for silicon and alumina. The tensile stresses resulting from this thermal mismatch in turn give rise to substrate warpage. The rigid low stress polyimides have TCE's that are matched to the substrate materials and as a result show little or no thermal mismatch. The substrate distortion resulting from the application and cure of a 4 μm thick film of Dupont's PI2525 to a 21 mil. thick 4" wafer is compared to that for the low stress PI2611D in Figures 4 and 5.

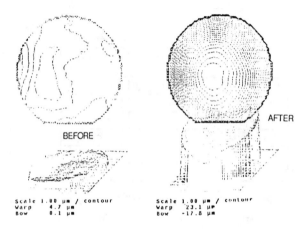

4" diameter wafer, 21 mils. thick; 4.4 μm dielectric thickness

Figure 4: PI2525 Polyimide Induced Wafer Deformation

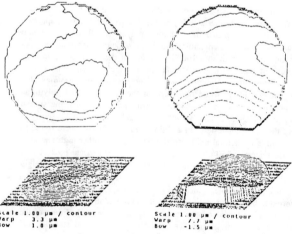

4" diameter wafer, 21 mils. thick; 3.8 μm dielectric thickness

Figure 5: PI2611D Polyimide Induced Wafer Deformation

390

The topographical plots shown in Figure 4 and 5 which were obtained on a GCA/Tropel Autosort show the initial bow and warp of a device quality wafer before and after polyimide coating. A beam displacement technique [25] was used to confirm the warp measurements obtained from this interferometric technique. Good correlation was obtained between measurement techniques for warpage values below 40 µm. Application of the PI2525 film to the substrate resulted in a net increase in warp of 18.4 µm. In the case of the low stress PI2611D a change of only 4.4 µm was observed.

The importance of minimizing dielectric induced substrate distortion to the processability of a multilayer thin film structure on silicon is highlighted by the graph in Figure 6. According to the MCNC design guidelines, a thickness of 8 to 10 µm per level of dielectric is required. A total dielectric thickness of 25 µm as shown in Figure 6 corresponds to 2 or 3 dielectric layers. From extrapolation of the test results to a dielectric thickness of 25 µm, a substrate warpage of >100 µm for PI2525 is anticipated, compared to <20 µm for the same thickness of PI2611D. Wafer warpage of >40 µm can result in hanging up of wafers on track systems used in an automated processing line. Increasing the thickness of the silicon wafers that are used in thin film processing (up to 100 mils. thick) has been used to reduce the warpage observed. If this approach is used, modifications of processing equipment may be necessary to be compatible with these non-standard substrate thicknesses.

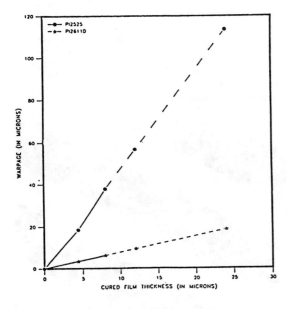

4" silicon wafer, 21 mils. thick

Figure 6: Cured Film Thickness vs. Substrate Warpage

Bow/Warp measurements were made for all of the materials evaluated using the same method described above. Stress values for the various dielectrics were calculated from the warp measurements using the following relationship where 'D' is the diameter of the wafer, 'd' is the warp measured, 'E_s' is Young's modulus for the substrate, 'v' is Poisson's ratio for the substrate, 'T_s' is the substrate thickness and 'T_f' is the film thickness [26].

$$\sigma = \frac{d}{(\frac{D}{2})^2} \frac{E_s}{3(1-v)} \frac{T_s^2}{T_f}$$

The stress values calculated for each of the dielectric classes are compared in Table 5. The low stress polyimides had a one order of magni-

tude lower stress than did the other materials studied. In comparison, blanket layers of evaporated metals show stresses in the 10^9 dyn/cm^2 range.

	TOTAL STRESS (dyn/cm^2)
STANDARD PI (DUPONT PI2525)	4.0×10^8
STANDARD PI (HITACHI PIQ13)	3.0×10^8
FLUORINATED PI (ETHYL XHP3119)	5.3×10^8
SILICONE/POLYIMIDE (GE SPI2000)	3.0×10^8
ACETYLENE TERMINATED PI (NATIONAL)	4.0×10^8
BENZOCYCLOBUTENE RESIN (DOW XU13005.02)	3.7×10^8
LOW STRESS POLYIMIDE (DUPONT PI2611D)	5.9×10^7
LOW STRESS POLYIMIDE (HITACHI PIQ L100)	5.2×10^7

STRESS WAS CALCULATED FROM WARP MEASUREMENTS ON 100 SILICON

Table 5: Calculated Stress for Dielectrics at 25°C

Adhesion

The long term reliability of a thin film interconnect is heavily dependent on the interlevel adhesion between its constituent layers. Consequently, polyimide/metal interfaces have been extensively investigated. Most of the work reported in the literature has involved metal on polyimide interactions [27-29]. More recently polyimide on metal interactions have also been studied [30-32]. Just as critical to the overall integrity of the substrate is the adhesion at the dielectric/substrate and dielectric/dielectric interfaces.

The adhesion of the various classes of dielectrics to a host of substrates was measured using the ASTM Tape Test Method [33] and the results compared in Table 6. Films of the materials under test were cast and cured according to the supplier's recommendations. When adhesion promoters were used, either the individual polyimide supplier provided their own proprietary adhesion promoter, or a 0.1% dilution of DUPONT's VM651 aminosilane adhesion promoter in deionized water was used. The polyimides were cured up to a maximum temperature of 400°C, and the BCB resin was cured up to a maximum temperature of 300°C in a N$_2$ purged quartz cage inside a high temperature Blue M oven.

SUBSTRATE	STD PI PI2525	FLUORINATED XHP3119	SILICONE/PI SPI2000	ACETYLENE TERMINATED EL5010	BCB DOW	LOW STRESS PI2611D
Cr	5	5	5	5	5	4
Al	5	5	5	5	4	4
Cu	0	0	0	0	4	0
self	5	-	4	4	5	3
SiO$_2$	5	5	5	5	4	0
SiO$_2$ + AP	5	5		5	-	2
Si$_3$N$_4$	4	4	5	5	5	0
Si$_3$N$_4$ + AP	5	5	-	5	-	3

ADHESION TAPE TEST AS PER ASTM D3359-83, METHOD B (0- POOR; 5- EXCELLENT)

Table 6: Adhesion of Dielectrics to Various Substrates

All of the materials evaluated with the exception of the low stress polyimides showed excellent adhesion to most of the test substrates. The low stress polyimides adhered well to chromium and aluminum but relatively poorly to either SiO$_2$ or Si$_3$N$_4$ even when adhesion promoters

391

were used. Recent reports in the literature suggest that adhesion between layers of low stress polyimide and metal on polyimide can be dramatically enhanced by O_2-plasma roughening of the underlying dielectric layer [34]. Due to the oxidation of copper by the polyimide precursors during curing, very poor adhesion was observed for all of the polyimide classes tested. Similar results for polyimide on copper have been published [30-32]. The BCB-1 adhered well to all of the substrate materials tested. In addition, this material does not appear to oxidize exposed copper.

Planarization

The major limiting factor in determining the total number of metal layers achievable in a given thin film process is the degree of planarization obtained with the interlayer dielectric. With conductor layers up to 4 µm thick and feature widths and spacings ranging from 10 to 100 µm, the roughness of the thin film structure increases rapidly unless the dielectric layer can adequately planarize the underlying metal layer. Recently, reports describing the enhancement of dielectric planarization using grinding/polishing [35] and etchback [36] techniques have appeared in the literature. The correlation between solids content of a coating casting solution for a given polyimide precursor and its degree of planarization (DOP) has been shown [37,38]. Table 7 compares the solids content, viscosity and degree of planarization over features of 10 µm width and 25 µm spacing (4 µm thick) for the classes of dielectrics evaluated.

MATERIAL	% SOLIDS	VISCOSITY (IN POISE)	D.O.P (%) (10/25)
BENZOCYCLOBUTENE	55	1.5	91
ACETYLENE TERMINATED PI	35	15	91-93
SILICONE/POLYIMIDE	26	12	22
FLUORINATED PI	15-19	15-25	25-30
STANDARD PI	14-17	11-19	24-34
LOW STRESS	12-13.5	20-110	25-28

PLANARIZATION OVER 10 um LINES AND 25 um SPACING

Table 7: Solids Content, Viscosity and Planarization of Dielectrics Evaluated

The polyamic acid precursors of conventional polyimides are relatively high molecular weight and show poor solubility in most organic solvents. Consequently, the maximum percent solids content achievable in an NMP based solvent system is typically less than 20% solids. The viscosities of these rather dilute solutions are also quite high (10 to 135 poise) due to their elevated molecular weights. The fluorinated and silicone polyimide precursors are slightly more soluble than the conventional polyimides, but are also limited in solubility due to their molecular weight. The acetylene terminated polyimides and the BCB are supplied in the form of low molecular weight oligomers. Since these materials in turn have much greater solubility in their proprietary casting solvent systems, higher percent solids solutions are possible (as high as 55% solids), while at the same time maintaining viscosities comparable to those of the other more dilute polyimide solutions.

The planarization data presented in Table 7 was obtained by applying a single coat of dielectric in the 6 to 10 µm thick range by spin coating, over a test structure consisting of 4 µm thick, 10 µm wide conductors with a spacing between adjacent conductors of 25 µm. The films were cured according to the specific material supplier's specifications. Planarization measurements were made on a DEKTAK 3030 profilometer with a 2.5 µm diameter stylus. The degree of planarization achieved for most of the dielectrics tested under the above conditions ranged between 20 and 35%. The acetylene terminated polyimides and the BCB with their high percent solids composition gave planarization results above 90%. Based on the DOP model for multiple identical coatings developed by Day et al. [37], a 10 µm thick layer of the BCB or acetylene terminated materials applied in double 5 µm thick coats, would result in essentially complete planarization over features of these dimensions.

Reactive Ion Etching

High etch rate, anisotropic and low contamination reactive ion etching through thick layers of dielectric is a critical element of MCNC's thin film interconnect technology. A split cathode magnetron reactive ion etching system designed and built at MCNC has been shown to be particularly well suited for the anisotropic etching of thick layers of dielectric (up to 100 µm thick films) with an etch rate of >1 µm per minute, and minimal etch residue [10]. A clean etch is of particular importance in technologies that use the selective via filling processes such as electroless nickel plating. Residue free via floors are required to etch, activate and then plate to the underlying metal layer.

The classes of dielectrics evaluated fall into 2 broad groupings: silicon containing and non-silicon containing materials. All of the non-silicon containing materials are readily etched in a N_2/O_2 gas mixture with minimal etch residue [10]. A PECVD SiO_2 or Si_3N_4 mask is used to define the etch. For the BCB and silicone polyimides, both of which contain silicon, an SF_6/O_2 gas mixture is required to effect the etch due to the passivating effect of the silicon in the dielectric to an oxygen plasma etch. Considerably greater amounts of etch residue were observed with the BCB and silicone polyimides. In addition, due to the different dielectric etch chemistry, a metal mask (Al or Ti) must be used to define the etched features. Representative etch results for a non-silicon containing dielectric is shown in Figure 7.

Figure 7: SEM Micrographs of RIE Etched PI2525 Polyimide

Conclusions

Of the classes of new materials evaluated, the low stress and acetylene terminated polyimides and the BCB resin appear to show the most immediate potential for multilayer thin film applications. The low stress polyimides exhibit some very desirable properties including their order of magnitude lower tensile stress than all of the other coatings evaluated (10^8 dyn/cm^2) and the resultant reduction in substrate distortion during the building up of these thick multiple coats of dielectric.

The rigid rod-like polymer chain structures characteristic of these polyimides also provide an order of magnitude reduction in the moisture absorption, typically less than 0.5%, and a dielectric constant of 2.9. Low stress materials show relatively much poorer adhesion to the common thin film surfaces than any of the other candidates unless additional processing steps are incorporated to enhance adhesion. For example, much improved interlayer adhesion (to self) is achievable using oxygen plasma to roughen up the surface of the underlying coat of low stress polyimide. Acetylene terminated polyimides are of interest due to their inherent differences in chemistry. These materials are supplied in low molecular weight acetylene terminated oligomers which exhibit much higher solubility in NMP based casting solvent systems than is observed with the much higher molecular weight polyimides. Consequently the acetylene terminated materials are available in high percent solids, low viscosity solutions which provide for improved planarization over the rough topographies characteristic of these multilayer structures yet have excellent thermal stability. Unfortunately, the fully cured acetylene terminated polyimides exhibit most of the other undesirable properties of the conventional polyimides such as considerable moisture uptake. The BCB material evaluated here has a dielectric constant around 2.6 and has a moisture absorption of <0.5%. It is however considerably more sensitive to thermal treatments above 300°C, and is therefore not recommended for processes involving high temperature steps such as metal annealing which is typically carried out at 400°C.

Fluorinated polyimides, of interest because of their lower moisture absorption and dielectric constant (2.6-2.7), are generally more susceptible to solvent crazing than are the other materials evaluated, and therefore care must be taken in selecting a fluorinated polyimide that is compatible with a multiple coat process. A relative ranking of the candidate material classes with respect to the specific material properties investigated is summarized in Table 8.

PROPERTY	STANDARD PI	FLUORINATED PI	SILICONE/ PI	LOW STRESS PI	ACETYLENE TERMINATED PI	BENZOCYCLO-BUTENE
1) SOLVENT COMPATIBILITY	1	1-5	1	1	1-5	1
2) THERMAL PROPERTIES	2	3	3	1	2	5
3) MOISTURE ABSORPTION	5	3	3	2	4	1
4) STRESS	5	5	5	1	5	5
5) ADHESION	1	2	2	5	2	2
6) PLANARIZATION	5	5	5	5	1	1
7) RIE ETCHABILITY	1	1	4	1	1	4

RELATIVE RANKINGS FOR EACH MATERIAL PROPERTY ON A 1 TO 5 SCALE (1-BEST; 5-WORST)

Table 8: Summary of Evaluation Results

Since an optimum polymer dielectric for thin film substrates is not commercially available at this time, great care must be taken in tailoring the selection of the material to the processing needs of the intended application. Recent work in the area of blending of rigid and flexible polyamic acid solutions promise some hope for achieving future dielectric formulations with enhanced properties for thin film applications [39], where the obvious advantages of the low stress rigid polyimides can be obtained with the lower dielectric constant and satisfactory adhesion properties of fluorinated polyimide chemistries.

Acknowledgements

The authors would like to thank the MCNC cleanroom staff for their assistance in the completion of this work, especially Drs. S. M. Bobbio, G. J. Dishon and P. K. Bhattacharya for their direct contributions to this effort, and, S. Rahim and R. Sheffield of Northern Telecom Electronics, for the thermal analyses of the materials in this study.

References

[1] R. Jensen, J. Cummings and H. Vora, "Copper/Polyimide Materials System for High Performance Packaging", IEEE Trans. Comp. Hyb. and Manuf. Technol. CHMT-7, 1984, pp. 383-393.

[2] T. Watanari and H. Murano, "Packaging Technology for the NEC SX Supercomputer", Proc. 35th Electronic Components Conf., 1985, pp. 192-198.

[3] C. Bartlett, J. Segelken and N. Teneketges, "Multichip Packaging Design for VLSI-Based Systems", Proc. 37th Electronics Components Conf., 1987, pp. 518-525.

[4] C. C. Chao, K. D. Scholz, J. Leibovitz, M. L. Cobarriviaz and C. C. Chang, "Multilayer Thin Film Substrate for Multichip Packaging", Proc. 38th Electronic Components Conf., 1988, pp. 276-281.

[5] L. F. Miller, "Controlled Collapse Reflow Chip Joining", IBM J. Res. Develop, vol. 13, pp. 239 1969.

[6] L-T Hwang, I. Turlik and A. Reisman, "A Thermal Module Design for Advanced Packaging", Journal of Electronic Materials 16, pp. 347-355, 1987.

[7] D. Nayak, L-T Hwang, I Turlik and A. Reisman, "A High-Performance Thermal Module for Computer Packaging", Journal of Electronic Materials 16, pp. 357-364, 1987.

[8] G. J. Dishon and R. C. Chapman, "Thick Copper Lift-Off Process for High Performance Computer Packaging Applications", presented at the 1988 Spring ECS Meeting, Atlanta, Georgia, Abstr. # 57, pp. 82.

[9] G. J. Dishon, "Electroless Nickel Via-Fill for High Performance Computer Packaging Applications", presented at the 1988 Spring ECS Meeting, Atlanta, Georgia, Abstr. # 60, pp. 86.

[10] G. J. Dishon, S. M. Bobbio, T. G. Tessier, Y. S. Ho and R. F. Jewett, "High Rate Magnetron RIE of Thick Polyimide Films for Advanced Computer Packaging Applications", accepted for publication in Journal of Electronic Materials, Dec. 1988.

[11] S. Numata, K. Fujisaki, D. Makino and N. Kinjo, "Chemical Structures and Properties of Low Thermal Expansion Polyimides", Proc. 2nd Int. Conf. on Polyimides Ellenville, NY, 1985, pp. 492-510.

[12] C. J. Araps, S. M. Kandetzke and M. A. Takacs, U. S. Patent # 4,749,621 (June 7th, 1988).

[13] R. A. Kirchhoff, C. E. Baker, J. A. Gilpin, S. F. Hahn and A. K. Schrock, "Benzocyclobutenes in Polymer Synthesis", 18th Int. SAMPE Tech. Conf. pp. 478-489, 1986.

[14] R. W. Johnson, T. L. Phillips, R. C. Jaeger, S. F. Hahn and D. C. Burdeaux, "Thin Film Silicon Multichip Technology", Proc. 38th Electronic Components Conf., pp 267-275, 1988.

[15] P. E. Garrou, "Cyanate Esters and Benzocyclobutenes: Dielectric Materials for Advanced Electronic Applications", accepted for presentation at the 1989 Spring ECS Meeting.

[16] J. F. McDonald, H. T. Lin, N. Majid, H. Greub, R. Philhower and S. Dabral, "New Systems for Fabrication of Wafer Scale Interconnections in Multichip Packages", Proc. 38th Electronics Components Conf. 1988, pp. 305-314.

[17] M. T. Pottiger, D. L. Goff and W. J. Lautenberger, "Photodefinable Polyimides: II. The Characterization and Processing of Photosensitive Polyimide Systems", Proc.38th Electronic Components Conf., 1988, pp. 315-321.

[18] K. P. Ackermann, R. Hug and G. Berner, "Multilayer Thin-Film Technology", Proc. 7th Int. Electronic Packaging Soc. Conf., 1987, pp. 519.

[19] T. G. Tessier, G. M. Adema, S. M. Bobbio and I. Turlik, "Low Temperature Etch Masks for High Rate Magnetron RIE of Polyimide Dielectrics in Thin Film Interconnects", accepted for presentation at the 3rd Int. SAMPE Electronic Materials and Processes Conference, LA, California, June 20-22, 1989.

[20] D. D. Denton, D. R. Day, D. F. Priore, S. D. Senturia, E. S. Anolick and D. Scheider, "Moisture Diffusion in Polyimide Films in Integrated Circuits", Journal of Electronic Materials, vol. 14, 1985, pp. 119-136.

[21] Don Hofer, "Polyimides in Packaging for Electronics", presented at the ACS Polyimide Symp. (Reno), 1987.

[22] R. J. Jensen, R. B. Douglas, J. M. Smeby and T. J. Moravec, "Characteristics of Polyimide Material for Use in Hermetic Packaging", Proc. VHSIC Packaging Conf., Houston, Texas, April 1987, pp.193-205.

[23] R. J. Jensen, "Recent Advances in Thin Film Multilayer Interconnect Technology for IC Packaging", in Electronic Packaging Materials Science III. Pittsburgh: Materials Research Society, 1988, pp.73-79.

[24] P. Geldermans, C. Goldsmith, and F. Bedetti, "Measurement of Stresses Generated During Curing and in Cured Polyimide Films" in Polyimides II. New York: Plenum Press, 1984, pp. 695-711.

[25] L. M. Mack, A. Reisman and P. K. Bhattacharya, "Stress Measurement of Thermally Grown Thin Oxides on (100) Silicon Substrates", accepted for publication in Electrochemical Soc. Journal.

[26] Equation from the Handbook of Thin Film Technology, Maisel and Glange, New York: McGraw Hill Inc., 1970.

[27] R. J. Jensen, J. P. Cummings and H. Vora, "Copper/Polyimide Materials System for High Performance Packaging", Proc. 34th Electronic Components Conf., 1984, pp. 73-81.

[28] N. J. Chou and C. H. Tang, "", J. Vac. Sci. Technol. A2, 751 (1984).

[29] P. S. Ho, B. D. Silverman, R. A. Haight, R. C. White, P. N. Sanda and A. R. Rossi, "Delocalized Bonding at the Metal-Polymer Interface", IBM J. Res. Dev. vol. 32, 1988, pp. 658-668.

[30] S. P. Kowalczyk, Y.-H. Kim, G. F. Walker, J. Kim, "Polyimide on Copper: The Role of Solvent in the Formation of Copper Precipates", Appl. Phys. Lett. vol. 52 (5), 1988, pp. 375-376.

[31] Y.-H. Kim, J. Kim, G. F. Walker, C. Feger and S. P. Kowalczyk, "Adhesion and Interface Investigation of Polyimide on Metals", J. Adhesion Sci. Technol. vol. 2 (2), 1988, pp. 95-105.

[32] S. Uchimura and D. Makino, "Investigation of the Reaction Between Polyimide and Copper by XPS, Proc. 3rd Int. Conf. Polyimides Ellenville, New York, 1988, pp. 251-253.

[33] ASTM Standard Test Methods for Measuring Adhesion by Tape Test (D3359-87), Method B, 1987.

[34] B. T. Merriman and J. D. Craig, "New Low Coefficient of Expansion Polyimide for Inorganic Substrates," presented at the 2nd Annual DUPONT High Density Interconnect Symp., Wilmington, DE, Oct. 4-5, 1988.

[35] T. Pan, S. Poon and B. Nelson, "A Planar Approach to High Density Copper Polyimide Interconnect Fabrication," Proc. 8th Int. Electronic Packaging Soc., 1988, pp. 174-189.

[36] D. Volfson and S. D. Senturia, "Polyimide Planarization Method for a Plated Via Post Multilayer Interconnect Technology," Proc. 3rd Int. Conf. on Polyimides, Ellenville, NY, 1988, pp. 180-181.

[37] D. R. Day, D. Ridley, J. Mario and S. D. Senturia, "Polyimide Planarization in Integrated Circuits," in Polyimides II. New York: Plenum Press, 1984, pp. 767-781.

[38] L. B. Rothman, "Properties of Thin Polyimide Films", J. Electrochem. Soc., vol. 127, 1980, pp. 2216-2220.

[39] D. Y. Yoon, M. Ree, W. Volksen, D. Hofer, L. Depero and W. Parrish, "Molecular Composites and Molecular Order of Rigid Polyimides," Proc. 3rd Int. Conf. on Polyimides, Ellenville, NY, 1988, pp. 1.

Evaluation Of Polyimides As Dielectric Materials For Multichip Packages With Multilevel Interconnection Structure

A. W. Lin

AT&T Bell Laboratories
Murray Hill, New Jersey 07974

Abstract

Using Temperature-Humidity-Bias (THB) screening test, nine commercial polyimides were evaluated for applications of multichip packages with a multilevel interconnection structure. Among these polyimides, DuPont PI2555 performed best under the test condition, i.e 85°C/85%RH (relative humidity)/180 Volts DC bias over 3 mils spacing. However, a slow increase of leakage current was observed in-situ on the TiPdAu triple track test samples coated with PI2555. This THB performance was improved by modifying PI2555 with a proprietary additive. However, the thermal stability of PI2555 was degraded by the additive. The dielectric constant of the modified PI2555 was determined as 3.4 at 1 KHz similar to that of PI2555.

1. INTRODUCTION

As microelectronic device's performance continues to excel, it becomes clear that a reliable and process-compatible organic polymer is needed for advanced VLSI multichip packaging applications. The polymer will serve as an interlevel dielectric on the packaging circuit with a multilevel interconnection structure. In order to meet the increasing demand of inter-chip connection density by VLSI and VHSIC, multilevel structure becomes essential for the packaging circuits. [1,2]

One consequence of the increasing interconnection density is high chip packaging density, which in turn results in high power dissipation. In addition, another increasing demand on the device packaging is high speed performance. Circuit design can be optimized to achieve the high speed performance. One aspect of the optimization is using a dielectric material with low dielectric constant to reduce parasitic capacitance, which can cause signal propagation delay. [1,2]

Among the organic polymers, polyimide has been known for its high thermal stability (>300°C), low dielectric constant (ca. 3.5) and excellent chemical resistance. Polyimide has been considered as a dielectric candidate for the multichip packages. However, there have been concerns about its degradation by hydrolysis and its water absorption. [3] Therefore, nine commercial polyimides have been studied using a Temperature-Humidity-Bias (THB) test. In this paper, the test results obtained at an accelerated aging conditions, i.e. 85°C, 85% relative humidity(RH) and 180 volts DC bias over 3 mils spacing, will be presented. Since Du Pont polyimide PI2555 performs best under the test conditions, more detailed studies of this polyimide also will be discussed.

2. EXPERIMENTAL

The THB test samples were 24-pin ceramic DIPs with thin film TiPdAu triple track meander patterns (3 mil line width and space) coated with various polyimides. The triple track test samples with TC (thermal compression)-bonded leads were cleaned using Freon TA vapor degreasing followed by a boiling in hydrogen peroxide and a rinse in deionized water. The test samples were then dried in a clean oven. The polyimides were coated on the test samples using spin coating or flow coating within 2 hours of the cleaning. All the polyimide solutions were commercially available. For some THB tests, Du Pont polyimide PI2555 was coated on the samples pre-treated with an adhesion promoter VM651. For some other THB tests, PI2555 was modified using a proprietary additive. Each of the polyimides was cured using the conditions recommended by its manufacturer. The cured polyimide films have a thickness in the range of 4 to 10 μm.

The THB test conditions were 85°C/85%RH/180 Volts DC bias, except for few experiments specified in later sections. In THB test, leakage currents were measured in-situ as a function of test time. Average leakage currents reported here are means of log values. Details of the test procedure and equipments have been described earlier. [4-8]

3. RESULTS AND DISCUSSIONS

3.1 THB Screening Test of Selected Polyimides

Table I shows a list of commercial polyimides available at the beginning of this study. The list, which by no means is complete, shows that there are four classes of polyimides: A) thermoplastic, preimidized, B) thermoset, preimidized with additional crosslinking, C) thermoset, non-preimidized (in polyamic acid form), and D) photosensitive polyimides. The thermoplastic polyimides were excluded from this study due to their poor chemical resistance. The photosensitive polyimides were evaluated in another study. Therefore, nine thermoset polyimides as marked by asterisks in Table I were evaluated using the THB screening test.

THB test results of the nine polyimides coated on TiPdAu triple tracks are shown in Figure 1 along with controls, i.e. TiPdAu triple tracks with and without Dow Corning 3-6550 RTV. The RTV has been recognized as one of the best encapsulants for integrated circuits. [7,8] Its leakage currents usually starts at 10^{-9} ampere and decreases with time as shown in Figure 1. In this study, the RTV-coated triple tracks were used to check the effectiveness of cleaning, since THB performance of the RTV had been well characterized. [4-8] The bare triple track samples were used for monitoring the contamination level in the test chamber. Normally the leakage current on the bare triple tracks starts at 10^{-7} to 10^{-8} amperes and decreases with time. However, if the test chamber is contaminated, the leakage currents of the bare triple tracks will increase with time as shown in Figure 1 during part of this test. The source of contamination for this transient leakage current was not determined.

The leakage currents of most of the nine polyimides increased with time. Although the leakage current of SIL-10 decreased initially, it increased sharply after 150 hours of the test. DuPont PI2555 showed

Reprinted from *Proc. 39th Electron. Components Conf. (ECC)*, pp. 148–154, 1989.

the best overall performance among the nine polyimides, because both its initial and final leakage currents (at ca. 1000 hours) were lowest.

Hitachi PIQ (polyimide isoindoloquinazolinedione) has been known for its improved thermal stability over other commercial polyimides.[9] THB tests of PIQ along with PI2555 and several polyimide-amides have been studied by Bowie and Gabriel.[10] Their results showed that PIQ did not perform as well as PI2555 under the 85°C/85%RH/180 V test conditions.

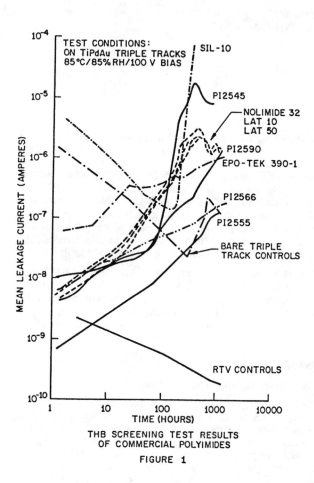

THB SCREENING TEST RESULTS
OF COMMERCIAL POLYIMIDES

FIGURE 1

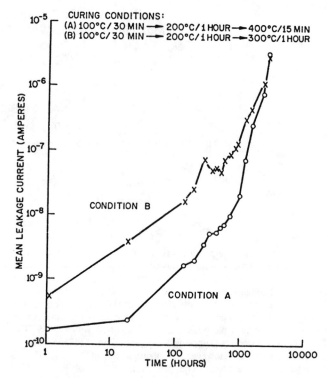

EFFECT OF CURING CONDITION ON THB PERFORMANCE
OF PI2555 COATED ON TiPdAu TRIPLE TRACKS

FIGURE 2

3.2 More Studies of DuPont Polyimide PI2555

Since PI2555 performed best among the nine polyimides in the THB screening test, it was selected for further studies.

3.2.1 Effect of Curing Conditions In the THB screening test, each polyimide was cured under the conditions recommended by its manufacturer. However, these recommended conditions, which typically are optimized for mechanical properties, may not yield the best THB performance. In order to find optimum conditions for curing THB samples, PI2555 was cured in air under the following conditions: A) 100°C/30 minutes -> 200°C/1 hour -> 400°C/15 minutes, and B) 100°C/30 minutes -> 200°C/1 hour -> 300°C/1 hour, and their THB performances were compared. Figure 2 shows that Condition A provides roughly an order of magnitude improvement in leakage current up to 1000 hours. This improvement probably is due to a higher degree of crystallinity and/or increase of polyimide film density as a result of the higher final curing temperature (400°C vs. 300°C).[11,12] This effect of curing conditions on THB performance has been confirmed repetitively in six other THB tests. A summary of these test results is shown in Figure 3.

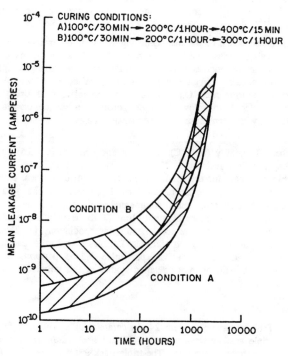

SUMMARY OF THB TEST RESULTS
OF TiPdAu TRIPLE TRACKS COATED WITH PI2555

FIGURE 3

3.2.2 Effect of Adhesion Promoter As shown in Figure 4, there are three possible paths for the leakage currents observed in THB tests. It has been widely believed that good adhesion of the coating material on substrate is a necessity for achieving good THB performance.[8] Since the RTV is very permeable to water, a strong chemical bonding between the RTV and substrate surface (alumina ceramic or silicon wafer) is believed to be the major factor responsible for the RTV's excellent THB performance.[8]

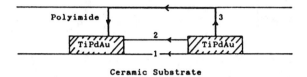

Possible Leakage Paths shown on the Cross-Section of
A Triple Track Test Sample

FIGURE 4

Adhesion of PI2555 films on Al_2O_3 ceramic substrate, TiPdAu and other thin film metallizations has been studied.[13] 90° peel strengths larger than 1.96 pounds have been measured on Al_2O_3 and TiPdAu using an Instron tester and one inch-wide PI2555 films (10 μm thick). However, the adhesion can be degraded by moisture. For example, PI2555 film completely lost adhesion on Al_2O_3 ceramic surface after 2 hours in boiling water. It was interesting to note that a heat treatment at 300°C/15 minutes restored the adhesion providing the PI2555 film remained in contact with the surface.

DuPont recommended VM651 (an organosilane coupling agent) as an adhesion promoter for improving adhesion of its polyimides on various substrates.[14] A 0.1% aqueous solution of VM651 was spun-dried at 3000 rpm on the triple track samples shortly after they were cleaned. PI2555 was then coated on the VM651-treated samples within 30 minutes. THB test results of these triple track samples along with controls, which were not treated with VM651, are shown in Figure 5. The adhesion promoter only provided slight improvement during the first 24 hours of the test. This result implies that the adhesion of PI2555 film on ceramic substrat may only survive 24 hours under the THB test conditions (as compared to the complete loss of adhesion in boiling water within two hours). Additional implication is that, contrary to common belief, adhesion may not be the primary factor responsible for the slowly degraded performance of PI2555. Instead, the hydrolysis of bulk polyimide could result in the slow increase of leakage current after long hours of the test. As shown in Figure 4, both path 2 and 3 can be affected by the hydrolysis.

3.2.3 Effect of Humidity In order to determine if temperature, humidity, bias or any combination of the three caused the increase of leakage current, a special experiment was conducted. Thirty TiPdAu triple track samples coated with PI2555, which were cured using condition B, were divided into three groups. Groups #1 and #2 were tested together in a humidity chamber at 85°C/85%RH, with and without continuous 180 volts DC bias, respectively. Bias was applied on group #2 only during the in-situ current measurements which lasted about 30 seconds per sample. Group #3 samples were aged in an air convection oven at 85±1°C/24±4%RH with no bias, and their leakage currents were measured at ambient conditions (23±2°C/46±24%RH).

As shown in Figure 6, groups #1 and #2 had similar leakage current increases in spite of the bias difference. While group #3 retained its low leakage currents for the entire 1500 hours of test. It should be noted that the leakage currents of group #3 samples were not measured at 85°C. However, the contribution to leakage current due to such temperature difference (85°C vs 23°C) is expected to be small, because PI2555's electric properties are stable over a wide temperature range including 85°C.

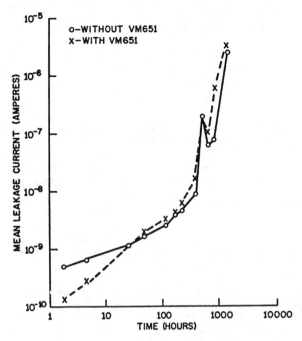

EFFECT OF ADHESION PROMOTER VM651
ON THB PERFORMANCE OF TiPdAu TRIPLE TRACKS
COATED WITH PI2555

FIGURE 5

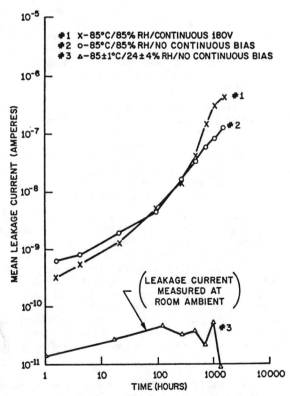

LEAKAGE CURRENTS OBSERVED ON TiPdAu
TRIPLE TRACKS COATED WITH PI2555
UNDER VARIOUS THB TEST CONDITIONS

FIGURE 6

A second THB test was conducted using the same group #1 and #2 samples. All the group #2 and majority of the group #1 samples regained their initial leakage currents (in 10^{-10} ampere range) after being baked at 300°C for 30 minutes. The leakage current increases of the samples were almost same as those observed in the first test. Only one sample from group #1 showed poorer performance than in the first test. Dendrites between the oppositely biased tracks were found on this sample at the end of the second test.

These results indicate that high humidity alone is responsible for the leakage current increases. It is less likely that the slow increase of leakage current is simply caused by the water absorption in PI2555. An equilibrium of the water absorption should not take hundreds of hours to be established, since diffusion coefficient of water in PI2555 film is expected to be in the range of 10^{-9} cm²/sec similar to Kapton film.[15]

It is possible that hydrolysis of PI2555 (in bulk and at the interface with the ceramic substrate) occurs under the THB test conditions, and the baking at 300°C recovers the hydrolysis damages.

3.2.4 Effect of Additive As part of the efforts to improve THB performance, a proprietary additive was mixed into PI2555 polyamic acid solution. Figure 7 shows that leakage currents of the TiPdAu triple track samples coated with the additive-modified PI2555 (0.2% percentage by weight) are less than those coated with PI2555 through the entire 10000 hours of the test. Both of the polyimides, modified and unmodified, were cured under the same condition (i.e. condition A described in next paragraph). As shown in Table II, in terms of the time required to reach certain leakage current level, an improvement of about 15 times was achieved by such modification.

Similar to the studies of unmodified PI2555, the effect of curing conditions on THB performance was examined for the modified PI2555. Four curing conditions were studied, i.e. A) 100°C/30 minutes -> 200°C/1 hour -> 400°C/15 minutes, B) 100°C/30 minutes -> 200°C/1 hour -> 300°C/1 hour, C) 100°C/30 minutes -> 200°C/1 hour -> 250°C/1 hour, and D) 100°C/1 hour -> 200°C/2 hours. All cures were done in air convection ovens. Figure 8 shows that conditions A) and B) yield similar THB performance, while conditions C) and D) result in poorer performance. The poorer THB performance could be attributed to incomplete curing of PI2555 (i.e incomplete imide linkage formation) and to solvent retained in the polyimide films. The main component of the solvent in PI2555 solution is NMP, which has low volatility and a boiling point of 202°C.

The reproducibility of the THB performance of the modified PI2555 cured under condition A or B has been demonstrated in a series of 10 experiments over a period of 22 months using 4 different lots of PI2555. Results of these experiments are summarized in Figure 9. Variations of leakage current among the experiments were observed, especially during the first one hundred hours. The cause of the variation is unknown. The THB performance of the modified PI2555 in each experiment, however, is more reproducible after 1000 hours.

Since the RTV likely will be used as a final encapsulant for the multilevel packages, its compatibility with the modified PI2555 was studied. Figure 10 shows that during the first 2000 hours of the test, the presence of Dow Corning 3-6650 gray RTV does not cause any significant performance differences on the samples coated with the modified PI2555. However, at the end of the test (8461 hours), one out of the twenty RTV-encapsulated samples showed a leakage current of 1.6×10^{-4} amperes, which is the maxium current limit set for the THB tests. Since polyimide has excellent chemical resistance, it is expected that the RTV encapsulant should not degrade polyimide and cause such high leakage current. Failure mode analysis of this sample was not done.

The effect of the additive on THB performance was also studied for a fluorinated polyimide, Du Pont PI2566. A similar but smaller beneficial effect was observed as shown in Figure 11. Also shown in Figure 11 is the compatibility of the RTV with the modified PI2566 (with 0.2% additive). In this case, the RTV-encapsulated samples had better THB performance than the ones without RTV.

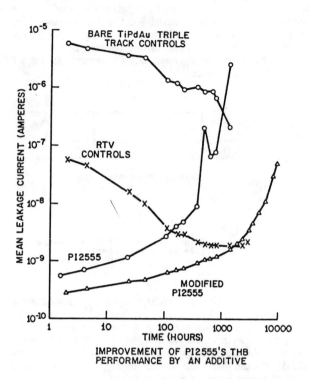

IMPROVEMENT OF PI2555'S THB PERFORMANCE BY AN ADDITIVE

FIGURE 7

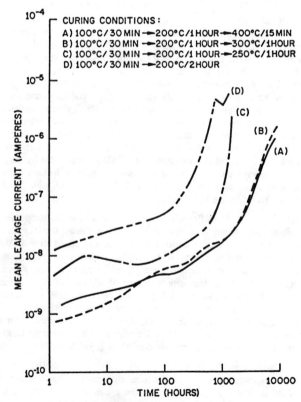

EFFECT OF CURING CONDITION ON THB PERFORMANCE OF TiPdAu TRIPLE TRACKS COATED WITH MODIFIED PI2555

FIGURE 8

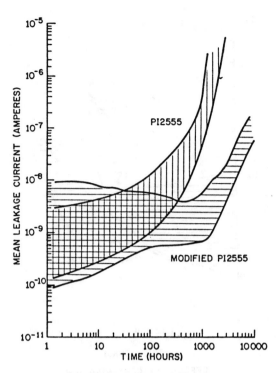

SUMMARY OF THB TEST RESULTS
OF TiPdAu TRIPLE TRACKS COATED
WITH PI2555 OR MODIFIED PI2555

FIGURE 9

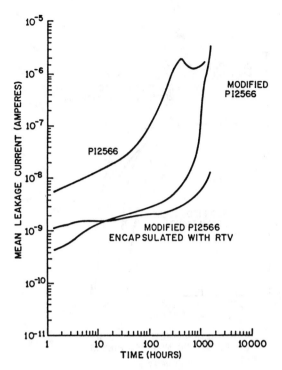

THB TEST RESULTS OF TiPdAu TRIPLE TRACKS
COATED WITH PI2566 OR MODIFIED PI2566

FIGURE 11

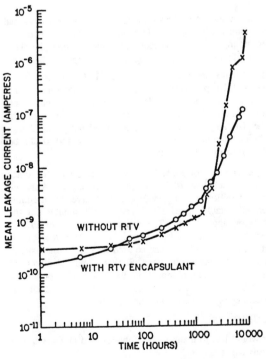

COMPARISON OF THB PERFORMANCES
OF MODIFIED PI2555 WITH/WITHOUT
RTV ENCAPSULANT

FIGURE 10

It has been demonstrated in the above discussion that the THB performance of the two polyimides can be improved by the additive. Similar beneficial effects also have been confirmed in other types of polymer, such as polysiloxaneimide, epoxy, polyurethane and polyester.[16,17] The magnitude of the effect varied depending on the types of polymer. The mechanism for this improvement, however, is yet to be identified.

As shown by the results of thermal gravimetric analyses in Figure 12, a degradation of 75 to 100°C in thermal stability of PI2555 is caused by the additive (0.2% by weight). Nevertheless, the modified PI2555 is still thermally stable enough for the multichip packaging applications. As for the high speed electrical performance, the dielectric constant of the modified PI2555 film was determined to be very similar to that of PI2555, e.g. 3.4 versus 3.5 at 1 KHz (Figure 13).

3.2.5 Effect of Contaminants In one experiment, the modified PI2555 was accidentally contaminated by using an epoxy nozzle for flow coating. The nozzle was found to be attacked by the modified PI2555 solution (presumably by NMP/aromatic hydrocarbons which are the solvents in PI2555). The THB performance of the triple track samples coated with the contaminated polyimide was much poorer than that of the uncontaminated ones (Figure 14). The identity of the contaminant was not determined.

In another experiment, ionic contaminants were purposely introduced onto the triple track samples by spin coating of a NaCl aqueous solution (4.42×10^{-3}M). Spotty discoloration was observed when the modified PI2555 was coated subsequently. The contaminated samples had an average leakage current of 7.93×10^{-12} amperes measured at room ambient before THB test. However, the leakage current jumped to 1.31×10^{-4} amperes after only 1.5 hours and remained high throughout the test.

Although the additive can improve PI2555's THB performance, the above two experiments indicate that contaminants, which are introduced before the modified polyimide is cured, can severely degrade its THB performance.

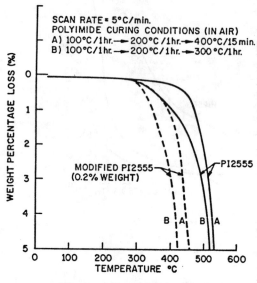

SCAN RATE = 5°C/min.
POLYIMIDE CURING CONDITIONS (IN AIR)
A) 100°C/1hr. ⟶ 200°C/1hr. ⟶ 400°C/15 min.
B) 100°C/1hr. ⟶ 200°C/1hr. ⟶ 300°C/1hr.

THERMAL GRAVIMETRIC ANALYSIS
OF PI2555 AND MODIFIED PI2555
FIGURE 12

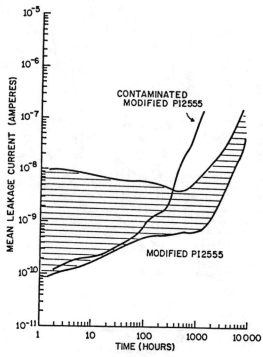

EFFECT OF CONTAMINATION
ON THB PERFORMANCE OF MODIFIED PI2555
FIGURE 14

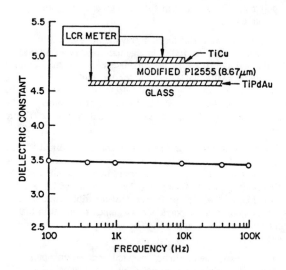

DIELECTRIC CONSTANT OF MODIFIED PI2555
FIGURE 13

4. SUMMARY

As part of the evaluation of polyimides as dielectric for advanced multichip packaging applications, DuPont PI2555 was identified as a top contender among the nine selected polyimides using the THB screening test. PI2555 was therefore subjected to in-depth studies.

The in-depth studies revealed the followings: 1) Curing conditions affected PI2555's THB performance. 2) The treatment on ceramic surface by the adhesion promoter VM651 had little improvement on the THB performance. 3) The slow increase of leakage current observed on the TiPdAu triple track samples coated with PI2555 was identified as being caused mainly by high humidity and could be attributed to the hydrolysis of the polyimide.

The THB performance of PI2555 was improved by ca. 15 times using a proprietary additive (0.2% weight percentage). Although PI2555's thermal stability was degraded by the additive, the modified PI2555 is thermally stable up to 300°C and is still suitable for the multichip packaging applications. The dielectric constant of PI2555 was not affected significantly by the additive. Therefore, it is concluded that the modified PI2555 can be used as a dielectric material for the multichip packaging applications. Similar beneficial effects of the additive have also been observed in the fluorinated polyimide PI2566, as well as in other types of polymers such as polysiloxaneimide, epoxy, polyurethane and polyester.

ACKNOWLEDGEMENTS

The author would like to thank C. N. Robinson and D. Gerstenberg for reviewing this paper.

REFERENCES

[1] C. C. Shiflett, et al, "High Density Multilayer Hybrid Circuits made with Polymer Insulating Layers (POLYHIC's)", p.481, Proceeding of 1986 ISHM Symposium.

[2] C. J. Bartlett, J. M. Segelken and N. A. Teneketges, "Multichip Packaging Design for VLSI-based Systems", Proceeding of 1987 Electronic Component Conference of IEEE.

[3] M. L White, "Encapsulation of Integrated Circuits", Proc. IEEE, 57(9), 1610-1616, (1969).

[4] R. G. Mancke, "A Moisture Protection Screening Test for Hybrid Circuit Encapsulants", IEEE Transaction, CHMT-4(4), 492, (1981).

[5] N. L. Sbar and R. P. Kozakiewicz, "New Acceleration Factors for Temperature,Humidity,Bias Testing", IEEE Transaction ED-26(1), 56, (1979).

[6] R. P. Kozakiewicz and N. L. Sbar, "Procedures for the Comparative Evaluation of the Moisture Protection of Coatings on Integrated Circuits", extended abstract, Electrochemical Society, Spring meeting, p.123, (1977).

[7] D. Jaffe, "Encapsulation of Integrated Circuits Containing Beam Lead Devices with a Silicone RTV Dispersion, IEEE Transaction PHP-12(3), 182, (1976).

[8] N. L. Sbar, "Temperature-Bias-Humidity Performance of Encapsulated and Un-encapsulated TiPdAu Thin Film Conductor in an Environment contaminated with Chlorine", IEEE Transaction PHP-12(3), 176, (1976).

[9] A. Saiki, S. Harada, T. Okubo, K. Mukai and T. Kimura, J. Electrochem. Soc. 124(10), 1619-1622, (1977).

[10] M. H. Bowie and W. Gabriel, "Temperature-Humidity-Bias Performance of Polyimide Coatings on Triple Track Conductors", private communcation.

[11] A. Endo and T. Yoda, "Thermal and Physical Properties and Etching Characteristics of PI Films", J. Electrochem. Soc., 132, 155, (1985).

[12] N. Takahashi, D. Y. Yoon and W. Parrish, Macromolecules, "Molecular Order in Condensed States of Semiflexible Poly(amic acid) and Polyimide", 17, 2583, (1984).

[13] R. H. Mills and A. W. Lin, "Adhesion Of Materials On Single And Multilayer HIC's With Polymer Dielectrics", unpublished results.

[14] DuPont, Technical information bulletin #PC-7, April, 1982.

[15] L. Iler, W. J. Koros, D. K. Yang and R. Yui, "Sorption and Transport of Physically and Chemically Interacting Penetrants in Kapton Polyimide", Proceeding of First Polyimide Technical Conference, November 10-12, 1982.

[16] I. M. Asher, private communication.

[17] R. G. Mancke, private communication.

TABLE I: Commercial Polyimides

COMPANY	PRODUCT NAME	CLASS+
DuPont	PI2540, PI2545*	C
	PI2550, PI2555*	C
	PI2560, PI2562	C
	PI2566*	C
	PI2590*	C
	Photosensitive PI	D
Hitachi	PIQ	C
	PAL1000 photosensitive PI	D
Rhone-Poulenc	Nolimid 32*	B
	Lat 10*, 50*	B
	Kerimid 601	B?
Ciba-Geigy	XU218,XU218-HP	A
Upjohn	Polyimide 2080	A
Epoxy Tech, Inc.	Epo-Tek 390-1*, SIL-10*	B
3M	Photosensitive PI	D
Siemens	Photosensitive PI	D
EM Chemical	HTR-2 photosensitive PI	D

*Materials studies in this work.
+Classification of the commercial polyimides is discussed in the text.

Table II: Improvement of THB Performance by a Proprietary Additive

Leakage Current Level	PI2555	Modified PI2555	Improvement*
5×10^{-8} amperes	700 hours	10000 hours	14.3
1×10^{-8}	400	6000	15.0
5×10^{-9}	225	3800	16.9

*The improvement was estimated using the time required to reach a same leakage current level.

Benzocyclobutene (BCB) Dielectrics for the Fabrication of High Density, Thin Film Multichip Modules

DAVID BURDEAUX, PAUL TOWNSEND and JOSEPH CARR

The Dow Chemical Company, Central Research, Midland, MI

PHILIP GARROU

The Dow Chemical Company, Central Research, Charlotte, NC 28210

A new class of organic dielectrics, benzocyclobutenes, 1, are described and their application to the fabrication of thin film multichip modules is detailed. Key properties for 3, a siloxy containing BCB derivative include low dielectric constant (2.7), low loss (0.008 at 1MHz), low water absorption (0.25% after 24 h water boil) and high degree of planarization (>90% from one layer coverage). All other properties meet the requirements necessary for fabrication of thin film MCM structures.

Key words: Benzocyclobutene, BCB, multichip module, dielectric

INTRODUCTION

Performance advantages inherent in advanced VLSI semiconductor devices will not be realized at the system level, during the 1990's, with conventional electronic packaging and interconnect.[1-3] Individually packaged chips have typically been interconnected on a multilayer circuit board. Although this approach has served the industry well for many years, it has become clear that new technology will be necessary to meet the performance demands that will be required in the '90s and beyond. One high density interconnect technology that has evolved to meet these needs has been dubbed the "Multichip Module" (MCM). As shown in Fig. 1, it has blended IC, hybrid and PWB technologies into a new interconnect methodology.

Described early on by IBM,[4] MCM technology has seen intense recent activity in the US,[5-7] Japan[8-9] and Europe.[10]

In addition to achieving far superior electrical performance, equally important gains are made in size and weight and significant advances are expected in reliability.[11-12] It is also becoming clear that MCM will show significant cost performance benefits over other technologies once available in a full production mode.[13-14] Thus, if MCM lives up to its billing, it will find applications not only in high performance mainframe computers but also in the workstation, laptop, military, telecom, automotive and consumer market segments.

Each active research group has its own modifications to the overall fabrication process, such as, substrate material (silicon or ceramic), conductor (aluminum vs copper), via interconnection techniques (unfilled vs filled vs plated up posts) or methodology to attach the bare IC's (wire bonding vs TAB vs flip chip), to name just a few. However, one commonality among nearly all the current variations is the use of polymeric insulators to separate the conductor traces. Polymer dielectrics allow the fabrication of the thick interlayer insulators necessary to produce matched impedance transmission lines with acceptable losses. Typical inorganic dielectrics used in conventional integrated circuits cannot exceed ca 1 micron thickness without serious fabrication problems.[15]

Material requirements for the ideal polymeric thin film dielectric can be described as follows:[16]

LOW DIELECTRIC CONSTANT
THERMAL STABILITY IN EXCESS OF SUBSEQUENT PROCESSING AND REPAIR STEPS
GOOD ADHESION TO SUBSTRATE, CONDUCTOR AND SELF
LOW STRESS FORMATION
LOW WATER ABSORPTION
GOOD PLANARIZATION
EASY TO PROCESS
    SPIN COATABLE   SPRAY COATABLE
    PIN HOLE FREE   ETCHABLE

Benzocyclobutenes, BCB, a new class of polymeric dielectric have been developed to meet these requirements for MCM fabrication.

BENZOCYCLOBUTENE POLYMER FAMILY

Benzocyclobutene polymers are derived from monomers of the generic form 1.[16] The polymerization of such molecules is based on thermal rearrangement of the cyclobutyl functionality to give the highly reactive intermediate o-quinodimethane, 2, which can polymerize with similar molecules or react with a variety of other unsaturated functional groups.

STRUCTURE 1   STRUCTURE 2

The polymerization is a purely thermal process. No catalysts are required and no volatiles and produced, in contrast to polyimides which produce water

(Received June 22, 1990; revised August 15, 1990)

Reprinted with permission from *J. Electron. Mater.*, vol. 19, no. 12, pp. 1357–1366, December 1990. Copyright © 1990 AIME.

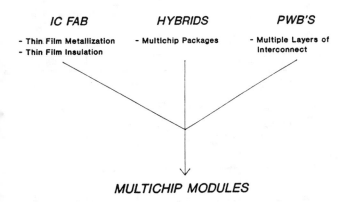

IC FAB
- Thin Film Metallization
- Thin Film Insulation

HYBRIDS
- Multichip Packages

PWB'S
- Multiple Layers of Interconnect

MULTICHIP MODULES

Fig. 1 — Multichip module technology.

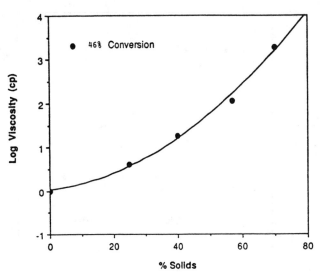

Fig. 2 — Viscosity vs % solids relationship for 3.

as a byproduct of their condensation polymerization. The properties of the polymers derived from 1 can be altered by appropriate choice of R group. The properties for three derivatives 3, 4, and 5, are shown in Table I and will be discussed later in the text. (5 is a formulation of 3 and 4.)

STRUCTURE 3

3

STRUCTURE 4

4

BCB is supplied "b-staged" (partially polymerized) to obtain materials with appropriate viscoelastic handling properties. B-staging is preformed by heating the monomers between 180 and 195° C for 2 to 8 hr. The b-staged materials are stored at room temperature (no refrigeration is required) without degradation. For coating applications, the b-staged materials are diluted with hydrocarbon solvents such as xylene or mesitylene. Solutions in the range of 25 to 70% solids have been used to cast polymeric films. Solids content and molecular weight control the solution viscosity and thus the thickness of the cured film. Figure 2 depicts the solution viscosity as a function of solids content for b-staged 3 (46% converted). By changing the degree of b-staging, different viscosity profiles can be obtained.

BCB coatings are fully polymerized, in a nitrogen environment, at 250° C, in 1 hr or less. The fully cured polymers are crosslinked thermosets.

ELECTRICAL PROPERTIES

Dielectric constant (ϵ') is the critical electrical parameter for the dielectric chosen to fabricate a multichip module. The lower the ϵ', the faster the basic wave propagation velocity, the lower the line capacitance per unit length and the higher the char-

acteristic impedance that can be obtained for a fixed conductor cross section. For a lower ϵ', transmission lines can be designed wider and the dielectric thickness decreased, to maintain the same characteristic impedance while lowering the line resistance and cross talk.

ϵ' of 3, 4, and 5 are shown in Table I. Capacitance measurements on free standing films, plaques or coatings typically yield ϵ' in the range between 2.6 and 2.8. As can be seen in Fig. 3, the ϵ' of 3 is flat from 1 kHz to 1 MHz, between 25 and 200° C. BCB 4 and 5 show nearly identical response. Measurements made on 3 at 10 GHz revealed an ϵ' of 2.58.[19]

Dissipation factor (ϵ'') or loss tangent is another important electrical parameter, although more so at high frequencies. ϵ'' at ca 1 kHz is shown in Fig. 4. Measurement of the ϵ'' of 3 at 10GHz revealed an increase to 0.002.[19] Breakdown voltage (or strength) measured on thin BCB films deposited on silicon substrates give values in the range of 4.0×10^6 V/cm.

Figure 5 shows the relationship between dielectric constant and line capacitance for various thicknesses of polymer dielectric. (For a conductor cross section of $10 \times 2.5 \ \mu m$.) It can be seen that for an

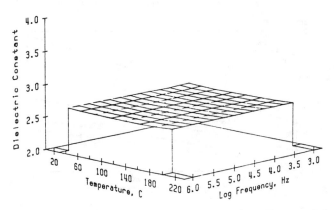

Fig. 3 — Dielectric constant variation with temperature and frequency for 1.

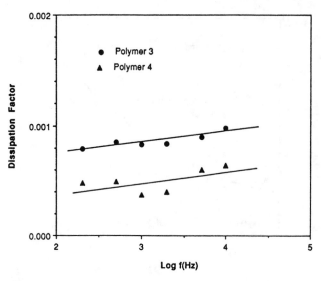

Fig. 4 — Dissipation factor vs frequency for **3** and **4**.

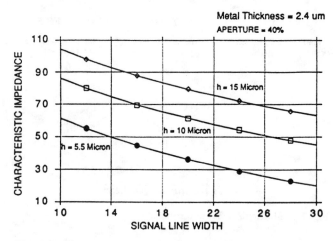

Fig. 6 — Characteristic impedance vs signal line width for several BCB layer thicknesses.

insulator thickness of 6 μm, use of BCB would result in 20% less line capacitance when compared to a PI with dielectric constant of 3.5. Figure 6 presents the relationship between characteristic impedance and signal line width for a 2.4 μm thick conductor located at various heights (BCB insulator thicknesses) above a perforated gound plane (40% aperture). It can be seen, for instance, that a 60 ohm impedance can be obtained for a 20 μm wide line, 10 μm above the ground plane.

BCB DEPOSITION TECHNIQUES

BCB can be applied by typical IC spin coating techniques. Figure 7 shows the variation of coating thickness (cured) with spin speed for solution of **3**. Coating thickness uniformity for a cured film of **3** on a 125 mm wafer is shown in Fig. 8. Details are given in the experimental section. For this typical experiment, 11 measurements taken across the wafer surface reveal a 1% standard deviation for a mean single coat thickness of 7.25 μm. Film shrinkage is not a problem since, in practice, most of the xylene solvent has evaporated during the spin procedure.

Since no volatiles are produced during the BCB cure, typical shrinkage of <5% can be expected vs 25% more typical of polyimides.[5]

CURE ATMOSPHERE

The cast films are cured by heating (ramp schedule differing for different derivatives) to 250° C for about one hr under a nitrogen atmosphere. The cure schedule is detailed in the experimental section. While all manufacturers recommend that their organic dielectrics be cured in the absence of air, to avoid high temperature oxidation, it is imperative that this recommendation be followed during the curing of BCB. A properly cured BCB film is clear and transparent to the eye. A yellowish or darker BCB film is indicative of oxidation. This is depicted by the infrared spectra shown in Fig. 9. A film cured in air reveals absorptions in the 1700–1850 1/cm region. Such absorptions are indicative of the oxidation of the benzylic CH_2 groups on BCB to anhydride and/or carbonyl species. The exact nature of these species, that are bound into the polymeric

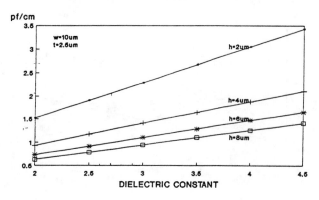

Fig. 5 — Line capacitance vs dielectric constant at various dielectric thicknesses.

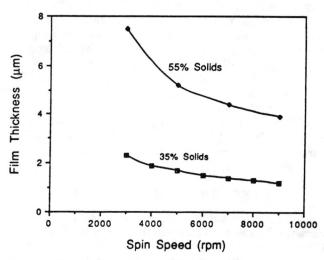

Fig. 7 — Variation in spin speed with film thickness for **3**.

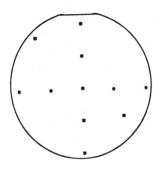

# Determinations	11
Mean Thickness (um)	7.25
% Standard Deviation	1.0
Spin Condition (10 sec. spread cycle)	30 sec./3000 rpm

Fig. 8 — BCB coating uniformity.

films, has not been determined. Oxidized films reveal a degradation in electrical and physical properties vs those shown in Table I. It is recommended that the furnace atmosphere be monitored by oxygen sensors. We have determined that ca 100 ppm of oxygen is allowable during cure.

THERMAL STABILITY

Thermal stability in the BCB family is determined to a large extent by the R group that links the thermally reactive units. As stated earlier thermal stability for a candidate MCM dielectric needs to be in excess of subsequent processing and/or repair steps.

Thermal stability is typically determined by analysis of TGA data (thermogravametric analysis) generated by ramping the sample temperature at

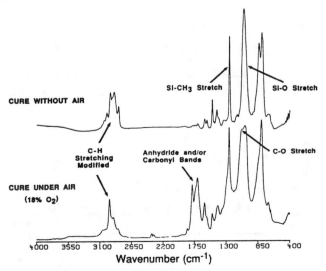

Fig. 9 — FTIR spectra of BCB 3 cured in the presence and the absence of air.

10° C per min and looking for the point at which weight loss occurs. The MCM practicioner is really interested in the isothermal stability of a polymer at MCM processing temperatures. In addition, thermal stability is also surface area dependent and thus the most informative data are those obtained on thicknesses of films that are similar to those used to fabricate the particular MCM structure. The typical TGA spectrum of 3 is shown in Fig. 10. Films of 3, 4 and 5 are compared to similar thickness films of typical PI's in Fig. 11. MCM processing, in general, does not require temperatures at or above 350° C for anywhere near two hr. We, therefore, feel that the stability of 3 is adequate for MCM fabrication rework and repair.

It should also be noted that polymer Tg is also an important thermal stability parameter. A thermoset polymer above its Tg becomes a rubbery "liquid." Cracking of SiO_2 and Si_3N_4 masking layers has been reported when using PI's with Tg's in the 300–320° C range[18]

ADHESION

The long term reliability of an MCM is heavily dependent on the adhesion between the given interfaces. A qualitative method that is routinely used is the ASTM tape.[19] A special adhesive tape is applied to the surface of the film and stripped. If the film detaches, poor adhesion is noted. In a modification of this test, the film is exposed to boiling water for 4 hr before the tape test is carried out. Another measure of adhesion is the stud pull test. In the stud pull test a metal pin is attached to the film surface with an epoxy adhesive. The stud is then attached to a pull tester and a force is applied normal to the surface of the film until either the film-substrate or film-epoxy bond fail. In practice, one sometimes obtains fractures that are not cleanly at one surface or the other. Even in these cases, however, one obtains values for the adhesion strength that must be equal to or less than the interface of interest, thus resulting in values that are lower limits to the bond strength of the interface.

Although published reports[15,20,21] on the adhesion of 3 to various surfaces such as itself, silicon dioxide, alumina, aluminum and to some extent copper have been reported as excellent, it became evident that adhesion was sensitive to surface preparation. To circumvent potential problems, an adhesion promoter has been developed for BCB.

BCB adhesion promoter is formulated as a spin-on solution, applied immediately prior to the application of the BCB prepolymer, after thoroughly cleaning the substrate surface. The coupling agent is designed to have one end of the molecule rigorously attach to the underlying metal or oxide surface while the other end crosslinks directly into the BCB polymer matrix. The adhesion promoter has proven to provide increased adhesion over a wide variety of surfaces under differing process conditions.

Table I.   Properties of BCB Based Resins.

Property	BCB XU13005	BCB XU130028	BCB XU130028
Glass Transition Temp. (TMA, DMA)	>350° C	>350° C	>350° C
Flexural Modulus	480 kpsi	540 kpsi	747 kpsi
Linear Coefficient of thermal expansion 25° C to 300° C, TMA	65–70 ppm/°C	42 ppm/°C	27 ppm/°C
Water Absorption 24 hour water boil	0.25%	0.52%	0.87%
Weight Loss at 350° C, Nitrogen, 2 hrs, 20 micron film on silicon wafer	2%	1%	0%
Dielectric Constant, 1 MHz (+/− 0.1)	2.7	2.7	2.7
Dissipation Factor 1 MHz	0.0008	0.0006	0.0004

Qualitatively, BCB coated wafers subjected to boiling water for 4 hr show no coating release after repeated tape pulls. Figure 12 shows stud pull data for **3** on A1 and SiO$_2$ with and without adhesion promoter after water exposure for 60 min at 95° C. The error on such data is typically 10–15%.

As all practitioners are aware, there are two metal/polymer interfaces; polymer deposited on metal and metal deposited on polymer. The latter interface has been notorious in its poor adhesion. It has been found that surface roughening of the BCB, by argon gas ion sputtering or oxygen plasma treatment, prior to deposition enhances the adhesion of deposited metal on the BCB. The adhesion of sputtered Al to BCB's **3** and **4** are shown in Table II. The surfaces of the polymer have been roughened by Ar sputtering in the metallization equipment just prior to metal deposition. Using this rougening technique, stud pulls in the 10,000 psi range are typically obtained (+/− 10%). As mentioned above, this sets a lower limit to the bond strength of the interface. This surface roughening is particularly important when

evaporating metal onto the BCB surface, since this technique offers significantly less inherent surface roughening than the sputtering technique.[22]

Polymer to polymer adhesion is also essential to fabricate a reliable structure. A "soft cure" procedure has been developed that increases BCB layer to layer adhesion. The cure schedule can be adjusted to leave from 50–95% of the BCB groups unreacted depending on subsequent process requirements. These unreacted groups crosslink into the subsequent layer to form a monolithic coating. Significant layer to layer adhesion is thus obtained.

PLANARIZATION

Among the many properties that make polymeric insulators attractive in MCM fabrication is their ability to planarize topographical conductor features. Relief existing on a surface prior to performing photolithography will effect the resolution and quality of the operation. Since the definition obtainable in photoresists is dependent on the UV exposure that the resist sees, thickness variations can compromise this procedure. Well planarized sur-

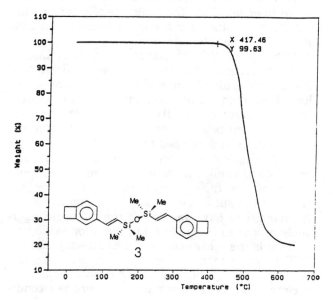

Fig. 10 — TGA of **3** (ramp rate 10° C/min).

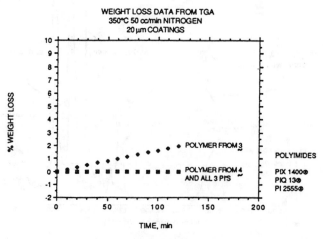

Fig. 11 — Isothermal stability of **3** and **4** compared to typical PI's at 350° C.

BCB Coating Adherence on Aluminum

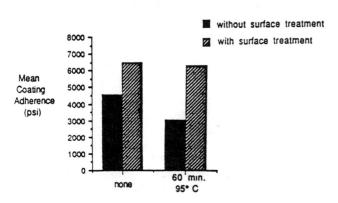

Mean Coating Adherence (psi)

■ without surface treatment
▨ with surface treatment

Water Exposure Prior to Testing

BCB Coating Adherence on Silicon Oxide

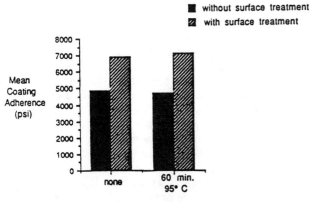

Mean Coating Adherence (psi)

■ without surface treatment
▨ with surface treatment

Water Exposure Prior to Testing

Fig. 12 — BCB **3** coating adhesion on Al and SiO_2.

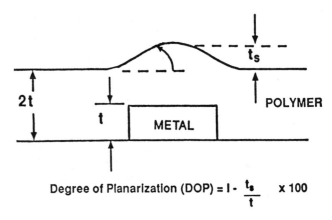

Degree of Planarization (DOP) = $1 - \dfrac{t_s}{t} \times 100$

Fig. 13 — Planarization of metal feature.

faces are also electrically and mechanically more stable due to fewer discontinuities.[15]

The degree of planarization (DOP) is defined as the step height resulting after dielectric deposition and cure vs the initial step height of the conductor pattern as shown in Fig. 13. This process has been studied in detail for polyimides[23] where it has been shown that the DOP of fully cured films is the composite of two processes, drying and curing. During the solvent loss process (drying) the film stays planar until a critical point after which the film shrinks due to solvent loss while the polymer can no longer flow leading to loss in perfect planarity. Further loss of planarity occurs during the curing process which

Table II. Adhesion of Sputtered Al to BCB.

	13005 (**3**)	31028 (**4**)
# of Determinations	9	10
Average Adherence (psi)	9,800	10,240
% Standard Deviation	12	18

Surface Preparation:
BCB Soft Cured to 220° C
Ion Gun Treat Prior to Deposit

reduces film thickness by a further ca 25% for PI's.[5,23] Typical DOP for commercial PI dielectrics ranges between 18 and 30%.[24]

BCB coatings exhibit outstanding topographical leveling. Table III details the planarization obtained over a 50 μm isolated line for spin-coated **3**. Figure 14 shows a surface profile of a 25 μm wide 2.4 μm thick Al conductor before and after spin coating with 7.5 μm of **5**. The DOP is 95%. One should recall that DOP is independent of dielectric thickness after a polymer to conductor thickness ratio of 2–3/1 is exceeded, *i.e.* with a 1 μm thick conductor pattern, 2–3 μm or 20 μm of polymer will give the same DOP if both are applied in one coat. The only way to increase DOP for a given polymer system (% solids and Mw) is to apply multiple coatings. Figure 15 shows the planarization obtained while building a 4 layer module with BCB **3**.[25]

ETCHING

BCB's (**3**, **4** and **5**) are highly resistant to wet chemical etching. They are however, dry (plasma) etchable. **3**, and **5**, due to their Si content, cannot be etched in a pure O_2 plasma, but rather need mixtures of SF_6/O_2 or CF_4/O_2. Johnson *et al.* have studied this process in detail[21,26] and shown that etch rates >1 μm per min are obtainable using either gas mixture in parallel plate or barrel reactor configurations. The etch rate for **3** as a function of SF_6/O_2 gas ratio is shown in Fig. 16. BCB **4** contains no Si, and therefore is etchable in a straight O_2 plasma.[22]

Defining features in BCBs **3** and **5** necessitates the use of a hard metal (usually Al) mask since the fluorine containing etch gas mixtures are not compatible with the use of typical photoresist or SiO_2 masks. When Al is used as the conductor material, the masking Al cannot be chemically stripped away without etching the Al conductor exposed at the bottom of the vias. A lift-off process using sandwiching layers of photoresist has been developed to alleviate this potential problem.[21] In our laboratories a copper hard mask has been used for the same purpose. In a similar fashion lift-off techniques were used to pattern **3** with a Al mask when Cr/Cu/Cr

Table III.   Planarization of BCB 3 Applied to Spin and Spray Coating.

BCB	Appl Tech	Conductor, t (μm)*	BCB (μm)	BCB/Line	ts (μm)	DOP
3	Spin	4.5	10	2.22	0.44	90.3

*50 μm Wide isolated line

metallurgy was used.[18] In our laboratories, a copper hard mask has been used for the same purpose. Some practitioners have noted that a 20% over etch or a 1:1 HF/H_2O rinse is necessary to completely remove any SiO_2 residue that could interfere with further processing.[22]

WATER ABSORPTION

Moisture absorption impacts both the electrical properties[5,27] and processability[5,28] of potential MCM dielectrics.

Figure 17 shows the moisture absorption of 3 vs relative humidity.[29] This <5% variation of with humidity results in a propagation delay variation of <2.2%. Figure 18 shows the effect of this water absorption on propagation delay and characteristic impedance (for the structure described). Moisture absorption of 4 and 5 are given in Table I.

Absorbed moisture can also cause severe damage to modules during fabrication.[28] Water vapor that has been absorbed during the storage or fabrication of MCMs will be driven off during processing steps that occur at temperatures >100° C. Delamination or blistering of layers will occur if the moisture is released more rapidly than it can diffuse out. This "steam" has also been known to adversely affect adhesion at adjacent interfaces. For all of the above processing reasons, numerous extended bakeout cycles are routinely employed during the fabrication of MCMs using polymer dielectrics with significant moisture absorptions. Such bakeouts are not necessary when fabricating MCM's with BCB 3 or 5.[15,30]

STRESS

Film stresses impact both the fabrication and long term reliability of MCM. Since the CTE of polymers are typically greater (25 to 65 ppm) than either of the prevalent substrate materials (silicon ca 2.5 or alumina ca 7) or metallizations (Al 22 or Cu 17), the stresses that are generated are tensile. Film stress develops during cure (film shrinkage) and as the polymer cools down after the modules highest temperature exposure. As discussed previously, for PIs the highest temperature exposure is the inherent cure temperature of the polymer, while for BCB it is the highest temperature of processing, either brazing, soldering or Al alloy anneal. Stress is also generated in sputtered metal films. It has been noted that process control is needed in both these areas to control the overall stresses in a structure.[15,31]

The tensile stresses that are generated in polymers during fabrication can result in substrate warpage especially when using thin (0.5 mm) silicon substrates.[20] Warpage causes problems during photolithography since the vacuum fixtures cannot hold the wafers down flat enough to allow sharp focusing over the entire wafer diameter during photolith-

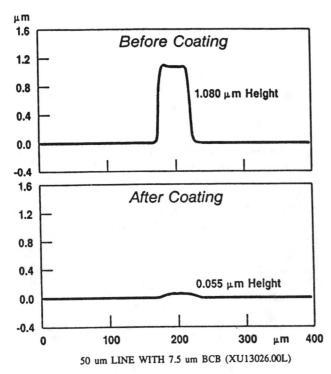

50 um LINE WITH 7.5 um BCB (XU13026.00L)

Fig. 14 — Typical BCB planarization.

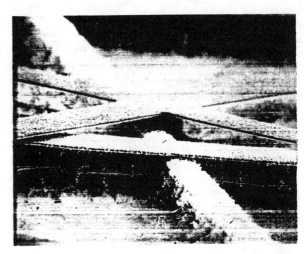

Fig. 15 — BCB planarization on a 4 metal layer structure (BCB etched away).

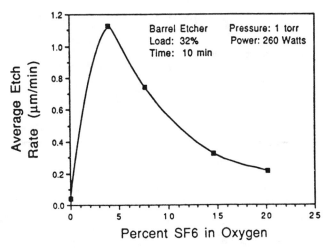

Fig. 16 — Etch rate of BCB 3 as a function of SF_6/O_2 ratio.

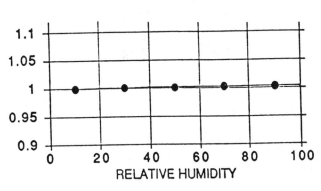

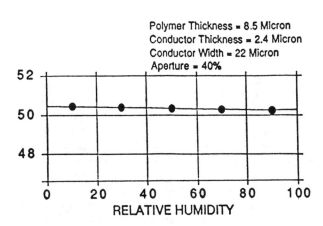

Fig. 18 — The effect of relative humidity on characteristic impedance.

ography. Problems will also be encountered in automated cassette to cassette wafer handling equipment, due to wafer sticking, if the bow becomes too severe. Warpage problems increase with the diameter of the substrate and the total number of dielectric layers. Thicker silicon substrates and ceramic substrates are much more resistant to such bowing.

Stress is also a reliability issue since excessive stress buildup can result in delamination or cracking of the dielectric layers. As it pertains to interface reliability, film stress must be studied in conjunction with the modulus and tensile strength of the polymer and the film adhesion at the specific interface in question.[15,26,32] Thermal stress, in a film, is defined as the product of the films biaxial modulus and the CTE difference between the film and the substrate over a given temperature range as shown below in Eq. 1.[33]

$$\Delta \sigma_f = \frac{E_f}{(1 - v_f)} (\alpha_s - \alpha_f) \Delta T \qquad (1)$$

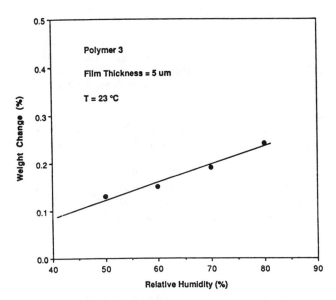

Fig. 17 — Water absorption vs relative humidity for BCB 3.

where:

$$\sigma_f = \text{film stress}$$
$$E_f(1 - v_f) = \text{biaxial modulus of film}$$
$$\alpha_s = \text{substrate CTE}$$
$$\alpha_f = \text{film CTE}$$
$$T = \text{temperature}$$

Film stresses are generally biaxial in the plane of the substrate, except near the film edges where the biaxial stresses are supported by shear stresses at the film substrate interface. It has been shown that the shear stresses between two layers is proportional not only to the CTE of the two interface materials and the temperature excursion they have seen, but also to the modulus of the materials and inversely proportional to the thickness of the layers.[34]

Lower modulus translates into lower shear stress. A thicker layer translates into lower shear stress since the thicker layer can be more "compliant" to such shear forces. It is important to note that film stress is independent of film thickness but shear stress at the interfaces is dependent on film thickness. Since interface shear stress is localized at the film edge, or more importantly, corners or edges of patterned features, this is where delamination is first observed. If the adhesive strength of the film to substrate interface is greater than the shear stress, delamination will not occur. Thus CTE, modulus, adhesion and film thickness all play important roles in determining the stability of any thin film interface.

Figure 19 shows the measured stress for a film of 3. This stress is much lower than that normally obtained from Al thik films.[35] We have not observed problems with BCB cracking or delamination on 4 layer parts with total dielectric thicknesses up to 35 μm *i.e.* 8.5 μm per layer.

TEST STRUCTURES FOR PERFORMANCE EVALUATION

Although the properties of polymer dielectrics can be measured on free standing films and the processing of these dielectrics can be observed by studying the individual process steps, it is even more important to assess the reliability of actual thin film circuits. A three level metal test structure fabricated using 5, to examine the performance of BCB polymers in thin film circuits.[30] The test pattern contains capacitors, via arrays, via chains, metal meanders and line and space patterns. A reliability study was performed on the 25 μm, 100 via "daisy chain" pattern by cycling 100 times between 30 and 140° C. The resistance of each stitch-through was measured before and after thermal cycling (cycling to a lower temperature was precluded by equipment on hand) Table IV summarizes the results. The stability of the resistance measurements indicates good continuity in the interconnect structure. More stringent high temperature storage, thermal shock, TBH (temperature/humidity under bias) and PCT (pressure and temperature) tests are underway in order to gather longer term reliability data.

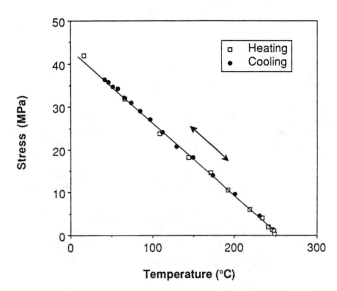

Fig. 19 — The effect of relative humidity on signal propagation delay.

Table IV. Resistance of Vias after Thermal Cycling of BCB 5 Structures.

Stitch Through Thermal Cycling 30 to 140° C, 5 Min Hold, 5 Samples Resistance Through Structures

Feature Size, μm	Zero Cycles, Ω	100 Cycles, Ω
25	47.7	48.0
50	15.3	15.2

EXPERIMENTAL SECTION

Wafer Pretreatment

Wafers were baked for 30 min at 200° C to remove any water absorbed in the thermal oxide layer.

Spin Coating Procedure

5 ml of BCB adhesion promoter are poured onto the wafer, spread at 500 rpm for 2 sec and spun at 5000 rpm for 15 sec on a Solitec 5100 spin coater. BCB (5 ml) is then poured onto the wafer and spread (2 sec at 500 rpm and spun (30 sec at 5000 rpm or 7000 rpm) to get 5.5 or 4.5 μm coatings respectively.

Curing of the Polymer

Curing was carried out in a Blue M oven (DCC-146) equipped with HEPA filtration and 100 scfh nitrogen purge. Oxygen content is monitored and kept below 100 ppm (<0.1%). Time/temperature plots for the soft cure and hard cure procedures are shown below in Fig. 20.

Polymer Etching

BCB etching was performed in a LFE 301C Barrell Plasma Etcher using 95sccm oxygen and 15 sccm SF6 at a power setting of 260 W and a pressure setting of approximately 1 Torr.

Aluminum Deposition and Etching

Metallization was performed in a Leybold-Heraeus L560 box coater equipped with dc magnetron sputtering in argon and an Ar ion gun.

The BCB surface is cleaned and roughened prior to metal deposition by ion gunning at 10 sccm argon, 40 mA, 500V, 0.0005 mbar for 1 min.

Al/1% Cu alloy is deposited at 229 sccm Ar at a pressure of 0.002 mbar and 1500 W. 2 μm of metal are deposited in 60 min.

Aluminum etching is performed using standard wet etch techniques in a phosphoric acid etchant comprised of: water (51%), phosphoric acid (41.8%), acetic acid (4.8%) and nitric acid (2.9%). 2 μm of Al are etched at 45° C in 13.5 min.

CONCLUSIONS

MCM technology has evolved to meet the needs of higher density interconnects. Although there are

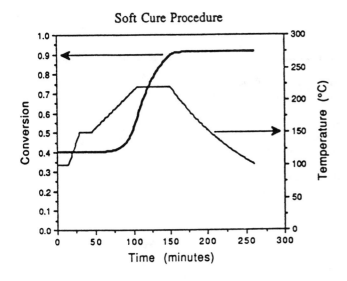

Soft Cure Procedure

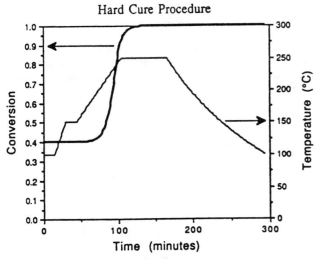

Hard Cure Procedure

Fig. 20 — Soft and hard cure schedules for BCB 3.

many modifications of thin film, MCM processing, it is becoming common practice to use polymeric insulators to separate the conductor traces. BCB's can be used to fabricate such thin insulator layers. They are of interest due to their low dielectric constant, low moisture uptake, good thermal stability and adhesion and the high degree of planarization that they offer.

ACKNOWLEDGMENTS

The authors wish to thank: J. Gilpin, A. Kennedy, T. Manial, C. Fulks, B. Pelon, S. Hahn and R. McGee for their valuable contributions to this work.

REFERENCES

1. C. Neubauer, R. O. Carlson, F. A. Fillion, T. R. Haller, Proc IEPS 1988, p. 163.
2. K. Shambrook and P. Trask, Proc. 39th ECC (1989), p. 656.
3. J. Balde, ASM Electron. Mater. Handbook, Vol. I, 1989, p. 297.
4. C. S. Ho, D. Chance, C. Bajoren and R. Acosta, IBM J. Res. and Dev. *36*, 286 (1982).
5. R. Jensen and H. Vora, IEEE Trans. CHMT, 1984, p. 384; R. Jensen, R. Douglas, J. Smeby and T. Moravec, Proc. VHSIC Pkging Conf., Houston, TX (1987), p. 193.
6. C. Bartlett, J. Segelken and T. Teneketges, IEEE ECC, (1987), p. 518.
7. R. P. Himmell and J. J. Licari, Proc. ISHM, 1989, p. 462.
8. T. Yamada, K. Otsuka, K. Okutani and K. Sahara, Proc. 5th IEPS, (1985), p. 551.
9. T. Watari and H. Murano, Proc. 35th ECC, (1985), p. 192.
10. J. Bailey, Proc. NEPCON West, 1989, 1308.
11. J. Hagge, Proc. 38th ECC, (1988), p. 282.
12. J. Hagge, Proc. NEPCON West, (1980), p. 271.
13. G. Messner, Proc. ISHM, (1988), p. 28.
14. M. Sage, Proc. ISHM, (1988), p. 1.
15. J. Reche, Proc. 3rd Int. SAMPE Elect. Conf., (1989), p. 948.
16. R. Kirchoff, USP 4,540,763 (1985).
17. D. Arthur, Rogers Corp., personal communication.
18. T. Tessier, G. Adema, S. Bobbio and I. Turlik, Proc. 3rd Int. SAMPE Elec. conf., (1989), p. 85.
19. L. Maissel and R. Glang eds., "Handbook of Thin Film Technology," McGraw Hill, (1970); J. Smith and D. Hensen Solid State Tech., (Nov. 1986), p. 135.
20. T. Tessier, G. Adema and I. Turlik, Proc. 39th ECC, (1989), p. 127.
21. R. W. Johnson, T. Phillips, S. Hahn, D. Burdeaux and P. Townsend, Proc. ISHM, (1988), p. 365.
22. M. Berry, T. Tessier, I. Turlik, G. Adema, D. Burdeaux, J. Carr and P. Garrou, Proc. 40th ECC, (1990), p. 746.
23. D. R. Day, D. Ridley, J. Mario, S. D. Senturia, Polyimides II, NY, Plenum Press (1984), p. 767.
24. B. Merriman, J. Craig, A. Nader, D. Goff, M. Pottinger and W. Lautenberger, Proc. 39th ECC, (1989), p. 5.
25. Photo courtesy of J. Reche, Polycon Corporation.
26. P. Townsend, S. Hahn, D. Burdeaux, M. Thomsen, R. McGee, J. Gilpin, J. Carr, R. W. Johnson and T. Phillips, Proc. 3rd Int. SAMPE Elec. Conf., (1989), p. 67.
27. D. Denton, D. Day, D. Priore, S. Senturia, E. Anolich and D. Schneider, J. Electron Mater. *14*, 119 (1985).
28. T. Tessier, I. Turlik, G. Adema, D. Sivan, E. Yung and M. Berry, Proc. 9th IEPS, (1989), p. 294.
29. D. Denton, Univ. Wisconsin, private communication.
30. D. Burdeaux, P. Townsend, S. Hahn, M. Thomsen, J. Gilpin, R. McGee and J. Carr, Proc. NEPCON (West), (1989), p. 937.
31. J. Reche, Proc. NEPCON (East), 1989, p. 1002.
32. R. Tummala and J. Rymaszewski eds., "Microelectronics Packaging Handbook," Van Nostrad Reinhold, New York, (1989).
33. P. Townsend, D. Barnett and T. Brunner, J. Appl. Phys. *62*, 4438 (1987).
34. W. Chen and C. Nelson, IBM J. Res. and Dev. *45*, 179 (1979).
35. A. Sinha and T. Sheng, Thin Solid Films *48*, 117 (1978).

NEW LOW COEFFICIENT OF THERMAL EXPANSION POLYIMIDE FOR INORGANIC SUBSTRATES

Burt T. Merriman, John D. Craig, Allan E. Nader,
David L. Goff, Michael T. Pottiger, and William J. Lautenberger

E. I. du Pont de Nemours & Company, Inc.
Electronics Department
Experimental Station Laboratory
P.O. Box 80334
Wilmington, DE 19880-0334

Abstract

Recent interest in using polyimides for high density interconnect applications (HDI) in which multilayer structures are formed has emphasized the need for polyimides which have matched coefficient of thermal expansion (CTE) with the underlying substrate. A polyimide chemistry has been identified which allows for a significant lowering of the CTE and subsequent reduction of stress on silicon and ceramic substrates while giving improved dielectric and water absorption properties. Both characterization of the materials properties and processing parameters will be discussed and compared with currently used materials.

Introduction

Recent advances in speed and density of integrated circuits used in high performance systems have created an increasing challenge for higher density packaging to ensure reduced interconnection delays. High density interconnects (HDI) offer a solution to this problem.

Typical HDI structures rely on thin film conductors such as copper or aluminum for signal, power, and ground layers. Polyimides are the preferred interlayer dielectric because of their low dielectric constant and processing flexibility. Underlying HDI substrates are typically silicon, ceramic, and co-fired multilayer ceramic.

Most commercially-available, aromatic polyimide chemistries have CTEs in the range of 20 to 40 ppm with dielectric constants in the range of 3.0 to 4.0 and absorb between 2.5-4.0% by weight moisture. When constructing multilayer structures on silicon or ceramic, in order to minimize stress a CTE between 3-5 ppm would be advantageous. The amount of water absorption will also effect the dielectric constant of a particular polyimide. For example, for Du Pont Kapton[R] polyimide film, the dielectric constant can vary from 3.1 to 3.8 depending upon the relative humidity. Moisture absorption will also effect the ease of processing during device fabrication.

New generation polyimides are targeted at providing a lower dielectric constant, lower stress, better CTE match with the substrate, and superior film properties. The use of different and improved monomers has yielded polyimides which help achieve these goals. One such polyimide is the biphenyl dianhydride (BPDA) and paraphenylene diamine (PPD) back-bone chemistry. This polyimide has been characterized and compared with Type I based on pyromellitic dianhydride (PMDA) and oxydianiline (ODA) and Type III based on benzophenone tetracarboxylic acid dianhydride (BDTA), metapheylenediame (MPD) and oxydianiline (ODA) and shows significant improvement in a number of properties including lower stress and thermal coefficient of expansion. This basic chemistry was first disclosed in patents filed by Du Pont in 1961 [1,2].

Experimental

Materials

Polyamic Preparation: Molecular structures of the aromatic diamine and dianhydride monomers used to prepare the polymers are shown in Figure 1. Pyromellitic dianhydride (PMDA), 3,3', 4,4'-benzophenone tetracarboxylic dianhydride (BTDA), biphenyl dianhydride (BPDA), 4,4'-oxydianiline (ODA), metaphenylenediamine (MPD) and paraphenylenediamine (PPD) were obtained from commercial sources.

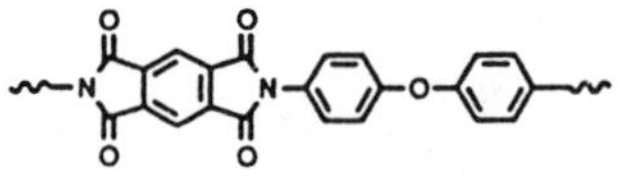

Type I - PMDA/ODA

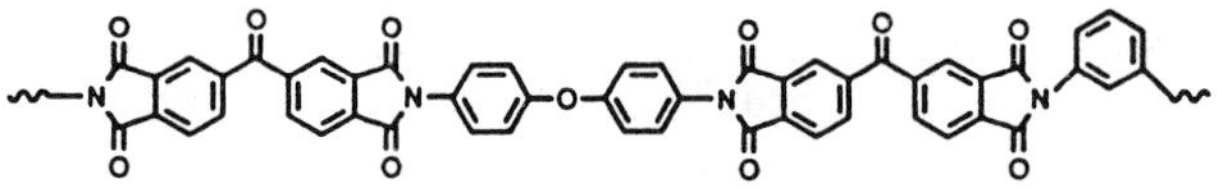

Type III - BTDA/ODA/MPD

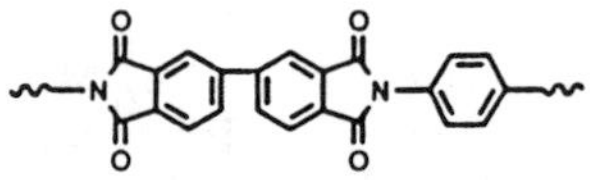

Type V - BPDA/PPD

Fig. 1 Polyimide Chemical Structures

Polyamic acid precursor solutions were prepared by mixing equimolar portions of diamine and dianhydride at room temperature in N-methyl-2 pyrrolidone (NMP) for 8-24 hours. One to two mil thick polyimide films were prepared by casting the polyamic acid solutions onto glass slides using a draw-down technique. Unless otherwise noted, the polyamic acid films were converted to polyimide by heating at 135°C for 15 minutes, 300°C for 30 minutes, and 400°C for 10 minutes in a convection (air) oven. The films were removed from the glass slides by placing in boiling water and then air dried.

Experimental Methods

Thermal Analysis and Mechanical Property Measurments:
Isothermal and dynamic thermogravimetry (TGA) studies were done with a Du Pont 990 Thermal Analyzer using a 951 TGA module. To determine the onset of rapid decomposition, a heating rate of 10°C/min. was used

Reprinted from *Proc. 39th Electron. Components Conf. (ECC)*, pp. 155–159, 1989.

up to 350°C where the 20 mg sample was held for 30 minutes and then a continued heating of 10°C/min. to the decomposition point. Weight loss at 500°C was done by taking a sample cured at 350°C and ramping 20°C/min. to 500°C and holding for two hours in air. Glass transition temperatures were determined by DSC using a 910 DSC module. Thermal coefficient of expansion measurements in the x-y plane were obtained with 2x 20 mm strips of cured 2 mil films which were secured in the clip jaws of a TMA 7 Perkin-Elmer Thermomechanical Analyzer. A 30 mN tension was maintained on the film, and a conditioning program run from 0-350°C at 30°C/min. was made. The sample was cooled to ambient conditions, and the run of record performed at 10°C/min. The direction of film draw was both the x and y direction. The instrument was calibrated with Pt foil (9 ppm/°C) and an expansion correction applied for the metal clips.

Stress-strain measurements were measured on an Instron tensile tester at a strain rate of 2 inches/min.

Electrical Property Measurements: Dielectric property measurements were made by vacuum depositing 500 $\mathring{A}$ thick gold electrodes on a 1 mil thick cured film. The film was sandwiched between brass point contacts well insulated by Teflon film and placed in an environmental chamber. An HP 4275A multi-frequency LCR meter was used to measure the capacitance of the film between the frequency range of 10 KHz to 10 MHz from which the dielectric constant was determined.

Stress Measurements: An Ionic Systems Stressgauge II was employed to determine wafer stress. Four inch silicon production wafers (Monsanto) with 1,1,1 orientation, were selected having initial (uncoated) readings of 1700 - 2300 units. These wafers were further segregated in lots within +/- 50 units. Sufficient wafers were selected to provide five wafers for each coating evaluated. Wafer thickness was determined with a digital micrometer. All operations were performed in a class 100 clean room.

Wafers were spin coated with resin and cured at 55°C for 90 minutes, 150°C for 30 minutes, then programmed to 300°C at 2°C/min., and after 20 minutes at 300°C, heated to 450°C at 2°C/min. and maintained at 450°C for 60 minutes in a convection oven with an air atmosphere. Wafers were allowed to cool to room temperature overnight. Two micron cured films were generally employed.

Stress measurements were made once each day over five days while maintaining the wafers in a constant temperature and humidity environment. Film thickness was determined with a Tencor instrument. The measurements were used to compute wafer tensile stress with the Ionicstress Gauge II program, which computes stress as a function of wafer deflection vs. the uncoated substrate. Stress values generally declined over 2-3 days to a plateau indicative of the relaxed film. An average of five wafers was employed for each coating examined. Stress σ was calculated using the equation:

$$\sigma = \pm \frac{1}{6} E \frac{h_s^2}{h_p} \frac{1}{R} \tag{1}$$

where E is the generalized Young's modulus for Si, and h_s and h_p are the thickness of Si and the polymer film respectively, and R is the radius of curvature.

Planarity Measurements: Three-inch silicon wafers with one micron of aluminum were patterned photolithographically and processed to make planarity templates with repeated patterns of 10 micron lines/10 micron spaces and 20 micron/20 micron spaces. Solutions of the polyamic acids were spin-coated and cured to give a final film thickness of 2 μm. The profiles of the aluminum templates were measured before and after coating with an Alpha Step profilometer. Planarity was measured as defined by the equation:

$$\% \text{ Planarity} = [1 - \frac{t_s}{t_m}] \times 100 \tag{2}$$

where t_s is the resulting feature height and t_m the original feature height.

Moisture Absorption Measurements: Moisture absorption measurements were conducted using a Du Pont 1090 Thermal Analyzer. Ten mg samples of cured film were placed in a platinum sample holder, heated to 120°C until dried, allowed to cool to 35°C under dry nitrogen, and then exposed to 25 to 65% RH. Moisture uptake was then monitored and recorded after one hour exposure.

Adhesion Measurements: Adhesion measurements were done on spin-coated and cured, three inch silicon wafers using the ASTM D-3359-83 Method B standard boiling water tape pull test.

Processing Conditions

Wafer Preparation: The wafers were cleaned prior to use by washing them in Freon[R] TMS and then scrubbing them on a Silicon Valley Group Model 8620SSC wafer scrubber. The wafers were then baked at 300°C for 30 minutes to remove any water.

Adhesion Promoter Application: An organosilane adhesion promoter (VM-651) was used to enhance the adhesion of the polyamic acid coatings to wafers. A 0.5 wt% solution of VM-651 in deconized water was prepared and puddled onto the wafer and spun at 5000 rpm for 30 seconds to air dry.

Spin Coating: Pinhole free coatings of between 1-12 microns were obtained by spin coating using a dynamic dispense technique. With the wafer spinning at approximately 500 rpm, a sufficient volume of material was dispensed into the center of the wafer to cover the wafer surface. The wafer was then accelerated up to the desired spin speed using an acceleration rate of 20,000 rpm/sec. Figure 2 compares the spin speed versus coating thickness curves for the Type V chemistry formulated at 10.5 and 13.5 percent solids in NMP. It was found that a volume of approximately 1.5 cm^3 adequately covered a bare 4 inch (150 mm) silicon wafer. All spin coating operations were performed on a Silicon Valley Group System 86 wafer track line.

	PI-2610	PI-2611
Solvents	NMP	NMP
Solids (%)	10.5	13.5
Viscosity (poise)	25-30	110-135
Density (g/cm^3 @ 25°C)	1.07	1.07
Filtration (μm)	1.0 absolute	1.0 nominal

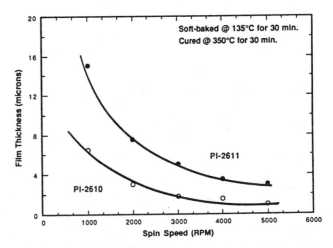

Fig. 2 Coating Thickness Versus Spin Speed for the Type V Chemistry.

413

Soft-Bake: The coated wafers were soft-baked in a forced air convection oven at 135°C.

Cure: The coated wafers were imidized using both a convection (VWR Model 1610D) and a vacuum drying oven (Heraeus Model VT5042 EKP). Curing schedules will depend upon whether a single coat or a multicoat process is required in order to obtain the desired film thickness and planarity.

Plasma Etching: Cured wafers were treated for two minutes in 100% oxygen using a Technics PE-II-A plasma etcher.

Results and Discussion

Comparison of Chemistry Type with Properties

Overall Property Comparisons: Evaluation of the thermal properties of the Type I and III polyimide chemistries compared to Type V shows that the Type V has a higher glass transition temperature, higher decomposition temperature, and less weight loss at 500°C in air for two hours (Table I).

Table 1: Cured Film Properties of Different Chemistries

Thermal Properties	Type I	Type III	Type V
Melting Point (°C)	None	None	None
Glass Transition Temperature (°C)	>400	>320	>400
Decomposition Temperature (°C)	580	560	620
Weight Loss (500°C in air; 2 hours)	3.6	2.9	1.0
Mechanical Properties			
Modulus (kg/mm^2)	300	245	900
Tensile Strength (kg/mm^2)	10.5	13.5	35
Elongation (%)	40	15	25
Electrical Properties			
Dielectric Constant (@ 1 kHz; 50% RH)	3.5	3.3	2.9
Dissipation Factor (@ 1 kHz)	0.002	0.002	0.002
Dielectric Breakdown Field [(V/cm)x10^{-6}]	>2	>2	>2
Volume Resistivity [(Ω/cm)x10^{-16}]	>1	>1	>1
Surface Resistivity (Ωx10^{-15})	>1	>1	>1

Comparison of the mechanical properties shows that the Type V chemistry has higher tensile modulus and tensile strength with % elongation values intermediate with those of the Type I and Type III chemistry.

Evaluation of the electrical properties shows a reduction in dielectric constant for the Type V chemistry. Due to the fact that the Type V chemistry significantly absorbs less water (Fig. 3), a corresponding lowered dependency of dielectric constant with R.H. is observed [3] (Fig. 4). These lower levels of water absorption will also affect the ease of processing multilayers during device fabrication.

Planarity measurements indicate planarity values for Type V lower than Type I and Type III (Table II). Previous studies have shown that wt% solids of the polyamic acid precursor solution plays a dominant role in controlling planarity [4]. Currently the Type V chemistry is available only in lower % solid formulations than Type I and Type III. Higher levels of planarity can be achieved by using multiple coats.

Relationship of Stress and TCE: When building multilayer structures which involve fabrication steps involving heating and cooling cycles, stress will develop at the interfaces because of differences in thermal expansion characteristics of the various layers of material.

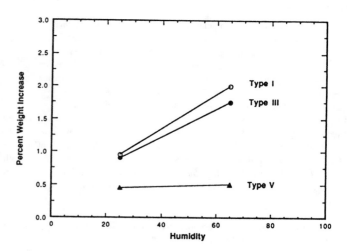

Fig. 3 Moisture Absorption Versus Chemistry Type

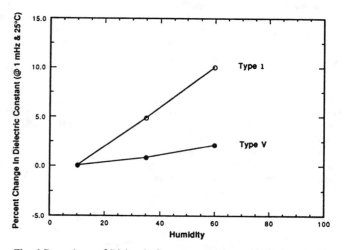

Fig. 4 Dependency of Dielectric Constant on Moisture Absorption Versus Chemistry Type.

Table II: Planarity Values Versus Chemistry Type and Solids Content.

Chemistry	Solids	Planarity*
I	14	21
III	19	30
V	11	18

*20 μm lines and 20 μm spaces. 1 μm high metal lines single coating. 2μm thick

A widely used approach for modeling the stress in polymer films is to treat the polymer as a linear elastic material (time independent). Neglecting the contributions from solvent loss, cure, and physical aging, and assuming that the tensile modulus of the substrate is nearly constant over the temperature range of interest, the stress can be given by

$$\sigma = \int_{T_r}^{T_c} E_f(T)[\alpha_s(T) - \alpha_f(T)]dT \qquad (3)$$

Where $E_f(T)$ is the tensile modulus of the film. $\alpha_s(T)$ and $\alpha_f(T)$ are the coefficient of thermal expansion (CTE) of the substrate and film

414

respectively. T_r is room temperature. and T_c is the final cure temperature. Another commonly used assumption is to assume that the polymer properties are independent of temperature

$$E \neq f(T) \text{ and } \alpha \neq f(T) \qquad (4)$$

Therefore. equation (3) reduces to

$$\sigma = E_f(\alpha_s\text{-}\alpha_f)\int_{T_r}^{T_c} dT = E_f(\alpha_s\text{-}\alpha_f)\Delta T \qquad (5)$$

When a polymer is elongated along an axis. the elongation is accompanied by a contraction along the remaining two axes. The relationship between tensile strain and volume change is defined by the Poisson ratio.

$$\mu \equiv \frac{1}{2}\left[1 - \frac{1}{V}\left(\frac{dV}{de}\right)\right] \qquad (6)$$

Since the tensile modulus is usually measured in a uniaxial direction while the strain occurs in all directions. a correction factor is included

$$\sigma = \frac{E_f}{1-\mu_f}(\alpha_s - \alpha_f)\Delta T \qquad (7)$$

where $\Delta T = T_c - T_r$. Equation (7) implies that if the coefficient of thermal exapnsion of the film matches that of the substrate. the coating will be low in stress regardless of how the film is processed.

Assuming that equation (7) is an adequate model for the stress in a film coated on a substrate. then is should be possible to calculate the CTE of the polymer based on measurements of the tensile modulus and the experimentally determined stress by using the rearranged equation (8)

$$\alpha_f = \alpha_s - \left[\frac{\sigma 1 - \mu_f}{E_f\Delta T}\right] \qquad (8)$$

Poisson's ratios for Type I. Type III. and Type V used were 0.42, 0.28. and 0.50 respectively [5] while the coefficient of thermal expansion of silicon was assumed to be 3.0 ppm/°C. Experimentally determined stress values for the various chemistry types are shown in Table III indicating a significant reduction in stress for the Type V chemistry. The calculated CTE results using equation (8) are summarized in Table IV. Comparison of calculated values with experimentally determined values indicate only fair agreement depending on the source of experimentally determined values [7,8]. These apparent differences may be due to the fact that equation 8 does not take into account residual stress due to factors other than the differences in CTEs between polymer and substrate. In previous studies. some intrinsic stress was claimed for the Type I chemistry while negligible for another polyimide chemistry [6]. Another consideration is the recognition that the CTEs are not independent of temperature as illustrated in Fig. 5 and that a more appropriate stress model should be considered to take this into account.

Table III: Stress Values Versus Chemistry Type.

Chemistry	Stress (dyne/cm^2 x10^8)
I	4.3
III	5.4
V	0.4

Table IV: Calculated Versus Experimentally Determined CTE Values.

Chemistry	Experimentally Determined CTE (ppm/°C)	Calculated CTE (ppm/°C)
I	35. 35*, 22**	22.5
III	40	47.4
V	19. 5*. 3**	3.5

* Reference 8
**Reference 9

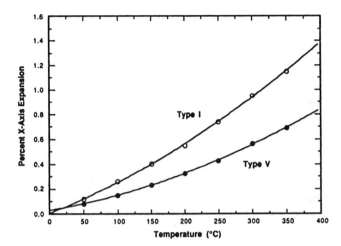

Fig. 5 CTE as a Function of Temperature

Adhesion Measurements: Using the ASTMD-3359-83 Method B test for adhesion. the Type V chemistry was found to have adhesion of silicon comparable to the Type III and better than Type I (Table V). For intercoat adhesion testing. a series of curing conditions for the base and top layer were conducted indicating a wide range of curing conditions available for the type III chemistry where good adhesion was obtained while a narrower processing window exists for the Type I and Type V chemistry (Table VI). For the Type V chemistry. it is recommended that B-stage of the innerlayers be done at 150-175°C with a final cure at 400°C. or alternately each layer be fully cured at 400°C but then be subject to O_2 plasma prior to subsequent coating.

Table V: Adhesion Versus Chemistry Type

Chemistry	Surface	No Boiling Water*	72 Hours in Boiling Water*
I	Si	5	3
III		5	5
V		5	5

*Standard boiling water tape pull test (ASTM D-3359-83 Method B).

5 = Excellent
4 = Very Good
3 = Good - No adhesion loss
2 = Partial Failure
1 = Failure

Table VI: Intercoat Adhesion Versus Chemistry Type.

Material	Cure Conditions			Treatment	Adhesion (Tape Test)
	Base (°C)	Top (°C)	Final (°C)		
Type I	300	300	400	---	Good
	350	350	400	---	20% Fail
	400	400	---	---	60% Fail
Type III	300	300	400	---	Good
	350	350	400	---	Good
	400	400	---	---	Good
Type V	150	150	400	---	Good
	175	175	400	---	Good
	200	200	400	---	Fail
	250	250	400	---	Fail
	300	300	400	---	Fail
	350	350	400	---	Fail
	400	400	---	---	Fail
	400	400	---	O_2 Plasma*	Good

*2 minutes in 100% oxygen

Conclusion

Evaluation of the physical and electrical properties of the Type V chemistry based on biphenyldianhydride (BPDA) and paraphenylene diamine (PPD) have been studied and compared with Type I and Type III chemistries. It has been found that significant reduction in film stress and CTE while lowering of the dielectric constant (2.9) and moisture absorption (<1 wt%) can be achieved. In order to obtain higher levels of planarity, multiple coats of material are needed with special processing required to maximize intercoat adhesion. Overall, these improved properties should make the type V chemistry an excellent candidate for consideration as a dielectric in HDI applications.

References

[1] United States Patent No. 3.179.614, April 20, 1965. Polyimide-Acids. Composition. Thereof and Process for their Preparation. W. M. Edwards.

[2] United States Patent No. 3.179.634, April 20, 1965. Aromatic Polymides and the Process for Preparing them. W. M. Edwards.

[3] A. K. St. Clair, T. L. St. Clair, and W. P. Winfree. "Low Dielectric Polyimides for Electronic Applications." presented at National ACS Meeting. Polymeric Materials for Electronic Packaging Symposium. Sept 25-30. 1988.

[4] B. T. Merriman. "High Planarity Polyimides: I. Synthesis and Applications in VLSI Multilevel Interconnections." presented at Joint Japanese/U.S. Electrochemical Society Meeting. Hawaii. October. 1987.

[5] F. Maseek, S. D. Senturia, "Elastic Properties of Polyimide Thin Films." Proceedings of Third International Conference on Polyimide. Ellenville. NY. November. 1988. pp. 131-134.

[6] P. Goldermans, C. Goldsmith, and F. Bedetti. "Measurement of Stresses Generated During Curing and in Cured Polyimide Films." Polyimides: Synthesis. Characterization, and Applications. vol. 2 pp. 695-711. 1982.

[7] S. Numata and N. Kinjo. "Chemical Structures and Propeties of Low Thermal Expansion Coefficient Polyimides." Polymer Engineering and Science. July 1988. Vol. 28. No. 14. pp. 906-911.

[8] D. C. Hofer. "Organic Packaging Materials. A Customer's Viewpoint." presented at National Research Council Meeting. January 7. 1988.

THE APPLICATION OF PHOTOSENSITIVE POLYIMIDE DIELECTRICS IN THIN-FILM MULTILAYER HYBRID CIRCUIT STRUCTURES

Guy V. Clatterbaugh and Harry K. Charles, Jr.

The Johns Hopkins University
Applied Physics Laboratory
Johns Hopkins Road
Laurel, MD 20707

ABSTRACT

The requirements that circuit wiring be dense and that dielectric permittivity be low to accommodate the high speed and high density of VLSI device technologies has created much interest in polyimide as an interlayer dielectric for multilevel hybrid circuits. Light sensitive polyimide precursors are now being investigated as an alternative to conventional precursors due to the reduced number of steps (i.e., processing time) required to form the complex interlayer via connections. The development of a direct photoimaging process methodology offers some unique challenges not encountered when using the conventional photoresist methods for etching via-holes in polyimide films. These include precise film exposure and development to optimize via-hole geometries and high-temperature curing resulting in extreme film shrinkage and residual stress. Two commercially available light-sensitive polyimide precursors were evaluated for use in thin-film multilayer circuits on ceramic. A three conductor level test circuit was designed and fabricated on Al_2O_3 which included multiple via structures, interdigitated capacitors, and transmission line structure such as microstrip, buried microstrip and coplanar waveguide. Mechanical tests were performed to determine both metal and polymer adhesion. Various aspects of processing are discussed including the response surface methodology approach to process development. Optimum cure schedules were determined through the use of differential scanning calorimetry (DSC) and thermogravimetric analysis (TGA).

INTRODUCTION

The advancement of microelectronics toward more complex and faster device technologies has resulted in a revolution in packaging. Todays packaging technologies must accommodate the high I/O capabilities as well as the increases in operational speed of todays VLSI devices. The printed circuit boards and the thick-film hybrids of yesterday are no longer adequate to meet the extremely dense wiring requirements of the state-of-the-art silicon devices. The multilayer thin-film hybrid with line geometries of one mil (on 99.6% alumina) or less (on silicon) is required to meet the challenge of VLSI technology and fully realize its potential. The thin-film multilayer hybrid permits the fabrication of resistors and ground planes without the problems of congestion associated with traditional thin-film hybrids containing crossovers.

Polyimide has recently been examined as an interlayer dielectric material for fabricating thin-film multilayered hybrid structures. The principle features of polyimide include, a low dielectric constant (e_r = 3.5), a smooth surface for the fabrication of narrow line widths, and resistance to high temperatures and solvents. Also, polyimides are compatible with existing thin-film processing methods and offer good planarization or leveling characteristics necessary for fine-line multilevel structures.

Using polyimide as an interlayer dielectric for multilevel circuits necessitates the fabrication of vias in the dielectric to connect different conductor layers. Fabricating via-holes in traditional non-photoimagable polyimides involved either wet or dry etching techniques. Wet etching is performed on either fully or partially cured polyimide using standard photoresist processing while dry etching is done on fully cured polyimide using either plasma etching, reactive ion etching (RIE) or reactive ion milling (RIM). The disadvantages of these traditional methods include; 1) poor resolution (wet chemical), and 2) process complexity and equipment cost (plasma, RIE, and RIM).

In photosensitive polyimide precursors, via-holes can be patterned directly by applying the precursor, exposing the substrate to UV light , developing away the unexposed portion of the precursor film followed by a high temperature cure. Until recently [1,2], their has been little information in the literature concerning the processing of thick layers of photosensitive polyimide precursors. Product vendors provide some information in making thin films for use as passivation and protection layers and alpha-barriers for integrated circuits and memory devices. This information is pertinent to circuit applications involving silicon substrates where only thin films are required but not relevant to applications involving Al_2O_3 (alumina) or AlN (aluminum nitride) where thicker films are necessary.

There is even less information concerning the subsequent processing of multiple dielectric layers.

This study addresses the use of photosensitive polyimide precursors for thin-film multilevel hybrid applications when thick dielectric interlayers are required. Two commercially available photosensitive polyimide formulations were investigated; Selectilux HTR 3-200 of Merck Electronic Chemicals distributed by EM Industries, Inc. and Probimide 348 manufactured by the Ciba-Geigy Corporation. The substrate material used in this study was the 99.6% Al_2O_3 typically used for thin-film hybrid applications. The particular topics discussed in this article include; the basic chemistry of the photosensitive polymers investigated, optimization of spin coating parameters, drying and curing of thick polyimide films, fabrication of reliable vias, and circuit parameters and design guidelines based on processing and reliability constraints. Results from thermal analysis methods such as TGA (thermogravimetric analysis) and DSC (differential scanning calorimetry) are also discussed. These results are then applied to the problem of determining an optimal cure schedule. The response surface methodology is discussed with respect to optimizing process parameters. This method is used to investigate the effect of spin coating parameters on film thickness and uniformity.

POLYIMIDE CHEMISTRY AND PROPERTIES

Chemical Properties

Developing a process for fabricating thin-film circuits using photosensitive polyimides requires a brief understanding of the basic chemistry involved. In non-photoimagable polyimides, polyamic acid , the polyimide precursor, is formed by a polycondensation reaction of an acid dianhydride with a diamine base as shown in Figure 1. Both Selectilux HTR 3 and Probimide 348 use the dianhydride, pyromellitic acid (PMDA) and the diamine, oxydianiline (ODA) to form the polyamic acid. These are the same constituent materials used in the manufacture of Dupont's KAPTON film. The polyamic acid is readily soluble in a polar organic solvent such as n-methyl pyrrolidone (NMP). The polyimide is converted from the polyamic acid at temperatures high enough to remove most of the NMP and initiate ring closure as illustrated in Figure 1. The ring closure or imidization reaction results in the release of two molecules of water per every polymer repeat unit.

In both the HTR 3-200 and the Probimide 348, the photosensitized polyamic acid contains photoreactive groups which may then lead to photocross-linking with adjacent polymer chains upon exposure to ultraviolet light. This leads to a differential solubility with respect to the exposed and unexposed portions of the film and thus can be developed much the same as a negative acting

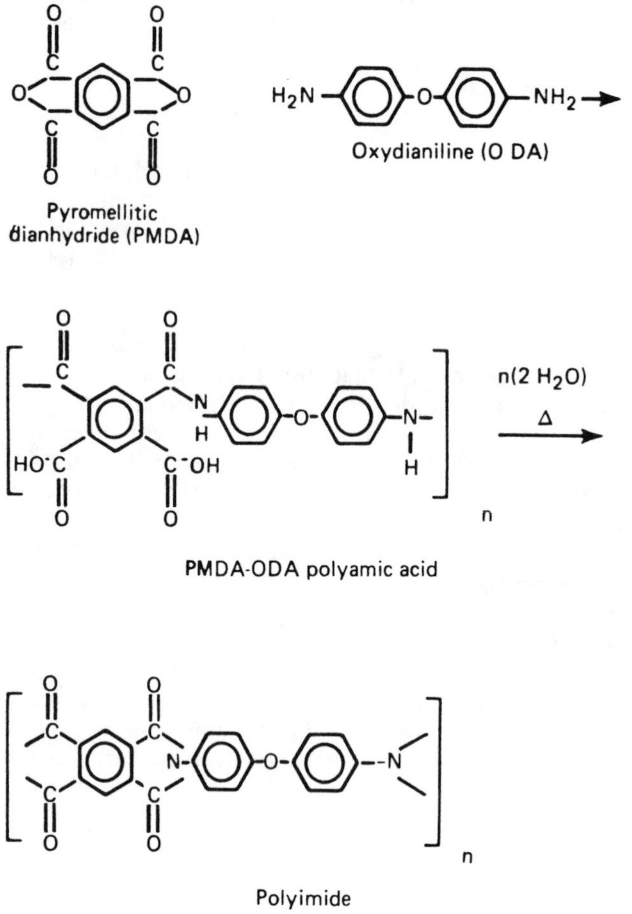

Figure 1. The chemistry of PMDA-ODA polyimide

photoresist. The chemical reactions of the photosensitive polyimides are shown in Figure 2.

The molecular weight of PMDA-ODA polyamic acids is known to increase as the weight percent solids (PMDA dianhydride) in NMP is reduced. This is accompanied with a reduction in the viscosity of the precursor. The viscosity of the Probimide 348 precursor was quoted by the manufacturer to be 4800 Centapoise [3] and the molecular weight approximately 10,000. The Selectilux HTR 3-200 is advertised to be 2350 centapoise [4] with a higher molecular weight near 30,000. The lower viscosity for the HTR 3-200 indicates the presence of a higher percentage of NMP solvent which will thus result in a significant reduction in film thickness upon high temperature cure. The manufacture reports an approximate film shrinkage due to solvent and water loss of about 57 percent. Ciba-Geigy reports a 33 percent reduction in film thickness when fully cured.

The photospeeds are approximately the same for both materials, however the UV wavelength at which the photoreactive groups are most sensitive is

418

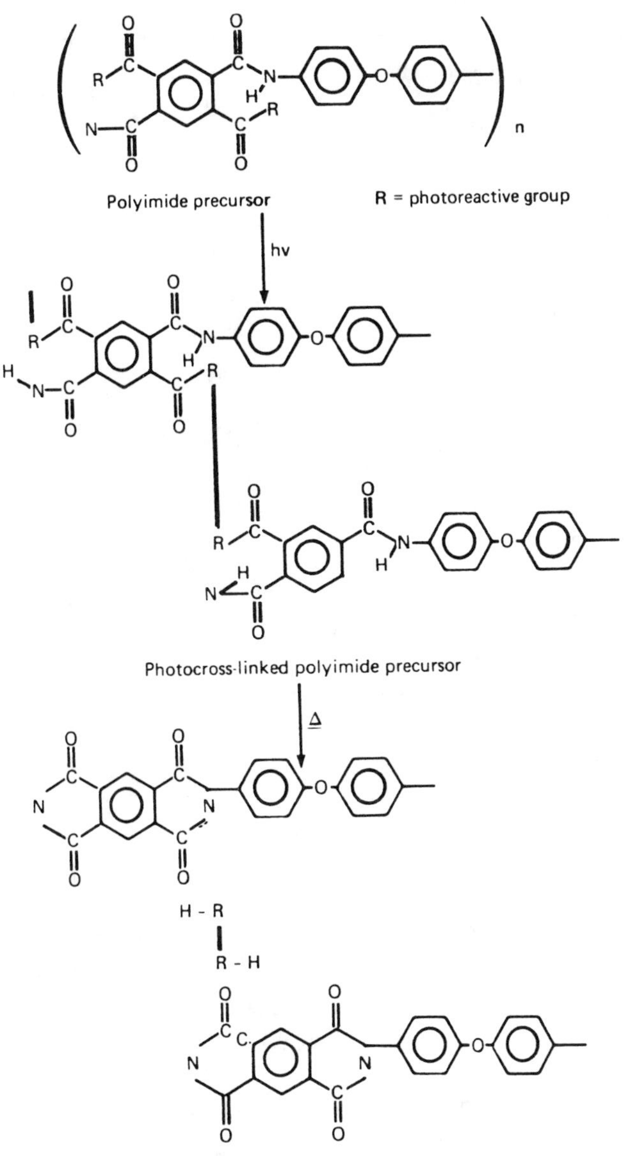

Polyimide precursor   R = photoreactive group

Photocross-linked polyimide precursor

Polyimide

Figure 2. The chemical reactions of light sensitive polyimide precursor

different. The HTR 3-200 has a photospeed of about 400 - 1000 mJ/cm^2 (30 microns-70 microns final thickness) in the 350-450 nm UV range with maximum sensitivity around 400 nm. The Probimide 348 has a sensitivity of (500 - 1200 mJ/cm^2 (30 microns - 70 microns final thickness) with a maximum sensitivity near 436 nm.

Thermal Analysis

Thermogravimetric Analysis (TGA)

The Perkin-Elmer Model TGS-2 thermogravimetric System was used to analyze uncured samples of the polyimide precursors.

Thermogravimetry is a technique whereby a sample is heated at a constant, preferably linear rate while the sample weight is continuously monitored. The weight change versus temperature is then plotted and some information concerning the thermal stability of the original sample, the composition and thermal stability of intermediate products, and the composition of the residue may be determined. Original sample weights varied from 10 mg to 20 mg and the heating rate was approximately 5°C/min. Prior to thermogravimetric analysis the samples were placed on a hot plate to dry and were exposed to ultraviolet light to achieve photocross-linking of the polyimide precursor. The percent weight loss vs. temperature for both samples is illustrated in Figures 3 and 4.

The thermogravimetric analysis indicates a significant weight loss occurring during the curing process for both of the samples tested. A weight loss after curing of about 53% was observed for the HTR 3-200 sample and 36% for the Probimide 348 sample. The HTR 3-200 exhibited a large solvent loss below 200°C due to the evaporation of the NMP solvent. The Probimide 348 samples exhibited little weight loss until the boiling point of the NMP was reached (205°C). The lower concentration of solvent in these samples require higher temperatures to initiate solvent evaporation. The bulk of the weight loss above 205°C is due to the water loss resulting from the imidization of the polyamic acid.

Differential Scanning Calorimetry (DSC)

Both of the photosensitive polyimides were also analyzed using a Perkin-Elmer DSC-2 Differential Scanning Calorimeter. The DSC method measures the change in the amount of heat (liberated or absorbed) with a change in the temperature. Heat or enthalpic changes can be either endothermic or exothermic and are a result of the following; phase transitions such as boiling, vaporization, etc., bond breaking reactions, changes in crystalline lattice, and chemical reactions. DSC is an ideal method for use in polymer chemistry to characterize such parameters as the glass transition temperatures, polymerization, thermal decomposition and oxidation type reactions.

Prior to analysis, the two specimens were dried on a hot plate at 80°C and then exposed to UV light to photocross-link the polyimide precursors. The results of the analysis are shown in Figures 5 and 6. The DSC spectra for the HTR 3-200 shows endothermic peaks at approximately 150°C and 185°C. The Probimide 348 DSC spectra also showed these two characteristic peaks but at somewhat higher temperatures, 185°C and 220°C, respectively and smaller in intensity. The higher temperatures correspond to the higher temperatures necessary to remove the smaller concentration of solvent present in the Probimide 348. The rapid change in slope for

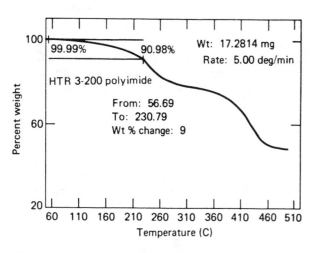

Figure 3. TGA of HTR 3-200 photosensitive polyimide precursor

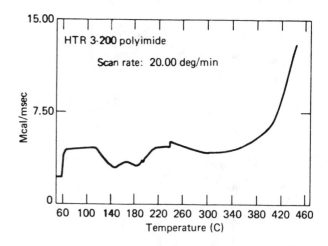

Figure 5. DSC of HTR 3-200 photosensitive polyimide precursor

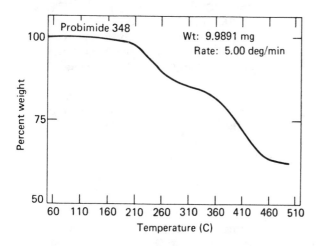

Figure 4. TGA of Probimide 348 photosensitive polyimide precursor

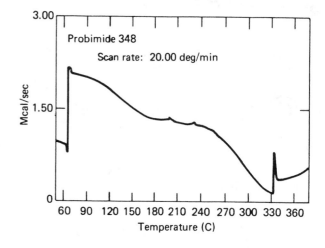

Figure 6. DSC of Probimide 348 photosensitive polyimide precursor

both samples above the 250°C temperatures is due to the evaporation of H_2O formed during the imidization reaction.

Mechanical Properties

Stresses in cured polyimide films are composed of "intrinsic" stresses due to film shrinkage and thermal stresses resulting from differences in the thermal coefficients of expansion (TCE) between the film and the substrate. Intrinsic stresses resulting from solvent loss alone may or may not be a significant factor depending on the cure cycle. If the film is partially cured and or cured to rapidly, such that there is a significant portion of solvent remaining in the specimen after the cure cycle, large stresses can accumulate in the film when it is returned to room temperature. These stresses are much higher than are present when the film is cured

slowly to 400°C and then brought back to room temperature. This is probably due to the looser structure present in the film with the unevaporated NMP present resulting in a higher TCE [5]. Additional drying steps which call for temperatures in excess of 200°C are not recommended as they may put the film under abnormally large strains.

The temperature coefficient of expansion for the PMDA-ODA has been reported to be 20 ppm/°C [4]. The glass transition temperature for this polyimide has been reported to be approximately 385°C [6]. The TCE for a common substrate material such as alumina is about 6.7 ppm/°C. Thus, cooling down from the glass transition temperature can result in large accumulated stresses. However, these stresses should be well below the fracture strength of the film [7] provided that there is sufficient adhesion to the substrate and there are no flaws in the film surface. The Young's modulus at room temperature has been reported at $.3 \times 10^6$ PSI [4,7]. One other researcher determined the Young's modulus for thin films to be 1×10^6 PSI at room temperature and about $.15 \times 10^6$ PSI at about $275^{\circ}C$ with a slight hysteresis depending on whether the film was being heated or cooled [8].

Film stresses should not increase significantly with film thickness, however adhesion may degrade if film thickness inhibits photocross-linking of polyimide near the substrate. Increased film thickness can also increase substrate warpage by introducing bending moments which are proportional to the cube of the largest substrate dimension [9].

ELECTRICAL AND CIRCUIT PARAMETERS

Electrical Properties

The electrical properties of cured PMDA-ODA have been fully characterized in the literature [4,10]. The relative dielectric constant for fully cured PMDA-ODA is about 3.6 near room temperature at 10 kHz. The dissipation factor has been measured to be about .004 at 1 MHz. It has been shown that the dissipation factor can vary depending on the cure schedule and final temperature of the cure [4]. The dissipation factor measures the amount of ring closure and at earlier stages in the cure, the amount of water and solvents present in the film, however it is generally insensitive to small amounts of absorbed water after curing [10]. The dissipation factor is also effected by the atmosphere during the cure cycle. The presence of an inert gas such as N_2 or vacuum during the final stages of the cure will decrease the dissipation factor as well as the dielectric constant and dielectric strength. This indicates that complete cyclization of the polyamic acid is not achieved unless the polyimide is vacuum cured during the high temperature portion of the cure cycle. The reason for this may result from the increase in the outgassing rate while vacuum curing. This in turn increases the molecular mobility [11] which otherwise slows down at or near the glass transition temperature.

Circuit Requirements

An evaluation of polyimide as a suitable interlayer dielectric for use in VHSIC and VLSI device application depends upon several factors; a complete characterization of the electrical properties of the polyimide (dielectric constant and dissipation factor), a knowledge of the specific characteristics of the logic family that is to be interconnected, and the determination of electrical parameters for a given multilevel configuration. The electrical parameters will be dependent upon the use and placement of distributed power and ground planes within the multilevel stack, the thickness of the dielectric layers and the geometry of conductor lines. The requirements of the device and the specific application will determine the range of acceptable electrical parameters.

Electrical parameters essential to good circuit performance include signal line capacitance and resistance for low speed applications and characteristic impedance, attenuation and crosstalk immunity for higher speed applications. Polyimide with it's inherently low dielectric constant should provide an excellent medium for moderate to high speed data transmission and the dissipation factor is sufficiently low (typically below .006 at $25^{\circ}C$ for fully imidized specimen) for packaging most high speed silicon devices.

Evaluation of a suitable substrate wiring technology depends primarily on the specific device characteristics for the logic family that is to be interconnected. These characteristics include the input and output impedances, noise margins and the unloaded rise and fall times of the device output drivers. The rise time (t_r) and the fall time (t_f) of the device output drivers are important in determining the maximum unterminated length of a signal line before reflections and ringing effects need to be considered. The rise and fall times are also useful for time-domain analysis of voltage crosstalk. The maximum unterminated or "critical" length of a transmission line can be determined by the relationship

$$l_c = \frac{(t_r \text{ or } t_f)}{2\,T_d.} \tag{1}$$

The unit length time delay, T_d, is simply

$$T_d = \frac{1}{v_p} = \frac{c}{\sqrt{e_r}} \tag{2}$$

where v_p is the propagation constant, c is the speed of light in vacuum and e_r is the effective dielectric constant of the medium. The time delay can be modified to include the effects of bonding wire, package leads, crossover capacitance, etc. through the relation

$$T_d' = T_d \sqrt{1 + C_l/C_o}, \tag{3}$$

where C_o and C_l are the isolated line and parasitic load capacitance, respectively.

The device output impedance is important in determining the optimum signal line impedance for the unterminated line. For example, ringing--stemming from reflections in improperly terminated lines--can be eliminated by matching the driver source impedance to the line impedance. Typically, the output impedances for high-to-low and low-to-

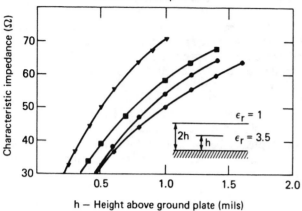

Figure 7. Characteristic impedance of covered microstrip lines in polyimide

high transitions are different, and an exact match is not possible. However, a good compromise can usually be found. Generally, the higher the output impedance, the less capacitance that a source can drive for a given propagation delay. For short lines ($l < l_c$) the total propagation delay can usually be represented by an equation of the form

$$t_p = t_{pi} + \alpha R_o C_l \qquad (4)$$

where t_{pi} is the internal delay, αR_o is the dynamic output impedance (R_o is the static output resistance), and C_l is the load capacitance.

The critical length of a signal line buried in polyimide with a dielectric constant of 3.5 for a high speed pulse with a rise or fall time of 1 ns would be approximately 6.3 inches. This compares to approximately 4 inches for a typical thick-film dielectric. Improvements in speed can be achieved with the ability to fabricate line with narrower traces which result in lower crossover capacitance and thus, a higher propagation velocity. The characteristic impedance for covered microstrip lines as a function of dielectric thickness for 1 mil, 1.5 mil and 2 mil wide signal lines is depicted in Figure 7. Since the relative dielectric constant is low, the impedance for the covered microstrip is only slightly different from that of the microstrip configuration. The microstrip line which has been well characterized in the literature should be adequate for design purposes below microwave frequencies. Today, the unloaded output impedances for most devices are in the 50 ohm to 65 ohm range and thus the ability to fabricate dielectric layers of 13 to 25 microns will be required. While the manufacturer should strive to achieve desirable values for the characteristic impedance of signal lines, thickness variations across the substrate due to

processing and material variables limit the ability to manufacture signal lines with tightly controlled impedances.

PROCESS DEVELOPMENT

Response Surface Methodology

Response surface methodology is a set of statistical procedures used by researchers to assist them in studying process-oriented problems particularly in cases where there are large numbers of variables which may effect some element of a process. The effected element is referred to as a response (such as film thickness and uniformity, percent imidization, slope of via sidewall, etc.). The independent variables or factors are those input or process parameters that may influence the response (such as the spin speed, cure cycle, exposure time, etc.). The response surface methodology or RSM approach can be made useful when only qualitative response information is available. This can be done by ascribing a numerical figure of merit to the response. This is done quite often in the food industry when the response has been a qualitative or subjective response such as taste. The RSM approach provides an experimenter a systematic approach for studying the salient features of a system or process and optimizing the independent variables such that the desired responses are realized.

The RSM method begins with the experimental design which is some variation of the full or fractional factorial experimental design method [12,13]. The factorial design eliminates the need for a one-factor-at-a-time experiment which may require several samples (perhaps 4 or more) at each level of a variable to estimate the mean value and variability of the results and test the significance of a particular factor on a given response. When many variables are present, this approach can be shown to be very inefficient. For example, a single factorial matrix design experiment that included three variables could be performed using only 8 experimental trials and the information obtained would include synergistic effects or interactions. An example of a two-level factorial design for three variables or factors is indicated in Figure 8. The two levels, (+1) and (-1) correspond to a high level for that variable and low one, respectively. A three dimensional representation of this experimental design is illustrated in Figure 9.

In order to estimate curvature effects a three-level experimental design is required. The two level model can be transformed into a three-level factorial design with the addition of a (0) level which represents a value for each factor midway between the (+1) and (-1) levels. An example of a three-level, two-factor experimental design used to study the effect of spin speed and spin time on film thickness and film

Trial	X_1	X_2	X_3	Response
1	−1	−1	−1	Y_1
2	+1	−1	−1	Y_2
3	−1	+1	−1	Y_3
4	+1	+1	−1	Y_4
5	−1	−1	+1	Y_5
6	+1	−1	+1	Y_6
7	−1	+1	+1	Y_7
8	+1	+1	+1	Y_8

Figure 8. A 2-level 3-factor full factorial design scheme

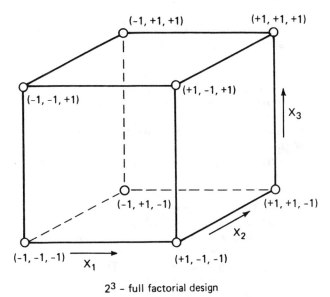

2^3 – full factorial design

Figure 9. A 3-dimensional representation of the experimental space for a full 2^3 factorial design

Table 1
Experimental design for polyimide thickness and uniformity study

Seq. No.	Plan No.	Variable levels (X1)	(X2)	Response values (Y1)	(Y2)
1	2	2500.00	15.00	———	———
2	4	2500.00	25.00	———	———
3	9	2000.00	20.00	———	———
4	6	2500.00	20.00	———	———
5	5	1500.00	20.00	———	———
6	7	2000.00	15.00	———	———
7	10	2000.00	20.00	———	———
8	3	1500.00	25.00	———	———
9	1	1500.00	15.00	———	———
10	8	2000.00	25.00	———	———

Definitions of variables	(−1.0)	(0.0)	(1.0)
(X1) spin speed	1500.00	2000.00	2500.00
(X2) spin time	15.00	20.00	25.00

Definition of responses

(Y1) thickness
(Y2) uniformity

Table 2
Number of experimental trials for RSM techniques

Factors	Trials in 3-level factorial design	Trials in box-Behnken experimental design
2	9	11
3	27	15
4	81	27
5	243	46
6	729	54

uniformity is shown in Table 1 (Box-Behnken design). The three level model allows for a fit of the experimentally determined responses to a 2nd order polynomial equation of the form

$$Y = a_0 + a_1X_1 + a_2X_2 + a_3X_1^2 + a_4X_2^2 + a_5X_1X_2.$$

The above equation can be plotted as a function of the response and is typically referred to as a response surface. For the 2-factor example above, the surface can be plotted in three dimensions or as is usually done, as a series of equal-response lines, or contours.

The number of trials required to use the Box-Behnken experimental design for two or more factors is indicated in Table 2. It is obvious from the table that this method provides an economical approach and a useful strategy for the development of processes involving large numbers of process parameters. This method was used quite extensively to obtain optimum processing parameters for the development of a polyimide thin-film hybrid and will be discussed further below.

Surface Preparation

Bare and metallized alumina substrates are cleaned using a solvent cleaning process consisting of tricholroethylene boil, followed by an acetone boil and then isopropyl boil. The substrates are then blown dry in nitrogen and dehydration baked for 30 minutes at 130°C. It is important to remove oils and other organic contaminants from the surface as the polyimide precursor tends to pull away from these areas.

Adhesion of polyimide films to bare alumina substrates is inadequate for multilevel circuit applications. Some researchers have reported large improvements in the adhesion of polyimide films to alumina with the use of a silane adhesion promoter. The commonly used adhesion promoter for alumina and silicon is y-aminopropyltriethoxysilane. The adhesion promoter is typical spin deposited and then dehydration baked for an hour at 130°C prior to spin coating the substrate with polyimide. Chrome metallization is also an effective method to promote adhesion of polyimide to alumina. Since ground and power planes are useful for high speed digital circuits, it can be beneficial to deposit a chrome/copper/chrome (or chrome/gold/chrome) metallization on the bare ceramic to serve as both distributed power and ground plane and an effective adhesion layer. Adhesion tests performed on fully cured polyimide films indicated that chrome offered excellent adhesion without the need for an adhesion promoter [7].

In this study, thick polyimide films were deposited and fully cured on alumina substrates with a silane adhesion promoter. These substrates were then diced into small .6 inch square wafers onto which small metal studs were epoxied. These studs were pulled with an Sebastian SMT Pull Tester. The film adhesion was so good that in every case the alumina shattered before loss of adhesion could occur. It has been reported elsewhere that the adhesion promoter does not prevent degradation of adhesion strength after prolonged aging at in an 85°C/80% RH environment [7]. It was also noted that films adhered to chrome metallization did not experience any adhesion loss after temperature and humidity aging.

Polyimide Application

There are several methods for applying thick polyimide coatings; the foremost being spin coating. Lower viscosity polyimide precursor solutions can be spray coated, however this technique is limited to thicknesses less than that required for circuits involving alumina substrates. Spin coating usually consists of an initial application of a sufficient quantity of the precursor at the center of the wafer. The substrate is then spun at a low speed (500 RPM) for about 5 seconds to spread the material evenly about the surface of the substrate. The substrate is then spun at higher speeds depending on the final thickness required. It is important to prevent the formation of bubbles when dispensing the material. This can be accomplished by pouring the precursor onto a teflon rod and allowing the viscous material to flow down the rod onto the substrate.

After spinning, the substrate is then placed on a level surface which is properly covered to protect the film from any particulate contamination. This dwell period is necessary to facilitate planarization around metallization features and along the edge of the substrate where the precursor tends to accumulate as a result of the spin coating. Most vendors recommend a dwell period of at least 45 minutes to 1 hour prior to soft-baking.

The primary process parameters involved in the spin coating process include; the viscosity, the volume of material dispensed, the speed used to spread the material, the final spin speed, the spin time and the dwell time. Other than the material viscosity, the two most likely parameters to effect the final thickness would include the final spin speed and spin time. A parameter study was conducted using the RSM methods discussed above to investigate the relationship between spin speed and spin time on film thickness and uniformity for the two photosensitive polyimides evaluated. The experimental plan is shown in Table 1. The thickness measurements were made using a Dektak Stylus Profilometer. The patterned test substrates consisted of a uniform array (20 x 20) of 25 mil x 25 mil square holes spread out across a 2" x 2" chrome metallized alumina substrate. The profilometer measures the step height encountered on a preselected edge for each hole. The mean thickness and the degree of uniformity (DOU) of the film thickness were measured for each substrate condition . The DOU parameter was simply the standard deviation of the film thickness divided by the mean film thickness (σ/μ). That is, the smaller the degree of uniformity, the more uniform is the film thickness.

The cured thickness as a function of spin speed for various spin times for both polyimide samples is shown in Figures 10 and 11. The thickness decreases monotonically as a function of both spin speed and spin time as would be expected. The effect of spin speed and spin time on the degree of uniformity of film thickness across the substrate is illustrated in Figures 12 and 13. The data indicates that for each spin time, there is a range of spin speeds such that the degree of uniformity parameter (σ/μ) is minimized. Thus, nonuniformity can result if the spin speed is either too slow or too fast. RSM contour plots indicated that the minimum degree of uniformity or equivalently, the most uniform film thickness was independent of both spin speed and spin time and was dependent only on the thickness. In the HTR 3-200 this thickness was approximately 9 microns (cured) while that for the Probimide 348 was near 14 microns (cured) independent of the method of obtaining these values. That is to say that from uniformity considerations alone, these two polyimides are tailored to a film thickness of 9 microns and 14 microns, respectively.

When exposing polyimide with small via structures using proximity methods (for sloped sidewalls) only a small amount of nonuniformity can be tolerated. The total thickness variation across a substrate can easily exceed 4 to 6 standard deviations mainly due to the edge bead resulting from spin

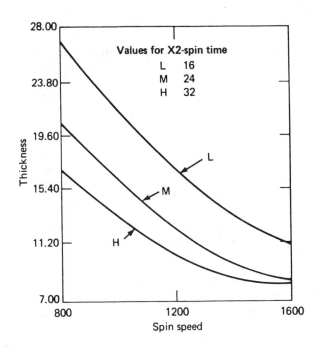

Figure 10. Fully cured film thickness for HTR 3-200 polyimide as a function of spin speed and spin time

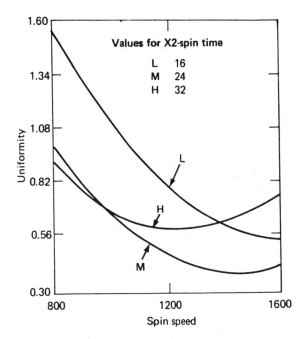

Figure 12. Degree of uniformity of film thickness for HTR 3-200 polyimide as a function of spin speed and spin time

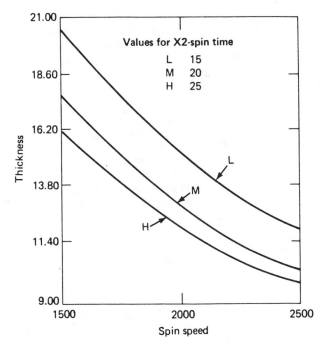

Figure 11. Fully cured film thickness for Probimide 348 polyimide as a function of spin speed and spin time

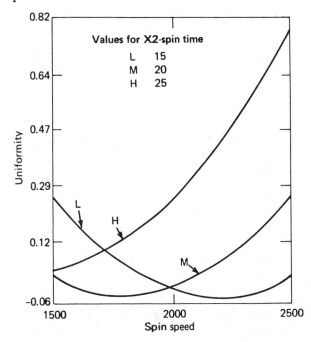

Figure 13. Degree of uniformity of film thickness for Probimide 348 polyimide as a function of spin speed and spin time

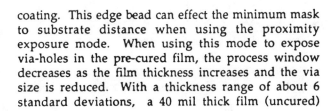

coating. This edge bead can effect the minimum mask to substrate distance when using the proximity exposure mode. When using this mode to expose via-holes in the pre-cured film, the process window decreases as the film thickness increases and the via size is reduced. With a thickness range of about 6 standard deviations, a 40 mil thick film (uncured)

exposed in the proximity mode can tolerate a maximum degree of uniformity of .025. Films this uniform are difficult to achieve. Larger via-holes will require a less stringent control of film uniformity. Typically, for predictable results, the diameter of via-holes should be approximately 2.5 times that of the final cured film thickness. This will allow for sloped sidewalls greater than or equal to 45 degrees with sufficient area at the bottom of the via-hole to maintain electrical resistance sufficiently low.

Soft-bake

The soft-bake is required to drive out some of the solvents needed in the precursor solution to permit spin coating. The film must be dried to reduce flow and permit contact printing during the exposure and to allow for proper development. The drying time depends on the thickness of the film and the percent of solvent in the precursor. The soft-bake can be accomplished either by heating in an infrared or convection oven or on a hot plate. All three methods were investigated, however soft-baking on a controlled temperature hot plate was determined to yield the most consistent results.

The HTR 3-200 was soft-baked on a hot plate at approximately 80°C. Several soft-baking tests were performed to determine an optimal soft-bake duration. These tests indicated that if the film is dried for too short a time, the film is too tacky for contact printing leaving some residue on the mask. If the film is dried for too long a period, then crazing of the substrate occurs. The soft-bake times were established by baking a series of coated substrates of a given thickness for various time durations, yielding some too tacky and some exhibiting crazing. A time intermediate between these two extremes was chosen. For 40 micron (uncured) films, the soft-bake time was about 2 hours and twenty minutes.

A second drying method was tried which utilized a prebake at 40°C followed by a bake at 80°C until the films were sufficiently dry for contact printing, however this resulted in a significant reduction in the photospeed. This reduction in the photospeed may have been due to the prolonged drying at a lower temperatures although this effect is not well understood.

The Probimide 348 was also soft-baked at 80°C on a hot plate. Soft-bake times were just long enough to insure that the film was no longer tacky. The tackiness was tested by applying a glass slide at selected intervals to the surface of the film while drying. If there was no residue remaining after removing the slide then this time period was used for soft-baking. The manufacturer recommends an initial bake at 80°C for 30 minutes followed by a higher temperature bake at 110°C for an additional 30 minutes [3]. This procedure was not found to be necessary and was not used in this study but may be useful when shorter soft-bake times are desirable.

Exposure

Via Formation

A critical step in the multilevel process is the fabrication of reliable vias. Via-holes must be small enough to permit the high density circuitry that thin-film processing offers. The sidewalls must be sloped to facilitate the adhesion of metal sputtered into the via-holes. Via-holes must be free of residual polyimide prior to subsequent metallization steps to prevent opens and maintain a low electrical resistance across the via.

Exposure was done with a OAI Hibralign mask aligner which uses a high-pressure mercury arc lamp. The emission spectrum for this ultraviolet light source is depicted in Figure 14. The three commonly used frequencies for UV-sensitive materials are the 365 nm, 405 nm and 436 nm lines. Polyimides have an absorption peak near the 365 nm and 405 nm lines and therefore do not transmit much light at these wavelengths. The transmittance is approximately zero for films thicker than 8 microns and 10 microns at the wavelengths 365 nm and 405 nm, respectively [14]. Films of photoreactive polyimides which are activated by light at these wavelengths and are thicker than the above values can not be properly photocross-linked. This can result in undercutting and cracking of sidewalls rendering via-holes unsuitable for layer-to-layer interconnections. Sidewall cracking will typically lead to catastrophic cracking after curing.

A via test pattern was generated with 2 mil, 3 mil, and 4 mil diameter round and square via-holes. The pattern included an underlying ground plane and second layer metallization which could be probed to measure via resistance. A Dektak Profilometer with an 8 micron stylus was used to profile the via-hole sidewalls prior to and after curing. Scanning electron microscopy was used to evaluate the via quality. Films approximately 1 mil thick were made using previously determined spin coating parameters.

The Selectilux HTR 3-200 was the first polyimide precursor formulation investigated. The HTR 3-200 polyimide precursor is primarily

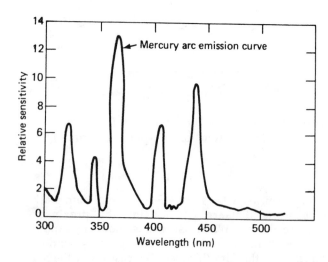

Figure 14. Emission spectrum for high-pressure mercury lamp

photoreactive near 400 nm [4]. During the exposure, a yellow filter (Corning CS-34) was used to block out the 365 nm (i)-line of the mercury arc lamp in order to photocross-link films greater than 10 microns with any consistency. Exposure times were also increased to account for the reduced intensity of the 405 nm and 436 nm wavelengths as well as the reduced sensitivity at these wavelengths. The intensity of the UV light source was 7 mW/cm^2 measured at 400 nm. The mask aligner was used in the proximity mode with N$_2$ purge to obtain sloped sidewalls.

Suitable via structures in the 32 micron (uncured) thick film could not be fabricated with HTR 3-200 polyimide formulation after several attempts even when using the recommended i-line suppression filter. Various exposure times (100 to 450 second exposures) and proximity settings (5 -25 microns) were used with filtering and without. SEM photomicrographs of vias made with and without i-line suppression are shown in Figures 15 and 16. The use of the filter did increase the depth of cross-linking. This eliminated most of the cracking around the via-holes but not the severe undercutting of via-holes after development. The undercutting, however rendered these via-holes unsuitable for interconnections. Exhaustive attempts at making acceptable via structures were not undertaken and further investigations on the application of this polyimide in the multilayer thin-film hybrid technology are still in progress.

Initial attempts at forming vias with the Ciba-Geigy Probimide 348 have yielded good results. The Probimide 348 polyimide precursor is highly photoreactive at the 436 nm peak in the mercury arc emission spectrum [3]. Relatively short exposure times using an i-line suppression filter were sufficient to fabricate via-holes in thick films of 20 mils and

better. Uniformity of development was quite good for both round and square via-hole geometries. Proximity setting were varied between 5 microns and 15 microns. At the 15 microns setting, the 2 mil vias were almost completely closed, however the degree of closure was fairly uniform across the substrate. Figures 17 and 18 illustrate a 2 mil round via and a 3 mil square via in a twenty micron thick Probimide 348 polyimide film using a ten micron proximity, 630 mJ/cm^2, filtered (l<400 nm absorption - 50% 0f 405 nm line and 80% 436 nm) exposure. The 10 micron proximity setting yields nicely sloped sidewalls and a sufficient opening at the bottom of the via-hole for electrical conduction.

Development

The two predominant methods for developing polyimide films are immersion development in an ultrasonic bath and spray development. Spray development is the recommended method for developing thick polyimide films when small isolated figures are present as this method is less aggressive than ultrasonic development. However, both methods should be suitable for developing openings in the exposed film.

In this study spray development was used exclusively. Two sprays were used, one containing the manufacturers recommended developer and the other containing a semiconductor grade isopropyl alcohol. The substrate was initially sprayed with the developer for a specified duration followed by a shorter period (about 10 s) during which time both the developer spray and the isopropyl alcohol were sprayed together followed by a rinse with the isopropyl alcohol. After the rinse, the substrate was spun at about 3000 RPM for 30 seconds.

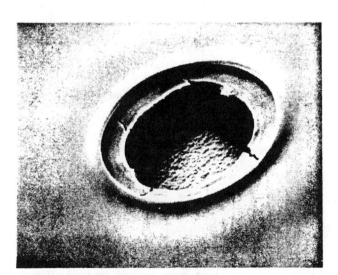

Figure 15. Via-hole developed from 36 micron (uncured) HTR 3-200 photosensitive precursor without i-line suppression filter

Figure 16. Via-hole developed from 36 micron (uncured) HTR 3-200 photosensitive precursor with i-line suppression filter

Figure 17. Via-hole developed from 30 micron (uncured) Probimide 348 photosensitive precursor

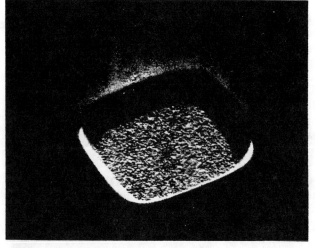

Figure 18. Via-hole developed from 30 micron (uncured) Probimide 348 photosensitive precursor

Cure Schedule

The results of the TGA and DSC thermal analyses were used to develop a cure schedule for both the HTR 3-200 and the Probimide 348 polyimides. The three important elements of the cure schedule include; the evaporation of solvents, the imidization of the polyamic acid, and the cool-down period. Although the NMP solvent has a boiling point around 205°C, most of the solvent can be evaporated at temperatures around 125°C. Between 130°C and 200°C the solvent concentration is low and remains constant. Therefore a bake at temperatures around 125°C will remove most of the NMP without initiating a significant amount of imidization. The residual solvent can be removed at the boiling temperature of NMP. It is important to remove most of the solvent to minimize residual stress build-up. The imidization reaction ceases as the softening temperature or glass transition approaches the reaction temperature thus freezing out molecular motion [15]. Therefore, long dwells at low temperature are probably unnecessary for purposes other than solvent removal. Since the glass transition temperature is near 380°C, a dwell time above this temperature for about 1 hour is recommended for a complete cure.

The high temperature potion of the cure should be done in a nitrogen purged atmosphere or preferably in vacuum. In this study, the cure was done in a nitrogen-purged convection oven. Curing under vacuum (or to a lesser degree, in N_2) can increase the molecular mobility by increasing the outgassing rate thereby enhancing imidization as well as preventing severe oxidation of the film surface. It can also increase the surface tension thereby improving the flow characteristics of the film necessary for planarization [11].

The two cure schedules for the HTR 3-200 and the Probimide 348 are illustrated in Figures 19 and 20. The two schedules are quite similar with the main exception that the solvent bakeout portion for the HTR 3-200 is somewhat longer due to the large amounts of NMP present. Also, the second dwell temperature is higher for the Probimide 348. This is because the DSC indicated that solvent evaporation was occurring at a higher temperature for this polyimide than for the HTR 3-200. Cool-down periods are long to allow for some stress relaxation while cooling.

The cure was monitored using a Dektak Profilometer to measure changes in polyimide thickness after curing. Cured samples were cured again at 400°C for 1 hour and then measured again for variations in thickness. No further reduction in thickness was observed indicating no further shrinkage due to H_2O loss through imidization. TGA analysis provides a simple way to qualitatively evaluate weight loss as a function of heating the sample. TGA analysis of fully cured samples indicated no significant weight loss up to 400°C.

Multilevel Polyimide Metallization

Via-hole Cleaning

Frequently, via-holes will contain small residual amounts of the polyimide material after development. Failure to remove this slight film residue will almost invariably render the via nonconductive. To help insure proper electrical conductivity through the via, the film surface containing the via-holes were plasma cleaned in a Branson O_2-plasma barrel etcher or back-etched in the sputtering apparatus prior to metallization. Residual films are typically much less than a micron and

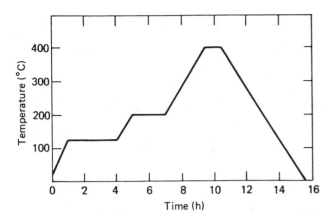

Figure 19. Cure schedule for HTR 3-200 polyimide

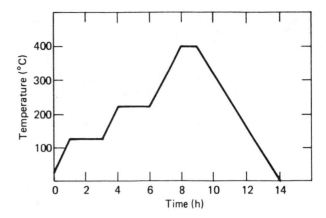

Figure 20. Cure schedule for Probimide 348 polyimide

plasma etching or back-etching parameters must be adjusted according to the actual residual film thickness.

Metallization

The bottom layer metallization is made up of sputtered chrome (.2 microns) to serve as an adhesion layer to the bare ceramic, followed by a sputtered copper conductor layer (1.4 microns), and a third layer of sputtered chrome serving as an adhesion layer for the first dielectric layer. Most of the ceramic is metallized since this layer will act as an interdigitated power and ground plane. Vias through the first dielectric layer will connect the power and ground to subsequent layers. The large areas of chrome offer an excellent intermediate surface for adhering the polyimide film to the ceramic substrate. The interdielectric metallization also consists of a sputtered chrome adhesion layer (.2 microns), a sputtered copper (1.4 microns) layer and a second chrome adhesion layer. The top layer metallization is a sputtered chrome adhesion layer followed by a sputtered copper or gold layer depending on the particular application. When conductor thicknesses greater than 3 microns are desirable, metal layers are typically plated-up to the desired thickness. Plated interlayer conductors can then be sputtered with a chrome adhesion layer prior to patterning. Metallizing the top layer with chrome/copper followed by plated nickel and gold yields a metallization scheme that is compatible with most soldering, bumping and wire bonding processes.

Prior to metallization, the substrates are cleaned by boiling in isopropyl alcohol and then blown dry in nitrogen. The substrates are then vacuum baked at 200°C for 2 hours. The vacuum baking is necessary to remove any absorbed moisture prior to metal deposition. Absorbed moisture trapped in the film prior to metallization can cause blistering in the metal film due to water vapor formed during the sputtering operation. Heat sinking the substrates

while sputtering is recommended to prevent overheating which may cause some film cracking if allowed to cool to fast.

Photolithography and Etching

The photolithography of the metallized polyimide surface is critical in assuring that via contacts are etched properly. Next to residual polyimide in left in via-holes after development, photoresist breakdown along the edge of the via-hole is the most important cause of electrical opens. Bulging--formation of a ridge approximately 2 micron high--near the perimeter of a via can result from film shrinkage while curing. In order to keep the photoresist from thinning significantly in this region, a more viscous resist should be used than that required for patterning the bare ceramic layer. Thinning of the photoresist becomes more acute as the depth of the via-hole is increased. A dry film photoresist may also be used but will be accompanied with some resulting decrease in feature resolution.

The metallized polyimide films are first cleaned using a standard solvent cleaning (acetone boil followed by isopropyl boil and blown dry with N_2) and a dehydration bake at 130°C for 30 minutes. Then Shipley 1822 positive photoresist is spun on at 4000 RPM's for 30 seconds. The photoresist is baked at 90°C for 30 minutes in a convection oven. The pattern is exposed at an energy of 120 mJ/cm^2 (365 nm). The pattern is developed in Shipley's Microposit 351 developer solution diluted in DI water (1:3.5 ratio) for approximately 45 seconds, rinsed in DI water and then blown dry in N_2. The pattern is then flood exposed at 240 mJ/cm^2 and baked an additional 30 minutes at 120°C to improve the resistance to the metal etchant solution.

The etching of the chrome\copper\chrome metallization was done using a 50/50 solution of H_2O and hydrochloric acid. The etchant is activated with the addition of a zinc rod. The substrate is then removed from the solution and rinsed in DI water and blown dry in N_2. The copper is then etched in a ferric chloride etchant solution. After completion of the copper etching , the substrate is rinsed in DI water and then returned to the chrome etchant to remove the remaining exposed chrome metallization. The chrome\gold\chrome system may also be used without any adjustments to the process with the exception that the copper etch be replaced with a gold etch.

DISCUSSION

Planarization

Planarization or leveling of the film around features (such as metal traces, etc.) after spin coating is essential for fine-line multilevel hybrid applications. Planarization is less of a problem when very thick films are involved. Film thickness has been shown to be independent of the step height (line edges) when the thickness of the film is greater than the step height [16]. Also, based on film shrinkage considerations, polyimide films with less solvents (i.e. low molecular weight) planarize better than polyimides containing high solvent concentrations (i.e. high molecular weights). Due to the thick nature of the films used in this study, planarization was not found to be much of a problem for either of the two polyimides evaluated.

Pattern Alignment

An important requirement for the soft baked polyimide film is that it be transparent enough to permit proper alignment to underlying features. The HTR 3-200 dries a dark blue and is very difficult to align the mask to an underlying metal pattern. The Probimide 348 is very transparent and eliminates the need for any special alignment fixturing.

Multiple Dielectric Layers

Presently, all test designs have been restricted to two dielectric layers. Results appear to confirm that polyimide has good adherence to itself. Tests on chrome metallized unpatterned (but exposed) substrates have indicated that at least four polyimide layers of about 15 microns each could be deposited and cured without any adhesion problems, however extensive testing on these samples has not yet been undertaken. All intermediate conductor layers have a chrome adhesion layer on both top and bottom. This assures adequate adherence of the polyimide film to the underlying metallization. The adhesion of sputtered copper alone to polyimide was found to be relatively poor.

CONCLUSIONS

A process was presented for the fabrication of multi-level thin-film structures. A three metal level thin-film test structure resulting from this method is illustrated in Figure 21. Two prospective photosensitive polyimide precursor materials were evaluated. Of the these, the Probimide 348 appeared to be the most promising for applications requiring thick dielectric layers. The uniformity is tailored to film thicknesses around 14 microns which should be suitable for use with 1 mil traces and 2 mil via-holes. The photospeed is excellent requiring about 35 mJ/cm^2 per micron to provide excellent resolution. The edge bead which formed around the edge of the substrate after spinning was markedly less than for the HTR 3-200. The soft-baked film is transparent even for relatively thick films (>30 microns) making it ideal for alignment to underlying patterned features and registration marks. The shelf life of the precursor is about 4 months when refrigerated at temperatures just above 0°C.

Additional process development must be performed to determine process parameter windows for optimizing the resolution of via-holes and fine-line geometries. Via-hole shapes must also be optimized to yield sloping sidewalls (preferably 45 degrees) to permit a more uniform sputtered sidewall metallization. These process parameters include; soft-bake time, exposure parameters such as proximity length and exposure energy and development method (spray or immersion) and time. In addition, improving the reliability and yield of multilevel dielectrics will require some improvements in the patterning of metal pads surrounding via-holes to circumvent photoresist breakdown particularly when the film thickness exceeds 10 microns and the via-hole diameter exceeds 4 mils.

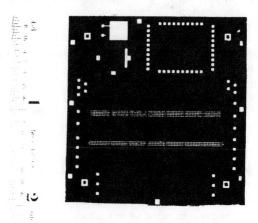

Figure 21. Three conductor level thin-film test pattern

ACKNOWLEDGMENTS

The authors gratefully acknowledge the efforts of J. A. Weiner in SEM microphotography, G. S. Mitchell, A. J. Milne, and A. Cutchember in substrate fabrication. A special thanks is extended to Paula Kelly and Sharon Kirkham for preparing this manuscript.

REFERENCES

1. Ackermann, K., Hug, R. and Berner, G, "Multilayer Thin-Film Technology," 1986 Proceedings of the International Symposium on Microelectronics, p. 519.

2. Rickerl, P.G., Stephanie, J. G., and Slota, P. Jr., "Evaluation of Photosensitive Polyimides For Packaging Applications," 37'th Electronic Components Conference, May, 1987, p. 220.

3. "Probimide 300 Photosensitive System," Ciba-Geigy Corporation, Microelectronic Chemicals Division.

4. "Selectilux HTR 3 Light Sensitive Polyimide Precursor," Merck Electronic Chemicals, Distributed by EM Industries.

5. Geldermanns, P., Goldsmith, C., and Bedetti, F., "Measurement of Stresses Generated during Curing and in Cured Polyimide Films," in "Polyimides: Synthesis, Characterization, and Applications," K.L. Mittal, Ed., Vol. II, p. 695, Plenum Press.

6. Bessonov, M.I., and Kuznetsov, N.P., "Softening and Melting Temperatures of Aromatic Polyimides," in "Polyimides: Synthesis, Characterization, and Applications," K.L. Mittal, Ed., Vol. I, p. 385, Plenum Press.

7. Jensen, R.J., Cummings, J.P., and Vora, H., "Copper/Polyimide Material System for High Performance Packaging," IEEE Trans on Components, Hybrids, and Manufacturing Technology, Vol CHMT-7, No. 4, Dec 1984.

8. Lacombe, R.H. and Greenblatt, J., "Mechanical Properties of Thin Polyimide Films," in "Polyimides: Synthesis, Characterization, and Applications," K.L. Mittal, Ed., Vol. II, p. 647, Plenum Press.

9. Riemer, D.E., "The Effects of Thick-Film Materials on Substrate Breakage During Processing," 1983 International Microelectronics Symposium, October, p. 599.

10. Day, D.R. and Senturia, S. D., "In-Situ Monitoring of Polyamic Acid Imidization with Microdielectrometry," in "Polyimides: Synthesis, Characterization, and Applications," K.L. Mittal, Ed., Vol. I, p. 249, Plenum Press.

11. Chao, C.C. and Wang, W.V., "Planarization Enhancement of Polyimides by Dynamic Curing and the Effect of Multiple Coatings," in "Polyimides: Synthesis, Characterization, and Applications," K.L. Mittal, Ed., Vol. II, p. 783, Plenum Press.

12. Hicks, Charles R., "Fundamental Concepts in the Design of Experiments," Holt, Rinehart and Winston, New York, 1973.

13. Meyers, Raymond, "Response Surface Methodology," Virginia Polytechnic Institute, 1976.

14. Kataoka et al., "Characteristics of a Highly Photoreactive Polyimide," in "Polyimides: Synthesis, Characterization, and Applications," K.L. Mittal, Ed., Vol. II, p. 933, Plenum Press.

15. Numata, S., Fujisaki, K. and Kinjo, N., "Studies on the Thermal Cyclization of Polyamic Acid," in "Polyimides: Synthesis, Characterization, and Applications," K.L. Mittal, Ed., Vol. I, p. 259, Plenum Press.

16. Day, D.R., et. al., "Polyimide Planarization in Integrated Circuits," in "Polyimides: Synthesis, Characterization, and Applications," K.L. Mittal, Ed., Vol. II, p. 933, Plenum Press.

Part 7
Cofired Glass/Ceramic Technology

NO REVOLUTION in any technology is without its response in changing and updating the previous technologies to meet the changing needs of the market. The shift to low dielectric constant organic coatings has been met by the ceramics industry with low dielectric constant ceramics. Thus the NEC glass–ceramic technology.

Bonding the semiconductor chips to the substrate imposes high stresses unless the coefficient of thermal expansion is a near match. The paper by Pence *et al.* examines low CTE glass–ceramic substrates.

Alternatively, ceramic substrates need higher conductivity interconnect leads, and easier processing. The papers on low-temperature cofireable ceramics permit the use of these more conductive interconnects.

Ceramic substrates, like silicon, can provide resistors for circuit terminations. They can provide thin film or thick film resistors on the surface. These can be cofired, as in the article by Sawhill *et al.*, or tantalum thin film, as in the paper from NTT (Ohsaki *et al.*).

LOW TEMPERATURE CO-FIREABLE CERAMICS:
A NEW APPROACH FOR ELECTRONIC PACKAGING

by

A. L. Eustice, S. J. Horowitz,
J. J. Stewart, A. R. Travis and H. T. Sawhill

E. I. du Pont de Nemours & Co., Inc.
Photosystems and Electronic Products Department
Electronic Materials Division
Wilmington, Delaware

Abstract

A new system for manufacturing multilayer interconnect hybrids which uses low temperature co-fired ceramic materials has been developed. The new low temperature co-fired ceramic materials system combines the benefits of high conductivity precious metal conductor metallurgy, design flexibility and cost effectiveness of the thick film process with punched vias and the multilayer capability of the co-fired green sheet process.

This paper describes a low temperature co-fired materials system consisting of a dielectric in green tape form, gold, silver/palladium and silver via fill and signal line conductor compositions, and compatible thick film conductors and resistors for post-firing.

Key processing steps, including via punching, via filling, lamination and firing are reviewed and design guidelines for making multilayer hybrids are presented. Performance of the dielectric/silver bearing conductor system under temperature-humidity bias and high temperature - high voltage conditions is described, and data for compatible materials including conductors and resistors are presented.

Introduction

Evolving integrated circuit technology is the driving force for development of new cost effective multilayer packaging. Innovative packaging designs are required to meet the demands for increased input/output connections, faster switching speeds and greater heat dissipation for new complex integrated circuit devices. In terms of materials and systems requirements include:

- conductor materials with high conductivity,
- fine line resolution,
- small via size,
- low dielectric constant,
- high thermal conductivity,
- hermeticity and reliability,
- controlled thickness and geometry,
- cost effective materials and processes.

Integrating passive components, such as resistors, into the package can contribute to increased design flexibility and circuit density; while the ability to manufacture in-house facilitates control of proprietary designs and speeds prototyping.

Many of these needs are currently being met with thick film hybrid technology[1] or high temperature co-fired alumina technology[2][3].

Features of thick film and high temperature co-fired technology are compared in Table 1. The thick film process is practiced in-house by many equipment and circuit manufacturers. Low firing temperatures and low investment combined with proven reliability, flexibility in implementing new designs and the availability of compatible high performance resistors account for high industry penetration. The thick film process is a sequential process; multilayers are built up layer by layer by consecutive printing and firing of conductive and insulating layers. As many as three dielectric printing and firing steps with corresponding via fill printing and firings may be required for each insulating layer. This process becomes less cost effective as the number of conductive layers increases. In addition, elaborate and labor-intensive inspection programs are required to detect pinhole defects in printed dielectric layers.

Most ceramic packages made using high temperature co-fired technology are purchased from a few specialized suppliers. High temperatures and the reducing atmosphere firing necessary to prevent oxidation of tungsten or similar refractory metals are factors contributing to a high investment process which is difficult to set up and operate in-house. In addition, there are no commercial suppliers of high quality alumina dielectric tape or compatible refractory metal conductors. Metals such as tungsten have high sheet resistance and do not easily meet the signal speed requirement for many new designs.

Recently, low temperature co-fired ceramic systems have been developed which combine the benefits of thick film technology with the processing advantages of co-fired ceramics. Tosaki[4], Shimada[5], Utsumi[6], Nishigaki[7], and others[8][9][10] have described low temperature co-fired ceramic systems which fire in air and incorporate precious metal conductors. Niwa et al.[11] have reported a low temperature co-fired system which is fired in a special atmosphere in a complex process and uses copper internal metallization. The above systems have been developed for in-house use and are not available as merchant materials.

A low temperature co-fired ceramic system incorporating precious metal conductors and allowing air firing in conventional thick film furnaces was described by Vitriol and Steinberg[12]. Pond and

Reprinted from *Proc. 36th Electron. Components Conf. (ECC)*, pp. 37–47, 1986.

Vitriol[13] described use of this low temperature ceramic system in a custom packaging application. More recently, Steinberg, Horowitz and Bacher[14] reviewed this low temperature co-fired process and provided key data on system performance. This paper will describe that low temperature co-fired tape system which is now commercially available, and detail reliability of silver-bearing inner layer conductors under temperature-humidity bias and high temperature - high voltage conditions and performance data for post-fired top conductors and resistors.

Materials System

The materials system consists of a dielectric in the form of a green tape, compatible conductor paste compositions for signal line and via fill printing and compatible thick film conductors and resistors for post-firing. (Figure 1)

Green Tape

The dielectric tape is cast on a Mylar® polyester film carrier coated with a release coating to facilitate stripping. Rolls are slit to a final width of 10 inches. Proprietary coating technology produces a high quality tape with few defects and reproducible properties. Table 2 lists properties of the dielectric tape.

TABLE 2

GREEN TAPE PROPERTIES

Thickness	0.114mm (.0045 inch)
Width	254mm (10 inch)
Carrier Tape Thickness	0.127mm (.005 inch)
Color	White
Density	2.08 g/cm^3
Tensile Strength	0.13 kg/mm^2 (186 psi)
Youngs Modulus	19.87 kg/mm^2 (28.2x10^3 psi)

The tape dielectric was developed using the extensive technology base existing for thick film multilayer dielectric compositions [15][16]. The ceramic part of this system is composed of glass phases with carefully engineered viscosity/temperature characteristics, and refractory filler materials selected to optimize the thermal expansion coefficient and dielectric constant. Relative amounts of the glass and filler phases are adjusted to produce dense hermetic structures after firing in a modified thick film firing profile. Glass content at too low a level results in poor hermeticity while glass content at too high a level can lead to entrapment of organics and blistering during firing. Sintering to high density is achieved at the relatively low firing temperatures and short firing times typical of thick film processing. Fired densities greater than 96% of theoretical are achieved.

A new polymer binder system was developed to give clean burnout during a short burnout cycle and meet rheological requirements for tape casting. Addition of a plasticizer controls properties during lamination.

Conductors

Inner layer conductors and via fill conductors have been developed to be compatible with the green dielectric tape. These conductors use the same types of fine metal powders found in standard thick film conductor compositions and are designed to shrink at the same rate as the dielectric tape to prevent warpage. Solvents were selected to be chemically compatible with the polymer binder and to meet drying specifications during manufacturing.

Top Layer Materials

Top layer conductors are printed and fired after the co-firing operation to optimize registration for automatic component placement. Post-fired top conductors are commercially available thick film conductor compositions and provide high adhesion and excellent soldering or bonding without plating. Commercially available thick film resistor compositions can also be post-fired to accommodate designs requiring resistors.

Processing

The low temperature co-fired materials system was developed to be compatible with standard thick film equipment and processes. Circuits can be fabricated using existing thick film equipment - printers, drying ovens and belt furnaces. Other equipment - laminating presses and numerically controlled via punches, are commercially available and can be purchased with moderate additional investment. Tooling required for mechanical registration can be readily built. Figure 2 is a flow chart describing the process.

Blanking

Tape is removed from the Mylar® backing and sheets are blanked to proper dimensions using a compound die. Registration holes and orientation marks can also be punched during blanking. Tooling for making 3 inch x 3 inch and 5 inch x 5 inch blanks has been built and used successfully in fabricating prototype circuits.

Via Punching

Vias are punched in each layer of dielectric tape using a computer-controlled sequential punch (soft tooling), or rigid array of punches (hard tooling). Laser drilling of vias has been demonstrated, but complete details have yet to be defined. Since shrinkage is slightly different in directions parallel to and normal to the tape casting direction, alternate layers are rotated 90° prior to via punching.

Via Filling and Conductor Printing

Vias are filled by screen printing a specially designed high solids conductor paste composition. The screen printer is operated in the on-contact mode and via filling is completed using a single pass. Sheets of tape are positioned using spring-loaded registration pins in the printing stage: a porous ceramic plate in the stage through which a vacuum is drawn holds down the tape and ensures complete via filling. Tape with filled vias is dried at 120°C for 5 minutes.

Conductor patterns are printed onto the tape using standard off-contact screen printing procedures. Print resolution is superior to that obtained on 96% alumina substrates due to the smooth porous tape surface. Tape with conductor patterns is dried at 120°C for 5 minutes.

Inspection, Collating and Registration

Punched, printed and dried tape dielectric layers are visually inspected for defects as they are collated and registered prior to lamination.

Lamination

Sets of collated and registered circuit layers are placed in a pressing die in which the die cavity has dimensions identical to those of the blanked sheet. The confined die prevents lateral distortion during pressing and registration pins assure mechanical registration. A standard press operated at 70°C and 3000 psi pressure is used for lamination.

Cutting

Laminated stacks are trimmed to pre-fired size using a hot razor, steel rule die set or dicing saw with a carbide blade. Cutting can also be completed after firing using a saw with a diamond blade or by laser scribing.

Burnout

Laminated parts are placed on alumina setters and pre-fired in a convection oven at 350°C for one hour. During burnout approximately 85% of the organics are volatilized and removed preventing blistering during subsequent firing. After burnout parts are not handleable but can be easily transported on the setters. Figure 3 contains data for weight loss and rate of weight loss as a function of temperature (TGA and derivative) and shows evolution of the plasticizer and binder components at approximately 200°C and 350°C respectively.

Firing

Burned out parts on setters are transferred to a conventional thick film furnace for firing. A firing profile of approximately 2 hours is required with a peak temperature of 850°C and a dwell time of 15 minutes. This profile is a standard 60-minute thick film profile run at approximately 75% of the normal belt speed. The combined burnout and firing profile is shown in Figure 4. The 2-step burnout/firing process allows maximum use of existing equipment. Future equipment and process development are expected to lead to a process which would combine burnout and firing into one continuous operation.

Burnout/firing studies have shown that weight loss is independent of heating rate and the number of layers for heating rates between 5 and 20°C per minute and up to eight layers[14]. The high glass content of the tape allows parts to relax and conform to the setter topology during firing. Thus a high quality setter will produce flat, smooth fired parts.

Results

Shrinkage

Control of shrinkage is critical for any co-fired system and is achieved by precise control of raw materials and the casting process. Figure 5 is a compilation of shrinkage data for 15 lots of tape. The average shrinkage in the plane of the tape is 12%, and variability (both lot-to-lot and within the same lot) is within + 0.2%. Average shrinkage in the direction normal to the plane of the tape is 17.5% + 0.5%. Shrinkage data refers to standard lamination conditions of 3000 psi and a temperature of 70°C. Multiple data points for a given tape lot represent measurements on samples taken from different rolls of tape cast from the same single lot of slip in a single casting run. Variations represent the combination of measurement and process effects. Studies are in progress to clarify contributions from measurement, process, and materials.

Conductor Properties

Properties of inner layer (co-fired) and top layer (post-fired) conductors are given in Table 3. Inner layer conductor compositions are formulated to print somewhat thinner than typical thick film conductor compositions.

TABLE 3

PROPERTIES OF CONDUCTORS

	Resistance mΩ/□	Fired Thickness μm	Resolution Lines/Spaces μm
INNER LAYER (CO-FIRED)			
Gold (5717D)	5	7	132/91
Silver (6142D)	5	8	109/114
Silver/Palladium (6144D)	20	8	132/88
TOP LAYER (POST FIRED)			
Gold (5715)	4	8	150/100
Platinum/Gold (4596)	80	15	150/100
Silver/Palladium (6134)	20	15	175/75

Figures 6, 7, and 8 and Table 3 show the line resolution capability of gold, silver and silver/-palladium inner layer conductors co-fired on top of 6 blank laminated layers of tape. A screen incorporating 125 μm lines and spaces was used for printing.

The initial and aged adhesion strength of leads soldered to post-fired top layer conductors on a pre-fired tape base are shown in Table 4. The adhesion test was a peel test[17], the pad size 2 x 2mm and the solder 63 Sn/37 Pb. Aging temperature was 150°C. Both the Pt/Au and Pd/Ag top conductor compositions show no decreases in adhesion when aged at 150°C. Adhesion values are similar to those measured for the same conductors when the substrate is 96% alumina or a pre-fired thick film dielectric.

TABLE 4

AGED ADHESION OF POST-FIRED CONDUCTORS

Time (hours)	0	48	142	500
6134 - Pd/Ag	26.0	24.5	25.0	25.0
4596 - Pt-Pd-Au	26.6	30.0	23.6	26.0

Aging Temperature - 150°C
Peel Test - 2mm x 2mm pads
Adhesion Units (Newton's)

1 mil gold wire was thermosonically bonded to gold 5715 post-fired on top of 6-layer tape laminates. Results are given in Table 5 and are equivalent to those expected from 5715 fired on a 96% Al_2O_3 substrate. Each data point represents the average of 50 wire pulls.

TABLE 5

BOND STRENGTH AS A FUNCTION OF AGING TIME (150°C)
AU(5715)

Time	Bond Strength (grams)	% Bond Failures
Initial Value	8.5 (7.7)	0
48 Hours	8.6 (7.7)	0

( ) - Minimum value

Bonding Conditions:

First Bond - power (0.2W), time (63 ms)
Second Bond - power (0.4W), time (63 ms)
.001 inch Au wire
Stage Temperature - 150°C
K&S Model 2401 Bonder

Electrical Properties

Table 6 summarizes electrical properties of the dielectric system. Properties are similar to those of conventional thick film materials.

TABLE 6

PROPERTIES OF FIRED DIELECTRIC

Dielectric constant (K)	8.0	(1kHz)
Dissipation factor (DF)	0.2%	
Insulation resistance (IR)	$>10^{12}$ Ω	

Physical Properties

The inorganic part of the tape was formulated such that fired laminates match the thermal expansion coefficient of the high alumina ceramics used in standard IC packages. This approach allows leadless chip carriers to be surface mounted to mother boards fabricated using the low temperature co-fired ceramic system. Table 7 summarizes physical properties; measurement procedures are given in Reference[14]. The fired density of co-fired laminates is greater than 96% of theoretical density.

TABLE 7

PHYSICAL PROPERTIES

Thermal Expansion (25-300°C)	
Fired dielectric	7.9 ppm/°C
96% alumina	7.0 ppm/°C
Fired density	
Theoretical	3.02 g/cm^3
Actual	>2.89 g/cm^3
Camber	
Fired	±75 μm (±3 mil)
68mm x 68mm (2.7x2.7 inch)	
Surface Smoothness	
Fired dielectric	0.8 μm/50 mm
50mm x 50mm (2x2 inch)	(Peak to peak)
Thermal conductivity	
Fired dielectric	15-25% of 96% alumina
Flexural strength	
Fired dielectric	$2.1x10^3$ kg/cm^2 ($3.0x10^4$ psi)
96% alumina	$3.8x10^3$ kg/cm^2 ($5.6x10^4$ psi)

Figures 9 and 10 are scanning electron micrographs of cross sections through fired laminates made using the low temperature ceramic system. These micrographs illustrate the dense, pore-free microstructure which occurs during firing. Dark regions in Figure 9 are from the ceramic fillers included in the tape formulation. Figure 10 is a composite from a part containing 7 metallized layers made with the 100% silver metallization. All 7 buried metal layers can be seen, as well as an internal via and a via leading to the surface. The part is free of delaminations.

Figure 11 is a micrograph of the top surface of a fired laminate. For comparison, the surface of a standard 96% alumina substrate is also shown. The fully dense surface of the tape dielectric is apparent.

Pond and Vitriol[13] reported that modules made using the low temperature co-fired ceramic system and sealed by solder reflow would pass leak testing at $1x10^{-10}$ cm^3/s after 100 thermal cycles from -65°C to 150°C.

Reliability of Silver Bearing Inner Layer Conductors

Silver and silver/palladium alloys can provide high conductivity buried layer metallizations with relatively low cost; however, industry concern for silver migration[18][19] need to be addressed. Silver bearing conductors have been shown to be reliable in many thick film applications and are widely used in the consumer, automotive and tele-communications industries. Silver is known to diffuse in glass during high temperature firing and to migrate along pore surfaces in the presence of an electrical field and moisture, causing shorts between circuit layers or closely spaced conductor lines[18][19][20].

In the low temperature co-fireable ceramic system silver diffusion is minimized by the low firing temperature, short firing time, and limited number of refires, while a defect-free dielectric tape, high fired density and pore-free microstructure prevent silver migration.

The reliability of silver bearing conductors in multilayer circuits was evaluated under temperature-humidity-bias (85°C / 85% RH / 7.5 Vdc) and high temperature-high voltage conditions (150°C/100 Vdc). Test parts were fabricated using processing conditions described above. Each sample consisted of 6 blank tape layers, for mechanical strength, and a seventh and eighth tape layer upon which electrodes were printed to form a parallel plate capacitor. After firing each capacitor had an area of 0.65 square centimeters.

Test conditions for temperature-humidity-bias and high temperature-high voltage tests are summarized in Table 8.

TABLE 8

TEST CONDITIONS FOR SILVER RELIABILITY TESTS

85°C / 85% RH / 7.5 Vdc

Conductor Metallurgy - 100% Ag, 20Ag/1Pd, 12Ag/1Pd, 8Ag/1Pd, 3Ag/1Pd

Current Limiting Resistor - 130 k Ω, 58 μA per part

Test intervals -   0, 78, 325, 912, 1200 hours

150°C / 100 Vdc

Conductor Metallurgy - 12Ag/1Pd, 3Ag/1Pd

Current Limiting Resistor- 240 k Ω, 417 μA per part

Test Intervals- 0, 24, 188, 240, 400, 820, 1350 hours

Parts were removed from the environmental chamber to equilibrate at ambient conditions approximately one hour prior to making measurements. Insulation resistance measurements were made at 100 Vdc approximately 20 seconds after applying the voltage.

Figures 12 and 13 show average insulation resistance results for both the temperature-humidity-bias and high voltage-high temperature studies. For all cases, average values were greater than 1×10^{12} ohms. No failures occurred and minimum values of insulation resistance, as shown in Figures 14 and 15, were greater than 1×10^{11} ohms for all conductor metallizations studied.

Reliability was examined using the electrolytic hermeticity test. Leakage current as determined in an electrolytic hermeticity test has been shown to correlate with accelerated life (temperature-humidity-bias) testing[21]. Thick film dielectrics with leakage currents below 19 μA/cm^2 did not fail when tested for 750 hours under 85°C/85% RH/20 Vdc conditions. Typical leakage current for air-fired thick film dielectric compositions was 1-10 μA/cm^2. Electrolytic hermeticity tests on multilayer structures made with the low temperature ceramic system gave leakage currents 2 orders of magnitude below those of thick film multilayer dielectrics (Table 9).

TABLE 9

ELECTROLYTIC HERMETICITY RESULTS

	LEAKAGE CURRENT
Thick Film Dielectric	
(air-fire)	1-10 μA/cm^2
Low Temperature Ceramic System	
Silver electrode	<.01 μA/cm^2
Silver/palladium electrode	<.01 μA/cm^2

POST-FIRED RESISTORS

Samples for evaluation of the properties of post-fired resistors were prepared by post-firing Ag/Pd 6134 terminations and Birox® 1900 series resistors[22][23] on a pre-fired base made up of 6 layers of tape. The base was prepared as described in the section on processing. 6134 was printed using a 200 mesh screen with an 0.5 mil emulsion, dried at 150°C for 10 minutes, and fired in a profile with a 10-minute dwell at 850°C and 60 minutes above 100°C. 1900 Series resistors were printed using a 325 mesh screen with an 0.5 mil emulsion, dried at 150°C for 10 minutes and fired in a profile with a 10-minute dwell at 850°C and 30 minutes above 100°C.

Resistance and temperature coefficient of resistance (TCR) was measured on 1.5mm x 1.5mm resistors. The hot temperature coefficient of resistance (HTCR) was measured by comparing resistance values at room temperature to those at 125°C. Similarly, the cold temperature coefficient of resistance (CTCR) was measured by comparing resistance values at room temperature to those at -55°C. Resistance and temperature coefficient data are given in Table 10. for comparison, data for Birox® 1900 series resistors post-fired over dielectric 5704 has also been measured.

Post laser trim stability was determined by monitoring resistance changes on 1mm x 1mm resistors trimmed to 150% of fired value with a single plunge cut. Data is given in Table 11, with results for resistors fired over dielectric 5704 measured for comparison.

TABLE 10

RESISTANCE AND TCR DATA

	1911 (10 Ω/□)		1921 (100 Ω/□)		1931 (1000 Ω/□)		1933 (3000 Ω/□)		1939 (10k Ω/□)		1949 (100k Ω/□)		1959 (1M Ω/□)	
	Tape	5704	Tape	5704	Tape	5704	Tape	5704	Tape	5704	Tape	5704	Tape	5704
Avg. Resistance (Ω/□)	10.8	9.5	136.4	100.3	1210.	923.	3.3k	2.5k	12.0k	12.0k	123.6k	109.0k	1.59M	1.10M
Hot TCR (+25°C - +125°C) ppm/°C	+133	+200	-32	+88	+50	+123	-10	+164	+80	+119	+62	+108	-65	16
Cold TCR (+25°C - -55°C) ppm/°C	+89	+191	-90	+58	-8	72	-48	+131	+23	+55	-14	+25	-190	-111

1.5mm x 1.5mm resistors
Terminations - Pd/Ag 6134

TABLE 11

POST LASER TRIM STABILITY

Birox® 1900 Series
Post-fired on Tape & on Dielectric 5704
48-Hour Exposure Data

	% $\triangle$R			
	25°C		150°C	
	Tape	5704	Tape	5704
1911	.24(.51)	.13(.30)	.40(.64)	.23(.41)
1921	.19(.41)	.06(.16)	.21(.45)	.14(.29)
1931	.02(.05)	.09(.27)	0.0(.02)	.02(.05)
1933	.19(.31)	.22(.41)	.20(.37)	.13(.42)
1939	.07(.15)	.02(.08)	.17(.23)	.12(.28)
1949	.09(.20)	.03(.18)	.13(.30)	.05(.15)
1959	.10(.23)	.25(.69)	.10(.23)	.24(.59)

Trim conditions
  power   - 0.7-1.0 W
  frequency - 3.0 kHz (1911, 1921 - 2.0 kHz)
  trim speed - 10 mm/sec.

( ) - maximum value

DESIGN CRITERIA

A prototype laboratory has been set up to determine design limits for using the low temperature co-fired materials system to make multilayer interconnect circuits. After approximately one year of in-house experience, and input from a limited number of customer test accounts, the design limits summarized in Table 12 have been developed.

TABLE 12

SYSTEM CAPABILITIES
LOW TEMPERATURE CO-FIRED CERAMIC SYSTEM

	Demonstrated μm (inch)	Possible μm (inch)
Minimum via Diameter	229 (.009)	125 (.005)
Minimum via Spacing (center to center)	458 (.018)	250 (.010)
Minimum Line Width	179 (.007)	125 (.005)
Minimum Line Spacing	179 (.007)	125 (.005)
Maximum Fired Dimensions	125mm x 125mm (5 x 5)	>125mm x 125mm (5 x 5)
Maximum Number of Layers	20	>40

CONCLUSIONS

A commercial low temperature co-fired ceramic system has been developed.

● The system consists of

(1) a dielectric in green tape form,

(2) gold, silver/palladium and silver via fill and signal line conductors,

(3) thick film conductor and resistor compositions for top layer customization.

- post-fired silver/palladium and platinum/gold conductions for high aged adhesion of soldered connections

- post-fired gold conductors for high bond strengths of 1 mil thermosonically bonded gold wires

- compatibility with post-fired Birox® 1900 series resistors has been demonstrated.

- Standard thick film equipment and processing can be used.

  - burnout and firing are completed within 3 hours in a modified thick film profile at a peak temperature of 850°C.

- Average shrinkage is 12%, and lot-to-lot control of + 0.2% has been demonstrated.

- Silver bearing conductors have been shown to be reliable in temperature-humidity bias and high temperature-high voltage tests.

ACKNOWLEDGEMENTS

The authors gratefully acknowledge the assistance of G. B. Mills, L. B. Straughn, J. E. Hickey, D. F. Rhodes, A. G. Mitchell and P. W. Kehn.

REFERENCES

[1] A compilation of references to multilayer thick film technology is given in R. Cadenhead, "Multilayer Thick Film Anthology", Int. J. for Hybrid Microelectronics I, No. 1, pp. 16-27 (1984).

[2] B. Schwartz, "Microelectronics Packaging: II", Ceramic Bulletin 63, No. 4, pp. 577-581 (1984).

[3] M. L. Stock, "Multilayer Co-fired Ceramic Technology for Hybrid Applications", Hybrid Circuit Technology, May 1985, pp. 15-18.

[4] H. Tosaki, N. Sugishita and A. Ikegami, "New Approaches to Multilayer Hybrid IC With Interlayered Resistors", 1981 ISHM Proceedings, pp. 100-105 (1981).

[5] Y. Shimada, K. Utsumi, M. Suzuki, H. Takamizawa, A. Nitta and S. Yano, "Low Firing Temperature Multilayer Glass-Ceramic Substrate", Proceedings 33rd Electronic Components Conference, pp. 314-319 (1983).

[6] K. Utsumi, Y. Shimada, H. Takamizawa, S. Fuji, S. Nanamatsu, "Application Monolithic Multi-components Ceramic (MMC) Substrate for Voltage Controlled Crystal Oscillator (VCXO)", Proceedings 34th Electronic Components Conference, pp. 433-440 (1984).

[7] S. Nishigaki, S. Yano, J. Fukuta, M. Fukaya and T. Fuwa, "A New Multilayer, Low Temperature - Fireable Ceramic Substrate", 1985 ISHM Proceedings, pp. 225-234 (1984).

[8] T. D. Thanh and N. Iwase, "Low Temperature Sintered Ceramics for Hybrid Functional Circuit (HFC) Substrates", Proceedings International Microelectronics Conference, Tokyo, Japan, pp. 220-223 (1984).

[9] K. Kawakami and M. Takabatake, "A Low Temperature Co-fired Multilayer Ceramic Substrate", presented at Electronics Division Meeting of American Ceramic Society, Orlando, Florida, October 14-15, 1985.

[10] K. Kondo, M. Oluyama and Y. Shibata, "Low Firing Temperature Ceramic Material for Multilayer Ceramic Substrates", presented at Electronics Division Fall Meeting of American Ceramic Society, Orlando, Florida, October 13-16, 1985.

[11] K. Niwa, Kurihara, Y. Yokouchi and N. Kamehara, "Multilayer Ceramic Circuit Board with Copper Conductor II", presented at Electronics Division Fall Meeting of American Ceramic Society, Orlando, Florida, October 13-16, 1985.

[12] W. A. Vitriol and J. I. Steinberg, "Development of a Low Temperature Co-fired Multilayer Ceramic Technology", 1983 ISHM Proceedings, pp. 593-598 (1983).

[13] R. G. Pond and W. A. Vitriol, "Custom Packaging in a Thick Film House Using Low Temperature Co-fired Multilayer Technology", 1984 ISHM Proceedings, pp. 268-271 (1984).

[14] J. I. Steinberg, S. J. Horowitz and R. J. Bacher, "Low Temperature Co-fired Tape Dielectric Materials Systems for Multilayer Interconnections", 5th European Hybrid Micro-electronics Conference 1985, Stresa, Italy, pp. 302-316 (1985).

[15] J. F. Sproull, D. J. Gerry and R. J. Bacher, "A High Performance Gold/Dielectric/Resistor Multilayer System", 1977 ISHM Proceedings, pp. 20-24 (1977).

[16] B. E. Taylor, R. R. Getty, J. Henderson and C. R. S. Needes, "Air and Nitrogen Fireable Multi-layer Systems: Materials and Performance Characteristics", 1983 ISHM Proceedings, pp. 43-53 (1983).

[17] "Method of Test for Wire Peel Adhesion of Soldered Thick Film Conductors to Ceramic Substrates", A74672, E. I. du Pont de Nemours and Co., Photosystems and Electronic Products Department, Electronic Materials Division, Wilmington, Delaware, 19898.

[18] D. E. Riemer, "Materials Selection and Design Guidelines for Migration-Resistant Thick Film Circuits with Silver-Bearing Conductors", Proceedings 31st Electronic Components Conference, pp. 287-292 (1981).

[19] H. M. Naguib and B. K. MacLaurin, "Silver Migration and the Reliability of Pd/Ag Conductors in Thick Film Dielectric Crossover Structures", IEEE Trans Components, Hybrids and Manufacturing Technology CHMT-2, No. 2, pp. 196-207 (1979).

[20] A. Hornung, "Diffusion of Silver in Borosilicate Glass", Proceedings 1968 Electronic Components Conference, pp. 250-255 (1968).

[21] C. R. S. Needes and D. P. Button, "Reliability Testing of Thick Film Multilayer Materials", Proceedings 35th Electronic Components Conference, pp. 505-511 (1985).

[22] T. R. Allington, H. E. Schmidt, L. H. Slack and B. E. Taylor, "A Low Cost Thick Film Multilayer Hybrid System", Proceedings 34th Electronic Components Conference, pp. 314-323 (1984).

[23] H. E. Schmidt and T. R. Allington, "Laser Trimming Resistors over Dielectrics", Hybrid Circuit Technology, October 1985, pp. 35-43.

TABLE 1

A COMPARISON OF THICK FILM AND HIGH TEMPERATURE CO-FIRED TECHNOLOGY

	Thick Film	High Temp. Co-Fired
Dielectric	Paste Composition Glass + Fillers	Tape 90-96% Al_2O_3
Diel. Constant	7-10	9.5
Conductor Metallurgy	Ag, Au, Pd/Ag Cu	W, Mo, Mo-Mn
Sheet Resistance m Ω/□/12.5 μ	2-4	16-30
Firing	800-950°C 45-60 Minutes Air (Precious Metals) N_2 (Cu)	>1500°C >24 Hours H_2 Atmosphere
Vias	Screen Printing .010 in. diam. .025 in. center	Punching .005 in. diam. .010 in. center
Line Resolution Line/Spaces	.006/.006 inch	.004/.004 inch
Multilayer Proc.	Sequential, many steps	Co-fired, fewer steps
Substrate Required	Yes	No
Lead Attach.	Direct	Plating Req.
Resistor Compatibility	Yes	No
Capital Investment	Low	High

CO-FIRED CERAMIC PROCESS FLOW

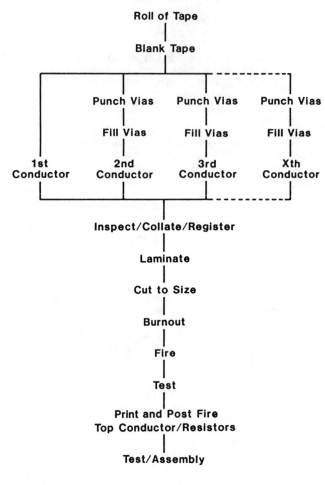

Figure 2 - Co-Fired Ceramic Process Flow

MATERIAL SYSTEM

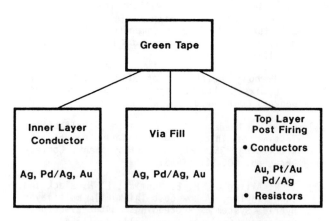

Figure 1 - Low Temperature Co-Fireable Ceramic Materials System

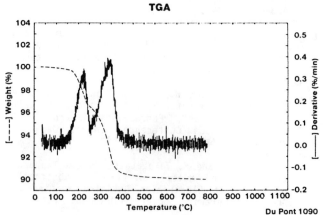

Figure 3 -
Weight Loss (TGA) as a Function of Temperature
(Heating Rate - 5°C/minute)

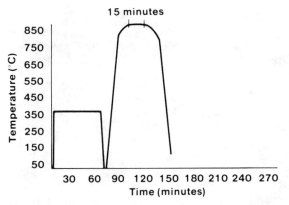

Preburnout
350°C 1 hour
85% weight loss

Firing
Modified Birox profile
.75 speed
15 minutes @ 850°C

Figure 4 - Burnout/Firing Profile for Low Temperature Co-Fired Ceramic

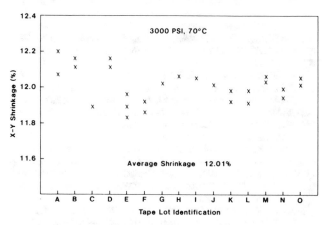

Figure 5 - Lot-to-Lot Shrinkage Control

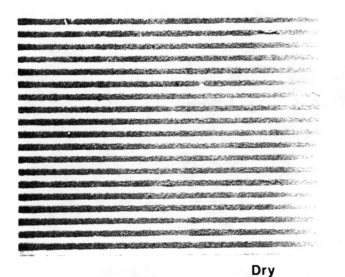

Dry   **Fired**

Figure 6 - Line Resolution (Dry and Fired) for Gold Inner Layer Conductor

Dry   **Fired**

Figure 7 - Line Resolution (Dry and Fired) for Silver Inner Layer Conductor

Dry   **Fired**

Figure 8 - Line Resolution (Dry and Fired) for Silver/Palladium Inner Layer Conductor

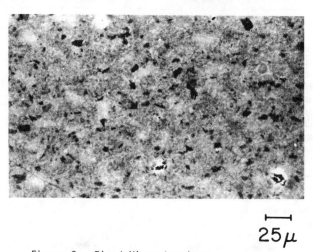

25μ

Figure 9 - Fired Microstructure

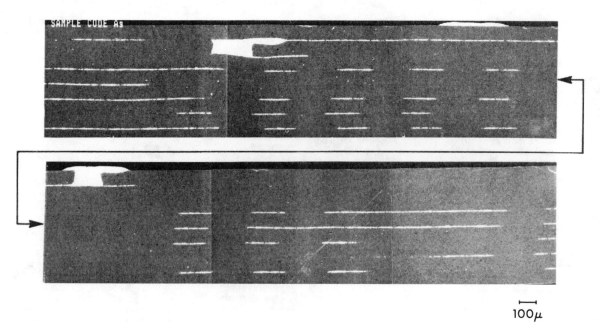

100μ

Figure 10 - Multilayer Structure, 7 Buried Layers, Silver Metallization

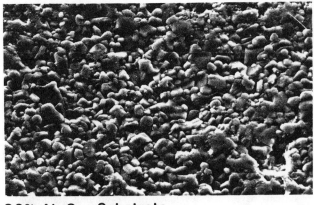

96% Al₂O₃ Substrate

Low Temperature Cofired Tape

$\overline{10\mu}$

Figure 11 -

Surface Density of Standard 96% Al$_2$O$_3$ Substrate

and Low Temperature Co-fired Ceramic

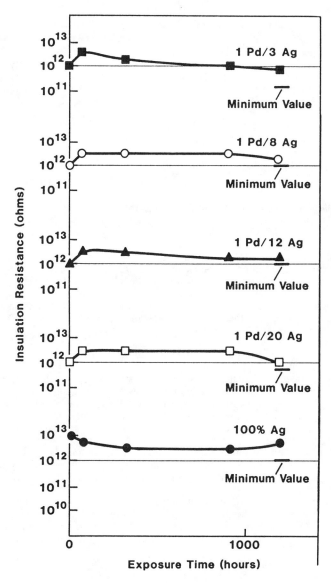

Figure 12 -

Insulation Resistance as a Function of Exposure
Time for Silver and Silver/Palladium Inner Layer
Conductors, Test Condition 85°C/85% RH/7.5 Vdc.

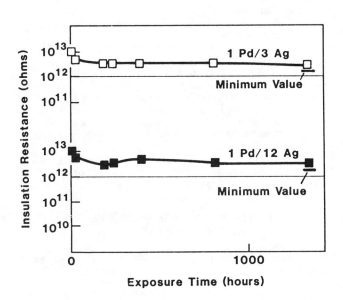

Figure 13 -

Insulation Resistance as a Function of Exposure
Time for Silver/Palladium Inner Layer Conductors,
Test Condition 150°C/100 Vdc.

445

A NEW LOW TEMPERATURE FIREABLE Ag MULTILAYER CERAMIC SUBSTRATE HAVING POST-FIRED Cu CONDUCTOR (LFC-2)

By
S. Nishigaki, J. Fukuta, S. Yano, H. Kawabe, K. Noda and M. Fukaya
Narumi China Corp.
Narumi Technical Lab.
Denjiyama, Midoriku, Nagoya, Japan

Abstract

This paper describes a new Cu conductor and RuO_2 resistor system compatible with a low temperature fireable ceramic substrate interlayered Ag conductor, which was presented at ISHM'85[1]

With current trends in hybrid microelectronics, thick film Cu multilayer systems have been developed on the grounds mainly of electrical conductivity, cost reduction, and high circuit density. The aim of our system is to develop a substrate, with the same function as a Cu multilayer substrate having thick film resistors.

The process is as follows: First; a ceramic green body [60 glass ($CaO\text{-}Al_2O_3\text{-}SiO_2\text{-}B_2O_3$) 40 alumina] containing a multilayered Ag conductor for internal conductor and RuO_2 resisters are co-fired in an air atmosphere at about 900°C (if necessary, Ag-Pd conductor for external conductor can be also formed). Secondly; a Cu conductor is formed on the top of the multilayered substrate by subsequent post firing at 600°C in an N_2 atmosphere. It was found that the top Cu conductor pattern can be throughly interconnected by post-firing through Ag-15Pd vias interconnecting with the internal Ag conductor pattern. (As a matter of course, these Ag-15Pd vias are formed by co-firing.) This temperature, namely 600°C, should be below the eutectic temperature (779°C) of Ag and Cu.

The resulting substrate is characterized as follows.
Conductor
1) High speed signal processing can be attained due to the high electrical conductivity of both internal (Ag) and external (Cu) conductors and also due to the low dielectric constant ($\varepsilon = 7.7$) of the ceramic.
2) High circuit density can be attained by forming a fine component-pattern of the Cu conductor onto the substrate by post-firing and by interlayering the signal patterns of the Ag conductor by co-firing.
3) Post-firing of the Cu conductor having high migration-resistivity has an advantage for fine patterning with accurate component registration.
4) Two kinds of top conductors, comprised of Cu and Ag-Pd can be implemented on the top of the substrate in response to the customer's requirement. The conductors should be selected by taking into account that migration in the top circuit pattern is severe and there is a necessity to use wire bonding pads (Ag-Pd) or flip chip pads (Cu).
Process
1) Multilayered co-fired high conductive conductor (Ag) patterns can be easily formed by air firing without any difficulty in burning out the tape binder. It is very efficient to have only the top Cu conductor formed by N_2 firing.
2) RuO_2 resistors can be formed by co-firing in an air atmosphere much more easily than in the case of N_2 fireable resistors. These RuO_2 resistors do not deteriorate at all because they are protected by the ceramic overcoat during post firing.

1. INTRODUCTION

With the rapid progress of LSI technology, it is obvious from current trends in hybrid microelectronics that requirements for increased hybrid circuit density, improved data processing rates, higher reliability, and cost reduction are becoming more important. Thick film copper multilayered substrates, under these circumstance, are being focused on, because of the lower cost of the conductor along with its high electrical conductivity. Cu conductor is also advantageous for fine patterning due to lower migration, which is caused by the lower ionization level of Cu oxide than Ag oxide. These oxides are formed by the reaction of the conductor and environmental humidity (water). In the Cu thick film process, it is also required to implement thick film resistors, which should be fireable in an N_2 atmosphere.

However, it is well known that the process of making Cu multilayer substrates requires strictly controlled firing in an N_2 atmosphere,[2],[3],[4],[5],[6] by which organic binders in insulator layers should be removed as decomposed or oxidized gas without the conductor being oxidized. In an N_2 firing, burning out the binder becomes more difficult as the thickness of insulator layers increases. In the Cu thick film process, another difficulty lies in forming thick film resistors with desired resistance from such materials[7],[8],[9] as LaB_6, SnO_2, La-Ta-B, $TaSi_2$.

Our new Cu conductor and RuO_2 resistor system interlayering a silver conductor is thought to overcome these difficulties allowing the same function as a thick film Cu conductor multilayer system having thick film resistors. This new system is of course different from the other Cu thick film processes reported, which are compatible to RuO_2 resistors (reduction method[10] and dual method[11],[12]).

Previously, we reported on a similar kind of low temperature co-fireable (880°C) ceramic substrate system, which is composed of 60 (by weight) glass ($CaO\text{-}Al_2O_3\text{-}SiO_2\text{-}B_2O_3$) 40 alumina ceramic, a Ag multilayer conductor (internal), a Ag-20Pd top conductor and RuO_2 resistors. This new system is characterized by replacing the Ag-20Pd top conductor with a Cu top conductor which is formed by post-firing at 600°C onto the co-fired substrate having a Ag multilayer conductor and RuO_2 resistors.

This paper focuses mainly on the process, reliability and characteristics of interconnection of Cu and Ag conductors with RuO_2 resistors.

2. OUTLINE OF OUR PREVIOUSLY REPORTED SYSTEM

In our previous system (abb. LFC-1), a mixture of an alumino-calcium borosilicate glass ($CaO\text{-}Al_2O_3\text{-}SiO_2\text{-}B_2O_3$)

powder and alumina powder in weight percent of 60 and 40 respectively was used as ceramic insulator material for green tape, printed insulator layer and printed overcoat. The resulting physical properties of this ceramic are summarized in Table 1. A low dielectric constant (7.7) and a low thermal expansion coefficient ($5.5 \times 10^{-6}/°C$) close to that of Si chip ($3.5 \times 10^{-6}/°C$) are its special features. Figure 1 shows a schematic cross section of LFC-1 substrate, in which the kind of materials of each layer are represented. For the internal multilayer conductor including vias, almost pure Ag was employed along with insulator layer having a high electrical resistivity ($>10^{14}\Omega \cdot cm$) and a high breakdown voltage ($>4KV$ at 30μm thick) after severe environmental tests. The internal conductor had high electrical conductivity (2.3 mΩ/□, surface resistivity at 8μm thick) almost the same as Cu conductor. This property is very beneficial for obtaining a high-speed processing rate. On the other hand, Ag-20Pd having higher migration-resistivity than pure silver was used for the external (top) conductor. External resistors with a surface resistivity ranging from 10Ω/□ to 1MΩ/□ along with good TCR of less than ±250ppm/°C were formed on the top of the substrate together with the insulator layer as overcoat. The resulting green component was co-fired at 880°C for 20 minutes in a modified conventional belt furnace. Finally, laser trimming of the RuO_2 resistors was carried out through the overcoat.

3. THE NEW LOW TEMPERATURE FIREABLE CERAMIC SUBSTRATE (LFC-2) HAVING Cu-Ag INTERCONNECTION

3-1. Process (including materials and structure)

3-1-1. Co-firing of the green component containing multilayer Ag conductor, RuO_2 resistors and Ag-15Pd vias for interconnection

A mixture of an alumino-calcium borosilicate glass (CaO-Al_2O_3-SiO_2-B_2O_3) powder and alumina powder in weight percent of 60 and 40 respectively is used for ceramic insulator materials. At first, a green sheet with the thickness of 0.1~0.4 mm is formed by a conventional doctor blade process employing a slurry consisting of the insulator material powder and organic vehicles comprising of solvent, plasticizer and binder. Next, through holes or via holes (with a diameter, for example, 0.3 mmφ for 0.3 mm thick green sheet) are formed by a punch and die method on the green sheet blanked to the required size.
Subsequent conductor printing for via hole filling and for patterning on green sheets is carried out using Ag-15Pd conductor paste and Ag paste (almost pure silver) respectively. These green sheets with printed conductors are stacked and then laminated at 110°C under a pressure of 70kg/cm². On this laminated body, RuO_2 resistors using an in-house RuO_2 paste system (100 Ω/□~1 MΩ/□) [mixture of RuO_2 powder and glass powder (CaO-Al_2O_3-SiO_2-B_2O_3)] are formed on Ag-15Pd electrodes. The resistors are also covered using overcoat paste containing the same insulator material. The resulting green component is co-fired in a modified conventional belt furnace at 880°C for 20 minutes.

3-1-2. Post-firing of the Cu top conductor and resistivity adjustment of RuO_2 resistors by laser trimming

On the co-fired green component, Cu top pattern-conductor paste (Du Pont 6001D was used in this study) is formed so as to interconnect with the top vias implemented on the co-fired substrate. The printed substrate is then fast-fired at 600°C for 10 minutes in an N_2 atomsphere (as shown in Fig. 2 along with the heat profile of co-firing). Finally, the co-fired RuO_2 resistors are laser-trimmed through the overcoat to adjust the resistivity. The whole process is shown in Fig. 3. The schematic cross section is also illustrated in Fig. 4.

4. RESULTS AND DISCUSSION

4-1. Physical and thermal properties of LFC-2 ceramic substrate

4-1-1. Good physical properties of LFC ceramic

Our sintered ceramic is composed of three phases including Al_2O_3 particles, the glass phase of alumino-calcium borosilicate and anorthite crystalline. The anorthite crystalline is formed as the result of a reaction between the glass phase and Al_2O_3 particles as shown in XRD patterns in Fig. 5. Figure 6 shows that the sintered ceramic has a sound microstructure having extremely few pores and a mechanical strength of 2,000 kg/cm². The low dielectric constant of the ceramic (7.7 at 1MHz) is caused mainly by the glass composition, which has a low dielectric constant of 7. The low thermal expansion coefficient ($5.5 \times 10^{-6}/°C$) is also attributed to the segregation of anorthite crystalline having a low thermal expansion coefficient of $4.5 \times 10^{-6}/°C$.

4-1-2. High deformation temperature of glass constituent

Our ceramic has the following further additional characteristics; ①The ability to perform a fast firing while leaving the substrate blister-free due to an easy burn out of organic materials in the multilayered green body by virtue of the high deformation temperature (730°C) of the glass constituent. Oxydized gas escapes through pores which are kept open up to the high deformation temperature (730°C) before sintering begins. It is explicit from figure 7 that the number of blisters generated are proportional to the firing speed and are inverse to the deformation temperature of the glass constituent. Our sintered ceramic has no blisters even in a firing ratio of 200°C/minute. ② A high heat deformation resistance in repetition of heat treatment by post-firing. It is clear from Fig. 8 that in LFC-2 substrate no bending, due to softening of the glass constituent, is observed even by repeating the firing 5 times. Such high heat deformation resistance is attributed to the crystallization of anorthite and results in high heat resistance in post-firing.

4-2. Insulation properties of LFC-2 ceramic

4-2-1. High breakdown voltage and high insulation resistance of insulator layers between Ag and Ag or Ag and Cu conductors

LFC-2 ceramic substrate composed of three phases namely Al_2O_3 powder, CaO-Al_2O_3-SiO_2-B_2O_3 glass and anorthite crystalline has excellent insulation properties such as high insulation resistance (IR), high breakdown voltage (BDV) and high migration resistivity. We can see from Fig. 9 that the insulator layers with a thickness of 30μm between Ag conductors, have high insulation resistance above $10^{14}\Omega \cdot cm$ and high break down voltage of above 4KV (equivalent to 133KV/mm) after long exposure to severe environmental conditions. These values are relatively higher than those of alumina insulator layers between W conductors. Since our insulator layers have a very high BDV, we tested the BDV using a thinner insulator layer by screen printing in place of a green sheet. We employed the test pattern shown in Fig. 10 for evaluations of insulation properties of both insulator layers between the Ag-Ag conductors and the Ag-Cu conductors.

It can be seen from Fig. 11 that the insulator layer between the Cu top conductor and the Ag conductor also has a high insulation resistance (above $10^{14}\Omega \cdot cm$) and high breakdown voltage (above 5KV/30μm). These excellent insulation properties of co-fired insulator layers seem to be due to the fine composite structure of the insulator layers, which exclude carbon, and to the fine microstructure with very few pores. Figure 6 shows this microstructure in comparison to an alumina substrate (co-fired). In addition, it is explicit from Fig. 12 that any Ag and Cu diffusion into the insulator layers can not be seen due to the good compatibility between the ceramic insulator material and the Ag and Cu conductors, and to the lower sintering temperature with short holding time.

4-2-2. High migration resistivity of Cu conductor compatible to LFC ceramic

On the other hand, the Cu top conductor has an excellent migration resistivity as shown in Fig. 13. It can be seen from the figure that the migration resistivity of the Cu conductor with a narrow electrode gap (0.1 mm) is 50 times superior to Ag-20Pd conductor formed on LFC substrate by co-firing. It should be however noted that the Ag-20Pd conductor itself has one digit higher migration resistivity (0.5 mm gap) than that which has been post-fired on alumina substrate. For reference, the migration test pattern and a magnified view of dendritic growth (caused by Cu^{++}) after the migration test (100μA) are given in Fig. 14. The higher migration resistivity of the Cu conductor compared to the Ag-Pd conductor is thought in general to be caused by the lower ionization of Cu oxide ($Cu \rightarrow CuO \rightarrow Cu^{++}$ in water) than that of the Ag oxide ($Ag \rightarrow Ag_2O \rightarrow Ag^+$ in water).[13] It is obvious from these results that migration resistivity depends on not only the kind of metals but on also the kind of insulator materials. LFC substrate seems to restrict the ion current related to ionization of Cu or Ag oxide due to the fine composite structure containing anorthite crystalline free of pores, and to the electrochemically stable property of the ceramic.

4-3. Electrical and physical properties of the conductor system

4-3-1. Highly conductive conductors (Ag and Cu) and good compatibility of the top conductor (Cu) with solder

Both conductors composed of a Ag multilayer inner conductor (co-fired) and a Cu top conductor (post-fired) have high electrical conductivity or low surface resistivity of 2.3mΩ/□ (8μm) and 2.5mΩ/□ (18μm) [ref. specific electrical resistivity of Ag and Cu are 1.63μΩ·cm and 1.69μΩ·cm respectively]. These conductor patterns, including interlayered Ag and top-layered Cu, are mainly used as signal lines and component pads, respectively. It is important for component pads to have high registration for positioning and to have high migration resistivity for adjacent fine lines. Cu top conductor pads require three kinds of performance including good solderability, solder leach resistance and peel strength. As shown in Fig. 15, it can be found that the Cu conductor test pattern, which was formed on the LFC substrate by post-firing, has an excellent solder wettability not only initially but also after 10 repeated solder dippings in 62Sn/36Pb/2Ag at 260°C. This means also that Cu conductor patterns have good solder leach resistance. On the other hand, conventional 80Ag-20Pd conductors on alumina substrates initially have an excellent solder wettability, but a considerable amount of solder leaching occures with only 5 repeated solder dippings. After 10 times, 90% of the solder is leached.

It is also very significant for component pads to have high

peel strength with a minimum value of 20N (2mm □) after ageing. In Fig. 16 it can be seen that the Cu top conductor formed by post-firing on LFC substrate has good peel strength (with saturation at 50~100 hours ageing) as observed in a Cu conductor in alumina substrate.[14]

4-3-2. Highly reliable interconnection between Cu and Ag conductors

We realized a highly reliable interconnection between the Cu and the Ag conductors by post-firing the Cu conductor at a temperature lower than 779°C (Cu-Ag eutectic temperature). For the Cu top conductor paste, we selected Du Pont 6001D that can be sintered at 600°C. Figure 17 shows the cross sectional view of the Cu-Ag interconnection. We can see from the microstructural photograph of Fig. 17 that the Cu top conductor pattern is soundly interconnencted with the Ag-15Pd via conductor, which is also further interconnected with the internal Ag multilayer pattern. For the via conductor material, Ag-15Pd was used taking into account shrinkage-matching with the green sheet (0.3mmt) and firing temperature.

In order to see the interdiffusion of these metals between the Cu top conductor and the 15Ag-Pd via conductor EPMA analysis was carried out. It is explicit from Fig. 18 that no significant interdiffusion can be observed. In order to verify the reliability of the interconnection system of the Cu external conductor and the Ag-15Pd via holes including the Ag internal conductor, the electrical resistance change of this interconnection system after heat cycles was measured using a test pattern having 804 via holes as shown in Fig. 19. The resistance change ($\triangle R/R$) is very small, as low as below 2% after 500 cycles as shown in Fig. 20. This means that two different kinds of metals, Cu and Ag, are being interconnected without any degradation even after many heat cycle tests. The small increase of resistance change after heat cycle tests seems to be caused by a small oxidation of the Cu conductor.

4-4. High Performance of RuO$_2$ resistors

It is well known that RuO$_2$ resistor systems has an advantage for obtaining a wide range of surface resistivity along with good TCR and high reliability with air firing. Our in-house RuO$_2$ resistor system has also good resistivity-TCR characteristics with a surface resistivity ranging from 100Ω/□ ~ 1MΩ/□ and with a TCR less than ±100 ppm/°C. Fig. 21 and Fig. 22 show that our RuO$_2$ resistor system which has an overcoat layer 10μm thick, is very stable even after post-firing at 600°C in an N$_2$ atmosphere, resulting in a small change of resistance (<1.2%) and allowable difference of TCR. For these studies, the test pattern as shown in Fig. 23 was used.

The fact means that RuO$_2$ resistors are well protected by the overcoat insulator layer (with 10μm thickness) for post-firing. The co-fired resistors with subsequent laser-trimming also have high reliability showing a change of less than 0.3% after severe ageing tests such as high temperature with high humidity, heat cycle, thermal shock and heat resistance for solder. Figure 24 is a cross sectional view of the sound microstructure of these resistors which were laser-trimmed through the overcoat.

4-5. Trial application

In order to realize this system for application, we fabricated the following three kinds of trial substrates. Figure 25 shows a Ag multilayer co-fired substrate (34 x 34 x 1.8 mm^3) on which the Cu top conductor is to be post-fired later. This substrate was produced to study such processing as green-multilayering (8 Ag conductor layers and 7 insulator layers using each 0.3mmt green tape) and multi-via-filling (804 vias for each two layers). As the 2nd trial substrate, a multichip mother board $\{90 \times 40 \times 0.8 \text{ mm}^3$, 3 Ag layers, top Cu layer, 200µm line width, 0.3mmt green tape for 3 insulator layers$\}$ for 64K bits frame memory was fabricated to study processing of interconnecting Ag and Cu (top) conductors. An HIC substrate ($30 \times 60 \times 0.8$ mm^3) containing two conductor layers (top layer: Cu, 2nd layer: Ag) and thick film resistors of 41 pieces was fabricated to study total process performance including RuO$_2$ resistors, as the third trial substrate. Electrical functions as good as those in conventional alumina substrates were obtained in the latter two kinds of substrates, shown in Fig. 26.

5. CONCLUSION

A new low temperature co-fireable (880°C) Ag multilayer ceramic substrate having a post-fired (600°C) Cu conductor on the top of the substrate was developed (LFC-2). This substrate can be easily obtained by co-firing in air together with RuO$_2$ resistors, which do not deteriorate in subsequent post-firing in N$_2$ for Cu conductor formation. In addition, high reliability of Cu (top)-Ag (inner) interconnections, insulator layers between the Ag multilayers and RuO$_2$ resistors covered with 10µm overcoat in the new substrate has been attained. This new substrate has an advantage for use with analog and digital co-existing microcircuits with high density and high speed in the circuits along with the following excellent features.

1) Low dielectric constant (7.7) of ceramic (60 glass CaO-Al$_2$O$_3$-SiO$_2$-B$_2$O$_3$-40 alumina) and high electrical conductivity (2.3~2.5 mΩ/□) of Ag (multilayer) and Cu (top) conductors realize high speed signal processing.
2) High migration resistivity and fine patterning with accurate registration of Cu top conductor make high surface component density possible.
3) A Cu top conductor system including a multilayer Ag conductor seems to be capable of replacing the thick film Cu multilayer system.
4) Easily formed RuO$_2$ resistors compatible with the Cu conductor appear capable of replacing N$_2$ fireable resistors.

6. ACKNOWLEDGEMENTS

The authors wish to thank Mr. S. Mayumi, and Mr. S. H. Tohyama of Naurmi China Corp. for helpful discussion and the people of the 3rd. section of Narumi Technical Lab. for their technical assistance in this program.

7. REFERENCE

1) S. Nishigaki, S. Yano, J. Fukuta, M. Fukaya and T. Fuwa, "A New Multilayered, Low-Temperature-Fireable Ceramic Substrate", ISHM (1985) 225–234.
2) P. T. Kitano and Dietrich E. Riemer, "The nitrogen furnace in large-volume thick-film production", ISHM (1985) 568–573.
3) C. R. Delotte, "Control of furnace atmosphere: Part I –furnace adjustment", ISHM (1985) 592–594.
4) J. P. Bradley, "Cu thick film nitrogen atmosphere furnace design and firing process considerations", ISHM (1985) 435–440.
5) R. J. Bacher and V. P. Siuta, "Firing process-related failure mechanisms in thick film Cu multilayers", ECC (1986) 471–480.
6) B. E. Taylor, R. R. Getty, J. Henderson and C. R. S. Needes, "Air and Nitrogen-Fireable Multilayer System: Materials and Performance Characteristics, Solid State Technology", 4 (1984) 291–295.
7) D. L. Hankey, M. X. Panzak and R. C. Sutterlin, "Introduction of a novel copper-compatible nitrogen firing resistor system", IMC (1986) 231–238.
8) C. R. S. Needs and P. C. Donohue, "Thick film materials for copper hybrid circuits", IMC (1986) 239–248.
9) O. Makino, H. Watanabe, T. Ishida, "Nitrogen-fireable base metal silicide thick film resistor", IMC (1986) 267–272.
10) C. Y. Kuo, "Thick Film Cu Conductor and Ruthenium-Based Resistor System for Resistor Circuits", ISHM (1983) 54–59.
11) T. Kaneko, M. Yotsuyanagi and M. Sakamaki, "Development of a Cu multilayer hybrid IC including thick film resistor", IMC (1986) 302–308.
12) C. R. S. Needs, "Base metal thick film materials a review of their technology and application", IMC (1982) 94–101.
13) D. E. Riemer, "Material Selection and Design Guidlines for Migration-Resistant Thick-Film Circuits with Silver-Bearing Conductors", ECC (1981) 287–293.
14) C. Y. Kuo, "Adhesion of Thick Film Copper Conductors", ISHM (1981) 70–78.

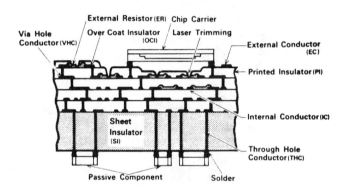

Fig. 1 Cross Sectional View of Representative LFC-1 Substrate having Ag multilayer Conductor and Resistors

ER: External Resistor (RuO$_2$)	THC: Through Hole Conductor (Ag)
OCI: Over Coat Insulator (Ceramic)	VHC: Via Hole Conductor (Ag)
PI: Printed Insulator (Ceramic)	EC: External (TOP) Conductor (Ag-20Pd)
SI: Sheet Insulator (Ceramic)	IC: Internal Conductor (Ag)

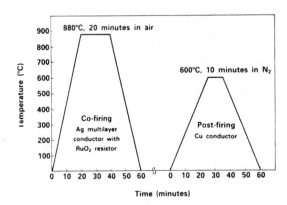

Fig. 2 Heat Profile for LFC-2 Substrate

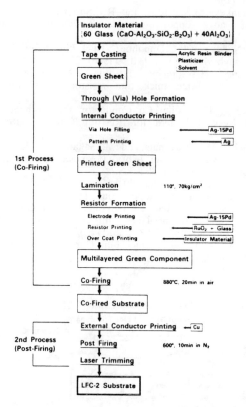

Insulator Material
[60 Glass (CaO-Al₂O₃-SiO₂-B₂O₃) + 40Al₂O₃]

Tape Casting ← Acrylic Resin Binder / Plasticizer / Solvent

Green Sheet

Through (Via) Hole Formation

Internal Conductor Printing

Via Hole Filling ← Ag-15Pd

Pattern Printing ← Ag

Printed Green Sheet

Lamination   110°, 70kg/cm²

Resistor Formation

Electrode Printing ← Ag-15Pd

Resistor Printing ← RuO₂ + Glass

Over Coat Printing ← Insulator Material

Multilayered Green Component

Co-Firing   880°C, 20min in air

Co-Fired Substrate

External Conductor Printing ← Cu

Post Firing   600°, 10min in N₂

Laser Trimming

LFC-2 Substrate

1st Process (Co-Firing)

2nd Process (Post-Firing)

Fig. 3 Fabrication Process of Co-firing (Ag Multilayer Conductor and RuO₂ Resistor) and Post-firing (Cu Conductor) of the New Low Temperature Fireable Ceramic Substrate (LFC-2)

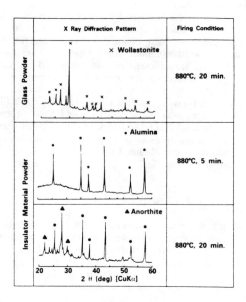

Fig. 5 Changes in X-ray Diffraction Pattern of Insulator Material and Its Glass Constituent

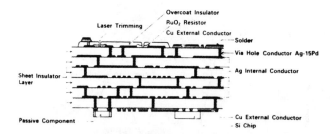

Fig. 4 Cross-sectional View of LFC Substrate with Post-fired Cu External Conductor

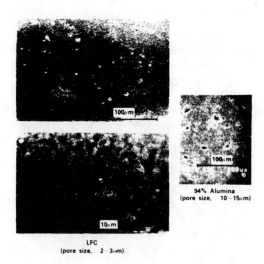

Fig. 6 Microstructure of LFC Substrate with Comparative Alumina Substrate (Co-fired) (Polished Surface)

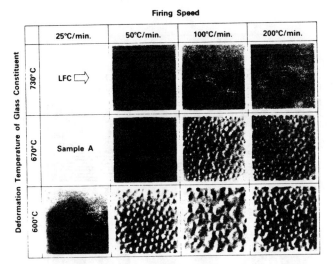

Firing Speed

	25°C/min.	50°C/min.	100°C/min.	200°C/min.
730°C	LFC ⇨			
670°C	Sample A			
600°C				

(Deformation Temperature of Glass Constituent)

Fig. 7 Observation of Blistering in LFC and Other
Substrates Showing the Relationship between De-
formation Temperature of Glass Constituent and
Firing Speed

Materials history 1) mixed ratio of glass and alumina: 60:40 by wt.
    2) kinds of glass: LFC: CaO-Al₂O₃-SiO₂-B₂O₃
                        Sample A: PbO-Al₂O₃-SiO₂-B₂O₃
                        Sample B: SiO₂-B₂O₃
    3) max. firing temp: 900°C
    4) furnace: IR experimental furnace

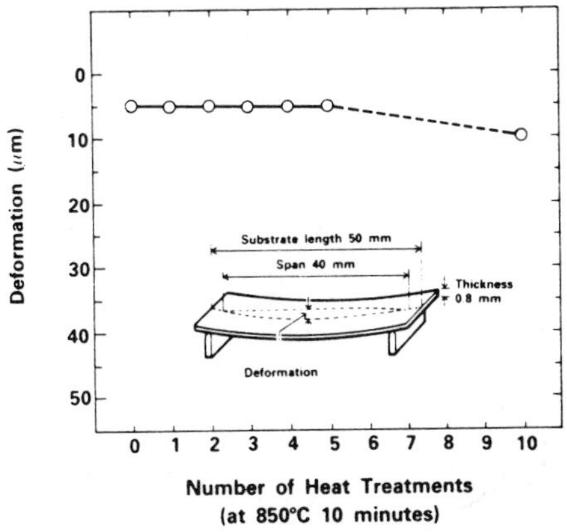

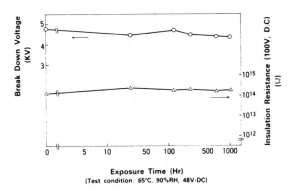

Fig. 8 Deformation of Substrate as a Function of Repeated Heat Treatments

Fig. 9 Insulation Properties of Insulator Layer (30µm by
Screen Printing) between Ag Multilayer Conductors as a Function of Exposure Time

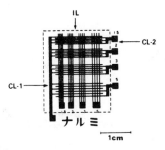

Fig. 10 Test Pattern for Measurement of Insulation Resistance and Breakdown Voltage

CL-1: Top conductor Layer (Cu or Ag)
IL: Insulater Layer
CL-2: Inner Conductor Layer (Ag)

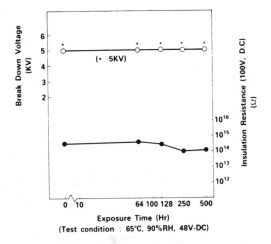

Fig. 11 Insulation Properties of Insulator layer (30 µm by
Screen Printing) between Cu and Ag Conductor as
a Function of Exposure Time

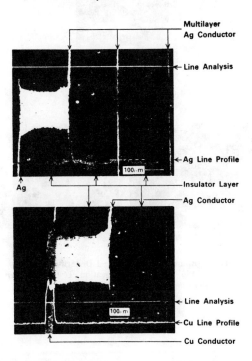

Fig. 12 No Significant Diffusion of Ag and Cu from Their
Conductors into Insulater Layer

451

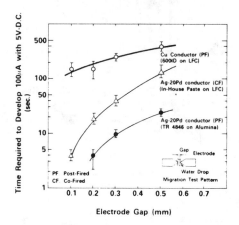

Fig. 13 Migration-Test Results for External Cu Conductor as a Function of Electrode Gap.

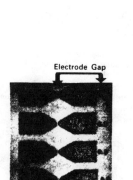

Cu Migration Test Pattern

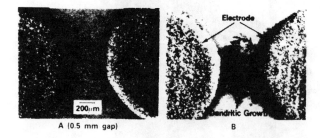

A (0.5 mm gap)   B

Fig. 14 Migration Test Pattern and Dendritic Growth of Cu Ion (A) Before Migration Test; (B) After Migration Test (100μA)

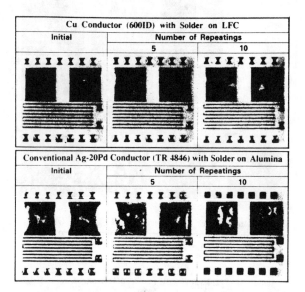

Fig. 15 Solder Leach Test as a Function of Repeating Solder Dipping

(Dipping Condition: 5 second immersion in 62Sn/36Pb/2Ag at 260°C)

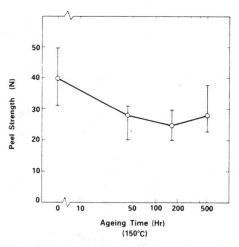

Fig. 16 Peel Strength of Cu Top Conductor (2mm × 2mm Pads) as a Function of Ageing Time with 62Sn/36Pb/2Ag Solder

452

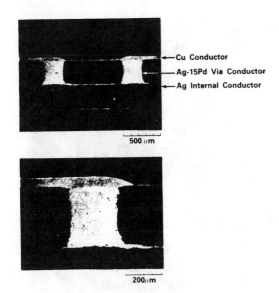

Fig. 17 Cross-sectional View of Interconnection between
Cu Top Conductor and Ag Internal Conductor

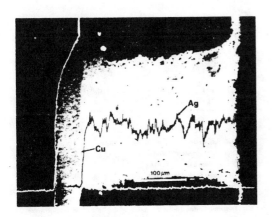

Fig. 18 No Significant Inter-Diffusion in Interconnection
of Cu and Ag

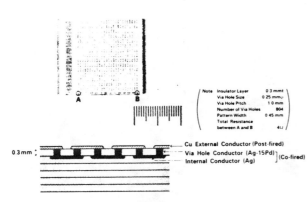

Fig. 19 Cross-sectional View of Test Pattern for Reliability
of Interconnections between Cu (top) – Ag-15Pd
(via hole) – Ag (internal)

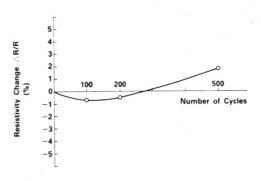

Fig. 20 Small Change of Electrical Resistivity of Cu-Ag
Interconnections after Heat Cycle Test
(Condition : −40°C ⇄ RT ⇄ 150°C, 1 Hr/cycle)

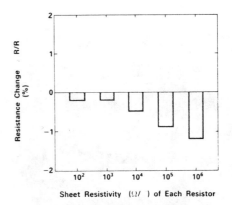

Fig. 21 Resistance Change of Various Resistors (Having a
10μm Thick Overcoat) after Post-firing
(Firing condition : 600°C 10 minutes
in N₂ atmosphere)

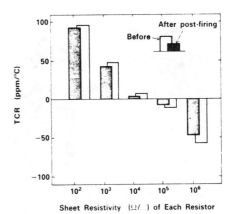

Fig. 22 TCR Change of Various Resistors (Having a 10μm
Thick Overcoat) Before and After Post-firing
(Firing condition : 600°C 10 minutes
in N₂ Atmosphere)

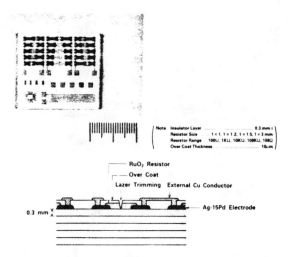

Note Insulator Layer .... 0.3 mm t
Resistor Size   1 × 1, 1 × 1.2, 1 × 1.5, 1 × 3 mm
Resistor Range   100Ω, 1KΩ, 10KΩ, 100KΩ, 1MΩ
Over Coat Thickness 10μm

— RuO₂ Resistor
— Over Coat
Lazer Trimming External Cu Conductor
0.3 mm — Ag-15Pd Electrode

Fig. 23 Cross-sectional View of Test Pattern for Reliability of RuO₂ Resistors

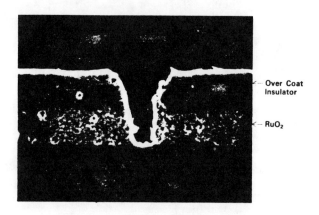

← Over Coat
Insulator

← RuO₂

Fig. 24 Laser Trimmed RuO₂ Resistor

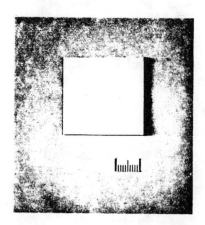

Fig. 25 Ag Multilayer Co-fired Substrate

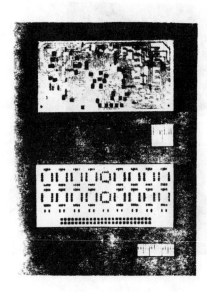

Fig. 26 Trial Substrate of LFC-2, (A) HIC Having Cu Top Conductor and (B) 64K Bits Frame Memory Having Cu Top Conductor and Multilayered Ag Conductor

Table 1.   Physical Properties of the LFC-Substrate

Bulk Density	2.9 g/cm³
Flexural Strength	2.000 kg/cm²
Thermal Expansion Coefficient (r.t. 250°C)	5.5 × 10⁻⁶/°C
Dielectric constant (1MHz)	7.7
Dissipation Factor (1MHz)	3 × 10⁻⁴
Insulation Resistance (100VD.C)	10¹⁴ Ω·cm
Young's Modulus	11,800 kg/mm²
Surface Roughness	0.2 μm Ra
Thermal Conductivity	0.006 cal/cm·sec·°C

NEW TCE-MATCHED GLASS-CERAMIC MULTI-CHIP MODULE:
I. ELECTRICAL DESIGN AND CHARACTERIZATION [†]

W.E. Pence [*], J.P. Krusius [*‡], R. Subrahmanyan [+], C.Y. Li [+]
Cornell University, Electrical Engineering,
[+] Materials Science and Engineering

G. Carrier, G.L. Francis, R.J. Paisley
Corning Glass Works, Research Laboratory

L.M. Holleran
Advanced Packaging Systems

Abstract

A glass-ceramic multi-chip packaging technology for high-I/O count, large-area silicon chips has been explored. The module substrate material is a new glass-ceramic (Corning 9641), which can be TCE (Thermal Coefficient of Expansion) matched to silicon (3.4 ppm/°C). In addition, this material possesses a low dielectric constant (5.2), resulting in reduced package signal propagation delays. A 3", 36-chip test vehicle has been designed, fabricated, and tested to assess the electrical and thermal properties of the new glass-ceramic and their usefulness in packaging applications. Chip and module sizes equal to or greater than 1 cm and 3", respectively, are targets for this new technology, referred to as the Hybrid Systems Module (HSM). The electrical design and characterization of the HSM test vehicle are described in this paper. Materials and thermal test results are discussed in a companion paper at this conference.

Introduction

The HSM (Hybrid Systems Module) test vehicle is designed to demonstrate the viability of Corning 9641 glass-ceramic as a substrate material for advanced packaging applications. To accomplish this in a convincing manner required an aggressive multichip package design, which is capable of testing the ability of the material to pass shrinkage control, surface planarity, and multilayer registration requirements, as well as demonstrating the unique properites of the material which distinguish it from competing ceramics. The unique properties of the 9641 glass-ceramic are (1) a low dielectric constant (5.2), and (2) a Thermal Coefficient of Expansion (TCE) which can be matched to Si (3.4 ppm/°C). The former is advantageous for reducing off-chip delays in the module, while the latter results in significantly reduced thermal expansion-induced stresses, especially critical in flip-chip die attach using area arrays of solder balls for electrical and mechanical contact. TCE-matching not only improves reliability, but results in the ability to mount larger chips, increasing module functionality.

The HSM test module technology is similar to the IBM Multilayer Ceramic (MLC) technology [1], except that it contains only 4 layers of cofired glass-ceramic and is 3"x3" in size. Three layers of buried Au metallization (power, ground, and signal), with minimum widths and spaces of 10 mils, are used (Fig. 1). The top surface carries an additional layer of metallization containing sputtered bonding pads for solder reflow and die attach. Chips are mounted via flip-chip bonding (95/5 Pb-Sn). An array of 6 x 6 chips, each 1 cm x 1 cm, are mounted on the sputtered pads with an array of 81 (9 x 9), 6 mil diameter solder bumps. The power and ground planes contain large-area bypass capacitors, and the signal plane contains a variety of signal-line test structures. The silicon chip set designed for this test vehicle carries signal lines, resistive heaters, and thin-film calibrated temperature-sensing resistors, for the evaluation of the electrical, thermal, and mechanical characteristics of the assembly.

DC, low frequency AC (< 10 MHz), small signal microwave (< 18GHz), and TDR (Time Domain Reflectometry) (t_{rise} = 45 ps) measurements have been performed on module signal lines. AC measurements have been used to extract parameters for interconnect transmission line models. Power and ground plane inductances have also been measured with AC techniques. TDR and two-port measurements have been made to determine propagation delays, impedance control, and crosstalk for the signal lines. Analysis of this data is made to determine the useful frequency range for the package.

Electrical Design

One of the main purposes of the HSM test vehicle is to provide information on the electrical properties of a typical interconnect system fabricated in a cofired 9641 glass-ceramic substrate. The electrical design goal for the test module is to simulate the electrical behaviour of a large multi-chip package. As a minimum, a single layer containing signal

[†] Work at Cornell University supported in part by the Semiconductor Research Corporation (SRC)

[‡] Currently on sabbatical leave at IBM Yorktown Heights

Reprinted from *Proc. 39th Electron. Components Conf. (ECC)*, pp. 647-651, 1989.

traces is required. In addition, a ground plane and via structure are also necessary for controlling signal line impedance and for feeding signals up to the top surface solder-joint pads or probe sites. The test module contains a single layer of signal lines, a ground distribution mesh, and a power distribution mesh. The power and ground layers are designed to supply current to the on-chip heaters and to provide some impedance control for the signal plane. Test structures are designed to provide electrical data for characterizing the module.

10 mil signal line test structures are designed into the signal planes to measure the following: (1) impedance control, (2) propagation delay, (3) reflections, (4) crosstalk, and (5) losses. The power and ground planes contain large-area (1 cm) capacitors for measuring layer-to-layer capacitances. Simple continuity structures are used to measure electrical resistance through the vias into each of the buried layers. These structures are used to check level-to-level registration as well as to measure via resistance. On-chip thin-film structures include an array of four serpentine heaters for simulating a chip power-on condition (either uniform or differential heating), and a calibrated, temperature sensing resistor. Signal lines are also placed on the chip for measuring signal propagation characteristics from the substrate signal lines through the on-chip interconnects. Measurements made on these test structures are presented in the next sections.

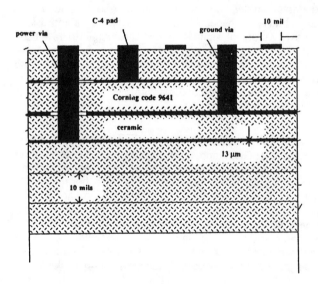

Figure 1. Sketch of the HSM test module cross section

DC Characterization

Electrical characterization of the HSM module is accomplished with several test structures which populate the module and the chip set. All data acquisition is automatic using an IBM PC-AT and Lotus Measure software to control an IEEE 488 bus. Instructions, coded as Lotus macros, are issued to the measuring equipment, and data is returned to the spreadsheet automatically for statistical analysis. DC continuity structures for measuring resistance from the top surface to the three buried layers (power, ground, and signal) provide data on layer-to-layer registration.

Measurements on the DC continuity test structures showed no opens or shorts in any of the 60 continuity test sites probed. DC measurements of the 10 mil wide Au signal lines provide resistance per unit length data on the interconnects as well as a check on the line thickness and Au ink post-fired resistivity. Calculated line thickness is found to be less than 1 mil. Via resistance and power/ground mesh series resistance are also measured. Final DC resistance measurements for the on-chip test structures, including thin-film signal traces, resistive heaters, and calibrated thermometers are made via Kelvin four-point probe techniques. A thermal chuck is used to determine resistance as a function of temperature for the thermal test structures. Table 1 summarizes the DC electrical parameters of the module.

parameter	measured value	comment
Line resistance	199 mΩ/cm	
Via resistance	102 mΩ	
Mesh resistance	< 500 mΩ	Power/Ground
Chip Heater	296 kΩ (RT)	chip power 2W
Temp. Sensor	28 kΩ (RT)	

Table 1. DC measurement results for the HSM module

Low-frequency AC Characterization

LCR measurements on the HSM test vehicle are carried out to determine signal line capacitance and inductance per unit length, as well as power-to-ground capacitance. Characteristic impedances for the signal lines are calculated from this data, to be compared with TDR measurements. Capacitance per unit length for the nominally 10 mil wide signal lines is 1.46 pF/cm (10 MHz), and inductance per unit length is 4.3 nH/cm (10 MHz). This results in a claculated nominal impedance of 54.3 Ω. Post-fired dielectric thickness is determined to be about 6 mils, resulting in a total power-ground capacitance in the module of about 1 µF. It should be noted that although the power-ground mesh was designed to provide some decoupling capacitance for each chip site, this value is far too low to serve as complete decoupling for the entire 36-chip module. The low dielectric constant of the 9641 material requires that (1) for integrated bypass capacitors to be useful, separate high-dielectric constant materials must be laminated into the module before firing, or (2) thin-film capacitors must be incorporated into the assembly after firing, or (3) discrete bypass capacitors must be mounted on the completed package.

Time Domain Reflectometry

TDR measurements are carried out with a Tektronix 7854 mainframe oscilloscope with a 7S12 TDR plug-in unit. This provides 45 ps risetime pulses and a time resolution of about 25 ps. Fig. 2 shows an HSM substrate sample in the high-frequency test jig. Impedance measurements are made in a 50 Ω system to determine the location of impedance discontinuities down the length of the signal test lines. Fig. 3 shows the measurement for a 3", 10 mil wide signal line running over the non-uniform ground plane. Impedance variations of about 15 Ω can be observed due to nonuniformity in the ground plane, neighboring signal

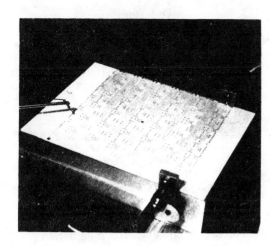

Figure 2. Photograh of an HSM test vehicle

lines, and vias and connectors. The observed variation in characteristic impedance gives rise to internal reflections along the path of the 10 mil wide, 3" long line. Ringing caused by these reflections is evident in the Time-Domain Transmission plot of Fig. 4. The left trace is the input signal (45 ps risetime), and the right trace is the extracted signal at the opposite end of the same 3" long trace measured in Fig. 3. Notice that the output signal is delayed by about 620 ps, the risetime

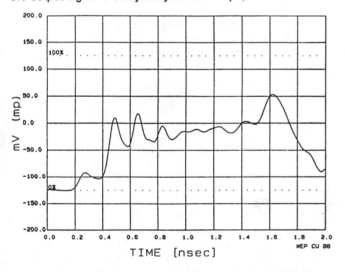

Figure 3. TDR Impedance measurement for 3" line

of the output signal is degraded, and ringing is clearly visible for several nanoseconds after the pulse is received. In Fig. 5, a meandering signal path 33 cm in length is injected with a 45 ps risetime pulse, and the signal is sampled at 6 regular distance intervals as it propagates down the line. There are 11 3" segments and 10 turns in this signal trace. The signal is sampled first after 3", and at intervals of 6" thereafter. Serious degradation in the signal risetime occurs with increasing length, and visible amplitude attenuation is also present.

Insertion loss plots can be extracted from Figs. 4 and 5 by FFT methods [2]. After measuring the insertion loss of the test jig and establishing that cutoff (3 dB) is well above the cutoff frequency for the test module, insertion loss is plotted as a function of frequency for the 10 mil wide, 3" long test line (Fig. 6).

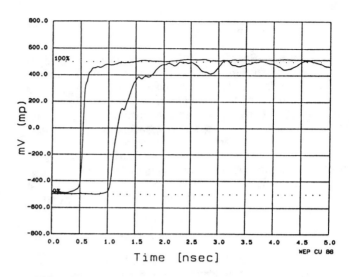

Figure 4. Time-domain transmission for the 3" line

Measured losses in this case are due to skin effect at higher frequencies and the fact that the impedance of the test line is not precisely matched to the 50 Ω cables and measuring equipment. Additionally, the internal impedance fluctuations which cause ringing of the test line can appear as losses in Fig. 6. Skin-effect losses are due to the rough surface properties of the thick-film signal lines.

Fig. 7 is a plot of the near-end crosstalk for two 3" long, parallel lines separated by 10 mils. Saturated crosstalk voltage is about 80 mV along 3". Table 2 summarizes the TDR test results measured for the HSM test module.

parameter	measured value
propagation delay	79.2 ps/cm
phase velocity	1.263×10^{10} cm/s
average impedance	61.9 Ω
impedance range	+/- 15 Ω

Table 2. TDR data for 3" long, 10 mil wide line

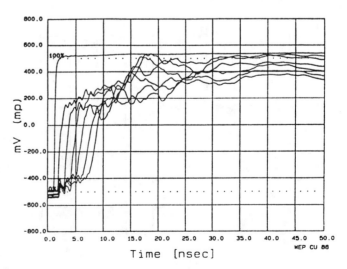

Figure 5. Time-domain transmission for 20 cm line, with sampling at regular distances

457

Frequency-Domain Measurements

A Hewlett-Packard 8510A network analyzer equipped with a set of Gore-Tex High-precision test cables is used to measure the frequency domain characteristics of the test module signal lines. Insertion and return loss measurements to 18 GHz can be performed, as well as impedance measurements and phase and dispersion measurements. Insertion and return loss measurements have been performed for the 3" test signal lines. In addition, 50Ω stripline structures, fabricated in 9641 and in Alumina, are measured to compare insertion and return losses for the two materials. Crosstalk test structures are also fabricated to compare Alumina to the 9641 material.

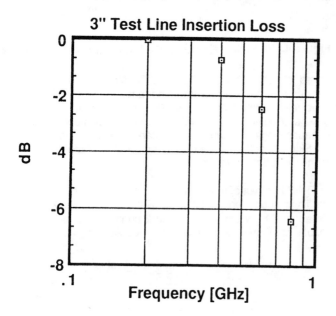

3" Test Line Insertion Loss

Figure 6. Insertion loss data determined from the measurement of Fig. 4 for the 3" test line

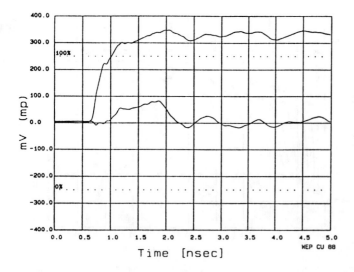

Figure 7. Near-end crosstalk for 3" parallel lines, separated by 10 mils.

Discussion

Electrical parameters for the HSM test module are summarized in Table 3. Thermal measurements are given in a companion paper at this conference. The electrical data set forth in Table 3 demonstrate the usefulness of the Corning 9641 material for advanced packaging applications.

parameter	value
impedance (calc)	54.3 Ω
impedance (meas)	61.9 +/- 15 Ω
cutoff frequency	620 MHz
loss tangent	5.76×10^{-4}
conductor losses	.014 dB/cm
skin depth frequency	250 MHz

Table 3. Signal line electrical parameters

The performance of a real multi-chip packaging application incorporating the 9641 glass-ceramic material can be extrapolated from the electrical test results described in this paper. For a real 36-chip, 3" package, signal routing with appropriate wiring rules requires 10-20 signal planes. Appropriate wiring rules guarantee that crosstalk, reflections, and losses do not result in false switching. The 9641 provides two unique advantages for use as a substrate material, a low dielectric constant and a TCE match to Si. These advantages will be discussed in the next two paragraphs.

Propagation delays in a module are proportional to the square root of the dielectric constant. The ratio of propagation delays in the 9641 glass-ceramic to the propagation delays in Alumina ($\varepsilon \sim 10$) are $(5.2/10)^{0.5} = 0.72$. Propagation delays are thus reduced by 38% in the 9641 material over Alumina. While this number may not seem very large, for processors with clock rates under 10 ns this can translate into a significant reduction in cycle time.

The advantages of the TCE match to Si provided by this glass-ceramic material are demonstrated with a simple example. Consider the differences between two 3" square multichip modules, one fabricated from Alumina and the other from the 9641 glass-ceramic. Due to the TCE mismatch between Alumina and Si, assume that reliability in the flip-chip solder joints requires that chips be no larger than 5 mm. With a matched substrate, chips up to 1 cm in size (or larger) can be mounted. Each of the latter IC chips has 4 times the Si area, and, for the sake of simplicity, is assumed to hold 4 times the number of circuits. Applying Rent's rule (I/O = α(circuits)$^\beta$) with $\alpha = 0.5$ and $\beta = 0.5$ gives twice the I/O count for the 1 cm chips than for the 5 mm chips. To accomodate the same number of circuits per module, four times as many smaller chips must be mounted in the same area. Since the smaller chips have only 1/2 as many I/Os, the total number of chip signal I/Os in the module (chipsxI/Os) is greater by a factor of 2 for the module with the small chips, i.e. the Alumina module. Module routing in this case requires more signal planes to accomplish. This results in a higher cost for the module itself, plus additional manufacturing costs due to the larger number of chips to mount. Finally, more of the critical signal paths would be chip-to-chip because of the lower level of integration in the chips, resulting in a longer cycle time. These calculations are summarized in table 4.

parameter	3" Alumnina module	3" 9641 module
chip size	0.5 cm	1.0 cm
circuits/chip	10^4	4×10^4
chip I/O	50	100
chip count	225	49
circuits/module	2.25×10^6	1.96×10^6
chip I/O x chip count	11250	4900
module signal planes	more	fewer
off-chip propagation delays	105 ns/cm	76 ns/cm
off-chip nets	more	fewer

Table 4. Sample calculations for 3" hypothetical modules fabricated with Alumina and 9641.

The striking differences between the two columns in Table 4 are the total chip count and the total signal port count (chips x chip I/O). The Alumina module requires 225 chips (vs. 49 for the 9641 module) to obtain a similar number of circuits in the module, and has over 6000 more chip I/O ports to interconnect! The matched substrate material provides a substantial reduction in routing problems and lowers the number of parts which could fail.

There are clear advantages to using a matched substrate, especially for applications invlolving large, high integration chips (such as CMOS), which have many I/Os. This materials technology would be especially useful for mid-size processors (i.e. high-end PCs, workstations, and minicomputers), which could be integrated into one or two 3" modules. VLSI CMOS, dissipating much less power than ECL chips, is an ideal candidate for the IC technology incorporated into the module, since cooling would not be as difficult. Significant improvements in size, weight, and speed could be obtained over conventional PC board technology for this type of system.

References

[1] A.J. Blodgett and D.R. Barbour, "Thermal Conduction Module: A High-Performance Multilayer Ceramic Package," *IBM Journal of Research and Development*, Vol. 26, No. 1, pp. 30-36, January 1982.

[2] C.J. Stanghan and B.M. MacDonald, "Electrical Characterization of Packages for High-Speed Integrated Circuits," *IEEE Transactions on Comp., Hybrids, and Manuf. Tech.*, Vol. CHMT-8, No. 4, pp. 468-473, December 1985.

NEW TCE-MATCHED GLASS-CERAMIC MULTI-CHIP MODULE:
II. MATERIALS, MECHANICAL, AND THERMAL ASPECTS[†]

G. Carrier, G.L. Francis, R.J. Paisley
Corning Glass Works, Research Laboratory

R. Subrahmanyan [+], C.Y. Li [+], W.E. Pence [*], J.P. Krusius [*‡]
Cornell University, Electrical Engineering,
[+] Materials Science and Engineering

L.M. Holleran
Advanced Packaging Systems

Abstract

A glass-ceramic multi-chip packaging technology has been explored. The module substrate material is a new glass-ceramic (Corning 9641), which can be TCE (Thermal Coefficient of Expansion) matched to silicon (3.4 ppm/°C) and possesses a low dielectric constant (5.2). The 9641 substrate permits large-area silicon chips (~ 1cm) to be mounted using flip-chip techniques. Multilayer metallization (Au) layers in the substrate are screen printed, and the laminated assembly is cofired to produce a 3", 36-chip test module. 1 cm size silicon test chips with 81 I/Os are mounted via flip-chip bonding (95/5 Pb-Sn). Chip and module sizes equal to or greater than 1 cm and 3", respectively, are targets for this new technology, referred to as the Hybrid Systems Module (HSM). Significant improvements in solder joint reliability can be obtained using the 9641 material as a matched substrate. This permits larger chips to be mounted on the module without a degradation in the mean time to failure of the solder joints. Processing, materials, and thermal characterization data is presented in this paper. Electrical characterization results are given in a companion paper at this conference.

Introduction

The materials, mechanical, and thermal aspects of the HSM test vehicle are described in this paper. The test module contains 4 layers of cofired glass-ceramic and is 3"x3" in size. Three layers of Au metallization (power, ground, and signal), with minimum widths and spaces of 10 mils, are used. The top surface carries an additional layer of metallization containing sputtered bonding pads for solder reflow and die attach (Fig. 1). An array of 6 x 6 chips, each 1 cm x 1 cm, are mounted on these pads with an array of 81 (9 x 9), 6 mil diameter solder bumps. The silicon chip set designed for this test vehicle carries signal lines, resistive heaters, and thin-film calibrated temperature-sensing resistors, for the evaluation of electrical, thermal, and mechanical characteristics of the test vehicle. The on-chip heaters simulate active chip operation to about 2 W per chip. Solder joints are stressed by thermal cycling the entire assembly via the ambient, or by powering the chip heaters. A complete assessment of the mechanical and reliability characteristics of the solder joints and their adhesion to the substrate will be presented.

Thermal cycling data for the test modules is currently being measured. Studies of the solder joint failure for large chips is in progress. A new damage integral method has been applied to the analysis of the solder joint failure. Based on measured data and this theoretical method, a projection of the mean time to failure for this materials system and bonding method is made.

Measured results for the 3"x3" HSM test vehicle demonstrate that this material possesses excellent mechanical and thermal properties for packaging applications. Significant improvements compared to competing materials technologies have been obtained. Data will be projected to real systems applications involving large-area IC mounting technologies, such as flip-chip and flip-TAB. System advantages obtainable from the new material will also be discussed.

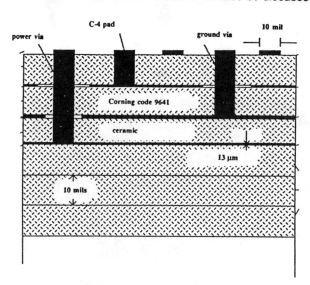

Figure 1. HSM Test Vehicle Cross Section

[†] Work at Cornell University supported in part by the Semiconductor Research Corporation (SRC)

[‡] Currently on sabbatical leave at IBM Yorktown Heights

Reprinted from *Proc. 39th Electron. Components Conf. (ECC)*, pp. 652–655, 1989.

Test Vehicle Design

The 3" HSM test vehicle is designed to demonstrate the viability of Corning 9641 glass-ceramic as a substrate material for advanced packaging applications. To accomplish this required an aggressive multichip package design, which is capable of testing the ability of the material to pass shrinkage control, surface planarity, and multilayer registration requirements, as well demonstrating the unique properties of the material which distinguish it from competing ceramics. The test vehicle is designed to simulate a large multi-chip package, similar in nature to the IBM Multilayer Ceramic (MLC) technology [1]. The unique properties of the 9641 glass-ceramic are (1) a low dielectric constant (5.2), and (2) a Thermal Coefficient of Expansion (TCE) which can be matched to Si (3.4 ppm/°C) (Fig. 2). The former is advantageous for reducing off-chip delays in the module, while the latter results in significantly reduced thermal expansion-induced stresses, especially critical in flip-chip die attach using area arrays of solder balls for electrical and mechanical contact. TCE-matching not only improves reliability, but results in the ability to mount larger chips, increasing module functionality. The following sections describe the processing and fabrication of the HSM test vehicle. Thermal test results are also presented.

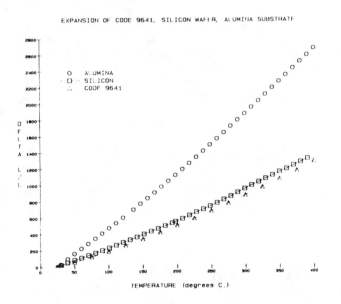

Figure 2. Thermal expansion of Alumina, Si, and the 9641 material

Fabrication

Corning Code 9641 glass-ceramic is a surface-nucleated glass-ceramic material developed at the Corning Glass Works research laboratory. The modulus of rupture (MOR) is 34,000 psi (ASTM test F-394). The expectation of a low dielectric constant and TCE match to Si initiated studies into the applicability of this material to advanced packaging. The HSM test vehicle program is one aspect of this work. The ceramic and test chip processing for this project are described in the next sections.

In preparation for fabricating cofired substrates from the 9641 material, the ground glass and a proprietary binder system are combined to form a slurry. This slurry is cast on a commercial casting table over a polyester support film. A doc-tor blade ensures that the tape is cast to a uniform thickness of 8-10 mils. In subsequent processing, the cast tape is left on the polyester backing for as long as possible.

Six different Au inks are used during fabrication. Power and ground layers are screened with a commercial Au ink. The signal plane is screened with an ink which has a lower post-fired resistivity than that used in the power and ground planes. This is done to provide low series resistance signal lines. A special Au via fill ink is used for the inner vias (power to ground, ground to signal), while the outer vias (signal to pad) are filled with a solder-resistant ink which contains Platinum. A special ink is also used for top-surface pads which are designated for later wire-bonding to a test fixture. Finally, a separate solder-resistant ink is used for the solder pads, which provides good adhesion of the reflowed solder joints to the ceramic substrate and any underlying vias. Wetting problems with this ink forced the use of post-fired sputtered pads for the solder joints.

Screening of the power, ground, and signal planes is accomplished using a 325 mesh stainless steel screen (22.5 °C) with 0.5 mils of emulsion. Green sheets are attached to a registration frame for use in aligning subsequent layers prior to firing. Via layers are punched using a computer-controlled vertical punch with a 10 mil bit. Punched and screened layers are cut into 6" squares and laminated using the registration frames at 25 psi and 75 °C for 1 minute. A 5 mil registration was maintained for the 4-layer laminate over the full 3" of the substrate using these methods.

Firing of the green sheet laminates followed a complex temperature-time profile, with peak temperature reaching above 900 °C for several hours. The lower firing temperature permits better conductors to be used (Alumina fires at 1625 °C) such as Au and Cu. The temperature scheduling during firing is dependent on the material and is custom-tailored to the 9641 material. During firing the glass particles will first sinter to a high density, then devitrify (ceram), resulting in a highly crystalline glass-ceramic whose primary phase is cordierite. Careful control of all process parameters is necessary to produce the phases which produce the desired materials properties. Finally, the cofired substrates are rough cut into 3.5" squares and prepared for reflow die attach and testing. Figures 3-7 show details of the substrate.

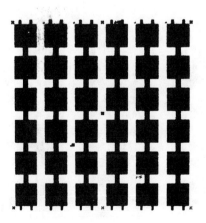

Figure 3. Screened green sheet (power mesh)

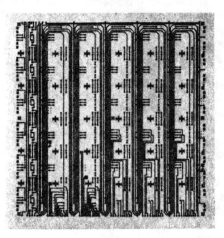

Figure 4. Screened green sheet (signal plane)

Figure 5. Cutaway view of the laminated module, showing the power, signal, and top surface layers.

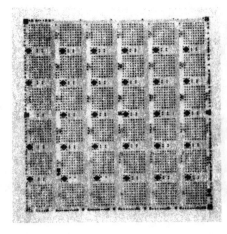

Figure 6. Top view of completed module.

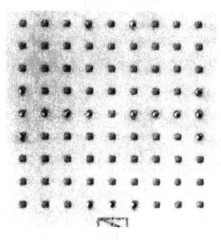

Figure 7. Closeup of one chip site, showing 9x9 array of solder pads.

The silicon test chips are fabricated by the authors. Artwork, glass mask generation, and all processing is accomplished on-site. Processing is done primarily using proprietary liftoff methods [2]. Feature sizes are nominally 20 μm lines and spaces. There are two separate metal depositions, a passivation oxide, and top surface pads. Ni and Cr metallization is used for the on-chip heaters and temperature-sensing resistor. Solder pads (Cr/Cu/Au) are reverse-bias sputtered to provide good adhesion and low contact resistance from the chip test structures to the substrate pads.

Solder is evaporated through a shadow mask onto the Si test chips at each pad site. The solder is reflowed in preparation for chip mounting. Solder is also deposited onto the substrate pads and reflowed. Chips are then mounted to the substrate and heated to complete the chip attachment. Completed modules are tested electrically for continuity and thermal cycled to determine mean time to failure for the solder joints.

Solder Joint Reliability

One of the objectives of the HSM test vehicle is to measure the number of cycles to failure, N_f, for 1 cm Si chips mounted on a 9641 substrate. The number of cycles to failure for the solder joints can be estimated using a damage integral approach [3]. By integrating a crack growth law which is assumed to be a function of the nominal stress intensity factor, the temperature and the environment, thermal fatigue life can be predicted. For the simplest case when the damage accumulation per cycle is constant, the number of cycles to failure N_f can be estimated by the following relation

$$\text{constant} = N_f \int_0^t dt A_0 \exp(Q/RT)\sigma^r$$

where t is the time period of a cycle, A_0 is the pre-exponential term in the rate constant expression, Q is the Arhenius activation energy, and r is a constant between 2.5 and 4. The following equation governs determines the shear stress in the solder joint:

$$\dot{\sigma}A_j/S = \Delta\alpha\dot{T}L - \dot{\varepsilon}h \, .$$

where $\dot{\sigma}$, $\dot{T}$, and $\dot{\varepsilon}$ are the time derivatives of stress, temperature, and strain, respectively. A_j is the area of the solder joint, L is the distance from the solder joint to the neutral point, and h is the solder joint height. S is an effective stiffness for the solder joint, and $\Delta\alpha$ is the thermal expansion mismatch. Clearly the time derivative of the stress is proportional to the TCE mismatch. Increasing the thermal expansion mismatch results in larger stresses in the solder joint. From the integral equation above, σ appears under the integral raised to the power r, which is generally anywhere from 2.5 to 4. Thus the larger stress drastically increases the value of the damage integral. Since the left-hand side of this equation is a constant, number of cycles to failure N_f must be reduced. Clearly a matched substrate ($\Delta\alpha \approx 0$) will result in a substantial increase in N_f.

Standard thermal cycle tests are established in mil std 883C. Cycling for both the bare substrate and the mounted chips is accomplished according to this standard. Failures in the solder joints can be measured in three ways: (1) electrically using the on-chip signal and thermal structures to measure resistance from substrate to chip, (2) visually by inspecting the joints after cross-sections are obtained, or (3) by a pull test. Thermal shock tests have been completed for bare substrates, using procedures similar to standards set forth in mil-std 883C. All substrates maintained full electrical continuity following thermal shock, and no evidence of cracking or other material damage was found.

The number of cycles to failure, N_f, is established from the thermal cycle test data. This experimental data is correlated with the theoretical predictions made using the damage-integral approach described earlier. From this theory and early test results, it is expected that N_f should be significantly increased for a module using the 9641 glass-ceramic. As discussed in the companion to this paper, the expansion match provides important systems benefits in addition to improvements in reliability. Final thermal cycle tests are under way at the time of this writing, and will be available at the presentation of the paper.

References

[1] A.J. Blodgett and D.R. Barbour, "Thermal Conduction Module: A High-Performance Multilayer Ceramic Package," IBM Journal of Research and Development, Vol. 26, No. 1, pp. 30-36, January 1982.

[2] T.C. Mele, A.H. Perera, J.P. Krusius, "High Resolution Trilayer Electron Beam Resist System Employing P[MMA/MAA] and Reliable Reactive Ion Etch Process," SPIE Vol. 923, pp. 217-223, 1988.

[3] R. Subrahmanyan, J.R. Wilcox, C.Y. Li, "An Integral Method for Mechanical Reliability Assessment in Solder Joint Design," Proceedings of the 1989 Nepcon West Conference, Los Angeles, CA, March 6-9, 1989.

Low Dielectric Constant Multilayer Glass–Ceramic Substrate with Ag–Pd Wiring for VLSI Package

YUZO SHIMADA, YOSHIHARU YAMASHITA, AND HIDEO TAKAMIZAWA

Abstract—VLSI package development continuously demands new packaging technology to meet high propagation speed and high wiring density requirements for a multi-chip mounted substrate. Important factors for the microelectronic packaging substrate are dielectric constant and electrical resistivity of conductors. New glass–ceramic materials with low dielectric constant, which can be sintered at about 900°C in air, have been developed. It has become possible to attain extremely low electrical resistivity on conductors due to silver and palladium particle shape control. The electrical design is a point to be considered for the packaging substrate. Pulse transmission properties are influenced greatly by the kinds of ground planes and wiring planes used. Therefore, basic pulse transmission properties, such as propagation delay, characteristic impedance, and crosstalk coupling noise, were measured for various multilayer structures. New technologies, which can precisely control substrate properties, have been established. Characteristics for the new multilayer glass–ceramic (MGC) substrate are summarized.

1) The new MGC substrate has a small propagation delay, due to the low dielectric constant material ($\epsilon \simeq 3.9$).
2) Low electrical resistivity and low cost is realized by using a silver-palladium conductor system ($< 4 \ \mu\Omega\cdot cm$).
3) Improved electrical design to control characteristic impedance and crosstalk coupling noise is used.
4) A highly accurate fabrication process required to take advantage of the new materials has been developed.

As a result, the low dielectric constant MGC substrate, with silver-palladium conductors, can be applied to the VLSI multi-chip packaging substrate for use in a large-scale computer system.

I. INTRODUCTION

VLSI PACKAGE development continuously demands new packaging style technology. Current approaches for silicon LSI microelectronic packaging are multilayered ceramic substrates [1] and thermal conduction modules [2]. These modules are fabricated from alumina ceramic layers using molybdenum or tungsten metallization as conductors. However, due to the increase in switching speeds and circuit density in VLSI chips, higher propagation velocity and interconnect densities are required for new multi-chip substrates.

A very important property of substrate materials, for microelectronic packaging with high switching speed, is dielectric constant. In conventional packaging technology with high dielectric constant alumina ($\epsilon \simeq 9$–10), the delay introduced in the package exceeds the chip delay. Therefore, it is necessary to reduce the dielectric constant for the substrate to below that for alumina [3].

Another important factor in microelectronic packaging is to minimize line resistance, to prevent any undesired voltage drop. In conventional packaging technology, the ability to reduce signal linewidth and line thickness is limited because of the relatively high resistivity of refractory metals, such as molybdenum or tungsten. For achieving high wiring density and fine patterns, it is necessary to use low electrical resistivity conductors for signal lines and interconnections.

A glass–ceramic material, a mixture of alumina and lead borosilicate glass powders, which can be sintered at about 900°C in air, had been reported [4], and low firing temperature multilayer glass–ceramic (MGC) substrates with gold wiring [5] and monolithic multicomponent ceramic (MMC) substrates [6] were realized. The dielectric constant for this material is 7.8, smaller than that for alumina. This substrate, therefore, is advantageous for high pulse transmission speeds. However, the VLSI demand will require even higher transmission speeds on the substrate.

Several new glass–ceramic material systems have been developed. Their dielectric constants are extremely small, such as 5.0 and 3.9. These material systems can also be sintered at around 900°C in air.

Electrical resistivity of interconnect metal controls the reduction of package dimensions. Gold performs well as an interconnect, but at a cost penalty. A silver–palladium conductor has been developed that achieves a lower resistivity than gold at a lower cost. Resistivity gains are possible as a result of silver and palladium particle shape control, and proper conductor composition selection. In this way, the new MGC substrate technology satisfies expectations for high propagation speed and high wiring density.

Electrical design is a point to be considered in substrate packaging. Pulse transmission properties are greatly influenced by the kinds of ground planes and wiring planes used [7]. Therefore, basic pulse transmission properties, such as propagation delay, characteristic impedance, and crosstalk coupling noise, were measured for various multilayer ceramic configurations. These properties depend a great deal on dielectric constant, layer thickness, spacing between conductors, linewidth, the kinds of reference planes used, etc. Propagation delay showed very small values, due to the low dielectric constant for the new glass–ceramic materials. Semiconductor circuits generally require line impedances

Manuscript received Februry 14, 1987; revised September 30, 1987. This paper was presented at the Electronic Components Conference, Boston, MA, May 1987.

Y. Shimada and H. Takamizawa are with the Materials Development Center, NEC Prefecture, 213 Japan.

Y. Yamashita is with the Electronic Device Department, NEC Hyogo LTd., 321 Sugasawa, Yamasaki-cho, Shiso-gun, Hyogo Prefecture, 671-25 Japan.

IEEE Log Number 8718634.

Reprinted from *IEEE Trans. Components, Hybrids, Manuf. Technol.*, vol. 11, no. 1, pp. 163–170, March 1988.

464

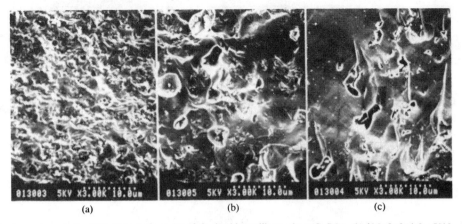

Fig. 1.   Electron micrographs of fracture surface. (a) Al₂O₃–lead borosilicate glass (G-C(I)). (b) 2MgO·2Al₂O₃·5SiO₂–borosilicate glass (G-C(II)). (c) SiO₂–borosilicate glass (G-C(III)).

ranging between 50 and 100 Ω. Characteristic impedances of 50 to 100 Ω were realized. Coupled noise considerations are very important, because the wiring density is restricted by limiting available spacing, and lengths of parallel conductors.

In addition, new technologies have been established for fabricating precisely the new MGC substrate. New technologies include via hole formation, fine-line metallization, highly accurate lamination, and I/O pin attachment. Consequently, high-density packaging and precise shrinkage control are realized.

The low dielectric constant MGC substrate, with silver–palladium wiring, will be applied to the VLSI multi-chip packaging substrate for use in a large-scale computer system.

This paper describes the new glass–ceramic material systems with low dielectric constant, the low electrical resistivity conductor material, the new fabrication technologies, the new MGC substrate properties, and the basic pulse transmission properties.

Low Dielectric Constant Materials

Dielectric materials used for the MGC substrate are a current Al₂O₃–lead borosilicate glass (G-C(I)), cordierite (2MgO·2Al₂O₃·5SiO₂)–borosilicate glass (G-C(II)), and SiO₂–borosilicate glass (G-C(III), which have controlled particle size. The dielectric material mixture compositions are 55 wt % Al₂O₃ and 45 wt % lead borosilicate glass for G-C(I), 45 wt % cordierite (2MgO·2Al₂O₃·5SiO₂) and 55 wt % borosilicate glass for G-C(II), and 35 wt % SiO₂ and 65 wt % borosilicate glass for G-C(III). The optimum sintering temperature for these materials is about 900°C in air. For G-C(I), when sintered, the fine structure becomes a continuous vitreous network of lead borosilicate glass, having crystals of Al₂O₃ and microcrystals. This glass–ceramic body has high flexural strength, 3000 kg/cm² or more. On the other hand, because G-C(II) and G-C(III) bodies have voids, flexural strengths are low, 1500 kg/cm² or less.

Fig. 1 shows electron micrographs of fracture surfaces for the three material systems, fired at about 900°C.

Frequency dependences of dielectric constants for the three material systems are shown in Fig. 2. Dielectric constants hardly change between 1 kHz and 10 MHz, and are 7.8, 5.0, and 3.9 at 1 MHz for G-C(I), (II), and (III), respectively.

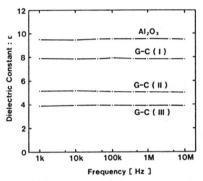

Fig. 2.   Dielectric constants frequency dependence.

The dielectric constants for ceramics generally will vary with their polarization and atomic numbers, and inversely with their bond strengths and densities. The technique using the multiporous structure, which increases the specific volume of a material, would be beneficial for low dielectric constants. G-C(II) and G-C(III) have extremely small dielectric constants, because of their physical properties and multiporous structures.

Low Electric Resistivity Conductor Materials

As a result of an investigation on silver and palladium particle shape control, optimum powders for low electrical resistivity have been found. Fig. 3 shows the silver and palladium powders used for the MGC substrate. This silver–palladium system has lower electrical resistivity than the conventional system. It features close packing in the paste phase and after sintering. The relationship between resistivity and silver–palladium composition ratio is shown in Fig. 4. The resistivity for 5 wt % palladium and 95 wt % silver composition is around 3.5 μΩ·cm. The resistivity for 100 wt % silver is around 1.8 μΩ·cm. Fig. 5 shows the resistivity variation for silver (I) and silver (II) compositions in the 95 wt % Ag–5 wt % Pd system.

Low conductor resistivity reduces voltage drop along the line length. Resistance at the utilization frequency is formulated as follows:

$$R = \frac{\rho l}{A} \qquad (1)$$

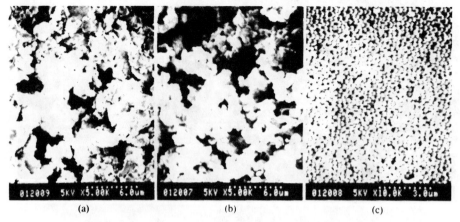

Fig. 3. Silver and palladium powders used for the MGC substrate. (a) Ag (I). (b) Ag (II). (c) Sphere Pd.

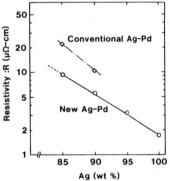

Fig. 4. Relation between resistivity and silver content for new Ag–Pd system.

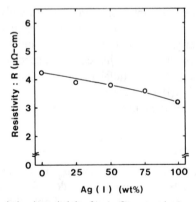

Fig. 5. Variation in resistivity for Ag(I) content in Ag$_{95}$–Pd$_5$ system.

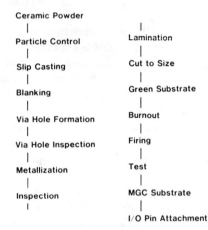

Fig. 6. MGC substrate fabrication process flowchart.

where ρ is specific resistivity, l is the length, and A is the cross-sectional area. The line resistance increases with decreasing linewidth and thickness. The higher line resistances produce voltage drops and localized heating, which can affect device switching behavior [8]. The MGC substrate is advantageous for the high wiring density, due to decreasing line dimensions, such as width and thickness, and for the high frequency pulse, because of decreasing waveform rounding on the rise pulse. A reduction in the processing temperature to around 900°C facilitates the use of the low-resistivity metal, such as Ag–Pd, Ag, Au, or Cu.

FABRICATION TECHNOLOGIES

A flowchart for the MGC substrate fabrication process is shown in Fig. 6. Ceramics and glass powder are controlled to a desired particle size, owing to electrical properties, mechanical properties, and shrinkage control. A slurry consisting of the mixed powders and organic liquid vehicles (solvent, plasticizer, butyral binder) is cast into 100 μm thickness ceramic green sheets on the polyester film coated silicone layer by a conventional slip casting process. Poly-vinyl-butyral binder is used in sufficiently low quantities to prevent deterioration of substrate characteristics. Deviation in green sheet thickness is less than 1 percent. Tape is removed from the polyester film and sheets are blanked.

Via Hole Technology

Via holes are formed through the green sheet simultaneously by a punching machine with a punch and die. A computer-controlled sequential mechanical punch system and a flight punch system using piezoelectric actuator, have been developed for via hole formation.

Fundamentally, via holes are formed mechanically by using die and punch. Fig. 7 shows a flight punch system. The puncher is basically driven by piezoelectric ceramic actuators, which are produced by ceramic green sheet technologies. These systems have achieved low production cost.

Since shrinkage is slightly different in directions parallel to and normal to the tape casting direction, alternate layers are rotated 90° prior to via hole formation. Hole inspection is performed by an optical tester. Accuracy is checked against the original design.

Fig. 7. Flight punch system.

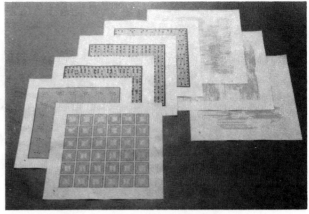

Fig. 8. Printed green sheets.

Fig. 9. Green substrate and sintered MGC substrate.

Metallization Technology

Low-resistivity silver–palladium conductor paste is applied onto the punched sheets by the semiautomatic printing equipment. In this process, the via holes are filled with paste to form contacts between signal lines. The via filled paste has a specially designed high solids paste composition. Print resolution is superior to that obtained on conventional alumina substrates for HIC, due to the smooth porous sheet surface. The printed green sheet uses an even thinner printed line, owing to the low electrical resistivity conductor system. Conductors and filled via holes are inspected visually and by optical testers. Printed green sheets are shown in Fig. 8.

Lamination Technology

The metalized green sheets are precisely stacked in a pressing die in sequence by the stacking machine. The die cavity has dimensions indentical to those of the green sheet, which is blanked into $100 \sim 150$-mm square dimensions with high precision. These sheets are next laminated together at 110°C for 30 min under about $200 \sim 250$-kg/cm^2 pressure by the press machine. The confined die prevents lateral distortion during pressing. Density inhomogeneities in the laminated samples influence any shrinkage in the sintered substrate. Therefore, this lamination process should be homogeneously carried out by means of correct die and punch with a flat surface. Further, the printed green sheets are put between flat organic sheets, coated with silicone, during the lamination to achieve homogeneous density and a smooth substrate surface. The green substrate is trimmed to pre-fired size using a dicing saw with a diamond blade.

The trimmed green substrate is placed on an alumina setter and pre-fired in a convection furnace at about 500°C for a few days, until the organic binder burns out. Sintering is performed in an electric furnace at about 900°C in air. The linear shrinkage for the co-fired substrate with conductor is approximately 13.0 percent. Electrical connections on the sintered substrate are checked using the tester tool. The green substrate and the sintered MGC substrate are shown in Fig. 9. MGC substrate size is 100 mm × 100 mm.

I/O Pin Attachement Technology

According to Rent's rule [9], increasing integration on the substrate tends to increase the I/O terminal numbers on the packaging substrate. It is necessary for the substrate package to have many I/O terminals, such as the pin grid array (PGA) type.

Pins are automatically inserted into holes in the substrate, and contact sputtered bonding pads. Pin attach uses 80 wt % Au–20 wt % Sn solder. Pins are weighted individually during solder attach. Bonding strength exceeds 7 kg. A substrate with I/O pins is shown in Fig. 10.

MGC Substrate Properties

Typical properties of MGC substrates and glass–ceramic materials are summarized in Table I.

Physical Properties

The sintered density for G-C(I) is close to the theoretical density value, since there are very few voids, but G-C(II) and G-C(III) have lower density values, compared with theoretical values, due to closed pores. For the G-C(I) system, each microcrystalline structure grows as a result of the Al$_2$O$_3$ chemical reaction with the lead borosilicate glass. The vitreous network grows concurrently with the chemical reaction and crystallization, until it is rendered integral with the crystals and the microcrystalline structures, thereby forming a dense crystalline phase having no voids. Therefore, this MGC substrate has high flexural strength about 3500 kg/cm^2. Other MGC substrates, using G-C(II) and G-C(III), have less flexural strength due to the presence of voids.

Camber and roughness are extremely small on individual substrates. These MGC substrates have good flatness.

Fig. 10.   Substrate with I/O pins.

TABLE I
TYPICAL MGC SUBSTRATE PROPERTIES

Property		Al$_2$O$_3$–Glass G-C(I)	2MgO·2Al$_2$O$_3$ 5SiO$_2$–Glass G-C(II)	SiO$_2$–Glass G-C(III)
Sintered density	[g/cm^3]	3.10	2.40	2.15
Flexural strength	[kg/cm^2]	3500	1500	1400
Camber	[/100 mm]	20 μm	40 μm	40μm
Roughness	[Ra]	0.3 μm	0.9 μmm	0.5 μm
Dielectric constant (1 MHz)		7.8	5.0	3.9
Dissipation factor (1 MHz)		0.3 percent	0.5 percent	0.3 percent
Insulation resistance (50 V dc)	[Ω·cm]	>10^{14}	>10^{13}	>10^{13}
Leakage current (20 V dc)	[μA]	<1	<1	<1
Thermal expansion coefficient	[deg^{-1}]	42 × 10^{-7}	79 × 10^{-7}	19 × 10^{-7}
Linear shrinkage		13.0 percent	13.7 percent	13.9 percent
Shrinkage tolerance		<0.3 percent	<0.3 percent	<0.3 percent

Electrical Properties

Dielectric constants for G-C(I), G-C(II), and G-C(III) kinds of MGC substrates are 7.8, 5.0, and 3.9, respectively, at 1 MHz. The MGC substrate for the G-C(III) has an extremely small dielectric constant, in comparison with conventinal ceramics. The dissipation factor is less than 0.5 percent.

The dielectric constant and dissipation factor were measured with a multi-frequency digital LCR meter, with 1 V rms in the frequency range from 1 kHz to 10 MHz.

The insulation resistance is more than 10^{13} Ω·cm, and the interlayer leakage current is extremely low.

Thermal Properties

The thermal expansion coefficient is one of the important factors which must be considered for the packaging substrate. The MGC substrate for the G-C(I) system has a 42 × 10^{-7} deg^{-1} thermal expansion coefficient, that is constant between room temperature and 250°C, and is close to that for the silicon semiconductor chips. It is necessary to improve G-C(II) and G-C(III) toward the optimum thermal expansion coefficient.

Shrinkage

Substrate shrinkage is influenced by the particle size of the low dielectric constant material powder. Linear shrinkage values are 13.0, 13.7, and 13.9 percent, respectively, for G-C(I), G-C(II), and G-C(III). It has become possible for co-fired substrates to achieve extremely small shrinkage tolerance of 0.3 percent or less, as a result of the new technologies.

Substrate Specifics

Typical MGC substate specifications are shown in Table II. The new MGC substrate can be used to achieve high wiring density and fine-line patterns.

PULSE TRANSMISSION PROPERTIES

The electrical design is a point to be considered for the MGC substrate, due to high wiring density. Pulse transmission properties are greatly influenced by the kinds of ground planes and wiring planes used. Accordingly, basic pulse transmission properties, such as propagatin delay, characteristic impedance, and crosstalk coupling noise, were measured for various multilayer structures. Fig. 11 shows estimation patterns for pulse transmission properties. Signal lines are formed in the substrate. The dielectric for G-C(I) is 7.8, that for G-C(II) is 5.0, and that for G-C(III) is 3.9. Signal conductor linewidths are 0.08 to 0.16 mm and thickness values range from 0.01 to 0.012 mm. The ground plane has a mesh pattern at the 45° angle for X–Y signal lines, and ground conductors are screened at 0.5-mm pitch. Pulse transmission properties were measured with Time Domain Reflectometry (TDR).

Propagation Delay

The dielectric constant, which is an important property for substrate material, controls the signal transmission speed. The propagation delay time (T_{pd}) is formulated as follows:

$$T_{pd} = \frac{\sqrt{\epsilon}}{c} \text{ [s/m]} \qquad (2)$$

TABLE II
TYPICAL MGC SUBSTRATE SPECIFICATIONS

Signal line: Width	100 μm
line-to-line	200 μm
Via hole: Diameter	100 μm
center-to-center	300 μm
Resistivity for conductor	1.8–10 $\mu\Omega\cdot$cm
I/O pin bonding strength	>7 kg
Green sheet thickness	100 μm
Number of layeres	34

Fig. 11.   Estimation patterns for pulse transmission properties.

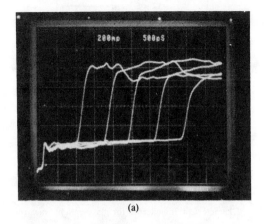

(a)

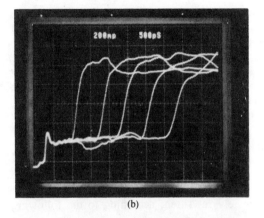

(b)

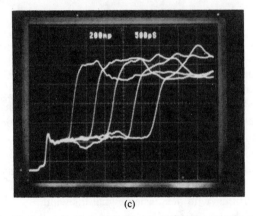

(c)

Fig. 12.   Propagation delay. (a) Al$_2$O$_3$–glass (G-C(I)). (b) 2MgO$\cdot$2Al$_2$O$_3\cdot$ 5SiO$_2$–glass (G-C(II)). (c) SiO$_2$–glass (G-G(III)).

where ϵ is the dielectric constant, and c is the speed of light. The propagation delay for many megabits of information, which have to travel through large computer systems and supercomputer systems [10], cannot be disregarded. Therefore, it is desirable to reduce the dielectric constant as much as possible. Any technique which increases the specific volume of a material, and any lower dielectric constant material would be beneficial.

Fig. 12 shows TDR propagation delay data for substrates made from G-C(I), G-C(II), and G-C(III). Data are available for five different signal line lengths on each substrate. Delay time is calculated from the transmission time differences for individual signal line lengths. The propagation delay for the new MGC substrates, with low dielectric constant, is shown in Fig. 13. It is very small, in comparison with that for a conventional alumina substrate. In the figure, the theoretical curve between propagation delay and dielectric constant, calculated using (2), is shown. The observed results agree well with the estimates.

Characteristic Impedance

Characteristic impedance depends on the dielectric constant, the geometry of the signal line, and the distance between the signal line and the ground plane. The characteristic impedance (Z_0) is formulated according to the following relationship:

$$Z_0 = \sqrt{\frac{L}{C}} \ [\Omega] \qquad (3)$$

where L is inductance and C is capacitance for the line.

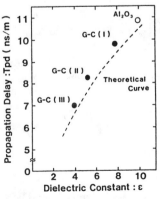

Fig. 13.   Relation between propagation delay and dielectric constant for multilayer ceramic substrate using various dielectric materials.

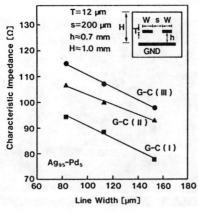

Fig. 14. Characteristic impedances for glass–ceramic materials.

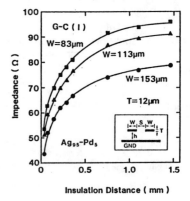

Fig. 15. Impedance versus insulation distance.

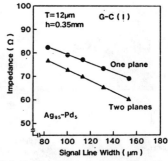

Fig. 16. Relation between impedance and ground structure.

Characteristic impedances for the new MGC substrates were measured. The observed results are shown in Fig. 14. An illustration of the test substrate configuration is shown in the figure, together with the test results. Basic parameters affecting characteristic impedances are:

W signal linewidth,
h distance between line and ground plane,
s line space,
H distance between ground plane and surface plane,
T line thickness.

In the case of this configuration, the characteristic impedance is approximately rewritten, according to the following relationship:

$$Z_0 = \sqrt{Z_{0e} \cdot Z_{0o}} \qquad (4)$$

where Z_{0e} is the even mode impedance and Z_{0o} is the odd mode impedance. These equations are approximately shown for strip lines, when the ground contains a solid plane all over the layer. The estimation substrate, used for pulse transmission properties, contains other conductors, such as orthogonal lines and via holes, in addition to the necessary pattern. Therefore, the observed results did not agree well with estimates. In Fig. 14, the characteristic impedance increases as the substrate material has a low dielectric constant, and increases further with decreasing signal linewidth.

Fig. 15 shows the relationship between the characteristic impedance and the distance between signal line and ground plane on the G-C(I) substrate. The characteristic impedance is also influenced by distance h, decreasing with shortening distance h.

The characteristic impedance values for various configurations, such as the coupled line and the line between ground planes, are shown in Fig. 16 for G-C(I). In the case of signal lines between ground planes, the characteristic impedance becomes small.

As mentioned above, the new multilayer substrate provides the ability to approach the optimum characteristic impedance and to realize 50–100 Ω.

Crosstalk Coupling Noise

Crosstalk coupling noise is a very important factor, because the packaging density is restricted by limiting spacing and length of parallel lines. Near-end crosstalk is quite significant, and far-end crosstalk is zero in a homogeneous medium. However, in the multilayer ceramic substrate, the presence of conductors in adjacent channels and orthogonal lines and via holes reduces mutual capacitance. Therefore, capacitive coupling decreases. On the other hand, the orthogonal lines and via holes have little effect on the mutual inductance. Therefore, the inductive coupling is larger than the capacitive coupling, contributing to the far-end crosstalk coupling noise.

Fig. 17 shows the relationship between the crosstalk coupling noise and the line space on the G-C(I) intralayer. The parallel line length is 77 mm. In this case, when the line space is wider than about 0.5 mm, the crosstalk coupling noise is less than 10 percent.

These results can be used to determine routing restrictions on the circuit design.

CONCLUSION

A new MGC substrate, which has a low dielectric constant and can be sintered at around 900°C, has been developed by means of green sheet technologies. A reduction in the processing temperature to around 900°C facilitates the use of low-resistivity metals, such as the silver–palladium system.

The electrical design is a point to be considered for the packaging substrate on a large-scale computer system. Basic pulse transmission properties, such as propagation delay, characteristic impedance, and crosstalk coupling noise, were measured for various multilayer structures.

The low dielectric constant MGC substrate, with Ag–Pd wiring for VLSI packages, has the following outstanding characteristics.

1) The new MGC substrate has a small propagation delay,

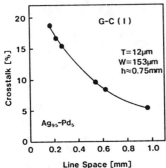

Fig. 17. Relationship between crosstalk coupling noise and line space.

due to the low dielectric constant material ($\epsilon \simeq 3.9$). It is advantageous in regard to high signal transmission speed.

2) Low electrical resistivity can be realized by using silver–palladium conductor system ($< 4\ \mu\Omega \cdot cm$). It is advantageous for achieving a desirable low voltage drop and low cost.

3) The low dielectric constant and resistivity of materials used in this system allows electrical design for controlling characteristic impedance and crosstalk coupling noise.

4) A highly accurate fabrication process has been established. Consequently, high-density packaging and precise shrinkage control are realized.

The low dielectric constant MGC substrate with silver–palladium wiring will be applied to the VLSI multi-chip packaging substrate for use in a large-scale computer system.

ACKNOWLEDGMENT

The authors are very grateful to T. Ohno, General Manager of the Material Development Center and K. Utsumi, Research Manager of the Material Research Laboratory, for their advice and guidance. They would also like to thank Y. Shiozawa, T. Ohtsuka, and H. Goto for their assistance with fabrication and measurements, as well as T. Sato, Supervisor of the Circuit Engineering Department, Computer Engineering Division, for his advice regarding pulse transmission properties.

REFERENCES

[1] A. J. Blodgett, "A multi-layer ceramic, multi-chip module," in *Proc. 30th Electronic Components Conf.*, pp. 283–285, 1980.

[2] A. J. Blodgett and D. R. Barbour, "Thermal conduction module: A high-performance multilayer ceramic package," *IBM J. Res. Develop.*, vol. 26, pp. 30–36, 1982.

[3] L. E. Cross and T. R. Gurnraja, "Ultra-low dielectric permittivity ceramics and composites for packaging applications," in *Mat. Res. Soc. Sympo. Proc.*, vol. 72, pp. 53–65, 1986.

[4] Y. Shimada, K. Utsumi, M. Suzuki, and H. Takamizawa, "Contribution of glass particle size on Al$_2$O$_3$-glass ceramics," in *Proc. Anu. Meeting Ceramic Soc. Japan*, p. 114, 1982.

[5] ——, "Low firing temperature multilayer glass-ceramic substrate," *IEEE Trans. Comp., Hybrids, Manuf. Technol.*, vol. CHMT-6, no. 4, pp. 382–388, 1983.

[6] Y. Shimada, K. Utsumi, T. Ikeda, and S. Nagasako, "Monolithic multi-components ceramic (MMC) substrate," in *Proc. 3rd Int. Microelec. Conf.*, pp. 227–234, 1984.

[7] P. N. Venkatachalam. "Pulse propagation properties of multilayer ceramic multichip modules for VLSI circuits," in *Proc. 33th Electronic Components Conf.*, pp. 130–134, 1984.

[8] B. Schwartz, "Microelectronis packaging: II," *Ceramic Bull.*, vol. 63, no. 4, pp. 577–581, 1984.

[9] B. S. Landman and R. L. Russo, "On a pin versus block relationship for partition of logic graphs," *IEEE Trans. Comput.*, vol. C-20, pp. 1469–1479, 1971.

[10] T. Watari, "The NEC SX supercomputer technology," in *Proc. IEEE Int. Conf. on Computer Design VLSI in Computers*, pp. 54–57, 1986.

LOW TEMPERATURE CO-FIREABLE CERAMICS
WITH CO-FIRED RESISTORS

By

H. T. Sawhill, A. L. Eustice, S. J. Horowitz,
J. Gar-El and A. R. Travis

E. I. du Pont de Nemours & Co., Inc.
Photosystems and Electronic Products Department
Electronic Materials Division
Wilmington, Delaware

ABSTRACT

Key processing parameters and performance characteristics of a commercial low temperature co-fireable ceramic materials system for the manufacture of multilayer interconnect hybrids are reviewed. Thick film resistor compositions for post- and co-firing on top layers are discussed and preliminary results on resistors for internal co-firing are presented.

INTRODUCTION

Existing high temperature co-fired ceramics [1] are a cost-effective way for manufacturing complex multilayer interconnects for electronic packaging. However, firing temperatures above 1500°C, reducing atmospheres and extended firing cycles limit widespread in-house adoption. In addition, available conductor metallurgies (W, Mo-Mn) limit applications to those without severe conductivity requirements. Thick film technology (compatible with high conductivity metallurgies) is practiced in-house by many equipment and circuit manufacturers, but yields are adversely effected ·and production becomes less cost-effective as the number of layers in the package increases.

Low temperature co-fired ceramic systems offer the benefits of existing thick film technology combined with the processing advantages of co-fired ceramics. These systems allow cost-effective production of high density multilayer interconnect hybrids with the benefits of in-house control.

Low temperature systems developed for in-house use and based on glass filled composites [2-7], crystallizable glasses [8-10] and crystalline phase ceramics [11-13] have been reported in the literature. These systems are designed for use either with precious metal conductors (for air firing) [2-6,8-11] or with Cu or Ni conductors (for nitrogen or reducing atmosphere firing) [7,12,13]. This paper discusses characterizations of processing parameters for a commercial low temperature ceramic materials system designed for air firing using conventional thick film processing equipment. The performance of thick film resistor compositions for post- and co-firing on top layers are described, and preliminary results for new resistor compositions for co-firing inner layer resistors are presented.

MATERIALS SYSTEM AND PROCESSING

The materials system consists of a cast dielectric tape, compatible signal line and via fill conductor compositions, thick film resistor compositions for post- and co-firing on top layers, developmental resistors for buried configurations and thick film conductor compositions for customizing top layers. The materials system (Table 1) was designed for compatibility with standard thick film equipment and processing. Thick film printers, drying ovens and belt furnaces can be used in circuit fabrication and additional equipment (laminating presses and via punching devices) is commercially available. Blanking and registration tooling can be readily built. The processing steps involved in making a package with this system include tape blanking,

TABLE 1

LOW TEMPERATURE CO-FIREABLE CERAMIC MATERIALS SYSTEM

	Gold	Silver	Silver/Palladium	Gold/Platinum
Conductors				
Inner Layer Conductors (Co-Fired)	5717D	6142D	6144D	-
Via Fill Conductors (Co-Fired)	5718D	6141D	6143D	-
Top Layer Conductors (Post-Fired)	5715	-	6134	4596

Resistors	
Top Layer (Post- or Co-Fired)	- Birox® 1900 Series Resistor Compositions
Internal (Co-Fired)	- Under Development

via punching, via filling, conductor and resistor printing, laminating, laminate cutting, burnout and firing. A flowchart of the processing steps is shown in Figure 1. Detailed descriptions of the individual processing steps can be found in earlier publications [14,15].

CO-FIRED CERAMIC PROCESS FLOW

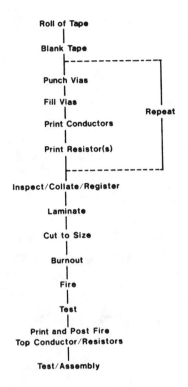

Figure 1 - Flowchart for Processing of Low Temperature Co-Fired Materials System

GREEN DIELECTRIC TAPE

Green dielectric tape (851AT) is cast on a Mylar® carrier treated with a release agent to facilitate stripping. Proprietary coating technology produces high quality green tape with minimal defects and thickness control within ± 2% of the nominal thickness (.0045 inch/.114 mm) across the width of the roll. The polymer binder system was developed to give clean burnout during a short burnout cycle and to meet rheological requirements for tape casting. Plasticizer additions control the physical properties of laminated parts. Physical characteristics and elastic properties of the green tape have been reviewed in previous publications [14,15].

The tape contains a glass phase with composition and particle size distribution engineered for viscosity and shrinkage control, and refractory fillers selected to match the thermal expansion of the substrate to alumina and to minimize the

dielectric constant. Careful control of these constituents gives uniform lateral shrinkage and flat fired parts without the need for confined sintering. Viscous sintering of the glass phase yields dense (>96% of theoretical) and hermetic structures after firing in a modified thick film firing profile. Firing is a 2-step burnout/firing process that allows maximum use of conventional thick film processing equipment. Burnout takes place at 350°C for 1 hour, removing approximately 85% of the organic constituents. The firing profile (Figure 2) is a modified thick film profile run at a reduced belt speed to yield a peak temperature of 850°C, dwell time at peak of 15 minutes and total cycle time of approximately 90 minutes. The lot-to-lot and inter-lot shrinkage reproducibility is shown in Figure 3. The

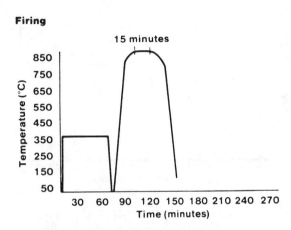

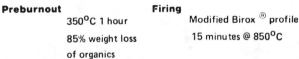

Preburnout	Firing
350°C 1 hour	Modified Birox® profile
85% weight loss of organics	15 minutes @ 850°C

Figure 2 - Burnout/Firing Profile for Low Temperature Co-Fired Materials System

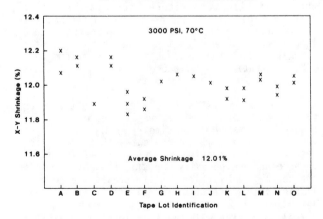

Figure 3 - Lot-to-Lot Lateral Shrinkage Control

473

average shrinkage in the plane of the tape is 12% and variability (both lot-to-lot and within the same lot) is within ± 0.2%. Average shrinkage in the direction normal to the plane of the tape is 17.5% ± 0.5%. Shrinkage data refers to standard lamination conditions of 3000 psi at a temperature of 70°C. The insensitivity of lateral shrinkage to lamination time is demonstrated in Figure 4. Error bars represent a total range of twice the standard deviation of the measured values based on double measurements of four samples.

Influence Of Laminating Time On X-Y Shrinkage Of Tape Dielectric

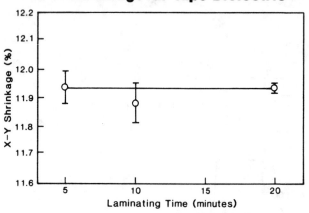

Figure 4 - Lateral Shrinkage During Firing of Tape Dielectric as a Function of Laminating Time (70°C @ 3 kpsi)

Laminates are fabricated by collating and registering 8 layers of tape into a pressing die possessing cavity dimensions identical to those of the blanked sheets. The confined die prevents lateral distortion and provides uniform uniaxial pressure during lamination. Registration pins assure precise mechanical registration. Tape shrinkage variation with lamination temperature is shown in Figure 5. For temperatures above 65°C, the lamination temperature does not appear to influence the lateral shrinkage during firing. At temperatures greater than 80°C, laminates become increasingly flexible but the potential for lateral distortion increases. The recommended 70°C laminating temperature allows for process latitude and easy handleability.

Lamination pressure has a direct influence on the lateral shrinkage during firing as shown in Figure 6. Mercury porosimetry analysis [16] of the samples in Figure 6 indicates that lamination results in a reduction of the volume fraction of interconnected porosity in the laminate. This change in green packing density causes the roughly linear correlation between laminating pressure and lateral shrinkage for pressures between 2.5 and 4 kpsi. Careful control of laminating conditions provides additional assurances of tight tolerances in the lateral shrinkage.

Influence Of Laminating Temperature On X-Y Shrinkage Of Tape Dielectric

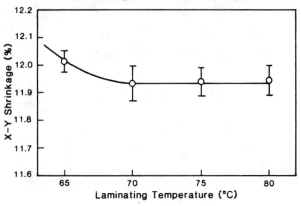

Figure 5 - Lateral Shrinkage During Firing of Tape Dielectric as a Function of Laminating Temperature (10 min. @ 3 kpsi)

Influence Of Laminating Pressure On X-Y Shrinkage Of Tape Dielectric

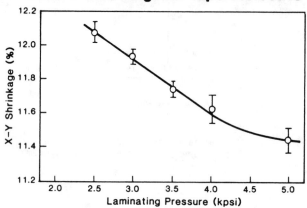

Figure 6 - Lateral Shrinkage During Firing of Tape Dielectric as a Function of Laminating Pressure (10 min. @ 70°C)

CONDUCTORS

Co-fired inner-layer and via fill conductors were developed to have sintering rates compatible with that of the tape dielectric. Top layer post-fired conductors are commercially available thick film compositions that provide high adhesion strengths and excellent soldering and bonding properties without plating. Post-firing of conductors requires an additional processing step, but the operation optimizes registration for automatic component placement. Properties of inner layer co-fired and top layer post-fired conductors are given in Table 2.

Table 2

Properties of Conductors

	Resist-ance mΩ/□	Fired Thick-ness μm	Resolution[1] Lines/Spaces μm
Inner Layer (Co-Fired)			
Gold (5717D)	5	7	132/91
Silver (6142D)	5	8	109/114
Silver/Palladium (6144D)	20	8	132/88
Top Layer (Post-Fired)			
Gold (5715)	4	8	150/100
Platinum/Gold (4596)	80	15	150/100
Silver/Palladium (6134)	20	15	175/75

(1) Co-fired conductors undergo 12% shrinkage

In an earlier report [14], it was shown that leads soldered to post-fired silver/palladium and gold/platinum conductor compositions have high aged adhesion, and .001 inch Au wires thermosonically bonded to post-fired gold conductors have high bond strengths and good reliability. Also, silver migration studies showed high insulation resistance for silver and Ag/Pd alloy conductor metallurgies tested under temperature-humidity-bias and high temperature-high voltage conditions.

RESISTORS

The addition of resistors to multilayer interconnect packages contributes to increased design flexibility and circuit density. Initial characterization studies were performed using commercial Birox® 1900 Series resistor compositions [17,18] for post-firing onto substrates made with tape dielectric. Favorable results led to characterization of 1900 Series resistors for co-firing on the top layer of tape laminates.

Studies were performed to ascertain the influence of lamination and tape shrinkage on the electrical properties of co-fired resistors. Finally, the 1900 Series resistor compositions were tested in co-fired internal configurations. Results prompted changes in resistor chemistry to accommodate increased interaction between the glass phases in the tape and in the resistor. The procedures and results from these different studies are discussed in the remaining sections.

POST-FIRED RESISTORS

Samples for the evaluation of post-fired resistors were prepared by post-firing Birox® 1900 Series resistors [17,18] with 6134 Ag/Pd terminations on pre-fired substrates. Terminations were printed using a 200 mesh screen with an 0.5 mil emulsion, dried for 10 minutes at 150°C and fired in a 60-minute profile with a 10-minute dwell at 850°C. The 1900 Series resistors were printed using a 325 mesh screen with an 0.5 mil emulsion, dried for 10 minutes at 150°C and fired in a 30-minute profile with a 10-minute dwell at 850°C.

Resistance and temperature coefficient of resistance (TCR) values were measured on 1.5mm x 1.5mm resistors. Hot and cold TCR values (HTCR, CTCR) are based on resistance changes between room temperature (25°C) and 125°C and -55°C respectively. Resistance and TCR data are presented in Table 3.

Post laser trim stability was determined by monitoring resistance changes for unencapsulated 1mm x 1mm resistors trimmed to 1.5x their fired value using a single plunge cut. Figures 7-9 show resistor changes under storage conditions of 25°C, 150°C and 40°C/90% RH respectively. Average resistance changes were less than 0.8% for 1000 hours storage under all three conditions tested. Re-testing is underway for 1931 resistors under 40°C/90% RH conditions because of anomalies in the initial measurements. Trim conditions for the post-fired resistor series were 0.7-1W, 3kHz (1911,1921-2kHz) and a 10 mm/s trim speed.

TABLE 3

Resistance and TCR Values for Top Layer Post- and Co-Fired
Birox® 1900 Series Resistors

Composition	Nominal Resistance Value (Ω/□)	Resistance (Ω/□)		HTCR (ppm/°C)		CTCR (ppm/°C)	
		Post-Fire	Co-Fire	Post-Fire	Co-Fire	Post-Fire	Co-Fire
1911	10	10.8	11.6	133	-27	89	-88
1921	100	136.0	132.0	-32	-41	-90	-90
1931	1000	1.21 k	1.05 k	50	-30	-8	-90
1939	10 k	12.0 k	11.8 k	80	60	23	-52
1949	100 k	123 k	97.0 k	62	100	-14	-13
1959	1 M	1.59 M	1.20 M	-65	97	-190	-32

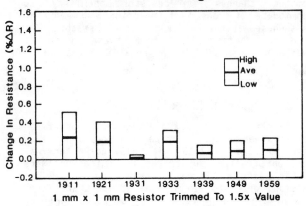

**1900 Series Resistors
Post-Fired On Tape Dielectric
Stability After Laser Trimming
1,000 Hours Storage @ 25°C**

Figure 7 –   Post-Trim Stability of Post-Fired
Birox® 1900 Series Resistors
After 1000 h at 25°C

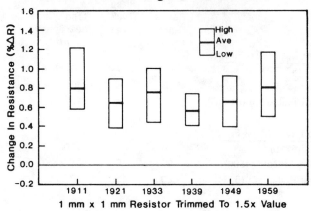

**1900 Series Resistors
Post-Fired On Tape Dielectric
Stability After Laser Trimming
1000 Hours Storage @ 40°C – 90% RH**

Figure 9 –   Post-Trim Stability of Post-Fired
Birox® 1900 Series Resistors
After 1000 h at 40°C/90% RH

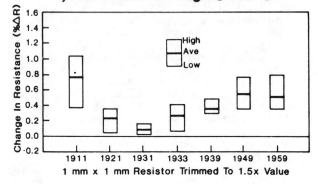

**1900 Series Resistors
Post-Fired On Tape Dielectric
Stability After Laser Trimming
1,000 Hours Storage @ 150°C**

Figure 8 –   Post-Trim Stability of Post-Fired
Birox® 1900 Series Resistors
After 1000 h at 150°C

CO-FIRED TOP LAYER RESISTORS

Samples for evaluation of top layer co-fired resistor properties were prepared by co-firing Birox® 1900 Series resistors terminated with 6144 Ag/Pd on an 8-layer tape laminate. Resistor and termination pastes were printed using 325 mesh screens with 0.5 mil emulsions, and dried at 125°C for 10 minutes after conductor and resistor printing. Standard laminating and burnout/firing conditions as described earlier were used.

Resistance and TCR data for 1900 Series resistors co-fired on the top layer are shown in Table 3, together with data for post-fired 1900 resistors. Differences in firing profiles and termination materials, combined with expected greater interactions of resistor and tape constituents, account for the observed differences. R. Pond [19] reported results of Birox® 1900 Series resistors printed on gold terminations co-fired on top of tape laminated substrates.

Post laser trim stability of selected 1900 Series resistors was determined by monitoring resistance changes of unencapsulated 1mm x 1mm resistors trimmed to 1.5x of their fired values using a single plunge cut. Trim conditions for the co-fired resistor series were 1W, 1kHz and a 5mm/s trim speed. Figures 10-12 show post trim stability under storage conditions of 25°C, 150°C and 40°C/90% RH, respectively. Average resistance changes were below 0.6% after 216 hours for all members and under all conditions tested.

CO-FIRED INTERNAL RESISTORS

1900 Series resistors co-fired in buried configurations showed systematic increases in resistance and large negative shifts in TCR compared with both post-fired and co-fired top layer 1900 Series resistors. The systematic shifts are attributed to the increased interaction between glass phases in the tape and resistor compositions due to the additional contact area and boundary conditions imposed by buried configurations. Substrate interactions with resistors are well-documented for the case of alumina substrates[20], and X-ray microprobe analysis has confirmed these findings.

Selected 1900 Series Resistors Co-Fired On Tape Dielectric Stability After Laser Trimming 216 Hours @ 25°C

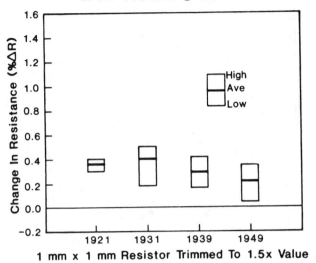

Figure 10 - Post-Trim Stability of Co-Fired Birox® 1900 Series Resistors after 216 h at 25°C

Selected 1900 Series Resistors Co-Fired On Tape Dielectric Stability After Laser Trimming 216 Hours @ 150°C

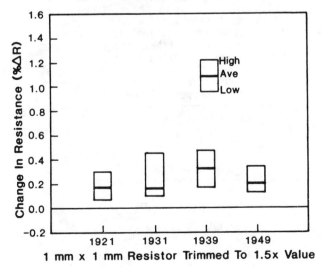

Figure 11 - Post-Trim Stability of Co-Fired Birox® 1900 Series Resistors After 216 h at 150°C

Selected 1900 Series Resistors Co-Fired On Tape Dielectric Stability After Laser Trimming 216 Hours @ 40°C -90% RH

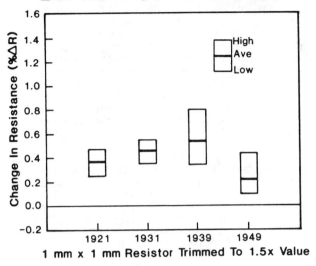

Figure 12 - Post-Trim Stability of Co-Fired Birox® 1900 Series Resistors After 216 h at 40°C/90% RH

Resistivity and TCR data for developmental resistor compositions are shown in Table 4. Developmental resistor compositions with resistivity between 1000 $\Omega/\square$ and 1M $\Omega/\square$ are based on Ru pyrochlore technology and show excellent compatibility with the tape dielectric and buried conductor components. The sintering rates and thermal expansion coefficients for these resistors have been engineered toward this purpose. Programs are in place to characterize performance of these compositions. Developmental low end resistors (R <1000 $\Omega/\square$) are based on RuO_2 technology and require additional development to match the sintering rate of the resistors (containing refractory RuO_2) to that of the tape laminate, and to improve TCR control.

TABLE 4

Resistance and TCR Values of Developmental Resistors for Internal Co-Firing

Resistivity $\Omega/\square$	HTCR (125°C) (ppm/°C)	CTCR (-55°C) (ppm/°C)
10	++	++
100	++	++
1000	- 38	- 101
10 k	- 41	- 113
100 k	- 88	- 130
1 M	- 91	- 196

++ Indicates large positive TCR.

RESISTOR PROCESSING

The last three sections have compared three different resistor firing configurations. In this section, the effects of key processing parameters including process sequence, laminating and tape shrinkage are discussed. A test was made to determine the influence of the lamination step on resistor properties. For each 1900 Series member, samples were made with (1) resistors printed on laminated but unfired substrates which received no subsequent lamination, and (2) resistors printed on tape and laminated together with 7 blank layers to form an 8-layer laminate. No differences were observed in either the resistivity or TCR between these samples. The only difference observed was in the small displacement of the resistor into the tape for samples with resistors receiving lamination.

In the case of co-fired top and buried resistors, laminating pressure was found to have no influence on either the resistivity (Figure 13) or TCR for the 100k $\Omega/\square$ member, 1949. In a parallel experiment, 'special' tape lots were produced to exhibit a wide range of lateral shrinkage values. Again, the value of the tape shrinkage had no noticeable influence on either the resistivity (Figure 14) or TCR values for the 1949 resistor composition. Additional work is required to demonstrate equivalent behavior for all 1900 Series members.

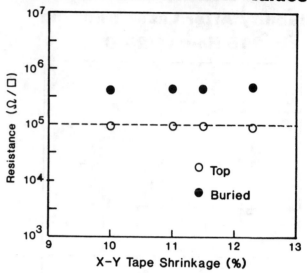

Figure 14 - Resistivity of Co-Fired Birox® 1949 Resistors as a Function of Lateral Tape Shrinkage

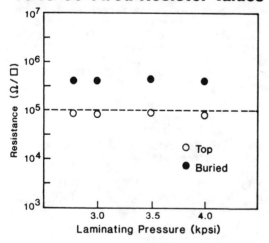

Figure 13 - Resistivity of Co-Fired Birox® 1949 Resistors as a Function of Laminating Pressure

SUMMARY

A low temperature co-fireable materials system is commercially available for the manufacture of multilayer interconnect hybrids. The system consists of a tape dielectric, gold, silver and silver/palladium signal and via fill conductors, co-fireable top layer resistors and developmental resistors for internal configurations, and thick film conductors and resistors for post-firing. Processing studies have shown that lateral shrinkage of the tape dielectric is independent of lamination time and lamination temperature (70°-80°C) but that a linear relation exists between shrinkage and laminating pressure.

1900 Series resistor compositions have been shown to be reliable for post-firing and co-firing on top layers. Developmental resistors designed for co-firing in internal configurations require modifications to accommodate greater levels of interaction between glass phases in the tape dielectric and in the resistor. Efforts are underway to complete development of a resistor series for co-fired internal applications.

ACKNOWLEDGEMENTS

The authors gratefully acknowledge the assistance of G. B. Mills, L. B. Straughn, J. E. Hickey, D. F. Rhoades, A. G. Mitchell, P. W. Kehn and P. G. McMonagle.

REFERENCES

[1] B. Schwartz, "Review of Multilayer Ceramics for Microelectronic Packaging", J. Phys. Chem. Solids, Vol. 45, No. 10, pp. 1051-1068 (1984).

[2] H. Tosaki, N. Sugishita and A. Ikegami, "New Approaches to Multilayer Hybrid IC With Interlayered Resistors", 1981 ISHM Proceedings, pp. 100-105 (1981).

[3] Y. Shimada, K. Utsumi, M. Suzuki, H. Takamizawa, A. Nitta and S. Yano, "Low Firing Temperature Multilayer Glass-Ceramic Substrate", Proceedings of 33rd Electronic Components Conference, pp. 314-319 (1983).

[4] K. Utsumi, Y. Shimada, H. Takamizawa, S. Fuji, S. Nanamatsu, "Application Monolithic Multi-components Ceramic (MMC) Substrate for Voltage Controlled Crystal Oscillator (VCXO)", Proceedings of 34th Electronic Components Conference, pp. 433-440 (1984).

[5] S. Nishigaki, S. Yano, J. Fukuta, M. Fukaya and T. Fuwa, "A New Multilayer, Low Temperature-Fireable Ceramic Substrate", 1985 ISHM Proceedings, pp. 225-234 (1984).

[6] K. Kondo, M. Oluyama and Y. Shibata, "Low Firing Temperature Ceramic Material for Multilayer Ceramic Substrates", presented at Electronics Division Fall Meeting of the American Ceramic Society, Orlando, Florida, October 13-16, 1985.

[7] K. Niwa, Kurihara, Y. Yokouchi and N. Kamehara, "Multilayer Ceramic Circuit Board with Copper Conductor II", presented at Electronics Division Fall Meeting of the American Ceramic Society, Orlando, Florida, October 13-16, 1985.

[8] IBM, U.S. Pat. 4413061 (1983).

[9] K. Kawakami and M. Takabatake, "A Low Temperature Co-Fired Multilayer Ceramic Substrate", presented at Electronics Division Meeting of the American Ceramic Society, Orlando, Florida, October 14-15, 1985.

[10] J. W. Lau, J. H. Enloe and R. W. Rice, "Effects of Powder Parameters on The Microstructure of a Low Temperature Glass-Ceramic Substrate", presented at the Annual Meeting of the American Ceramic Society, Chicago, Illinois, April 27-May 1, 1986.

[11] T. D. Thanh and N. Iwase, "Low Temperature Sintered Ceramics for Hybrid Functional Circuit (HFC) Substrates", Proceedings of International Microelectronics Conference, Tokyo, Japan, pp. 220-223 (1984).

[12] S. Tosaka, S. Hirooka, N. Nishimura, K. Hoshi and N. Yamaoka, "Low Temperature Fired Multilayer Ceramic Substrate", Proceedings of 1st IEEE: CHMT Symposium, Tokyo, Japan, Oct. 1-3, 1984, pp. 29-32, published 1985.

[13] H. Mandai, K. Sugoh, K. Tsukamoto, H. Tani and M. Murata, "A Low Temperature Co-Fired Multilayer Ceramic Substrate Containing Copper Conductors", IMC 1986 Proceedings, Kobe, Japan, May 28-30, pp. 61-64.

[14] A. L. Eustice, S. J. Horowitz, J. J. Stewart, A. R. Travis and H. T. Sawhill, "Low Temperature Co-Fireable Ceramics: A New Approach for Electronic Packaging", 36th Electronic Components Conference, Seattle, Washington, May 5-7, pp. 37-47, 1986.

[15] A. L. Eustice, S. J. Horowitz and A. R. Travis, "Low Temperature Co-Fireable Ceramics for Multilayer Interconnect Hybrids", IMC Proceedings, Kobe, Japan, May 28-30, pp. 49-60, 1986.

[16] W. D. Kingery, H. K. Bowen and D. R. Uhlmann, Introduction to Ceramics, pp. 531-532, © 1976, J. Wiley & Sons, NY.

[17] T. R. Allington, H. E. Schmidt, L. H. Slack and B. E. Taylor, "A Low Cost Thick Film Multilayer Hybrid System", Proceedings of 34th Electronic Components Conference, pp. 314-323 (1984).

[18] H. E. Schmidt and T. R. Allington, "Laser Trimming Resistors Over Dielectrics", Hybrid Circuit Technology, October 1985, pp. 35-43.

[19] R. Pond, "Processing and Reliability of Resistors Buried in Low Temperature Co-Fired Ceramic Structure", presented at the 1986 ISHM Symposium, Oct. 6-8, Atlanta, GA.

[20] W. S. Machin and R. W. Vest, "Reactivity of Alumina Substrates with High Lead Glasses", Processing of Crystalline Ceramics, pp. 243-51, ed. H. Palmour, III, R. F. Davis and T. Hare, Plenum Press, NY, 1977.

Part 8
Thermal Analysis of Multichip Modules

THERMAL phenomena are among the least understood and most complicated phenomena encountered in electronic systems. A potential physical limitation for making reliable electronic systems is the ability to remove the generated heat from the active devices. With the extreme high density and speed of multichip packages, heat removal becomes a critical factor in assuring high system reliability.

One of the best overviews of the thermal managements of multichip modules is in the paper by Bar-Cohen. When compared to the analysis of thermal stresses in the MCNC multichip constructions, some understanding of the problems should be possible. Add the paper from AT&T Bell Laboratories on the polyimide-on-silicon flip chip technology, and a look at the low-temperature ceramic constructions, and most of the considerations have been covered.

Thermal Management of Air- and Liquid-Cooled Multichip Modules

AVRAM BAR-COHEN, MEMBER, IEEE

Abstract—The state-of-the-art in air- and liquid-cooled multichip modules is examined. An effort is made to identify the salient features of eight distinct modules, define their thermal characteristics, and establish a consistent basis for comparing and evaluating their thermal performance.

INTRODUCTION

Chip Parametric Trends

THE Electronic Numerical Integrator and Computer (ENIAC), invented nearly 40 years ago by Mauchly and Eckert at the University of Pennsylvania, represents a critical inflection point in electronic technology and the gateway to the programmable high-speed computing devices of today. The ENIAC was a magnificent example of vacuum tube technology, employing 18 000 tubes in cabinetry weighing 30 tons and stretching 80 ft along three walls of a room [1], and it served to demonstrate the utility of multipurpose computing machines. However, ENIAC's 140-kW dissipation and mean time between failures (MTBF) of 0.5 h, while performing approximately 5000 instructions/s (0.005 MI/s), inadvertently established the need for a new class of electronic device.

The requisite device was provided in 1958 by Kilby in the form of the silicon integrated circuit [2], which has constituted the basic building block for the electronic industry ever since. Moreover, Kilby's integrated circuit spawned an unprecedented improvement in the reliability, speed of data processing and storage capacity of computing machines. When compared to ENIAC's performance, the technological achievements underpinning AT&T's Electronic Switching Systems (ESS) computer controller's 2 min/yr down time (!) [3] and the 24 MI/s scalar processing provided by a single processor of Control Data Corporation's Cyber 990, as well as the 1.3 billion floating point operations (GFLOPS) 6-ns machine cycle time of NEC Corporation's recently announced SX-2 supercomputer [4] and the 7.0-GFLOPS 10-ns clock of the first ETA-10, delivered at the end of 1986 [5], appear totally miraculous.

Much of the improvement in these system parameters can be traced to reductions in the IC feature size from the initial several hundred microns down to approximately 10 μm in 1975, 2 μm in 1983 and essentially 1 μm today. As suggested in Table I (derived from [6]), this miniaturization has facilitated a dramatic increase in the scale of circuit integration from Kilby's one transistor per circuit in 1958 and less than

Manuscript received December 9, 1985; revised February 5, 1987.
The author is with Corporate Research and Engineering, Control Data Corporation, 8100 34th Ave. S., P.O. Box 0, Minneapolis, MN 55440-4700.
IEEE Log Number 8714432.

TABLE I
EVOLUTION OF INTEGRATED CIRCUITS[a]

Year	Type	Components	Gates	I/O
1960	SSI—small-scale integration	1–40	1–10	14
1965	MSI—medium-scale integration	40–400	10–100	24
1970	LSI—large-scale integration	0.4–4.5k	0.1–1k	48
1975	VLSI—very large-scale integration	4.5–300k	1–80k	64
1982	ULSI—ultra large-scale integration	>300k	>80k	160

[a] Table derived from [6].

100 components (transistors, diodes, resistors or capacitors) per chip in the early 1960's to approximately 100 000 in 1980. More recently, chip gate density has taken a further jump to 450 000 components in the 1982 Hewlett-Packard 32-bit CPU chip [7] and to 460 000 in Texas Instruments 1984 VHSIC static RAM chip [8]. By the year 1990, the Semiconductor Research Corporation expects a 10 million component, 1.25 μm chip to be available [9].

Thermal Requirements

This five-order-of-magnitude increase in circuit integration, in the past 25 years, has been associated with successive revolutions in device technology, proceeding from TTL to ECL to NMOS and, most recently, to CMOS, and has been accompanied by almost a three-order-of-magnitude reduction in feature size, an order-of-magnitude increase in the characteristic chip dimension, and, perhaps most importantly, a percipitous drop in transistor switching energy from more than 10^{-9} J in 1960 to nearly 10^{-13} J in present-day devices [9]. Not surprisingly, much of the potential benefit of this enormous decrease in switching energy has been used to increase circuit speed and chip heat removal requirements have actually risen from the 0.1 to 0.3 W typical of the SSI devices, used in the early 1960's, to 1 to 5 W, commonly encountered in today's LSI ECL components and VLSI CMOS devices, and to values in excess of 10 W for the 10 000-gate ECL integrated circuits beginning to appear in advanced technology computers [10].

Based on short-term extrapolations of present trends in chip technology, it may well be anticipated that power dissipation in a leading-edge bipolar chip, 1 cm on a side, will reach 25 W by 1988 and a somewhat larger 10 million component chip, dissipating 100 W, could be encountered by 1990 [9]. Longer

Reprinted from *IEEE Trans. Components, Hybrids, Manuf. Technol.*, vol. CHMT-10, no. 2, pp. 159–175, June 1987.

TABLE II
STATE-OF-THE-ART IN CHIP MODULE THERMAL PARAMETERS

Technology	Chip Size (mm)	Maximum Power Dissipation (W)	Maximum Chip Flux (W/cm²)	Thermal Resistance (K/W)
Single chip modules				
Mitsubishi Alumina HTCP [29]	8 × 8	4	6.25	5.2
Mitsubishi SiC [36]	8 × 8	4	6.25	4.7
Hewlett-Packard Finstrate [7]	6.3 × 6.3	4	10.10	<8.7
Hitachi S-810 [22]	1.9 × 4	1	13.1	7.0
Fujitsu M-380 [40], [41]	4.5 × 4.5	3	14.8	8.0
Fujitsu M-780 [41], air-cooled	9.3 × 9.3	6.5	7.5	3.5
Fujitsu M-780 [41], water-cooled	9.3 × 9.3	9.5	11.0	2.5
Burroughs PGA [42]	4.5 × 4.5	5	24.3	12
Motorola MCA-2 [23]	7 × 7	12	24.5	3.3
Sperry Compact HX [43]	5 × 5	10	40	4.9
Multichip modules				
Mitsubishi HTCM [30]	8 × 8	4	6.25	7.3
NEC SX liquid-cooled module [34]	8 × 8	>5.4	>8.4	5 (water cooled)
Hitachi RAM [24]	1.9 × 4	1	13.1	34.7
IBM 4381 [26]	4.6 × 4.6	3.8	17.0	17.0
IBM 3090 TCM [12]	4.85 × 4.85	7	29.8	8.7 (water cooled)
NTT grooved substrate [36]	8 × 8	15.1	23.6	3.3 (water cooled)

term projections, which differentiate between high density and high-speed integrated circuits and address allowable current densities and on-chip delays, suggest that power dissipation for a 1×1-cm chip may reach approximately 125 W (for 50 million components, switching at 10 pS or a billion components, switching at 200 pS) for switching energies of 10^{-16} J at the end of the present century. Interestingly, a close approach to the theoretical switching energy of approximately 10^{-19} J [9], would make it possible in the next century to increase the product of switching speed and gate density by a further two or three orders-of-magnitude, without a significant rise in chip power dissipation.

Nevertheless, it must be noted that the heat fluxes encountered in today's "cutting edge technology" chips already pose a very significant challenge to the thermal packaging engineer. Table II reveals these to be in the range of 5–40 W/cm² and thus comparable, at the upper end, to the thermal loading experienced by re-entry vehicles and, even at the lower end, to heat fluxes imposed on rocket motor cases. The anticipated 1990 VLSI heat fluxes of approximately 100 W/cm² are seen in Fig. 1 to be in the range of thermal loadings associated with nuclear blasts. While rocket heat shields and targets of nuclear blasts can be expected to reach temperatures in excess of 1000°C, successful thermal management of microelectronic components, operating in this same heat flux range, requires that maximum chip surface temperatures be maintained

between 50 and 100°C. These thermal requirements and the crucial need to reduce off-chip time delays, as well as the need to provide significantly longer system MTBF's at substantially lower prices, have recently focused the efforts of the thermal packaging community on multichip modules. The high packaging density associated with these multichip modules has further exacerbated the thermal management task and placed these modules at the forefront of thermal control technology.

Much of the discussion which follows will, therefore, be devoted to examining both air-cooled and liquid-cooled state-of-the-art multichip modules. An effort will be made to identify the salient technological features of such modules, as Hitachi's Silicon Carbide six-chip RAM module [11] and IBM's 3090 100-chip 500-W Thermal Conduction Module [12], define their thermal characteristics, and establish a consistent basis for comparing and evaluating their thermal performance. The paper will conclude with a brief evaluation of the impact of the current trends in computer system architecture and packaging on multichip module thermal control requirements.

Thermal FOM's

The severe thermal requirements described in the previous section have not only resulted in the application of ever more sophisticated thermal design techniques to electronic equipment but have made it essential that chip and module designers

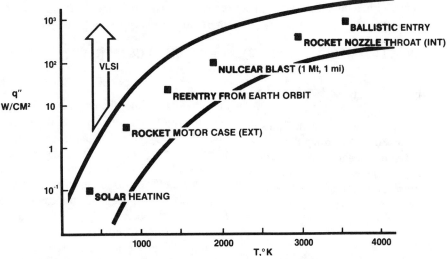

Fig. 1. Perspective on microelectronic heat fluxes.

address the thermal control requirement early in the design process. As a consequence, electronic designers have made use of a variety of figures-of-merit (FOM's) which have attempted to address the limitations on device performance imposed by chip heat dissipation.

Keyes was, perhaps, the first to examine the fundamental limits on integrated circuit performance systematically, and in his pioneering paper [13] he suggested that 20 W/cm² be taken as a realistic maximum attainable heat transfer rate, based on nucleate boiling of R-113 and the expectation that the contact resistance between the chip and any extended surface would lie in the range of 1–10°C/(W/cm²).

Unfortunately, this particular choice of chip heat flux failed to anticipate the considerable reductions that have been achieved in interface resistances during the past decade [14] and neglected the cooling potential of both compact heat exchangers [15] and flow boiling of dielectric refrigerants [16]. As clearly shown in Table II, maximum chip heat fluxes are already in excess of 20 W/cm² and a careful reading of the thermal literature fails to reveal any absolute limit on chip heat removal. Consequently, neither the FOM's proposed by Keyes [13], [17] nor more recent modifications [18], [19] can be expected to order various packaging strategies correctly, and other alternatives must be sought.

The venerable chip thermal resistance R_T, based on the chip power dissipation and chip-to-coolant temperature difference, is perhaps the most common basis for comparing thermal packaging technologies. However, as may be seen in Table II, R_T appears to vary widely and, often, in an unanticipated manner, providing significantly different values for apparently similar packaging approaches and vice versa. It would, therefore, appear equally unwise to rely exclusively on R_T for a thermal packaging FOM though, as will be shown, this parameter may be used to establish an appropriate FOM when related to the packaging density.

CHIP MODULE THERMAL RESISTANCE

Definition

Due to the significant variations in the ambient conditions, power dissipations and allowable chip or junction tempera-

tures associated with the various electronic system configurations and/or imposed by the manufacturers of computers, the thermal performance of chip packaging techniques is commonly compared on the basis of overall (junction-to-coolant) thermal resistance R_T. This packaging figure-of-merit is generally defined in a purely empirical fashion to equal

$$R_T = (T_j - T_f)/q_c \qquad \text{K/W} \qquad (1)$$

where T_j and T_f are the junction and coolant (fluid) temperatures, respectively, and q_c the chip heat dissipation.

Unfortunately, however, most measurement techniques are incapable of detecting the actual junction temperature, i.e., the temperature of the small volume at the interface of p-type and n-type semiconductors, and hence this term generally refers to the average temperature or a representative temperature on the chip. Since the failure rate of integrated circuits has long been known to be accelerated by an increase in junction temperature [20], the lowest value of R_T is to be preferred.

Examination of various packaging techniques reveals that the junction-to-coolant thermal resistance is, in fact, composed of an internal, largely conductive, resistance and an external, primarily convective, resistance. The internal resistance R_{jc} is encountered in the flow of dissipated heat from the active chip surface through the materials, used to support and bond the chip, and on to the case of the integrated circuit package. The flow of heat from the case directly to the coolant, or indirectly through a fin structure and then to the coolant, must overcome the external resistance R_{ex}.

Internal Resistance

Conductive thermal transport is governed by the Fourier equation which, in one-dimensional form, is expressible as [21]

$$q = kA dT/dx \qquad \text{W} \qquad (2)$$

where q is the heat flow, k is the thermal conductivity of the medium, A is the cross-sectional area for heat flow, and dT/dx the temperature gradient in the direction of heat flow. For composite, rectilinear structures, as encountered in many chip modules, the Fourier equation (with temperature and time invariant properties), takes the form

$$q = (T_i - T_0) / \sum_p (\Delta x / kA) \qquad (3)$$

where T_i and T_0 are the temperatures internal and external to the composite structure, respectively, Δx the thickness of the material in the direction of heat flow, and the summation sign pertains to p distinct layers of material. Assuming that power is dissipated uniformly across the chip surface and heat flow to be largely one-dimensional, (3) can be used to provide a first-order approximation for the internal chip module resistance, as

$$R_{jc} = (T_j - T_c) / q_c = \sum_p (\Delta x / kA) \qquad K/W. \qquad (4)$$

When expressed in this form, the summed terms are seen to represent the thermal resistances of the individual layers of silicon, solder, adhesives, etc. As the thickness of each layer decreases and/or the thermal conductivity and cross-sectional area increase, the resistance of the individual layers decreases. Although two-dimensional conduction effects are often of significance in chip packages, especially in the presence of nonuniform power dissipation in the chip, (4) does suggest that R_{jc} can be altered, not only by the choice of support materials and bonding technology (i.e., choice and thickness of bonding materials), but also by a change in geometry and, particularly, the cross-sectional or "footprint" area of the chip and case. An evaluation and comparison of chip packaging technologies on the basis of R_{jc}, which ignores this geometric influence, may well lead to erroneous conclusions.

External Resistance

The resistance to thermal transport from a surface to a fluid in motion, i.e., the convective resistance, varies inversely with the wetted area and the heat transfer coefficient h. For a particular geometry and flow regime, h may be found from available empirical correlations and/or theoretical relations. For flow along plates and in the inlet zones of parallel-plate channels, as may well be encountered in electronic cooling applications, the low velocity, or laminar flow, average convective heat transfer coefficient is given by [21]:

$$h = 0.664(k/l) \, (\text{Re})^{0.5} \, (\text{Pr})^{0.333} \qquad W/m^2K$$

$$\text{for Re} < 2 \times 10^5 \qquad (5)$$

where k is the fluid thermal conductivity, l the characteristic dimension of the surface, Re the Reynolds number (equal to $\rho V l / u$), and Pr the Prandtl number ($c_\rho u / k$). Inserting the various parameters associated with the Re and Pr in (5), the laminar heat transfer coefficient is found to be directly proportional to the square root of fluid velocity and inversely proportional to the square root of the characteristic dimension. Furthermore, increases in the thermal conductivity of the fluid and in the Pr, as are encountered in replacing air with a liquid coolant, can be expected to result in higher heat transfer coefficients.

In higher velocity, turbulent flow, the dependence of the convective heat transfer coefficient on the Reynolds number increases and is typically given by [21]

$$h = 0.036(k/l) \, (\text{Re})^{0.8} \, (\text{Pr})^{0.333} \qquad W/m^2K$$

$$\text{for Re} > 3 \times 10^5. \qquad (6)$$

In this flow regime, the convective heat transfer coefficient is thus found to vary directly with the velocity to the 0.8 power and inversely with the characteristic dimension to the 0.2 power. The dependence on fluid conductivity and Pr remains unchanged.

Applying (5) or (6) to the transfer of heat from the case of a chip module to the coolant, the external resistance $R_{ex} = 1/hA$ is found to be inversely proportional to the wetted area and to the coolant velocity to the 0.5–0.8 power and directly proportional to the length scale in the flow direction to the 0.5–0.2 power. It may thus be observed that the external resistance can be strongly influenced by the fluid velocity and package dimensions and that these factors must be addressed in any meaningful evaluation of the external thermal resistances offered by various packaging technologies.

For a fixed value of the convective heat transfer coefficient, the external resistance can be reduced by enlarging the surface area in contact with the coolant. Since it is also generally desirable to minimize the projected area or footprint of the chip module, this extended area is best provided by a fin structure or compact heat exchanger attached to the module case. However, the presence of the fin structure and an additional bonding layer (needed to attach the fin to the case), introduces new thermal resistances which must be incorporated in the expression for the external resistance.

In the context of the present approximate formulation for chip module thermal resistance, the fin-to-case interface can be treated as an additional material layer with a resistance of $\Delta x / kA$, and use can be made of the "fin efficiency" concept to deal with the conductive resistance of the fin structure. Fin efficiency η is generally defined as the ratio of the average temperature difference between the fin and the coolant to the temperature difference between the fin base and the coolant [21]. Fin efficiency can thus be expected to range from 0 to 1, with high thermal conductivity, short thick fins providing the highest values of η.

Using this approach, heat transfer by a fin or fin structure can be expressed in the form

$$q_f = hA \, [\eta (T_0 - T_f)] \qquad (7)$$

where q_f is the fin heat dissipation, η is the fin efficiency, $T_0 - T_f$ the temperature difference between the base and the coolant, and A the full wetted area. Since the external resistance of a chip module is defined in terms of the temperature difference between the case and the coolant, (7) can be used to modify R_{ex} to reflect the contribution of a fin structure or compact heat exchanger, as

$$R_{ex} = (T_c - T_f) / q_c = (\Delta x / kA)_b + (1/\eta hA). \qquad (8)$$

In an optimally designed fin structure, η can be expected to fall in the range of 0.5–0.7 [21]. Relatively thick fins in a low velocity flow of gas are likely to yield fin efficiencies approaching unity. This same unity value would be appropriate, as well, for an unfinned surface and thus serve to generalize the use of (8) to all package configurations.

Total Resistance

To the accuracy of the assumptions employed in the preceding development, i.e., uniform heat dissipation and one-dimensional conduction, the overall chip module resistance can be found by summing the internal and external resistances given by (4) and (8) to yield

$$R_T = R_{jc} + R_{ex} = \sum_p (\Delta x / kA) + (1/\eta hA). \qquad (9)$$

In evaluating the thermal resistance by this relation, care must be taken to determine the effective cross-sectional area for heat flow at each layer in the module. For single-chip modules, the requisite areas can be readily obtained, though care must be taken to consider possible voidage in solder and adhesive layers. The determination of the appropriate areas in multichip modules is far more difficult but, for modules involving chips of identical geometry and power dissipation, this task can generally be performed, to an acceptable level of accuracy, by defining a "unit cell" around each chip in the module.

As previously noted in the development of the relations for external and internal resistances, (9) shows R_T to be a strong function of both the convective heat transfer coefficient and geometric parameters (thickness and cross-sectional area of each layer). Thus the introduction of a superior coolant, use of thermal enhancement techniques which increase the local heat transfer coefficient, or selection of a heat transfer mode with inherently high heat transfer coefficients (e.g., boiling) will all be reflected in appropriately lower external and total thermal resistances. Similarly, improvements in the thermal conductivity of and reduction in the thickness of the relatively low conductivity bonding materials (e.g., soft solder, epoxy, silicone) would act to reduce the internal and total thermal resistances.

However, frequently, even more dramatic reductions in the total resistance can be achieved simply by increasing the cross-sectional area for heat flow, within the chip module (e.g., chip, substrate, heat spreader) as well as along the wetted, exterior surface. The implementation of this "scale-up" approach generally results in a larger module footprint and/or lower volumetric packaging density, both of which are highly undesirable, and yet is rewarded with an apparently better packaging figure-of-merit.

Evidence for this difficulty can be found by comparing the Hitachi S-810 air-cooled single-chip package [22] and the Motorola MCA-2 air-cooled single-chip package [23] listed in Table II. While the Hitachi package offers a resistance which is twice that of the Motorola package, six of the Hitachi single-chip modules are mounted on a 2.5 cm² ceramic substrate and extend approximately 1 cm high. By contrast, although both packages are cooled by ambient air at comparable velocities, 5 m/s for the S-810 and 3.8 m/s for the MCA-2, one MCA-2 PGA package is 15 cm² in area and nearly 2 cm high. Further support for this concern can be found by comparing the dimensions and thermal resistances of the multichip modules, shown in Tables II and III, as will be discussed in a later section.

In evaluating packaging approaches, it must therefore be understood that the thermal resistance is not a true figure-of-merit, and if R_T is to be used, care must be taken to determine the specific reason(s) for a change in chip-to-coolant thermal resistance. Alternatively, a more consistent thermal packaging figure-of-merit must be sought.

MULTICHIP MODULES

In recent years most of the leading manufacturers of computers and other microelectronic equipment have begun to develop the technology and packaging strategies needed to insert multichip modules into their product families. Several companies, notably Hitachi [11], [24], have chosen to emphasize "mother chips," or relatively small modules containing four to six small chips on a common substrate, while many others have pursued the development of substantially larger modules containing 30 to as many as 133 chips, as in IBM 3080's TCM [25].

Table III displays the salient thermal features of a selection of air- and water-cooled multichip modules based on descriptions in the open literature. Although most of these modules have been used in operating commercial equipment, the absence of uniformity in the veracity and depth of the available detail significantly constrains the accuracy of any comparison of thermal characteristics. Nevertheless, in succeeding sections, an effort will be made to define the primary thermal packaging features of this representative sample of multichip modules.

Air-Cooled Modules

IBM 4381 Module: The IBM 4381 midrange processor, elements of which are described in [26]–[28], consists of a single board containing 22 modules. Each of the modules shown in Fig. 2 is 64 × 64 mm and approximately 40 mm high and houses up to 36, though typically 31, logic chips of approximately 7000 elementary components, or 704 circuits, in an area of 4.6 × 4.6 mm [28]. In each 882 input/output pin module the chips are solder-bumped on a multilayer ceramic (MLC) substrate and separated from a ceramic cap by a layer of thermal paste, 0.1–0.35 mm (4–14 mils) thick. The ceramic cap, tin–lead soldered over a 1.5-mm-wide seal band, supports a 25-mm-high array of 256 hollow-pin aluminum fins cooled by a wide jet of air exhausting from a nearby vertical plenum. The design of this plenum provides for the parallel flow of air jets onto each module and thus provides an identical ambient air temperature for each fin array, regardless of location on the board.

Based on information provided in [26], it appears that 55 percent of the heat released by the chips conducts down into the MLC substrate and 45 percent up through the thermal paste to the cap. However, more than half of the heat flowing into the MLC conducts back up through the solder seal to the cap. Thus, in total, nearly 75 percent of the dissipated heat exits the module via the fin array and the remainder is removed at the exposed surfaces of the board.

TABLE III
MULTICHIP MODULE PARAMETERS—GENERAL

Technology	Number of Chips	Number of I/O's	Number of Transistors[a] ($\times 10^3$)	Area (mm²)	Chip Area Fraction[b] (mm)	Height
Mitsubishi HTCM [30]	9	624	108	66 × 66	0.13	22
Hitachi RAM [24]	6	108	?	27.4 × 27.4	0.06	16
Honeywell SLIC [33]	110	240	?	80 × 80	?	50
NEC SX [34]	36	2177	144	125 × 125	0.15	60
IBM 4381 [26]	36	882	252	64 × 64	0.19	40
IBM 3090 [12]	100	1800	600	150 × 150	0.10	60
NTT [36]	25	900	?	85 × 105	0.18	5[c]

[a] Based on an average of four transistors/gate.
[b] Chip-to-module area ratio.
[c] Assumed value for module-to-module spacing.

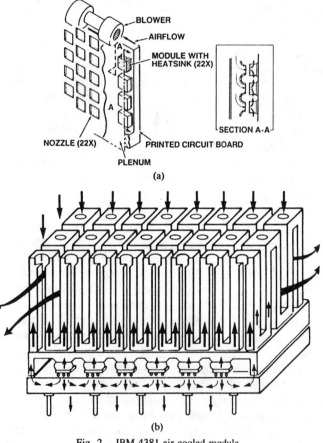

(a)

(b)

Fig. 2.   IBM 4381 air-cooled module.

This thermal design, with an air flow of 211 l/s at 20 mm of water pressure head, has been found to facilitate the transfer of up to 3.8 W per chip, 90 W per module, and 1.3 kW per board, while maintaining all the chips below the maximum specified temperature of 90°C [26]. It may be noted that while 3.8 W represents the maximum chip dissipation, the average value for a high dissipation module ranges from 2.5 to 2.9 W and the average chip dissipation for the 22-module system is only 1.65 W. The IBM 4381 module has been shown to provide an external resistance, based on total module power,

of approximately 0.23°C (or K) per watt. Alternately, in a module containing 36 identical chips, the chip-to-air thermal resistance has been found to equal 17 K/W, divided nearly equally between the external (8 K/W) and internal (9 K/W) resistances [26].

Due to the heat spreading effect of the cap and substrate, a realistic, though still conservative, assessment of chip temperature can be obtained by using the module-based external resistance to determine the cap temperature and the chip-based internal resistance to calculate the chip-to-cap temperature difference. Following this procedure, a 3.8-W chip in a 90-W 4381 module, cooled by ambient air at 25°C, would attain a temperature no greater than 80°C rather than the 89.6°C attained if each and every chip dissipated 3.8 W.

A close examination of the 4381 module reveals that its thermal performance is intimately related to three interface resistances: the thermal paste layer, the cap/MLC seal, and the fin array/cap attachment. Due to the significant thermal resistance of the paste layer, approximately 8 K/W for a 0.25-mm (10 mil) layer with $k = 1.25$ W/mK [27], failure to control manufacturing tolerances closely on the ceramic cap and solder-bumped chips could induce unacceptably large chip-to-chip temperature variations and excessive temperatures at high-power large-gap chips. Similarly, since some 30 percent of the dissipated heat flows through the cap/MLC seal and 75 percent across the bond between the cap and the fin array, these resistances must be minimized by design and maintained close to design values during assembly, if chip temperatures are to remain at acceptable levels.

Hitachi Silicon–Carbide RAM Module: In an attempt to provide a reliable, densely packaged, and thermally acceptable IC package, Hitachi's Device Development Center has focused much of its recent effort on mother chips, or small multichip modules [11]. The 27.4 × 27.4-mm and approximately 16-mm-high, 108-lead silicon carbide module, described in [11], [24] and shown in Fig. 3, epitomizes this approach.

The Hitaceram 101 SiC RAM module contains six 1-W ECL chips, each 1.9 × 4 mm and providing 1 kbit of memory, solder-bumped to a silicon substrate which is, in

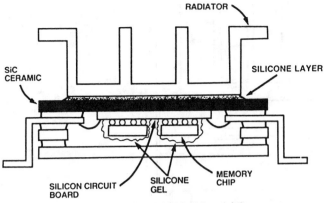

Fig. 3. Hitachi SiC RAM module.

turn, gold/tin eutectic bonded to the SiC. Both the aluminum fin structure and the lid are attached to the module with a layer of silicone rubber (with filler), approximately 50 μm thick. The chips are encapsulated in silicone gel for protection from humidity and from alpha particles emanating from the solder.

The heat released by each chip is conducted through 77 solder bumps (52 of which are purely thermal in function) to the silicon substrate, or 'mother chip' and then through the low resistance gold eutectic bond to the SiC. An aluminum heat sink, approximately 8 mm high and 20 × 20 mm at the base, with four longitudinal fins, serves to transfer the dissipated heat to the ambient air blown past the RAM module.

The results reported in [24] reveal the theoretical thermal resistance of the RAM module, at an air velocity of 3 m/s, to be 34.7 K/W, based on the heat dissipation of a single chip, or nearly 5.8 K/W, based on total module dissipation. This latter value compares most favorably with a measured value of 5.5 K/W. The theoretical value corresponds to a chip-to-silicon substrate resistance of 1.44 K/W, calculated on the basis of module dissipation, negligible resistance through the gold eutectic bond, an additional 0.25 K/W imposed by heat conduction through the silicone rubber bonding the heat sink to the module and, finally, the resistance of the fin structure equaling 4.1 K/W. The air-cooled heat sink and the solder-bump structure are, therefore, the two primary thermal resistances in the Hitachi RAM module.

Using the stated theoretical thermal resistance values and a module dissipation of 6 W for a 25°C inlet air temperature and an assumed 10°C rise in the air flowing past the modules, the maximum chip temperature can be expected to approach 70°C. The relatively modest temperature rise at the chip would appear to allow adequate thermal control of the Hitachi RAM module at the stated air velocity of 3 m/s. However, it should be noted that, assuming laminar flow in the fin passages and using (5), at a more typical velocity of 6 m/s, the external module resistance could be expected to decrease to approximately 3 K/W. As a result, the total resistance from chip-to-air for the 6 m/s air velocity is likely to approach 4.5°C/W of module dissipation or 27 K/W based on the dissipation of an individual chip.

As previously noted, the thermal resistance of the Hita-ceram substrate, shown in [24] to equal approximately 0.025 K/W, results in a negligible temperature rise of 0.15°C for a

module dissipation of 6 W. While at first glance this result appears to provide a thermal justification for the use of SiC, it must be noted that if alumina were used as the substrate material (with a thermal conductivity 13.5 times lower than SiC), this resistance would still be relatively negligible at 0.34 K/W and a resulting temperature difference through the substrate of less than 2°C. Reduced lateral conduction in the low conductivity substrate may result in a higher heat flux through the silicone rubber layer, which bonds the heat sink to the substrate, and produce a consequent rise in the temperature difference through the rubber. However, in the present design this resistance (at 0.14 K/W) accounts for less than 1°C and is thus unlikely to exceed 2°C even with a low-conductivity substrate.

These calculated values appear to suggest that, contrary to some reports in the literature, the high thermal conductivity of SiC may be of limited significance in determining the thermal performance of, at least, air-cooled multichip modules. Nevertheless, the use of such substrates may well be justified by the near-equality of the thermal coefficients of expansion of silicon and SiC, leading to significantly lower thermal stress in the solder joints and/or bonding layers between these two materials.

Mitsubishi High Thermal Conduction Module

Mitsubishi's concern over the high thermal resistance of flip-chip bonded devices has led to the development of a packaging technology for LSI chips which relies on heat transfer from both the top and solder-bump sides of each chip to achieve a relatively low junction-to-ambient thermal resistance [28], [29]. A 66 × 66 × 22-mm-high 624-I/O multichip pin-grid array module represents one articulation of this packaging concept and is described in detail in [29].

This Mitsubishi module, shown in Fig. 4, contains nine 3-kgate ECL chips, each 8 × 8 mm with 223 I/O's and a maximum dissipation of 4 W, which are solder-bumped to the substrate (apparently alumina) and soldered on the top to a 13 × 13 × 0.25-mm-thick copper plate. During assembly the copper plate is pressed up against the ceramic cap enclosing the chips, the module is filled with hydrogen, and the assembly is heated to melt the solder, sealing the module and attaching the copper heat spreaders. When the module has cooled, a longitudinal-fin heat sink is epoxied to the module gap. The cooling of the structure and solidification of the solder were found to produce a nominal 30 μm gap between the copper plate and the cap.

In operation, heat released by the chips flows to the heat sink along two parallel paths, one passing through the solder bumps and the substrate and then on to the cap and heat sink through the module seal and the other path going directly to the cap and heat sink via the thermal spreader and the gas gap. A detailed thermal analysis of this multichip module, reported in [29], revealed that, in the absence of heat transfer from the pins to the air (a conservative assumption), a chip located at the edge of the substrate could be expected to dissipate approximately 18 percent of its heat through the solder bumps and 82 percent across the gas gap, while only 13 percent of the heat would go through the solder bumps of the center chip.

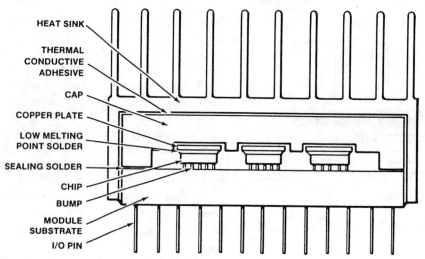

HEAT SINK

THERMAL CONDUCTIVE ADHESIVE

CAP

COPPER PLATE

LOW MELTING POINT SOLDER

SEALING SOLDER

CHIP

BUMP

MODULE SUBSTRATE

I/O PIN

Fig. 4. Mitsubishi high thermal conduction module.

The resulting chip-to-ambient thermal resistance, defined in terms of the chip power dissipation, was calculated to be 6.8, 7.0, and 7.3 K/W for a corner chip, edge chip, and the center chip, respectively, at an imposed air velocity of 6 m/s. Measured values in the velocity range of 2–6 m/s, with an inlet air temperature of 25°C, were generally found to agree with the calculated values to within five percent.

The overall thermal resistance of the Mitsubishi multichip module was found to include a 3.0-K/W chip-to-heat-sink resistance for the central chip and an internal resistance of approximately 2.5 K/W for the peripheral chips. Test results shown in [29] indicate that the conduction resistance of the gap between the heat spreader and the ceramic cap, when the module is filled with pressurized hydrogen, accounts for less than 0.5 K/W (though calculations would suggest nearly 0.9 K/W). For the dimensions shown thermal conduction through the silicon chip, copper heat spreader, and ceramic cap appears to contribute approximately 0.6–1.2 K/W to the internal thermal resistance of the central chip. It may thus be surmised that the three interface resistances, offered by the solder used to attach the heat spreader, by the hydrogen layer and by the epoxy bonding the heat sink to the module cap, account for more than 50 percent of the internal resistance. Assuming an air inlet temperature of 25°C and a 10°C rise in air temperature across several modules, the maximum chip temperature can be expected to lie below 65°C and average approximately 63°C for the stated conditions.

It must be noted, however, that to achieve this thermal performance the module must remain hermetically sealed, the gas gap dimension must be kept to approximately 1 mil, and the quality of the mating surfaces and/or the thickness of the bonding layers must be carefully controlled to the conditions attained in the prototype module described in [29]. Alternately, the authors of [29] suggest that the temperatures of the chips could be lowered by several degrees Celsius by taking advantage of heat transfer to the printed circuit board on which the module is mounted.

Liquid-Cooled Modules

IBM 3081 Thermal Conduction Module: At the heart of the 9 MIP, IBM 3081 processor complex is the hermetically

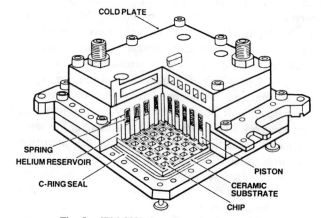

COLD PLATE

SPRING

HELIUM RESERVOIR

C-RING SEAL

PISTON

CERAMIC SUBSTRATE

CHIP

Fig. 5. IBM 3081 thermal conduction module.

sealed, water-cooled thermal conduction module (TCM) shown in Fig. 5. The 1800-pin TCM is approximately 150 × 150 × 60 mm high and contains up to 133 chip sites arrayed on a 90 × 90 mm 33-layer ceramic substrate [25]. Each 4.6 × 4.6 mm TTL chip, containing up to 704 circuits for a peak heat dissipation of 4 W is attached to the substrate via solder bumps. Up to 300 W can be removed from each TCM, and as many as nine TCM modules may be mounted on a single 700 × 600-mm printed circuit board [31].

Heat released by each chip in the TCM is conducted, via a spring-loaded aluminum piston in a helium atmosphere, to the water-cooled heat exchanger, constituting the cap of the module. In the TCM design it was desired to bring the water-cooled surfaces as close as possible to the chip heat sources while, at the same time, allowing for variations in chip height and location, arising from manufacturing tolerances. Additionally, allowances had to be made for nonuniform thermal expansion and contraction along the primary thermal path [32].

While thermal control of the chips in the TCM involves heat flow along several parallel paths, its developers found it convenient to describe the performance of the TCM in terms of the resistances along the primary thermal path from a single chip [32]. Since the heat dissipating devices are on the solder-bump side of the chip, heat flow to the piston will encounter a conduction-constriction resistance of approximately 1.15 K/W

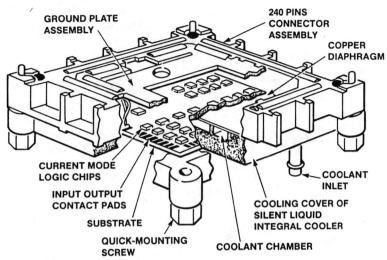

Fig. 6. Honeywell silent liquid integral cooler module.

in crossing the chip. Although the piston contacts the chip, most of the heat flow occurs across the helium-filled gap surrounding the piston/chip contact zone and engenders a 3-K/W resistance.

Heat flow through the metallic piston and the helium gap separating the piston from the housing results in an additional resistance of 3.2 K/W and conduction through the housing to the water-cooled heat exchanger imposes a 1.6-K/W penalty. Interestingly, nearly 50 percent of the internal resistance, or 4.2 K/W can be traced to the two helium-gap conduction resistances, emphasizing once again the importance of controlling interface resistance in multichip modules.

Thermal transport in the heat exchanger, by conduction through the heat exchanger walls and by convection to the water flowing in the channels, is constrained by a final 1.5–3 K/W 'external' resistance. The total chip-to-coolant resistance for the IBM TCM, defined in terms of the maximum chip dissipation, is thus, approximately, 11.2 K/W, including an internal resistance of nearly 9 K/W and a typical external resistance of 2.25 K/W (at approximately 40 cm^3/s water flow rate). Based on these reported values, it would appear that the temperature of a 4-W chip in a 300-W 3081-configuration TCM with a water inlet temperature of 24°C is approximately 69°C and is thus considerably below the stated design requirement of 85°C.

This performance margin has apparently made it possible to use an essentially unmodified TCM in the IBM 3090 Series 25 MIP machine. In this application the TCM is reported to dissipate approximately 500 W, generated by 100 enhanced-ECL chips, 4.85 × 4.85 mm in size with a peak dissipation of 7 W from 612 circuits [12]. To achieve the desired chip temperature limit of 85°C with an inlet water temperature of 24°C, it is thus necessary for the 3090 TCM to provide an overall thermal resistance of approximately 8.7 K/W. An increase in the water flow rate through each TCM and optimization of the water distribution system, as well as thermal optimization of the piston and housing design for the slightly larger and more widely spaced 3090 chips, may very well have been sufficient to reduce the TCM resistance to this value.

Honeywell (HIS) Silent Liquid Integral Cooler (SLIC): Packaging of the Honeywell Information Systems DPS-88 computer is based on the use of an 80 × 80 mm multilayer ceramic substrate, housed within an aluminum frame which is termed a micropackage. Each such leadless package, with 240 I/O's and 110 sites for CML LSI chips, is connected, together with seven other micropackages, to a 534 × 318-mm mother board [33]. As seen in Fig. 6, cooling of the DPS-88 components is provided via the micropackage cover, the SLIC, which serves as a liquid-cooled cold plate. A flexible copper diaphragm, which conforms to the back surface of the substrate on one side and is wetted by the circulating water (or possibly in contact with a water-cooled plastic membrane) on its other side, constitutes the bottom of that cover. Each micropackage is reported to dissipate 60 W, for an average chip dissipation of approximately 0.55 W [33].

Unfortunately, while Honeywell was one of the first companies to develop water cooling for mainframe computers, little additional information is available in the open literature on the thermal performance of the SLIC. For an assumed water inlet temperature of 24°C, as well as a peak allowable chip temperature of 70°C and a maximum chip dissipation of 0.75 W, the overall thermal resistance of the SLIC can be calculated to equal approximately 60 K/W.

In the absence of published data, it is difficult to evaluate the veracity or significance of this value and one can only speculate as to whether the 0.55-W chip dissipation and 60-W module dissipation represent a mature technology or are to be associated with a nonoptimized cooling system. Note, however, that a first-cut analysis of the SLIC package suggests that the interface resistance between the copper diaphragm and the ceramic substrate can be expected to contribute as much as one-third of the overall thermal resistance of the micropackage.

NEC SX Liquid Cooling Module: Nippon Electronic Corporation's (NEC) latest supercomputer, reported to achieve 1.3 gigaflops and 6-ns machine cycle time, implements several new packaging technologies and liquid cooling systems [34], [35]. Both logic and RAM chips, TAB'd and packaged in ceramic chip carriers which are in turn solder-

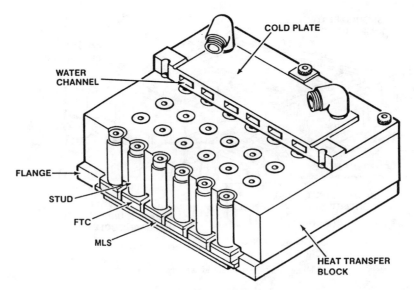

Fig. 7.   NEC SX liquid-cooling module.

bumped onto an alumina/polyimide multilayer ceramic substrate, are thermally controlled by adjustable metal studs which conduct the heat dissipated by the chips to a water-cooled cold plate. Using this approach, shown schematically in Fig. 7, NEC claims to have achieved significant improvements in gate density, signal "flying time," wiring density, I/O density, and heat removal capability relative to its highest class general purpose computer, the S-1000 [34].

NEC's water-cooled multichip package (MCP) is approximately 125 × 125 × 60 mm high and can accommodate up to 36 flipped TAB carriers (FTC's), 12–14 mm on a side, solder-bumped onto a 100 × 100 × 2.75-mm-thick alumina/polyimide substrate. Twelve such modules can be mounted on a single 541 × 457 × 4.9-mm-thick printed wiring board. Each MCP includes 2177 I/O pins brazed on its bottom surface and is capable of dissipating 250 W through the water-cooled heat exchanger on its top surface. Each 12 × 12-mm FTC contains a 1000-gate 5.7-W (typical) CML logic chip which is solder-bumped to the ceramic substrate and TAB'd to the copper/tungsten FTC cap. Four 1-kbit bipolar RAM chips, typically dissipating a total of 5.4 W, can be packaged in a similar manner in the slightly larger (14 × 14 mm) FTC.

Thermal control of the FTC's in the multichip module is provided by the liquid cooling module (LCM), which consists of a heat transfer block (HTB), a cold plate, and 36 studs. As reported in [34], the HTB removes heat by fine-gap contact between the stud and the FTC. The stud is placed in a machined hole in the HTB and has a unique shape to ensure fine-gap contact with the FTC. To establish this contact, each stud is adjusted individually before its position is fixed in the HTB. After the stud location is secured, the HTB is detached, a thermal compound applied to the cap of each FTC and the HTB reassembled. To complete the LCM, the cold plate is bolted to the HTB and liquid lines are attached to the inlet and outlet ports on the cold plate.

While no detailed test results have been released for the LCM, the authors of [34] claim to have achieved a chip-to-water thermal resistance of 5 K/W. This relatively low value

of thermal resistance would appear to justify the pains taken in assembling the LCM to minimize the gap between the stud and the chip carrier and thereby limit the thermal interface resistance at the surface of the FTC. In the absence of a detailed thermal analysis of this thermal packaging configuration, it is difficult to evaluate either the FTC's contribution to the overall thermal resistance or the potentially deleterious effect of differential thermal expansion and stud-to-HTB clearance tolerances on the as-built thermal performance.

NTT Liquid-Cooled Substrate: In contrast to the other modules described in this section, the NTT water-cooled substrate approach [36] has yet to be commercialized and, as a consequence, the reported performance characteristics are not directly comparable to those of the preceding modules. Nevertheless, since the NTT module may well presage a major thrust in the direction of integrally cooled multichip substrates and since NTT has already built and tested a prototype module of this design, it appears appropriate to include the NTT module in this discussion.

NTT's interest in liquid-cooled substrates resulted from the desire to improve the volumetric packaging density of computer CPU's so as to reduce the vertical interconnect length between high planar-density modules. The technique involves mounting VLSI chips on a multilayered alumina substrate which incorporates very fine coolant channels between via holes. Substrate fabrication includes formation of a coolant distributor and collector, located on opposite ends of the multilayered substrate, so that, in operation, the coolant is distributed uniformly to the channels and the heated fluid is mixed in an internal plenum before flowing on to the next module. As explained in detail in [36], the fine channels, the distributor, and the collector for each substrate are formed by punching the green sheet prior to, but in identical fashion to, the process used in the formation of via holes. The prepared substrate is then cofired, and the I/O pins brazed on to the bottom of the substrate after cooldown.

In the NTT prototype module, shown in Fig. 8, a 5 × 5 array of 8-mm square VLSI chips was mounted on a 85 ×

Fig. 8.   NTT water-cooled substrate.

105×1.2-mm six-conductor layer alumina substrate containing 29 0.8×0.4-mm channels and 900 I/O's, both on a 2.54-mm pitch. The coolant distributor and collector were 7×77 mm in external dimension. In this prototype module the NTT researchers sought to maintain the maximum chip junction temperature below 85°C with an inlet water temperature, to the rack, of 25°C.

Based on preliminary experimental results and extensive finite-element modeling, a water flow rate of 17 cm³/s at a pressure drop of 0.2 atm was selected for this system. This flow rate was found to result in less than a 5°C coolant temperature rise across each module and a convective resistance in the liquid channels of approximately 0.45 K/W on a single chip basis. In the NTT design the chips can be epoxied or soldered to the substrate. Use of the thermally less desirable epoxy die bond was found to produce a bonding resistance 0.7 K/W per chip higher than obtained with Sn/Pb solder.

The bonding and convective resistances together with the conductive resistances in the alumina and chip, as well as the sensible temperature rise in the coolant flowing across the module, combine to produce an overall, worst case thermal resistance of 3.3 K/W for the prototype module. The NTT integrally cooled substrate can thus accomodate 25 identical VLSI devices, each dissipating 15.1 W, with a flow rate of 17 cm³/s of 35°C water. Furthermore, NTT's modeling studies suggest that the allowable chip dissipations could be increased by 20 percent by replacing the epoxy die bond with a solder bond, and by approximately 25 percent as a result of enlarging the die size to 10×10 mm. Alternately, increasing the substrate thickness to 6 mm, as might be appropriate for 35 conductor layers, can be expected to reduce the allowable chip dissipation by some 20 percent.

TECHNOLOGY COMPARISON

General Trends

No evaluation of the state-of-the-art in the thermal control of multichip modules would be complete without a direct comparison of the various strategies which have been implemented and the thermal performance which has been achieved in these modules. Unfortunately, such comparison is made most difficult by variations in the thermal specifications, associated with competing device technologies and differing operating modes, and by incompleteness in the reporting of both empirical and analytical results. Nonuniformities in heat dissipation, frequently encountered in multichip modules, can further obscure such a comparison by significantly altering the apparent thermal resistance values.

Conductive heat flow in the module cap and base, as well as in the fin structure employed for convective heat dispersion, can be expected to "smooth" temperature variations which would otherwise result from the chip-to-chip variation in heat generation. Consequently, the temperature of the convectively cooled surfaces, i.e., heat sink, can generally be determined on the basis of average chip dissipation (or, conversely, total module dissipation) while the "local" temperature difference between the active chip surface and the heat sink can best be determined by addressing the actual heat dissipation of a specific chip.

While a correct calculation of the maximum chip temperature anticipated for a particular module should thus be based on the appropriate combination of average and peak power dissipation, an upper bound estimate of the maximum chip temperature can be obtained by multiplying the chip-to-coolant thermal resistance by the peak power dissipation. The former approach has been used throughout this paper to determine the relevant maximum chip temperatures but, in the interest of simplicity, several of the comparisons—presented in Figs. 9–12—are based on this latter procedure.

In examining the multichip modules described in the previous chapter, it is apparent that much of the thermal design activity, devoted to these units, has focused on the reduction of air-gap conduction and thermal interface resistances at both the chip and module surfaces. Gaps in the conduction path from the chip to the heat sink have generally been minimized by material selection, to reduce thermal differential expansion, by high-tolerance manufacturing processes, to limit the variation from chip-to-chip, and by adjustment during assembly to neutralize the variations that have accumulated during module assembly. High thermal conductivity "greases" and high conductivity gases, notably helium and hydrogen, have been used to reduce the resistance of the internal gaps, and wherever possible, bare interfaces have been coated with thermal grease, soft solder, or silicone rubber to enhance interface conductance while providing a stress-reducing compliant layer. No less importantly, many designs appear to incorporate high thermal conductivity "thermal spreaders," in close proximity to or bonded to the chip, to reduce the high heat flux dissipated on the chip surface to more manageable levels.

Overall Thermal Resistance

The overall or chip-to-fluid thermal resistance, as well as the division between internal (chip-to-heat-sink) and external (heat-sink-to-fluid) resistance, associated with each of the multichip modules examined herein is displayed in Fig. 9. It may be noticed that in the air-cooled modules, notably the IBM 4381, Mitsubishi's HTCM, and Hitachi's RAM, the external resistance is roughly comparable to or greater than the internal resistance, while in the water-cooled modules the external resistance is generally less than one-third of the total chip-to-fluid resistance. While it is also apparent in Fig. 9 that

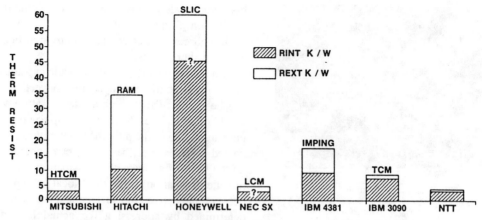

Fig. 9. Technology comparison for multichip modules—thermal resistance.

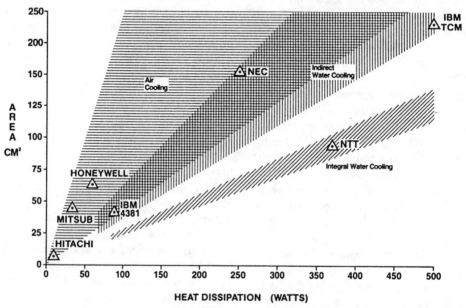

Fig. 10. Multichip module length versus total heat dissipation.

NTT's and NEC's liquid-cooled modules offer the lowest resistances at 3.3 and 5 K/W, respectively, somewhat surprisingly the third lowest value is provided by Mitsubishi's air-cooled high thermal conduction module and the highest thermal resistance is associated with Honeywell's water-cooled SLIC. If this latter module is excluded from the comparison, liquid cooling is seen to provide a generally lower thermal resistance than has been achieved by air cooling.

However, as noted in an earlier section, the evaluation of the chip-to-fluid thermal resistance, without regard for the chip area, the module area/chip, and/or the module volume/chip, can be expected to obscure the significant contribution of area to the conductive and convective transport of heat. Thus, while both the NEC and IBM 4381 modules contain 36 chips, the NEC chips and module are significantly larger than in the IBM 4381 (8 × 8-mm chips versus 4.6 × 4.6-mm chips, 125 × 125-mm module versus 64 × 64-mm module). This difference, rather than a true packaging technology advantage, would appear to be responsible for the significantly lower thermal resistance of the NEC module. In fact, as shown in Tables III and IV, the heat flux and heat density, referenced to

the module dimensions, of the NEC water-cooled module is substantially below the values of the IBM air-cooled module.

Furthermore, since the resistance defined in terms of a single chip and chip-to-chip spacing, or packaging density, can vary from one module to another, the heat dissipation capability of a "high resistance" module may exceed the capability of a "low resistance" module. This is one of the factors associated with the higher dissipation of the 36-chip 64 × 64-mm IBM 4381 module relative to the nine-chip 66 × 66-mm Mitsubishi HTCM module. Thus, in these cases as in others, the thermal resistance fails to serve as a useful figure-of-merit by which to classify and evaluate thermal packaging technology.

Module Heat Dissipation

In view of the "geometric" limitation encountered in the use of the thermal resistance, it may be appropriate to examine the thermal capability of the various modules strictly in terms of the heat flux and volumetric thermal density. The variation in reported heat dissipation with module area, as well as the

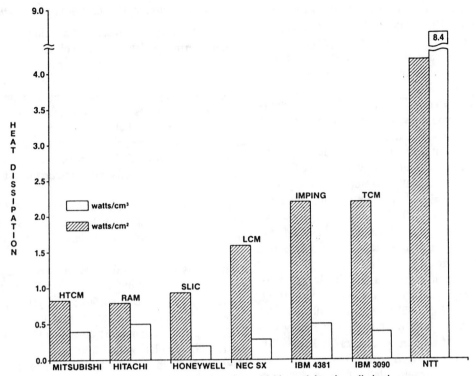

Fig. 11.   Technology comparison for multichip modules—heat dissipation.

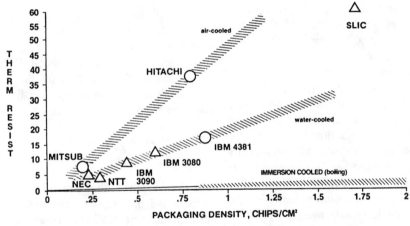

Fig. 12.   Thermal resistance versus packaging density for multichip modules.

TABLE IV
MULTICHIP MODULE PARAMETERS—THERMAL

Technology	Total Dissipation (W)	Max Chip Q (W)	Heat Flux (W/cm²)	Heat Density (W/cm³)	RINT[a] (K/W)	REXT[a] (K/W)
Mitsubishi HTCM [30]	36	4	0.83	0.4	3.0	4.3
Hitachi RAM [24]	6	1	0.8	0.5	10.1	24.6
Honeywell SLIC [33]	60	>0.5	0.9	0.2	(60)	
NEC SX LCM [34]	250	>5.4	1.6	0.3	(5)	
IBM 4381 IMPNG [26]	90	3.8	2.2	0.5	9.0	8.0
IBM 3090 TCM [12]	500	7.0	2.2	0.4	7.2	1.5
NTT [36]	377	15.1	4.2	8.4	2.8	0.5

[a] Based on the chip heat dissipation. RINT = internal thermal resistance; REXT = external thermal resistance.

heat flux and volumetric heat transfer density, across the sample of multichip modules described in this paper, is presented in Figs. 10 and 11.

For identical technologies and allowable temperature difference between the chip and ambient fluid, module heat dissipation could be expected to vary directly with module area. Examination of Fig. 10 reveals that, to a first approximation, module heat dissipation does, indeed, appear to depend directly on the area of the module, almost independently of the technology used. However, it may also be noted that, with the exception of the Honeywell module, the water-cooled packages appear to provide substantially greater heat dissipation capability than could be achieved with air cooling. Interestingly, the heat dissipation of IBM's impingement-cooled module falls along the line associated with the IBM-TCM and NEC's liquid modules, when it is extrapolated towards the origin for small module sizes.

The difficulties involved in relying on this simple comparison of thermal packaging capability can be seen by comparing the performance of the IBM 4381 module with that of Mitsubishi's HTCM. While both modules are of nearly the same area, with the HTCM six percent larger, the IBM-4381 module dissipates 2.5 times as much heat. While at first, this advantage would appear to result totally from a superior thermal design, which allows IBM to achieve a far larger packaging density, on closer examination much of this advantage is found to result from the larger chip-to-ambient temperature difference for the IBM module (55°C versus 35°C for the HTCM). At an identical temperature difference, the impingement-cooled module would appear to provide only a 60-percent higher heat dissipation capability.

The values of module heat flux and volumetric heat density, shown in Fig. 11, display these same limitations in, perhaps, a more direct way. In the figure, the IBM 4381 air-cooled module is shown to provide the same heat flux capability as the water-cooled TCM and both IBM modules fall significantly above the NEC module. This latter module, however, operates at nearly one-half the temperature difference of the IBM modules. Somewhat surprisingly, while the volumetric heat removal rate of the 4381 module is only marginally higher than that of the Mitsubishi HTCM, the heat flux capability is nearly three times higher than that reported for both the Hitachi and Mitsubishi units. In this representation, as opposed to the simple comparison of thermal resistances or total heat dissipations, both the Hitachi and Mitsubishi modules are seen to offer nearly identical heat transfer capability. On the other hand, as in previous representations, Honeywell's SLIC is again seen to lie in close proximity to the values associated with air-cooled modules, while the integrally cooled NTT substrate provides a significantly higher capability than the other water cooled configurations.

The comparisons shown in Figs. 10 and 11 suggest that, while an examination of the heat flux and volumetric heat dissipation capability of multichip modules does offer some insight into the similarities and differences among these modules, neither criterion is either sufficiently general or sufficiently consistent to serve as a thermal packaging figure-of-merit.

Thermal Resistance Versus Packaging Density

The preceding has revealed that a simple comparison of the overall thermal resistances, as well as of heat dissipation flux and density, fails to embrace the critical salient features of the thermal packaging technologies in use in multichip modules. To obtain a meaningful comparison, it appears necessary to relate the thermal resistance to some measure of the packaging density. While several choices can be explored, including chip area, module volume per chip, and module projected area (or footprint) per chip, this latter parameter is thought to offer the best basis for a consistent comparison of state-of-the-art technology.

This particular choice is partially motivated by the observed importance of heat spreading in the structure, bridging between the chip and the coolant, in many of the multichip modules and the emergence of single-board computers. Heat spreading is not only effective in providing a large convectively cooled surface but can significantly reduce the heat flux at the "softly" bonded surfaces and gas gaps and, thereby, lower the thermal resistance across these critical junctures. While this can be seen most clearly in the Mitsubishi HTCM, where a copper plate with 2.5 times the chip area is interposed between the chip and the hydrogen gap, published analyses confirm the importance of spreading the dissipated heat into the heat sink structure in both IBM modules. Since, in a module with an array of identical chips, heat can only be distributed across the "area of influence" of each chip, the module area per chip appears to be the best measure of packaging density.

Returning to the representation (9) of R_T and multiplying both sides of the equation by the module area per chip, henceforth referred to as A_c, the thermal resistance relation takes the form

$$R_T A_c = \sum_p (\Delta x/k)(A_c/A) + (1/\eta h)(A_c/A). \quad (10)$$

The left side of the equation is of a form frequently encountered in compact heat exchangers, where conduction through metal structures and layers of corrosion products must be combined with convection at the wetted surfaces to define a "composite" heat transfer coefficient and is recognizable as the inverse of this composite heat transfer coefficient, U [21].

By analogy to such compact heat exchangers, it might be anticipated that similarly configured chip modules, fabricated of identical materials and maintaining the same area ratios, would display essentially equal values of the composite heat transfer coefficient and that, in addition, various distinct combinations of materials and dimensions could also result in identical U values, signifying equivalence in thermal performance. Alternately, true breakthroughs in thermal packaging could be expected to result in significantly and unequivocally higher values of U.

Re-expressing (10) in terms of this composite heat transfer coefficient and dividing both sides of the equation by A_c, the overall module resistance is found to equal

$$R_T = (1/U)(1/A_c). \quad (11)$$

Consequently, a plot of R_T versus the reciprocal of the normalized chip area should display a linear relationship for either identical or equivalent thermal packaging technologies. Conveniently, the reciprocal of A_c is the packaging density, i.e., chips per unit area, and thus a plot of the type shown in Fig. 12 possesses physically meaningful coordinates and a slope which is the reciprocal of the composite heat transfer coefficient.

As anticipated, IBM's 3081 and 3090 TCM's are seen in Fig. 12 to lie along the same composite heat transfer coefficient locus, which includes, as well, the LCM by NEC. This approach thus correctly identifies the inherent similarities between the TCM and LCM technologies (both are based on heat transfer from the chips to water-cooled pistons or studs) and properly relates the lower thermal resistance achieved by the NEC module to the significantly larger chip size and lower chip packaging density chosen by the NEC developers.

Despite the apparent differences in the thermal paths and packaging details between the Hitachi RAM module and the Mitsubishi HTCM module, and the vastly different thermal resistances of these two units, earlier comparisons of heat removal capability in Figs. 10 and 11 had already established a surprising degree of similarity, associated, no doubt, with a similar sequence of conductive, interface and heat sink resistances, which together yield nearly identical values of the composite heat transfer coefficient. With nearly equal heat fluxes, these two modules do, in fact, provide similar maximum chip temperatures (29°C above ambient for the HTCM and 35°C above ambient for the RAM) and would yield exactly the same chip temperatures at a common air velocity of 6 m/s. These two modules can, therefore, be expected to lie along the same locus on a map of thermal resistance versus packaging density, as seen in Fig. 12.

The same figure displays once again the two thermal anomalies among the multichip modules. Honeywell's Silent Liquid Integral Cooler is found to lie between the air- and water-cooled technology streams, in relative proximity to the Japanese air-cooled modules and displays relatively poor performance for a water-cooled module. It must be recalled, however, that this evaluation of the SLIC module is based on relatively incomplete information which may not do justice to the thermal design.

Alternately, IBM's 4381 air-cooled module is seen to fall precisely along the water-cooled technology stream. There can thus be no doubt that, by the use of a dense array of fins and careful thermal design and optimization, the IBM developers have succeeded in obtaining a composite heat transfer coefficient in an air-cooled module that is identical to that usually associated with water-cooling. It may be anticipated that a similarly optimized water-cooled TCM would display a considerably higher value of U and a lower slope on the coordinates of Fig. 12. The high potential of the water-cooled modules can be seen most clearly in the NTT module, which lies considerably below the IBM and NEC technologies but still above a theoretical line for direct immersion of the chips in a dielectric fluid.

While this was by no means the primary purpose of this paper, it is to be noted that the composite heat transfer coefficient U defined earlier and shown in Fig. 12, does appear to offer significant capability for comparing and classifying thermal packaging technologies. An evaluation of this potential figure-of-merit for a far wider sample of single and multichip modules would be needed to establish its viability.

FUTURE REQUIREMENTS

As remarkable as the hardware achievements of the electronic industry are, they are far from sufficient to meet the requirements of the new generation of computers taking form in simulators and board rooms around the world. Japan's Fifth Generation Computer Project aims at providing the capability for image recognition, verbal and written information processing, and direct machine translation of Japanese into English, as well as supporting highly sophisticated "expert systems" in a wide variety of disciplines. The United States' Microelectronic and Computer Technology Corporation, operating in the private sector, and the Department of Defense's Project for Strategic Computing and Survivability, as well as the European Strategic Program for Research on Information Technology (Esprit), all have similar aims.

It has been estimated [38] that the achievement of DOD's and Japan's goals will require a capability of as much as 1000 gigaflops by 1990–1992 and Reddy, Director of the Robotics Institute at Carnegie–Mellon University, has recently asserted that, "...billion transistor superchips would be barely adequate for performing the computations required for artificial intelligence applications..." and that such applications may require as many as 100 trillion operations per second [39].

Fortunately, perhaps, Meindl of Stanford University believes that the requisite "gigascale" integration will be achieved before the end of the century [38]. However, many experts believe, to the contrary, that only modest improvements in IC performance can be achieved by reducing feature size below 0.5 μm. Furthermore, while the growth in speed and memory of the largest computers has been approximately exponential since the days of ENIAC, the performance of today's machines may well be as much as a factor of six below historical projections which were previously thought to be conservative [36].

The confluence of these three factors—burgeoning demand for computational capacity, "maturation" of silicon chip technology, and the apparent transition from exponential to asymptotic growth in machine performance—has placed the computer industry at a "tri-via" along its development path. It stands at this tri-via, poised for either a new revolution in chip technology (GaAs FET's or perhaps HEMT's), or a more rapid evolution of existing technology (reduced feature sizes, cryogenic operating temperatures, optical interconnects, wafer scale or three-dimensional packaging), or, as a third alternative, more effective utilization of existing technology by reliance on coarse- and fine-grain parallelism.

Regardless of the particular course taken by the computer industry in the future, it appears that packaging, in general, and thermal packaging, in particular, are destined to play a pivotal role. At the present time, packaging technology lags seriously behind IC technology and, as suggested by Whalen, Packaging

Program Director at MCC, a "revolution in packaging" is needed just to support the demands of 1-μm devices [39]. This assertion is supported by data showing that, even at the IC level, the package for 2-μm technology already costs twice as much as an LSI chip. This situation will, no doubt, be further exacerbated as the industry moves in any one or a combination of the three indicated directions.

A heavy reliance on parallel processing can be expected to impose temperature uniformity requirements, across modules, boards and CPU's, far in excess of current practice and necessitate the use of inherently adjustable or physically adjustable thermal control techniques.

Operation of silicon CMOS devices at cryogenic temperatures, though advantageous in reducing cycle times, will require extreme care in assembling materials of distinctly different thermal expansion characteristics and massive insulation of the components and/or CPU to reduce the flow of heat from the environment into the cryogenic enclosure. Similarly, the use of on-chip or on-module light sources and optical fibers and/or the use of holograms and optical switches is likely to impose considerably more stringent thermal control requirements than currently in use.

The thermal management of wafer-scale integration devices would appear to demand a far greater degree of thermal interface management than in evidence in today's products and the removal of heat fluxes at the module level that are as much as three to five times higher than presently encountered. Furthermore, three-dimensional packaging, or chip stacking, to reduce signal transit times, is likely to increase the volumetric heat dissipation rate by a similar factor while geometrically constraining the thermal control system.

Finally, while new device technologies promise to lower the heat dissipation per gate significantly relative to today's silicon devices, the lower junction temperatures necessitated by these technologies—near 0°C for optimum GaAs FET's performance and −200°C for HEMT's—along with the more stringent temperature uniformity requirements necessitated by the brittleness of GaAs—can be expected to demand continued development of thermal packaging techniques.

REFERENCES

[1] W. H. Desmonde, *Computers and Their Uses*. Englewood Cliffs, NJ: Prentice-Hall, 1969.

[2] J. S. Kilby, "Invention of the integrated circuit," *IEEE Trans. Electron. Devices*, vol. ED-23, pp. 648–654, 1976.

[3] G. Zorpette, "Computers that are 'never' down," *IEEE Spectrum*, pp. 46–54, Apr. 1985.

[4] W. Toshihiko and M. Hiroshi, "Packaging technology for the NEC SX supercomputer," in *Proc. 1985 IEEE Electronic Components Conf.*, 1985, pp. 192–198.

[5] J. Connolly, "ETA delivers first field-test single-CPU supercomputer," *Computerworld*, p. 2, Jan. 12, 1987.

[6] E. C. Blackburn, "VLSI packaging reliability," *Solid State Technol.*, pp. 113–116, Jan. 1984.

[7] J. W. Beyers, E. R. Zeller, and S. D. Seccombe, "VLSI technology packs 32-bit computer system into a small package," *Hewlett-Packard J.*, pp. 3–6, Aug. 1983.

[8] P. J. Klass and B. M. Elson, "New circuits expected to exceed projections," *Aviation Week Space Technol.*, pp. 46–51, July 30, 1984.

[9] R. M. Burger *et al.*, "The impact of IC's on computer technology," *IEEE J. Comput.*, pp. 88–95, Oct. 1984.

[10] K. Ohno *et al.*, "Semiconductor technologies for the FACOM M-780," *Fujitsu*, vol. 37, no. 2, pp. 108–115, 1986.

[11] K. Otsuka *et al.*, "Considerations of VLSI chip interconnection methods," presented at the IEEE Computer Society Spring Workshop, Palm Desert, CA, May 1985.

[12] E. Davidson, "Packaging technology of the IBM 3090 series systems," presented at the IEEE Computer Society Spring Workshop, Palm Desert, CA, May 1985.

[13] R. W. Keyes, "Physical limits in digital electronics," *Proc. IEEE*, vol. 63, pp. 740–767, May 1975.

[14] W. Nakayama, "Thermal management of electronic equipment: A review of technology and research topics," *Appl. Mech. Rev.*, vol. 39, no. 12, 1986.

[15] D. B. Tuckerman and R. F. W. Pease, "High performance heat sinking for VLSI," *IEEE Electron Device Lett.*, vol. EDL-2, no. 5, 1981.

[16] K. R. Samant and T. W. Simon, "Heat transfer from a small high-heat-flux patch to a subcooled turbulent flow," presented at the *AIAA Thermophysics Conf.*, Boston, MA, 1986.

[17] R. W. Keyes, "A figure of merit for IC packaging," *IEEE J. Solid-State Circuits*, vol. SC-13, pp. 265–266, Apr. 1978.

[18] O. G. Folberth, "The interdependence of geometrical, thermal and electrical limitations for VLSI logic," *IEEE J. Solid State Circuits*, vol. SC-16, pp. 51–52, Feb. 1981.

[19] H. Kanai, "Low energy LSI and packaging for system performance," *IEEE Trans. Components, Hybrids, Manuf. Technol.*, vol. CHMT-4, pp. 173–180, June 1981.

[20] A. J. Wager and H. C. Cook, "Modeling the temperature dependence of integrated circuit failures," in *Thermal Management Concepts in Microelectronic Packaging*. ISHM Technical Monograph Ser. 6984-003, 1984, pp. 1–43.

[21] A. D. Kraus and A. Bar-Cohen, *Thermal Analysis and Control of Electronic Equipment*. New York: McGraw-Hill, 1983.

[22] F. Kobayashi *et al.*, "Packaging technology for the supercomputer Hitachi S-810 array processor," in *Proc. 1984 IEEE Electronic Components Conf.*, pp. 379–382.

[23] L. M. Mahalingham, J. A. Andrews, and J. E. Drye, "Thermal studies on pin grid array packages for high density LSI and VLSI logic circuits," *IEEE Trans. Components, Hybrids, Manuf. Technol.*, vol. CHMT-6, pp. 246–255, Sept. 1983.

[24] K. Okutani, K. Otsuka, K. Sahara, and K. Satoh, "Packaging design of a SiC ceramic multi-chip RAM module," in *Proc. 1984 Int. Electronic Packaging Society Conf.*, pp. 299–304.

[25] S. Oktay and H. C. Kammerer, "A conduction cooled module for high performance LSI devices," *IBM J. Res. Develop.*, vol. 26, no. 1, pp. 55–66, 1982.

[26] R. G. Biskeborn, J. L. Horvath, and E. B. Hultmark, "Integral cap heat sink assembly for the IBM 4381 processor," in *Proc. 1984 Int. Electronic Packaging Society Conf.*, pp. 468–474.

[27] J. Brady and M. Courtney, "Hermetic tin/lead solder sealing for the air-cooled IBM 4381 module," presented at the IEEE VLSI Packaging Workshop, Santa Clara, CA, 1984.

[28] G. G. Werbizky and F. W. Haining, "Circuit packaging for large-scale integration," in *Proc. 1985 IEEE Electronic Component Conf.*, pp. 187–191.

[29] M. Kohara, S. Nakao, K. Tsutsumi, H. Shibata, and H. Nakata, "High thermal conduction package technology for flip chip devices," *IEEE Trans. Components, Hybrids, Manuf. Technol.*, vol. CHMT-6, pp. 267–271, 1983.

[30] ——, "High thermal conduction module," in *Proc. 1985 IEEE Electronic Components Conf.*, pp. 180–186.

[31] R. F. Bonner, J. A. Asselta, and F. W. Haining, "High performance printed circuit board for the IBM 3081 processor," presented at the 31st IEEE Electronic Components Conf., Atlanta, GA, 1981.

[32] A. D. Kraus, A. Bar-Cohen, and R. C. Chu, "Thermal management of microelectronics: Past, present and future," *Computers in Mech. Eng.*, vol. 1, no. 2, pp. 69–79, 1982.

[33] J. Lyman, "Special report—Supercomputers demand innovation in packaging and cooling," *Electronics*, pp. 136–143, Sept. 22, 1982.

[34] T. Watari and H. Murano, "Packaging technology for the NEC SX supercomputer," in *Proc. 1985 IEEE Electronic Component Conf.*, pp. 192–198.

[35] H. Matsuo, "Packaging technology for NEC high speed computers (ACOS 1500 and SX-2)," presented at the IEEE Computer Society Spring Workshop, Palm Desert, CA, May 1985.

[36] T. Kishimoto and T. Ohsaki, "VLSI packaging technique using liquid-cooled channels," in *Proc. 1986 IEEE Electronic Component Conf.*, May 1986, pp. 595–601.

[37] M. Kohara, M. Hatta, H. Genjo, H. Shibata, and H. Nakata, "Thermal-stress-free package for flip-chip devices," in *Proc. 1984 IEEE Electronic Components Conf.*, pp. 388–393.

[38] J. P. Martino, "Looking ahead with confidence," *IEEE Spectrum*, pp. 76–81, Mar. 1985.

[39] R. Reddy, Int. Solid State Circuits Conf., New York, Feb. 1985.

[40] J. D. Meindl, quoted in *Business Week*, p. 83, June 10, 1985.

[41] B. Whalen, Int. Solid State Circuits Conf., New York, Feb. 1985.

[42] T. Murase, H. Hirata, and S. Ueno, "High density three dimensional stack packaging for high speed computer," in *Proc. 1982 IEEE Electronic Component Conf.*, pp. 448–455.

[43] H. Yamamoto, Y. Udagawa, and T. Okada, "Cooling and packaging technology for the FACOM M-780," *Fujitsu*, vol. 37, no. 2, pp. 124–134, 1986.

[44] T. E. Lewis and D. L. Adams, "VLSI thermal management in cost driven systems," *IEEE Trans. Components, Hybrids, Manuf. Technol.*, vol. CHMT-5, pp. 361–367, Dec. 1982.

[45] N. Goldberg, "Narrow channel forced air heat sink," *IEEE Trans. Components, Hybrids, Manuf. Technol.*, vol. CHMT-7, pp. 154–159, Mar. 1984.

INTERNAL THERMAL RESISTANCE OF A MULTI-CHIP PACKAGING DESIGN
FOR VLSI-BASED SYSTEMS

Y. C. Lee, H. T. Ghaffari, and J. M. Segelken

AT&T Bell Laboratories
600 Mountain Avenue
Murray Hill, N.J. 07974

Abstract: A heat transfer study is conducted for the steady state internal thermal resistance of a multi-chip packaging technology for VLSI based systems. This technology, which is known as Advanced VLSI Packaging (AVP), has chips flip-chip soldered and interconnected on a silicon substrate. AVP's thermal management approach is to dissipate chip power through the silicon substrate to a heat sink or other packaging levels. We need to control the chip-to-substrate interface; therefore, the present study using a three-dimensional heat conduction analysis characterizes this interface. We simulate thermal performance of typical AVP assemblies affected by thermal vias, solder bump heights, high-power I/O drivers, and chip sizes. We also analyze and measure the internal resistances of an experimental package consisting of three WE32100 chips. These resistances are predicted as 3.7, 4.7, and 5.0°C/Watt respectively; they are confirmed by the experimental data.

Differed from existing techniques, our cooling approach is designed for general-purpose multi-chip packaging technologies. The present study demonstrates the achieved low thermal resistance: 3.0°C/Watt for a 1 cm square chip and 10°C/Watt for a 0.25 cm square chip. The study also provides the insight into the roles of different conduction paths involved; this understanding assures the success of our new approach and will benefit other similar packaging technologies as well.

1. Introduction

A new packaging technology interconnecting flip-chip solder attached chips on a silicon substrate is being developed at AT&T Bell Laboratories (Ref. 1,2,3). This technology, which is referred to as Advanced VLSI Packaging (AVP), will impact electronic system designs by decreasing size, weight, and power, while increasing performance and reliability. It is a general-purpose technology that employs Integrated Circuits (chips) having substantially different sizes, I/O arrangements, and power dissipation densities.

To manage the power dissipation of flipped chips, we generally have two choices. One is to follow existing approaches which conduct the heat from the chip back side to a heat sink through a piston, conductive paste, or other mediums (Ref. 4). The other is to dissipate the heat from the chip active side to a heat sink through the silicon substrate. The first choice provides a good thermal path regardless of interconnection technology on the substrate. However, it needs a sophisticated thermal contact design, and this design usually is not flexible to handle substantially different chips. Alternatively, the second choice provides the needed flexibility with a simple design; therefore, it is the preferred method used for the general-purpose AVP technology described in this paper.

Dissipating the heat through the silicon substrate needs a good control of the chip-to-substrate interface. This interface consists of several major dissipation elements such as solder bumps, thermal vias, and the air gap. They affect not only heat dissipation, but electrical interconnection density, soldering assembly yield, and solder bump reliability. Therefore, every AVP design needs a trade-off evaluation. In order to effectively conduct this evaluation, a simulation tool is required to characterize the AVP package thermal performance with respect to different interface structures.

To establish such a tool, a numerical heat transfer study using a three-dimensional finite-difference method was conducted. This study simulated steady-state internal thermal resistances affected by thermal vias, solder bump heights, high-power I/O drivers, and chip sizes. In parallel, an experimental package consisting of three WE32100 chips was also analyzed and the internal resistances were measured. The measured data confirmed the predicted resistances.

AVP's thermal mangement approach represents a new strategy for cooling general-purpose multi-chip packages. The present study established the needed simulation tool providing design guidelines. These guidelines assure the success of our new approach, and will also benefit other similar packaging technologies as well.

2. Configuration and Governing Equations

Figure 1 shows a typical cross-section of the AVP structure. A chip is flip-chip soldered onto a silicon substrate. This figure illustrates all three major thermal dissipation elements: solder bumps, the air gap, and thermal vias.

The heat conducts from the chip to the multilayer structure through solder bumps and the air gap. It further transfers to the silicon substrate through polyimide layers and thermal vias. Plated nickel thermal vias mechanically connect solder bumps and the silicon substrate; they provide good thermal paths through low thermal conductivity polyimide layers. As mentioned earlier, the designer needs to know the effects on the thermal resistances of the size, number, and distribution of thermal vias and of the solder bump height.

To analyze these effects in detail, we carried out a numerical analysis for a configuration consisting of only one quarter of a chip bonded to a large silicon substrate being attached to a heat sink. This configuration assumed the power dissipation of the chip and the distribution of thermal dissipation elements are symmetric to the X and Y axes passing through the center of the chip. The validity of this assumption will be shown later in several sensitivity analyses.

Figure 2 shows the top view of the grid systems along with the distribution of signal, P/G solder bumps. P/G solder bumps are specifically marked to represent bumps having thermal vias underneath because these usually serve as via connection to power and ground planes in AVP. The substrate size is much larger than the chip size in order to avoid the influence from artificially assumed adiabatic boundary conditions around the substrate.

The side view of the heat transfer model configuration is shown in Figure 3. The materials and their thermal conductivities of all layers are listed in Table 1.

The governing equation of a three-dimensional heat conduction problem in terms of Cartesian coordinates is

$$\frac{\partial^2 T}{\partial X^2} + \frac{\partial^2 T}{\partial Y^2} + \frac{\partial^2 T}{\partial Z^2} = 0 \qquad (1)$$

The boundary conditions of the studied problem, see Figures 2 and 3, are

1. constant external surface temperature of the heat sink (T=0°C in present analysis), and

Reprinted from *Proc. 38th Electron. Components Conf. (ECC)*, pp. 293–301, 1988.

500

2. adiabatic (or symmetric) conditions of the remaining five surfaces.

The constant temperature condition will be verified in the experimental study. The adiabatic conditions were assumed for the worst-case analysis.

In addition, simulating one watt power dissipation, a uniform heat source (5.9×10^7 W/m^3) is specified in the chip bottom layer. This layer is named the active layer in the present study and has a thickness of 169 μm (see Figure 3).

A general purpose computer program has been developed and written to solve this three-dimensional heat conduction problem. The program was developed employing a control volume finite-difference approach. Details of the approach can be found in Patanker's book (Ref. 5).

3. Results and Discussion

Before studying the internal resistance, we will describe the temperature distributions within example AVP packages. A case having a 1 cm square chip consisting of 104 signal solder bumps and 32 P/G solder bumps as shown in Figure 2 is to be discussed. The bump height is 25 μm and the area of thermal via is 10^4 μm^2.

Figure 4 shows the temperature distribution across the active layer of a one Watt chip. The relatively large temperature gradients around the P/G solder bumps indicate that these bumps are more effective heat sinking elements than signal solder bumps. Thermal paths passing through P/G solder bumps connect to the silicon substrate through thermal vias; their overall resistances are much lower than those passing through signal solder bumps. As a result, P/G solder bumps are more important elements for heat dissipation.

Figure 4 also shows a non-uniform temperature distribution across the chip. The maximum temperature difference is about 0.6°C. This difference is small, so the non-uniform distribution should have no substantial effect on any temperature-dependent chip behavior.

Figure 5 shows the temperature distribution along a vertical line passing through the chip center. The temperature difference between the silicon substrate and the heat sink is very small (<0.2°C). A thin layer of silicone gel, 12.5 μm thick over the entire silicon area, is used here for the silicon attachment. The small temperature difference shows that this relatively new attachment method is a very effective means for heat transfer. Alternatively, the most substantial temperature jump occurs across the air gap. This jump controls about 80% of the total internal resistance. Since the air gap is part of the chip-to-substrate interface, this result confirms our expectation: to effectively control the internal resistance, we have to control the interface.

Studying the interface, the proportional power dissipation through major elements were calculated. They are 53% through the air gap, 7% through the signal solder bumps, and 39% through P/G solder bumps. Another negligible 1% is transferred from the chip edges through the air to the substrate. These results indicate the importance of the air gap and P/G solder bumps. The role of the air gap is controlled by the solder bump height, and that of the P/G solder bumps is controlled by the thermal vias. The following sections focus on understanding these two controlling factors.

3.1 Effect of Thermal Via

Referring to Figure 2, each thermal via occupies a calculation cell (100 μm by 100 μm in X-Y planes) the same as that for solder bumps. The effective thermal conductivity of that cell is calculated by

$$K_{eff} = RK_{via} + (1-R) K_{polyimide} \qquad (2)$$

where

$$R = A_{real\ via\ area}/A_{cell\ area}$$

If the via size is 10^4 μm^2 area, the effective thermal conductivity of the cell is exactly the same as that of the via material. If the via size is reduced, the effective cell thermal conductivity is also decreased. The calculation of effective thermal conductivity here was based on one-dimensional heat conduction through the cell in Z-direction. Since the lateral conduction of each thermal via is negligible, this assumption was believed reasonable.

Figure 6 shows the internal thermal resistances with respect to different nickel via sizes. The internal thermal resistance becomes stablized as the via size is increased beyond 2500 μm^2. This stablization can be explained as follows. Major conduction paths affected by the thermal vias are from junction to the heat sink through the chip, P/G solder bumps and thermal vias. The main thermal barrier of each path shifts from the thermal via to solder bump as the via size increases. Therefore, increasing the via area beyond a critical size, 2500 μm^2 in this case, contributes little to a change in the overall internal thermal resistance.

Another interesting observation is that the internal resistance is 6.9°C/Watt if the thermal vias are eliminated. It is about 1.5 times higher than that having 10^4 μm^2 thermal vias. This increase is not critical for the 1 watt chip power dissipation, but it will cause a 22°C difference if the power level goes to 5 watts.

Figure 6 also shows the individual contribution to the total power dissipation through the air gap, P/G solder bumps and signal solder bumps. The air gap dominates the heat transfer (84%) when the thermal via size is zero, and it is reduced to a stablized value 52% when the thermal vias are large (>10^4 μm^2). The contribution of P/G solder bumps increases substantially when the via size changes from zero to 2500 μm^2, then it is stablized as 41%. The contribution of signal solder bumps is always small; however, it is not negligible. These bumps control 7-10% of the total power dissipation.

Similar to the via size, the via distribution also affects the thermal resistance. A case with the worst distribution having all eight vias on one side was simulated. Figure 7 shows the temperature distribution in the active layer of this case. Comparing Figures 7 and 4, we find that the temperature pattern changes substantially with this worst case distribution, but the maximum temperature is still not changed. As a result, we conclude that the via distribution effect is not significant.

So far, we have only discussed a chip having 32 P/G solder bumps; i.e. with 32 thermal vias. This number varies with different chips; therefore, we are interested in knowing the effect of the number of thermal vias. Figure 8 clearly indicate that the internal resistance monotonically decreases when the number of thermal vias increases. There is about 18% change by halving or doubling the number of P/G solder bumps. This change will increase to a 40% reduction of the thermal resistance if we connect all the solder bumps to the silicon through thermal vias. In this case, the internal resistance is as low as 3.0°C/Watt.

3.2 Effect of Solder Bump Height

The height of solder bumps studied in the above cases is 25 μm. This height may vary depending on design requirements. As shown in Figure 8, the internal resistance increases 75% if the bump height changes from 25 μm to 75 μm. This significant thermal penalty is caused by the substantial increase of the air gap dimension.

However, this significant penalty is not generally true for every design. When the importance of the air gap is reduced, this penalty should be also decreased. A dashed curve shown in Figure 8 presents the effect of bump height in a case having 128 thermal vias. Increasing the number of thermal vias make the thermal paths passing through the solder bumps be dominant paths. Because solder bumps have low resistances, they are not the major thermal barrier in those paths; as a result, the variation of bump height causing resistance changes will not substantially affect the thermal performance.

501

3.3 Effect of High-Power I/O Drivers

Usually, each chip has a non-uniform heat generation because different functional units operate at different frequencies and capacitive loadings. In particular, drivers close to the I/O pads consume much more power than any other integrated circuit elements. However, to model this time and space dependent heat generation problem is very difficult, and requires examining the differences between uniform and non-uniform power dissipation cases.

Figure 9 shows the chip temperature distribution in the worst case having all the heat sources specified near I/O pads. These pads are represented by solder bump cells as those marked in Figure 2. Differing from Figure 4, these high-power drivers result in a reversed temperature pattern. The maximum temperature occurs close to the I/O pads instead of the chip center. Nevertheless, the maximum temperature is not changed. In addition, the roles of major dissipation elements are also not changed. The calculated power dissipation contributions are 49% through the air gap, 43% through the P/G solder bumps, and 8% through the signal solder bumps. The P/G solder bump contribution increases only 4%, from 39% to 43%, compared to that in the previous uniform heat source case.

Studying the effects of high-power I/O drivers and the distribution of thermal vias is a sensitivity analysis. The results show that these two effects are not significant for predicting the maximum internal resistance. Because these effects are the only two major space dependent ones, the results also validate our assumption made for Figure 2; i.e. only one quarter of a chip is needed for simulation due to symmetry.

3.4 Effect of Chip Size

From the above studies, we obtained a thorough understanding of a 1 cm square chip. In addition to this understanding, we need to study another important consideration: the effect of chip size.

We simulated a small chip having a 0.25 cm square size with 20 signal solder bumps and 16 P/G solder bumps. This chip's temperature distribution across the active layer is shown in Figure 10. Comparing Figure 10 with Figure 4, we observe two major differences caused by chip size changes. One is the very high internal resistance of 18.4°C/Watt for the small chip. This increase of the internal resistance is expected because the power density increases 16 times. Actually, we expect an even higher value if another factor is not involved. This factor is the other feature also shown in the Figure 10; the area ratio between P/G solder bumps and the chip is significantly increased.

To understand this area ratio effect, we calculated the contributions of thermal paths to the total power dissipation. They are 12% through the air gap, 6% through the signal solder bumps, and 81% through the P/G solder bumps. Clearly, the P/G solder bumps dominate the power dissipation and the air gap conduction becomes less important. The P/G solder bumps are more conductive than air and result in more efficient heat transfer.

Because the roles of the three elements are changed, we reviewed their roles carefully. Thermal vias are the dominant thermal dissipation elements, so we expect a sharper increase of the resistance when the size of thermal vias decreases toward zero. Figure 11 confirms this expectation. The stablized internal resistance is 17.8°C/Watt, and the resistance jumps to 68.5°C/Watt when the via size is zero.

Figure 11 also shows the effect of solder bump height. As mentioned earlier, the variation of solder bump height is not effective when the thermal paths passing through P/G solder bumps become dominant ones. Compared to Figure 8, the thermal penalty having increased bump height is substantially reduced for this small chip.

In the same figure, the effect of number of thermal vias is also presented. Reducing the number below 16 causes a large thermal penalty. The thermal resistance is 28°C/Watt with eight vias, and is 68.5°C/Watt when the vias are eliminated. Toward the other direction, increasing the number from 16 to 32 reduces the thermal resistance from 17.8 to 10°C/Watt. This 10°C/Watt resistance is very impressive considering the 16 Watt/cm^2 high power density. If needed, we can further decrease this resistance by adding more solder bumps and thermal vias.

One watt power dissipation is not unusual for small size and high power chips -- e.g NMOS, bipolar or GaAs chips. It is very possible that AVP packages may carry these small chips along with other large chips. The size effect should be considered carefully.

4. Experimental Study

In parallel to the numerical study, we also made measurements on an experimental package consisting of WE32100 Central Processing Unit (CPU), Memory Management Unit (MMU), and Math Accelerator Unit (MAU). As shown in Figure 12, this 160 I/O pin grid array package has the silicon substrate attached to an aluminum heat sink using a thin layer of silicone gel. This attachment is the same as that previously simulated (See Figure 3). More details describing the package can be found in Ref. 1.

Figure 13 shows the experimental setup carrying out the heat transfer experiment. A heater with a copper spreader was attached using thermally conductive epoxy to each chip. The copper spreader has a K-type thermocouple embedded. All package surfaces except the heat sink are insulated as stipulated by the adiabatic boundaries assumed in the numerical analysis.

Table 2 lists the chip and heat sink measured temperatures taken from a representative case. In high package power conditions, the whole assembly was cooled by impinging an air jet instead of natural convection. In either case, the heat sink temperature is very uniform confirming our assumptions made in the numerical analysis. The three chips, however, show different temperatures because they have different thermal resistances to the heat sink. Their resistances were calculated by the formula

$$R = \frac{(T_{chip} - T_{heat\ sink})}{P_{chip}} \quad (3)$$

where the chip power is one third of the total power input. These resistances have to be reduced by 0.3°C/Watt that is contributed from the thermally conductive epoxy attaching the copper spreader and the chip.

Along with the measurement, these chips resistances were also analyzed. The CPU and MAU have sizes very close to 1 cm square but have different number of thermal vias. In the experimental package, there are 130 vias for the CPU and 80 vias for the MAU. MMU is smaller, about 0.8 cm square, and has 120 thermal vias. The solder bump height of the assembly is around 35 μm. Based on these configurations, we calculated the thermal resistances as 3.5°C/Watt for CPU, 4.2°C/Watt for MMU, and 4.5°C/Watt for MAU.

These resistances were further corrected because the experimental package differs from our simulated one with a much smaller area ratio between the chip and the substrate. This smaller ratio limits the spreading in the substrate and the heat sink and causes thermal penalties. These thermal penalties were calculated with numerical analysis, and they were 0.5°C/Watt for MAU and MMU and 0.2°C/Watt for CPU. After correcting the calculated resistances, we obtained the predicted resistances for CPU, MMU, and MAU to be 3.7, 4.7, 5.0°C/Watt respectively.

Figure 14 compares the predicted and measured results for those three chips. The measured resistance range from 2.5 to 3.7°C/Watt for CPU, 3.8 to 5.6°C/Watt for MMU, and 3.5 to 6.0°C/Watt for MAU. Their averaged resistances are 3.4, 5.1, and 5.4°C/Watt, respectively. As shown in the figure, the experimental and numerical results match well. Good agreement between these two group of results confirms the accuracy of the numerical analysis.

The calculation for thermal penalties also illustrates a procedure to extend the simulated single-chip results to those in a multi-chip package. These penalties are small in the studied case; however, sometimes, they are significant because of very non-uniform chip power distribution or heat sinking. In these cases, we should analyze the whole package with discrete heat sources representing multi-chips, and add the calculated resistances with the corresponding chip-to-substrate ones to obtain the overall internal resistances. The chip-to-substrate resistances can be estimated from the present analysis.

5. Summary

A computer simulation tool has been established to provide design guidelines of a new multi-chip packaging technology for VLSI based systems. This technology uses a new cooling approach that dissipates the power of the flipped chip to a heat sink through the silicon substrate. Using the established tool, this paper studied this approach's steady-state internal resistances. Several conclusions can be drawn from the study.

1. Providing thermal vias to all the solder bumps, the resistance was shown to be only 3.0°C/Watt for a 1 cm square chip, and to be only 10°C/Watt for a 0.25 cm square chip having a 16 Watt/cm^2 high power density. This performance is similar to or even better than that of existing approaches.

2. The air gap and solder bumps having thermal vias underneath are the critical thermal dissipation elements for the chip-to-substrate interface.

3. The size and the number of thermal vias are important factors. They affect the internal resistances as well as the roles of the air gap and solder bumps. Another factor, the distribution of thermal vias, however, is not influential.

4. The solder bump height affects the internal resistance with various degrees. Enlarging the bump height substantially increases the resistance when the area ratio between the solder bumps having thermal vias and the chip is very small. This resistance increase is not influential when the ratio is large.

5. The high-power I/O drivers cause non-uniform heat generation across the chip. This effect changes the chip's temperature pattern; however, it does not significantly influence the maximum chip temperature.

6. Reducing chip size increases the power density and results in a high internal resistance. This increased resistance can be substantially decreased by effectively using thermal vias. Solder bumps having thermal vias become the dominant dissipation elements when the chip size is very small, e.g. 0.25 cm square.

7. The experimental study measuring a real package consisting of three WE32100 chips confirmed the accuracy of the numerical prediction. The numerical results were fine-tuned by correcting the heat spreading effect of the experimental configuration. This kind of correction is not critical in this study but may be needed for other multi-chip packages where the substrate has large temperature gradients. These gradients may be caused by very non-uniform chip power distribution or heat sinking.

6. Acknowledgement

The authors wish to thank N. A. Sanders of Hypertherm Inc., Hanover, NH, and David Kniep for their earlier contributions to this study while they were at AT&T Bell Laboratories.

References

1. C. J. Bartlett, J. M. Segelken, N. A. Teneketges, "Multi-Chip Packaging Design for VLSI-Based Systems," ECC Conference Proceedings, May 1987.

2. H. J. Levinstein, C. J. Bartlett, W. J. Bertram, Jr., "Multi-Chip Packaging Technology for VLSI-Based Systems," ISSCC Conference Proceedings, Feb 1987.

3. A. C. Adams, R. S. Bentson, W. J. Bertram, H. J. Levinstein, W. Q. McKnight, J. J. Rubin, B. A. terHaar, "High Density Interconnect for Advanced VLSI Packaging," Proceedings of Electrochem. Soc., May 1987.

4. A. Bar-Cohen, "Thermal Management of Air and Liquid-Cooled Multi-Chip Modules," ASME National Heat Transfer Conference, Denver, 1985.

5. Patankar, S.V. "Numerical Heat Transfer and Fluid Flow", Hemisphere Pub. Co., 1980.

Table 1 : Materials and Thermal Conductivities in the AVP Structure		
Layer	Material	Thermal Conductivity
IC Device	Doped Si	98.4
Solder Bump	60Sn/40Pb	53.4
Polymer	Polyimide	0.25
Thermal Via	Nickel	90.
Substrate	Doped Si	98.4
Gel	Silicone Gel	0.4
Heat Sink	Aluminum	216.5
Air	Air	0.03
		Watt/m·°C

Table 2: Measurement Data							
V	I	MMU	CPU	MAU	TC1	TC2	TC3
Volt	Amp	°C	°C	°C	°C	°C	°C
5.9	0.4	69	68	68	65	65	65
10.3	0.265	76	75	76	72	72	71
14.6	0.2	80	78	80	75	75	75
20.2	0.157	83	82	84	78	78	78
25.3	0.13	86	84	87	81	81	81
30.7	0.111	89	87	89	83	84	83
40.4	0.089	93	91	93	87	87	87
3.3*	0.491	35	34	35	32	32	32
5.1*	0.645	47	46	48	42	42	42
8.6*	0.594	62	60	63	54	54	54
10.3*	0.531	66	63	67	57	57	57
15.3*	0.373	68	66	70	59	59	59
20.2*	0.297	72	69	73	61	62	61
29.9*	0.221	78	74	79	66	67	66
40.0*	0.176	82	78	83	69	70	69
50.3*	0.146	86	81	86	72	73	72
60.5*	0.126	89	84	89	74	75	74
69.7*	0.112	91	86	92	76	77	76
80.6*	0.100	95	89	95	79	80	79
89.5*	0.092	97	91	98	81	82	81
100.*	0.085	100	94	101	83	84	83
110.*	0.079	102	96	103	85	86	85
120.*	0.075	104	98	106	87	88	87
130.4*	0.070	107	101	109	89	90	89
138.0*	0.067	109	102	110	90	91	90

Note: 1. Room temperature is 23°C.
2. "*" means the package is cooled by an impinging jet.
3. Data is taken from a representative case measured.

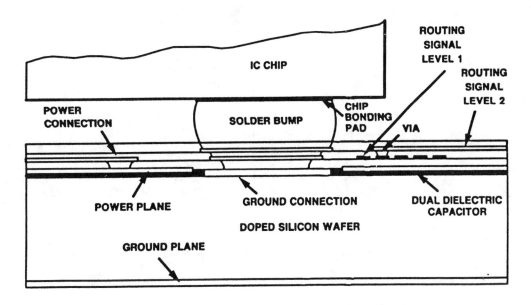

Figure 1   AVP Cross section.

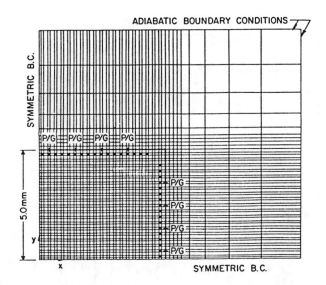

Figure 2   Top view of 1/4 simulated chip with grid system.

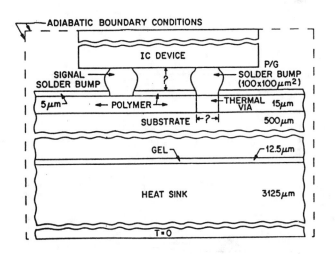

Figure 3   Cross-sectional view of the heat transfer model configuration.

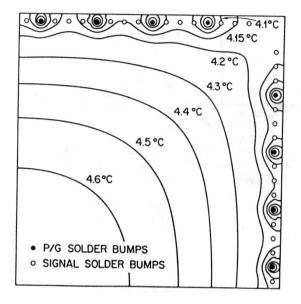

Figure 4   Temperature distribution in active layer (1 cm square chip with 1 watt power).

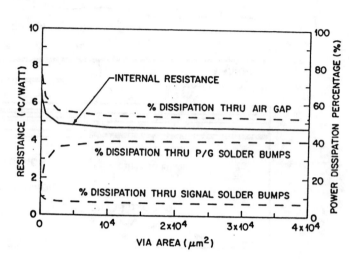

Figure 6   The effect of thermal via area (1 cm square chip with 32 thermal vias and 25 μm bump height)

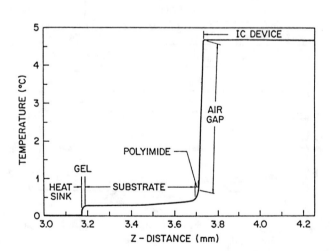

Figure 5   Temperature distribution along a line passing through chip center.

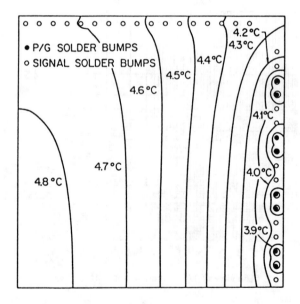

Figure 7   Temperature distribution in active layer with respect to a different via distribution. (1 cm square chip with 1 watt power).

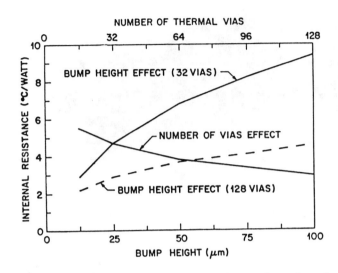

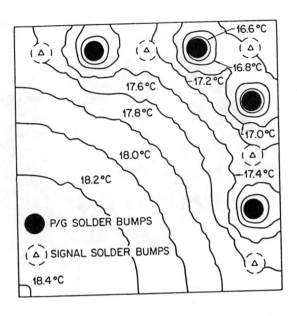

Figure 8　The effects of solder bump height and number of thermal vias (1 cm square chip).

Figure 10　Temperature distribution in active layer (0.25 cm square chip with 1 watt power).

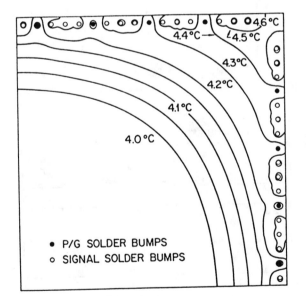

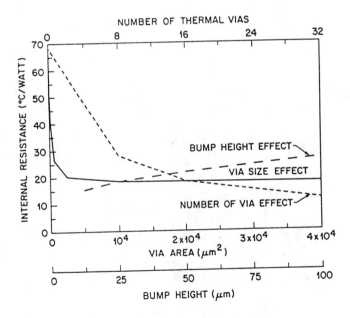

Figure 9　The effect of high-power I/O drivers (1 cm square chip with 32 thermal vias and 25 μm bump height)

Figure 11　The effects of thermal via area, solder bump height and number of thermal vias (0.25 cm square chip).

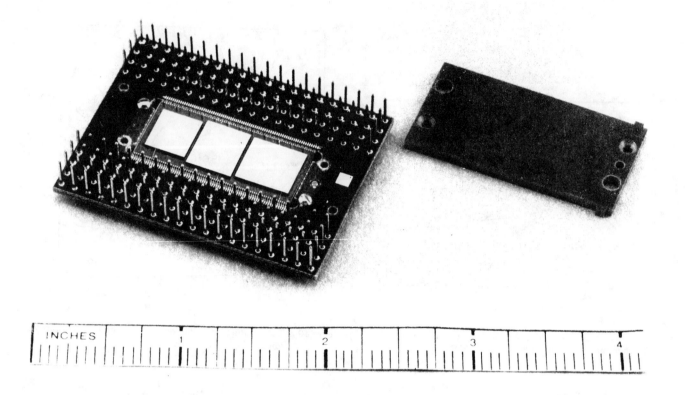

Figure 12   The AVP experimental package.

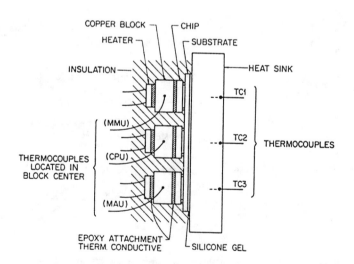

Figure 13   Experimental setup for temperature measurement.

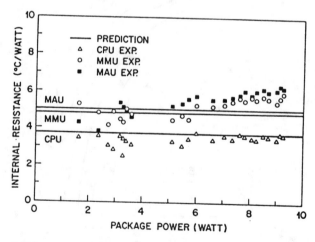

Figure 14   Measured and predicted thermal resistances for the experimental package.

THERMAL / STRESS ANALYSIS OF A MULTICHIP PACKAGE DESIGN

Robert Darveaux[1,2], Lih-Tyng Hwang[1], Arnold Reisman[1,3], Iwona Turlik[1,4]

[1]Microelectronics Center of North Carolina, PO Box 12889, Research Triangle Park, NC 27709

[2]Dept of Materials Science and Engineering, North Carolina State University, PO Box 7907, Raleigh, NC 27695

[3]Dept of Electrical and Computer Engineering, North Carolina State University, PO Box 7911, Raleigh, NC 27650

[4]Bell Northern Research, PO Box 13478, Research Triangle Park, NC 27709

Abstract

In this paper, we provide a thermal analysis of our package design, with emphasis on thermally induced stress in the most critical package components. The package employs flip-chip solder bonding and thin film interconnections between chips. Indium was chosen as the die attachment medium between each chip and the water cooled heat sink. A methodology is given to estimate the stress on a solder bump array when the indium die bond deforms during a power-up. It has been shown that stress can be reduced by decreasing the die bond thickness. However, very thin die bonds will have a reduced fatigue life due to increased plastic strain per power cycle. Therefore, the package design can be optimized by using the thinnest possible indium die bond which has an adequate fatigue life.

Introduction

The density and speed of today's devices are the driving forces behind packaging technology development. A package suitable for VLSI/ULSI devices demands high density interconnections and the capability to supply current at tight voltage-drop tolerances. Furthermore, heat removal becomes a critical factor in assuring high system reliability. A thin film, multichip module (shown in Figure 1) is currently under development at the Microelectronics Center of North Carolina (MCNC). The design is targeted for high performance VLSI/ULSI applications. In this paper, we provide a thermal analysis of our package design, with emphasis on thermally induced stress in the most critical package components.

The heat removal capability of a package is usually characterized by its thermal resistance, defined as $R_T = \Delta T/P$, where ΔT is the device temperature rise, and P is the chip power. Thermal resistance is a function of the package design, and the thermal conductivity of the die attachment and heat sink materials. Details on the heat sink design and performance for the MCNC module can be found in references [1,2]. As die size increases, a compliant die attach material is needed to prevent fracture of the die during thermal or power cycling of the package. In a flip package, compliance can be achieved through mechanical means (e.g. spring loaded contacts to the heat sink), or through material design (i.e. using a soft, deformable bonding material).

In the MCNC package, indium has been chosen as the die attachment medium between each chip and the water cooled heat sink because of its favorable properties, and its compatibility with the package process sequence. Due to the high thermal conductivity of indium (0.81W/cm°C), a device-to-water thermal resistance of 0.4°C/W can be achieved with the MCNC design [3]. However, due to the lack of mechanical compliance in the heat sink or in the interconnection substrate, the indium die bond must deform to accommodate all displacements from thermal expansion and warp in the structure. As the indium deforms, the solder bump connections of each chip will be subjected to stress. The maximum stress on the solder bumps was estimated for our module, which contains twenty five, 1μm technology CMOS gate array chips, with 100,000 gates each.

In this paper, we provide a methodology to estimate stress on the solder bumps, which are the most critical components of our design. The following sequence of calculations was used for estimation of the stress: 1) predict the heat flux from each chip and from the lossy transmission lines on the substrate, 2) calculate the steady state temperature distribution using 2-D finite difference computer simulations, 3) calculate the worst case thermal expansion displacements between the substrate and the heat sink, 4) calculate the time required to reach the yield point of the indium die bond, and 5) estimate the stress on the solder bumps from yield stress data on indium die bond joints. This methodology can be applied to any given package design to estimate thermal stresses on the critical components.

Package Configuration

A cut-away view of the multichip package design is shown in Figure 1. The package contains a 5 x 5 array of 1cm x 1cm chips on a 1.8cm pitch. The chips are reflow bonded to thin film interconnections on the substrate with 95Pb5Sn solder. The nominal dimensions of the solder bumps are 125μm dia x 75μm high on a 250μm pitch. Each chip has a 37x37 bump array for a total of 1356 connections (three bumps removed from each corner, and one from the interior for registration). The 10.2cm x 10.2cm (4in x 4in) aluminum nitride substrate has molybdenum conductors between the thin film interconnections and the I/O pins. The interconnection structure has six or more layers of copper conductors in a polyimide dielectric. Electroless nickel vias are used to connect signal lines and to bring power directly to the chips. The heat sink is made of densified, reaction-sintered silicon carbide with 15% silicon to fill interstices from the firing process. The heat sink has 15, 1mm x 5mm water channels designed to accommodate a total flow rate of 500cc/s (7.9gpm).

The indium bond is made by reflowing to metallization on the heat sink and on the chip backside. The nominal die bond thickness is 650μm, but due to thickness tolerances on the chips, solder bumps, and support ring, and flatness tolerances on the heat sink and substrate, the die bond thickness might vary from 500μm to 800μm across the module. The support ring around the package periphery prevents deflections due to mechanical loading, but all thermally induced displacements must be absorbed by the indium die bond.

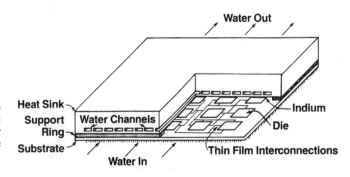

Figure 1. MCNC Multichip Package.

Reprinted from *Proc. 39th Electron. Components Conf. (ECC)*, pp. 668–671, 1989.

Package Power Dissipation

Heat generated in the substrate interconnections, in addition to each chip, should be considered in very high speed applications of a thin film module. The heat sources on each chip are internal logic gates, on-module output drivers, and off-module output drivers. In the substrate, heat will be generated in the lossy transmission lines and in the power distribution system. When pulse rise time approaches signal propagation time in a transmission line, termination is necessary to reduce ringing. In this case, terminating resistors would be an additional source of heat in the substrate. In the present analysis, it was assumed that termination would not be used. However, if terminating resistors are required to guarantee electrical performance, and are integrated into the substrate, the additional heat source has to be incorporated into the analysis. The average heat flux on each chip and on the substrate is shown in Table 1.

The amount of heat generated by output drivers and transmission lines was calculated from the number of interconnections on- and off-module, the distribution of line lengths on the module, and the average driver switching frequency. The number of lines required to interconnect chips on a module can be predicted from the number of I/O's on each chip and the average fan out (number of receivers per driver). A Rent's Rule type relationship which predicts chip I/O's as a function of gate count for CMOS logic gate array chips is given by Turlik et.al. in reference [4]. For 100,000 gates on a chip, approximately 474 signal I/O's will be required. Assuming an average fan out of two, 7900 lines are needed to interconnect the 25 VLSI CMOS chips on the module.

Based on an empirical wiring algorithm developed by Heller et.al. [5], the average line length on the module will be 3.26cm. Therefore, 257m of wire will be required to interconnect the chips. The distribution of line lengths which make up this 257m was assumed to be similar to the distribution given in reference [6], Table 2-7.

The power dissipation in the lossy transmission lines was calculated using SPICE computer simulations. The average power dissipation was found to be about 0.1mW/cm, so 2.6W will be dissipated on the entire module by interconnections. If parallel termination is used for the interconnection lines, power dissipation could be much higher.

The power dissipation by the output drivers was calculated from

$$P_{driver} = \frac{1}{2}CV^2f \qquad (1)$$

where the line capacitance off-module was assumed to be 125pF, and the capacitance on-module was determined from the transmission line length. The number of on-module and off-module drivers allowed to simultaneously switch will ultimately be determined by ΔI noise considerations, so the average power due to these heat sources will depend on the inductance of the module power distribution system and the driver current slew rate. For the purposes of this analysis, we have assumed that 10 off-module and 53 on-module drivers will switch per clock cycle on each chip. The heat generated by the power distribution system on the substrate was calculated from the total current supplied to the module and the resistance of the power distribution lines and vias.

The heat generated by internal logic gates on each chip was calculated from the number of gates and their average switching frequency. For three levels of metal, the gate power is approximately 0.006mW/MHz for 1μm technology CMOS gate array chips [7]. It was assumed that 10% to 30% of the gates toggle per clock cycle on average [8], so the effective switching frequency will be 10MHz to 30MHz for a 10ns system cycle time. For this analysis, 20MHz is assumed to be the nominal effective frequency, and 10MHz and 30MHz are assumed to be the lower and upper extremes.

Thermal Analysis

A 2-D finite difference approach was employed in the thermal analysis. The cross section used in the model is shown in Figure 2. From symmetry considerations, the chip centerline and the centerline between chips can be modeled as insulating boundaries; i.e. there is no net heat flow across the boundary. Also, the air gap between chips was assumed to be insulating (natural convection was neglected).

Heat was assumed to be generated in the device layer of the silicon chip and in the thin film layers on the substrate. Heat was assumed to be removed by forced convection to water in the channels, and by natural convection to air off the substrate bottom surface. The ambient water and air temperatures were assumed to be 25°C.

The heat transfer coefficients were calculated using the relations in reference [9]. At the channel surfaces, h=3.57W/cm²°C, and on the substrate surface, h=0.0012W/cm²°C. Radiative heat transfer and lateral conduction to the package support structure, piping, etc., were neglected (these are conservative assumptions).

A thermal contact resistance of 0.025cm²°C/W was assumed at the soldered interfaces [10], and thermal constriction of heat flow through the solder bump array was accounted for by using the relation given in ref [11]. Details on the thermal analysis methodology is provided in ref [3].

An isothermal plot for the module is shown in Figure 3. In the simulation, the effective logic gate frequency was 20MHz (power = 13W), the water temperature was 25°C, and the indium die bond thickness was 800μm. The highest chip temperature is 30.5°C (5.5°C above water temperature). The average temperature rise at the top of the substrate is 5.4°C, and the average temperature rise at the bottom of the heat sink is 2.1°C.

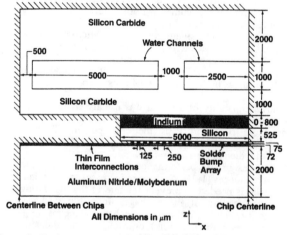

Figure 2. Package cross section for 2-D finite difference model.

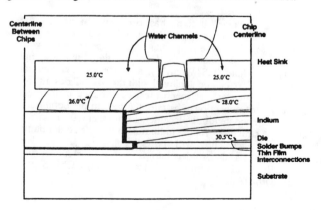

Figure 3. Isothermal plot at 0.5°C intervals. Heat flux = 13W/cm² on chip and 0.049W/cm² on substrate, indium thickness = 800μm.

Thermal Stress Analysis

In general, there will be stress on each solder bump array whenever the module undergoes a thermal transient. This stress is due to thermal expansion and warp in the package, which result in significant displacements between the heat sink and the substrate. As depicted in Figure 4, there will be both normal and shearing forces on a chip. Normal forces will arise due to localized thermal gradients, and bowing of the substrate from expansion in the polyimide/copper interconnection layers. Shearing forces will arise from the in-plane expansion mismatch between the substrate and the heat sink. Normal forces will be greatest for chips in the package interior, and shear forces will be greatest for the chips at the corners of the package. As a first order approximation, the stress in the solder bumps will be 6x the stress in the indium, because the load bearing area on the backside of a chip is 6x that on the device side of the chip.

As stress levels increase during a thermal transient, all materials will deform elastically until either the indium, polyimide, or solder bumps yield, or until brittle fracture occurs at the bump interfaces. A good estimation of the peak stress can be obtained once the indium flow stress and the rate of the substrate temperature change are known.

Since indium is at 0.70 of its melting point at room temperature, its deformation characteristics are very strain rate sensative, i.e. the flow stress increases dramatically with strain rate. Therefore, thermal transients which occur very rapidly will produce the highest stress in the package. In general, there will be four types of thermal transients: 1) die attachment, 2) shipping and storage, 3) power-up or power-down during testing or service, and 4) chip address during testing or service. The power-up and power-down transients result in the highest thermal stress because they produce the largest rate of temperature change.

The approach used to calculate the peak stress on the solder bumps during a power cycle was to 1) calculate the worst case thermal expansion displacement between the substrate and heat sink, 2) calculate the time required to reach the yield point of the indium die bond, which determines the strain rate of the material, and 3) estimate the stress on the solder bumps from flow stress data on indium die bond joints. The stress level will reach a peak quite rapidly, then decay exponentially with time as the indium undergoes creep. The expansion mismatches, transient times, indium strain rates and flow stresses, and solder bump peak stresses are shown in Table 2 for three chip power levels. The methodology used to obtain these values is outlined below.

The worst case shear displacement between the heat sink and substrate is given by

$$\delta = DNP\,(\alpha_{hs}\,\Delta T_{hs} - \alpha_{subs}\,\Delta T_{subs})\qquad(2)$$

where DNP is the distance from package neutral point, ΔT is the average temperature change, and α is the thermal expansion coefficient.

The worst case displacement in the normal direction will depend on 1) localized thermal gradients, 2) the nature of the substrate/heat sink attachment around the package periphery (e.g. rigid, sliding, free), and 3) the bending moment exerted on the substrate by the interconnection layers. Since this displacement is difficult to calculate without a finite element analysis, it was assumed that the normal displacement would be no more than 10x the shear displacement of the package. The simplest way to ensure that normal displacements are indeed within this range is to increase the substrate thickness.

The time required to heat or cool the substrate is given by [12]

$$Q_{in} - Q_{out} = Q_{stored}\qquad(3)$$

where Q is the rate of energy change. For cooling, Q_{in} is zero and Q_{stored} is negative. Substituting in the appropriate parameters yields

$$Q_{in} - \frac{\theta}{R_T} = \rho V c\,\frac{dT}{dt}\qquad(4)$$

where ρ, V, and c are the substrate density, volume, and specific heat, respectively, R_T is the composite thermal resistance between the substrate and the water, $\theta = T(t) - T_{water}$, and t is time.

Solving for time yields

$$t = R_T \rho V c\; ln\,((\theta_i - Q_{in} R_T)/(\theta - Q_{in} R_T))\qquad(5)$$

where $\theta_i = T_{initial} - T_{water}$. By rearranging equation (5), it can be seen that the rate of temperature change decreases exponentially as the substrate approaches its final temperature. This implies that the indium strain rate (and flow stress) will be highest at the beginning of the transient. The time to reach 0.07% strain in the die bond was used to calculate the maximum indium strain rate.

Constant strain rate loading was applied to indium die bond joints to determine the flow stress as a function of strain and strain rate. Shear, tensile and compressive loading were applied to 1cm x 1cm specimens, which were 500μm to 800μm in thickness. The indium flow stress values listed in Table 2 correspond to a strain level in the die bond for the worst case heat sink / substrate expansion mismatch. The peak stress on the solder bumps during a power transient will be about 6x the indium flow stress, assuming that the solder bump and indium bonding processes are 100% effective. If there are non-bonded areas on either side of the chip, the solder bump stress will change proportionately.

Since the indium creep rate is inversely proportional to grain size, the flow stress of the die bond increases with grain size. This effect is shown by the data in Table 2, where the 500μm die bond has a lower flow stress than the 800μm die bond, due to the smaller indium grain size. On the other hand, the thinnest die bonds are subjected to nearly twice as much plastic strain in a power cycle, which results in a reduced fatigue life. Therefore, the package design can be optimized by using the thinnest possible die bond which has an adequate fatigue life.

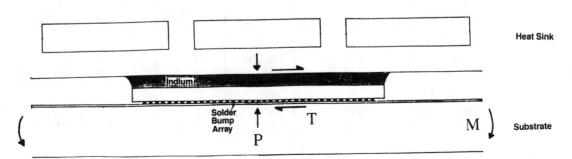

Figure 4. Thermally induced loading on chip during package power-up. Normal loads due to substrate bowing and localized temperature gradients, shear loads due to in-plane expansion mismatch between heat sink and substrate.

Conclusions

We have presented a thermal analysis of our multichip package design, with emphasis on thermally induced stress in the most critical package components (the solder bumps). The package design is such that each indium die bond must deform to accommodate displacements due to thermal expansion and warp in the structure. A methodology was provided to estimate the stress on a solder bump array when the die bond deforms during a power-up. For a die bond thickness range of 500 - 800μm and a power dissipation of 13W/chip, the worst case stress was estimated to be 6.6 - 7.2 MPa in shear, and 22 - 44 MPa in tension or compression. The solder bump stress can be reduced by decreasing the nominal die bond thickness. However, very thin die bonds will have a reduced fatigue life due to increased plastic strain per power cycle. Therefore, the package design can be optimized by using the thinnest possible die bond which has an adequate fatigue life.

TABLE 1

Power Dissipation in a Thin Film, Multichip Package

% Gates Switching per Clock Cycle	10%	20%	30%
Chip Power	(W)	(W)	(W)
Logic Gates	6.0	12.0	18.0
Off-Module Drivers	0.6	0.6	0.6
On-Module Drivers	0.4	0.4	0.4
Total	7.0	13.0	19.0
Average Heat Flux (W/cm^2)	7.0	13.0	19.0
Substrate Power	(W)	(W)	(W)
Transmission Lines	2.6	2.6	2.6
Power Distribution System	0.5	1.4	3.0
Total	3.1	4.0	5.6
Average Heat Flux (W/cm^2)	0.038	0.049	0.069

References

[1] L-T. Hwang, I. Turlik, A. Reisman, "A Thermal Module Design for Advanced Packaging," J Electronic Materials, Vol 16, No 5, 1987.

[2] D. Nayak, L-T. Hwang, I. Turlik, A. Reisman, "A High Performance Thermal Module for Computer Packaging," J Electronic Material, Vol 16, No 5, 1987.

[3] R. Darveaux, L.T. Hwang, A. Reisman, I. Turlik, "Thermal Analysis of a Multichip Package Design," Submitted to J Electronic Materials, 1988.

[4] I. Turlik, A. Reisman, R. Darveaux, L-T. Hwang, "Multichip Packaging for Supercomputers," Proc NEPCON West, Anaheim, CA 1989.

[5] W.R. Heller, W.E. Donath, W.F. Mikhail, Proc 14th Design Automation Conf, New Orleans, LA, 1977.

[6] R.R. Tummala, E.J. Rymaszewski, "Microelectronics Packaging Handbook," Van Nostrand Reinhold, New York 1989.

[7] VLSI Systems Design - 1988 Semicustom Design Guide, CMP Publications, Inc.

[8] E. Evans, Motorola Inc. Chandler, AZ, Private Communication, February, 1989.

[9] W.M. Kays, M.E. Crawford, "Convective Heat and Mass Transfer," McGraw-Hill, New York, 1980.

[10] M.M. Yovanovich, M. Tuarze, "Experimental Evidence of Thermal Resistance at Soldered Joints," J Spacecraft, July, 1969.

[11] V.W. Antonetti, M.M. Yovanovich, "Thermal Contact Resistance in Microelectronic Equipment," The International Journal for Hybrid Microelectronics, Vol 7, No 3, September, 1984.

[12] F.P. Incropera, D.P. Dewitt, "Fundamentals of Heat Transfer," Chapter 5, Wiley, New York 1981.

TABLE 2

Peak Thermal Stress on Solder Bumps During a Power Transient

	Worst Case Heat Sink / Substrate Expansion Mismatch (μm)	Transient* Time (s)	Maximum Indium Strain Rate ($10^{-4}s^{-1}$)	Indium Flow Stress (MPa)	Peak Solder Bump Stress (MPa)
Die Bond Thickness	650	500 - 800	500 - 800	500 - 800	500 - 800
7.0 W/chip					
Shear	0.42	7.56 - 30.0	0.93 - 0.23	0.8 - 0.9	4.8 - 5.4
Normal	4.20	0.37 - 0.61	19 - 11	3.1 - 4.6	18.6 - 27.6
13.0 W/chip					
Shear	0.76	2.60 - 5.74	2.7 - 1.2	1.1 - 1.2	6.6 - 7.2
Normal	7.60	0.20 - 0.33	35 - 21	3.7 - 7.3	22.2 - 43.8
19.0 W/chip					
Shear	1.10	1.62 - 3.04	4.3 - 2.3	1.2 - 1.3	7.2 - 7.8
Normal	11.0	0.14 - 0.22	50 - 32	3.8 - 8.1	22.8 - 48.6

*Time to reach 0.07% strain in indium die bond.

THERMAL MANAGEMENT CONSIDERATIONS FOR A
LOW-TEMPERATURE, CO-FIREABLE CERAMIC SYSTEM

Edward L. Rich, III, Scott K. Suko
Angela J. Martin, Brian H. Smith

David G. Onn, Andrew J. Whittaker
Ryan E. Giedd

Westinghouse Electric Corporation
Defense and Electronics Center
Baltimore, MD 21203

Applied Thermal Physics Laboratory
University of Delaware
Newark, DE 19716

Frank K. Patterson

E.I. Du Pont deNemours and Co., Inc.
Electronic Materials Division
Wilmington, DE 19898

ABSTRACT

Higher system performance represents one of the major driving forces for advances in packaging technology. High density and higher speed are both associated with higher performance. Both trends drive up the total dissipated power and power density for multichip and multilayer packages. A potential physical limitation for producing reliable systems is the ability to remove the generated heat from the active devices, especially in power applications. Thermal management, therefore, has become an important and challenging consideration for determining system performance. Thermal management at device level packaging must involve efficient and cost-effective removal of dissipated heat from the device to ensure reliable device and system performance.

This paper will describe the thermal properties and thermal management considerations associated with a low-temperature, co-fireable Green Tape^{TM} dielectric and conductor system from Du Pont. This material system was developed for fabricating high-density multilayer hybrids and multichip modules.

The intrinsic thermal conductivity of the Green Tape^{TM} dielectric which will be presented was derived from measurements of the thermal diffusivity and the specific heat as a function of temperature between 20 and 150°C. Samples with a variety of processing furnace profiles with peak cycle temperatures between 800 and 925°C will be compared. The thermal diffusivity and specific heat values used to determine the thermal conductivity were measured using a novel "electronic flash" technique which will be described. Selected samples were studied by conventional laser-flash and longitudinal bar techniques in order to confirm the accuracy of this new high-speed method for obtaining the thermal parameters of materials.

As a means of increasing the thermal conductivity of the Green Tape^{TM} dielectric for practical applications, test circuits with eight dielectric tape layers were fabricated incorporating thermal vias 25 mils in diameter filled with a dielectric-compatible gold composition. The contiguous vias were oriented parallel to the direction of the projected heat flow from active devices. These test coupons also included bonding pads for device attachment. The test coupons without devices were submitted to MIL-STD-883 tests, including thermal shock. The results of these tests and interpretation of same will be presented. Other test configurations incorporating the thermal vias were also evaluated using the "electronic flash" technique to determine the enhancement of the effective thermal conductivity of the Green Tape^{TM} dielectric. These measurements were used to determine the thermal conductivity of the composite structure including vias.

Finally, a power MOSFET was soldered to the metallized pad above the via array of the test circuits and temperatures were recorded at known power loadings both with and without the gold thermal vias. Design guidelines, including optimal thermal via size and location, will also be discussed.

1.0 INTRODUCTION

The design of hybrid electronic devices has utilized thick-film interconnects on monolithic alumina, and traditionally recognized a separation between power and small signal circuits. Highly dissipative devices requiring insulation have been mounted on separate substrates of BeO, AlN and SiC; while more moderate heat producers have been directly included on the alumina substrate with small signal devices. The thick-film on alumina processes are a mature art. Packing densities are limited, however, because much of the surface real estate is utilized for interconnection and printed resistors.

A design density improvement results from placing the interconnecting circuitry within the layers of a co-firable multi-layer substrate. The Du Pont Green Tape™ system offered an alternative to the serial print/dry/fire sequence of multi-layer thick film and to the reliability high resistance conductors and high shrinkage variation of high temperature co-fired alumina.

Accompanying the circuit design densification has been a requirement for distributed power conditioning circuits to regulate, store, and switch energy to adjacent loads. A thermal management technique for use with the Green Tape™ system was borrowed from multi-layer printed circuit practice, to include thermal vias (solid metallic plugs) within the laminate to lower the thermal impedance from a power chip to the underlying package. The thermal vias tend to shunt the relatively high thermal impedance of the Green Tape™ substrate and are formed by mask printing on a standard screen printer.

In this paper methods and measurements of the thermal conductivity of the Green Tape™ with and without various configurations and compositions of thermal vias will be described, including the acid test of thermal performance, junction temperature measurement of an actual power device.

2.0 MATERIALS SYSTEM AND PROCESSING

The standard Du Pont Green Tape™ Materials Systems consists of a low K ceramic filled glass tape (851AT) and co-fireable thick film conductor paste compositions all designed for compatibility with conventional thick film processing equipment. The dielectric is specifically formulated for multilayer applications and as such, it has a relatively low dielectric constant (7.9 at 1 MHz) and a high insulation resistance of $>10^{12}$ ohms.

Individual layers of the standard as-cast dielectric tape are 0.0045" thick which shrink 12 $\pm$0.2% during firing (0.0037") to form a hermetic, pore free composite structure with glass as the continuous phase. The fired density of this composite structure is >96% of theoretical. In addition to this standard dielectric, we also used a new dielectric tape with the same chemical composition but with an as-cast thickness of 0.0125". This tape was also designed to shrink 12% during firing.

The conductors used in this thermal management study consisted of both gold and silver metallurgies (See Table I). For the co-fireable thick film via fill conductors, four different compositions were used. Three of

these compositions were gold and consisted of the standard 5718D gold and two experimental gold pastes designated A and B. The latter two conductors are related to 5718D but have different sintering rates and paste rheologies. The fourth conductor was a silver via fill composition, 6141D.

The co-firing and post firing conductor compositions for line traces, ground, chip attach, and bonding pads were gold. The product numbers and characteristics of these conductors are shown in Table II.

TABLE I

VIA FILL CONDUCTOR COMPOSITIONS

Product	Metallurgy	Description
Du Pont 5718D	Gold	---
Du Pont A	Gold	Modified 5718D with sintering inhibitors
Du Pont B	Gold	Same as above
Du Pont 6141D	Silver	---

TABLE II

TOP LAYER CONDUCTOR COMPOSITIONS

Product	Metallurgy	Description
Du Pont 5715	Gold	Post firing conductor for wire bonding, etc.
Du Pont 9191	Gold	Post firing mixed bonded conductor
Du Pont 5731D	Gold	New co-fireable top layer gold for Green Tape™ System

The processing steps used in fabricating all of the multilayer structures used in this thermal management study included blanking out layers of the dielectric tape with a 3.0" x 3.0" precision die. All thermal and interconnecting vias plus holes were punched out with a Tam Model 159 numerically-controlled punch using cylindrical tooling with different diameters depending upon the application.

Via filling was carried out using an MPM printer and Mylar stencils. The balance of the fabrication steps and equipment description including conductor printing, laminating, burnout and firing have been well documented in previous publications.[1,5,6]. A flow chart of the general processing steps for multilayer fabrication is shown in Figure 1. Several variations to the sequence of processing steps used in the fabrication of thermal management structures were used and will be described in more detail for each different structure type.

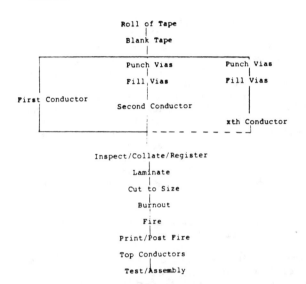

FIGURE 1

CO-FIRED GREEN TAPE™ MATERIAL SYSTEM PROCESS FLOW

3.0 THERMAL PROPERTIES OF THE 851AT DIELECTRIC

3.1 Measurement Techniques

The thermal properties of three Green Tape samples fired at 800, 850 and 925°C respectively, were determined using a variety of measurement techniques.

An estimate of the z-axis thermal conductivity (normal to the plane of the layers of the tape precursor) was obtained using a calibrated single probe thermal comparator (1). The probe tip which is brought into contact with the sample forms one junction of a differential thermocouple. Both junctions are integrally mounted on a temperature controlled copper block. Prior to making a measurement both junctions are isothermally matched. When the probe tip is placed in contact with a test material the resulting heat flow alters the tip temperature generating a differential voltage which is calibrated to give an estimate of the sample thermal conductivity.

The z-axis thermal diffusivity was determined at room temperature using the laser flash technique[3]. The front face of a small, plane parallel slab of material is subjected to an instantaneous pulse of energy delivered by a 100 J Nd/glass laser. The temperature history of the opposite face is monitored and recorded. Use of a digital data acquisition system permits the resulting temperature transient to be analyzed in a microcomputer to yield the thermal diffusivity (α) from the equation:

$$\alpha = 0.1388 \times 1^2/t^{1/2} \qquad \text{Eq. (1)}$$

where 1 = the length of the sample and $t_{1/2}$ = the time taken for the rear face to achieve one half its maximum temperature rise.

Globally, this technique is responsible for 70% of the thermal diffusivity data accumulated to date. It is widely regarded to be the most rapid and accurate technique available for the determination of thermal diffusivity.

The z-axis thermal diffusivity and the specific heat were measured in the temperature range 20 to 120°C using the newly developed electronic flash technique[2]. This novel apparatus is a modification of the laser flash method, devised specifically for dielectric materials. The laser is replaced as the source of the energy pulse by a thin film of resistively heated graphite applied to one surface of the sample. Sophisticated electronic circuitry permits a brief, high energy electrical pulse to be applied across the heater. Since the amount of energy actually absorbed by the sample can be precisely quantified, a direct measure of both the thermal diffusivity (α) and specific heat (Cp) can be obtained. A knowledge of the mass density (ρ) of the sample allows the derivation of the thermal conductivity (k) from the equation:

$$k = \alpha \times \rho \times Cp \qquad \text{Eq. (2)}$$

The thermal conductivity in the x-y direction (parallel to the plane of the layers of the precursor tape) was measured in the temperature range 100 to 500 °K using a micro-computer controlled shielded longitudinal bar apparatus. A heater attached to one end of a bar shaped sample is used to establish a steady temperature gradient along the central portion of the bar. The cross-sectional area (A) of the bar, the temperature gradient (dT/dx) and the rate of energy supplied (q) are all measured and the thermal conductivity (k) is obtained directly from the Fourier equation:

$$q = kA \ dT/dx \qquad \text{Eq. (3)}$$

3.2 Thermal Properties

3.2.1 Sample Configuration and Preparation.
Two different sample configurations were used for the three different test methods. These test methods included longitudinal bar, laser flash, and electronic flash.

Figure 2 shows a picture of the completed structure of the dielectric for the longitudinal bar technique described previously. This particular structure was fabricated using 37 blanked out 3.0" x 3.0" layers of the 851AT dielectric tape. Four holes were punched in each layer using a 0.0937" diameter punch for two of the holes and a 0.024" punch for the other two. In order to provide enough samples,

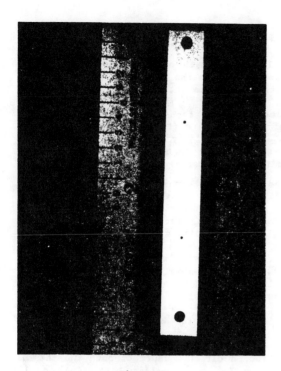

Figure 2
Longitudinal Bar

eight individual patterns were generated on each 3.0" x 3.0" tape layer with a step and repeat pattern. All thirty-seven layers were then collated with good hole pattern alignment and laminated in a confining die using a Pasedena Uniaxial Press at 3000 psi, 70°C for 10 minutes. The 3.0" x 3.0" laminates were burned out at 350°C and then fired at three different peak temperatures, 800°C, 850°C and 925°C. These were then cut into 2.0" x 0.250" x 0.125" bars for thermal conductivity measurements.

For both the laser flash and electronic flash measurements, rectangular bars without holes were fabricated in the same manner as above. Twenty-three of the 3.0" x 3.0" blanked out dielectric tape layers were laminated, burned out and fired at the three different peak temperatures. These were than cut into 0.472" x 0.472" x 0.074" pieces.

3.2.2 Thermal Property Results

The room temperature thermal properties of the three samples tested are summarized in table III. Thermal diffusivities in the range 8.3 x 10^{-3} to 9.6 x 10^{-3} cm^2/sec were recorded in the z direction using the electronic flash and laser flash techniques. Excellent agreement is observed between the data obtained using the two techniques the difference being no more than 5% and in two cases <1%. No systematic trend in the diffusivity data with the variation in processing temperature is noted. A maximum value (averaged over the two techniques) of 9.5 x 10^{-3} cm^2/sec was recorded for the sample processed at 850°C.

There appears to be a steady increase in the specific heat capacity from 0.74 to 0.91 J/gK as the processing temperature is increased from 800 to 925°C. Since the composition of each of the samples is nominally identical this is a surprising result. The determinate error in the specific heat measurement is of the order of +/-5%, so that the 20% increase observed lies outside the bounds of experimental error. No explanation is at present available for this observation.

Using the measured values of thermal diffusivity and specific heat combined with measured mass densities in equation (2) enables derivation of thermal conductivity data. The values shown in Table III were derived using the

TABLE III

ROOM TEMPERATURE THERMAL PROPERTIES OF
GREEN TAPE^{TM} PROCESSED AT DIFFERENT TEMPERATURES

Process Temp. °C	α_z cm^{2}sec^{-1} x 10^{-3}	α_z cm^{2}sec^{-1} x 10^{-3}	C_p (Jg^{-1}K^{-1})	P (g(cm^{-3})	Kz (W/mK)	Kz* (W/mK)	Kz** (W/mK)	K(x-y) (W/mK)
800	9.23	9.15	0.740	3.07	2.1	2.08	2.7	---
850	9.45	9.55	0.850	2.74	2.2	2.22	3.0	3.8
925	8.73	8.34	0.910	3.02	2.4	2.29	2.6	---

* - denotes low flash measurements

** - denotes single probe thermal comparator measurements

diffusivity results obtained from the electronic and laser flash experiments combined with the same specific heat and mass density data. Values in the range 2.1 to 2.4 W/mK were recorded. A 10% increase in the thermal conductivity is observed as the processing temperature is increased from 800 to 925°C. It is difficult to assess the significance of this apparent trend, given the uncertainty associated with the specific heat data.

Measurements made in the z direction using the single probe thermal comparator do not follow the same trend and in fact reflect the unsystematic behavior of the diffusivity data. If all the thermal conductivity data accumulated in the z direction at room temperature is combined, an average value of 2.39 W/mK is obtained with a standard deviation of +/-12%.

It seems reasonable to conclude that the ultimate z-axis thermal conductivity of the Green Tape^{TM} is not critically dependent on the processing temperture.

The typical temperature dependence of the thermal properties obtained using the electronic flash system is indicated in Figures 3 to 5. These data represent the sample processed at 800°C. A slight decrease in thermal diffusivity is observed as the temperature increases from from 20 to 120°C, while a slight increase in the specific heat is noted. These competing trends result in a thermal conductivity which appears to remain invariant with temperature over the temperature range.

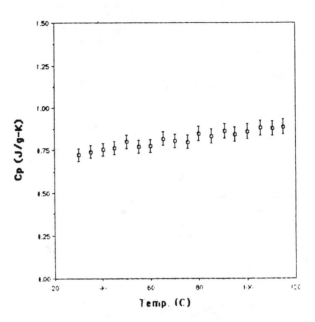

Glass/Alumina A Tp = 800 C

Figure 4
Specific Heat vs Temperature

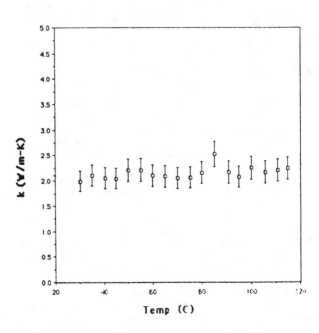

Glass/Alumina Tp = 800C

Figure 5
Thermal Conductivity vs Temperature

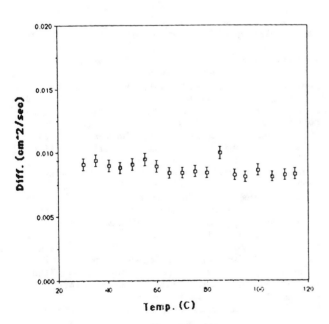

Glass/Alumina A Tp = 800 C

Figure 3
Thermal Diffusivity vs Temperature

517

Measurement of thermal conductivity made in the x-y direction on the shielded longitudinal bar system show an almost linear increase with temperature, rising from a value of 1.9 W/mK at 100 K to 6.0 W/mK at 490°K. These measurements were made on the sample processed at 850°C. It is interesting to note that the room temperature value of 3.8 W/mK listed in table III is substantially higher than the values obtained in the z direction.

This apparent anisotropy is perhaps due to internal structural modulation arising during processing. Segregation of a small proportion of the glassy phase to the surface of the individual layers of tape, during the casting process, could lead to the formation of a low thermal conductivity layered interface interphase. This would serve to drastically reduce the z-axis thermal conductivity, while having a minimal effect in the x-y plane. This hypothesis has yet to be confirmed by microstructural investigation.

4.0 INCREASING THERMAL CONDUCTIVITY WITH THERMAL VIAS

4.1 Design Considerations

The object of thermal vias is to effectively conduct heat from power dissipating devices mounted on a substrate with high thermal impedance. The thermal vias are large diameter holes punched into layers of Green Tape™ and are aligned vertically. The resultant via stack is a solid metallic plug. The size of the via should be the largest possible taking into account that the via must be filled with paste at each layer. A via diameter of .0255" was chosen for availability of the punch, although a .020" diameter via may be just as appropriate. The .0255" via has an area 6.5 times greater than that of the recommended .010" diameter via used for normal interconnections. The optimum array of thermal vias would be the maximum number of vias that could be placed under a particular device and still maintain structural integrity, ie., cracking between vias punched in a row or cracking as a result of thermal cycling. The actual substrate design will dictate the maximum numbers vias that can be used. An array of 3 x 5 .0255" vias was successfully punched with .055" centers for the samples discussed in this paper. The limits have not been defined as to the smallest centers (recommended 1 diameter between edges of vias) and the largest diameter via that can be filled.

4.2 Materials and Substrate Fabrication

For the HEXFET™ 4 temperature measurements, two different multilayer configurations were assembled with thermal vias. The via diameter for all of the thermal vias independent of configuration was 0.0255" as punched in the green state.

Three gold compositions and one silver composition were used to fill the thermal vias for cofiring. The gold conductors included the Du Pont standard via fill gold 5718D plus two compositional modifications described earlier and designated A and B. These modifications were formulated to enhance adherence of the via fill metallurgy to the dielectric sidewalls during firing. The major effort in this program was focused on these gold conductors and especially the A modification. The silver conductor was the standard 6141D via fill composition.

The gold conductors for IC chip mounting, wire bonding tracks and ground plane layers were screen printed with a post firing 5715 gold, 9791 gold or a newly developed cofireable top layer gold conductor 5731D.

4.2.1 Eight Layer Structure

Two series of these eight dielectric layer structures were fabricated. The first series designated X consisted of 0.0255" thermal vias punched out in each tape layer using 15 vias in a 3 x 5 matrix which would be associated with the HEXFET™ device. Adjacent to these were ten additional thermal vias in a 2 x 5 matrix. These functioned solely as a control for optical inspection before and after thermal cycling.

The 3 x 5 via matrix for the X series had the thermal vias spaced such that vias in the long direction were on 0.075" centers and vias in the short direction on 0.067" centers. The second series designated Y had the 3 x 5 matrix of these thermal vias all on 0.055" centers.

Individual layers of the dielectric tape for both X and Y series were via filled with gold formulation A or silver composition 6141D. It was difficult during the printing operation to prevent overfilling the vias with either composition. After collating and laminating the contiguous vias, overfilling was not evident. However, after firing, the thermal vias did protrude above the dielectric on both top and bottom surfaces. This was compensated for in the the short term by leaving the vias for layer 1 and 8 (see Figure 6a) unfilled. After cofiring, the vias still protruded slightly above the surface. These were abrasively reduced prior to post firing either 5715 or 9791 gold bonding pads on both bottom and top surfaces of the eight layer structure. For some structures in Series X, all of the thermal vias were left unfilled until lamination was complete. For these structures, all eight laminated layers were filled with one printing step using gold conductor compositions 5718D, gold conductor A and silver conductor 6141D (see Figure 6b). The laminates and via fills were then cofired followed by post firing of the gold chip attach and wire bonding pads.

(a)   (b)

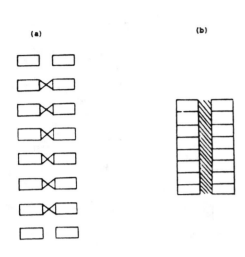

Figure 6
Via Filling for Eight Layer Structure

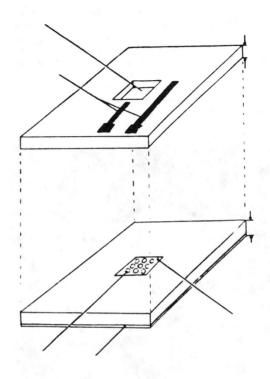

Figure 7
Cavity Part with Thermal Vias

4.2.2 Two Layer Structure

For this structure only two dielectric layers were used. Each dielectric tape layer was 0.0125" thick in the as cast state compared to the standard 0.0045" 851AT dielectric tape. A 3 x 3 matrix of 0.0255" vias on 0.075" centers were punched out in the tape layer. These were then via filled with either gold composition 5718D, gold composition A or gold composition B. After drying the via fills at 120°C, a chip attach pad for the HEXFET™ 4 with dimensions 0.320" x 0.270" was screen printed over these thermal vias with a new cofireable gold conductor 5731D. A second dielectric layer with an open window pattern (punched out using a nibbling punch) was placed with the open window pattern around the HEXFET™ 4 chip attach pad. The two dielectric layers were than laminated together at 2000 psi, 70°C for 10 minutes to create a recessed cavity for the HEXFET™ 4 chip.

Laminates with each of three different thermal via compositions were then burned out and cofired in the standard manner.

After firing, a ground plane of 9791 gold was screen printed on the back side of the two layer structure and in contact with the thermal vias. The two layer structure was then fired using a one hour cycle with ten minutes at a peak temperature of 850°C. Two wire bonding pads were then printed with 9791 and fired on the top surface of the dielectric. A representation of the completed assembly in exploded form prior to chip attach is shown in Figure 7.

4.3 Test Package Preparation and Configuration

The Green Tape™ substrates used for the thermal evaluation included a post-fired gold conductor pattern. Substrates with and without thermal vias (control group) were produced. A chip attach pad on one side and full backside metallization was used. On the substrates containing vias, this metallization coated both surfaces of the via array. Two line traces terminating in a bond pad provided the necessary gate and source tracks from the chip. A HEXFET™ type silicon transistor (Size 4) measuring .170 x .220 in. was positioned on the chip pad and soldered in place. Gold wirebonds (.001 in diameter to minimize thermal conduction) connected the gate and source of the chip to their respective line traces and from these traces to the package terminals. The drain connection was made through the soldered chip base and utilized the thermal vias to conduct to the backside metallization and ultimately the package which houses the substrate. A single, small diameter (.010 in.) via was placed under the pad of the control substrates in order to accomplish this drain connection.

The package used was a standard gold plated copper PHP (Power Hybrid Package) with KOVAR terminals through the side walls (see Figure 8). This package acts as an isothermal heat sink as well as providing terminals for the thermal test circuit. The chip-to-substrate and substrate-to-package soldering was accomplished with AuGe (88% Au, 12% Ge) solder. The soldering was performed in a three zone hydrogen belt furnace profiled for AuGe. The resulting test package was x-rayed for voids in the solder joints, especially under the chip. The joints did exhibit a certain percentage of voiding, which affects the thermal results.

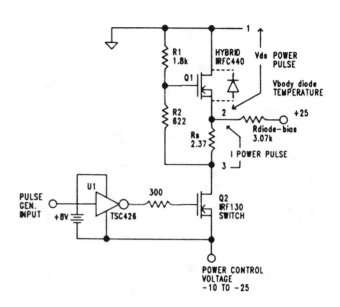

Figure 9
Thermal Impedance Test Circuit

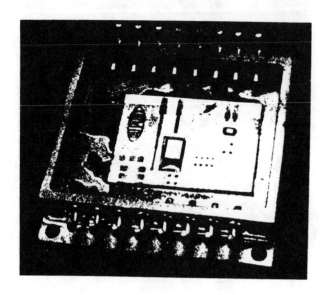

Figure 8
Power Hybrid Test Package

In addition to the aforementioned test substrates, a cavity part version was produced which contains the chip attach pad sunken into a window (see Figure 7). It is believed that the shortened thermal path resulting from this configuration will improve heat transfer. The test results of this approach will be available at the time of the conference.

4.4 Test Method
To evaluate the performance of various types of thermal vias, power MOSFETs were soldered to a Green Tape™ substrate in a power hybrid package as described before. A testing circuit (Figure 9) was constructed which provides a controlled amount of power to the MOSFET. The circuit periodically interrupts the applied power and forward biased the intrinsic body diode of the MOSFET. The forward drop of this diode, which was previously calibrated against temperature, is used as a temperature sensor.

The power pulses were 4.8 ms in duration and the interruptions for reading temperature were 400µs long. The forward diode drop was read on a Tektronics 7A13 plug-in by adjusting the differential comparator offset. Approximately fifteen minutes passed before each measurement to ensure steady state conditions.

A thermal resistance in degrees C per watt as seen by the power MOSFET was calculated for each sample. The temperature difference is between the MOSFET die and the copper hybrid package. Package temperature was monitored with a thermocouple.

4.5 Results
Test results are given for a HEXFET™ size 4 soldered onto metallization covering an array of thermal vias. Measured power inputs were approximately five and ten watts. After thermal stabilization, the power input, junction temperature and case temperature were measured.

The test was configured with a pattern (22.4 mil diameter vias on 50 mil centers) of eighteen (18) effective vias (the power chip does not cover all vias in a pattern). The results are shown in Table IV.

An additional sample with fewer vias, type A gold and standard lamination correlated well with the value for similar composition and lamination shown in Table IV.

5.0 THERMAL CONDUCTIVITY OF COMPOSITE STRUCTURES
- DIELECTRIC WITH THERMAL VIAS

5.1 Sample Configuration and Fabrication
A limited study was carried out to evaluate the thermal conductivity enhancement of the 851AT dielectric using thermal vias. The technique utilized for this measurement was the laser flash.

TABLE IV

Thermal Impedances Measured with
HEXFET™ Size 4

Via Fill	$\theta_{J-C}(°C/W)$ (5w)	$\theta_{J-C}(°C/W)$ (10w)
Stand. Gold (5718D)	3.56	3.73
Modified Gold (A) Stand. Lamination	4.41	4.46
Modified Gold (A) Modified Lamination & Fill	3.55	3.58
Stand. Silver (6141D) Stand. Lamination	3.95	3.99
Stand. Silver (6141D) Modified Lamination & Fill	2.43	2.57
Aluminum Nitride (Shapal) (No Thermal Vias)	2.65	2.67
Green Tape™ (851AT) (No Thermal Vias)	12.6	(Too hot)

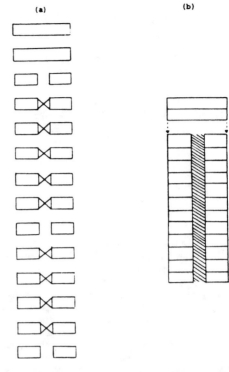

Figure 10
Via Filling for Fourteen Layer Structure

A fourteen dielectric layer structure was constructed as a test vehicle. Twelve layers contained thermal vias in the two matrix layouts as described in Section 3. In order to increase the thermal impedance on the side exposed to the laser beam, the last two layers of the dielectric were deliberately left blank so as to cover the thermal vias with an insulating layer to accommodate the thin film graphite heater.

The thermal via pattern for each of the twelve layers consisted of a 5x5 matrix of 0.0255" diameter vias on both 0.075" and 0.055" centers. The conductor compositions used to fill these vias were gold conductors 5718D, A and B plus silver conductor composition 6141D.

In order to minimize the effect of overfilling the vias during the screen printing operation, three of the twelve thermal via layers were not filled. These were layers 1, 6 and 12. This is diagrammatically represented in cross-section shown in Figure 10a. A second approach to constructing this multilayer was to laminate all twelve layers without via fills. This was followed by a single via filling step for all twelve layers followed by laminating the top blank dielectric layers (Figure 10b).

5.2 Test Method

Despite the precautions taken to insure that the electronic flash technique could be applied to these samples, the enhanced thermal diffusivity of the composite structures coupled with the reduced length of the samples gave rise to half rise times of the order of 100 ms. The lowest accurately measurable half rise time for the electronic flash technique is 150 ms, due to the finite duration of the electrical pulse and associated delays incurred in quantifying its magnitude. Consequently all the measurements were made using the laser flash technique.

The rear surfaces of the samples were coated with a thin layer of copper allowing an intrinsic thermocouple to provide point detection of the temperature transient. By locating the thermocouple precisely on a thermal via and then on the dielectric between the vias, upper and lower bounds for the effective thermal diffusivities of the composites were determined.

5.3 Results

The room temperature thermal properties of the Green Tape™ composite samples are presented in Table V.

TABLE V

ROOM TEMPERATURE THERMAL PROPERTIES

OF GREEN TAPE™ COMPOSITE

Via Filler	α_U (cm^{2}sec^{-1}) x 10^{-3}	α_L (cm^{2}sec^{-1}) x 10^{-3}	k_{EU} (W/mK)	k_{EL} (W/mK)
Std gold	46	11.4	59.07	3.36
Mod gold	25.0	16.2	9.56	4.94
Silver	18.6	15.8	4.21	3.58

Clearly the addition of metal filled thermal vias enhances the thermal transport properties of the monolithic Green Tape™. The highest upper bound thermal diffusivity (corresponding to the temperature sensor being placed directly on the end of a thermal via) was obtained for the sample filled with standard gold. The value of 46 x 10^{-3} cm^2/sec is approximately a factor of two higher than either of the upper bound values obtained for the other two samples. The lowest upper bound value was recorded for the sample filled with silver. Due to the disagreement of the silver result with that obtained with an actual power semiconductor, further testing is intended.

All the lower bound values were higher than the values obtained for the monolithic material. It is however interesting to note that the value obtained on the standard gold filled sample was the lowest recorded for the three composite samples tested.

Upper and lower bounds on the effective thermal conductivity of the composite samples were determined by assuming that the sample averaged specific heat and mass density could be used in equation (1) in conjunction with the upper and lower bound values of thermal diffusivity. This is an over simplification, but it certainly permits the comparative performance of the three samples to be assessed. A correction was also made for the additional layers of the Green Tape™ dielectric.

The lower bound values for all three samples lie in the range 3.4 to 5 W/mK and all are greater than for the monolithic material. The upper bound values lie in the range 4.2 to 59 W/mK, the highest value being obtained on the sample filled with standard gold. Clearly the standard gold filler offers the greatest potential enhancement of the thermal conductivity of the Green Tape™, by a substantial margin. Unfortunately, the non-uniform distribution of the vias in the samples measured makes it difficult to precisely evaluate the effective composite conductivity.

Due to the microscopic structural modulation of the sample, large differences are observed between the effective upper and lower bound composite conductivities. The column containing values of k_{EU}, shown in Table V, represents the effective thermal conductivity in the region of the via material (and not, it should be noted, the intrinsic conductivity of the via filler).

TABLE VI

FILLER DENSITY

Via Filler	Theoretical ρ(gcm^{-3})	Measured ρ(ycm^{-3})	T.D.
Std gold	18.8	15.1	80.3
Mod gold	18.8	10.2	54.3
Silver	10.5	7.0	66.6

The much higher upper bound thermal conductivity obtained for the standard gold filled material is almost certainly due to the relatively high density of the gold fillter phase. Table VI indicates that the standard gold is 80% dense while the modified gold is only 54% dense. However the silver, which has a higher theoretical thermal conductivity than gold, is 66% dense yet performs comparatively poorly. This is possibly due to its tendency to oxidize much more rapidly than the gold.

One point worthy of note is that a single value of thermal diffusivity of 40 x 10^{-3} cm^2/sec was obtained with the temperature sensor placed on a modified gold via. However, subsequent measurements reproducibly gave a value half this magnitude. This is perhaps the result of mechanical degradation of the via, given its relatively low density, under the load of the point temperature sensor.

In the as supplied composite materials the thermal vias occupied only 4% of the total volume. Even with the correction for the added Green Tape™ layers the volume fraction was still only 10%. It seems reasonable to conclude that a relatively modest increase in the volume loading of standard gold filled thermal vias would significantly enhance the effective thermal conductivity of the Green Tape™ composite.

6.0 RELIABILITY TESTING

Thirty-seven (37) parts containing thermal vias in the standard test pattern (Figure 11) have been subjected to 2000 thermal shock cycles (-55 to +125°C), liquid-to-liquid, according to

MIL-STD-883, without any problem. No micro-cracking of thermal or electrical vias was observed. The testing is continuing to 5000 cycles. Results of ongoing testing under humidity and temperature will be presented at ISHM.

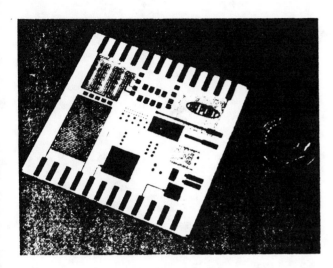

Figure 11
Standard Test Pattern

Thermal cycling was initially considered but has been deferred due to the excellent results of the thermal shock tests.

7.0 SUMMARY

The thermal testing of Green Tape[TM] dielectric has determined a design conductivity of 2.2w/mK. Addition of thermal vias allows enhancement of the thermal performance exceeding that of 96% alumina and equalling or exceeding that of aluminum nitride, while obtaining the high density advantages of buried interconnects.

*Research supported by E.I. du Pont de Nemours and Co., Inc., the State of Delaware Research Partnership Fund, and Westinghouse Independent Research and Development Funds.

**HEXFET[TM] is a trademark of International Rectifier Corporation.

References

1. A.L. Eustice, S.J. Horowitz, J.J. Stewart, A.R. Travis, and H.T. Sawhill, 36th Electronics Component Conference (1986), Proceedings, pp 37-47 "Low Temperature Co-fireable Ceramics: A New Approach For Electronic Packaging".

2. Giedd R.E., Onn D.G.
   To be published.

3. Parker W.J., Jenkins R.J., Butler C.P., Abbott G.L.
   J. Appl. Phys. 32 No. 9 p1679-1684 (1961).

4. Powell R.W.
   "Thermal Conductivity", Vol. 2, Chapter 6, Tye R.P. (ed.) (1969) Academic Press (London).

5. J.I. Steinberg, S.J. Horowitz and R.J. Bacher, Solid State Technology, January (1986), pp 97-101, "Low Temperature Co-fired Tape Dielectric Material Systems for Multilayer Intraconnections".

6. W.A. Vitriol and J.I. Steinberg, International Society for Hybrid Microelectronics (ISHM), Philadelphia, PA, 1983, Proceedings, pp 593-598, "Development of a Low Temperature Co-fired Multilayer Ceramic Technology".

Part 9

Electrical Analysis and Testing of Multichip Modules

ALL THIS TECHNOLOGY has to be perceived by the system user as advantageous for circuit performance as well as cost. Most of the arguments are presented within the specific technology papers. The three papers selected here give some indication of the electrical performance to be expected using these new technologies.

One of the best electrical analyses was that presented at the IEPS by Schacham and others of the MCNC. While its intent was to compare the performance of conductive interconnects with the possibilities of optical interconnections, this paper gives a good indication of the ultimate performance of multichip modules, and is worth reviewing.

Electrical Characteristics of Copper/Polyimide Thin-Film Multilayer Interconnects

THOMAS A. LANE, FRANK J. BELCOURT, MEMBER, IEEE, AND RONALD J. JENSEN

Abstract—A multilayer substrate containing two-chip emitter-coupled logic (ECL) ring oscillator circuits was fabricated and tested to determine the static and dynamic electrical characteristics of the copper/polyimide thin-film multilayer (TFML) interconnections between chips. The circuit oscillated at 122 MHz without delay lines; delay lines in the circuit enabled measurement of 88 ps/cm propagation delay and -28 dB crosstalk between adjacent parallel lines. The measured electrical characteristics agreed with predictions obtained from static electrical models and SPICE simulations of the interconnections. The ring oscillator circuit has general applications for determining dynamic electrical characteristics of a variety of packaging structures, such as co-fired ceramic single-chip packages and tape-automated-bonding (TAB) leadframes.

INTRODUCTION

TO TAKE advantage of the increased speed and density of VLSI circuits, multichip packages are being developed to reduce signal delays, power requirements, and the physical size of electronic systems. A number of IC's are mounted on substrates having very dense interconnect structures, sandwiched between voltage or ground planes. High-speed logic circuits require short interconnections with high propagation velocity and low loss. Signal rise times have decreased to such an extent that the interconnects must be treated as transmission lines with large bandwidths and controlled characteristic impedance. Sources of noise, such as reflections from impedance discontinuities, and crosstalk between adjacent signal lines must be minimized. CAD tools are becoming increasingly important for simulating electrical performance prior to package fabrication. It is, therefore, essential that the high-speed electrical characteristics of the interconnect technology are accurately known.

A high-density interconnect technology being developed at Honeywell and a number of other electronics companies is based on multiple thin-film layers of copper, aluminum, or gold conductors and polyimide dielectrics [1]–[3]. Although a number of investigators have reported on the fabrication and physical implementation of this technology, there has been relatively little discussion of its high-speed electrical performance. This paper addresses the measurement and prediction of the static and dynamic electrical characteristics of thin-film multilayer (TFML) copper/polyimide interconnections using a strip-line test structure and a high-speed ring oscillator circuit.

Manuscript received February 11, 1987; revised July 22, 1987. This paper was presented at the 37th Electronic Components Conference, Boston, MA, May 11–13, 1987.

The authors are with the Honeywell Physical Sciences Center, MN09-1300, 10701 Lyndale Avenue South, Bloomington, MN 55420.

IEEE Log Number 8717178.

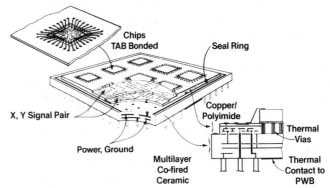

Fig. 1. Typical thin-film multilayer package structure.

THIN-FILM MULTILAYER PACKAGING APPROACH

A conceptual design for a multichip package using thin-film multilayer interconnections is shown in Fig. 1. Multiple layers of thin-film copper separated by a polyimide dielectric are patterned on a co-fired ceramic substrate up to 100 mm square. The co-fired ceramic substrate can contain internal metal layers for power and ground distribution, a pin grid array for connection to a printed wiring board (PWB), metal strips for thermal contact to the PWB, and a seal ring for hermetic sealing. Packages of this type can accommodate up to 1000 pins and still leave room for adequate heat sinking to the PWB.

The Cu/polyimide interconnect structure typically consists of five metal layers: two layers of orthogonal signal lines sandwiched between ground or voltage planes in an offset stripline configuration, and a top layer for the attachment of chips and other components, such as termination resistors and decoupling capacitors. For typical 50-Ω signal lines, the conductor lines are 25 μm wide and 5 μm thick and dielectric layers are 25 μm thick. Center-to-center spacing of signal lines is usually less than 125 μm to achieve high interconnect density. The chips can be electrically bonded to the top layer using wire bonding or tape-automated bonding (TAB). A dense array of thermally conductive copper vias are fabricated beneath each chip site to improve the heat transfer through the polyimide layers to the substrate.

ELECTRICAL CHARACTERISTICS

Electrical characteristics that are important for the design of a high-speed multichip package include the fundamental transmission line properties of the interconnect as well as the dynamic characteristics of signals propagated between chips. The fundamental transmission line properties include the line resistance (R), capacitance (C), inductance (L), and dielectric

Reprinted from *IEEE Trans. Components, Hybrids, Manuf. Technol.*, vol. CHMT-12, no. 4, pp. 577–585, December 1987.

conductance (G), all per unit length, and the characteristic impedance (Z_0), and propagation constant (γ). These characteristics are functions of the cross-sectional geometry of the interconnect, the electrical properties of the conductor and dielectric, and the frequency of the signal. The transmission line properties, combined with the electrical characteristics of the driver and receiver circuits and other interconnect structures, such as wire bonds, vias, and stubs, affect the dynamic behavior of signals reaching the receiver circuits. Important dynamic characteristics include the propagation delay, the rise and fall times of the signals, attenuation, noise due to signal reflections, and crosstalk between adjacent lines.

Measurements of transmission line properties can be made with standard instruments, such as the universal bridge and network analyzer. Measurements of dynamic characteristics are best made using functional circuits which provide signal characteristics and impedances corresponding to the actual package application. Honeywell has developed a ring oscillator circuit that uses 100K emitter-coupled logic (ECL) chips to propagate signals through the interconnects at frequencies over 100 MHz with rise times of 1 ns. Interconnects selected for test can be switched into the oscillator circuit with selective wire bonding. Parallel interconnects can be measured for crosstalk coupling. The following section describes the experimental approach in greater detail, and will be followed with a presentation of the measurements as compared to expected values.

EXPERIMENTAL

Stripline Test Structure

Because the interconnect delay lines in the ring oscillator test package (described later) were quite long and exhibited a high dielectric conductance due to process imperfections, an offset stripline test structure was fabricated to measure process geometries and their effects on static electrical characteristics. The offset stripline is a transmission line structure containing two adjacent layers of signal lines sandwiched between ground or voltage reference planes, resulting in the same characteristic impedance on both signal layers [4]. The reference planes in the test structure were a mesh with 62-μm square holes on a 125-μm pitch to permit outgassing from the polyimide. The striplines varied from 1 to 8 cm in length and were designed to a characteristic impedance of 50 Ω using the cross-sectional geometries given in the first column of Table I. See also Fig. 2.

The width and thickness of the conductor lines and the dielectric thickness were measured at each step during the fabrication of the striplines using a calibrated microscope to measure linewidths and a profilometer to measure film thicknesses. Static electrical characteristics R, C, and G were measured at frequencies of 10 kHz to 10 MHz with an HP 4275A LCR meter. The characteristic impedance and attenuation of the striplines were measured with an HP 8754A network analyzer.

Ring Oscillator Circuit

A ring oscillator circuit was used to obtain the dynamic electrical characteristics of TFML interconnections. Ring

TABLE I
MEASURED DIMENSIONS OF OFFSET STRIPLINE TEST STRUCTURE

	Design Value (μm)	Measured Average (μm)	Standard Deviation Between Substrates (μm)
Dielectric thickness			
layer 1 (a_1)	22.5	24.4	1.2
layer 2 (a_2)	25.0	27.3	2.3
layer 3 (a_3)	27.5	26.2	2.4
Conductor thickness (t)			
x metal	5.15	6.2	—
y metal	5.15	5.2	—
Conductor linewidth (w)			
x metal	25.4	26.0	1.6
y metal	25.4	28.4	1.0

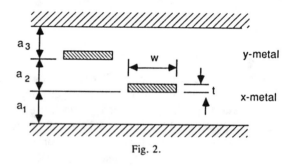

Fig. 2.

oscillator circuits are widely used for evaluating propagation delay and power dissipation since they provide their own signal source [5]. The circuit is constructed from an odd number N of inverting gates connected in series to form a ring. The measured frequency of oscillation is related to the intrinsic gate delay by $f = 1/(2 \cdot N \cdot T_d)$, where T_d is the internal gate delay. In this study the ring oscillator was used to investigate the interconnects, rather than the intrinsic gate delay, by providing a stable oscillating signal that could be propagated through a variety of interconnect structures.

The ring oscillator circuit in this study consisted of five ECL gates on two Fairchild F100102 IC's, connected so that the signals passed between the chips after each gate. The nominal oscillating frequency of the circuit was 120–125 MHz. Signal swings were from 0.3 to 1.1 V, with driver rise and fall times of approximately 1 ns.

A schematic diagram of the circuit is shown in Fig. 3. A number of optional interconnect lines can be inserted into the circuit between the fifth and first gates: 1) a short jumper (1-cm) wire which provides the minimum delay path, 2) a 40-cm-long serpentine delay line which permits measurement of propagation delay and rise time, and 3) a 15-cm-long delay line to which are tapped 2-cm and 6-cm stubs connected to ECL gates to measure the effects of fanout. The delay lines may be terminated at the far end with a parallel resistor to ground, or at the near end with a series resistor. Two lines, 40 and 15 cm long, terminated at both ends in the characteristic impedance, are routed parallel to the serpentine delay line to permit measurement of crosstalk.

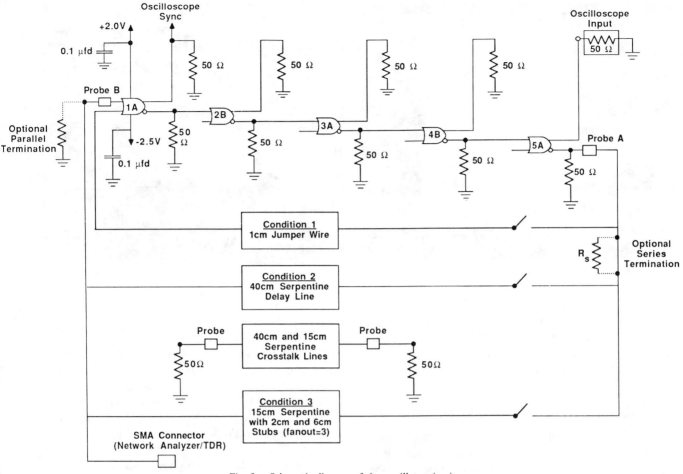

Fig. 3. Schematic diagram of ring oscillator circuit.

Ring Oscillator Test Package

The physical layout of the ring oscillator test package is shown in Fig. 4. Three two-chip circuits are fabricated on a 2 × 2-in substrate. The circuits differ in their delay and crosstalk lines. From top to bottom in Fig. 4, these circuits are 1) a 25-μm-wide delay line with adjacent parallel crosstalk lines on a 125-μm pitch, 2) a 50-μm-wide delay line with crosstalk lines on a 250-μm pitch, and 3) a delay line with fanout as described earlier. The test package is fabricated in four metal layers: a bottom ground plane, two layers of signal interconnect lines and delay lines, and a top voltage plane and chip attach/bonding layer. The serpentine delay lines are patterned on adjacent orthogonal layers between reference voltage planes, forming an offset stripline as described in the previous section, with a large number of crossovers. Both the delay lines and the crosstalk lines have probe points on the line, permitting direct measurement of the line with a high-impedance probe and sampling oscilloscope. The top voltage plane has a 0.2-mm holes on 0.38-mm centers to facilitate outgassing from the polyimide during processing.

The ring oscillator test packages were fabricated using standard TFML processes described previously [1]. The substrate was a smooth tape-cast ceramic (99-percent Al_2O_3), 2 × 2 × 0.025 in thick. Conductor layers are nominally 5 μm thick and consist of sputtered copper sandwiched between thin adhesion layers of Cr or TiW. The polyimide dielectric layers

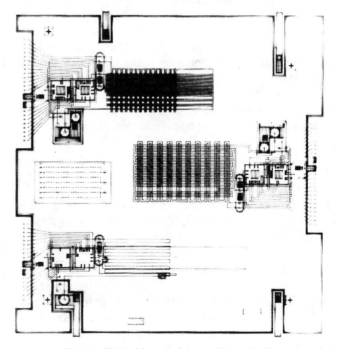

Fig. 4. Physical layout of ring oscillator circuit.

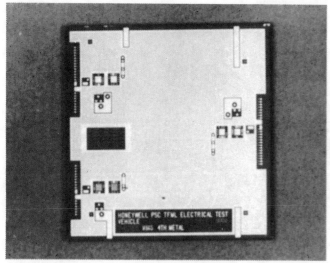

Fig. 5.   Completed TFML ring oscillator package.

Fig. 6.   Test fixture connected to network analyzer.

are nominally 25 μm thick, creating 50-Ω characteristic impedance for 25-μm-wide lines. Staggered vias are used for connection through more than one dielectric layer. The ECL chips were attached with epoxy and bonded with 1-mil aluminum wire. A completed package with wire-bonded chips is shown in Fig. 5.

Test Approach

The test setup consisted of test instruments, a test fixture in which the ring oscillator package was mounted, and a probe station consisting of a microscope and various types of probes, coaxial cabling, and power supplies. This setup provided a constant 50-Ω transmission line environment for making accurate tests on high-frequency (100–400 MHz) signals. An important factor in maintaining the 50-Ω environment was the test fixture, which consisted of a multilayer printed wiring board mounted in a copper frame. This fixture provided mechanical support for the TFML substrate as well as decoupling capacitors, termination resistors, and 50-Ω lines feeding SMA connectors for interfacing the TFML substrate to the instruments and power supplies.

The test setup also included the two power supplies needed for the ECL circuits. The voltage levels were skewed such that $V_{CC} = +2.0$ V, $V_{tt} = GND$, and $V_{EE} = -2.5$ V [6]. These voltage levels were chosen to support oscilloscope measurements using a sampling head where the 50-Ω input impedance was used to terminate the line.

The instruments used for static testing were a Kerr Universal Bridge, an HP 4275 *LCR* meter, and an HP 8754A Network Analyzer. The bridge was used to measure capacitance, resistance, and conductance at 1.6 kHz by probing delay lines directly. The HP 4275A was used to make the same measurements at frequencies of 10 kHz to 10 MHz. The network analyzer was used to measure insertion loss and input impedance over a frequency range of 50–400 MHz, from which the attenuation and characteristic impedance of the line were determined as discussed in the next section. Fig. 6 shows the test fixture connected to the network analyzer. Characteristic impedance was also determined from time domain reflecto-

metry measurements using a Tektronix 7S12 General Purpose Sampler with a S-52 Pulse Generator Head.

Measurements of dynamic waveforms in the ring oscillator were made by connecting a Tektronix 7854 Oscilloscope, with sampling plug ins, to various test points in the circuit. Two types of probes were used: high-frequency high-impedance active probes, and direct connections from the test fixture SMA connectors to the 50-Ω inputs of the oscilloscope. The TFML substrate was designed for high-frequency probing by providing all test points with an adjacent ground pad. Fig. 7 shows the test fixture containing the substrate being measured, with high-frequency scope probes applied to the substrate.

RESULTS AND DISCUSSION

Fabrication Geometries

The design values and measured dimensions of the offset stripline test structure are given in Table I and Fig. 2. Dielectric layer thicknesses, obtained by spray coating polyimide, are within 2 μm of target values and have a standard deviation of less than ten percent between substrates. The sputtered conductor thickness is usually within 0.2 μm of target values (although a systematic error on the x layer resulted in a deviation of 1.0 μm for the structure in Table I). Conductor linewidths vary by as much as 3 μm from target values, with less than seven-percent standard deviation between substrates. All of these geometrical variations have minor effects on the predicted electrical characteristics of TFML interconnects. We are continuing to modify equipment and processes to improve the reproducibility and uniformity of the conductor and dielectric geometries, as well as to improve the yield of package fabrication.

Static Electrical Measurements

Measured values for the resistance (R), capacitance (C), and dielectric conductance (G) of the offset stripline test structure at frequencies of 10 kHz to 10 MHz are given in Table II. The dissipation factor, tan δ, which is the ratio of conduction current to displacement current in the dielectric, was calculated from the measured values of G and C based on

Fig. 7. High-frequency probing of ring oscillator package in test fixtures.

TABLE II
MEASURED AND CALCULATED STATIC ELECTRICAL PARAMETERS OF
STRIPLINE TEST STRUCTURE

Frequency (MHz)	R (Ω/cm)	C (pF/cm)	G (nS/cm)	tan δ
0.01	1.09	1.32	0.26	0.003
0.1	1.12	1.31	5.4	0.007
1	1.14	1.30	57	0.007
10	1.06	1.32	570	0.007
Calculated value (from Table I)	1.04	1.21		

the equation

$$\tan \delta = \frac{G}{\omega C} \qquad (1)$$

where ω is the angular frequency of the measurement ($\omega = 2\pi f$). It can be seen that the derived value of tan δ is constant (0.007) from 0.1 to 10 MHz, and drops to 0.003 at 10 kHz. This is consistent with the manufacturer's published value of 0.002 for the dissipation factor of DuPont 2555 polyimide, and with published values of tan δ, which vary from 0.007 to 0.01 [1], [7].

Calculated values of R and C are also given in Table II, based on the measured conductor and dielectric geometries

TABLE III
MEASURED STATIC ELECTRICAL PARAMETERS OF DELAY LINE IN RING
OSCILLATOR PACKAGE

Frequency (kHz)	R (Ω/cm)	C (pF/cm)	G (nS/cm)
1.6	1.93	1.61	79.0

given in Table I. (Values for the x and y metal layers differed by less than four percent and are, therefore, averaged.) Resistance per unit length was calculated from the simple expression for dc resistance:

$$R = \frac{\rho}{w \cdot t} \qquad (2)$$

where ρ is the conductor resistivity (1.6×10^{-6} Ω·cm for copper) and w and t are the conductor width and thickness, respectively. Skin effects are unimportant here since the skin depth of copper at the highest frequency of 10 MHz is 20 μm, which is much greater than the conductor line thickness. The capacitance of the offset stripline was calculated from a model which includes the effects of fringing and parallel capacitance [4, eqs. 2-7 to 2-12]. A dielectric constant of 3.5 was assumed for polyimide, based on past measurements [1]. Note that this model assumes solid ground planes, whereas the actual ground planes were a mesh, which would result in lower capacitance.

Very good agreement was obtained between the measured and calculated values of R and C. The measured R is slightly higher than the calculated value, possibly because actual lines have a trapezoidal cross section whereas the calculations assume a rectangular cross section. The measured C was also slightly higher than predicted values, probably because the added capacitance of the end pads and the crossovers between the two adjacent layers of signal lines were included in the measured capacitance. The predicted value of C, taking the mesh planes into account, would be lower than the value in Table II, making the discrepancy even larger.

Measured values of R, C, and G for the 40-cm-long delay line on the ring oscillator test package are given in Table III. These values were measured at 1.6 kHz with the Kerr Universal Bridge. The design values for the cross-sectional geometry were similar to those of the stripline test structure (Table I), although the linewidth and layer thickness were not measured during processing. Compared to design values, the ring oscillator substrate had high values of R (by a factor of 2), possibly because of large via resistance or an incorrect conductor thickness or linewidth. The measured capacitance is also about 20 percent above the design value, due to the added capacitance of many crossovers, plus the mutual capacitance of adjacent lines. Finally, the measured value of G is very large. We believe this is due to a measurable conductive residue left on the dielectric after patterning the metal lines, which caused large leakage currents to ground. The processing problems which resulted in the large deviation from design values have since been corrected, as evidenced by the more recent data in Tables I and II. However, the actual measured values of R, C, and G for the ring oscillator were required for

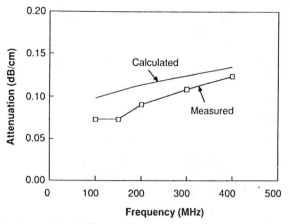

Fig. 8. Attenuation of offset stripline, measured with network analyzer and calculated from low-frequency measurements of resistance, capacitance, and conductance.

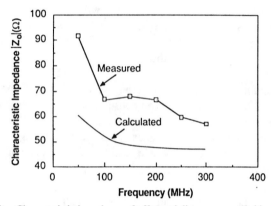

Fig. 9. Characteristic impedance of offset stripline, measured with network analyzer and calculated from low-frequency measurements of resistance, capacitance, and conductance.

further characterization of the dynamic behavior of the ring oscillator circuit.

Characteristic Impedance and Attenuation

The characteristic impedance and attenuation of a 7-cm long line in the offset stripline test structure were measured using the network analyzer over a frequency range of 50–400 MHz. The attenuation was determined by directly measuring the insertion loss of the unterminated line (in dB) and dividing by the line length. As shown in Fig. 8, the attenuation increases with frequency and has a small magnitude (less than 0.15 dB/ cm) over the full frequency range.

The characteristic impedance of a line is difficult to measure directly, while the input impedance can be measured easily with the network analyzer. A method commonly used to determine Z_0 is to measure the input impedance with the line open and shorted, from which Z_0 can be calculated. We used an improved version of this approach described in [8]. In this method, the input impedance of the line is measured twice, each time with a different known termination value. From the results of these measurements the characteristic impedance can be calculated. Fig. 9 shows a plot of the measured values of Z_0 obtained by this method over a frequency range of 50–300 MHz. Time domain reflectometry (TDR) using a 45-ps rise time pulse was also used to obtain a value of approximately 62

Ω for the characteristic impedance of the line at high frequency. This value confirms the network analyzer measurements shown in Fig. 9.

The characteristic impedance and attenuation were also calculated from the measured-low frequency parameters R, C, and G in Table II, using the defining equations for the characteristic impedance and attenuation of a transmission line propagating signals in a quasi-TEM (transverse electromagnetic) mode [9], [10]:

$$Z_0 = \sqrt{\frac{R + j\omega L}{G + j\omega C}} \tag{3}$$

and

$$\alpha = \text{Real } \{\gamma\}$$
$$= \text{Real } \{\sqrt{(R + j\omega L)(G + j\omega C)}\} \tag{4}$$

where the attenuation α is the real part of the propagation constant γ. The parameters R, C, and G for the offset stripline were measured directly, as discussed in the preceding section. The measured dc value of R was corrected for frequency dependence due to the skin effect. Since it is difficult to measure the inductance of a line directly, L was calculated from the general relationship for the phase velocity of a TEM wave:

$$\sqrt{LC} = \frac{\sqrt{\epsilon_r}}{c}$$

or

$$L = \frac{\epsilon_r}{c^2 C} \tag{5}$$

where c is the speed of light in vacuum and ϵ_r is again assumed to be 3.5. The derived value of L (2.97 nH/cm) and the measured values of R, C, and $\tan \delta$ ($G = \tan \delta \omega C$) from Table II were substituted into (3) and (4) to obtain the calculated curves for α and Z_0 as shown in Figs. 8 and 9.[1]

It can be seen from Figs. 8 and 9 that the measured values of α and Z_0 have the same frequency dependence as that predicted by (3) and (4). However, the measured attenuation is lower and the measured Z_0 is higher than the values calculated from the measured R, C, and G. We believe this is because the measured value of C was artificially high, due to the additional capacitance of end pads and cross overs which are included in the capacitance per unit length calculation. Furthermore, the derived value is of L is lower than that of the actual line which contains the added inductance of vias. In fact, if we substitute a value of C that is 15 percent lower than the measured value and a value of L that is 15 percent higher than the derived

---

[1] The data presented in Figs. 8 and 9 differ from the data presented in the *Proceedings of the 1987 Electronic Components Conference*. Figs. 8 and 9 in the *Proceedings* paper showed measured data for the ring oscillator test vehicle whereas the calculated curves were based on R, C, and G for the stripline test structure.

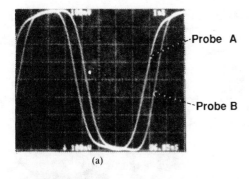

(a)

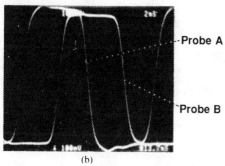

(b)

Fig. 10. Oscilloscope waveforms from operating ring oscillators. (a) Without delay line. (b) With 40-cm delay line.

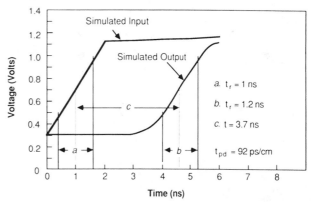

Fig. 11. SPICE simulation of input and output signals on 40-cm delay line in ring oscillator.

$$\tau_{pd} = 1/v_p$$

$$= \sqrt{LC}$$

$$= Z_0 C. \qquad (6)$$

value, we obtain almost perfect agreement with both the measured attenuation and characteristic impedance. Because of the short length of the measured line (7 cm), the parasitic capacitance and inductance of pads and vias can contribute significantly to measurement error. Therefore, we feel that the agreement between direct measurement of α and Z_0 and the values calculated from measured R, C, and G is reasonably good. We are continuing to design and fabricate new test structures to resolve the discrepancy between high-frequency and low-frequency electrical characteristics.

Dynamic Ring Oscillator Characteristics

The measurement results from dynamic testing are taken from photographs of signal waveforms in an operating ring oscillator [11]. By analyzing the photographs we have derived the interconnect parameters, such as propagation delay and rise time. The signal waveforms of an operating ring oscillator are shown in Fig. 10. Fig. 10(a) displays the driver and receiver waveforms of the operating ring oscillator without the 40-cm line in the ring (Condition I), measured at an unused gate output through an SMA connector. Fig. 10(b) displays the driver and receiver waveforms with the 40-cm line in the ring (Condition II), measured at the same points. The propagation delay is calculated from the difference in frequency and the length of the delay line and is 88 ps/cm. The rise times at the driver and receiver end of the line were also measured by probing the line directly at points A and B is Fig. 3. The values at Condition II were 1.0 ns at the driver end (point A) and 1.5 ns at the receiver end (point B).

The propagation delay of a lossless transmission line is the inverse of the propagation velocity of the electromagnetic waves in the dielectric medium and is given by

Substituting the measured value of C (1.6 pF/cm from Table III) and the Z_0 measured at high frequency (52 Ω), we obtain a lossless propagation delay of 83 ps/cm, which is close to the measured value of 88 ps/cm. (This is significantly higher than the theoretical propagation delay of 62 ps/cm for lossless lines in polyimide with an $\epsilon_r = 3.5$; however, we believe that the conductive residue combined with some oxidized residue may have increased the effective dielectric constant of polyimide.)

The high values of R and G for the ring oscillator delay line introduce additional signal delay by causing a decrease in the signal rise time. No simple closed-form equations exist that describe the time-domain behavior of a lossy transmission line. Therefore, we modeled the dynamic behavior through the use of a lumped-element model in which the transmission line is approximated by multiple segments consisting of a series resistance and inductance and a parallel resistance and capacitance to ground [12]. We inserted the measured transmission line parameters of R, C, and G (Table III) and the derived value of L into the model, and performed SPICE simulations using simplified driver and receiver equivalent circuits. Fig. 11 shows the result for a SPICE simulation of the 40-cm delay line. The simulated propagation delay of 92 ps/cm agrees well with the measured value of 88 ps/cm.

The crosstalk ratio was determined by measuring the voltage coupled onto the ends of an adjacent 40-cm line terminated at both ends with 50 Ω. The maximum amplitude of the coupled voltage was 30 mV, or an attenuation of 28 dB for the 800-mV signal. Crosstalk was simulated using a SPICE model which included two transmission lines with coupling parameters (mutual inductance and capacitance) obtained from design values [12]. The SPICE simulation of the crosstalk circuit indicated a coupled voltage of 1.1 mV, or an attenuation of 57 dB. The crosstalk is expected to be minimal in a package using these geometries because the signal lines are close to the ground planes and the line pitch is five times the linewidth. The measured crosstalk is larger than predicted due to the presence of other noise sources, such as power supply noise.

TABLE IV
MEASURED AND CALCULATED DYNAMIC ELECTRICAL
CHARACTERISTICS OF RING OSCILLATOR PACKAGE

Parameter	Observed Value	SPICE Simulation Results
Ring oscillator frequency without 40-cm interconnect	122 MHz	*
Ring oscillator frequency with 40-cm interconnect	64 MHz	*
Propagation delay	88 ps/cm	92 ps/cm
Driver end rise time	1.0 ns	1.0 ns
Receiver end rise time	1.5 ns	1.2 ns
Crosstalk amplitude	−28 dB (30 mV)	−57 dB

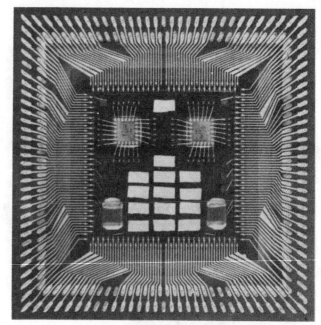

Fig. 13. Populated hybrid carrier for ECL ring oscillator circuit (320 × 320 mil), TAB-bonded in single-chip package.

Fig. 12. Ring oscillator circuit mounted in chip cavity of ceramic multilayer package.

Table IV is a summary of the dynamic measurements and predicted results. The SPICE simulations for propagation delay and rise time agree very well with the measured results. The measured crosstalk was negligible, although larger than predicted.

OTHER APPLICATIONS OF RING OSCILLATOR CIRCUITS

The ring oscillator circuit has been used to characterize the electrical performance of other packaging configurations. Fig. 12 shows a ring oscillator circuit mounted in the chip cavity of a 240-pin single-chip multilayer ceramic package. Chip decoupling capacitors and load resistors were mounted in the cavity along with two ECL chips. The interconnect to be tested was connected into the ring oscillator circuit through wire bonds between pads on the oscillator and the package. This arrangement permitted observation of the signal propagated through the interconnect, including the pads, pins, vias, and lead bonds. Using the longest package interconnect (approximately 3.5 cm), the ring oscillated at a frequency of 100 MHz with a 0.8-ns rise time. The total propagation delay was approximately equal to the calculated value of 102 ps/cm for a lossless interconnect in a ceramic dielectric.

We also fabricated a hybrid version of the ring oscillator circuit containing thin-film multilayer interconnects on a small ceramic carrier. The carrier is 0.320 × 0.320 in and contains all the necessary interconnections and bonding sites for the ring oscillator circuit, including two ECL IC's, chip resistors, and decoupling capacitors. A large number of these hybrid carriers can be fabricated in TFML on a ceramic substrate which is then diced. Fig. 13 shows a photograph of a completed hybrid carrier populated with IC's and discrete components, and TAB-bonded into the cavity of a single-chip multilayer ceramic package. This hybrid ring oscillator circuit provides an oscillating signal that can be propagated through various interconnect elements such as TAB leads, pins, vias, or interconnect lines within the ceramic package, permitting the measurement of propagation delay and the effects of impedance discontinuities.

CONCLUSION

The transmission line properties and dynamic electrical characteristics of a high-density thin-film multilayer (TFML) copper/polyimide interconnect technology were measured and compared with calculated or design values. Transmission line properties (characteristic impedance, attenuation) were measured at frequencies up to 400 MHz using a network analyzer; static electrical characteristics (R, C, and G) were measured at lower frequencies using a universal bridge or LCR meter. The static electrical characteristics of a simple offset stripline test structure agreed well with values calculated from the measured geometries of the fabricated structure, while the measured transmission line properties agreed fairly well with values calculated from the measured R, C, and G.

The dynamic performance of TFML interconnects was measured using an ECL ring oscillator circuit which generates signals with characteristics approximating those for the actual package application. The measured propagation delay of 88 ps/cm and rise time of 1.5 ns agreed well with the performance simulated on SPICE using a lumped-element model of the transmission line. Crosstalk between lines spaced on a 125-μm pitch was negligible, both for measured and predicted values. The methods discussed in this paper, as well as the quantitative results, will facilitate further CAD analysis of TFML package designs.

ACKNOWLEDGMENT

The authors would like to thank the technical staff of the Honeywell Physical Sciences Center, especially Dr. R. Sainati for technical assistance and insight, D. Saathoff and M. Propson for fabrication of TFML test packages, and B. Jacobsen, M. Campbell, P. Koehn, and C. Meyer for testing.

REFERENCES

[1] R. J. Jensen, J. P. Cummings, and H. Vora, "Copper/polyimide materials system for high-performance packaging," *IEEE Trans. Components, Hybrids, Manuf. Technol.*, vol. CHMT-7, pp. 384–393, 1984.

[2] T. Watari and H. Murano, "Packaging technology for the NEC SX supercomputer," *IEEE Trans. Components, Hybrids, Manuf. Technol.*, vol. CHMT-8, pp. 462–467, 1985.

[3] H. Tsunetsugu, A. Takagi, and K. Moriya, "Multilayer interconnections using polyimide dielectrics and aluminum conductors," *Int. J. Hybrid Microelectron.*, vol. 8, pp. 21–26, 1985.

[4] H. Howe, Jr., *Stripline Circuit Design.* Dedham, MA: Artech House, 1979, ch. 2.

[5] R. C. Eden and B. M. Welch, in *Very Large Scale Integration*, D. F. Barbe, Ed. Berlin: Springer-Verlag, 1980, ch. 5, p. 163.

[6] *F100K ECL Data Book.* Mountain View, CA: Fairchild Semiconductor, 1982, pp. 3–7.

[7] L.B. Rothman, "Properties of thin polyimide films," *J. Electrochem. Soc.*, vol. 127, pp. 2216–2220, 1980.

[8] J. E. McKay, "Measure the characteristic impedance," *Electron. Des.*, vol. 24, pp. 136–137, Nov. 22, 1976.

[9] R. Matick, *Transmission Lines for Digital and Communications Networks.* New York: McGraw-Hill, 1969.

[10] R. Sainati, "High speed packaging analysis and modeling," in *Proc. 35th Electronic Components Conf.*, 1985, pp. 365–371.

[11] T. Lane, "TFML ring oscillator test results," Internal Honeywell Rep., Oct. 1986.

[12] F. Belcourt and T. Lane, "Electrical CAD analysis for multilayer package design," in *Proc. 1986 ISHM Symp.*, 1986, pp. 802–808.

PERFORMANCE CHARACTERISTICS OF THIN FILM MULTILAYER INTERCONNECTS

IN THE 1-10 GHz FREQUENCY RANGE

by

D. J. Schwab, R. L. Thompson, and B. K. Gilbert
Mayo Foundation
Rochester, Minnesota 55905

and

K. Jayaraj, T. J. Moravec, R. J. Jensen, and R. Sainati
Honeywell Sensors and Signal Processing Laboratory
10701 Lyndale Avenue South
Bloomington, Minnesota 55420

Abstract

Multichip packaging approaches which provide short chip-to-chip interconnects are required to fully utilize the high speed potential of Silicon ECL and GaAs ICs. To maintain signal fidelity at GHz clock speeds, interconnects must be designed as impedance controlled transmission lines with proper concern for discontinuities, cross-talk and attenuation. At present, there is very little quantitative data for frequencies greater than 1 GHz on the effects of the typical situations encountered in hybrid and multichip packaging, i.e. wire bonds, bond pads, vias, stubs, changes in line widths, package feedthroughs, etc. Thus the design of a high speed system is left to guess work. This paper presents initial results from a program to develop much of this data.

The platform chosen to quantify high frequency multichip packaging effects is the copper/polyimide based Honeywell thin film multilayer (TFML) technology which is representative of the developing high density thin film approaches. Mayo foundation designed and tested a TFML substrate which was fabricated by Honeywell. Measurements of return loss, insertion loss, and characteristic impedance of interconnects of various lengths were made up to 9.045 GHz. Signal attenuation and crosstalk of 1 and 2 GHz digital signals were measured and simulated using Mayo Foundation's EM modeling software package. In general, the loss and crosstalk values obtained were low enough to permit the transmitting of GHz digital signals over typical interconnect lengths encountered in a multichip module.

Introduction

Systems are being designed to operate at clock frequencies greater than 1 GHz using the latest in GaAs and silicon ECL digital IC technology. Chip-to-chip transmission delays can be minimized in these systems by using multichip packaging (MCP) approaches using cofired ceramic, thin film and thick film interconnect technologies. Compared to ceramic and thick film materials, polymeric materials have much lower dielectric constants. Since the propagation delay of a lossless transmission line is directly proportional to the square root of the dielectric constant of the insulating medium, thin film packaging technologies that use polymeric dielectric materials are well-suited for multi-GHz systems.

Honeywell has pioneered the development of a thin film multilayer (TFML) technology [1,2] to interconnect GaAs ICs as well as DoD's Very High Speed Integrated Circuits (VHSIC). There are now several companies [3-6] involved in developing high-density interconnect technologies with polyimide as the dielectric medium and Al and/or Cu as the conductor. Honeywell's TFML approach consists of sequential deposition and patterning of copper conductor and polyimide dielectric materials on rigid substrates using thin film techniques. Polyimide with its low dielectric constant (~3.1), excellent processing characteristics, and extreme resistance to temperature, radiation, and chemical etchants is an ideal dielectric material. Polyimide is applied by either spin or spray coating, and vias are etched with excellent geometrical control by reactive-ion-etching. Copper is sputtered and patterned photo-lithographically. Copper is the conductor material of choice because of its high electrical conductivity. However, interface materials such as chromium and TiW must be used to promote adhesion to polyimide and insure reliable operation over military standard test conditions.

The baseline TFML design consists of a pair of orthogonal signal layers sandwiched between two ground/power planes, and a die-attach/bond-pad layer (Figure 1). Both the X and Y conductor lines are designed as offset transmission lines with a 50-ohm characteristic impedance. As with single chip packaging, the MCP must be designed to electrically interconnect the ICs, protect the chips, provide power and ground, and remove the heat. At multi-GHz frequencies, interconnect lengths become a significant fraction of the wavelength of the highest frequency harmonics, and therefore the interconnects must be designed as transmission lines with proper concern for impedance, crosstalk and attenuation [7]. Impedance mismatch must be minimized to reduce reflections and prevent ringing which can cause false switching. The narrow conductor line geometries and close spacing between conductors present additional problems [8], especially for signals with sub-nanosecond risetimes. Crosstalk between narrowly spaced signal lines can cause noise and false switching on quiet lines. Significant attenuation and rise-time degradation can be caused by losses in the transmission line. The transmission line loss is the sum of conductor loss and dielectric loss, both of which are dependent on frequency. At clock frequencies less than a few hundred MHz, dielectric losses are usually negligible for most materials. Dielectric losses are

Reprinted from *Proc. 39th Electron. Components Conf. (ECC)*, pp. 410–416, 1989.

very difficult to model, and experimental data are unavailable for the frequencies and geometries of interest for thin film packaging [7,9]. Conductor resistance consists of two components: a d.c. resistance and an a.c. resistance which increases with frequency. At frequencies greater than several GHz, conductor losses can be accurately modeled assuming a well-developed skin effect. However, no closed form solutions for conductor loss exist in the entire frequency range.

Several components of a multichip package have complex geometries with effects on high speed signal propagation that are difficult to model. These include vias, via pads, bond pads, wire-bonds, and stubs. Usually these components are considered to be "electrically short" and are represented by lumped element models. Their effect is usually negligible for thin film geometries, but this should be established by experimental measurements.

Honeywell and Mayo Foundation undertook the task of collecting the much needed experimental data on the performance characteristics of TFML interconnects in the 1-10 GHz frequency range, and establishing guidelines for designing multi-GHz digital systems. A test coupon, fabricated by Honeywell using the TFML technology, was designed and tested by Mayo Foundation. This paper presents a detailed description of the test coupon design, high-speed calibration and measurement techniques, and some initial results.

Honeywell/Mayo TFML Test Structure

The Honeywell/Mayo TFML test structure is a multilayer copper/polyimide substrate designed for microwave testing at frequencies up to 10 GHz. The substrate has two major sections, a passive section and an active section. The passive section consists of transmission lines of varying lengths and inter-lead spacings for measuring impedance, loss, reflection and crosstalk. The active section consists of eight GaAs die that can be used to demonstrate various types of chip-to-chip communication at up to 2 GHz clock rates.

The substrate was designed for Honeywell's TFML process, which uses copper conductor and a polyimide dielectric. A cross-sectional view of this structure is presented in Figure 1. At the bottom of the figure is a 40 mil-thick alumina substrate which serves as the base layer. The substrate surface is metalized, thus serving as the first metal layer; this layer is designated M1, and is used as a ground plane. A dielectric layer (designated D1) is applied to the metalized surface and patterned with vias; the next metal layer (M2) is then patterned on top of the first dielectric layer D1. This process is repeated until a total of 5 metal layers and 4 dielectric layers are in place. The 5-metal-layer configuration allows two stripline layers (M2 and M3) and one microstrip layer (M5) for signals, and two layers for power and ground (M1 and M4). Layer M4 is fabricated as a mesh plane as opposed to a solid plane to allow outgassing from the lower layers during the high temperature processing steps such as curing of polyimide layers.

The metal layers are 0.2 mils thick and the signal lines are 1 mil wide. The dielectric layers are 0.8, 0.6, 0.8, and 0.6 mils thick for dielectric layers D1, D2, D3, D4 respectively. The vias that pass through the dielectric layers are 1.5 mils in diameter, and require a 3 mil-square cover pad on each layer to insure proper contact. With these dimensions, all signal lines should exhibit a characteristic impedance of 50 ohms. For the crosstalk measurements, the signal line center-to-center spacing is 5 mil (the recommended design rule limit); thus, the expected crosstalk should be quite low, since the line-to-line spacing is about 4 times the line-to-ground plane spacing.

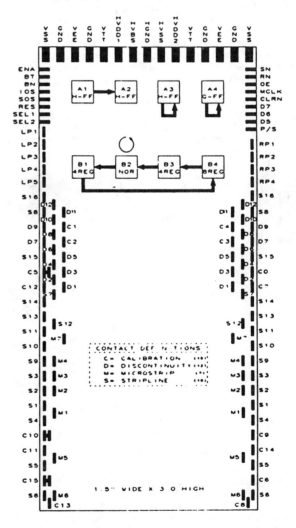

Figure 2. Block diagram of a 1.5" x 3.0" multilayer test structure fabricated with the TFML technology.

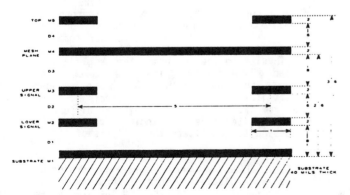

Figure 1. Cross section of a thin film multilayer (TFML) substrate designed to have 50 ohm signal (stripline and microstrip) lines.

Figure 2 is a block diagram of the Honeywell/-Mayo test structure showing the I/O and power contacts. The substrate is 1.5 inches wide and 3.0 inches high. The active section occupies the upper one-third of the structure, while the passive section occupies the lower two-thirds. The relative position of the eight GaAs die in the active section are shown, and the black arrows depict the main data flow through the GaAs integrated circuits. Power and control I/O for the active section are located along the edges of the upper part of the substrate, and will be contacted using ribbon wire and solder. The active section will not be described in further detail since at this time measurements have only been taken on the passive section. Results from the active section will be reported in a future paper.

In the passive section of the substrate, the design of the test and probe pads was critical. Since tests are made at frequencies up to 10 GHz, a method to insert signals into and extract signals from the substrate without introducing major discontinuities had to be identified. The Cascade Microtech Model 105-250 probes were selected as the best available for this application. These probes are shown in position above the TFML substrate in Figure 3. The probe body is mounted on a micropositioner and can be moved in the x, y and z directions. The probe consists of a tapered ceramic blade with coplanar lines on its under-surface. A plated-up signal pad at the end of each line and an adjacent plated-up ground pad contact the device under test. These probes have a characteristic impedance of 50 ohms, and are delivered with a calibration substrate so the calibration takes place at the probe tip, a very important feature. The calibration substrate contains all of the appropriate structures (short circuit elements, open circuit elements, 50 ohm loads, through sites and various reactive elements of different values) to perform a complete HP 8510 network analyzer calibration at frequencies up to 26.5 GHz.

Figure 3. Photograph of a TFML test substrate positioned in the microwave probe station contacted by two microwave probes.

The probes can be used for single-line transmission and reflection measurements with one probe positioned at each end of the line. In order to perform crosstalk measurements, a signal must be applied to one of the lines, with the coupled signal recorded from either end of the quiet line. In order to conduct such a test, triple signal contact probes are employed. Figure 4 shows the arrangement of the triple probe contacts for various measurements. The

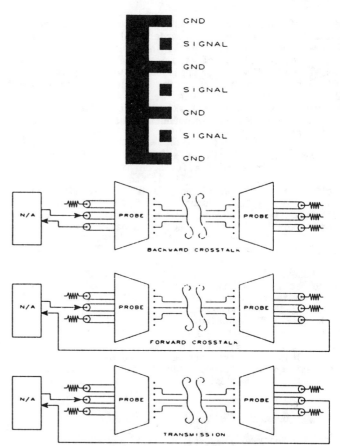

Figure 4. Schematic diagram of the triple probe configurations for crosstalk and transmission line measurements.

probe contacts are in an alternating ground-signal-ground sequence. The ground contact trace on either side of the signal trace minimizes crosstalk in the probe and also provides a ground contact in close proximity to each signal contact. For the backward crosstalk measurements a signal may be transmitted down one line, while the return signal will propagate back through an adjacent line in the same probe. Note that a probe is in place at the far end of the line under test, and that all of the signal lines are terminated. This same principle applies to the other measurements, with all unused probe ports terminated by means of SMA coaxial terminators.

Before measurements can take place on the TFML substrate, the "launch" effects from the probe onto the substrate must be understood and factored out. These "launch" effects occur because there is a change from a coplanar environment on the probe to either a microstrip or stripline environment on the substrate. The launch effects are further complicated by the fact that the probe site geometries are equivalent in size to the measured item, e.g., a via or a via pad discontinuity. Thus the launch site could generate an equal or even larger reflection than the test item. The launch site effects must be subtracted from each measurement. Fifteen different types of launch sites have been included on the TFML substrate design to facilitate the analysis of these effects.

The first five of these specially designed launch sites are dedicated to microstrip structures. The first site consists of an open pad the same size as all of the other probe site pads on the substrate. This open pad will have a fixed capacitance, which will cause a localized decrease in the impedance value. The second site is a pad that has been short circuited to ground, which will have an associated series inductance. The third site is a pad with an attached 25 mil line, with the far end left open. The fourth site has the same structure but with the far end shorted to ground. The fifth site is a line with a pad on each end, to be used for a through measurement.

The next five launch sites are associated with the upper stripline layer and the last five are associated with the lower stripline layer. These sites are similar to the microstrip sites except that they are buried and will have different values of associated parasitics. The signal path to the buried stripline layers is longer and contains more vias than the path for the microstrip layer, which is on the top surface. The stacked pads and vias will have larger associated inductance and capacitance than will the microstrip pads.

Figure 5. Symbolic descriptions of microstrip transmission lines and tests designed into the TFML substrate. The line lengths shown are nominal. Refer to the text for actual line lengths.

A listing and symbolic description of the microstrip lines are shown in Figure 5. The first section (M1) contains microstrip lines 1.36, 3.8, and 6.36 inches long (which are referred to as nominally 1.5", 4.5" and 7.5" in Figure 5). The impedance, loss and reflection parameters of these lines will be measured with the longer lines providing the most accurate loss measurements. The 1.36 inch lines make one pass across the substrate. The 3.8 and 6.36 inch lines are folded to make three and five passes across the substrate respectively.

Structure M2 in Figure 5 is a line triplet to be used for crosstalk measurements. The line pitch (i.e., the center-to-center spacing) is 5 mils and the coupled length is 1.36 inches. Both forward and backward crosstalk will be measured. Lines M3 and M4 are also on a 5 mil pitch with 3.8 and 6.36 inch coupled lengths, respectively. Structure M5 is a set of lines with a 10 mil pitch and 6.36 inch coupled lengths. Shorter lines with this interline spacing would not be expected to have measurable crosstalk in this type of copper/polyimide substrate. Structure M6 in Figure 5 is a set of lines with designed impedance changes. The top line contains a 50 ohm segment, which changes abruptly to a 25 ohm impedance through a change in linewidth, and returns abruptly to a 50 ohm segment. The center line changes from 25 ohms to 50 ohms, and back to 25 ohms. The bottom line begins at 25 ohms, changes abruptly to 50 ohms, then 25 ohms, and is completed with a 50 ohm segment. This type of structure was incorporated into the substrate to demonstrate that known impedance

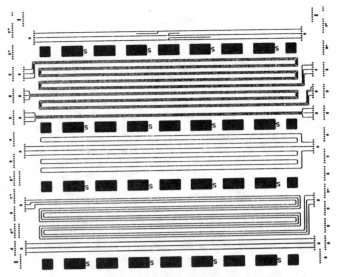

Figure 6. Layout of a portion of the passive test structure on the TFML substrate, showing isolated and coupled transmission lines.

discontinuities could be accurately measured. Structure M7 in Figure 5 consists of another set of lines with stubs of length 50 mils, 100 mils, and 200 mils, which present a different type of impedance discontinuity.

Figure 6 shows the microstrip portion of the substrate artwork at five times normal size. The pitch of the isolated lines is 20 mils to prevent electrical interactions. Note also that all of the line corners are curved to prevent reflections. The radius of all corners is 5 mils, which is five times the line width, and well above the microwave guideline of three times the line width. The 5 mil square pads at the ends of the lines are the contact sites for the high frequency probes, with neighboring pads for ground contacts. Stripline signal lines on this substrate are identical to the microstrip lines depicted in Figure 6, except that they are buried and thus the connections to these lines are made through vias.

Electromagnetic Simulations of TFML Substrate

The passive section of the TFML substrate was simulated with Mayo's Electromagnetic (EM) Modeling software package (part of the Mayo MagiCAD Integrated Circuit Design System [9,10]) to verify the performance of the TFML design prior to fabrication. The Mayo EM modeling package contains two sections: MMTL and Waveform Simulation. The Multilayer Multiconductor Transmission Line (MMTL) programs determine the distributed parameters (C, L, G, and R), characteristic impedance, and signal propagation velocity of multilayer, multiconductor transmission line structures (including microstrip and stripline). The Waveform Simulation programs determine the actual waveform responses at all line ends of a transmission line structure when an excitation signal is applied to any one or more of the lines. The MMTL programs can be applied to two-dimensional loss-free, two-dimensional lossy, or three-dimensional loss-free structures. The Waveform Simulation programs can be applied to two-dimensional loss-free or lossy structures. The results of these simulations of the TFML coupon will be discussed after the experimental results in the following section.

Results

In order to obtain early results, a simpler, three-layer microstrip version (ground-dielectric-top signal) of the TFML test structure was fabricated and tested first. This microstrip version is shown in Figure 7. DC resistance measurements were made on each of the lines of this three layer structure. The 1.36 inch lines displayed a resistance of 4.86 ohms. The 3.8 and 6.36 inch lines had resistances of 13.4 and 22.2 ohms, resulting in an average d.c. resistance of 3.5 ohms/inch. This measured value agrees very well with the expected value of 3.4 ohms/inch based on measured geometries and resistivity of pure copper.

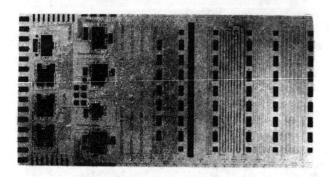

Figure 7.   Photograph of the three-layer microstrip version of the 1.5" x 3.0" TFML test substrate.

All of the lines on the microstrip version of the test structure were tested in three modalities, i.e., frequency domain, time domain and "digital" domain. Representative data from all three types of measurements are shown in Figures 8-12. For the frequency domain measurements, the Hewlett Packard Model 8510 Vector Network Analyzer was used to test for return and insertion losses over a range of 45 MHz to 9.045 GHz. Figure 8 shows the return loss and insertion loss for the 1.36 inch line. The return loss was less than -20 dB throughout the frequency range. This indicates that the microstrip line has a controlled impedance of 50 ohms without any major

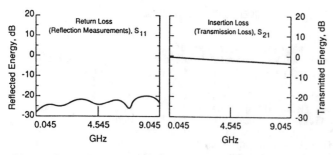

Figure 8.   Network Analyzer measurements showing the return and insertion losses for a 1.36" microstrip line fabricated with Honeywell's TFML technology.

sources of reflection. The transmission loss or insertion loss was linear with frequency, changing from -1.3 dB at 2 GHz to -3.4 dB at 9.045 GHz. The length of this microstrip line is representative of interconnect lengths in a typical multichip package designed for operation in multi-GHz frequency range. Assuming 10% of the cycle time of a 1 GHz system is devoted for propagation delays, the interconnect must be kept shorter than 0.8" for the Cu/PI system. From

the data in Figure 8, the transmission line loss for a 0.8" microstrip at 1 GHz may be calculated to be 0.4 dB which corresponds to less than 5% attenuation of the signal voltage.

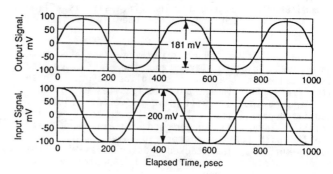

Figure 9.   The output waveform at the far-end of a 1.36" microstrip line. The lower curve corresponds to the input waveform which has a 2 GHz period, 140 psec edge rate and 200 mV amplitude.

Figure 9 shows the input and output "digital" waveforms applied to the same 1.36 inch microstrip line. A 2 GHz, 200 mV peak-to-peak wave with 140 psec edge rate was applied at one end of the line with a Colby Pulse Generator Model PG 1000A. At the other end of the line, the signal amplitude was measured at 181 mV (a 10% reduction or -1.0 dB insertion loss) with an HP 54120T oscilloscope. This data agrees very well with the network analyzer measurements (shown in Figure 8), and shows that the losses in a 1.36" line have a minor effect on a 2 GHz signal. Clearly, the TFML interconnects are suitable for transmitting high speed signals over typical interconnect lengths encountered in a MCP.

FREQUENCY DOMAIN MEASUREMENTS
(S Parameters)

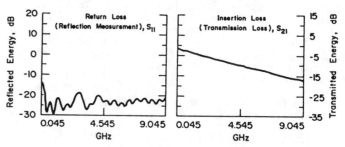

Figure 10.   Network analyzer measurements showing the return and insertion losses for a 6.36" microstrip line fabricated with Honeywell's TFML technology.

Figure 10 shows the return loss and insertion loss for the 6.36 inch line. As with the 1.36" line, the return loss was less than -20 dB throughout the frequency range. The higher return loss at 45 MHz may indicate a "contact" problem which was observed only on some measurements. The transmission loss or insertion loss was linear with frequency, increasing from -1.5 dB at 45 MHz to -17dB at 9.045 GHz. Not shown here are the insertion losses for the 3.8 inch line, which were -3.5 dB at 2 GHz and -9.5 dB at 9 GHz. The loss per unit length was independent of the line length. It increases linearly with frequency from -1 dB/inch at 2 GHz to -2.5 dB/inch at 9 GHz.

ANALOG TIME DOMAIN MEASUREMENTS
(Time Domain Reflectometry)

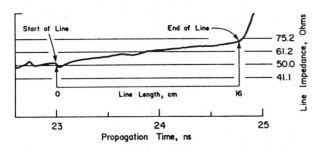

Figure 11. Time domain reflectometry measurements on a 6.36" microstrip line fabricated with Honeywell's TFML technology. Note that the line impedance begins at 50 ohms as designed, and increases to 73 ohms due to the series resistance of the line.

Figure 11 depicts a time domain measurement taken with a Hewlett Packard Model 54120T time domain reflectometry (TDR)/oscilloscope. The line impedance for the 6.36" line began at 50 ohms as designed and ended at 73 ohms. This impedance increase was caused by the series resistance (22 ohms) of the line. The smooth TDR trace confirms the lack of reflections along the line.

DIGITAL DOMAIN MEASUREMENTS
2 GHz Symmetric Pulse Train, 140 psec Input Signal Risetime

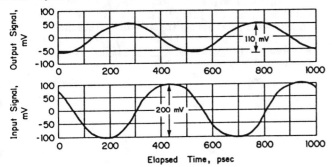

Figure 12. The output waveform at the far end of a 6.36" microstrip line. The lower curve corresponds to the input waveform which has a 140 psec edge rate, and a 200 mV amplitude. The attenuation of the output waveform is due to the resistive loss in the very long line.

Figure 12 shows the input and output "digital" waveforms applied to the same 6.36 inch microstrip line. A 200 mV peak-to-peak wave with 140 psec edge rate was applied at one end of the line. At the other end of the line, the signal amplitude was measured at 110 mV (a 45% reduction). The applied signal is essentially sinusoidal. Note that the output signal reduction agrees well with the measured insertion loss at 2 GHz in Figure 10. The length of this line coupled with its small cross-section are the primary reasons for the signal attenuation. Therefore, in typical multi-GHz systems, interconnects must be kept short to minimize propagation delays as well as signal attentuation.

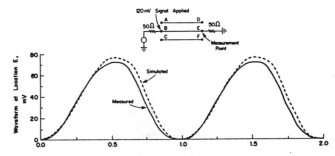

Figure 13. Measurements and lossy-line simulations (using Mayo's EM modeling package) of voltage waveforms at the far end of a 6.36" microstrip line. Input is a 1 GHz, 120 mV, 300 psec risetime pulse train.

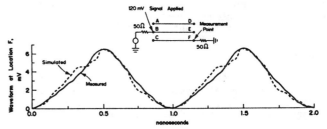

Figure 14. Measurements and simulations (using Mayo's EM modeling package) of crosstalk at the far end of a microstrip line adjacent to a 6.36" actively driven line. The input pulse train to the driven line has a GHz period, 120 mV amplitude and 300 psec risetime.

Figures 13 and 14 show comparisons between the measured and simulated waveforms for the far end of an actively driven and passively coupled 6.36 inch microstrip line. The lossy version of the Mayo Waveform Simulator was used for this simulation. The input signal, which was applied to the middle line, was a pulse train with a frequency of 1 GHz, an amplitude of 120 mV, and a risetime of 300 psec. The voltage response at the far end of the active line is shown in Figure 13. As expected, the amplitude of the signal has been reduced from 120 mV to approximately 70 mV due to resistive loss. The Mayo Waveform Simulator accurately models the amplitude of the output signal.

Figure 14 shows the far-end crosstalk of one of the passive lines adjacent to the driven line. The amplitude of the crosstalk is 6 mV, (less than -20 dB), even though the coupled line length is over 6 inches. This is one of the advantages of the TFML technology. The small dielectric thicknesses significantly reduce line-to-line coupling. Note that the model accurately predicts the magnitude of cross-talk. The discrepancy between the shapes of the measured and simulated lines may be attributed to either noise in the measurement, error introduced by using Fast-Fourier Transforms to convert between the time and frequency domains in the Waveform Simulation algorithm, or a combination of both.

Conclusion

The preliminary frequency, time, and digital domain testing and the electromagnetic simulations indicate that the Honeywell TFML technology is suitable for high performance digital systems that require operation at frequencies greater than 1 GHz.

The return loss was typically -20 dB which indicates the ability to design and fabricate controlled impedance lines with Honeywell's TFML tehcnology. The 1 and 2 GHz digital domain test measurements did not appear to be affected significantly by the reflections. The insertion loss of approximately -2.5 dB/inch at 9 GHz is acceptable for relatively short signal interconnect lines which, in any event, are required to minimize interchip propagation delays in a functioning system. Predictions from the Mayo EM modeling package show excellent agreement with measured signal and crosstalk waveforms. The crosstalk between adjacent lines is less than -20 dB even when the coupled line length is very long (>6").

Acknowledgement

The authors wish to acknowledge the help of J. Peterson who fabricated the test substrates. This research was supported in part under F33615-86-C110 from U.S.A.F. Wright-Aeronautical Lab.

References

1. R. Jensen, J. Cummings, and H. Vora, "Copper/-Polyimide Materials Systems for High Performance Packaging", IEEE Trans. Comp., Hybrids, and Manuf. Technol., Vol. CHMT-7, pp. 384-393, 1984.

2. R. J. Jensen, "Recent Advances in Thin Film Multilayer Interconnect Technology for IC Packaging", in Proceedings of the Symposium on Polymeric Materials for Electronic Packaging and High Technology Applications, J. R. Susko, R. W. Snyder, and R. A. Susko, eds., Proceedings Vol. 88-17, Electrochemical Society: Pennington, N.J., 1988.

3. C. J. Bartlett, J. M. Segelken, and N. A. Teneketges, "Multichip Packaging Design for VLSI-Based Systems", IEEE Trans. Comp., Hybrids, and Manuf. Technol., Vol. CHMT-12, pp. 647-653, 1987.

4. L. M. Levinson, C. W. Eichelberger, R. J. Wojnarowski and R. O. Carlson, "High Density Interconnects Using Laser Lithography", International Society for Hybrid Microelectronics Proceedings, pp. 301-306, 1988.

5. T. Watari, and H. Murano, "Packaging Technology for the NEC SX Supercomputer", IEEE Trans. Comp., Hybrids and Manuf. Technology, Vol. CHMT-8, pp. 462-467, 1985.

6. J. T. Pan, S. Poon and B. Nelson, "A Planar Approach to High Density Copper-Polyimide Interconnect Fabrication", International Electronics Packaging Society Proceedings, pp. 174-189, 1988.

7. R. Sainati, S. L. Palmquist, and T. J. Moravec, "Packaging High-Speed Multigigahertz GaAs Digital Integrated Circuits", Microwave Systems News and Com. Tech., Vol. 18, Part 1: No. 1, 94-96 Part 2: No. 2, 68-76 (1988).

8. K. Jayaraj, T. Moravec, R. Jensen, F. Belcourt, and R. Sainati, "High Speed, High Density, Low Loss Interconnections for VLSI, VHSIC, and GaAs on a Variety of Substrates", Electronic Materials and Processing, Proceedings of the First Electronic Materials and Processing Congress, Chicago, Il, Sept. 1988, ASM International: Metals Park, OH, 1988; pp. 111-118.

9. K. S. Olson, G. W. Pan, and B. K. Gilbert, comments on "Transient Analysis of Single and Coupled Lines with Capacitively-Loaded Junctions", IEEE Transactions on Microwave Theory and Techniques, MTT-35 (10), pp 929-930, 1987.

10. B. K. Gilbert and G. W. Pan, "The Application of Gallium Arsenide Integrated Circuit Technology to the Design and Fabrication of Future Generation Digital Signal Processors; Promises and Problems", Proceedings of the IEEE, 76(7), pp 816-836, 1988.

TRANSMISSION LINES FOR WAFER SCALE INTEGRATION

H. Stopper

Mosaic Systems, Inc.
1497 Maple Lane
Troy, MI 48084, USA

Abstract

Lossy transmission lines are as fast as loss-less transmission lines if they do not exceed a critical distance. The critical distance is larger for non-homogeneous lines than for homogeneous lines. Lines for wafer scale integration can be designed such that they fall into the high-speed regime and still have excellent noise characteristics.

Introduction

One of the many intriguing aspects of wafer scale integration (WSI) is the signal transmission delay reduction which can be gained from system size reduction. However, this advantage would be lost if the signal delay per line length on a wafer would be permitted to significantly exceed the delay per line length attained in more conventional packaging technologies, or if the physically available density were inextricably coupled with unmanageable levels of cross-talk [1]. It is the purpose of this paper to analyze the magnitude of these potential problems and to propose designs which actually realize the latent speed performance of WSI.

A general investigation of homogeneous and non-homogeneous, serially lossy TEM transmission lines will lead to the establishment of key parameters which govern transmission performance. Optimization of these parameters by transmission line design will be developed. Experimental verification of the basic conclusions will be given.

Serially Lossy Transmission Lines

It is assumed with justification pending that the transmission lines in the wafer scale environment can be treated as TEM transmission lines with frequency-independent series resistance, series inductance, and parallel capacitance. The analysis starts with the two-wire transmission line (micro-strip conductor over ground plane) and is directed first at the signal delay along a homogeneous line.

Delay of the Homogeneous Line

Homogeneous means that the parameters resistance, inductance, and capacitance normalized per length do not vary over the length of the line. Universal analyses of this type of line can be found in textbooks, but the resulting equations are complex and usually still need to be expanded from the special case of a voltage step applied to an infinitely long line to the realistic case of a finite transition time pulse applied to a finite length line. Simulation, on the other hand, proves to be very effective for individual cases but not too helpful to develop insight unless there is some guidance toward the most interesting cases to simulate. A suitable compromise appears to be a simplified general analysis whose inevitable gaps are filled by simulation results.

Presupposing familiarity with the loss-less transmission line, a lossy line can be approached by inserting discrete resistors between subsections of a loss-less line (Fig. 1). The number of subsections of the finite-length line be n, the value of each inserted resistor be R/n, the total line inductance be L, and the total line capacitance be C. The

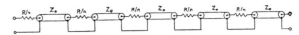

Fig. 1: Cable made lossy by discrete resistors

characteristic impedance of the (loss-less) line is known to be $Z_o = \sqrt{L/C}$, and the total signal delay is known to be $t_o = \sqrt{LC}$. As a signal travels down this line, it will be quasi-amplified at the end of each subsection according to the law of reflection for line discontinuities and attenuated at each subsection input due to the voltage divider consisting of the inserted resistor and the subsection input impedance which is equal to the characteristic impedance. What reaches the end of the line this way first shall be called the primary component of the output signal. It obviously travels with the speed characteristic for the loss-less line (speed of light) and has its amplitude multiplied by a factor which is the product of all the quasi-amplifications and attenuations, namely,

$$A_n = \left[\frac{2(R/n + Z_o)}{(R/n + Z_o) + Z_o}\right]^{n-1} \cdot [2] \cdot \left[\frac{Z_o}{R/n + Z_o}\right]^n \qquad (1)$$

With the definition of an attenuation constant

$$\alpha = \frac{R}{2 Z_o} \qquad (2)$$

Eq. (1) can be rewritten as

$$A_n = \frac{1 + \alpha/n}{1 + \alpha/2n} \cdot \frac{2}{(1 + \alpha/n)^n} \qquad (3)$$

This equation will be used later to determine the accuracy of finite-subsection approximation. The amplitude factor for the lossy line with truly distributed resistance can now be seen to be

$$A = \lim_{n \to \infty} A_n = 2 e^{-\alpha} \qquad (4)$$

The summated effect of all secondary reflections on the line output signal shall be called the secondary component. For digital signals, the secondary component facilitates the transition from a level established by the primary component to the final level. The primary component will be dominant if A is close to unity, but the secondary component will be dominant if A is close to zero.

There must be a specific value for **A** such that A=1:

$$\alpha_c = \ln 2 \qquad (5) \qquad \left(\frac{R}{Z_o}\right)_{\alpha = \alpha_c} = 2\ln 2 \approx 1.4 \quad (6)$$

Since R and α are proportional to the length of the line, Eq. (5) and (6) imply a specific length which shall be defined as the critical distance d_c. With R as the absolute resistance of a line with the arbitrary length d_o, or R' as the relative resistance per length,

$$d_c = \frac{2(\ln 2)Z_o}{R/d_o} = \frac{2(\ln 2)Z_o}{R'} \approx 1.4 \frac{Z_o}{R'} \qquad (7)$$

This means that an open-ended lossy line can carry a signal over a critical distance d_c with maximum speed and without loss of amplitude, which also means without transition time degradation. This fact can be interpreted as an alternate method of "termination": An (originally) loss-less line can not only be operated conventionally with a parallel destination

Note: The author is presently with the Image Processing Technology Laboratory of ERIM in Ann Arbor, MI.

Reprinted from *Proc. 1st Int. Conf. Computer Technol., Systs., Appl. (COMPEURO)*, pp. 551-554, 1987.

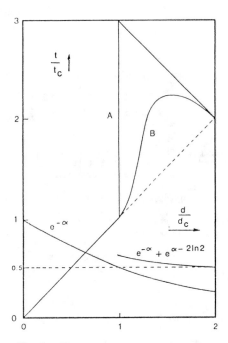

Fig. 2: Hypothetical delay at zero transition time. Primary (A) and complete (B) wave form

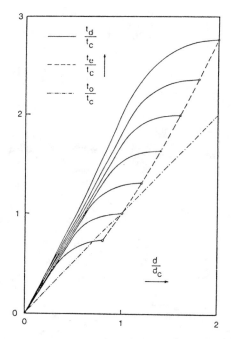

Fig. 3: Normalized digital delays (short range) obtained by simulation

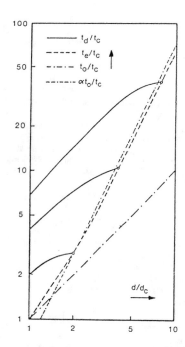

Fig. 4: Normalized digital delay (long range) obtained by simulation

terminator (maximum speed, but d.c. power dissipation) or a serial source terminator (no d.c. power dissipation, but maximum speed only for the end of the line), but also with a "distributed terminator", combining the advantages of both. Lumped terminators are ideally equal to Z_o, distributed terminators to $1.4\ Z_o$.

This surprising potential of the lossy line has been recognized before in ref. [2], and it has been reported there that such lines worked well in a certain thin film module over the range of $0.5\ d_c \leq d \pm 1.4\ d_c$. In the WSI case, however, longer lines cannot be avoided and specific attention must be paid to the effects ot the delay definition and of the finite pulse transition time t_T.

Fig. 2 curve A shows the digital delay (time which a binary signal requires to cross the 50% level) for a distance d along a line which is exactly $2d_c$ long, provided that only the primary component is considered and that $t_T=0$. t_c is defined as the delay of a line which is d_c long (lossy or loss less). Adding the secondary component changes curve A to curve B. While the relative delay of the lossy line appears to increase rapidly for $d>d_c$, it reaches again unity at $d=2d_c$.

Curve B shows only the underlying characteristic of the real delay curves for $t_T>0$. The influence of t_T can not be conveniently normalized. Delays generally increase with t_T but the curves are flat within 10% for commonly used transition times. Fig. 3 shows the simulation results for a number of lines from $0.8d_c$ to $2d_c$ total length which have been obtained for $t_T=t_e$, where t_e is the delay to the end of the lossy line. It appears that t_e can be approximated as

$$t_e \approx t_c \left(\frac{d_o}{d_c}\right)^{1.46}, \quad d_c \leq d_o \leq 2d_c \tag{9}$$

Lines shorter than d_c appear in total to be faster than light, and all other lines have at least a tail where signals travel faster than light. This phenomenon is nothing but the impact of gain variation on the digital delay for finite transition times. Beyond $2d_c$, the primary component vanishes quickly and the secondary component can be well approximated by the RC-line (Thomson cable). An exact

analysis of the open-ended RC-line can be found in textbooks and from that analysis the following simple approximation of the digital delay can be derived (details omitted):

$$t_e \approx \frac{RC}{2} = \alpha\ t_o = t_c\ (2\ln 2)\left(\frac{d_o}{d_c}\right)^2 \tag{10}$$

Eq. (10) refers only to end-of-line delays t_e. Intermediate delays t and end-of-line delays t_e as obtained from simulation are shown in Fig. 4, again for $t_T=t_e$. The overall results have been summarized in Fig. 5 as the ratio of t_e over the corresponding delay t_o of the loss-less line ($s_c=1$).

Delay of the Non-homogeneous Line

The piece-wise approximation of the lossy line in Fig. 1 can be modified such that Z_o of the first subsection be called Z_A and that both R/n and Z_o for the following subsections are increased from subsection to subsection by a factor k, which implies that the attenuation factor per subsection remains constant:

$$\frac{\alpha}{n} = \frac{k^i \cdot R/n}{2\ k^i Z_A} = \frac{1}{n} \cdot \frac{R}{2Z_A} \tag{11}$$

Eq. (1) and (3) change now to Eq. (12) and (13) respectively:

$$A_n = \left[\frac{2(R/n+Z_A)k^i}{(R/n+Z_A)k^i+Z_A k^{i-1}}\right]^{n-1} \cdot [2] \cdot \left[\frac{Z_A k^i}{(R/n+Z_A)\ k^i}\right]^n \tag{12}$$

$$A_n = \frac{1/2+1/2k+\alpha/n}{1+2\alpha/n} \cdot \frac{2}{(1/2+1/2k+\alpha/n)^n} \tag{13}$$

The relations become clearer if one substitutes

$$\frac{1}{k} = 1 - \frac{2\beta}{n} \tag{14}$$

and obtains

$$A_n = \frac{1+(\alpha-\beta)/n}{1+2\alpha/n} \cdot \frac{2}{(1+(\alpha-\beta)/n)^n} \tag{15} \quad A = \lim_{n\to\infty}A_n \simeq 2e^{-(\alpha-\beta)} \tag{16}$$

The difference between the non-homogeneous and the

homogeneous line is then that the attenuation factor α is reduced by an amount β. If Z_o of the last subsection is called Z_B, Eq. (14) can be transformed into

$$\frac{Z_B}{Z_A} = k^{n-1} = \lim_{n \to \infty} k^{n-1} = \lim_{n \to \infty} \frac{1}{(1 - 2\beta/n)^{n-1}} = e^{2\beta} \quad (17)$$

This means that the characteristic impedance grows exponentially over the length of the line from Z_A to Z_B with a growth factor

$$2\beta = \ln\frac{Z_B}{Z_A} \quad (18)$$

The critical distance can now be redetermined such that A=1 by Eq. (16) instead of Eq. (4), and a "stretch factor" s_c can be obtained by dividing the new critical distance over the old one:

$$s_c = 1 + \frac{\ln Z_B / Z_A}{2 \ln 2} \quad (19)$$

With $Z_B/Z_A=4$, for instance, $s_c=2$. This means that ideal transmission conditions are now found for lines with the length $2d_c$ rather than d_c.

In practical designs, it may be desirable to grow Z_o not exponentially but rather in one or two discrete steps, which will reduce the stretch factor slightly. Skipping the details of the derivation, Table 1 gives some examples of stretch factors for lines with discrete Z_o-steps.

Since β subtracts from but does not divide into α, the stretch factor decreases with increasing line length but not as drastically and as far as suggested by Eq. (16) because of the not yet considered secondary component. Fig. 5 shows stretch factors obtained by simulation and their effect on t_e as a function of d_o. The overall result is that lossy lines can be made quite effective up to at least $3d_c$ by suitable impedance control.

In order to provide proper distributed termination for very short lines, the above process can be reversed: inverse impedance ratios shrink d_c.

$\dfrac{Z_B}{Z_A}$	s_c		
	One Step	Two Steps	Exponential
2	1.42	1.46	1.50
3	1.59	1.69	1.80
4	1.70	1.83	2.00

Table 1: Stretch factors for discrete impedance steps

* only 1.63 for ratios 1:1.5:3 versus 1:1.73:3

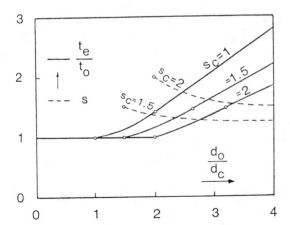

Fig. 5: Stretch factors and relative delays of homogeneous and non-homogeneous lossy lines versus the homogeneous loss-less lines

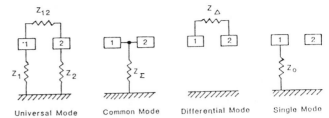

Fig. 6: Partial characteristic impedances and operating modes of a 3-conductor line

Cross-talk

Cross-talk between parallel lines can be analyzed by looking at the aggregate of talking and listening lines as a single multi-wire transmission system, in which the characteristic impedances of the several two-wire transmission lines have been replaced by partial characteristic impedances strung between the several wires of the multi-wire transmission line [3]. Certain complications in this analysis which stem from velocity dispersion among the partial waves in the multi-wire system do fortunately not apply in the WSI environment. If one approximates the general cross-talk problem by looking only at a three-wire transmission system, the above method can be explained in terms of the more familiar common/differential mode line analysis.

The relationships between voltages and impedances in this model (Fig. 6) are as follows:

$$u_1 = u_\Sigma + \frac{u_\Delta}{2} \quad (20) \qquad u_2 = u_\Sigma - \frac{u_\Delta}{2} \quad (21)$$

$$u_\Sigma = \frac{u_1 + u_2}{2} \quad (22) \qquad u_\Delta = u_1 - u_2 \quad (23)$$

$$Z_1 = Z_2 = 2 Z_\Sigma \,(24) \quad Z_{12} = \frac{4 Z_\Sigma Z_\Delta}{4 Z_\Sigma - Z_\Delta} \,(25) \quad Z_o = Z_\Sigma + \frac{Z_\Delta}{4} \,(26)$$

Reference [3] shows that the maximum cross-talk from conductor 1 to conductor 2 for loss-less lines is irrespectively of pulse rise time and line length solely determined by the voltage divider Z_{12}/Z_2. With the abbreviation

$$\gamma = \frac{4 Z_\Sigma - Z_\Delta}{2 Z_\Delta} \quad (27)$$

one derives from this voltage divider that the maximum cross-talk is

$$\frac{Z_2}{Z_{12} + Z_2} = \frac{2 Z_\Sigma}{4 Z_\Sigma Z_\Delta / (4 Z_\Sigma - Z_\Delta) + 2 Z_\Sigma} = \frac{1}{1 + 1/\gamma} \approx \gamma \quad (28)$$

The more complex case of the lossy lines can again be approached by inserting resistors R/n between n subsections of loss-less lines and by then letting n grow to infinity, except that this method has now to be applied separately to the common and differential mode waves. Since the resistors with values R/n are inserted into the conductors 1 and 2, their effective value for the common mode is only half as large, and their effective value for the differential mode is twice as large. Hence, the resulting attenuation constants are

$$\alpha_\Sigma = \frac{1}{2} \frac{R}{2 Z_\Sigma} = \frac{R}{4 Z_\Sigma} \quad (29) \qquad \alpha_\Delta = 2 \frac{R}{2 Z_\Delta} = \frac{R}{Z_\Delta} \quad (30)$$

With the help of Eq. (26) and (27), these attenuation constants can be related to the original attenuation constant and to each other:

$$\alpha_\Sigma = \frac{1 + \gamma}{1 + 2\gamma} \alpha \quad (31) \qquad \alpha_\Delta = (1 + \gamma) \alpha \quad (32)$$

$$\alpha_\Delta = (1 + 2\gamma)\, \alpha_\Sigma \qquad (33)$$

Since the cross-talk voltage is the difference between the common and differential mode voltages according to Eq. (21), the attenuation difference of Eq. (33) means that the lossy line generates additional cross-talk over and above the cross-talk of the comparable loss-less line. However, this addition is usually only a fraction of γ. Example: Let conductor 1 be driven by the signal source, while conductor 2 is shorted to ground (at the line input). At the critical distance d_c the maximum cross-talk generated by attenuation difference will be

$$\frac{U_2}{U_1} = \frac{1}{2}e^{-\ln 2} - \frac{1}{2}e^{-(1+2\gamma)\ln 2} = \frac{1}{4}(1 - e^{-2\gamma \ln 2}) \approx 0.35\gamma \qquad (34)$$

Thus, if one succeeds to make γ a very small number by physical design, parallel line cross-talk can be ignored in WSI even though lossy lines are somewhat inferior to loss-less lines in this regard.

WSI Transmission Line Design

If the relative permittivity of the insulator is 4 (SiO_2) and the resistivity of the conductors is 30 m$\Omega\mu$m (Al), the electrical parameters of a microstrip-and-groundplane line (Fig. 7) can be computed as follows:

$$Z_o = 189\frac{gh}{w}\ \Omega \qquad (35) \qquad\qquad C' = 0.035\frac{w}{gh}\frac{pF}{mm} \qquad (36)$$

$$\frac{1}{g} \approx 1 + \frac{2h}{\pi w}\left[1 + \ln(1 + \frac{\pi w}{2h})\right]\ (37) \qquad R' = \frac{30}{\frac{w}{\mu m}\frac{b}{\mu m}}\frac{\Omega}{mm} \qquad (38)$$

A correction for Z_o and C' is necessary if the capacitance versus crossing lines adds significantly to the basic line capacitance. In wafer designs as used in ref. [4], C' increases typically by 50%, and Z_o decreases accordingly by 18%.

It should be remembered that Z_o is the characteristic impedance of the underlying loss-less line and therefore not equal to the load on the line driver, which is much smaller.

The nomogram in Fig. 7 represents Eq. (35) and (36). w and h have been limited to the range which has actually been used on wafers designed per ref. [4]. In a given design, w can be varied extensively, but h is usually limited to the two values of an orthogonal wiring matrix.

With the above tools, various line designs can now be tried. The overall result is that the critical

distance can be pushed to at least 35mm, and that the impedance ratio can be expanded up to 4 where needed. Therefore, distances up to 100mm can be traversed almost with the speed of light (see Fig. 5).

Space does not permit to develop the cross-talk factor from Eq. (28), which is a function of w, h, and the spacing between the micro-strips. As a rule of thumb, γ is smaller than 3% if Z_o is smaller than 50Ω. Therefore, noise control can easily be reconciled with the above method of designing for speed.

Experimental Verification

Macro Experiments

A simple macro experiment was executed with ordinary 50Ω coax cable which had been made artificially lossy by insertion of discrete resistors as in Fig. 1. The length of the cable was 1m, the number of sections was five. Computing A from Eq. (3) with n=5 versus Eq. (4) shows that the inserted resistors have the same attenuation effect as truly distributed resistance if their total resistance R is made 10% smaller. Thus, the cable length could be made equal to the critical distance by making R=1.25Z_o=62.5Ω. It was observed that a pulse with t_T=10ns applied to the line input arrived indeed at the end of the line without noticeable distortion and with exactly the same delay (5.7ns) which was measured for an end-terminated but otherwise unmodified cable of the same length.

In order to demonstrate the stretch factor, a line with three sections of 1m length each was built such that the first section had three, the second section had two, and the last section had only one modified coax cable connected in parallel. R was chosen such that the length of the total structure (3m) represented 1.63 d_c, which means 34Ω per cable per meter. Again, delay equality (17ns) and transition time preservation (10ns) could be verified.

Micro Experiments

A 44mm long line was built on a wafer. Its resistance and capacitance were measured at low frequencies to be 42Ω and 11pF, respectively. With no cross-over capacitance involved, Z_o followed from Eq. (35) and (36) or directly from the nomogram with 26.4Ω. Eq. (7) delivers d_c=38.3mm, and, hence t_o=$Z_o C' d_c$ =0.255ns. The total line delay per Eq. (9) should then be 0.31ns if t_T=0.3ns. The actual delay was measured to be 0.34ns with t_T=1ns.

These results mean that at least down to 1ns the resistance of the microstrip (b=2μm) was not raised by the skin effect, that no significant losses occurred in the dielectric, and that the ground plane which was also 2μm thick did not contribute to the line resistance. A second experiment with a 1μm ground plane, however, raised the delay to 0.48ns.

Numerous other experiments with t_T=2ns and d>2d_c showed no deviations from predicted delay values, even with a thinner ground plane of 1μm.

References

[1] J.F. McDonald et al, "The Trials of Wafer Scale Integration," IEEE Spectrum, October 1984, pp. 32-39.

[2] C.W. Ho et al, "The Thin-Film Module as a High-Performance Semiconductor Package," IBM Journal of Research and Development, Vol. 26, No. 3, May 1982, pp. 286-296.

[3] H. Stopper, "Leitungen in Digitalrechnern," Dissertation, Technische Universität Berlin, 1968, D83.

[4] H. Stopper, "A Wafer with Electrically Programmable Interconnections," Digest of the ISSCC, 1985, pp. 268-269.

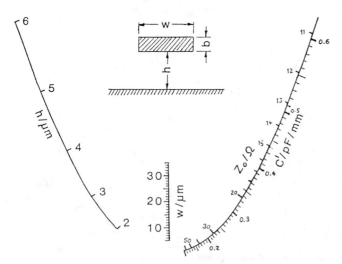

Fig. 7: Nomogram for characteristic impedance and capacitance per length as function of line geometry

WAVEGUIDES AS INTERCONNECTS FOR HIGH PERFORMANCE PACKAGING

BY

S.E. Schacham[1,4], T.G. Tessier[2], H. Merkelo[3], L.-T. Hwang[1], and I. Turlik[1,2]

[1] Microelectronics Center of North Carolina (MCNC),
P.O. Box 12889, Research Triangle Park, NC 27709.

[2] MCNC Resident Professional, Bell Northern Research
P.O. Box 13478, Research Triangle Park, NC 27709.

[3] Dept. of Electrical and Computer Engineering,
University of Illinois, Urbana, IL 61801-2991.

[4] Department of Electrical Engineering,
Technion- Israel Inst. of Technology, Haifa 32000, Israel.

ABSTRACT

The well known advantages of optical interconnects include high carrier frequency, low attenuation, high noise immunity, and low crosstalk. In this paper, a detailed comparison between optical and electrical interconnects is presented, with the emphasis on advantages and drawbacks of utilization of optical links. The impact of attenuation, crosstalk, and fanout on signal integrity is discussed. According to our results, at frequencies below 10 GHz there is no obvious advantage of using optical interconnects for signal distribution. Therefore, in high speed digital systems, a detailed design analysis has to be performed before a decision is made as to which interconnection scheme is superior for a given application. Similarly, the reduction of the number of lines and ports by multiplexing and demultiplexing is limited by the electronics, for both electrical and optical approaches. Retention of pulse shape eventually becomes the key parameter that forces the transition to the optical channel at ultra high frequencies. Approaching 10 ps rise-time, dispersion and attenuation of electrical signal is detrimental, while the optical waveform is basically unaffected, making the transition to optical interconnects essential. The implementation of a polyimide optical waveguide in the MCNC package is discussed.

Introduction

As both silicon and GaAs based technologies drive device speeds into the gigahertz frequency range, and rise-times down to small fractions of a nanosecond, interconnections assume ever increasing importance. Digital systems supported by these technologies are designed to operate in the multigigabit data regime, where synchronous clock and data distribution are extremely difficult to maintain. This means that the interconnection lines have to be treated as high frequency transmission systems. Skin effect, discontinuities, and dielectric imperfections cause severe pulse distortions and attenuation in addition to randomize propagation delays. Thin film multichip package technology offers high density interconnection capability and design trade-off flexibility. However, conductor conductivity, cross section, length, and driver/ receiver characteristics must be coordinated in order for the package interconnections to function properly. The increased density of components within integrated circuits requires hundreds of electrical interconnects for each chip, and thousands of I/O ports for the entire package. Optical interconnections offer the potential of alleviating many of these obstacles. The advantages of optical links are common knowledge, the most relevant ones are: low attenuation, noise immunity, low crosstalk, extremely high carrier frequencies (multiplexing capability) and propagation delays independent of fanout. Optoelectronics offers an important supplement to the electrical signal transmission, but poses its own set of technological challenges.

In the following sections a comparative analysis of conventional conductors versus optical interconnects is presented. The discussion is focused on limitations of electrical transmission lines, and applicability of the short range links in packages of high performance digital systems, operating well above clock-rates of 1 GHz. An effort is presented to define the conditions at which the "optical alternative" benefits the system performance. Finally, the implementation of optical waveguides with polyimides, as part of the MCNC high performance package is described.

Attenuation

Optical techniques promise interconnects that overcome important limitations of conventional electrical interconnects [1-4]. The attenuation as a function of frequency for a typical thin film microstrip line is shown in figure 1. The line attenuation per unit length is [5,6]

$$\alpha = \frac{R(\omega)}{2 \cdot Re(Z_o)} \tag{1}$$

with $Re(Z_o)$ is the real part of the characteristic impedance. Since R is a function of frequency, this attenuation incorporates the conductor series resistance loss factor including the skin effect, but not the dielectric loss. The characteristic impedance of a transmission line is given by

$$Z_o = \sqrt{\frac{R + jX + j\omega L_{ext}}{j\omega C_{ext}}} \tag{2}$$

where R and X are the real and imaginary parts of the internal impedance of the line, and L_{ext} and C_{ext} are the external inductance and capacitance of the line. At high frequencies, Z_o approaches $R_o = \sqrt{L_{ext}/C_{ext}}$. The attenuation is calculated for a polyimide microstrip homogeneously capped by the same material, with a 4 μm thick and 8 μm wide copper conductor of nominal $R_o = 58\ \Omega$. For frequencies below 0.2 GHz the attenuation increases due to decreasing characteristic impedance caused by an increasing capacitive reactance. With further increase in frequency, the capacitive reactance is compensated by the inductive reactance, and the signal attenuation is determined by the line resistivity. Above 1 GHz, there is a substantial increase in attenuation due to the skin effect as well as a decreasing internal inductance. Comparing this result to the optical alternative, it is obvious that the attenuation of communication quality optical fibers is lower by many orders of magnitude. However, the use of such fibers in a high performance package is highly impractical. The density and complexity of thin film multichip packages rule out the insertion and coupling of optical fibers, but allow the fabrication of optical waveguides directly in the package. The lowest losses for these waveguides are about 0.01 dB/cm, and are obtained with laser annealed polycrystalline ZnO. However, typical losses are in the range of 0.1-5 dB/cm [7]. Recently transparent polyimides have been introduced for waveguide applications, with attenuations in the range of 0.3-4 dB/cm [8,9]. In addition to the attenuation, optical interconnects suffer from various coupling losses (i.e. coupling of the light source to the waveguide and the waveguide to the detector, or coupling/ splitting between waveguides). It is important to point out that the losses of optical links are *not* frequency dependent throughout the whole signal frequency spectrum . On the other hand, the attenuation of electrical interconnects increases with frequency, and therefore eventually becomes larger than its optical counterpart.

Crosstalk

The effects of another deterrent, namely crosstalk, are investigated . For optical waveguides, if the cladding is properly designed, the crosstalk practically does not exist. Figure 2 shows the crosstalk between two transmission lines with the cross section of 8×4 μm^2 and 24 μm apart, embedded in an insulator with a relative permittivity of 3.5. Both series (figure 2a) and parallel (figure 2b) terminations are equal to the nominal high frequency characteristic impedance (58 Ω). For this particular geometry the crosstalk as a function of frequency reaches a plateau at a level of about -20 dB.

Sensitivity and Fanout

The effect of fanout on electrical interconnect was analyzed by Hangen et al. [1]. The increased fanout results in a lower effective characteristic impedance, e.g. a 50 Ω line is effectively reduced to a 37 Ω line in a ceramic package or to 34 Ω in polyimide for a fanout of 5. Larger fanout requires a larger driving power, and it results in slower signal response. The dependence of maximum data rate on fanout is a function of the line length and termination. Hansen et al. found (in their theoretical analysis) that for lossless lines the fanout is independent of frequency, and the longer the line- the larger the fanout it can sustain [1]. In practice, line losses result in drastic deterioration of signal with increased fanout and in severe attenuation for such long line lengths (50 cm). Figure 3a shows the signal waveform derived for a 2 cm copper line of the above cross section with a fanout of 1, while figure 3b shows the same for a fanout of 2, both waveforms were calculated for 8 GHz. The capacitance of each load was taken to be 3 pF, following Ref. 1. The peak level of the output waveform is below 50% of the input signal, and it spreads out to more than 5 times its original width for both parallel and series terminations. Already at a frequency of 1 GHz the additional load causes a noticeable signal distortion, and the attenuation of a 10 cm long line is such that even for a single load the output signal at 8 GHz is below 50% for both terminations.

The derivation of the effect of optical fanout on the maximum operating rate of an optical interconnect is based on the reduction of the output power, equally split among N loads. Since the noise increases with bandwidth, maintaining a specified bit error rate (BER) at a reduced output power implies a reduction in data rate. Assuming a Gaussian distribution, the BER is given by [10]

$$P(E) = \frac{1}{\sqrt{2\pi}} \int\limits_{Q}^{\infty} e^{-x^2/2} dx \tag{3}$$

where

$$Q = \frac{|D - i|}{<i_n^2>^{0.5}} \tag{4}$$

with D defined as the decision level (current level above which the state is "1" and below which it is "0"), i is the expected signal current and $<i_n^2>^{0.5}$ is the rms noise.

For long distance communications a BER of 10^{-9} is usually assumed for a frequency of 1 GHz. When manipulating words rather than single bits a much lower BER must be imposed, typically 10^{-17} at 1 GHz [2]. Figure 4 shows the BER as a function of Q for the 10^{-9} to 10^{-19} range. Note that due to the exponential nature of equation (3), this 10 order of magnitude change in the BER is obtained by an increase in Q of only 50%.

The analysis of the various noise sources involved in the detection of optical signals is elaborated in reference [10]. It is shown that in digital receivers the noise is dominated by circuit noise, $<i^2>_c$, while noise due the detected signal is negligible. For a p-i-n photodiode, the optical power required to achieve a desired error rate (i.e. the receiver sensitivity) is determined by Q as

$$\eta \overline{P} = \left(\frac{h\nu}{q}\right) Q <i^2>_c^{0.5} \tag{5}$$

where $h\nu$ is the photon energy. For a wavelength of 825 nm, $h\nu/q = 1.5$ V, and Q=8.5 for BER of 10^{-17} .

For a well designed FET front end circuit $g_m R_L \gg 1$ where g_m is the transconductance of the FET, and with a small gate leakage current (provided that its shot noise is negligible), the minimum circuit noise is given by reference [10] as

$$<i^2>_c = 4k \, T\Gamma \frac{(2\pi C_T)^2}{g_m} I_3 \cdot B^3. \tag{6}$$

The total capacitance C_T is the sum of the FET gate capacitances, the detector and stray capacitances. The numerical factor Γ is a noise factor associated with channel thermal noise and gate induced noise of the FET. It is about 0.7 for Si and 1.1 for GaAs FETs. The definite integral I_3 depends only on the pulse shape, and for a rectangular pulse it is equal to 0.03.

The sensitivity of p-i-n photodiodes as a function of data bit rate, up to 50 GHz, is shown in figure 5. As can be derived from equations 4 and 5, the minimal optical power required at the detector is proportional to $B^{1.5}$. In addition we introduced a frequency dependent BER. Assuming a constant number of errors per unit time, the BER is linearly proportional to the inverse of the bandwidth. Since the change in Q is minimal, this effect is secondary. The parameters used in equation 5 for deriving the noise generated by the FET were based on data of state-of-the-art GaAs transistors, capable of operating at these high frequencies. These were g_m =50 mmho and C_T=1.5 pF, rendering

$$\eta \bar{P} = 4.68 \cdot 10^{-8} \cdot Q \cdot B^{1.5} \tag{7}$$

when B is expressed in GHz. The quantum limit curve reflects the minimum average number of photons per pulse required to assure that the probability of zero photoelectrons generated is below the dictated BER (e.g. 39 for BER= 10^{-17} as compared to 21 for 10^{-9}). We also added a third curve representing the improved sensitivity attainable with integration of the detector and the FET on a single chip, thereby reducing the total capacitance to 0.5 pF.

The mathematical process of evaluating the effect of fanout on the maximum bit rate is quite intricate since the relationships between the detected power, Q, and the BER are not explicit. The computation is based on the assumption that the system is operating at its maximum data rate for a one to one link (N=1). The optical power is now split between N (>1) receivers equally. Knowing the reduced power, the data bit rate is derived. Figure 6 presents the maximum rate as a function of fanout, starting with 0.1, 1 and 10 GHz for N=1, and assuming no excess losses, neither through the waveguide nor due to coupling and splitting. A similar analysis is demonstrated in figure 7, assuming 4 different levels of total losses, starting with 10 GHz for N=1. The effect of these losses is quite devastating. When the excess losses (including coupling, splitting and total attenuation) are 4 dB per link, the maximum operating data rate is reduced more than 3 orders of magnitude for a fanout of 10.

Maximum Bit Rate

As discussed in the previous section, the only limitation on the maximum operating data bit rate of an optical interconnect is the power reaching the receiver. Improving coupling efficiency, reducing waveguide attenuation and development of more powerful sources will enable the use of higher and higher bandwidths. The receiver circuitry can be replaced by a bipolar front end, in which the circuit noise increases as the square of the bandwidth rather than its cube for FETs [10]. Avalanche photodiodes also provide better signal to noise ratio [10,11], however they need power levels (tens of volts) unaffordable in a package.

For electrical transmission line, high frequencies become an ever increasing obstacle. The two major limiting parameters are signal attenuation and signal dispersion. Figure 8 shows the propagation distance of two short Gaussian pulses (5 ps and 10 ps FWHM) until the positive peak amplitude is reduced to 50% of the input level. The conductor is 100 μm wide, the relative permittivity of the dielectric is ε_r=3.5, and the nominal characteristic impedance is 50 Ω. In figure 8a the signals are propagating on a dispersion dominated (lossless) microstrip. The propagation distance is 70 cm for the 10 ps signal and 18 cm for the 5 ps one. The distortion of the output signal is significant. These distances reduce to 13.3 cm and 3.5 cm respectively when the substrate relative permittivity is ε_r=10. When conductor losses are introduced, the propagation distance is substantially reduced. Figure 8b shows this effect for a copper (σ=5.8$\cdot$10^7 mho/m) microstrip of the same structure embedded in a dielectric material with ε_r=3.5. As is demonstrated, the propagation distance is below 16 cm even for the 10 ps signal and 8.5 cm for the 5 ps signal. Practically this indicates that maximum length of electrical interconnects is limited to a few centimeters at frequencies well below 100 GHz.

Polyimide Waveguides

The implementation of optical interconnects into a high performance multichip module was studied. Since polyimide materials are used as the dielectrics for the thin film interconnects, various combinations of polyimides are being explored to establish their suitability as optical links. Figure 9 shows the cross section of one such waveguide. The core material is a DuPont PI2525, with refractive index n=1.70. The cladding is either SiO_2 or DuPont PI2566 with a refractive indices of n=1.48 and n=1.58. These materials and the related processing schemes are compatible with processes involved in the manufacturing of the MCNC high performance multichip package. Various cross sectional dimensions were tested, starting from 5×10 μm^2 and up to 100×20 μm^2. Typical losses measured for these waveguides were of 3 dB/cm level. Fluorinated polyimides, known to have lower attenuation [8], are currently under investigation as core material for these waveguides.

Conclusions

The comparison between optical and electrical interconnects involves a large number of parameters, the key ones are discussed in depth in this paper. We have shown that the attenuation of a transmission line is increasing with frequency due to the skin effect, and approaches 1 dB/cm at 10 GHz. The attenuation of an optical link also depends on the material used for implementing the waveguide, however it is independent of frequency. Fanout reduces the maximum bit rate throughout the spectrum for optical interconnects, while its effect is noticeable only in the GHz regime for a terminated electrical line. Other parameters such as power consumption, depend on the specific technology and devices employed [1,3,4]. Presently, the consumption for both techniques is comparable. Optical waveguides are definitely superior as far as crosstalk is concerned. However, we have shown that crosstalk levels in electric transmission lines are usually below -20 dB for the few GHz range.

The reduction of the number of I/O ports by multiplexing the signal over a single link seems to be a very attractive approach. Still, the technology is limited and the fastest multiplexers commercially available in the few GHz range, could be used equally well in electric connections. The realization of optical sources and detectors along with the waveguide for optical interconnects, along with all the supporting electronics, is complex and is not likely to be widely implemented presently. In spite of that, when the operating frequencies reach 10 GHz, it will be necessary to implement optical links to preserve signal integrity. In these frequency ranges, the retention of pulse shape over significant distances becomes the key parameter that forces the transition from electrical interconnects to the optical channel. The electrooptical components must be integrated in order to minimize the required area and to assure performance enhancement. At these high frequencies additional electronics will be required to generate and detect optical signals, maintaining an extremely high signal to noise ratio. At that point, the entire package configuration will be driven by the tradeoff between core components and supporting electronics. In the meantime, dielectric materials with low relative permittivity must be employed for electrical signal distribution, not only because of the shorter propagation delays but also since they yield the maximum length of electrical connections.

References

1. P.H. Hangen, S. Rychnovsky, A. Husain, and L.D. Hutcheson, Opt. Eng., 1986, pp. 1076-1085, vol. 25.

2. D.H. Hartman, *ibid* pp. 1086-1102.

3. R.K. Kostuk, J.W. Goodman, and L. Hesselink, Appl. Opt., 1985, pp. 2851-2858, vol. 24.

4. M.R. Feldman, S.C. Esner, C.C. Guest, and S.H. Lee, Appl. Opt., 1988, pp. 1742-1751, vol. 27.

5. W.T. Weeks, L.L. Wu, M.F. McAllister, and A. Singh, IBM J. Res. Develop., 1979, pp. 652-660, vol. 23.

6. C.S. Yen, Z. Fazarinc, and R.L. Wheeler, Proc. IEEE, 1982, pp. 750-757, vol. 70.

7. F.S. Hickernell, Solid State Tech., Nov. 1988, pp. 83-88, vol. 31.

8. R. Reuter, H. Franke and C. Feger, Appl. Opt., 1988, pp. 4565-4571, vol. 27.

9. R. Selvaraj, H.T. Lin, and J.F. McDonald, J. Lightwave Tech., 1988, pp. 1034-1044, vol. 6.

10. R.G. Smith and S.D. Personick, <u>Semiconductor Devices for Optical Communications</u>, Springer-Verlag, 1980, pp. 89-160.

11. T.V. Muoi, J. Lightwave Tech., 1984, pp. 243-267.

ATTENUATION

8 BY 4 µm² LINE

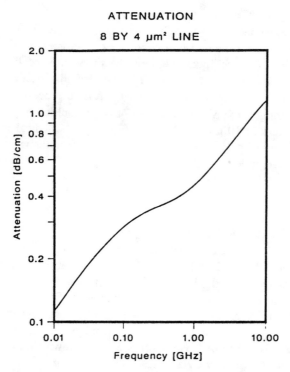

Fig. 1   Attenuation vs. data rate for a 8×4 μm^2 copper
microstrip, ε_r =3.5 dielectric, R_o =58 Ω.

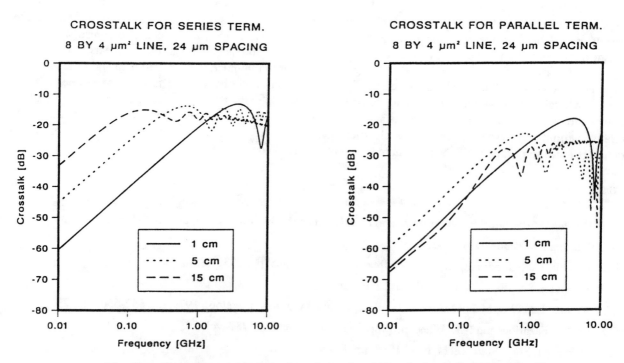

Fig. 2   Crosstalk vs. data rate for microstrip of Fig. 1, for 1, 5, 15 cm long lines.
(a) With 58 Ω series termination. (b) With 58 Ω parallel termination.

EFFECT OF FANOUT ON PULSE SHAPE

Length = 2 cm, Bit Rate = 8 GHz

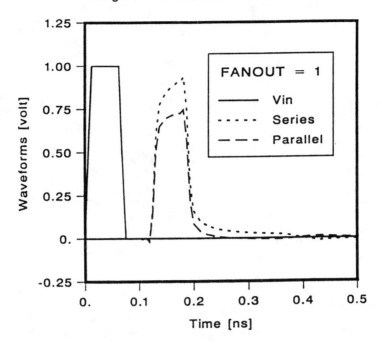

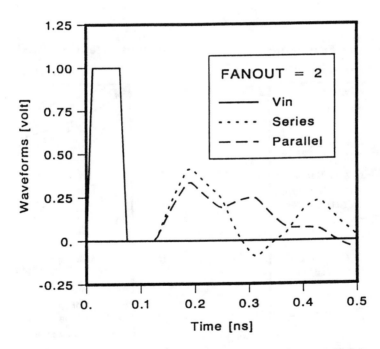

Fig. 3   Waveform distortion due to fanout for a 2 cm long microstrip of Fig. 1. Both series and parallel terminations. (a) Fanout of 1. (b) Fanout of 2.

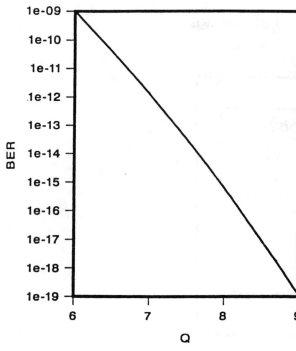

Fig. 4   Bit Error Rate vs. Q.

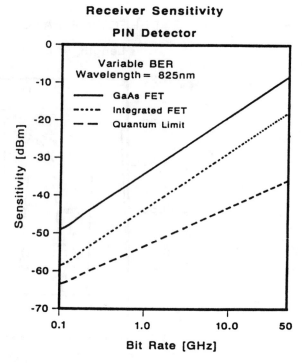

Fig. 5   Receiver sensitivity vs. data rate. BER (10^{-17} at 1 GHz) varies with rate. C_T=1.5 pF for GaAs FET; C_T=0.5 pF for integrated FET.

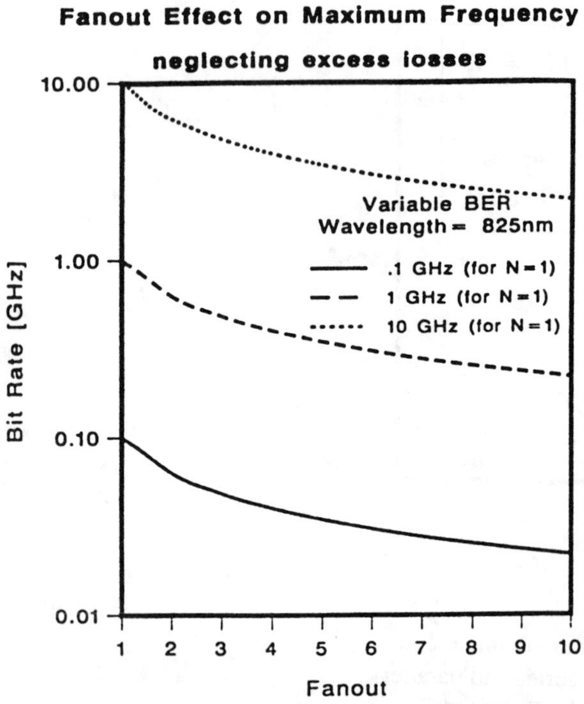

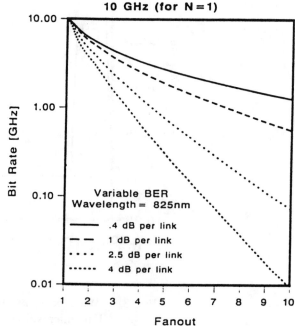

Fig. 6   Maximum bit rate as a function of fanout for various frequencies of lossless links.

Fig. 7   Maximum bit rate as a function of fanout for various levels of overall losses.

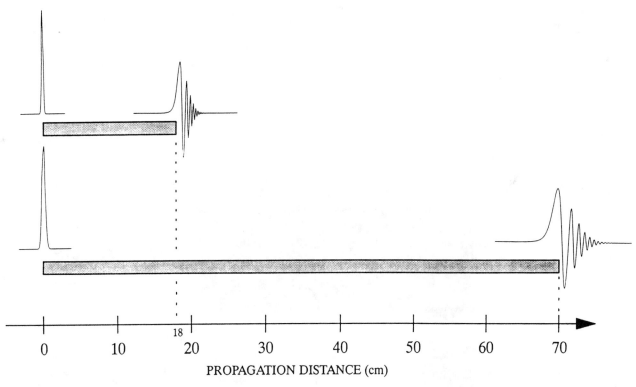

Fig. 8a   Propagation distance until amplitude diminishes to 50% of launched value, for a 5 ps FWHM Gaussian (top) and 10 ps (bottom) input signals, in a *lossless* microstrip 100 μ*m* wide, $\varepsilon_r = 3.5$, $R_o = 50\,\Omega$.

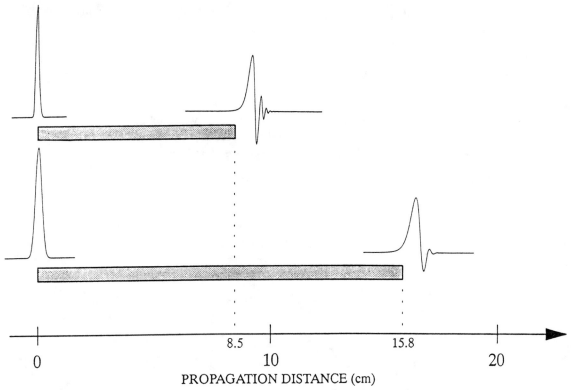

Fig. 8b   Same as (a) for *lossy* (copper, $\sigma = 5.8 \cdot 10^7$ mho/m) line.

Fig. 9   Cross-section of a polyimide waveguide.
Core material- DuPont PI2525, cladding-
PI2566. Core dimensions are 20×10 μm^2.

A COMPARATIVE STUDY OF INTERCONNECT TECHNOLOGIES

By

V K Nagesh, Dan Miller & Larry Moresco
Hewlett-Packard Co., Palo Alto, CA 94304

Abstract

Recent advances in VLSI and the increasing system cost/performance demands have challenged packaging technologies. There are many possible alternatives and it is important to consider different parameters like density, performance, cost, reliability etc., in choosing the appropriate packaging technologies for high performance digital systems. In this paper, both the interconnect and the substrate technologies are compared with a focus on the density and performance aspects.

Introduction

With the advances in VLSI technology and increased demand for high performance computers at lower costs, the role of electronic packaging of ICs and systems are becoming increasingly important. Chip bonding technologies invented to package simple 2-3 terminal transistors two to three decades ago can still be used to individually package ICs with several hundred terminals. However, considerations should now be given as to how the different chip bonding techniques and substrate technologies would affect the system performance, I/O densities, chip packing densities and overall system cost. In this paper an attempt is made mainly to compare chip bonding and substrate interconnect technologies in terms of density and performance.

Chip Bonding Technologies

There are essentially three main chip bonding techniques that are common in the industry: (a) Wire Bonding, (b) Tape Automated Bonding (TAB), and (c) Flip Chip Solder bump bonding (also known as C-4)[1]. Some of the characteristics of these technologies are presented in Table 1. While the first two are peripheral in nature (TAB is projected to have the area array

Reprinted with permission from *Proc. 9th Int. Electron. Packaging Conf.*, vol. I, pp. 199–208, 1989. Copyright © 1989 by the International Electronics Packaging Society, Inc.

possibility), Flip Chip bonding can be both peripheral as well as area array in nature. The number of chip I/Os possible in a peripheral scheme varies linearly as the chip size and inversely as the bonding pad pitch. In the area array, however, the number of I/Os possible increases as the square of the chip dimension. With an effective bond pitch of 4 mils, Wire Bond or TAB technologies will allow 400 pads for a one centimeter square chip and Flip Chip bonding will allow about 1600 pads on a 10 mil area grid array for the same chip size. This factor is important for chips with large number of I/Os (in the region of 400 - 600) as the peripheral bonds may make the IC chip to be unusually large in order to accomodate the required number of pads and thereby increasing the chip cost tremendously. Figure 1 plots the number of I/Os as a function of the chip size for the three different technologies. Another implication, of course, is that peripheral bonding schemes increase the footprint of the package reducing the number of chips packed per unit area. Even though this may help in the thermal management aspects, it increases the delay between chips and thus having detrimental effects on the system performance. Another factor related to the performance is the inductance of the interconnect; the lower the inductance, the lower would be the simultaneous switching noise. Out of the three bonding schemes, flip chip bonding has the lowest inductance at 50-100 pH compared to 2000-6000 pH for wire bonds and TAB leads. While a detailed cost comparison study between the three technologies is yet to be done, priliminary indications are that flip chip bonding, being a gang bonding scheme, is potentially the lowest cost alternative among the three for multichip module applications. Further, flip chip bonding will allow true reworkability which is very important in fabricating multichip modules.

Substrate/board Interconnects & Circuit Packing Densities

In this study, three basic types of technologies are compared. They are: (a) Pin Grid Array (PGA) packages on PC boards and TAB on PC boards, (b) Multilayer Ceramic (MLC) module technology, and (c) Copper-Polyimide (Cu-PI) thin film module technology. For both MLC and Cu-PI technologies, flip chip bonding of chips is assumed. Table 2 gives the pertinent characteristics of the PC boards, MLC substrates and Copper-Polyimide substrates. An important parameter to consider in comparing the different interconnect technologies may be the number of electrically accessible circuits per nanosecond as it takes into account the chip complexity in the form of number of circuits per chip, chip packing densities and the signal propagation time in the dielectric medium[2,3]. Actual accessibility will depend on the routability in the substrate technology chosen and will provide the upper bound on the calculated accessible circuits per nanosecond; actual accessibility will, of course, also depend on the hardware architecture parameters like logic delays, number of chip crossings etc.. Figure 2 shows the number of accessible circuits per nanosecond for the different technologies as a function of chip complexity. Based on the calculated number of accessible circuits, Cu-PI thin film technology ranks the highest followed by MLC-low dielectric constant, MLC and PC boards. Typical mainframe systems have 200K-1M logic gates in them. To pack the 200K circuits within a nanosecond (assume 40 - 10K gate array chips), following the calculations mentioned in the next section, it would require approximately 12 conductor layers (including power and ground) with Cu-PI technology using 4"x4" substrates and 2 track routing on a 10 mil pitch. Capability of up to 8 conductor layers has been reported using Cu-PI technology[3]. With MLC, it would require as many as 52 layers (including, signal, power & Ground, and space expansion layers) with a 4"x4" substrate and 1 track routing on 20 mil pitch to pack 200K circuits within a nanosecond. Capability to process MLCs upto 60 layers exists in the industry. With PGAs and TAB on PC board technology, with the current day PC board design rules, it is not possible to meet the 1 nS accessibility requirement due to both the number of vias required and the total layers that would be required (> 60 layers).

Connectivity Comparisons

In comparing different substrate/board technologies, it is very helpful to know the ability of each technology in terms of routing. In other words, it will be helpful if the information related to line width and spacings, number of possible layers, via pitches etc. can be integrated, so as to know how many chips with a given number of I/Os can be interconnected (even if it is on a first approximation basis). Using the empirical relations given by D.P.Seraphim[4] and G.Messner[5] (Figure 3), we have tried to compare the various interconnect technologies mentioned earlier. We have also added a boundary condition of 1 nS accessibility to normalize the signal propagation speeds in the various dielectric mediums. Substrate dimensions corresponding to the nS propagation time for the different mediums as well as dimensions corresponding to 12000 vias per layer for the different substrate technologies are given in Table 3. Table 4 presents the interconnect requirement data for a hypothetical test example of interconnecting 9 VLSI chips each with 625 I/Os. Average interconnect length is assumed to be 1.5 times the chip pitch and for every 4 terminals 3 interconnections are asssumed. The number of layers required to interconnect all the I/Os using the different technologies are calculated based on the design guidelines of each technology with the substrate dimension being limited by the nanosecond boundary condition. Also, every X,Y pair is assumed to be shielded between power/ground planes and space expansion layers required for MLC and Cu-PI substrates are included. Both MLC and Cu-PI substrates are assumed to use flip chip bonding for chip to substrate interconnections. As can be seen from the table, it requires a total of 6 layers with Cu-PI (2 tracks on 0.010" pitch) and 25 layers with MLC. Both the Cu-PI and MLC requirements calculated are within the capability of the respective technologies. Due to package and via placement limitations, 1nS boundary condition cannot be met by TAB/PGA on PC board technology. If this issue is ignored and number of layers required are calculated for PC boards with 4-tracks on 0.100" pitch, results indicate that it would require a total of 60 conductor layers.

Comparison of Chip Crossing Delays

To compare the performance of the different interconnect technologies, SPICE simulations were performed. Schematic used for this analysis is shown in Figure 4, and Figure 5 is a plot of the chip crossing delays as a function of the fanout for the different technologies simulated. ECL drivers were used, and two chip pitches per fanout was assumed in all cases. The delays plotted are total delays which include both the package delay and the IC delays. Flip chip solder bump interconnection was assumed for both Cu-PI and MLC technologies and a chip pitch of 1.5 cms was used. For TAB on board and PGA on board, a chip pitch of 6.5 cms was assumed. Table 5 provides the package delays and total delays for the different package implementations. In the case of Cu-PI structures total delay for a fanout of three was 2.3nS out of which package contribution to the delay was only 1.030nS (45% of the total delay). In the case of alumina MLC, total delay was 2.76nS and package contribution was 1.47nS (56% of the total delay). In the case of PGA on PC board, for the same fanout of three, total delay was 6.7nS out of which 5.4nS (81% of the delay) was due to packaging. TAB on PC board delay numbers are about 15-20% better than PGA on PC boards assuming the same chip pitch. Overall, Cu-PI structures had the best result with MLC results being a close second. Total delays in the case of TAB/PGA on PC boards were worse by a factor of three compared to Cu-PI technology.

Cu-PI and MLC module technologies

Both these technologies provide high density of interconnect and low package delays needed for packaging high performance digital systems. While the Cu-PI technology is a thin film additive technology, MLC is a thick film based technology with the different conductor layers processed on separate dielectric sheets, inspected and combined together. This is an important difference as it will have an effect on the process costs, yields and the total number of layers possible. With Cu-PI technology, however, feature sizes are much finer than MLC and hence will take a fewer layers to interconnect the same number of I/Os. Approximate cost of fabricating Cu-PI structures vs MLC implementing the same interconnect function can be 1.5-2.5 : 1. Integrating the Cu-PI and MLC technologies provides, however, ample opportunity to balance the performance and cost, in addition to having high module pin-outs possible with industry standard area array pins (like PGAs)[3]. Cost comparison between Cu-PI/MLC modules vs TAB/PGA on PC board implementation is being currently studied. Priliminary indications are that there will be significant benefits both in increasing the system performance and lowering the overall system cost with the module approach. Benefits will be greater if the high speed interconnects of the module are utilized with chip sizes optimized for overall system performance and cost.

Acknowledgements

The authors are indebted to the members of the Advanced Packaging technology department, and the management of Circuit Technology R&D and Hewlett-Packard Laboratories without whose encouragement and support this study would not have been possible.

References

1. N.G.Koopman et.al., p366, "Microelectronics Packaging Handbook", Van Nostrand Co., N.Y., 1989

2. D.Balderes & M.L.White, p351, Proc. of Electronic Components Conf., IEEE, 1985

3. C.C.Chao et.al., p76, 1988 IEEE Intl. Conf. on Computer Design (ICCD), 1988

4. D.P.Seraphim, p90, Proc. of Electron. Insulation Conf., 1977

5. G.Messner, p143, IEEE Trans. on Components, Hybrids, and Mfg. Tech., CHMT-10, No.2, June 1987

INTERCONNECTS

	WIRE BONDS (PGA on PCB)	TAB (on PCB)	FLIP CHIP (on MCM)
I/O DENSITY	400/sq.cm.	400/sq.cm.	1600/sq.cm.
CHIP DENSITY on PC Board / MCM (400 I/Os per Chip)	0.25 / sq.inch	0.25 / sq.inch	6 / sq.inch
INDUCTANCE	2–6 nH	2–6 nH (shielded)	5–10 pH
REWORK	Poor	Poor	Good
FAILURE RATE*	10^{-5}	N/A	10^{-7}

* *Percent failure/1000 hours of testing at 40,000 hours.*

Table 1

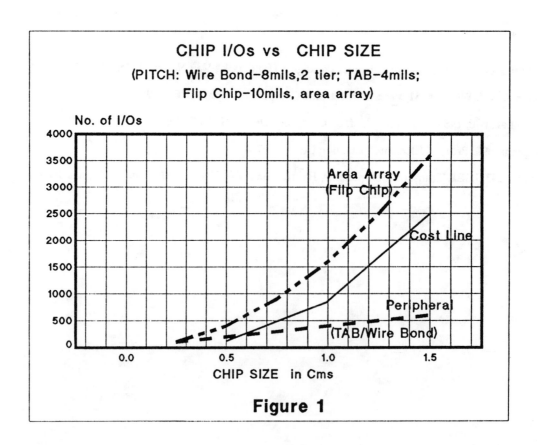

CHIP I/Os vs CHIP SIZE

(PITCH: Wire Bond–8mils,2 tier; TAB–4mils;
Flip Chip–10mils, area array)

Figure 1

SUBSTRATE/BOARD TECHNOLOGY

TECHNOLOGY	CONNECTION DENSITY IN/IN^2/LYR	SIGNAL LYR.S (Total Lyrs)	TOTAL LENGTH IN/IN^2	VIA DENSITY per IN^2
1. PC Boards (4 Track, 0.100" pitch)	40	12 (20)	480	400 (vias on 50 mil pitch)
2. Multilayer Ceramic Substrate (1 Track, 0.020" pitch)	50	20 (42)	1000	2500
3. Cu/PI Thin Film (2 Tracks, 10 mil Pitch)	200	4 (8)	800	10,000

Table 2

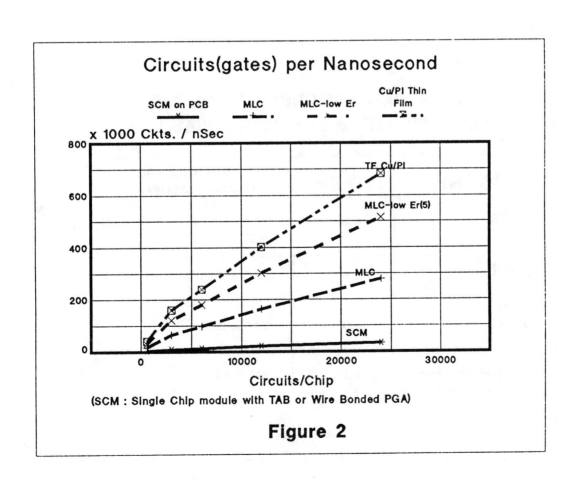

Figure 2

SUBSTRATE/BOARD CONNECTIVITY

Connectivity = 1.5 P (3/4 x # OF I/O'S)

Assumptions: AVG. INTERCONNECT LENGTH = 1.5 x Pitch

FOR EVERY 4 TERMINALS THERE ARE 3 INTERCONNECTIONS

50% EFFICIENCY IN ROUTING

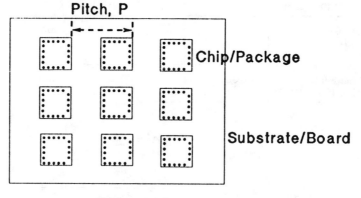

Figure 3

SUBSTRATE DIMENSIONS FOR 1nS ACCESSIBILITY & FOR 12,000 VIAS/LYR

Technology	1nS dia.	Dimensions for 12,000 vias/lyr.
1. PCB	4.5"	10.9" X 10.9" (0.100"pitch) 5.5" x 5.5" (0.050"pitch)
2. MLC	3.4"	2.2" X 2.2"
3. MLC – LOW Er	4.6"	2.2" X 2.2"
4. Cu–PI Thin Film	5.3"	1.1" X 1.1"

Table 3

INTERCONNECTING 9 VLSI CHIPS EACH WITH 625 I/O'S, WITH 1nS ACCESSIBILITY CONDITION:

Technology	Pitch	Wiring Length Required	No. of Signal Layers (& Total Layers)	
1. MLC (4"X4")	0.5"	6330"	8 (25**)	
2. Cu-PI Thin Film* (4"X4")	0.5"	6330"	2 (6**)	
3. TAB/PGA ON PCB*** (5"x5", 16mil OLB for TAB)	3"	38,000"	(3 track) 51 (80) (4 track) 39 (60)	

* Two tracks on 0.010" via pitch used
** Includes space Expansion Layers

*** Does not meet 1 nS criterion due to Pkg. & via placement limitations. No.s are hypothetical.

Table 4

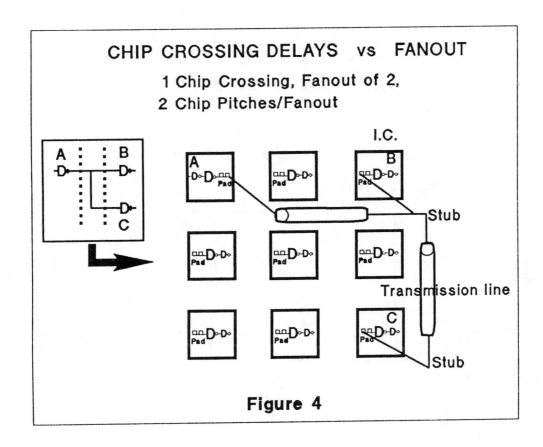

CHIP CROSSING DELAYS vs FANOUT

1 Chip Crossing, Fanout of 2,
2 Chip Pitches/Fanout

Figure 4

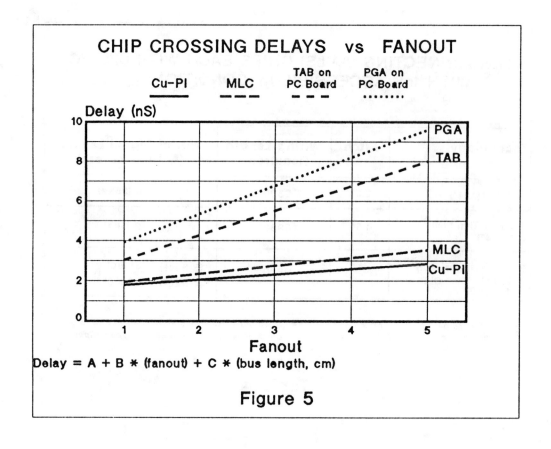

CHIP CROSSING DELAYS vs FANOUT

Delay = A + B * (fanout) + C * (bus length, cm)

Figure 5

CHIP CROSSING DELAYS vs FANOUT
Fanout of 3, 2 Chip Pitches/Fanout

	Pitch	Package Delay	IC Delay	Total Delay
PGA/PCB	6.5cm	5.4nS	1.3nS	6.7nS
TAB/PCB	6.5cm	4.2nS	1.3nS	5.5nS
MLC	1.5cm	1.47nS	1.3nS	2.77nS
Cu/PI	1.5cm	1.03nS	1.3nS	2.33nS

Table 5

DESIGN AND TESTING OF HIGH DENSITY INTERCONNECTION SUBSTRATES

by

Claude Hilbert and Claude Rathmell
Microelectronics and Computer Technology Corporation (MCC)
Austin, Texas

Multi-chip modules offer the potential advantages of a higher chip density, smaller interconnect delays and improved electrical performance. However the transition from printed-circuit-boards to high-density interconnection substrates increases the line losses by scaling down their cross-section. This paper discusses the design constraints imposed on high density substrates by lossy interconnect. We describe the trade-offs between line width, line length and number of interconnection layers as a function of line resistivity and driver impedance. The miniaturization of the interconnection substrates also increases the testing challenge. Fast test methods providing high resolution over a large field are required. We describe and compare the state-of-the-art test technologies for high density interconnection substrates.

Introduction

Traditionally printed-circuit boards (PCB) have been used to interconnect chips in single-chip packages (SCP). State-of-the-art PCBs can achieve down to 4 mil line widths and spaces. The number of metal layers can reach up to 40 in the most advanced applications [1]. As the interconnect requirements keep increasing, driven by the quest for increasingly powerful processors and faster cycle times, the limits of PCB technology are fast approaching. An increasing number of high performance machines are starting to use multi-chip modules consisting of (bare) chips on high density interconnect substrates [2]. These substrates may be multilayer structures of copper or aluminum with ceramic and/or polyimide as a dielectric. The typical line dimensions on these substrates are presently an order of magnitude smaller than the ones in PCBs. The line widths range from 15 to 25 μm and the pitches from 50 to 75 μm. The chips are usually attached by controlled-collapse-chip-connect (C4) [3] or tape-automated-bonding (TAB) [4] technology.

The advantages of this scale-down are multiple. (1) Thin film technology can provide higher interconnect densities in fewer layers because of the improved line width capabilities and smaller inter-layer vias. (2) Bare chips can be mounted almost "wall-to-wall" onto the substrate and provide 50-100% silicon coverage. On traditional PCBs the single-chip package, used as a space transformer to adapt the chip pad pitch to the much larger PCB pad pitch, often uses up an area ten times larger than the chip itself. This reduces the silicon coverage on PCBs to less than 10% typically. (3) Because of the increased chip density, signals need to propagate over smaller distances resulting in a speed-up of the machine. (4) At the same time, the signal cross-talk and power supply noise decrease because of the improved impedance control. (5) Because an intermediate level of packaging is eliminated, the number of interconnect layers is reduced and the system is miniaturized, the reliability will improve and the packaging costs will decrease as this technology matures.

The Packaging/Interconnect Program at MCC has developed high density interconnect substrates based on copper/polyimide technology as well as tape-automated-bonding (TAB) for chip attachment. Figure 1 shows a photo of a typical substrate fabricated at MCC for a dual 1750A processor on a 4 inch wafer. The actual processors (consisting of 6 SOS VLSI and 12 memory chips) are mounted on the 2.25" square center portion of the wafer. The peripheral part of the wafer is filled with electrical test and process monitoring structures. Figure 2 illustrates a typical cross-section of the substrate. It consists of 5 copper layers (2 signal layers, 1 power, 1 ground and 1 bonding layer) separated by polyimide on a ceramic substrate. The impedance is controlled close to 50 Ω. MCC uses a plated copper pillar and a mechanical polyimide polish technique which yields excellent planarity, allows the stacking of vias, and does not require a photo-imagable dielectric. Line widths and spaces down to 10 μm can be fabricated for a 5 μm-thick line. Figure 3 shows a photo of chips TAB bonded face-up onto a substrate. MCC TAB technology allows down to 100 μm pitch of the tape leads.

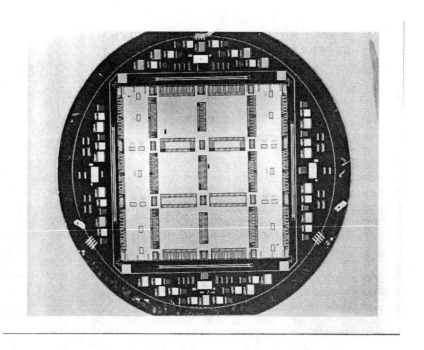

Figure 1:   Photo of copper/polyimide interconnect substrate fabricated at MCC.

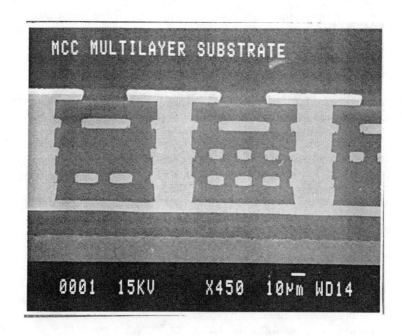

Figure 2:   Scanning electron micrograph of cross-section of MCC copper/polyimide substrate.

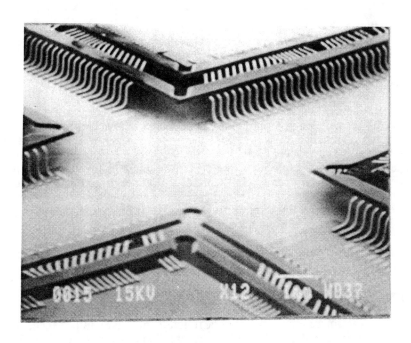

Figure 3: Photo of chips tape-bonded face-up onto a copper/polyimide substrate.

Design of high density interconnection substrate

Even though the substrate fabrication technology allows line widths and spacings to shrink to 10 μm, electrical performance requirements impose certain design restrictions. Cross-talk imposes a minimum line spacing, typically at least 2-3 times the line width, and the signal attenuation because of the line losses makes it necessary to design with a minimum line cross-section. We will analyze the design restrictions due to the line resistance, and describe the trade-offs between line width and cross-section, line length, number of interconnect layers as a function of the line resistivity, the dielectric material and the driver impedance.

Many variables are needed to describe and model the configuration of a transmission line and the circuit driving it, so it is necessary to make some assumptions in order to reduce the problem to a manageable size and to extract the main features. In the following discussion we choose a system where the circuit drives a controlled impedance interconnect of 50 Ω, feeding into an open circuit at the receiver end [5]. This assumption is a convenient starting point and is characteristic of many systems that are presently manufactured. The results that we derive can be easily scaled to other assumed impedances and termination conditions, and we give guidelines for doing this.

A controlled impedance is necessary to propagate fast signals and to minimize cross-talk. Consider an off-chip driver that has a 50 Ohm source impedance. A low characteristic impedance line ($Z_0 \ll 50\ \Omega$) compared to the output impedance of the driver, yields an unfavorable voltage division at the source which increases the number of reflections needed to build up the signal to a given amplitude. A high characteristic impedance line ($Z_0 \gg 50\ \Omega$) compared to the output impedance of the driver, increases cross-talk between neighboring lines for a fixed line pitch. In addition, practical considerations limit the maximum characteristic impedance that can be achieved: line geometries that can be manufactured are typically limited to aspect ratios less than or on the order of unity (thickness-to-width ratio). Such aspect ratios yield a maximum impedance of about 100 Ω for a dielectric constant of 3.5. A characteristic line impedance of 50 Ω is therefore a reasonable compromise for our study.

For the purpose of the following analysis we define the delay as the time it takes the signal to reach 90% of the input step amplitude at an open termination. There are a number of criteria that decide the method and impedance of the termination of the off-chip signal net: reflections, net topology, resources for termination (where are the resistors), active vs. passive, etc. In our study we assume an approach that is increasingly being used for high performance systems; an approach that is common in slow speed systems - namely not to terminate the net, but rather leave it open (subject to the input impedance of the receiver circuit, which we can assume to be much higher than the line characteristic impedance).

A high impedance termination is advantageous because the signal amplitude will be doubled by the reflection at the end; in this manner the required signal amplitude can be more easily achieved after a single time-of-flight on the first reflection. Additional practical considerations that make an open termination attractive are that in the steady-state the signal voltage is independent of the lossy line length and the power dissipation is zero. A matched termination would result in different steady-state voltages for different lengths of lossy line and increase the power dissipation. If the interconnect is to be operated in the time-of-flight regime, it is preferable to have a matching termination at the source which will eliminate any further reflections. Such an arrangement is also best suited to drive a 'star' or 'tree' network of several lines with different loads. The advantages of an open termination are to be weighed against an increase in cross-talk and a longer time for the system to settle due to reflections at the load.

We discuss later the sensitivity of our results to the above assumptions. The general conclusions reached with this model can be easily adapted to different configurations and circumstances.

Interconnect simulations

We have investigated the interconnect delay as a function of driver impedance, line width, length and resistivity using SPICE [6,7]. Figure 4 is a schematic representation of the interconnect circuits as used for these simulations. To achieve good fidelity in the simulations the interconnect line of resistance R is partitioned into 40 sections. Each section consists of a small piece of ideal (lossless) transmission line in series with one or several lumped elements which represent the losses in that small segment. When the skin effect is neglected, there is a single series resistor in each section whose value is given by the dc resistance of that element (R/40), the small resistance contribution of the ground plane being neglected. When the normal skin effect is taken into account, the frequency dependent resistance and inductance of each element is represented by a series of parallel R-L combinations with various time constants. The accuracy of this representation and the valid frequency range increase with the number of serial R-L elements. The values of these discrete resistors and inductors are obtained from a finite element modelling program developed at MCC [9]. We studied simulations with different numbers of elements and found with > 20 sections the simulated results were independent of the number of sections.

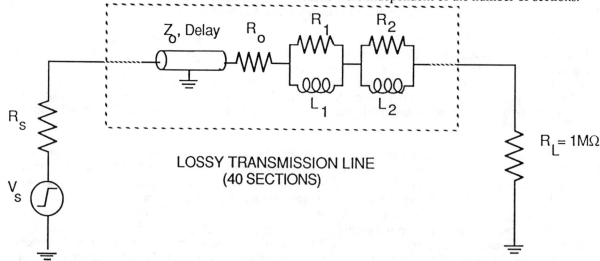

Figure 4: Schematic SPICE representation of unterminated lossy transmission line.

In all the cases we analyze, the real part of the controlled impedance $Z_0 = \sqrt{\dfrac{L}{C}}$ of the interconnect is assumed to be 50 Ω. The complex part of the characteristic impedance which is due to the losses is taken into account by the added resistors in our circuit model.

The thickness of the interconnect line is always taken to be half of the line width. The resistivity r we assume for copper at room temperature is 1.7 μΩ.cm; at 77 kelvins it is 0.22 μΩ.cm [6]. The dielectric constant e of the insulating material is taken to be 3.5, a typical number for polyimide or printed circuit board materials. This yields a velocity $v = \dfrac{1}{\sqrt{LC}} = \dfrac{c}{\sqrt{e}}$ of 1.6 x 10^8 m/s for the propagation of electromagnetic waves along the lines, or a time-of-flight of 62 ps/cm, assuming a TEM mode. The corresponding capacitance per unit length $1/vZ_0$ is 1.25 pF/cm. We assume the driver is a voltage source of 1V amplitude at time zero with a rise time which equals 10% of the time-of-flight for the line length under consideration.

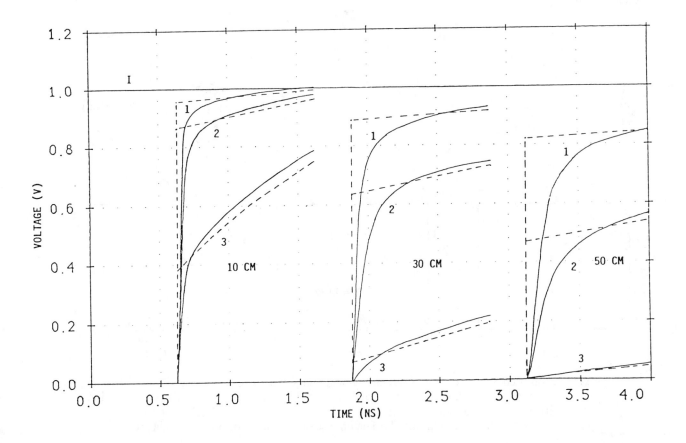

Figure 5:   Calculated attenuation and delay for 2, 5, and 10 μm-wide copper lines at 77 K whose lengths are 10, 30 and 50 cm. The dotted lines represent the behavior of the lines when the skin effect is neglected.

Figure 5 shows the results of calculations with and without skin effect for copper lines of various lengths d and widths w at 77 K. The voltage at the open end of the line is plotted. The source and the line impedances are

50 Ω. We see that for any line width, the signal arrives at the load after one time-of-flight (1.24 ns). There are two different portions to the arriving signal: first, the signal rises rapidly to a value determined by the voltage division at the line input and the exponential attenuation of the line [6]; second, a slow RC-like tail follows.

The skin effect rounds off the transition between the two regimes, because the higher frequencies experience a greater attenuation. However, even at this low temperature, it appears that the skin effect makes only a small difference. At signal levels up to 90%, the difference in delay between the two cases is less than 15%. The actual skin depth scales like the square root of the resistivity and the skin effect increases in importance as the skin depth decreases with respect to the line size. Hence the skin effect is even less important in the study of interconnect delay for smaller lines, higher temperatures, and more resistive materials. We conclude therefore that the normal skin effect can safely be disregarded for digital applications at or above 77 kelvins without risk of major error for rise times larger than about 100 ps. Finally, the anomalous skin effect only starts to be significant below 77 K when the skin depth and the electron mean free path become similar [6]. A further discussion of skin effects, and the conditions for which they are important, is included in Reference [10]. We will neglect the skin effect in the remainder of this paper.

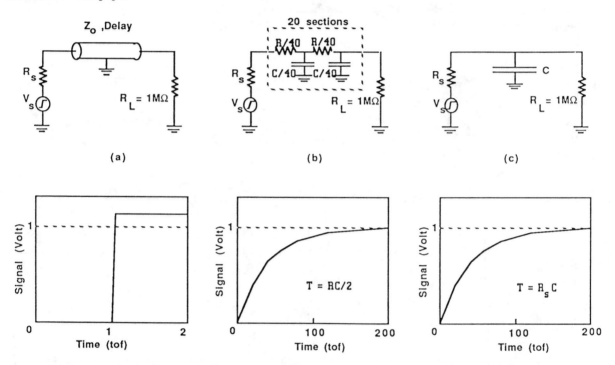

Figure 6: Simplified SPICE circuits (top) and typical signal response at the load (bottom) for the limiting propagation regimes. (a) Time-of-flight regime ($R_s < Z_0$), (b) lossy line diffusion, and (c) source resistance limited regime. (b) and (c) have an RC-type behavior with a charging time T. The time scale is in units of time-of-flights (tof).

SPICE simulations demonstrate the existence of 3 different limiting regimes for interconnect performance: (1) the time-of-flight regime, (2) the lossy line diffusion regime, and (3) the source limited RC regime. These 3 limiting regimes can be represented by the simplified SPICE circuits shown in Figure 6. In the time-of-flight regime (Figure 6a) the equivalent SPICE circuit is a lossless transmission line and the signal arrives at the load after one time-of-flight without attenuation. In the diffusion regime (Figure 6b) the line can be represented by a distributed RC circuit and the signal diffuses slowly along the line according to the equation

$$\frac{d^2V}{dx^2} = rc\frac{dV}{dt} \quad (3)$$

which is analogous to the equation for heat diffusion. V is the voltage amplitude as a function of distance x and time t. r and c are the resistance and capacitance per unit length, respectively. In the source limited regime (Figure 6c) the line can be represented by a single lumped capacitor. Multiple signal reflections are required to build it up to the desired amplitude at the load.

Isochronal delay surfaces

Numerous SPICE simulations were run in order to study the dependence of the delay on the source resistance, line width, length and resistivity. In order to better visualize and understand the results we chose to produce three-dimensional graphs of surfaces of equal delay in the length-width-source resistance space. These isochronal delay surfaces indicate the distance the signal can propagate in a given amount of time. Once again a conservative 90% point was taken on the wave form. These graphs were obtained from interpolation of SPICE runs with assumptions identical to the ones described previously.

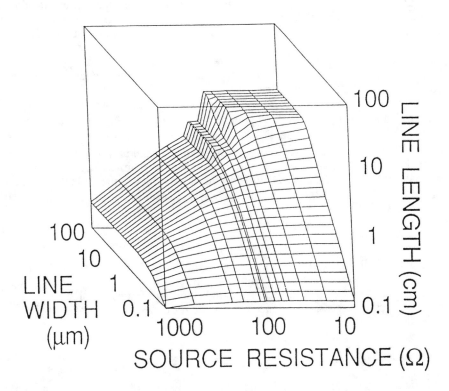

Figure 7:   Isochronal delay surface for 700 ps for room temperature copper, a dielectric constant of 3.5 and a characteristic line impedance of 50Ω. The lines on the surface correspond to line widths spaced equally on this logarithmic scale by 1 dB and to source resistances of 50, 65, 70, 85, 90, 100, 200, 500 and 1000 Ω respectively.

Figure 7 shows one such isochronal surface corresponding to a constant delay of 700 ps calculated for copper interconnect at 300 K. All scales are logarithmic. The three regimes previously discussed for the scaled interconnect can again be found on this three-dimensional figure. The flat plateau on top of the "mountain shell" corresponds to the time-of-flight regime; it is defined as the regime where the signal voltage reaches at least 90% of its final value on the first reflection. It is located in the low source resistance, large line width corner of the graph. The delay of 700 ps in the time-of-flight regime is roughly equivalent to a line length of 10 cm with a dielectric constant of 3.5. As one moves on the surface at low source impedance toward the smaller line widths, there is a transition into the lossy line, diffusive RC-regime. Here the distance the signal can travel in 700 ps decreases linearly with line width.

If one moves off the plateau in the direction of increasing source impedance, keeping a constant line width, one encounters a descending staircase as soon as the source impedance exceeds the characteristic line impedance by about 20%. At this point the voltage put on the line at time zero is less than 45% and voltage doubling at the load cannot achieve 90% of the signal at the first reflection, even for a lossless line. The exact location of the steps depends on the signal percentage that is required at the receiver. One needs to wait for the signal reflected off the load to travel back to the source, be reflected off the source, and then reach the load after three times-of-flight. This sudden tripling of the delay in the time domain for a given line length means that the length a signal can travel within a given amount of time (isochron) is abruptly reduced by a factor of three. As we increase the source impedance used in the simulation, more and more reflections are needed to reach the assumed 90% threshold of the signal at the receiver. Thus, the steps in the isochronal surface represent odd multiples of the time-of-flight. They are spaced increasingly closer together along the source resistance axis and their height also diminishes on this logarithmic length scale. Sufficient SPICE circuits were run only to resolve the first two steps in detail.

As the lines get narrower the location and nature of the steps changes because of the increasing line attenuation; the ridge moves toward lower source resistances and is less sharp. For very high source impedances (i.e. $R_S \gg Z_O$) multiple reflections are needed; in the time domain simulations the envelope of the multiple reflections tends asymptotically towards a simple exponential $R_S C$ charging curve. On the isochronal chart this corresponds to the steps gradually fading into the $R_S C$-limited slope. As can be seen again on this figure, the line losses do not contribute substantially to the signal delay in this source-limited regime, until narrow line widths less than a few μm are reached. This graph is obtained by interpolation from SPICE data at each given source impedance so that a small error can exist, especially along the ridge which represents the transition between the time-of-flight regime and the diffusion regime. Even in the worst case, however, the relative error in length is expected not to exceed 20% or 1 dB.

Universal scaling

This three-dimensional isochronal graph is "universal" in the sense that it can be scaled in order to describe the performance of any controlled impedance line with an open termination for any delay. Neglecting the skin effect there are only 3 parameters determining the propagation behavior of a transmission line:
- the source-resistance to characteristic impedance ratio or the reflection coefficient,
- the time-of-flight, and

- the attenuation coefficient $e^{-\left(\frac{R}{2Z_0}\right)}$.

How do they scale with the design parameters?
   (i)   The characteristic impedance affects the line performance by the voltage division on signal launch and by its reflection coefficient. Both factors can be kept constant by scaling the source resistance linearly with any changes in characteristic line impedance.
   (ii) The distance that a signal can travel in the time-of-flight is proportional to the permissible delay t. Also, the velocity of propagation of electromagnetic waves v, and hence the distance, depend on the dielectric constant e. Therefore the length axis scales as $t/\sqrt{e}$.

(iii) Lastly, the line attenuation is a function of both the line resistance and its characteristic impedance. The relevant factor is $R/2Z_0$ or (r d / w^2 Z_0). Therefore, we can take into account the varying line attenuation by scaling the line width axis proportionally to $(r$ t / $\sqrt{e}$ $Z_0)^{1/2}$.

In general, we can therefore deduce the performance of any controlled impedance interconnect with an open termination from one isochronal graph (or a single set of SPICE simulations) using a three-dimensional shift of the isochronal surfaces in the logarithmic line length, line width and source resistance space.

Design guide lines

The isochronal surfaces and the scaling relations described in the previous sections reflect the following general conclusions. For time-of-flight operation at a minimum line width, the characteristic line impedance needs to be equal to or greater than the driver impedance. In order to reduce the time-of-flight or increase the distance a signal can travel within a fixed amount of time, the dielectric constant must be minimized. Line widths can be reduced and line densities increased by going to lower resistivity conductors. These design recommendations must of course be weighed against process, cost, and reliability considerations.

The isochronal delay surfaces can be scaled and sectioned at any point and along any axis to provide specific information for a given parameter value set. Table I gives an overview of the minimum line width required for 10 Ω and 50 Ω drivers using 50 Ω copper conductor lines on a polyimide-type dielectric ($e = 3.5$).

Distance (cm)	Time-of-Flight (ps)	Min. Cu line width for time-of-flight (μm) $R_S=10\Omega$	Min. Cu line width for time-of-flight (μm) $R_S=50\Omega$	Max. Cu line density for T.o.F. operation (lines per cm) $R_S=10\Omega$	Max. Cu line density for T.o.F. operation (lines per cm) $R_S=50\Omega$
100	6,900	24.0	57.0	140	58
50	3,450	17.0	40.0	200	83
20	1,380	10.5	25.0	320	130
10	690	7.4	18.0	450	180
5	345	5.2	12.7	640	260
1	69	2.4	5.7	1,400	580

Table I: Minimum line width requirements for copper (300K), 50 Ω lines in order to achieve time-of-flight (plus 10% rise-time) for 10 Ω and 50 Ω drivers. The last column indicates the maximum line density for which copper interconnect operates in the time-of-flight regime. We assume a line thickness which equals half the line width and a line pitch which equals three times the line width.

Consider the case of a 10 cm x 10 cm copper/polyimide substrate. According to Table I the maximum line length of 20 cm requires a minimum line width of approximately 10.5 μm (or equivalently a cross-section of 55 μm^2) for 10 Ω drivers. This corresponds to a line density of 320 lines per cm or 810 lines per inch. A single x-y layer pair can provide, under these conditions, 640 cm/cm^2 or 1,620 in/in^2. This capability is amply sufficient to interconnect any of today's mainframe or supercomputer modules. For instance, the IBM 3081 thermal conduction module with a multi-layer ceramic substrate contains about 820 in/in^2. Similarly, the 44-layer printed circuit board in the ETA-10 contains only 680 in/in^2.

The above examples demonstrate the tremendous capabilities of high density interconnect substrates. Even though the resistance per unit length of the interconnect wires increases as their cross-section is scaled down, the increased silicon coverage decreases the required line length and thus mitigates the effects of a reduced cross-section. Table I gives first order design guide lines for a particular system. The minimum line widths and potential line densities for different systems can be obtained from the isochronal surface using the scaling relations.

Isochronal graphs can similarly be developed for terminated transmission lines and different signal percentage requirements. The same scaling relations will hold true. A matched termination will impose more stringent requirements on the line attenuation, whereas a lower signal percentage requirement will allow smaller line cross-sections and higher line densities for a given distance.

As the complexity of multi-chip modules increases, the interconnect requirements will continue rising. Wiring density as well as substrate size will keep increasing. These developments will make it necessary to go to more than a single x-y layer pair and/or to operate the module at lower temperatures. The choice between these options is partially an economic one: will it be cheaper to develop, produce and operate a multiple layer substrate at room temperature, or a substrate with fewer layers cooled to, say, 77 kelvins? Eventually it may even become necessary to use superconducting interconnect at liquid nitrogen temperatures.

Test Requirements

Testing of these high density, high performance substrates is an integral part of their manufacture. At various levels during the manufacture of the interconnect substrate, substrates are electrically tested for dc parametric information which is fed back to process engineering. This testing is performed on a set of parametric test structures which have been included in the mask sets for this purpose. We will briefly discuss the requirements for this testing. After the substrates have been completed, a final test to verify the network interconnections is required. We will consider three possible approaches to performing this testing, examining the performance and cost of each method.

Parametric Process Monitors

A variety of electrical monitors are used to test for process variations. These tests are conducted in order to find opens and shorts in the test structures, and to determine the electrical characteristics of the interconnects being manufactured. Serpentine structures, set in alternating directions on alternate substrate conductor layers, allow monitoring of metal deposition for opens in metal lines, as well as provide information on resistivity. Interdigitated structures, also set in alternating directions on alternate substrate conductor layers, allow monitoring for shorts and leakage between lines. Via strings provide a method of verifying the interconnections between conductor layers and measuring the average resistance of those connections. Also included are capacitor monitors, via alignment monitors, thickness uniformity monitors, and a battery of high frequency test structures which allow measurement of impedance, propagation delay, and attenuation per unit length at any frequency. Time domain reflectometry and spectral measurements are also used to reveal any discontinuities in impedance or resonances. These tests are performed at several steps during the fabrication of the interconnect substrates.

Interconnect Network Test

Testing the interconnect networks of a completed substrate implies verification of the electrical path for each connection specified in the design, both for continuity and isolation from other networks. Two of the approaches we will consider are taken directly from the printed circuit board test regime: flying probe systems and network capacitance probing. In addition, we will consider a third method for testing, Voltage Contrast Electron Beam (VCEB), which promises complete verification and high throughput. Wafer handling and alignment time are not considered for each system in this paper, but are crucial considerations for purchasing any production test system.

Flying Probes

Flying probe testers employ two or more single needle probes which are mounted on moving mechanical systems, positioning each probe at a programmed position and contacting the substrate at a bond pad for connection to an IC. By moving the probes to each bond pad in a network, this type of tester verifies the continuity of each interconnect network in a substrate. After a network has been verified in this manner, one of the probes may be moved to a ground connection, and then power connection(s), to check for isolation from power and ground. Checks for isolation from other networks is usually not performed with this type of tester on PC boards due to the time involved in making continuity checks from each network to every other network, although it is certainly possible. Some vendors use capacitance checks for this application, since the occurrence of internet shorts can be a yield concern. This type of testing is highly desirable for substrates for two reasons: assembled multi-chip module costs are high, and repair of a substrate network is not always possible after bonding IC's to the surface.

There are some clear disadvantages to this type of system, although some vendors have done a commendable job of making production worthy flying probe test systems for PC boards. These systems are mechanically complex, and require many constantly moving parts to perform their job. They also rely on sophisticated software and control systems for accurate placement of probes and collision avoidance. Therefore, long term reliability must be considered before committing to this type of system for testing substrates. Another principal consideration is wafer alignment accuracy and probe placement accuracy. These systems were originally developed for bare board testing, where the board is placed in a self-aligning fixture. Currently no manufacturer we contacted has a tester of this type configured with wafer alignment and handling provided for, although at least one is developing such a system. Also, the probing ability of these systems is typically intended for point-to-point spacing of 0.020" to 0.010", with placement accuracy of ±0.002" or more. This probe placement capability is inadequate for high-density substrate testing, where point-to-point spacing of 0.004" and placement accuracy of ±0.0005" are required. It remains to be seen whether a tester utilizing this technology, which is capable of testing high density substrates, reaches the market.

Capacitance Probes

A second method of testing these substrates is network capacitance probing, where a single probe is used to contact each bond pad on the surface of the wafer and measure the capacitance to a ground reference plane. These systems rely on either a single moving probe or a fixed probe with a moving wafer stage system. In either case, they are mechanically less complicated than multiple moving probe systems, and are well suited to handling and testing wafers. The capacitance of any given network is determined by the length of the network lines, and the dielectric material between the network and the reference plane. Measuring the capacitance value at each node of the network should yield the same result as the value measured at any other node within the measurement accuracy of the tester. Therefore the measurement of different capacitance values within any network is an indication that the network is open. The average capacitance for each network should be known prior to testing, so that the values obtained for any network may be compared to the expected values. Measurements exceeding the expected value by a significant amount is an indication that the network is shorted to some other structure. The capacitance values for each network are dependent on the dielectric thickness, and variations in that thickness result in variations of measured capacitance values. Because of this dependence, measured values may vary by as much as ±10% from expected values, meaning that the window of "passed" measurement for good networks must be quite large. This means that some defects may escape the test (i.e. two very small nets shorted). There are also some unlikely faults possible that could give node capacitance measurements which are still in the "pass" window, but the occurrence of such faults are quite infrequent.

Despite these limitations, capacitance probing is an inexpensive and effective method of verifying interconnect network continuity and detecting shorts. A capacitive probe test system configured suitably for production can be purchased for between $60K-$80K. The throughput of such a system could be extrapolated from our experience with testing a sample substrate with 400 networks connecting 1350 nodes on a high performance commercial test system. Not considering the time required for manual alignment of each wafer, several substrates were tested in an average time of 555 seconds (translating to a time of 0.412 seconds/node for stepping and measurement). The time required for testing increases linearly with the number of nodes to test. A rough estimate of the throughput of a multiple flying probe continuity measurement system would be 40% longer to perform the same verification, due to the fact that more stepping is being performed, and each network is also being measured for capacitance.

Voltage Contrast Electron Beam Test

A third method of verifying networks is to use a Voltage Contrast Electron Beam test system. With this approach, an electron beam performs all of the testing and no mechanical probes ever touch the substrate surface. An electron beam is used to read the voltage present at each node of a network, then inject a charge in the network, then read the other nodes of the network again. Each node of the network should show the charge voltage present, verifying electrical continuity. Each network is charged in turn until all of the nodes have been tested. If a charge is detected during the initial read, it indicates a short to an already charged network. Power and ground networks are treated as any other network. When complete, an electron flood gun is used to discharge the substrate. This test method provides a direct method of verifying the electrical continuity and isolation of each node of each network with a very high degree of confidence. In addition, this entire operation is performed in just seconds, rather than minutes. Provided that the entire substrate may be placed in the scan area of the VCEB, the test time for the above substrate would be less than 17 seconds. If stage movement is required for complete coverage of a substrate, test times increase accordingly. Still, the test time for a complete substrate is less than a minute in the worst case. The cost of a production version of a VCEB substrate test system is currently unknown, as only a prototype currently exists, but it should be in the range of $250K-$300K.

Conclusion

Multi-chip modules offer the potential advantages of a higher chip density, smaller interconnect delays and improved electrical performance. However the transition from printed-circuit-boards to high-density interconnection substrates increases the line losses by scaling down their cross-section. On the other hand, the increased silicon coverage and the corresponding decrease in line lengths mitigate the effect of the increased attenuation. In this paper we have discussed the design constraints imposed on high density substrates by lossy interconnect. They are visualized with the help of three-dimensional isochronal graphs, which can be combined with universal scaling relations in order to determine the propagation performance of any transmission line configuration. Examples of minimum line widths and maximum line densities are given.

The miniaturization of the interconnection substrates also increases the testing challenge. Fast test methods providing high resolution over a large field are required. We describe and compare flying probe systems, capacitance probes, and voltage contrast electron beam testers. The advantages of the VCEB system in throughput, certainty of results, and not probing the substrate surface make this system an attractive choice, but it is not yet a commercial product. The most useful, low-cost, production tool currently available is the capacitive probing system.

Acknowledgments

The authors wish to thank U. Ghoshal for his permission to use Figure 5, and F. Hartnett for the information on VCEB.

References

[1]   R.R. Tummala and E.J. Rymaszewski, "Microelectronics Packaging Handbook", Van Nostrand Reinhold (New York), 1989.

[2]   L.N. Smith, "High Density Copper Polyimide Interconnect", Proceedings of 3rd International SAMPE Electronics Conf., 939, June 1989.

[3]   L.S. Goldmann and P.A. Totta, "Area Array Solder Interconnections for VLSI", Solid State Technology, June 1983.

[4]   B. Nelson, et al., "Performance and Reliability of Tape Automated Bonding of High Lead Count, High Performance Devices", Proceedings of IPC, April 1988.

[5] C.W. Ho, D.A. Chance, C.H. Bajorek, and R.E. Acosta, "The Thin-Film Module as a High-Performance Semi-conductor Package", IBM Journal of Research and Development, Vol.26, Number 3, pp.286-296, May 1982.

[6] C. Hilbert, D. Gibson, and D. Herrell, "A Comparison of Lossy and Superconducting Interconnect for Computers", IEEE Trans. Electron Devices, Vol. ED-36, pp.1830, September 1989.

[7] H. Kroger, C. Hilbert, D. Gibson, U. Ghoshal, and L.N. Smith, "Superconductor-Semiconductor Hybrid Devices, Circuits and Systems", Proc. IEEE 77, 1287, 1989.

[8] R.E. Matick, "Transmission Lines for Digital and Communication Networks", McGraw-Hill (New York), 1969.

[9] U. Ghoshal, T.C. Wang, and L.N. Smith, "Modeling of Advanced Multichip/Wafer Scale Interconnects", Proc. 4th IEEE VLSI Multilevel Interconnection Conf., pp.265-272, June 1987.

[10] U. Ghoshal and L.N. Smith, "Finite Element Analysis of Skin Effect in Copper Interconnects at 77 K and 300 K", Proc. 1988 IEEE MTT-S International Microwave Symposium, Volume 1, pp.773-776, May 1988; and "Skin Effects in Narrow Copper Microstrips at 77 Kelvins", to be published in Trans. on Microwave Theory and Techniques, December 1988.

Application Of Automatic Optical Inspection (AOI)
in the
Manufacturing Technology Of Thin Film Multichip Module (MCM)

by

Giora Dishon and Oded Hecht

ORBOT Systems Ltd. ,
New Industrial Zone,
Yavne 70651, Israel

Abstract

The thin film multichip module (TF-MCM) packaging approach for high performance systems is being intensively developed and implemented throughout the electronic industry. Overcoming low yield barriers of the thin film multilayer process is essential for successful implementation of this technology in production. Yield enhancement can be obtained by applying a non-destructive testing of the multilayer structure at various stages of the manufacturing process. Automated optical inspection, which is widely used in manufacturing of printed circuit boards (PCB's), multilayer ceramics (MLC), and integrated circuits (IC), can be applied also to the TF-MCM manufacturing process. The various materials, processes, and design rules used in constructing the TF-MCM structure and the need to perform 100% quality inspections, dictate the development of a dedicated, high speed, versatile and intelligent AOI system. In addition to yield enhancement, applying automatic optical inspection through the manufacturing process can contribute to: improved long term reliability of the finished part, process control and quality assurance, manufacturing capacity increase, and evaluation of new processes and materials.

Introduction

The multichip module (MCM) packaging approach is being developed and implemented throughout the electronic industry. This technology uses a hybrid approach to achieve high density, high speed, and high I/O number for VLSI and GaAs based systems.

Multichip module technology for high performance systems has evolved from thick film hybrids, to co-fired ceramics (MLC's), to the present intense development efforts in thin film multilayers. This thin film approach was first detailed by C. Ho et al. [1], developed by numerous groups [2-4] and recently included in advanced new mainframe computers [5-7]. Extending semiconductor techniques to packaging through the wafer scale integration (WSI) approach has been under investigation for a long time, but yield problems and the need for elaborate redundancy schemes have so far prevented the implementation of the WSI approach. It appears that the "interim" technology of the thin film multichip module is a practical way to achieve many of the advantages of WSI and more (e.g. mixed IC types on the same module).

A large number of technological approaches are being explored in order to obtain reliable thin film multichip modules suitable for manufacturing [8]. Substrate materials for building the thin film structure include: silicon, alumina, aluminum nitride, silicon carbide, multilayer ceramics and glass ceramics. The materials practiced for the thin film interconnection structure differ in metalization (Cu, Au, Cr, Ti, Al, Ni, Sn and Pb), dielectrics (polyimides, BCB, parylene, SiO_2 and Si_3N_4), and photoresists. A large variety of processing techniques are being employed for the deposition and patterning of the layers. Conventional thin film processes such as vacuum deposition, spin coating, photolithography, wet and dry etching and liftoff are widely used, while thick film processes like electroplating, electroless plating and solder dipping are also practiced. All these processes are being developed and modified to accommodate thicker thin films and higher aspect ratios, which differ from both IC and thick film hybrid dimensions [9].

As the design complexity of the thin film interconnection drives towards larger substrate size, finer line resolution, and larger number of layers, overcoming low yield barriers of the multilayer processes is essential for a successful implementation of the TF-MCM approach in production. In addition to built-in redundancy, yield enhancement and therefore the cost justification of this approach to a wide range of applications (from mainframe computers to workstations, telecommunication and microcomputers) is needed. This goal can be accomplished by applying non-destructive inspection methods of the layers at different production stages. Yield and reliability can be further improved by adding different repair techniques. Electrical probing, while useful as a screening tool for defective parts, would not provide the necessary information for inspection/repair applications since the exact defect location cannot be determined [10].

Automatic Optical Inspection (AOI) is widely used today in manufacturing printed circuit boards (PCB)[11], multilayer ceramic packages (MLC)[12], and integrated circuits (IC). It is a non-destructive technique and potentially can be applied at most process steps. In the PCB industry, AOI has become the major inspection tool and both 100% quality inspection as well as process control are carried out[13]. In the IC industry only a low sample rate inspection is done. There are two major reasons for this:

1. The speed of today's IC AOI systems is far too low to allow 100% inspection.

2. A typical wafer contains 50 - 300 chips and even with the existing defect densities, the yield is reasonable and thus does not justify the cost of slow and expensive 100% inspection.

Therefore, AOI in the IC industry, is used mainly for process control and photomask inspection to help reduce the number of area defects.

The TF-MCM is a combined challenge. On one hand, its optical requirements approach those of IC (resolution, 3-D, transparent materials, etc.). On the other hand, a large substrate contains, in many cases, a single circuit and with the current defect densities, the yield is expected to be very low without 100% inspection and repair techniques. Therefore, the two challenges for an AOI system dedicated to TF-MCM are: speed and the quality of the acquired image.

Achieving high inspection speed (throughput) is critical to achieving the 100% inspection requirement. The factors that affect high speed are: high quality image acquisition and short image processing time, while maintaining system versatility and intelligence. These factors impose more difficulties when the fine line and tight tolerance geometries of the thin film multilayer structures are to be inspected. Additional constraints, that strongly affect the optical properties of the inspected layer, which result from the materials, processes, and structure of the TF-MCM, have to be considered when an inspection system is being developed.

Automatic Optical Inspection

The basic functional operation of an automatic optical inspection (AOI) system can be divided into four phases :

- Image acquisition
- Image processing
- Image analysis
- Flaw identification and reporting.

Fig. 1 provides a block diagram of an AOI system. The inspected surfaces are illuminated and the light reflected from the different materials, representing the image of the surface, is transmitted to the optical head. The scan table moves the inspected module under the optical head in both X and Y directions to obtain an image of the entire area.

The image is divided into picture elements (pixels) where each pixel has a grey level value according to the reflectivity of the surface material. The acquired image is then transferred to the image processing unit, which converts the image into digital form. *Fig. 2* shows a grey level histogram derived from an ideal surface. Setting the threshold level (the arbitrary dividing line between conductor and substrate) is critical in obtaining an accurate binary representation of the digitized image. The image analysis unit uses the binary image to identify features and applies defect detection operators to determine both design rule and reference violations. Different sources can be used to build the image for the reference comparison mechanism: CAD data, photomask, or "golden" part. Position and type of defects, as identified by the image analysis unit, are stored in a database and can be retrieved and displayed, either on the inspection system or on a verification station for further analysis and rework.

In order to carry out a design rule inspection, the system must acquire the basic logic design of the layer[11], such as linewidth and spacing, and consequently detect line and space violations, islands, pinholes, open ends and mousebites. For the reference comparison process, two basic concepts can be used: comparing to a bit map of the perfect image, and using the critical feature approach. The bit map approach is a computationally intensive solution - a 4x4 inch module inspected at a 1 μm pixel size contains 10^{10} pixels, thus making this approach very difficult to implement in a high speed and efficient manufacturing system. The critical feature approach, which uses only meaningful data such as feature type and coordinates to represent the image of a pattern, is much more efficient, even though much more complex and hard to develop. It uses sophisticated artificial intelligence algorithms to represent different features such as pads, vias, lines not ending in pads and junctions of lines. Implementing these algorithms in dedicated hardware maintains system versatility and intelligence without compromising on speed, while achieving very efficient storage of reference data.

The most important variable determining the system speed is the pixel size, as the throughput of an inspection system is actually evaluated by the rate of pixels per seconds. In the Variable Pixel approach[14] one can maximize pixel size used within the limits dictated by the combination of the application's physical feature dimensions and the inspection criteria and maintain the best throughput.

Thin Film MCM Technology and Optical Inspection

Fig. 3 provides a schematic cross section of a typical thin film MCM[15], and *Fig. 4* provides a block diagram of the process flow of the thin film multilayer structure fabrication. In manufacturing the thin film multilayer structure, the resultant yield is gated by the size and number of defects introduced into the structure by less than ideal manufacturing environments and processes[1]. There are three major kinds of fatal defects: intralevel metal line opens and shorts, interlevel metal line shorts, and defective interlevel vias. The defects are either process dependent (e.g. over-etch or under-etch), or caused by contamination and particulates. While the defect densities may be reduced by controlled clean-room environment and tight process control, they cannot be totally eliminated, thus leading to lower yield levels. In addition, nonfatal defects like near shorts, near opens, islands, pinholes and mousebites can reduce yield and strongly affect long-term reliability. *Figures 5 to 8* show some thin film typical defects[16]. Yield enhancement can be obtained by the combination of 100% quality inspection, photomask inspection and quality control inspection.

The materials, processes and design rules used in the TF-MCM interconnections impose significant challenges that need to be met by the automatic optical inspection system[17].

These difficulties arise for the following reasons:

1. A minimum feature of about 10 μm requires both line width measurement tolerances and minimum defect detections of about 3 μm.

2. High aspect ratios of line width to line thickness, thick dielectric layers, and unfilled vias, require a large depth of field in order to accommodate the non-planarity of the inspected surfaces and requires wide angle illumination to overcome these 3-D patterns.

3. Most of low dielectric constant polymers used as the dielectrics in the multilayer structure (polyimides, BCB) are transparent, therefore separation of top and lower metal layer is not simple. Photoresists impose the same type of constraint.

4. Due to the large range of different metals and deposition processes, different optical characteristics of the surface exist. These variations dictate the use of extremely flexible and elaborate illumination and optical systems.

5. Different shapes and angles of lines and pads require intelligent and dedicated hardware and software.

Fig. 9 shows a typical "raw" histogram of the surface of a thin film multilayer structure that demonstrates the poor contrast resulting from some of the difficulties discussed above. Contrast enhancement is required to set an adequate threshold for accurate inspection. Accurate inspection is characterised by low escape rates (undetected defects), and low rates of false alarms (pseudo-defects).

Effective inspection of fine line, thin film, geometries requires flexible, wide angle, illumination, and customized lenses that will allow adequate resolution down to 10 μm lines and large enough depth of field to accommodate the non-planarity of the inspected area. An advanced AOI system[17], with a highly sophisticated and dedicated hardware, will provide a better contrast between the different layers and thus enabling reliable optical inspection. Cost justification of optical inspection of thin film MCM's at every stage of the manufacturing process can be obtained only with high inspection speeds. An inspection speed of 72 mm²/sec, at 1 μm pixel size, which corresponds to an inspection rate of about 5 min/module for a 4"x4" inspected area, has been demonstrated[17]. Increasing the pixel size will increase inspection speed by a power of 2, e.g. from 72 mm²/sec at 1 μm pixel size to 450 mm²/sec at 2.5 μm pixel size.

AOI and Manufacturing of TF-MCM

It has been stated [1] that the yield of a multilayer structure is really determined by the total number of defects that occur on the total critical area in a multilevel thin film structure. Assuming a simple random defect model, derived from Poisson statistics, for a defect density of .1/cm² one finds the predicted yield for an area of 100 cm² approaching 0%[10]. This simplified analysis suggests that achieving practical yields in building TF-MCM's requires defect detection and repair techniques for each layer. AOI provides the defect detection technique thus needed. *Fig. 4* shows the metalization and dielectric repetitive process steps at which inspection can be introduced.

An AOI system can be utilized for: the detection of fatal and nonfatal defects, to facilitate dimensional control of controlled impedance signal line s, process control, and routine photo mask inspection.

The overall benefits of automatic optical inspection of thin film layers through the manufacturing of TF-MCM are:

1. Yield increase - through non-destructive detection and reporting of the location of all fatal defects for each metal layer, dielectric vias and photoresist patterns.

2. Improved long-term reliability - through early detection of nonfatal defects that cannot be detected by other testing methods.

3. Process control - through continuous monitoring of all process steps, both recurrence of defects and statistical process control can be maintained. AOI of each process step will promote the quality assurance of the finished product and increase the overall process latitude.

4. Manufacturing capacity increase - through better utilization of process equipment by early rejection of defective parts.

5. Technological advancement - AOI can be used to evaluate manufacturability of new processes and materials, helps interaction between development and production and between manufacturing and customers. Correlation between electrical testing of final product and the accumulated data of optical inspection, can be useful for better understanding of the technologies.

Conclusions

Automated optical inspection can be a powerful tool in the manufacturing process of thin film multichip modules. The characteristics of constructing the multilayer thin film interconnection structure - materials, processes, and design rules - dictate the use of a high resolution, fast, versatile and intelligent AOI system. The major reason for implementing 100% optical inspections, throughout the fabrication process, is the very much needed yield enhancement. Other benefits include improved long term reliability, process control, capacity increase, and easier and better implementations of new processes and materials. The overall outcome is higher efficiency and cost justification of this packaging approach for a wide range of applications from mainframe computers to communications, workstations and microcomputers.

Obtaining these features in an AOI system is based on sophisticated image processing techniques, using advanced and dedicated optical system, software algorithms and state of the art hardware.

References

1. C.W. Ho, D.A. Chance, C.H. Bajorek, R.E. Acosta, "The Thin Film Module as a High-Performance Semiconductor Package", IBM J. Res. Develop., Vol. 26, 3, May, 1982.

2. R. Wayne Johnson, "Thin Film Multichip Hybrids: An Overview", Proceedings of Technical Program Nepcon West '89, March 1989, pp. 655-672.

3. Maurice Sage, "Future of Multichip Modules in Electronics", Proceeding of Technical Program Nepcon West '89, March 1989, pp. 673-678.

4. George Messner, "Benefits of Multi-Chip Modules and Their Impact on PWB Markets", Joint T/MRC HMRC Technology Impact Seminar, ISHM, June 1989.

5. Digital Equipment Corporation, "Advanced Technologies for Use in VAX Mainframe Future Systems", Press Release, Sept. 25, 1989.

6. Shinichi Sasaki, Taichi Kon, and Takaam Ohsaki, "A New Multi-Chip Module Using a Copper Polyimide Multi-Layer Substrate", Proceedings of the 39th ECC Meeting, May 1989, pp. 629-635.

7. David Lammers, "SX-3: Paper tiger or for real?", Electronic Engineering Times, April 1989, pp. 35-37.

8. John K. Hagge, "Ultra-Reliable HWSI with Aluminium Nitride Packaging", Proceeding of Technical Program Nepcon West '89, March 1989, pp. 1271-1283.

9. G.J. Dishon, "Electroless Ni Via-Fill Process for High Performance Computer Packaging Application", Proceeding of the Symposium on Packaging of Electronic Devices, ECS Vol. 89-3, 1988 pp. 171-178.

10. T.G. Tessier, I. Turlik. G.M. Adema. D. Sivan, E.K. Yung and M.J. Berry, "Process Considerations in Fabricating Thin Film Multichip Modules", Proceedings of IEPS Meetings, Sept. 1989, pp. 294-313.

11. Douglass M. Domres, and James Macfarlene, "Automatic Optical Inspection Techniques for PWBs - Image Acquisition and Analysis Remain Areas of Challenge", Test & Measurement World, May 1983, pp. 71-74.

12. Lo-Soun Su, Thomas E. Wohr amd James J. Leybourne, "Automatic Pattern Inspection for Multilayer Ceramic Package", Proceedings of the 39th ECC Meeting, May 1989, pp. 616-622.

13. Nat Tinnerino, "Automatic Optical Inspection as a Process Control Tool", Electronic Packaging & Production, August 1989, pp. 40-45.

14. Shimon Ullman, "Automating the PCB Optical Inspection Process", Test & Measurement World, Sept. 1984, pp. 87-88.

15. Clinton C. Chao, Kenneth D. Scholz, Jacques Leibovitz, Maria Cobarruvias and Cheng C. Chang, "Multi-Layer Thin Film Substitute for Multi-Chip Packaging", IEEE Trans on CHMT, Vol. 12, 2, June 1989.

16. J.J.H.Reche, Polycon, Ventura, CA.

17. TF-501 Product Description and User Manual, Orbot Systems Ltd., Yavne, Israel.

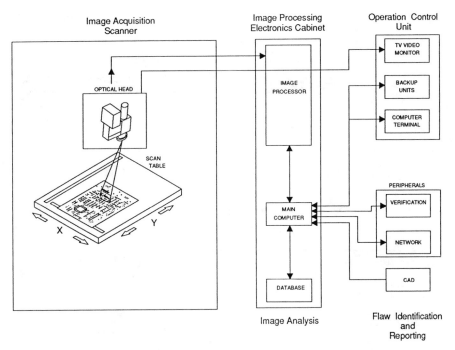

Fig. 1: Block Diagram of an Automatic Optical Inspection System

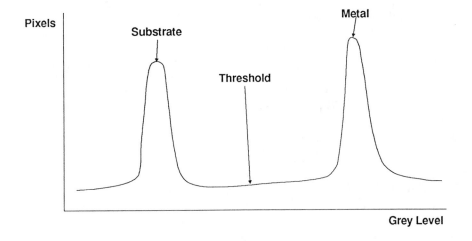

Fig. 2: A Grey Level Histogram derived from an ideal surface

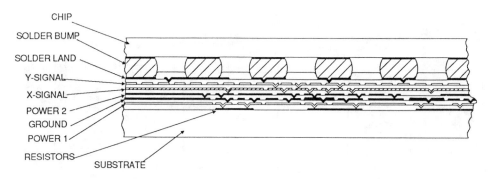

CHIP
SOLDER BUMP
SOLDER LAND
Y-SIGNAL
X-SIGNAL
POWER 2
GROUND
POWER 1
RESISTORS
SUBSTRATE

Fig. 3: A schematic cross section of a typical thin film MCM[15]

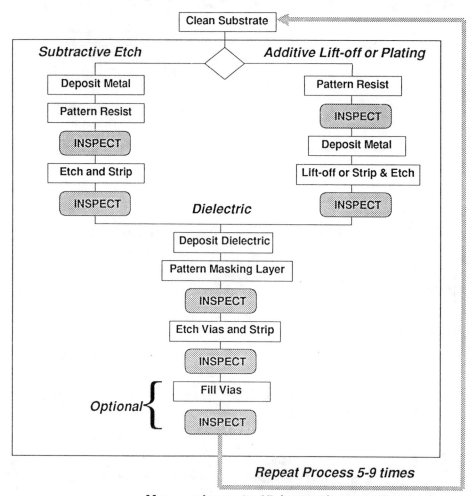

Clean Substrate

Subtractive Etch   **Additive Lift-off or Plating**

Deposit Metal   Pattern Resist

Pattern Resist   INSPECT

INSPECT   Deposit Metal

Etch and Strip   Lift-off or Strip & Etch

INSPECT   INSPECT

Dielectric

Deposit Dielectric

Pattern Masking Layer

INSPECT

Etch Vias and Strip

INSPECT

Optional { Fill Vias

INSPECT

Repeat Process 5-9 times

May require up to 45 inspections per finished part

Fig. 4: Block diagram of the process flow of thin film module fabrication, including suggested inspection steps

588

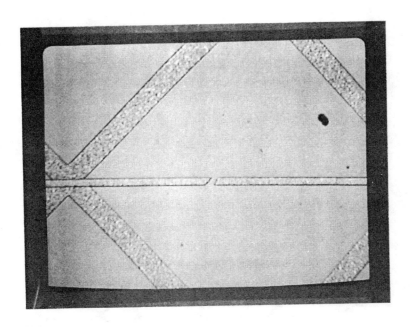

Fig. 5: Typical Defect - Line Open
(15 μm line)

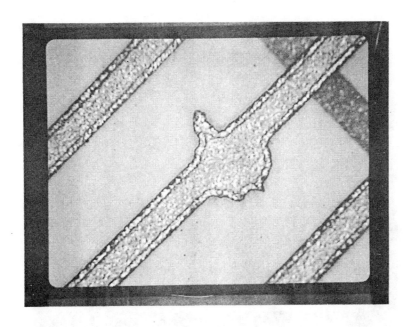

Fig. 6: Typical Defects - Large Protrusion with Space Violation
(25 μm line)

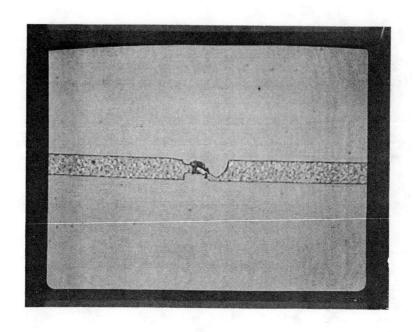

Fig. 7: Typical Defect - Mousebite (reliability defect)
(25 μm line)

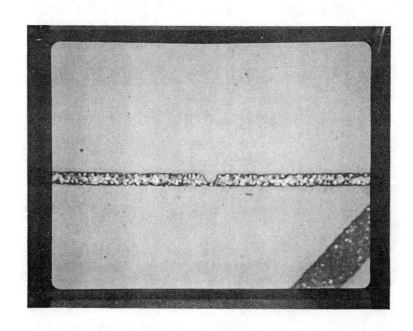

Fig. 8: Typical Defects - Mousebite (reliability defect)
(15 μm line)

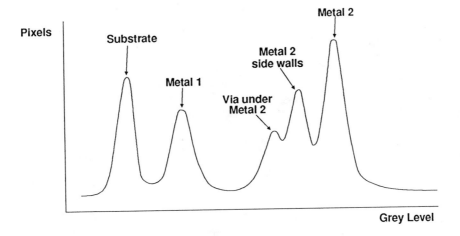

Fig. 9: A typical "raw" histogram of a thin film multilayer structure

Epilogue
What Is Missing?

MULTICHIP modules would have severe yield problems if testing were delayed until after final assembly. It is both customary and necessary to inspect the deposited circuits layer by layer. The defects, if any, are repaired before the next interconnect layer is deposited.

Similarly, the interconnect assembly is tested and the chips are tested before the assembly is made, and the final assembly is tested by both inspection means and functional performance tests.

At this time there is no full set of papers available that describe all the testing technology applied to multichip modules. Information on automated optical inspection, acoustic microscopy, X-ray and laser inspection, glow discharge, and voltage contrast SEM utilization is not all available in published papers. Boundary scan electrical testing can be used for assembly testing but also for board diagnosis. Repair procedures are under development.

The choices in this area will be left for a subsequent volume, as well as the reporting of the developing ideas already reported in workshops and yet to appear in print. This technology is in a rapid development phase, and we can but capture only what is presently public knowledge.

Author Index

Subject Index

597

Differential scanning calorimetry (DSC), analysis of photosensitive
 polyimides using, 419–420
DuPont Polyimide PI2555, testing of, 396–399

E

Electrically programmable silicon circuit board, 156–161
 amorphous silicon "antifuse," separation by, 157–158
 description, 156
 matrix of lines, 158–159
 mounted on silicon, 156
 power distribution system, 157
 "sea of bonding pads," 159–160
Electroless copper plated conductors, fine-line multilayer substrate with,
 311–316
Electroless nickel via–filling process flow, 309
Electroplating conductor, high-density multilayer interconnection with,
 144–149

F

Fireable ceramic substrate, using Cu–Ag interconnection, 447–449
Flip chip interconnection for MCMs, 130–138
 assembly process, 133
 bump processing, 130–132
 demonstration vehicle, 136–137
 design considerations, 132
 modeling/design verification, 133–136
 substrate processing, 132–133
Fluoropolymer composite MCMs, 330–338
 combining fluoropolymer composites and additive process, 333
 design rule objectives, 338
 fluoropolymer composite materials, 332–333
 Mictrotec process, 332
 processing advantages, 337
 reliability testing, 337

G

Green Tape(TM) dielectric and conductor system, 513–523
 increasing thermal conductivity with thermal vias, 518–520
 materials system/processing, 514–515
 reliability testing, 522–523
 thermal conductivity of composite structures, 520–522
 thermal properties of 851AT dielectric, 515–518

H

HDI, *See* High-density interconnects (HDI)
High density/high speed multichip packages, 323–325
 electrical performance, 325
 fabrication, 323–324
 moisture, 325
 properties, 324–325
 temperature, 324
High-density interconnection substrates
 capacitance probes, 577
 design guide lines, 575–576
 design/testing, 567–579
 flying probes, 576–577
 interconnect network test, 576
 interconnect simulations, 570–573
 isochronal delay surfaces, 573–574
 parametric process monitors, 576
 test requirements, 576
 universal scaling, 574–575
 voltage contrast electron beam test, 578

High-density interconnects (HDI)
 definition, 317
 fabrication, 317–319
 test results, 319–322
 using laser lithography, 317–322
High-density multichip interconnection, 123–129
 base/package, 126
 CAD and testing, 126–127
 comparison of approaches, 50–52
 interconnect and vias, 125–126
 package, 123
 packaging density, 127
 signal dielectric, 125
 substrate, 123–125
High-density multichip module by chip-on-wafer technology, 150–155
 AIN package, 152
 bumps, 152
 substrate, 151–152
High-density multilayer interconnection with photo-sensitive polyimide
 dielectric/electroplating conductor, 144–149
 application, 148–149
 dielectric layer/via hole formation, 146–147
 electrical via connection, 147–148
 fabrication, 144–148
 fine line conductor formation, 145–146
 multilayer structure, 148
 tests/evaluations, 148–149
High-speed digital components, 36–45
 approach, 36
 component reliability, 43
 electrical simulation and verification, 40–41
 interconnection/packaging challenges, 36
 interconnection requirements, 38–39
 mechanical, 43
 mechanical outlines, 36–37
 substrate technology development, 37–38
 thermal management, 42–43
Hitachi silicon-carbide RAM module, 488–489
Honeywell (HIS) Silent Liquid Integral Cooler (SLIC), 491
Honeywell/Mayo TFML test structure, 537–539
Hybrid wafer scale integration (HWSI)
 with aluminum nitride packaging, 201–210
 aluminum nitride properties, 203
 industry status, 201–202
 miniaturization potential, 202
 vs. surface mount technology, 201
 See also Aluminum nitride packaging

I

IBM 3081 thermal conduction module, 490–491
IBM 4381 module, 487–488
IBM, 1
Integrated circuit applications, MCMs, 19
Integrated circuits (ICs), vs. PWBs, 23–25
Interconnection costs, 11–13
 comparisons, 11–12
 cost measure, 11
 interconnection capability, 11
 interconnection substrates, hierarchy of, 11
 potential vs. usable wire, 12
 wiring demand measure, 12–13
Interconnection density, vs. line technology, 20
Interconnections
 cost–density analysis of, 21–29
 density requirements, 23
 trends in technology, 46
Interconnection on silicon, 1, 2

601

Editors' Biographies

R. Wayne Johnson (S'77–M'79–S'80–M'82–S'85–M'87) received the B.Sc. and M.Sc. degrees in 1979 and 1982, respectively, from Vanderbilt University, Nashville, TN, and the Ph.D. degree in 1987 from Auburn University, Auburn, AL, all in electrical engineering.

He is presently Assistant Professor of Electrical Engineering at Auburn University. He has established there teaching and research laboratories for thick- and thin-film hybrid technology. The thin-film research efforts are focused on materials, processing, and modeling for multichip modules. Power hybrids research is being directed in the thick-film laboratory. He has published and presented numerous papers at workshops and conferences and in technical journals. He has also worked for DuPont, Eaton, and Amperex.

Dr. Johnson is a member of the IEEE Components, Hybrids, and Manufacturing Technology Society and is President-Elect for the International Society for Hybrid Microelectronics.

Robert K. F. Teng (S'86–M'86) received the B.S. degree (magna cum laude) in electrical engineering from Mississippi State University, Starkville, in 1983, and the M.S. and Ph.D. degrees in electrical engineering from Purdue University, West Lafayette, IN, in 1985 and 1986, respectively.

In 1986, he joined the Department of Electrical Engineering, Mississippi State University as an Assistant Professor. He was in charge of the Microelectronics Fabrication Laboratory and the Hybrid Microelectronic Laboratory in the school. In 1989, he joined the Department of Electrical Engineering, California State University, Long Beach, as an Associate Professor. For the past seven years, he has been involved in research on fabrication of semiconductor devices, thin-film and thick-film hybrids, photovoltaic solar cells, reactive ion-beam etching and deposition, superconductive thin films, and computer-controlled manufacture. He has presented numerous papers at various conferences, and is the author of more than 25 technical articles, including 13 journal papers. He is currently working on nonvolatile static RAM, VLSI designs, and multichip modules.

Dr. Teng is a member of the International Society for Hybrid Microelectronics, Tau Beta Pi, and Eta Kappa Nu.

John W. Balde (SM'64–F'89) received the B.E.E. from Rensselaer Polytechnic Institute, Troy, NY.

He was a Research Leader in Interconnection Technology at AT&T. He has been a consultant on electronic packaging to over 40 international companies, and is now Senior Consultant at Interconnection Decision Consulting, Inc. in Flemington, NJ. He has contributed textbook chapters in the ASM Handbook, and has written papers on multichip modules in the conference proceedings of the Materials Research Society, IEEE Electronic Components Conference, International Electronics Packaging Society, IPC, International Society for Hybrid Microelectronics, NEPCON, INTERNEPCON, and ISHM NORDIC. He has also published articles in *Electronics*, *Electronic Packaging and Production*, *Circuits Manufacturing*, *Electronic Design*, *Journal of Electronic Materials*, *Computer Magazine*, and *Circuits World*. He holds 16 patents in system interconnection, flat cables, and tantalum thin-film technology.

Mr. Balde has received special awards from the IEEE for chip carrier development, from the IPC for contributions to packaging knowledge, and was chosen Man of the Year by the International Electronics Packaging Society for innovations and activities in packaging. He is former Chairman and Vice-Chairman of the IEEE Technical Committee on Packaging of the IEEE Computer Society, one of the 12 founders of the International Electronics Packaging Society and former Chairman of the Board, and is an active member of the Program Planning Boards of various technical societies, including the IEEE Components, Hybrids, and Manufacturing Technology Society, VLSI/GaAs Workshop, the Japanese, European, and U.S. Workshops of the IEEE Computer Society, and the International Society for Hybrid Microelectronics.